(Continued)

BIOLOGY
A Guide to the Natural World

ACADEMIC ADVISORS

Tania Baker
Massachusetts Institute of Technology

Anthony Ives
University of Wisconsin

Leon W. Browder
University of Calgary

Greg Podgorski
Utah State University

Anu Singh-Cundy
Western Washington University

Erica Suchman
Colorado State University

Leanne Field
University of Texas

Sara Via
University of Maryland

Nicholas Wade
The New York Times

BIOLOGY
A Guide to the Natural World
Third Edition

David Krogh
Berkeley, California

PEARSON

Prentice
Hall

Upper Saddle River, New Jersey 07458

Library of Congress Cataloging-in-Publication Data

Krogh, David.
    Biology : a guide to the natural world / David Krogh.-- 3rd ed.
        p. cm.
    ISBN 0-13-141449-6 (pbk.) — ISBN 0-13-141450-X (case) — ISBN
0-13-144956-7 (Instructor's ed.)
    1.   Biology.   I. Title.
    QH308.2.K76 2005
    570—dc22

                                        2003026478

Executive Editor: *Gary Carlson*
Editor-in-Chief, Science: *John Challice*
Production Editor: *Shari Toron*
Art Development Editor: *Kim Quillin*
Text Development Editor: *Annie Reid*
MediaLabs Development Editor: *Peggy Brickman*
Editorial Assistant: *Susan Zeigler*
Project Manager: *Crissy Dudonis*
Vice President of Production and Manufacturing: *David W. Riccardi*
Executive Managing Editor: *Kathleen Schiaparelli*
Assistant Managing Editor, Science: *Beth Sweeten*
Assistant Managing Editor, Media: *Nicole M. Bush*
Senior Marketing Manager: *Shari Meffert*
Marketing Manager: *Andrew Gilfillan*
Editor-in-Chief, Development: *Carol Trueheart*
Manufacturing Manager: *Trudy Pisciotti*
Manufacturing Buyer: *Alan Fischer*
Director of Creative Services: *Paul Belfanti*

Manager of Electronic Composition and Digital Content: *Jim Sullivan*
Assistant Formatting Managers: *Allyson Graesser, Bill Johnson*
Electronic Composition: *Bill Johnson, Beth Gschwind, Judith R. Wilkens*
Managing Editor of AV Management and Production: *Patricia Burns*
AV PE Chemistry and Science A Titles: *Connie Long*
Creative Director: *Carole Anson*
Art Director: *John Christiana*
Interior Designer: *Joseph Sengotta*
Cover Designer: *Luke Daigle/John Christiana*
Media Developer: *Mike Guidry/Lightcone Interactive*
Media Editor: *Andrew Stull*
Media Production Editor: *Aaron Reid*
Photo Research Administrator: *Melinda Reo*
Photo Researcher: *Jerry Marshall*
Copy Editor: *Chris Thillen*
Proofreader: *Michael Rossa*
Image Permission Coordinator: *Joanne Dippel*
Art Studio/Illustrator: *Imagineering, Steven Graepel*

Cover Photo Credits: Bison, © D. Robert Franz/Bruce Coleman Inc.; Fungus mushroom, maple leaf ©John Shaw/Bruce Coleman Inc.; Spider web in dew © Robert Falls Sr./Bruce Coleman Inc.; Detail head shot of captive jaguar, Belize, Central America ©Michael L. Peck/Imagestate; Little bee-eaters (Merops pusillus) on branch, Botswana © GettyImages/Stone; Nautilus seashell © Lester Lefkowitz/Corbis; Bird-of-paradise (strelitzia reginae), Molokai, Hawaii, close up © Getty Images/ImageBank

© 2005, 2002, 2000 Pearson Education, Inc.
Pearson Prentice Hall
Pearson Education, Inc.
Upper Saddle River, New Jersey 07458

Pearson Prentice Hall® is a trademark of Pearson Education, Inc.

Printed in the United States of America

10 9 8 7 6 5 4 3 2 1

ISBN 0-13-141449-6 (Paper)
ISBN 0-13-141450-X (Case)
ISBN 0-13-144956-7 (Instructor's Edition)

Pearson Education LTD., *London*
Pearson Education Australia PTY, Limited, *Sydney*
Pearson Education *Singapore, Pte. Ltd*
Pearson Education North Asia Ltd, *Hong Kong*
Pearson Education *Canada, Ltd., Toronto*
Pearson Educación de *Mexico, S.A. de C.V.*
Pearson Education—*Japan, Tokyo*
Pearson Education *Malaysia, Pte. Ltd.*

Essays

Forty-four essays appear in the book, most of them having an applied slant. They deal with such topics as acid rain, dietary fats, DNA fingerprinting, and the nature of human sexuality. In the How Did We Learn? essays, students can come to understand the inventiveness and the plain hard work that generally are prerequisites to scientific discovery.

MediaLabs

There are sixteen MediaLabs throughout the book. The topics were carefully chosen not only for student interest but also because they highlight issues that students may come across in their daily lives. Each MediaLab takes the reader on a journey of discovery through Web Tutorial activities and investigations.

Brief Contents

Contents

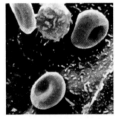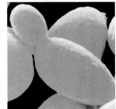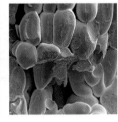

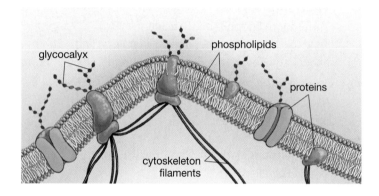

Unit 2   Energy and Its Transformations

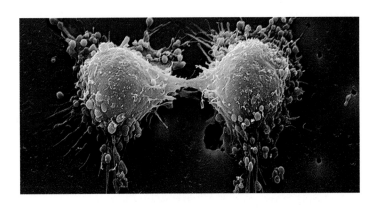

Unit 3 How Life Goes On: Genetics

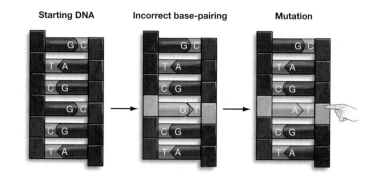

Starting DNA   Incorrect base-pairing   Mutation

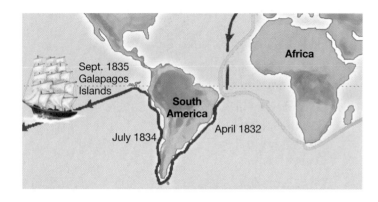

Chapter 27   Defending the Body: The Immune
System   **590**

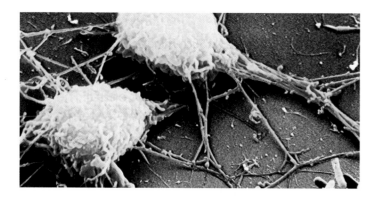

Chapter 28   Transport and Exchange: Blood, Breath,
Digestion, and Elimination   **608**

Preface

From the Author

Book titles may be the first thing any reader sees in a book, but they're often the last thing an author ponders. Not so with *Biology: A Guide to the Natural World*. The title arrived fairly early on, courtesy of the muse, and then stuck because it so aptly expresses what I think is special about this book.

Flip through these pages, and you'll see all the elements that students and teachers look for in any modern introductory textbook—rich, full-color art, an extensive study apparatus, and a full complement of digital learning tools. When you leaf slowly through the book and start to read a little of it, however, I think that something a little more subtle starts coming through. This second quality has to do with a sense of connection with students. The sensibility that I hope is apparent in *A Guide to the Natural World* is that there's a wonderful living world to be explored; that we who produced this book would like nothing better than to show this world to students; and that we want to take them on an instructive walk through this world, rather than a difficult march.

All the members of the teams who produced the three editions of *A Guide to the Natural World* worked with this idea in mind. We felt that we were taking students on a journey through the living world and that, rather like tour guides, we needed to be mindful of where students were at any given point. Would they remember this term from earlier in the chapter? Had we created enough of a bridge between one subject and the next? The idea was never to leave students with the feeling that they were wandering alone through terrain that lacked signposts. Rather, we aimed to give them the sense that they had a companion—this book—that would guide them through the subject of biology. *A Guide to the Natural World*, then, really is intended as a kind of guide, with its audience being students who are taking biology but not majoring in it.

Biology is complex, however, and if students are to understand it at anything beyond the most superficial level, details are necessary. It won't do to make what one faculty member called "magical leaps" over the difficult parts of complex subjects. Our goal was to make the difficult comprehensible, not to make it disappear altogether. Thus, the reader will find in this book fairly detailed accounts of such subjects as cellular respiration, photosynthesis, immune-system function, and plant reproduction. It was in covering such topics that our concern for student comprehension was put to its greatest test. We like the way we handled these subjects and other key topics, however, and we hope readers will feel the same way.

What's New in the Third Edition?

The third edition of the *Guide* has been substantially revised. Readers of this edition will find:

- Increased coverage of the diversity of the living world. Where once we had two chapters devoted to this subject, we now have three: one on animals, one on fungi and plants, and a third on microorganisms. This change has significantly increased our coverage of fungi and microorganisms.

- A new stand-alone chapter on the immune system.

- A general revision to the human anatomy and physiology unit that makes its coverage clearer and more relevant to the lives of students. All the major senses are now covered, whereas previously only vision was.

- A revamping of our biotechnology chapter, such that it now focuses on four sharply defined areas: transgenic biotechnology, reproductive cloning, forensic biotechnology, and personalized medicine.

- Substantially expanded coverage of human evolution.

- Information that has been updated, or that is new altogether, on such news-related subjects as oncogenes, dietary fats, human population growth, and global warming.

As is apparent from this list, the relevance of biology to students was never far from our minds as we shaped the subject matter for the third edition. Beyond the subjects mentioned, the book now contains coverage of such topics as the physiology of acne, the dangers of suntans and loud music, and the reasons for menstruation. Even where coverage isn't strictly "applied," we have tried, wherever possible, to note the broader importance of basic biological processes. As an example, the *Guide* has always covered the subject of meiosis, but in this edition students will find information on how meiosis has helped shape the living world by making it more diverse.

After the 2nd edition came out, faculty requested that we add coverage on two subjects that are heavy on scientific detail: functional groups in chemistry and the Hardy-Weinberg Principle in evolution. We now review both topics (in chapters 3 and 17, respectively), but we do so with the kind of approach you might expect from the *Guide*. Read the section on Hardy-Weinberg and you'll see that we explain not only what it is, but why it works as it does and how it is used in society. Further, since many faculty do not review Hardy-Weinberg, we placed the material on it in a location (the back of the chapter) that made it available to those faculty who teach it, while keeping it unobtrusive for those faculty who don't. Along similar lines, faculty will notice that the *Guide's* coverage of cellular respiration (in Chapter 7) has become more flexible in that faculty can choose to have students walk through its metabolic steps or not. How can it work both ways? This has to do with a more general change in the third edition.

The Third Edition Art Program

If you look at Figure 7.5 on page 139, you can see that all the steps of glycolysis have now been moved into the illustration. The main text tells students to walk through these steps as required in their class, but the only place the steps are actually reviewed is the figure. This same thing then happens with the Krebs cycle. Thus, faculty can have their choice: assign reading in Chapter 7 with or without the metabolic details.

This movement of information into illustrations is not limited to the *Guide's* Chapter 7, however. It takes place throughout the book. In dozens of figures, captions have now been integrated with drawings, whereas before captions existed at the bottom or side of a figure. Likewise, a substantial amount of main-text information has been moved into the figures. Where these changes have been made, students can stay focused on a small visual space, rather than moving back and forth between drawings and captions or drawings and main-text.

Beyond this change, the colors used in the third-edition illustrations are more vibrant and the figures have a greater sense of three-dimensional perspective. Lots of new photos have been added, and there is a better integration of photos with illustrations. The upshot is a book whose art is more appealing, but our goal was not simply to produce a prettier book. The complicated figures that appear in biology textbooks can be intimidating to students. To the extent that such figures can be made more inviting, they should do a better job of capturing and holding student attention. Also, notice the small "hand-pointers" that have been added to numerous figures. (See, for example, Figure 7.5 on page 139.) In scientific drawings, students often have trouble picking out what's important from what's not. The pointers should help.

Guide Design

As in the previous editions of the *Guide*, each chapter in this edition is divided into numbered modules (1.1, 1.2, and so forth), so that faculty can easily assign selected parts of a given chapter. The chapter sections are listed at the start of each chapter, and end-of-chapter summaries are indexed by section. On the first page of each chapter is a visual "filmstrip" that offers an intriguing preview of what's to come.

Flip through the pages of the *Guide*, and you'll note another useful design element right away: Text almost always occupies the top left of a page, with illustrations at the bottom. As a result, text continued from one page to the next is rarely broken up by a photo or illustration. Students reading text will not have their concentration broken by graphics when they turn to new pages.

The chief architect of the *Guide's* art continues to be artist and biologist Kim Quillin. Kim and I now have to communicate through electronic files, whisking them from one coast to the other, whereas when the book started out we communicated at a Berkeley Starbuck's. (Kim moved back to her native coastal Maryland after finishing her Ph.D. in biomechanics at U.C. Berkeley.) But our method of working has remained the same: We revise chapters at an early stage, based on the illustrations that Kim comes up with, thus ensuring a tight integration between text and illustrations. Put another way, the figures in the book aren't just adjuncts to the text. Rather, figures and text have shaped each other in a back-and-forth process.

Coverage of the Process of Discovery

One of the priorities for the third edition was to continue to impart to students a sense of how research results are arrived at in biology. Many of the book's chapters weave information on the process of discovery into explanations of what has been discovered. See, for example, Chapter 13 on Watson, Crick, and the DNA molecule; or Chapter 33 on proximate and ultimate causes in animal behavior. Each edition of the book has had series of stand-alone "How Did We Learn?" essays, and these have been updated and expanded for the third edition. (See the box on the discovery of penicillin in Chapter 20.) We have also kept in mind that, while faculty and students like these essays, they don't like them interrupting the flow of a chapter's main text. Thus, "How Did We Learn?" boxes continue to appear at the end of chapters.

Electronic Media and the Third Edition

The Guide continues to set a very high standard in its use of electronic media. For the third edition, the biggest change in media is that the primary entry point for it will now be the Worldwide Web, rather than the CD-ROM. A CD is included with each book, as was the case in the previous editions, but this CD now serves as a technical aid to the extensive set of materials that can be found at **www.prenhall.com/krogh3**. Students who have high-speed access to the Internet might not need to use the CD at all. But for students who lack such access, the CD will serve as an "accelerator" that will integrate with the book's web pages; it will quickly bring to the screen animations and other information-heavy modules.

The book's media offerings for students can be conceptualized as falling into two categories. First, there are the Web Tutorials—well-named because collectively they function as a kind of book-length tutor. Each of them leads students through a series of related biological concepts with the help of animations. If, upon reading Chapter 14 on genetic transcription and translation, a student isn't able to visualize how transfer RNA and messenger RNA work together at ribosomes, he or she can turn to the chapter's Web Tutorial and see this process laid out, step by step, with all the kinetics presented in animations. This story, of manufacturing proteins, is a Web "learning module" for Chapter 14—one of four contained in the web pages for that chapter. Each module walks students through a key chapter concept; each contains an interactive activity or exercise; and each ends with its own summary and mini-quiz.

All the Web Tutorials were developed by Mike Guidry and his colleagues at LightCone Interactive. Mike's team produced a tutorial for every chapter in the book, each one identified in the text with an icon like the one shown above.

Of course, students can turn to tutorial animations simply to make a given book illustration come to life; but they can also use the tutorials as just that—as learning sessions that employ interactive, step-by-step progressions. The proof here is in the pudding; take a look at some of the tutorials, and I think you'll agree they work very well.

Apart from the Web Tutorials, the *Guide* continues to have an extensive roster of the MediaLabs that proved so popular in the previous two editions. Produced by Peggy Brickman of the University of Georgia, these MediaLabs are aimed at making plain the linkage between biological concepts and real-world issues, and at fostering critical thinking about this linkage. A given lab starts by having students review, through a Web Tutorial, certain key concepts in a chapter. Then students are asked to investigate real-world issues connected to these concepts by going to suggested websites. (The cell cycle, covered in Chapter 9, may be intimately involved in the initiation of cancer, but what environmental factors are most important in getting cancer going? A web link in the MediaLab tells the tale.) Having done this digging, students are then asked to communicate what they have learned by writing brief essays on questions that are put to them. The book now has 16 MediaLabs, each integrated with the content of a specific chapter. Each MediaLab begins within the book itself (at the end of selected chapters), but then broadens out to the wide world of the Internet.

Many more web-based digital tools are available to students in this third edition of the *Guide*. Thanks to the work of media editor Andrew Stull, students looking at any chapter can click on a "Destinations" hyperlink at the website and be presented with a rich roster of chapter-specific Internet links. Self-quizzes for each chapter also are posted on the website, with quiz questions divided into "basic" and "challenge" sets. Beyond this, there is a set of audio files—18 of them new with this edition—that can be launched from the website. These are National Public Radio *Biocast* programs that have been integrated by their author, Bruce Hofkin, into each chapter in the book. Upon launching the *Biocasts* for Chapter 10, for example, a student can listen to a short program on a new technology that helps parents choose the gender of their child. This technology is connected to a basic concept covered in Chapter 10, sex determination in meiosis. Hofkin then brings the basic and applied science together in questions he poses at the end of the program.

All of these digital resources (and more) are available to students on the website, but faculty have additional resources at their disposal. The Instructor's CD-ROM contains all of the key animations on the student website; these are in turn part of a bank of images, known as the Instructor's Resource Center, containing every illustration and most of the photos in the book. The Instructor's Resource Center makes all the figures available in several formats, including PowerPoint slides that can be mixed and matched as desired, with figure parts, labels, and captions that can be edited. In addition, the Instructor's Guide and test-item file are embedded as a Word document in the CD-ROM, so that faculty can cut and paste what they need. Beyond these things, all the traditional media, such as transparencies, are available to faculty.

The *Guide to the Natural World* Team

Given all the names I've mentioned so far, it may go without saying that production of this book has been a team effort. It is my good fortune to have been given great teams for all the editions of *A Guide to the Natural World*. So large is an effort such as this that there are many people I've never met who have put in long hours on the book. I've noted Kim Quillin and her role in the book's art program, but this time around Kim was also ably assisted by an artist from nearby Virginia, Linda Berry. Annie Reid did her usual fine job as the book's developmental editor—the person who looked over everything Kim and I came up with and said whether it worked, after which she put the revised product together in a package that could be made into a book. Chris Thillen copyedited the manuscript, patiently making sure that the English language was used correctly. Shari Toron has been a great production editor, bringing together all the pieces of the project into the final product you see before you. Peggy Brickman not only developed the MediaLabs but also contributed greatly to the Web Tutorials produced by Mike Guidry and his co-workers. Jennifer Warner, Allison Wiedemeier, and Michelle Priest contributed tremendously to their revision for the third edition. I am grateful for the efforts of Michelle Priest and the team of people who contributed to new end of chapter questions and to the test bank and quizzes. These include Patricia Cox, Sherri Gross, Rick Heskett, Greg Podgorski, Daniel (Yunqiu) Wang, Jennifer Warner, and Allison Wiedemeier. Greg Podgorski deserves special mention because he so ably helped with so many facets of the book. Thanks needs to go out in advance to Andrew Gilfillan and Shari Meffert, who are just beginning to get the word out about the new edition of the *Guide* and to Crissy Dudonis who is finalizing the supplements package. Finally, we had great support at the top from Prentice Hall Editor Gary Carlson, who managed the project on its largest scale.

Apart from these team members, more than ninety faculty have now carefully critiqued every word and image you see in the *Guide*. (Is any written work more carefully reviewed than a textbook? Peer-reviewed scientific papers are the only other contenders that come to mind.) The names of reviewing faculty can be found beginning on page xxvii. Of these faculty, I need to make special note of the team of academic advisors who have provided advice not only on the details of the book, but on its overall structure and coverage. These advisors are listed across from the title page.

Finally, my thanks to all the faculty who used the first and second editions of *A Guide to the Natural World* in their courses and then let us know how they worked. Some of these faculty were reviewers, but some were instructors who sent in comments by e-mail just because they thought their feedback might be helpful. If they said the book needed some tweaking, we listened—the result being what you see in front of you. The main message from these faculty, however, was gratifying indeed. From them, we learned that we had done what we originally intended with *A Guide to the Natural World*: We had created a book that their students could understand. Moreover, they said, we did this not by leaving out the hard parts, but by thinking carefully about how all the parts should be presented.

I hope, of course, that the third edition will be just as effective. As a means of learning whether it is, let me extend an invitation. Whether your reactions to the book are positive or negative, send them to me via e-mail. My address is rdkrogh@pacbell.net. I'd like to hear from you.

David Krogh
Berkeley, California

Reviewers

We express sincere gratitude to the expert reviewers who worked so carefully with the author in reviewing the final pages to ensure the scientific accuracy of the text and art.

Gregory Podgorski—*Utah State University*
Sara Via—*University of Maryland*
Carla Bundrick-Benejam—*California State University, Monterey Bay*
Erica Suchman—*Colorado State University*
Robert Boyd—*Auburn University*
Judy M. Nesmith—*University of Michigan, Dearborn*
Leanne H. Field—*University of Texas, Austin*
Ray Ochs—*St. John's University*
James Cahill—*University of Alberta*
Mary Pat Wenderoth—*University of Washington*
Rick Relyea—*University of Pittsburgh*
Robert Curry—*Villanova University*

Media Reviewers

Robert S. Boyd, *Auburn University*
Sherri Gross, *Ithaca College*
Carolyn Glaubensklee, *University of Southern Colorado*
Gregory J. Podgorski, *Utah State University*
David A. Rintoul, *Kansas State University*
Ron Ruppert, *Cuesta College*
Brian Sailer, *Sam Houston State University*
Rebekah J. Thomas, *Saint Leo University*
Jennifer M. Warner, *University of North Carolina, Charlotte*
Jamie Welling, *South Suburban College*

Third Edition Reviewers

Ian M. Bird, *University of Wisconsin, Madison*
Robert Boyd, *Auburn University*
Peggy Brickman, *University of Georgia*
Carla Bundrick-Benejam, *California State University, Monterey Bay*
Warren Burggren, *University of North Texas*
Chantae Calhoun, *Lawson State Community College*
John Capeheart, *University of Houston-Downtown*
Kelly Sue Cartwright, *College of Lake County*
Marnie Chapman, *University of Alaska Southeast Sitka*
Richard Copping, *Lane College*
Lee Couch, *University of New Mexico*
Patricia Cox, *University of Tennessee*
Garry Davies, *University of Alaska, Anchorage*
Lee C. Drickamer, *Northern Arizona University*
Douglas Eder, *Southern Illinois University*
Elyce Ervin, *University of Toledo*
Rebecca Ferrell, *Metropolitan State College at Denver*
Leanne H. Field, *University of Texas, Austin*
Ralph Fregosi, *University of Arizona*
George W. Gilchrist, *College of Williams and Mary*

Tejendra S. Gill, *University of Houston*
Elliott Goldstein, *Arizona State University*
Sherri Gross, *Ithaca College*
James Hampton, *Salt Lake Community College*
Christopher Harendza, *Montgomery County Community College*
Barbara Harvey, *Kirkwood Community College*
Rebecca Jann, *Queens University of Charlotte*
Thomas W. Jurik, *Iowa State University*
Arnold J. Karpoff, *University of Louisville*
Harvey Liftin, *Broward Community College*
Ann S. Lumsden, *Florida State University*
Andrea Mastro, *Pennsylvania State University*
Mary Victoria, McDonald, *University of Central Arkansas*
Hugh A. Miller III, *East Tennessee University*
Leslie R. Miller, *Iowa State University*
Lee Mitchell, *Mt. Hood Community College*
Robert Mitchell, *Community College of Philadelphia*
Jeremy Montague, *Barry University*
Christopher B. Mowry, *Berry College*
Leann Naughton, *University of Wyoming*
Judy M., Nesmith, *University of Michigan, Dearborn*
James A. Nienow, *Valdosta State University*
Hunter O'Reilly, *University of Wisconsin, Milwaukee*
Eckle L. Peabody, *Tulsa Community College*
Kathleen Pelkki, *Saginaw Valley State University*
Gregory Podgorski, *Utah State University*
Michelle Priest, *University of Southern California*
Ameed Raoof, *University of Michigan Medical School*
Michael H. Renfroe, *James Madison University*
Todd Rimkus, *Marymount University*
Darryl Ritter, *Okaloosa-Walton Community College*
Carolyn Roberson, *Roane State Community College*
Jay Robinson, *San Antonio College*
William H. Rohrer, *Union County College*
Ron Ruppert, *Cuesta College*
Linda Smith-Staton, *Pellissippi State Technical Community College*
Salvatore A. Sparace, *Clemson University*
Steven Tanner, *University of Missouri*
Todd Templeton, *Metropolitan Community College*
Rani Vajravelu, *University of Central Florida*
William A. Velhagen, Jr., *Longwood College*
Edward L. Vezey, *Oklahoma State University, Oklahoma City*
Janet Vigna, *Grand Valley State University*
Stephen Wagener, *Western Connecticut State University*
Carol Wake, *South Dakota State University*
Charles Walcott, *Cornell University*
Yunqiu Wang, *University of Miami*
Jennifer Warner, *University of North Carolina, Charlotte*
Cheryl Watson, *Central Connecticut State University*
Richard Weinstein, *University of Tennessee, Knoxville*
Richard Barry Welch, *San Antonio College*
Elizabeth Forston Wells, *George Washington University*
Cathy Donald-Whitney, *Collin County Community College*
Allison Wiedemeier, *University of Missouri-Columbia*
Lorne Wolfe, *Georgia Southern University*
Calvin Young, *Fullerton College*
Martha C. Zúñiga, *University of California, Santa Cruz*
Victoria Zusman, *Miami-Dade Community College*

First and Second Edition Reviewers

Dawn Adams, *Baylor University*

John Alcock, *Arizona State University*

David L. Alles, *Western Washington University*

Sylvester Allred, *Northern Arizona University*

Gary Anderson, *University of California, Davis*

Marjay A. Anderson, *Howard University*

Michael F. Antolin, *Colorado State University*

Kerri Armstrong, *Community College of Philadelphia*

Mary Ashley, *University of Illinois, Chicago*

Jessica Baack, *Montgomery College*

Kemuel Badger, *Ball State University*

Tania Baker, *Massachusetts Institute of Technology*

Peter Bednekoff, *Eastern Michigan University*

Michael C. Bell, *Richland College*

William J. Bell, *University of Kansas*

David Berrigan, *University of Washington*

Lois A. Bichler, *Stephens College*

A.W. Blackler, *Cornell University*

Andrew Blaustein, *Oregon State University*

Robert S. Boyd, *Auburn University*

Bonnie L. Brenner, *Wilbur Wright College (City College of Chicago)*

Mimi Bres, *Prince George's Community College*

Peggy Brickman, *University of Georgia*

Leon Browder, *University of Calgary*

Arthur L. Buikema, *Virginia Polytechnic Institute and State University (Allegheny College)*

Warren Burggren, *University of North Texas*

Steven K. Burian, *Southern Connecticut State University*

Janis K. Bush, *University of Texas, San Antonio*

Linda Butler, *University of Texas, Austin*

W. Barkley Butler, *Indiana University of Pennsylvania*

David Byres, *Florida Central Community College, South Campus*

Van D. Christman, *Ricks College*

Deborah C. Clark, *Middle Tennessee State University*

William S. Cohen, *University of Kentucky*

Karen A. Conzelman, *Glendale Community College*

Tricia Cooley, *Laredo Community College*

Patricia Cox, *University of Tennessee, Knoxville*

John Crane, *Washington State University*

Garry Davies, *University of Alaska, Anchorage*

Paula Dedmon, *Gaston College*

Miriam del Campo *Miami Dade Community College*

Brent DeMars, *Lakeland Community College*

Llewellyn Densmore, *Texas Technical University*

Jean DeSaix, *University of North Carolina, Chapel Hill*

Jean Dickey, *Clemson University*

Christopher Dobson, *Front Range Community College*

Deborah Dodson, *Vincennes University*

Matthew M. Douglas, *Grand Rapids Community College (University of Kansas)*

Lee C. Drickamer, *Southern Illinois University*

Charles Duggins, Jr., *University of South Carolina*

Susan A. Dunford, *University of Cincinnati*

Ron Edwards, *University of Florida*

Douglas J. Eernisse, *California State University, Fullerton*

Jamin Eisenbach, *Eastern Michigan University*

George Ellmore, *Tufts University*

Patrick E. Elvander, *University of California, Santa Cruz*

Michael Emsley, *George Mason University*

David W. Essar, *Winona State University*

Richard H. Falk, *University of California, Davis*

Michael Farabee, *Estrella Mountain Community College*

Rita Farrar, *Louisiana State University*

John Philip Fawley, *Westminster College*

Eugene J. Fenster, *Longview Community College*

Christine M. Foreman, *University of Toledo*

Carl S. Frankel, *Pennsylvania State University*

Lawrence Friedman, *University of Missouri, St. Louis*

John L. Frola, *University of Akron*

Larry Fulton, *American River College*

Gail E. Gasparich, *Towson University*

Matt Geisler, *University of California, Riverside*

Claudette Giscombe, *University of Southern Indiana*

Carolyn Glaubensklee, *University of Southern Colorado*

Jack M. Goldberg, *University of California, Davis*

Judith Goodenough, *University of Massachusetts, Amherst*

Glenn A. Gorelick, *Citrus College*

Melvin H. Green, *University of California, San Diego*

G.A. Griffith, *South Suburban College*

Edward Hale, *Ball State University*

Gail Hall, *Trinity College*

Linnea S. Hall, *California State University, Sacramento*

Madeline Hall, *Cleveland State University*

Kelly Hamilton, *Shoreline Community College*

Steven C. Harris, *Clarion University*

Steve Heard, *University of Iowa*

Walter Hewitson, *Bridgewater State College*

Jane Aloi Horlings, *Saddleback College*

Eva Horne, *Kansas State University*

Michael Hudecki, *State University of New York, Buffalo*

Michael Hudspeth, *Northern Illinois University*

Terry L. Hufford, *The George Washington University*

Catherine J. Hurlbut, *Florida Community College*

Carol A. Hurney, *James Madison University*

Andrea Huvard, *California Lutheran University*

Martin Ikkanda, *Los Angeles Pierce College*

Rose M. Isgrigg, *Ohio University*

Anthony Ives, *University of Wisconsin, Madison*

Tom Jurik, *Iowa State University*

Anne Keddy-Hector, *Austin Community College*

Kathleen Keeler, *University of Nebraska*

Nancy Keene, *Pellissippi State Technical Community College*

Kevin M. Kelley, *California State University, Long Beach*

Jeanette J. Kiem, *Guilford Technical Community College*

Tom Knoedler, *Ohio State University, Lima Campus*

Don E. Krane, *Wright State University*

Jocelyn Krebs, *University of Alaska, Anchorage*

Kate Lajtha, *Oregon State University*

Erika Ann Lawson, *Columbia College*

Mike Lawson, *Missouri Southern State College*

Ann Lumsden, *Florida State University*

Paul Lurquin, *Washington State University*

James Manser, *Harvey Mudd College*

Michael M. Martin, *University of Michigan, Ann Arbor*

Paul Mason, *Butte Community College*
Michel Masson, *Santa Barbara City College*
Mary Colleen McNamara Albuquerque *T-VI A Community College*
Lee H. Mitchell, *Mount Hood Community College*
Scott M. Moody, *Ohio University*
Janice Moore, *Colorado State University*
Joseph Moore, *California State University*
Jorge A. Moreno, *University of Colorado*
Michael D. Morgan, *University of Wisconsin, Green Bay*
David Mork, *Saint Cloud State University*
Deborah A. Morris, *Portland State University*
Allison Morrison-Shetlar, *Georgia Southern University*
Richard Mortensen, *Albion College*
Michelle Murphy, *University of Notre Dame*
Courtney Murren, *University of Tennessee, Knoxville*
Royden Nakamura, *California Polytechnic State University*
Harry Nickla, *Creighton University*
Jane Noble-Harvey, *University of Delaware*
Marcy P. Osgood, *University of Michigan*
Andrea Ostrofsky, *University of Maine*
Maya Patel, *Ithaca College*
Patricia A. Peroni, *Davidson College*
Rhoda E. Perozzi, *Virginia Commonwealth University*
Carolyn Peters, *Spoon River College*
John S Peters, *College of Charleston*
Kim M. Peterson, *University of Alaska, Anchorage*
Raleigh K. Pettegrew, *Denison University*
Gary W. Pettibone, *State University of New York, College at Buffalo*
Holly C. Pinkart, *Central Washington University*
Barbara Pleasants, *Iowa State University*
John M. Pleasants, *Iowa State University*
Gregory J. Podgorski, *Utah State University*
Lynn Polasek, *Los Angeles Valley College*
F. Harvey Pough, *Arizona State University, West*
Don Pribor, *University of Toledo*
Louis Primavera, *Hawaii Pacific University*
Paul Ramp, *Pellissippi State and Technical Community College*
Regina Rector, *William Rainey Harper College*
Dennis Richardson, *Quinnipiac University*
Sonia J. Ringstrom, *Loyola University*
David A. Rintoul, *Kansas State University*
Laurel Roberts, *University of Pittsburgh*
Rodney A. Rogers, *Drake University*
Leslie Ann Roldan, *Massachusetts Institute of Technology*
Leslie Ann Roldan, *Massachusetts Institute of Technology*
Heidi Rottschafer, *University of Notre Dame*
John Rueter, *Portland State University*
Ron Ruppert, *Cuesta College*
Nancy Sanders, *Northeast Missouri Sate University*
Gary Sarinsky, *City University of New York, Kingsborough Community College*

Julie Schroer, *Bismarck State College*
Edna Seaman, *University of Massachusetts, Boston*
Ralph W. Seelke, *University of Wisconsin, Superior*
Prem P. Sehgal, *East Carolina University*
C. Thomas Settlemire, *Bowdoin College*
Robert Shetlar, *Georgia Southern University*
Mark A. Shotwell, *University of Slippery Rock*
Linda Simpson, *University of North Carolina, Charlotte*
Anu Singh-Cundy, *Western Washington University*
Peter Slater, *University of St. Andrews, UK*
Ellen Smith, *Arizona State University, West*
Philip J. Snider, *University of Houston*
Nancy G. Solomon, *Miami University*
Frederick W. Spiegel, *University of Arkansas*
Kathleen M Steinert, *Bellevue Community College*
Allan R. Stevens, *Snow College*
Donald P Streubel, *Idaho State University*
Erica Lynn Suchman, *Colorado State University*
Gerald Summers, *University of Missouri, Columbia*
Christine Tachibana, *University of Washington*
Rebekah J. Thomas, *Saint Leo University*
Joanne Tornow, *University of Southern Mississippi*
Todd T. Tracy, *Colorado State University*
Robin W. Tyser, *University of Wisconsin, LaCrosse*
Joseph W. Vanable, Jr., *Purdue University*
Sara Via, *University of Maryland*
Tanya Vickers, *University of Utah*
Janet Vigna, *Southwest State University*
Allan Hayes Vogel, *Chemeketa Community College*
Dennis Vrba, *North Iowa Area Community College*
Nicholas Wade, *The New York Times*
Jyoti R. Wagle, *Houston Community College*
John H. Wahlert, *Baruch College, The City University of New York*
Timothy S. Wakefield, *John Brown University*
Charles Walcott, *Cornell University*
Gene Walton, *Tallahassee Community College*
Sarah Ward, *Colorado State University*
Jennifer M. Warner, *University of North Carolina*
R. Barry Welch, *San Antonio College*
Jamie Welling, *South Suburban College*
John Whitmarsh, *University of Illinois at Urbana-Champaign*
Susan Whittemore, *Keene State University*
Sandra Winicur, *Indiana University, South Bend*
William Wischusen, *Louisiana State University*
Deborah Wisti-Peterson, *University of Washington*
Rachel Witcher, *University of Central Florida*
Mark A. Woelfe, *Vanderbilt University*
Lorne Wolfe, *Georgia Southern University*
Wade B. Worthen, *Furman University*
Robert Yost, *Indiana University Purdue University Indianapolis*

BIOLOGY
A Guide to the Natural World

1 *A Guide to the Natural World*

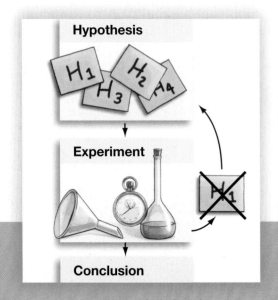

How scientists think about the world—the scientific method.
(Section 1.2, page 7)

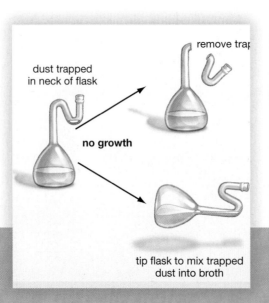

A famous experiment by a famous scientist.
(Section 1.2, page 8)

From atom to rain forest, the hierarchy of life.
(Section 1.3, page 13)

Science has great impact on our lives now and stands to have greater impact on them in the future. Science is both a body of knowledge and a means of acquiring knowledge. Biology, a branch of science, is the study of life.

1.1 How Does Science Impact the Everyday World?

Late in 2002, a Canadian chemist created quite a stir when, at a press conference, she announced that her company had cloned a human being. Now, set aside for the moment the fact that no evidence was ever offered for this claim, and just consider the announcement. Average citizens may not have understood exactly how the company carried off its supposed feat, but they had a rough idea of what the company was claiming: that it had produced one human being who was a genetic copy of another. Moreover they understood that this person was conceived not through sex, but through genetic manipulation in the laboratory.

Though this announcement may have been surprising, it was not startling, for by the time it was made people had some context in which to place cloning. It would not have been unusual in 2002 for a couple of office workers to note, at the watercooler, that it would be great if a particularly bright employee could be "cloned." Likewise, neighbors might observe, across the backyard fence, that one child seemed like a "clone" of another. The remarkable thing is how far the concept of cloning had to come to reach this common-knowledge status. Five years before the Canadian chemist made her announcement, the word clone was a technical term known to a handful of molecular biologists and a few science-fiction enthusiasts. What made the difference was, first, the cloning of Dolly the sheep in 1997 and then a steady tick of announcements: cloned pigs, cloned mice, and the fight over the possibility of a cloned human being. In a sense, however, the transition that cloning made is par for the course in the modern world. Here at the start of the twenty-first century, scientific innovations are moving with breathtaking speed from the laboratory to everyday life.

To appreciate this swift pace of change, consider electronic technology. The fundamental breakthrough that brought about modern electronics, including the computer, was the invention of the transistor at Bell Labs in 1947. Average American

Nature's finery—it's evolutionary.
(Section 1.4, page 14)

Poison dart frog—also evolutionary.
(Section 1.4, page 14)

You don't know what this insect is, but it scares you. Why?
(Section 1.4, page 15)

workers then *heard* about computers for 30 years, but not until the mid-1980s were they *using* computers right on their desktops, at home and at work. Once this revolution in circuits got going, however, things came in a torrent. VCRs and CD players were new-fangled devices in the mid-1980s; now people are "burning" their own CDs and throwing out their VCRs as antiquated relics of the era before DVDs (**see Figure 1.1**).

In a similar vein, the breakthrough that brought about the biotechnology industry was the description of the DNA molecule in 1953 by James Watson and Francis Crick. People heard more about genes in

(a)

(b)

Figure 1.1
Then and Now

(a) A technician enters data into the world's first programmable computer, run initially in 1948. Called "Baby," the computer was more than 2 meters (6 feet) tall and almost 5 meters (15.5 feet) wide, but had a total memory of only 128 bytes—less than one-ten-thousandth the amount of data that can be put on a common floppy disk today.

(b) An employee of Japan's Fujifilm corporation holds a tiny, mobile printer that can receive wireless signals from cell phones, like the one on the right, and print on card-sized, instant color film.

subsequent years, but it took another 25 years for the first genetically engineered medicine (human insulin) to appear. Now one-quarter of the corn grown in the United States is genetically modified; criminals are regularly convicted (and innocent people freed) through use of DNA "fingerprints," salmon are grown to eating weight in half the normal time thanks to gene splicing; and the announcement that a human being has been cloned is a claim that has to be taken seriously.

It may go without saying that these various innovations and discoveries entail value judgments on the part of society. Before the online music service Napster was shut down in 2001, it was an operation that facilitated large-scale music sharing or large-scale music theft, depending on your point of view. President Bush had to think long and hard in 2001 about whether to allow federal funding for embryonic "stem cell" research—a research avenue that holds out great promise, but entails the destruction of human embryos. For nearly a decade, people have argued passionately for and against the use of genetically modified foods.

Note that all these issues have been brought to society by science, we might say. But things also work the other way around. Scarcely a week goes by without society calling on scientists to provide advice on issues of the day. Think of such charged environmental questions as global warming or the logging of public lands. Is global warming being caused by human activity? Does thinning out forests reduce the risk of catastrophic forest fires? Then there is health. Does a high-fat diet lead to heart disease? Should women undergoing menopause undertake hormone replacement therapy? In all these instances, average citizens look to science to provide answers.

If we take a step back from all this, the message is that society is being driven by science and technology to a greater extent than ever before. Accordingly, to fully participate in the workforce, to make informed choices at the ballot box, or simply to make routine decisions, the average person must now be more scientifically literate than at any time in the past. To get a sense of how this plays out over the course of everyday life, let's look at some biology-related news that came to Americans through one magazine (*Time*) during one short period (January through May 2003). More than a score of biology-related stories were featured in *Time* during this period. Six are noted here.

Cracking the Ice said the headline in the February 3 issue of *Time*, in a story about why portions of enormous ice shelves that jut out from the Antarctic continent have started "calving" or breaking off into the sea with greater frequency. "Within the past year, scientists watched in awe as a giant ice shelf disintegrated in the Antarctic Peninsula in just over one month's time, and in a remote region of West Antarctica, satellites have detected an expanse where glaciers are worrisomely speeding up their transport of ice to the sea," wrote correspondent J. Madeleine Nash. One suspected cause of this series of events is global warming. But what is global warming? *Time* didn't have the space to delve into this question, but readers of this textbook who want to know can turn to page 725 for an explanation.

The Secret of Life was the cover story for *Time*'s February 17 issue, which celebrated the fiftieth anniversary of Watson and Crick's discovery of the structure of DNA (see **Figure 1.2**). Some human genes, *Time* said, "trace back to a time when we were fish; more than 200 come directly from bacteria. Our DNA provides a history book of where we come from and how we evolved. It is a family Bible that connects us all; every human being on the planet is 99.9% the same." But what are the lengths of DNA called genes? And how is DNA something that connects people not only with each other but with all living things? To find out, see Chapter 9, beginning on page 170.

Botox without the Needle was a March 3 story regarding new variants on the Botox injections people get as a means of getting rid of those pesky, middle-aged forehead lines. But did you know that Botox works by paralyzing muscles and that the substance it is purified from is a deadly bacterial toxin? How could a bacterial toxin paralyze muscles? And why should something that is a toxin in one context work as a cosmetic in another? For an account, see "The Muscular System" in Chapter 25 on page 551.

AIDS Vaccine: A Successful Failure was a March 10 story on the results of an AIDS vaccine trial that showed once again why AIDS is such a formidable enemy. AIDS is, of course, caused by HIV—the human immune deficiency virus—which is devastating because it attacks the very immune system that is meant to kill microbial invaders. How does HIV operate? To find out, see "Viruses: Making a Living by Hijacking Cells," in Chapter 20 on page 403. For an account of what AIDS does to the immune system, see "AIDS: Attacking the Defenders" in Chapter 27 on page 600.

The No. 1 Killer of Women was the title of an April 28 *Time* cover story. "No, it's not breast cancer," said

Time. "More women die of heart disease than of all cancers combined." The magazine also noted that "the common belief that premenopausal women are immune to heart problems is just plain wrong. Heart attacks strike 9,000 women younger than 45 each year." But what is a heart attack? What leads to one, and what does it do to the heart? You can find out in Chapter 28 on page 613.

Healthy Junk Food was a story that appeared in the May 12 issue. *Time* said such food is increasingly available, as evidenced by Frito-Lay's new "natural" varieties of such long-time favorites as Tostitos and Cheetos. What's the difference between the old and new varieties? For one thing, the natural versions are cooked in oil that has no trans fats in it. But what's a trans fat, and what does it have to do with health? You'll find an answer in "From Trans Fats to Omega 3's" in Chapter 3 on page 54.

What Do Americans Know about Science?

If you knew the scientific background to most of the *Time* stories, that probably puts you in a pretty select group of Americans. As you can see in

Figure 1.2
Science in the News
The importance of science to everyday life is reflected in the large number of news stories that focus on science. Pictured is a cover story on DNA that *Time* magazine ran in February 2003.

Figure 1.3, the average American has what might be called an uneven knowledge of science and the scientific process. Eighty-seven percent of a sample of American adults questioned in 2001 knew that the oxygen we breathe comes from plants. Yet one-quarter of this same group said that the sun goes around the Earth, rather than Earth around the sun. Almost 80 percent understood that "the continents on which we live have been moving their location for millions of years." Yet more than 50 percent believed that "the earliest human beings lived at the same time as the dinosaurs." (In fact, dinosaurs died out more than 60 million years before the first human beings appeared.) What factors go into making people scientifically literate? An important one is the number of science courses they have had in high school or college: as the number of courses goes up, so does the degree to which people feel well informed about scientific issues.

1.2 What Is Science?

Having looked a little at the impact that science has, it might be helpful to consider the question of what science *is*. The point here is to give you some sense of the underpinnings of science—to review something about the how and why of it before you begin looking at the nature of one of its disciplines, biology.

Science as a Body of Knowledge

Science is in one sense a process—a *way* of learning. In this respect, it is an activity carried out under certain loosely agreed-to rules, which you'll get to shortly. **Science** is also a body of knowledge about the natural world. It is a collection of unified insights about nature, the evidence for which is an array of facts. The unified insights of science are commonly referred to as *theories*.

It's unfortunate but true that *theory* means one thing in everyday speech and something almost completely different in scientific communication. In everyday speech, a theory can be little more than a hunch. It is an unproven idea that may or may not have any evidence to support it. In science, meanwhile, a **theory** is a general set of principles, supported by evidence, that explains some aspect of nature. There is, for example, a Big Bang theory of the universe. It is a general set of principles that explains how our universe came to be and how it developed. Among its principles are that a cataclysmic explosion occurred 13–14 billion years ago; and that, after it, matter first developed in the form of gases that then coalesced into the stars we can see all around us. There are numerous facts supporting these principles, such as the current size of the universe and its average temperature.

As you might imagine, with any theory this grand some *pieces* of it are in dispute; some facts don't fit with the theory, and scientists disagree about how to interpret this piece of information or that. On the whole, though, these general insights have withstood the questioning of critics, and together they stand as a scientific theory.

The Importance of Theories

Far from being a hunch, a scientific theory actually is a much more valued entity than is a scientific fact, for the theory has an *explanatory* power, while a fact is generally an isolated piece of information. That the universe is at least 13 billion years old is a wonderfully interesting fact, but it explains very little in comparison with the Big Bang theory. Facts are important; theories could not be supported or refuted without them. But science is first and foremost in the theory-building business, not the fact-finding business.

Figure 1.3
What Do Americans Know about Science?
Some results published by the National Science Foundation (*Science & Engineering Indicators—2002*).

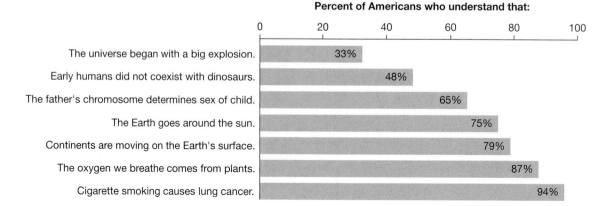

Percent of Americans who understand that:

The universe began with a big explosion.	33%
Early humans did not coexist with dinosaurs.	48%
The father's chromosome determines sex of child.	65%
The Earth goes around the sun.	75%
Continents are moving on the Earth's surface.	79%
The oxygen we breathe comes from plants.	87%
Cigarette smoking causes lung cancer.	94%

Science as a Process: Arriving at Scientific Insights

So how does a body of facts and theories come about? What is the process of scientific investigation, in other words? When **science** is viewed as a process, it could be defined as a means of coming to understand the natural world through observation and the testing of hypotheses. This process is referred to as the **scientific method**. The starting state for scientific inquiry is always *observation*: A piece of the natural world is observed to work in a certain way. Then follows the *question*, which broadly speaking is one of three types: a "what" question, a "why" question, or a "how" question. Biologists have asked, for example, What are genes made of? Why does the number of species decrease as we move from the equator to the poles? How does the brain make sense of visual images?

Formulating Hypotheses, Performing Experiments

Following the formulation of the question, various hypotheses are proposed that might answer it. A **hypothesis** is a tentative, testable explanation for an observed phenomenon. In almost any scientific question, several hypotheses are proposed to account for the same observation. Which one is correct? Most frequently in science, the answer is provided by a series of *experiments*, meaning controlled tests of the question at hand (**see Figure 1.4**). It may go without saying that scientists don't regard all hypotheses as being equally worthy of undergoing experimental test. By the time scientists arrive at the experimental stage, they usually have an idea of which is the most promising hypothesis among the contenders, and they then proceed to put that hypothesis to the test. Let's see how this worked in an example from history.

The Test of Experiment: Pasteur and Spontaneous Generation

Does life regularly arise from anything *but* life, or can it be created "spontaneously," through the coming together of basic chemicals? The latter idea had a wide acceptance from the time of the ancient Romans forward, and as late as the nineteenth century it was championed by some of the leading scientists of the day. So how could the issue be decided? The famous French chemist and medical researcher Louis Pasteur formulated a hypothesis to address this question (**see Figure 1.5**). He believed that many purported examples of life arising spontaneously were simply instances of airborne microscopic organisms landing on a suitable substance

and then multiplying in such profusion that they could be seen. Life came from life, in other words, not from spontaneous generation. But how could this be demonstrated? In 1860, Pasteur sterilized a meat broth in glass flasks by heating it, while at the same time heating the glass *necks* of the flasks, after which he bent the necks into a "swan" or S-shape. The ends of the flasks remained open to the air, but inside the flasks there was not a sign of life. Why? The broth remained sterile because microbe-bearing dust particles got trapped in the bend of the flask's neck. If Pasteur broke the neck off before the bend, however, the flask soon had a riot of bacterial life growing within it. In another test, Pasteur tilted the flask so that the broth *touched* the bend in the neck, a change that likewise got the microbes growing.

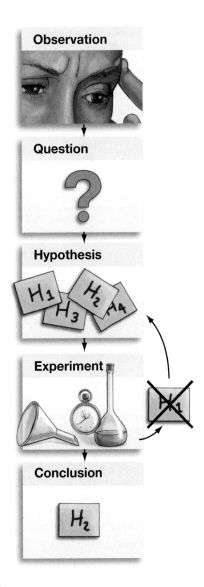

Figure 1.4
Scientific Method
The scientific method enables us to answer questions by testing hypotheses.

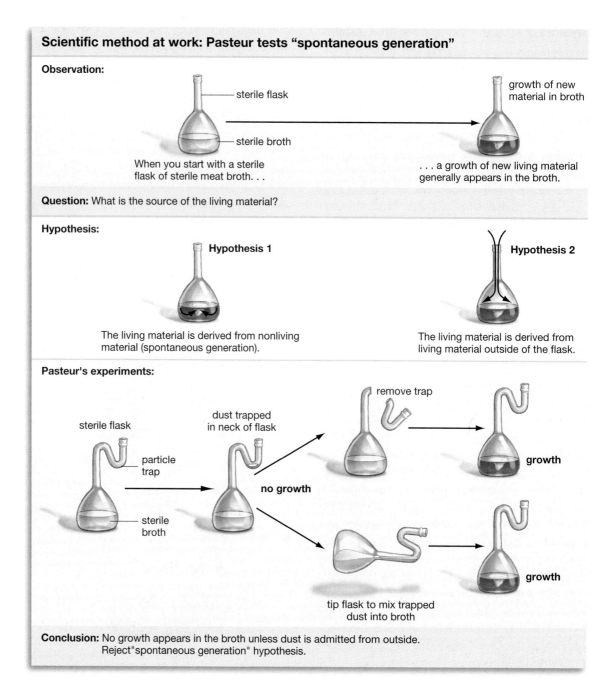

Scientific method at work: Pasteur tests "spontaneous generation"

Observation:

sterile flask

sterile broth

When you start with a sterile
flask of sterile meat broth...

growth of new
material in broth

...a growth of new living material
generally appears in the broth.

Question: What is the source of the living material?

Hypothesis:

Hypothesis 1

The living material is derived from nonliving
material (spontaneous generation).

Hypothesis 2

The living material is derived from
living material outside of the flask.

Pasteur's experiments:

sterile flask

particle
trap

sterile
broth

dust trapped
in neck of flask

no growth

remove trap

growth

tip flask to mix trapped
dust into broth

growth

Conclusion: No growth appears in the broth unless dust is admitted from outside.
Reject "spontaneous generation" hypothesis.

Figure 1.5
Pasteur's Experiments
Pasteur's spontaneous generation experiments and the scientific method. Nineteenth-century observation made clear that life
would appear in a medium, such as broth, that had been sterilized. But what was the source of this life? One hypothesis was that it
arose through "spontaneous generation," meaning it formed from the simple chemicals in the broth. Conversely, Pasteur hypothe-
sized that it originated from airborne microorganisms. He was able to design an experiment that offered evidence for this hypothesis. The device he used was an S-shaped flask, which enabled air to enter the flask freely while trapping all particles (including
invisible microorganisms) in a bend in the neck.

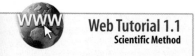

Web Tutorial 1.1
Scientific Method

Elements in Pasteur's Experiments

Now, note what was at work here. Pasteur had a preconceived notion of what the truth was, and de-
signed experiments to test his hypothesis. Critically,
he performed the same set of steps several times in
the experiments, keeping all the elements the same
each time—except for one. The nutrient broth was
the same in each test; it was heated the same
amount of time and in the same kind of flask. What
changed each time was one critical **variable**, meaning
an adjustable condition in an experiment. In this
case, the variable was either the shape of the flask
neck, or the tilt of the flask. Given that all other elements of the experiments were kept the same, the

experiments had rigorous controls: All conditions were held constant over several trials, except for a single variable. A **control** can be defined as a comparative condition in an experiment in which no variables are introduced. Pasteur's finding that no life grew in the bent-necked flask is interesting, but tells us very little by itself. We learn something only by comparing this finding to the result in the control condition: that life did grow when the flask neck was straight.

Note also that the idea of spontaneous generation was not banished with this one set of experiments—nor should it have been. Pasteur's experiments provided one of the *facts* mentioned earlier, in this case the fact that flasks of liquid will remain sterile under certain conditions. The idea that life arises only from life is, however, one of the scientific *theories* noted earlier, meaning that it requires the accumulation of many facts pointing in the same direction.

Other Kinds of Support for Hypotheses

Some scientific questions are difficult or impossible to test purely through experiment. For example, there currently is a controversy over whether birds are the direct descendants of dinosaurs. What kind of experiment could be run to test this hypothesis? Certain modern-day evidence is available to us—the DNA of living birds, for example—but examining DNA does not amount to an experiment. Instead it is observation, which is another valid way to test a hypothesis. Evidence from the past can also be observed, of course, which in this case means the observation of dinosaur and bird fossils. Indeed, fossils have been the key evidence in convincing most experts that birds are the descendants of dinosaurs. Experiment and observation are enhanced with the tool of statistics, which frequently is used in science, as you can see in "Lung Cancer, Smoking, and Statistics in Science" on page 10.

From Hypothesis to Theory

When does an idea move from hypothesis to theory? One of the ironies of the orderly undertaking called science is that there's nothing orderly about the change from hypothesis to theory. No scientific supreme court exists to make a decision. Scientists aren't polled for their views on such questions, and even if they were, at what point would we say something had been "proven"? When more than 50 percent of the experts in the field assent to it? When there are no dissenters left?

Provisional Assent to Findings: Legitimate Evidence and Hypotheses

This lack of finality in science fits in, however, with one of its central tenets, which is that nothing is ever finally proven. Instead, every finding is given only *provisional* assent, meaning it is believed to be true for now, pending the addition of new evidence. This principle is so deeply embedded in science that scientists rarely have reason to think about it (just as drivers would seldom contemplate why they are driving on the right-hand side of the road). Yet it is profoundly important because it is the thing that most starkly separates science from belief systems, such as those that operate in culture, politics, or religion. Every principle and "fact" in science is subject to modification, based solely on the best evidence available. There are no immutable laws and no unquestioned authority figures. This means there is a paradox in science: its only bedrock is that there is no bedrock; everything scientists "know" is subject to change.

In practice, this is a difficult ideal to live up to. Even when a body of evidence starts to point in a new direction, scientists—like anybody else—may be reluctant to give up old ways of thinking. Recognizing this tendency in human nature, Charles Darwin's friend Thomas Henry Huxley, writing in 1860, gave a beautiful description of the attitude scientists should have when investigating nature:

> Sit down before fact as a little child, be prepared to give up every preconceived notion, follow humbly wherever and whatever abysses nature leads, or you will learn nothing.

This principle of science's openness to revision is one of three important scientific principles having to do with hypotheses and evidence. Here are all three, stated briefly:

- Every assertion regarding the natural world is subject to challenge and revision, based on evidence.

- Results obtained in experiments must be *reproducible*. Different investigators must be able to obtain the same results from the same sets of procedures and materials.

- Any scientific hypothesis or claim must be *falsifiable*, meaning open to negation through means of scientific inquiry. The assertion that "UFOs are visiting the Earth" does not rise to the level of a scientific claim, because there is no way to prove that this is *not* so.

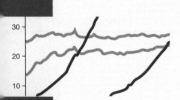

Lung Cancer, Smoking, and Statistics in Science

Valuable as they are, experimental and observational tests often are not enough to provide answers to scientific questions. In countless instances, scientists employ an additional tool in coming to comprehend reality—a mathematical tool—as you'll see in the following example.

The evidence that cigarette smoking causes lung cancer (and heart disease and emphysema and on and on) has been around for so long that most people have no idea why smoking was looked into as a health hazard in the first place. You might think that scientists were suspicious of tobacco decades ago and thus began experimenting with it in the laboratory, but this wasn't the case. Instead, the trail that led to tobacco as a health hazard started with a mystery about disease.

When the lung-cancer pioneer Alton Ochsner was in medical school in 1919, his surgery professor brought both the junior and senior classes in to see an autopsy of a man who had died of lung cancer. The disease was then so rare that the professor thought the young medical students might never see another case during their professional lifetimes. Prior to the 1920s, lung cancer was among the rarest forms of cancer, because cigarette smoking itself was rare before the twentieth century. It did not become the dominant form of tobacco use in the United States until the 1920s. This made a difference in lung-cancer rates because cigarette smoke is inhaled, while pipe and cigar smoke generally are not.

If you look at **Figure 1**, you can see the rise in lung-cancer mortality in U.S. males and females from 1930 forward. Note that women show a later rise in lung-cancer deaths; this is because women started smoking en masse later. (Also note that in the 1990s, lung-cancer rates finally began to level off—or drop in the case of men. This was a direct result of a decline in smoking that began in the 1970s.)

Given the lung-cancer trends that were apparent in males by the 1930s, the task before scientists was to explain the alarming increase in this disease. What could the cause of this scourge be, the medical detectives wondered? The effects of men being gassed in World War I? Increased road tar? Pollution from power plants? Through the 1940s, cigarette smoking was only one suspect among many.

Laboratory experiment eventually would play a part in fingering tobacco as the lung-cancer culprit, but the original indictment of smoking was written in numbers—in statistical tables showing that smokers were contracting lung cancer at much higher rates than nonsmokers.

It has sometimes been said that "science is measurement," and the phrase is a marvel of compact truth. For centuries, people had an idea that smoking might be causing serious harm, but this information fell into the realm of guessing or of *anecdote*, meaning personal stories. The problem with anecdote is that there is no measurement in it; there is no way of judging the validity of one story as opposed to the next. Related to anecdote is the notion of "common sense," which is valuable in many instances, but which also had us believing for centuries that the sun moved around the Earth. In the case of smoking, it took the extremely careful measurement provided by a discipline called *epidemiology*—the study of disease distributions—to separate truth from fiction.

1.3 The Nature of Biology

Let us shift now from an overview of science to a more narrow focus on **biology**, which can be defined as the study of life. But what is life? It may surprise you to learn that there is no standard short answer to this question. Indeed, the only agreement among scholars seems to be that there is not, and perhaps cannot be, a short answer to this question. The main impediment to such a definition is that any one quality common to all living things is likely to exist in some nonliving things as well. Some living things may "move under their own power," but so does the wind. Living things may grow, but crystals and fire do the same thing. Therefore, biologists generally define life in terms of a group of characteristics possessed by living things. Looked at together, these characteristics are sufficient to separate the living world from the nonliving. We can say that living things:

- Can assimilate and use energy
- Can respond to their environment
- Can maintain a relatively constant internal environment
- Possess an inherited information base, encoded in DNA, that allows them to function
- Can reproduce, through use of the information encoded in DNA
- Are composed of one or more cells
- Evolved from other living things
- Are highly organized compared to inanimate objects

Probability in Science

Note that "measurement" in this instance was a matter of calculating *probability*, which is often the case in science. Epidemiologists found a linkage between smoking and lung cancer, in the sense that those who smoked were more likely to get the disease. But having seen this, scientists then had to ask: Could this result be a matter of pure chance? A person tossing a coin might get heads five times in a row, and it might be written off to chance. But would it be the same if the person came up with heads *fifty* times in a row? No; at that point there would be justification for assuming that some force other than chance was in operation (such as a rigged coin). When the epidemiologists looked at their statistical tables and saw so many more smokers than nonsmokers getting lung cancer, they had to ask whether this result fell into the realm of five heads in a row, or fifty. Even in the earliest studies they concluded that more than chance was at work in the results. After many studies, they concluded that smoking was *causing* lung cancer. But how did they judge what was probable and what was not in an issue as complicated as this one? The researchers relied on techniques developed in the branch of mathematics called *statistics*.

The importance of probability and statistics to science can hardly be overstated. These tools are used frequently in nearly every scientific discipline. Imagine that 10 experimental plots of land are being compared, five with fertilizer added to them, the other five without. The plots with the added fertilizer end up with more growth but fewer kinds of plants. Could the differences between the two kinds of plots be a matter of chance? Here, as in so many other tests, scientists would use the tools of statistics to get at the truth.

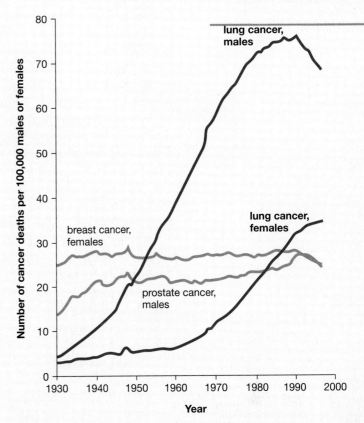

Figure 1
Rise in Lung-Cancer Mortality in U.S. Males and Females from 1930 Forward

Every one of these qualities exists in all the varieties of Earth's living things. The simplest bacterium needs an energy source no less than any human being. Our energy source is the food that's familiar to us; the bacterium's might be the remains of vegetation in the soil. The bacterium responds to its environment, just as we do. You would take action if you smelled gas in your house; the bacterium would move away if it encountered something it regarded as noxious. Humans maintain **homeostasis** or a relatively stable internal environment by, for example, sweating when they get hot. When the bacterium's external environment gets too hot, it has certain genes that will switch on to keep it functioning. Both humans and bacteria use the molecule DNA as a repository of the information necessary to allow them to live. Bacteria and human beings both reproduce—bacteria by simple cell division, human beings through the use of two kinds of reproductive cells (egg and sperm). A bacterium is a single-celled life-form, while humans are a 100-trillion-celled life-form. Bacteria and humans both evolved from complex living things and ultimately share a single common ancestor.

There are some exceptions to these "universals." For example, the overwhelming majority of honeybees and ants are sterile females; they can't reproduce, but no one would doubt that they're alive. In the main, however, if something is living, it has all these qualities.

11

Life Is Highly Organized, in a Hierarchical Manner

One item on the list of qualities requires a little more explanation. It is that living things are highly organized compared to inanimate matter. More specifically, they are organized in a "hierarchical" manner, meaning one in which lower levels of organization are progressively integrated to make up higher levels. The main levels in this hierarchy could be compared to the organization of a business. In a corporation, there may be individuals making up an office, several offices making up a department, several departments making up a division, and so forth. In life, there is one set of organized "building blocks" making up another (**see Figure 1.6**).

Actually life is not just "highly" organized. Nothing else comes *close* to it in organizational complexity. The sun is a large thing, but it is an uncomplicated thing compared to even the simplest organism. Consider that you have about 100 trillion cells in your body and that, with some exceptions, each of these cells has in it a complement of DNA that is made up of chemical building blocks. How many building blocks? Three billion of them. Now, you probably know that most cells divide regularly, one cell becoming two, the two becoming four, and so on. Each time this happens, each of the 3 billion DNA building blocks must be faithfully *copied*, so that both of the cells resulting from cell division will have their own complete copy of DNA. And this copying of the molecule—before anything is actually done with it—is carried out in almost all the varieties of the 100 trillion cells we have. Complex indeed. Let's see what life's levels of organization are.

Levels of Organization in Living Things

The building blocks of matter, called *atoms*, lie at the base of life's organizational structure. (See Chapter 2 for an account of them.) Atoms come together to form *molecules*, meaning entities consisting of a defined number of atoms that exist in a defined spatial relationship to one another. A molecule of water is one atom of oxygen bonded to two atoms of hydrogen, with these atoms arranged in a very precise way. Molecules in turn form what are called *organelles*, meaning "tiny organs" in a cell. Each of your cells has, for example, hundreds of organelles in it called *mitochondria* that transform the energy from food into an energy form your body can use. Such an organelle is not just a collection of molecules that exist close to one another. It is a highly organized structure, as you can tell just from looking at the rendering of it in Figure 1.6.

Figure 1.6
Levels of Organization in Living Things

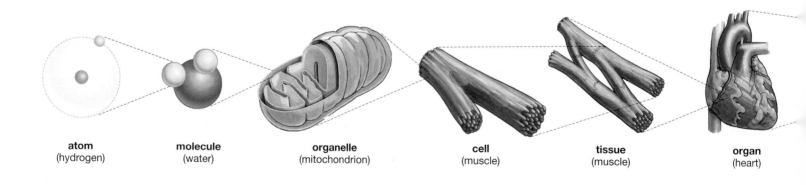

| atom (hydrogen) | molecule (water) | organelle (mitochondrion) | cell (muscle) | tissue (muscle) | organ (heart) |

At the next step up the organizational chain are entities that are actually *living*, as opposed to entities that are components of life. *Cells* are units that can do all of the things listed earlier: assimilate energy, reproduce, react to their environment, and so forth. Indeed, most experts would agree that cells are the only place that life exists. You may say: But isn't there a lot of material in between my cells? The answer is yes; it's mostly water with a good number of other molecules in the mix. But if all the cells were removed from this watery milieu, there would be nothing resembling life left in it.

The next step up is to a *tissue*, meaning a collection of cells that serve a common function. Your body contains collections of muscle cells that serve the same function (contraction). Each concentration of these cells constitutes muscle tissue. Several kinds of tissues can come together to form a functioning unit known as an *organ*. Your heart, for example, is a collection of muscle tissue and connective tissue, among other types. An assemblage of cells, tissues, and organs can then form a multicelled *organism*. (Of course, back down at the cell level, a one-celled bacterium is also an organism; it's just not one with organs and so forth.)

From here on out, life's levels of organization all involve *many* organisms. Members of a single type of living thing (a species), living together in a defined area, make up what is known as a *population*. When you look at *all* the kinds of living things in a given area, you are looking at a *community*. Finally, all the communities of the Earth—and the physical environment with which they interact—make up the *biosphere*.

1.4 Special Qualities of Biology

Biology traces its origins to the ancient Greeks. In the work of such Greeks as Hippocrates and Galen, we can find the origins of modern medical science. In the work of Aristotle and others, we can find the origins of "natural history," which led to what we think of today as mainstream biology and the larger category of the **life sciences**: a set of disciplines that focus on varying aspects of the living world. Apart from biology, the life sciences include such areas of study as veterinary medicine and forestry.

Despite its ancient origins, biology is, in a sense, a much younger science than, say, physics, which is one of the **physical sciences**, meaning the natural sciences not concerned with life. Western Europe's revolution in the physical sciences probably can be dated from the sixteenth century, when Nicholas Copernicus published his work *On the Revolution of Heavenly*

organism
(human)

population
(Yanomamö tribe)

community
(Amazon rain forest)

biosphere
(Earth)

Spheres, which demonstrated that Earth moves around the sun. Meanwhile, biology did not come into its own as a science until the *nineteenth* century. Prior to the 1800s, biology was almost purely "descriptive," meaning that the naturalists who we would today call biologists largely confined themselves to describing living things—what kinds there were, where they lived, what features they had, and so forth. Beginning in about the 1820s, however, biologists began to formulate biological theories as that term was defined earlier. They began to postulate that all life exists within cells, that life comes only from life, that life is passed on through small packets of information that we now call genes, and so forth. To put this another way, biologists in the nineteenth century began describing the *rules* of the living world, whereas before they were largely describing *forms* in the living world.

This change moved biology closer to the same scientific footing as physics. But biology was then, and remains now, a very different kind of science from any of the physical sciences, with physics being a clear case in point. One reason for this difference is that the constituent parts of physics are very uniform and far fewer in number than is the case in biology. Physics deals with only 92 stable elements, such as hydrogen and gold, and to a first approximation, if you've seen one electron, you've seen them all.

Meanwhile, in biology, if you've seen one species you've seen just that—one species. Each species is at least marginally different from another, and many are greatly dissimilar. Moreover, there are thought to be at least 4 million species on Earth. And each of these species has all the organizational levels of elements in physics *and more*. (They not only have electrons and atoms, they have organelles, cells, tissues, and so on.) Biology is concerned with the rules that govern all species, and you've seen that there are some biological "universals." But when cancer researchers are looking for the principles that underlie cell division, they are likely to be looking at only one of two main kinds of cells; when ecologists are looking at what causes dry grassland to turn into desert, their findings are likely to have little relevance to the rain forest. Put simply, the living world is tremendously diverse compared to the nonliving world, and such diversity means that universal rules in biology are likely to be few and far between. Biology is concerned with the *particular* to a greater degree than is the case in the physical sciences. Note also that "universals" in biology may not apply beyond Earth; we don't know if life even exists anywhere else, much less what its rules are. Meanwhile, the rules of physics truly are universal in that they are equally applicable on Earth or in the farthest reaches of the cosmos.

Biology's Chief Unifying Principle

Almost all biologists would agree that the most important thread that runs through biology is **evolution**, meaning the gradual modification of populations of living things over time, with this modification sometimes resulting in the develop-

Figure 1.7
Evolution Has Shaped the Living World

(a) A peacock displaying his plumage

(b) A poison dart frog in Colombia

(c) Pine trees in the South Pacific

ment of new species. Evolution is central to biology, because every living thing has been shaped by evolution. (There are no known exceptions to this universal.) Given this, the explanatory power of evolution is immense. Why do peacocks have their finery, or frogs their coloration, or trees their height (**see Figure 1.7**)? All these things stand as wonders of nature's diversity, but with knowledge of evolution they are wonders of diversity that *make sense*. For example, why do so many unrelated stinging insects look alike? Evolutionary principles suggest they *evolved* to look alike because of the general protection this provides from predators. Think of yourself for a moment as a bee predator. Having once gotten stung, would you annoy *any* roundish insect that had a black-and-yellow-striped coloration? You probably learned your lesson about this in connection with one species, but many species of insects are now protected from you simply by virtue of the coloration they share with the others (**see Figure 1.8**). Thus, there were reproductive benefits to individuals who, through genetic chance, happened to get a slightly more striped coloration: They left more offspring, because they were bothered less by predators. Over time, entire populations moved in this direction. They evolved, in other words.

The means by which living things can evolve is a topic this book takes up beginning in Chapter 16. Suffice it to say for now that a consideration of evolution is never far from most biological observations. So strong is evolution's explanatory power that, in uncovering something new about, say, a sequence of DNA or the life cycle of a given organism, one of the first things a biologist will ask is: Why would evolution shape things in this way?

1.5   The Organization of This Book

This book has something in common with the levels of organization you looked at earlier, in that it too goes from constituent parts to the larger whole. It begins with atoms, moves on to the biological molecules that atoms make up, and then goes to cells. The end of the book covers the highest levels of biological organization, which is to say natural communities and Earth's biosphere. In between, however, are tours of such facets of life as energy, DNA-encoded information, and reproduction. Even in these sections, however, you'll be moving in a general way from the small to the large, because much of the first part of the book is given over to **molecular biology**—meaning the study of individual molecules (such as DNA) as they affect living things. Then you'll move into evolution, which touches on **organismal biology**, meaning the study of whole organisms within biology. Next is the **physiology** or physical functioning of plants and animals, which largely concerns tissues and organs. Finally there is **ecology**, which is the study of the interactions of organisms with each other and with their physical environment. And so, let's begin to look at biology—as a body of knowledge and a way of learning.

Figure 1.8
Similar Enough to Yield a Benefit
These are two of the many stinging insects that have the black-and-yellow-striped coloration that warns away predators.

(a) Golden northern bumblebee

(b) Sandhills hornet

Chapter Review

Summary

1.1 How Does Science Impact the Everyday World?

- Science is playing an increasingly important role in the everyday lives of Americans. Scientific advances regularly confront society with choices that have an ethical dimension to them. Society frequently turns to scientists to answer questions about health, the environment, and other domains of life.

- Americans have an uneven knowledge about science. Almost 80 percent of adult Americans know that the continents are moving about the face of the Earth, for example, but one-quarter think the sun goes around the Earth.

1.2 What Is Science?

- In one of its facets, science is a body of knowledge, a collection of unified insights about nature, the evidence for which is an array of facts.

- The unified insights of science are known as theories. A theory is a general set of principles, supported by evidence, that explains some aspect of nature.

- Science can also be defined as a way of learning: a process of coming to understand the natural world through observation and the testing of hypotheses.

- Science works through the scientific method, in which an observation leads to the formulation of a question about the natural world. Then comes a hypothesis—a tentative, testable explanation that has not been proven to be true. The hypothesis may be tested through observation or through a series of experiments, as aided by statistical procedures. An example of hypothesis testing is Louis Pasteur's experiment on the spontaneous generation of life.
Web Tutorial 1.1: Scientific Method

- In science, every assertion regarding the natural world is subject to challenge and revision; results obtained in experiments must be reproducible by other experimenters; and scientific claims must be falsifiable, meaning open to negation through means of scientific inquiry.

1.3 The Nature of Biology

- Biology is the study of life. Life is defined by a group of characteristics possessed by living things. These are that living things can assimilate energy, respond to their environment, maintain a relatively constant internal environment, and possess an inherited information base, encoded in DNA, that allows them to function. Living things can also reproduce, are composed of one or more cells, are evolved from other living things, and are highly organized compared to inanimate objects.

- Life is organized in a hierarchical manner, running in increasing complexity from atoms to molecules and then in sequence to organelles, cells, tissues, organs, organisms, populations, communities, and the biosphere.
Web Tutorial 1.2: Hierarchical Organization of Life

1.4 Special Qualities of Biology

- Until the early nineteenth century, biology was largely a descriptive science, meaning it largely catalogued and described the Earth's living things. Beginning about the 1820s, however, life science researchers began to formulate biological theories, such as that life comes only from life and exists only within cells.

- Biology's subject matter—the living world—is notable for its complexity and diversity compared to other aspects of the natural world (such as stars and atoms). Because of this, biology does not deal in universal rules to the extent that a discipline such as physics does; instead, biological research may focus on particular species, processes, or portions of the living world.

- Biology's chief unifying principle is evolution, which can be defined as the gradual modification of populations of living things over time, with this modification sometimes resulting in the development of new species. Evolution provides the means for making sense of the forms and processes seen in living things on Earth today.

Key Terms

biology 10	**organismal biology** 15
control 9	**physical sciences** 13
ecology 15	**physiology** 15
evolution 14	**science** 6
homeostasis 11	**scientific method** 7
hypothesis 7	**theory** 6
life sciences 13	**variable** 8
molecular biology 15	

Understanding the Basics

Multiple-Choice Questions (answers in the back of the book)

1. Which of the following statements best describes the nature of a scientific hypothesis?
   a. A hypothesis is an idea that is widely accepted as a description of objective reality by a majority of scientists.
   b. A hypothesis must stand alone, and not be based on prior knowledge.
   c. A hypothesis must be testable through experimentation, observation, or mathematical demonstration.
   d. A hypothesis must deal with an aspect of the natural world never dealt with before.
   e. A hypothesis when accepted becomes a scientific law.

2. A *theory*, as used in scientific discourse, is
   a. An established fact, such as the distance from the Earth to the sun
   b. A long-accepted belief about the natural world
   c. A concept that is in doubt among most scientists
   d. A set of principles, supported by evidence, that explains some aspect of nature
   e. An initial guess about how some aspect of nature works

3. Pasteur's experiments on spontaneous generation made correct use of a variable in that Pasteur
   a. varied the bacteria he employed with each experiment
   b. used statistics to prove his hypothesis
   c. observed the bacteria as they were growing in the flasks
   d. held all conditions constant in each test except one
   e. was willing to vary to the extent necessary from the standard hypotheses of his day

4. Evolution is a central, unifying theme in biology because
   a. It is not a falsifiable hypothesis.
   b. Humans have evolved from ancestors we share with present-day monkeys.
   c. It has occurred in the past, even though it no longer operates today.
   d. The enormously diverse forms of life on Earth have all been shaped by it.
   e. Almost all biologists believe in it.

5. Biologists generally define life in terms of a group of characteristics possessed by living things. Which of the following is *not* a characteristic of living things?
   a. All living things possess an inherited information base, encoded in DNA, that allows them to function.
   b. All living things can respond to their environment.
   c. All living things can maintain a relatively constant internal environment.
   d. All living things evolved from other living things.
   e. All living things are composed of two or more cells.

Brief Review

1. What is science? In what ways is science different from a belief system such as religious faith?

2. What is a controlled experiment? Why is it important to keep all variables but one constant in a scientific experiment?

3. How did Louis Pasteur cast doubt on the idea of spontaneous generation?

4. Describe the defining features of life as we know it on the Earth.

5. Living things are organized in a hierarchical manner. List all the levels of the biological hierarchy that you can think of.

Applying Your Knowledge

1. Would you agree that it is valuable for a nation to have a citizenry that is reasonably well versed in science? Give reasons for your answer. Would you say this need has become especially urgent in the last two decades? If so, why?

2. Is it harder to prove a hypothesis than to disprove it? Imagine you wanted to establish that cheetahs are the fastest land animals, and assume you have the ability to clock any animal moving at its top speed. Now, what would it take to disprove the idea that cheetahs are the fastest land animals? What would it take to prove that cheetahs are the fastest land mammals, meaning no other land mammal could run faster than they?

3. If you were sent on an interplanetary mission to investigate the presence of life on Mars, what would you look for? Would you explore the land and the atmosphere? Imagine you discover an entity you suspect is a living being. Realizing that life elsewhere in the universe may not be organized by the same rules as on Earth, which of the features of life on Earth, if any, would you insist that the entity display before you would declare it living?

2 *Chemistry and Life*

Pure gold. It's elementary.
(Section 2.1, page 22)

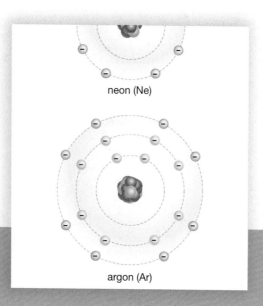

neon (Ne)

argon (Ar)

Filled shells mean stability.
(Section 2.2, page 24)

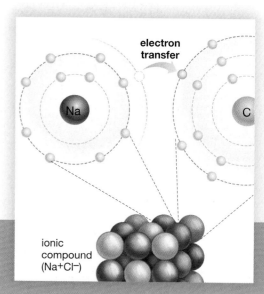

electron transfer

Na

C

ionic compound (Na^+Cl^-)

You put it on popcorn.
(Section 2.2, page 28)

Cities are made of buildings and buildings are made of bricks; bricks are made of earth and earth is made of . . . ? To answer this question, in this chapter we will look at what the material world is made of. Biology is our subject, but to fully understand it, you need to learn a little about what underlies biology. You need to learn a little about what *biology* is made of, in a sense. And to do this, you need to understand some of the basics in two other fields: chemistry and physics.

How are these disciplines relevant to biology? Well, consider chemistry. The average person probably is aware that living things are made up of individual units called cells. But beyond this bit of knowledge, reality fades and a kind of fantasy takes over. In it, the cells that populate people or plants or birds carry on their activities under the direction of their own low-level consciousness. A cell *decides* to move, it *decides* to divide, and so on. Not so. By the time our story is finished, many chapters from now, it will be clear to you that the cells that make up complex living things do what they do as the result of a chain of chemical reactions. Repulsion and bonding, latching on and reforming, depositing and breaking down; what makes people and plants and birds function at this cellular level is *chemistry*.

What is chemistry concerned with? Look around you. Do you see a table, light from a lamp, a patch of night or daytime sky? Everything that exists can be viewed as falling into one of two categories: matter or energy. You will learn something about energy in this chapter, but we are most concerned here with *matter and its transformations,* which is the subject of chemistry. Matter can be defined as anything that takes up space and has **mass**. This latter term is a measure of the *quantity* of matter in any given object. How much space does an object occupy—how much "volume," to put it another way—and how *dense* is the matter within that space? These are the things that define mass. For our purposes, we may think of mass as being equivalent to weight, though in physics they make a distinction between these two things.

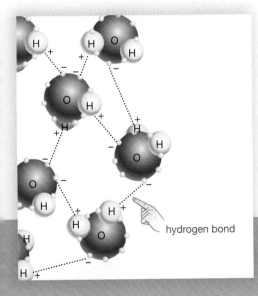

The ties that bond water molecules together.
(Section 2.2, page 29)

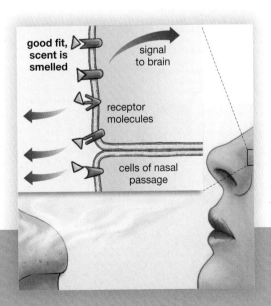

Do I smell bread in the oven?
(Section 2.3, page 30)

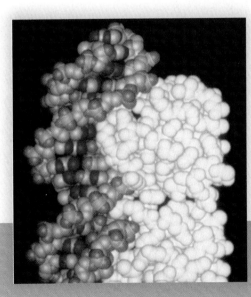

Many molecules are large and complex.
(Section 2.3, page 31)

Web Tutorial 2.1
**Structure of Atoms,
Elements, Isotopes**

2.1   The Nature of Matter: The Atom

Beholding matter all around us, it is natural to ask, what is its nature? A child sees a grain of sand, pounds it with a rock, sees the smaller bits that result and wonders: What is this stuff like at the end of these divisions? Not surprisingly, adults too have wondered about this question— for centuries. About 2,400 years ago, the Greek philosopher Plato accepted the notion that all matter is made up of four primary substances: earth, air, fire, and water. A near-contemporary of his, Democritus, believed that these substances were in turn made up of smaller units that were both invisible and in*div*isible—they could not be broken down further. He called these units atoms (**see Figure 2.1**).

Well, at least one cheer for Democritus, because he had it partly right. Centuries of painstaking work lying between his time and ours has confirmed that matter is indeed composed of tiny pieces of matter, which we still call atoms. But these atoms are not indivisible, as Democritus thought. Rather, they are themselves composed of constituent parts. A superficial account of *all* the parts scientists have discovered to date would go on for pages and still be incomplete. Physicists are continually slamming together parts of atoms with ever-greater force in an effort to determine what *else* there may be at the heart of matter. This is what the machines called "atom smashers" do. The physicists who run them could be compared to people who, in trying to find out what parts a watch has, throw it on the ground and record the way its various mechanisms fly out upon impact (**see Figure 2.2**).

Interesting stuff, but it is purely the business of physics, with little relation to biology. We are not concerned here with what's at the very end of these divisions. We do care a good deal, however, about what's *nearly* at the end of them.

Protons, Neutrons, and Electrons

For our purposes there are three important constituent parts of an atom: **protons**, **neutrons**, and **electrons**. These three parts exist in a spatial arrangement that is uniform in all matter. Protons and neutrons are packed tightly together in a core (the atom's **nucleus**), and electrons move around this core some distance away (**see Figure 2.3**). The one variation on this theme is the substance hydrogen, the lightest of all the kinds of matter we will run into. Hydrogen has no neutrons, but rather only one proton in its nucleus and one electron in motion around it.

These three "subatomic" particles have mind-bending sizes and proportions. As P. W. Atkins has pointed out, an atom is so small that 100 million carbon atoms would lie end to end in a line of carbon about this long: —————— (3 centimeters). Things are just as disorienting when we consider the size of the atom as a whole, relative to the nucleus. The whole atom, with electrons at its edge, is 100,000 times bigger than the nucleus. If you were to draw a model of an atom *to scale* and began by sketching a nucleus of, say, half an inch, you'd have to draw some of its electrons more than three-quarters of a mile away.

Although the nucleus accounts for very little of the *space* an atom takes up, it accounts for almost all

Figure 2.1
The Building Blocks of Life
Viewing this idealized feather at different levels of magnification, we eventually arrive at the building block of all matter, the atom. The atom selected here, from among a multitude that make up the feather, is a single hydrogen atom, composed of one proton and one electron.

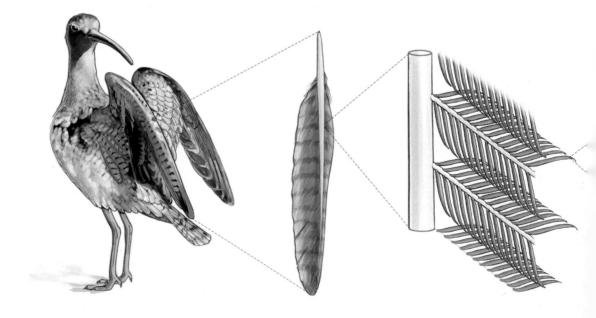

of the *mass* an atom has. So negligible are electrons in this regard, in fact, that all of the mass (or weight) of an atom is considered to reside with the nucleus' protons and neutrons.

The components of atoms have another quality that interests us: electrical charge. Protons are positively charged and electrons are negatively charged. Meanwhile, neutrons—as their name implies—have no charge, but are electrically neutral. Because all these particles do not exist separately, but *combine* to form an atom, as a whole the atom may be electrically neutral as well. The negative charge of the electrons balances out the positive charge of the protons. Why? Because in this state the number of protons an atom has is exactly equal to the number of electrons it has (though we'll see a different, "ionic" state later in this chapter). In contrast, the number of *neutrons* an atom has can vary in relation to the other two particles.

With this picture of atoms in mind, we can begin to answer the question that has been handed down to us through history: What is matter? We certainly have a commonsense answer to this question. Matter is any substance that exists in our everyday experience. For example, the iron that goes into cars is matter. But what is it that differentiates this iron from, say, gold? The answer is that an iron atom has 26 protons in its nucleus, while a gold atom has 79.

Fundamental Forms of Matter: The Element

Gold is an **element**—a substance that is "pure" because it cannot be reduced to any simpler set of component substances through chemical processes.

Figure 2.2
Getting to the Heart of Matter
Understanding what the tiny objects called atoms are made of requires the use of extremely large particle accelerators or "atom smashers," such as this one at Fermilab in Batavia, Illinois. Subatomic particles are accelerated around Fermilab's magnetic ring, some 6.3 kilometers or 3.8 miles in circumference, and then slammed into each other or into a fixed target. How big is this operation? The detectors that record the collisions weigh 5,000 tons apiece and are three stories high.

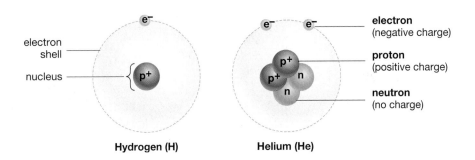

Figure 2.3
Representations of Atoms
One conceptualization of two separate atoms, hydrogen and helium. The model is not drawn to scale; if it were, the electrons would be perhaps a third of a mile away from the nuclei. The model also is simplified, giving the appearance that electrons exist in track-like orbits around an atom's nucleus. In fact, electrons spend time in volumes of space that have several different shapes.

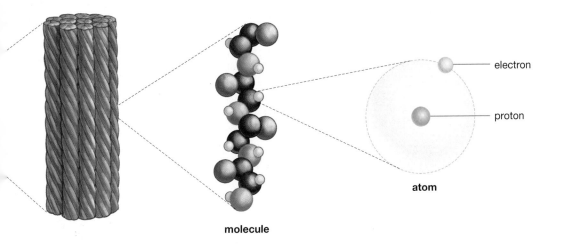

And the thing that defines each element is the number of protons it has in its nucleus. A solid-gold bar, then, represents a huge collection of identical atoms, each of which has 79 protons in its nucleus (**Figure 2.4**). In making gold jewelry, an artist may combine gold with another metal such as silver or copper to form an alloy that is stronger than pure gold, but the gold atoms are still present, all of them retaining their 79-proton nuclei.

Given what you've just read about protons, neutrons, and electrons, you may wonder why gold—or any other element—cannot be reduced to any "simpler set of component substances." Aren't protons and neutrons components of atoms? Yes, but they are not component *substances,* because they cannot exist by themselves as matter. Rather, protons and neutrons must *combine* with each other to make up atoms.

Assigning Numbers to the Elements

In the same way that buildings can be defined by a location, and thus have a street number assigned to them, these elements, which are defined by protons in their nuclei, have an **atomic number** assigned to them. We have observed that hydrogen has but one proton in its nucleus, and it turns out that scientists have constructed the atomic numbering system so that it goes from smallest number of protons to largest. Thus, hydrogen has the atomic number 1. The next element, helium, has two protons, so it is assigned the atomic number 2. Continuing on this scale all the way up through the elements found in nature, we would end with uranium, which has an atomic number of 92.

Given this view of the nature of matter, we are now in a position to answer the question posed at the beginning of the chapter: What is a handful of earth—or anything else—made of? The answer is one or more elements. If you look at **Figure 2.5**, you can see the most important elements that go into making up both the earth's crust and human beings.

Isotopes

All this seems like a nice, tidy way to identify elements—one element, one atomic number, based on number of protons—except that we're leaving out

Figure 2.4
Pure Gold
Gold is an element because it cannot be reduced to any simpler set of substances through chemical means. Each gold bar and nugget is made up of a vast collection of identical atoms—those with 79 protons in their nuclei.

Figure 2.5
Constituent Elements
The major chemical elements found in Earth's crust (including the oceans and the atmosphere) and in the human body.

Earth's crust

other 8%
oxygen (O) 50%
silicon (Si) 26%
aluminum (Al) 8%
calcium (Ca) 3%
iron (Fe) 5%

Human body

7% other
10% hydrogen (H)
65% oxygen (O)
18% carbon (C)

something. Recall that atoms also have neutrons in their nuclei, that these neutrons add weight to the atom, and that the number of neutrons can vary independently of the number of protons. What this means is that in thinking about an element in terms of its weight, we have to take neutrons into account. Furthermore, because the number of neutrons in an element's nucleus may vary, we can have various *forms* of elements, called **isotopes**. Most people have heard of one example of an isotope, whether or not they recognize it as such. The element carbon has six protons, giving it an atomic number of 6. In its most common form, it also has six neutrons. However, a relatively small amount of carbon exists in a form that has *eight* neutrons. Well, the element is still carbon, and in this form the number of its protons and neutrons equals 14, so the *isotope* is carbon-14, which is used in determining the ages of fossils and geologic samples.

Most elements have several isotopes. Hydrogen, for example, which usually has one proton and one electron, also exists in two other forms: deuterium, which has the proton, electron, and one neutron; and tritium, which has one proton, one electron, and two neutrons (**see Figure 2.6**). **Figure 2.7** shows you how isotopes are used in medicine.

The Importance of Electrons

In our account so far of the subatomic trio, we have had much to say about protons and neutrons, but little to say about electrons. This was necessary because we needed to go over the nature of matter, but in a sense you can regard what has been set forth to this point as so much stage-setting, because what's most important in biology is the way elements *combine* with other elements. And in this combining, it is the outermost electrons that play a critical role. Just as you come into contact with the world through what lies at your surface—your eyes, your ears, your hands—so atoms link up with one another through what lies at their outer edges. The interior of an atom is very quiet in a sense, while the atom's outer electrons exist in a world that can be one of continual forming and breaking of alliances.

2.2   Matter Is Transformed through Chemical Bonding

The process of chemical combination and rearrangement is called **chemical bonding**, and for us it represents the heart of the story in chemistry. When the outermost electrons of two atoms come into contact, it becomes possible for these electrons to

reshuffle themselves in a way that allows the atoms to become attached to one another. This can take place in two ways: One atom can *give up* one or more electrons to another, or one atom can *share* one or more electrons with another atom. Giving up electrons is called ionic bonding; sharing electrons is called covalent bonding. A third type of bond, which we'll get to shortly, also is important for our purposes: the hydrogen bond.

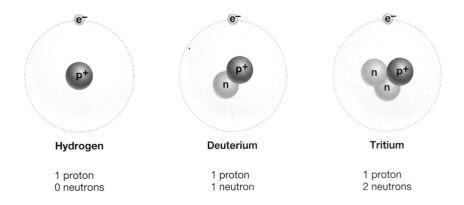

Hydrogen	Deuterium	Tritium
1 proton	1 proton	1 proton
0 neutrons	1 neutron	2 neutrons

Figure 2.6
Same Element, Different Forms
Pictured are three isotopes of hydrogen. Like all isotopes, they differ in the number of neutrons they have.

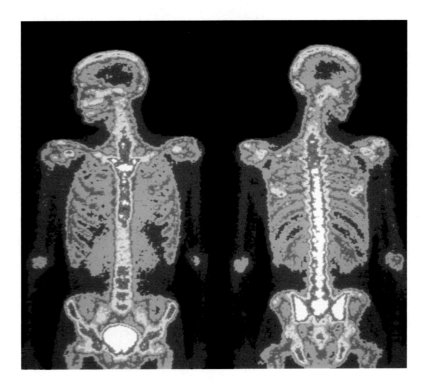

Figure 2.7
Imaging the Body with Isotopes
The radioactive isotopes technetium 99 was used in a nuclear medicine scan to provide these images of skeletal portions of the human body.

Web Tutorial 2.2
Chemical Bonding

Figure 2.8
Electron Configurations in Some Representative Elements
The concentric rings represent energy levels or "shells" of the elements, and the dots on the rings represent electrons. Hydrogen has but a single shell and a single electron within it, while carbon has two shells with a total of six electrons in them. Helium, neon, and argon have filled outer shells and are thus unreactive. Hydrogen, carbon, and sodium do not have filled outer shells and are thus reactive—they readily combine with other elements.

Energy Always Seeks Its Lowest State

Atoms that undertake bonding with one another do so because they are in a more *stable* state after the bonding than before it. A frequently used phrase is helpful in understanding this kind of stability: Energy always seeks its lowest state. Imagine a boulder perched precariously on a hill. A mere shove might send it rolling toward its lower energy state—at the bottom of the hill. It would not then roll *up* the hill, either spontaneously or with a light shove, because it is now existing in a lower energy state than it did before—one that is clearly more stable than its former precarious perch. When we turn to electrons, the energy is not gravitational, but electrical. Atoms bond with one another to the extent that doing so moves them to a lower, more stable energy state. The critical thing for our purposes is that atoms move to this more stable state by filling what is known as their outer shells.

Seeking a Full Outer Shell: Covalent Bonding

What are these "outer shells"? As it happens, electrons reside in certain well-defined "energy levels" outside the nuclei of atoms. The number of these energy levels varies depending on the element in question. Here we need only note the practical effect of these levels

on bonding: *Two* electrons are required to fill the first energy level (or shell) of any given atom, but *eight* usually are required to fill all the levels thereafter. If you look at the electron configurations pictured in **Figure 2.8**, you can see that two elements—hydrogen and helium—have so few electrons in orbit around them that they have nothing *but* a first energy level, while the other elements pictured have two or three energy levels. This means that hydrogen and helium require only two electrons in orbit around their nuclei to have filled outer shells, but that the other elements pictured require eight electrons to have this kind of complete outer electron complement.

How Chemical Bonding Works in One Instance: Water

To see how chemical bonding works in connection with this concept of filled outer shells, take a look at the bonding that occurs with the constituent parts of one of the most simple (and important) substances on Earth, water. In so doing, you'll see one of the kinds of bonding we talked about—covalent bonding.

The familiar chemical symbol for water is H_2O. This means that two atoms of hydrogen (H) have combined with one atom of oxygen (O) to form water. (See "Notating Chemistry" on page 00 for an explanation of symbols in chemistry.) Recall that hydrogen has but one electron running around in its single energy level. Also recall, however, that this first level is not completed until it has *two* electrons in it. Hydrogen could fill this shell in any number of ways. It might, for example, come into contact with another hydrogen atom. These two atoms can then form a **covalent bond** by sharing a *pair* of electrons with each other—one electron from each atom. These electrons can now be found orbiting the nuclei of both atoms. As a result, both have two electrons in their outer energy levels, which makes them filled.

Our hydrogen atom might also, however, come into contact with an oxygen atom, which has eight electrons. Looking at **Figure 2.9**, you can see what this means: Two electrons fill oxygen's first energy level, which leaves six left over for its second. But remember that the second shells of most atoms are not completed until they hold *eight* electrons. Thus oxygen, like hydrogen, would welcome a partner. Only it needs *two* electrons to fill its outer shell, which means that two atoms of hydrogen would do. Once again, pairs of electrons are shared. The oxygen atom and first hydrogen atom donate one electron each for the first pair; and the oxygen and second hydrogen atom each donate one electron for the second pair. The result? H_2O: Two atoms of hydrogen and one atom of oxygen, covalently bonded together and all of them

Unstable, very reactive atoms

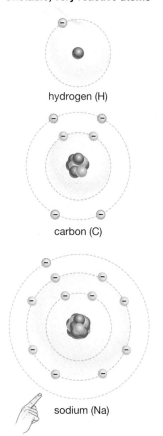

hydrogen (H)

carbon (C)

sodium (Na)

Outermost electron shells unfilled

Stable, unreactive atoms

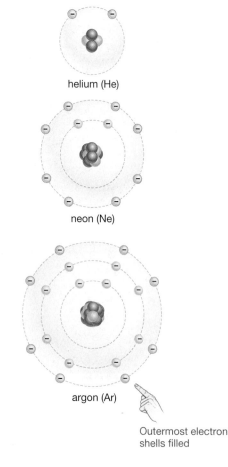

helium (He)

neon (Ne)

argon (Ar)

Outermost electron shells filled

"satisfied" to be in that condition. (Occasionally in nature, covalent bonding will take place in a way that leaves one atom with an unpaired electron, a potentially harmful phenomenon you can read about in "Free Radicals" on page 00.)

Matter Is Not Gained or Lost in Chemical Reactions

Note that when this pairing up of electrons happens, no matter has been gained or lost. We started with two atoms of hydrogen and one atom of oxygen, and we finish that way. The difference is that these atoms are now bonded. This points up an important principle, known as the **law of conservation of mass**, which states that matter is neither created nor destroyed in a chemical reaction.

What Is a Molecule?

When two or more atoms combine in this kind of covalent reaction, the result is a **molecule**: a compound of a defined number of atoms in a defined spatial relationship. Here, one atom of oxygen has combined with two atoms of hydrogen to create *one water molecule*. (What we commonly think of as water, then, is an enormous, linked collection of these individual water molecules.) A molecule need not be made of two different elements, however. Two hydrogen atoms can covalently bond to form one hydrogen molecule. On the other side of the coin, a molecule could contain many different elements bonded together. Consider sucrose, or regular table sugar, which is $C_{12}H_{22}O_{11}$ (12 carbon atoms bonded to 22 hydrogen atoms and 11 oxygen atoms).

Reactive and Unreactive Elements

The elements considered so far all welcome bonding partners, because all of them have incomplete outer shells. This is not true of all elements, however. There is, for example, the helium atom, which has two electrons. It thus *comes equipped*, we might say, with a filled outer shell. As such, it is extremely stable—it is unreactive with other elements. It is so unreactive that it is part of a family of elements that at one time were known as the inert gases, because it was thought that these elements never combined with anything. (In more recent times, however, all but two have been coaxed into such combining.) At the opposite end of the spectrum are elements that are extremely reactive. Look again at the representation of the sodium molecule in Figure 2.8. It has 11 electrons, two in the first shell and eight in the second, which leaves but one electron in the third shell—a very unstable state. Between the extremes of sodium and helium are elements with a range of outer (or *valence*) electrons.

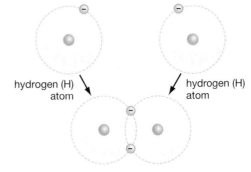

(a) Hydrogen molecule

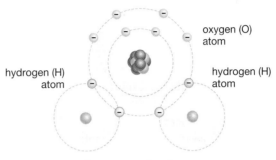

(b) Water molecule

Thus, there is a spectrum of stability in the chemical elements, based on the number of outer-shell electrons each element has—from 1 to 8, with 1 being the most reactive and 8 being the least reactive.

Polar and Nonpolar Bonding

Not all covalent bonds are created alike. When two hydrogen atoms come together, the result is a hydrogen molecule (H_2). Now, in the hydrogen molecule, the electrons are shared *equally*. That is, the two electrons the hydrogen atoms are sharing are equally attracted to each hydrogen atom. This is not the case, however, with the water molecule.

Look at the representation of the water molecule in **Figure 2.10a**. As it turns out, the oxygen atom has a greater power to attract electrons to itself than do the hydrogen atoms. The term for measuring this kind of pull is **electronegativity**. Because the oxygen atom has more electronegativity than do the hydrogen atoms, it tends to pull the shared electrons away from the hydrogen and toward itself. When this happens, the molecule takes on a **polarity** or a difference in electrical charge at one end as opposed to the other. Because electrons are negatively charged, and because they can be found closer to the oxygen nucleus, the oxygen end of the molecule becomes slightly negatively charged, while the hydrogen regions become slightly positively charged. We still have a covalent bond, but it is a specific type: a **polar covalent bond**. Conversely, with the hydrogen molecule—where electrons are being shared equally—we have a **nonpolar covalent bond**.

Figure 2.9
Covalent Bonding
A covalent bond is formed when two atoms share one or more pairs of electrons.

(a) Two atoms of hydrogen have come together, and each shares its lone electron with the other. This gives both atoms a filled outer shell—and stability.

(b) Two hydrogen atoms have linked with one oxygen atom; in this case, two pairs of electrons are shared, one pair between each of the hydrogen atoms and the oxygen atom.

gaining one electron. That's just how this encounter goes: Sodium does in fact lose its one electron, chlorine gains it, and both parties become stable in the process (**see Figure 2.11**).

What Is an Ion?

But this story has a postscript. Having lost an electron (with its negative charge), sodium (Na) then takes on an overall *positive* charge. Having gained an electron, chlorine (Cl) takes on a negative charge. Each is then said to be an **ion**—a charged atom—or, to put it another way, an atom whose number of electrons differs from its number of protons. We denote the ionized forms of these atoms like this: Na^+, Cl^-. Were an atom to gain or lose more electrons than this, we would put a number in front of the charge sign. For example, to show that the magnesium atom

has lost two electrons and thus become a positively charged magnesium ion, we would write Mg^{2+}.

Note that we now have two ions, Na^+ and Cl^-, with differing charges in proximity to one another. They are thus attracted to one another through an *electrostatic attraction,* and have an ionic bond between them. This hardly ever happens with just *two* atoms, of course. Many billions of atoms are bonded together in this way, up, down, and sideways from each other. This whole collection is, likewise, called an ion; or, if two or more elements are mixed together this way, an **ionic compound**. The particular ionic compound just described actually is very familiar. Sodium and chlorine combine to create sodium chloride, which is better known as table salt. The notation should properly be written Na^+Cl^-, but it is usually denoted as just plain NaCl.

**Figure 2.11
Ionic Bonding**

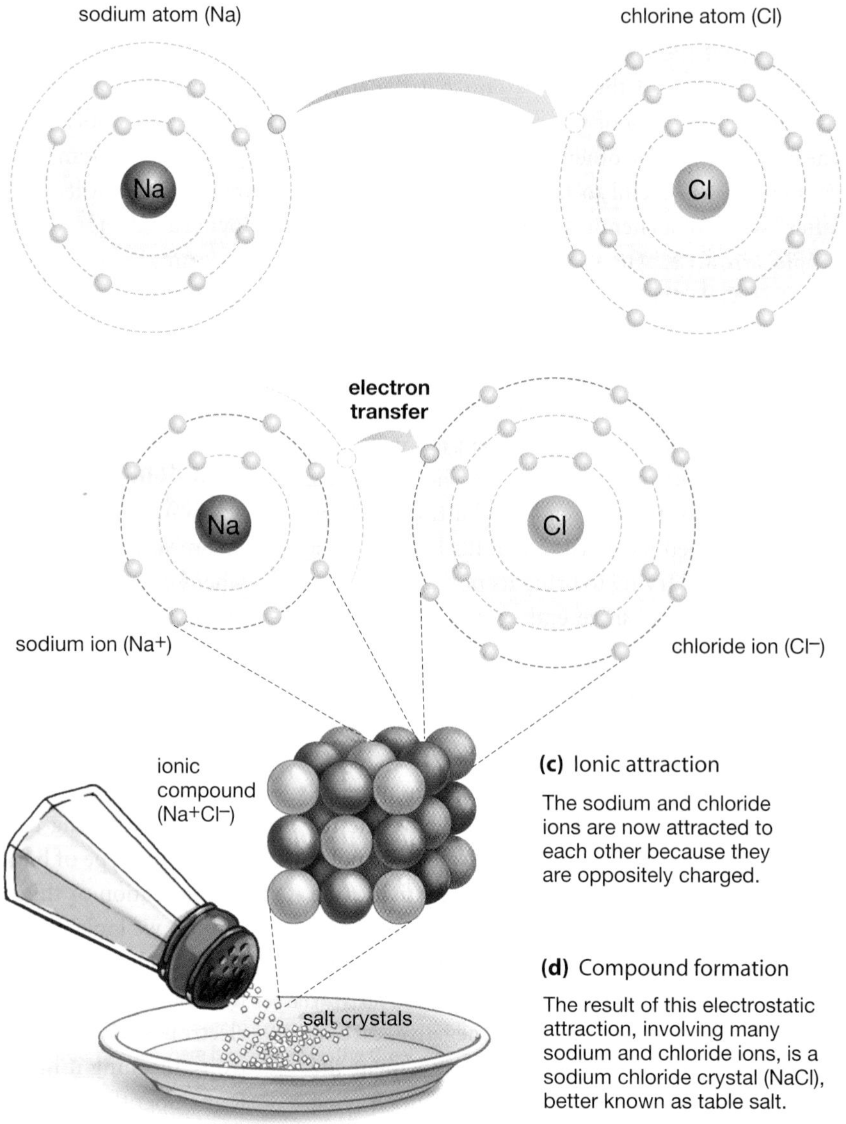

(a) Initial instability

Sodium has but a single electron in its outer shell, while chlorine has seven, meaning it lacks only a single electron to have a completed outer shell.

sodium atom (Na)

chlorine atom (Cl)

electron transfer

(b) Electron transfer

When these two atoms come together, sodium loses its third-shell electron to chlorine, in the process becoming a sodium ion with a net positive charge (because it now has more protons than electrons.) Having gained an electron, the chlorine atom becomes a chloride ion, with a net negative charge (because it has more electrons than protons).

sodium ion (Na+)

chloride ion (Cl–)

ionic compound (Na+Cl–)

(c) Ionic attraction

The sodium and chloride ions are now attracted to each other because they are oppositely charged.

salt crystals

(d) Compound formation

The result of this electrostatic attraction, involving many sodium and chloride ions, is a sodium chloride crystal (NaCl), better known as table salt.

Is it apparent how an ionic compound differs from a molecule? In an ionic compound, there is no fixed number of atoms linked up in a defined spatial relationship, as in H_2O. Rather, an undefined number of charged atoms are bonded together, as in NaCl.

To take a step back for a second, recall that bonding runs a gamut from the nonpolar covalent bonding (where electrons are shared equally) to the slightly charged polar covalent bonding (where they are shared somewhat unequally) to the charged ionic bonding (where electrons are gained or lost altogether). It's important to recognize that there is a *spectrum of polarity,* and that within it, some bonds are almost completely ionic (as with sodium chloride) while others are completely nonpolar (as with the hydrogen molecule).

A Third Form of Bonding: Hydrogen Bonding

We need to look at one more variant on bonding, called hydrogen bonding. Recall that in any water molecule, the stronger electronegativity of the oxygen atom pulls the electrons shared with the hydrogen atoms toward the oxygen nucleus, giving the oxygen end of the molecule a partial negative charge and the hydrogen end of the molecule a partial positive charge. So what happens when you place several water molecules together? A positive hydrogen atom of one molecule is weakly attracted to the negative, *unshared* electrons of its oxygen neighbor. Thus is created the **hydrogen bond**, which links an already covalently bonded hydrogen atom with an electronegative atom (in this case with oxygen; **see Figure 2.12**). Hydrogen bonding is a linkage that, for our purposes, nearly always pairs hydrogen with either oxygen or nitrogen. These relatively weak bonds are important in linking the atoms of a single molecule to one another, but they are just as important in creating bonds *between* molecules, as in the example. The hydrogen bond, indicated by a dotted line, exists in many of the molecules of life—in DNA, proteins, and elsewhere.

2.3   Some Qualities of Chemical Compounds

Molecules Have a Three-Dimensional Shape

We now need to make more explicit what has been noted only by implication in our diagrams of water molecules: that molecules and ionic compounds have a three-dimensional shape. It is useful to depict them as two-dimensional chains and rings and such, but in real life a molecule is as three-dimensional as a sculpture. A fair number of shapes are

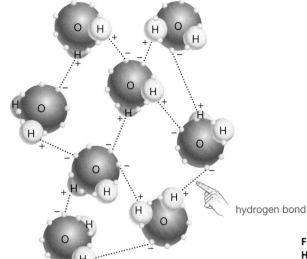

possible, even in simpler molecules. Atoms may be lined up in a row, or in triangles or pyramid shapes. As an example, look at another representation of the water and methane molecules (**see Figure 2.13**). You can see that in water there is a very definite spatial configuration: Its hydrogen atoms are splayed out from its oxygen atom at an angle of 104.5°.

Molecular Shape Is Very Important in Biology

Why does molecular shape matter? It is critical in enabling biological molecules to carry out the activities they do. This is so because molecular shape determines the capacity of molecules to latch onto or "bind" with one another. When, for example, you smell the aroma of fresh-baked bread, gas molecules wafting off the bread bind with receptor molecules in your nasal passages, thus sending a message to the brain about the presence of bread. It is the precise shape of the gas

Figure 2.12
Hydrogen Bonding
The hydrogen bond, in this case between water molecules, is indicated by the dotted line. It exists because of the attraction between hydrogen atoms, with their partial positive charge, and the unshared electrons of the oxygen atom, with their partial negative charge.

Web Tutorial 2.3
Geometry, Chemistry and Biology

Figure 2.13
Three-Dimensional Representations of Molecules

(a) In the case of water there is an angle of 104.5° between hydrogen atoms.

(b) Methane is a molecule with an angle of 109.5° between hydrogen atoms.

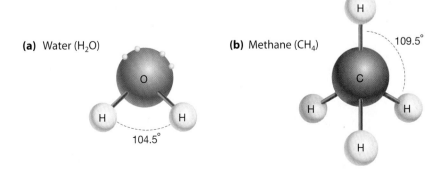

(a) Water (H_2O)   104.5°

(b) Methane (CH_4)   109.5°

Free Radicals

Their name makes them sound like a group of sixties activists set loose after years in jail, but **free radicals** aren't people at all; they are atoms or molecules. They *have* been set loose, however—to damage human bodies in illnesses that may range from cancer to coronary heart disease.

The way the public normally hears about these culprits is through the recommended means of limiting their harm: a diet rich in vitamins C, E, and the substance beta carotene. Looking at free radicals from another angle, however, we can see how they represent a damaging exception to the rules of chemical bonding.

You know that atoms "seek" to have a full outer energy shell, which in most cases means eight electrons. In covalent bonding, atoms achieve this state by sharing *pairs* of electrons with one another, one electron of each pair coming from each of the atoms

involved. Occasionally in nature, however, atoms come together to create a molecule in which one of the component atoms has an *unpaired electron* in its outer shell. Nitrogen can come together with oxygen, for example, to form nitric oxide (NO). Recall that oxygen has six outer electrons (and thus needs two for stability), while nitrogen has five outer electrons (and thus needs three). When oxygen and nitrogen hook up, they can share only two electrons with one another before oxygen's outer shell is filled. This, however, leaves nitrogen with an unpaired electron.

unpaired electron

$\cdot\ddot{N}:: \ +\ :\ddot{O}: \longrightarrow \ \cdot\ddot{N}::\ddot{O}:$

nitrogen   oxygen   nitric
                        oxide

free radical

Figure 2.14
The Importance of Shape
Gas molecules wafting off from bread (the triangles) bind with specific receptors on the surface of the cell, thus acting as signaling molecules that set a cellular process in motion. For this binding to take place, the gas and nasal receptor molecules must fit together; this fit is governed by the shape of each molecule.

molecules and nasal receptor molecules that allows them to bind with one another. Look at **Figure 2.14** to see how this works. If you look at **Figure 2.15**, you can get an idea of how large some biological molecules are, relative to the simple molecules considered so far.

Having learned a little about atoms and molecules, we now need to review a few final concepts that will aid us in understanding chemistry on a slightly larger scale.

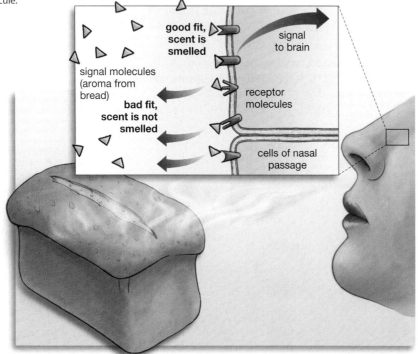

Solutes, Solvents, and Solutions

Take a glass of water and pour a little salt in it. When you stir that up, the salt quickly disappears. It has not actually gone anywhere, of course. It has simply mixed with the water. Now, if it has mixed uniformly, so that there are no lumps of salt here or there, you have created a **solution**—a homogeneous mixture of two or more kinds of molecules, atoms, or ions. The salt is what's being dissolved here, so it is the **solute**. The water is doing the dissolving, so it is the **solvent** (**see Figure 2.16**).

Molecules vary greatly in the degree to which they are *soluble*—are able to be dissolved—in different solvents. This is because, for something to act as a solvent, it must be able to form chemical bonds with the solute. The more bonding a solvent can do with a solute, the greater capacity that solvent will have to break down the solute.

A general rule is that like dissolves like. Substances that are nonpolar dissolve best in nonpolar solvents; substances that are polar dissolve best in polar solvents. Salt and water are both polar, and the one dissolves in the other. Conversely, the ingredient in soap that actually breaks up nonpolar greases and dirt is a long, nonpolar molecular chain composed of hydrogen and carbon atoms.

On to Some Detail Regarding Water

In biology, you will often see casual references to a molecule being *soluble* in a certain way. And you

Unstable molecules like this usually exist only briefly, as intermediate molecules in chemical reactions. And in people, that's just the problem. Human beings are among the many species that use a terrific amount of oxygen to extract energy from food. In this process, oxygen is constantly picking up electrons. Through the many steps in metabolism, oxygen may come together with other substances to create a type of free radical called reactive oxygen. Though any one reactive oxygen molecule is short-lived, the damage comes by way of a destructive chain reaction: Seeking partners, one free radical begets more, which beget more.

Where's the harm? Well, for one thing, free radicals may irritate or scar artery walls, which invites artery-clogging fatty deposits around the damage. They also may have a mutation-causing or "mutagenic" effect on human DNA, which can be a factor leading to cancer. Some primary sites of free-radical generation and damage are the "powerhouses" in cells—structures called mitochondria—which are primary sites at which energy is extracted from food. Indeed, a growing body of evidence supports a long-standing theory of human aging, which holds that many of the things we associate with getting older—memory loss, hearing impairment—can be traced to the cumulative effects of free radicals damaging DNA in the mitochondria, thus diminishing the body's energy supply.

Free radicals are the natural product of metabolism in human beings; they are the price we pay for being alive. However, they can be created in us in *greater* numbers in accordance with our behavior. Some of the usual suspects seem to be involved here—cigarette smoking, alcohol consumption, and sunlight exposure. Radiation provides a good example of how free radicals can be produced in us. Medical workers guard against getting excessive exposure to X-rays, because, like other forms of "ionizing" radiation, X-rays cause water molecules in living tissue to break down in such a way that they yield free radicals.

Against this production of free radicals, however, nature has also provided its own set of free-radical scavengers, among them the aforementioned beta carotene and vitamins C and E. We can control these, in the sense of making sure we have plenty of them in our diets. We can buy them in pill form, of course, but the jury is still out on the effect of very large doses of these substances. The best bet is to eat the right kinds of foods—meaning a lot of citrus fruits, whole grains, and vegetables of the green leafy, orange, and yellow variety.

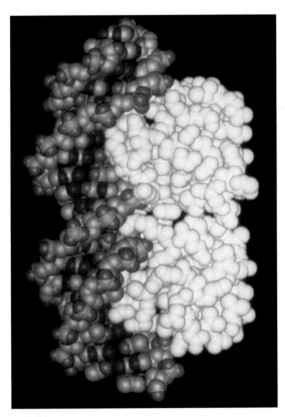

Figure 2.15
Complex Binding
A computer model of some real-life molecular binding. In this case a protein (the yellow atoms) binds to a length of DNA. (The protein is a "repressor" that turns off the activity of this section of DNA.) This space-filling model provides some idea of the enormous number of atoms and the complicated shapes that make up some of the molecules employed by living things.

will see this most commonly mentioned in two ways: as something being fat-soluble or water-soluble. Does it dissolve in fat or does it dissolve in water? Both things are important to life, but the importance of water as a solvent cannot be overstated. Water accounts for about 75–85 percent of a cell's weight, and most cells are surrounded by it. It is not difficult to see why water should be so prominent in this way. Living things existed solely in water for billions of years, much life lives in it today, and our own ancestors made the transition from sea to land essentially by carrying their wet environment with them—inside themselves. So important to life is this simple, common molecule that we end with it here, only to take it up again in our next chapter in a more detailed way.

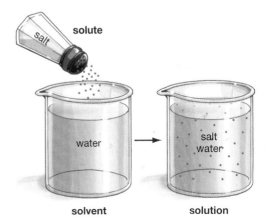

Figure 2.16
A Solute Dissolved by a Solvent Makes a Solution
When we pour a small amount of table salt (a solute) into water (a solvent), the salt crystals dissolve into the water, yielding a solution.

Chapter Review

Summary

2.1 The Nature of Matter: The Atom

- The fundamental unit of matter is the atom. The three most important constituent parts of an atom are protons, neutrons, and electrons. Protons and neutrons exist in the atom's nucleus, while electrons move around the nucleus, at some distance from it. Protons are positively charged, electrons are negatively charged, but neutrons carry no charge.

- An element is any substance that cannot be reduced to any simpler set of constituent substances through chemical means. Each element is defined by the number of protons in its nucleus.

- The number of neutrons in an atom can vary independently of the number of protons. Thus a single element can exist in various forms, called isotopes, depending on the number of neutrons it possesses.
  Web Tutorial 2.1: Structure of Atoms, Elements, Isotopes

2.2 Matter Is Transformed through Chemical Bonding

- Atoms can link to one another in the process of chemical bonding. Among the forms this bonding can take are covalent bonding, in which atoms share one or more electrons, and ionic bonding, in which atoms lose and accept electrons from each other.

- Chemical bonding comes about as atoms "seek" their lowest energy state. An atom achieves this state when it has a filled outer electron shell. Hydrogen and helium require two electrons in orbit around their nuclei to have filled outer shells, while most other elements require eight electrons to have filled outer shells.

- A molecule is a compound of a defined number of atoms in a defined spatial relationship. For example, two hydrogen atoms can link with one oxygen atom to form one water molecule.

- Atoms of different elements differ in their power to attract electrons. The term for measuring this power is electronegativity. Through electronegativity, a molecule can take on a polarity, meaning a difference in electrical charge at one end compared to the other. Covalent chemical bonds can be polar or nonpolar. A polar covalent bond exists when shared electrons are not being shared equally among atoms in a molecule, due to electronegativity differences.

- Two atoms will undergo a process of ionization when the electronegativity differences between them are great enough that one atom loses one or more electrons to the other. This process creates ions, meaning atoms whose number of electrons differs from their number of protons. The charge differences that result from ionization can produce an electrostatic attraction between ions. This attraction is an ionic bond. When atoms of two or more elements bond together ionically, the result is an ionic compound.

- Hydrogen bonding links a covalently bonded hydrogen atom with an electronegative atom. In water, a hydrogen atom of one water molecule will form a hydrogen bond with an unshared oxygen electron of a neighboring water molecule.
  Web Tutorial 2.2: Chemical Bonding

2.3 Some Qualities of Chemical Compounds

- Three-dimensional molecular shape is important in biology because this shape determines the capacity molecules have to bind with one another.
  Web Tutorial 2.3: Geometry, Chemistry and Biology

- A solution is a homogeneous mixture of two or more kinds of molecules, atoms, or ions. The compound being dissolved in solution is the solute; the compound doing the dissolving is the solvent.

- A general rule in chemistry is that like dissolves like: Substances that are nonpolar dissolve best in nonpolar solvents, while substances that are polar dissolve best in polar solvents.

Key Terms

atomic number 22	molecular formula 26
ball-and-stick model 27	molecule 25
chemical bonding 23	neutron 20
covalent bond 24	nonpolar covalent bond 25
double bond 26	nucleus 20
electron 20	polar covalent bond 25
electronegativity 25	polarity 25
element 21	product 27
free radical 30	proton 20
hydrogen bond 29	reactant 27
ion 28	single bond 26
ionic bonding 27	solute 30
ionic compound 28	solution 30
isotope 23	solvent 30
law of conservation of mass 25	space-filling model 27
	structural formula 26
mass 19	triple bond 26

Understanding the Basics

Multiple-Choice Questions (answers in the back of the book)

1. Carbon is an element with an atomic number of 6. Based on this information, which of the following statements is true? (More than one may be true.)
   a. Carbon can be broken down into simpler component substances.

b. Carbon cannot be broken down into simpler component substances.

c. Each carbon atom will always have 6 neutrons.

d. Each carbon atom will always have 6 protons.

e. Protons + electrons = 6.

2. Suppose that you are reviewing for a test, and some fellow students say that the equation for photosynthesis is $6CO_2 + H_2O \rightarrow C_6H_{12}O_6 + 6O_2$. How would you reply?

a. They are right.

b. The way this equation is written violates the law of the conservation of matter.

c. The CO_2 is held together by ionic bonds.

d. A hydrogen bond holds the two hydrogens to the oxygen in the water molecule.

e. There are 12 carbons in sugar.

3. Neon used to be called an inert gas. Thus it

a. easily forms perfect covalent bonds

b. easily forms ionic bonds

c. has a filled outer shell

d. is polar

e. all of these

4. Oxygen and hydrogen differ in their electronegativity. Thus

a. They can share electrons, but unequally.

b. Sometimes oxygen takes electrons completely away from hydrogen.

c. They can share electrons equally.

d. Hydrogen is attracted to oxygen, but does not bond with it.

e. They have the same number of protons.

5. A molecule that does not have a net electrical charge at one end as opposed to the other is:

a. an isotope

b. a polar molecule

c. a reactant

d. a nonpolar molecule

e. a solvent

6. You add sugar to black coffee, and the sugar dissolves. Thus the coffee is the _____ and the sugar is the _____.

a. solute … solvent

b. solvent … solute

c. polar covalent bond … nonpolar covalent bond

d. nonpolar covalent bond … polar covalent bond

e. ionic bond … hydrogen bond

7. The two strands of a DNA molecule are held together because large numbers of hydrogens that are covalently bonded to oxygen or nitrogen in one strand are weakly attracted to oxygens or nitrogens in the opposite strand. Therefore the two strands of DNA are held together by

a. polar covalent bonds

b. nonpolar covalent bonds

c. ionic bonds

d. inert bonds

e. hydrogen bonds

8. While baking cookies for a friend, you're having a hard time keeping your roommate from eating them because the smell is driving her wild. She is able to smell the cookies because the molecules wafting from them

a. have a shape that allows them to bind to receptors in her nose

b. form permanent attractions to nerves in her nose by ionic bonding

c. form temporary attractions to nerves in her nose by covalent bonding

d. dissolve in her blood and travel to her brain

e. travel along her nerves—causing changes as they go—until they reach the brain, where they are recognized

Brief Review

1. As with most elements, carbon comes in several forms, one of which is carbon-14. What are these forms called, and how does one differ from the other?

2. Draw a line and label one end "complete + or − charge" and the other end "no charge" to indicate the charges on the molecules or ions after bonding has occurred. Along the line, indicate where polar covalent bonds, nonpolar covalent bonds, and ionic bonds should be placed.

3. Why are free radicals so dangerous when they are produced in our bodies?

4. Compare the size of an atom with the size of its nucleus. Where are the electrons? In light of this, what occupies most of an atom?

5. Why are atoms unlikely to react when they have their outer shell filled with electrons?

Applying Your Knowledge

1. In the Middle Ages alchemists labored to turn common materials such as iron into precious metals such as gold. If you could journey back in time, how could you convince an alchemist that iron cannot be changed into gold?

2. Why does a balloon filled with helium float? Hydrogen can make balloons float, but it is not used for this purpose today, because it is flammable. Based on chemical principles reviewed in the chapter, can you see why helium is not flammable? (*Hint:* Think what you are adding to a fire when you blow on it.)

3

Water, pH, and Biological Molecules

Water's cohesive power.
(Section 3.1, page 37)

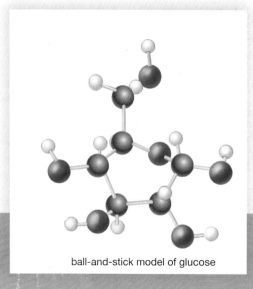

ball-and-stick model of glucose

The most important energy source for our bodies.
(Section 3.3, page 45)

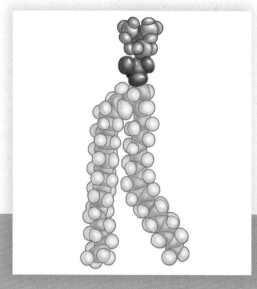

A dual-natured molecule with head and tails.
(Section 3.6, page 55)

Imagine that it's raining outside as you read this, the rhythm of individual drops combining to create a sound that is like no other. It's a sound that most people find comforting. And isn't that the way with other sounds water makes when it's in motion? With ocean waves or a running stream? It may be that these things are soothing purely *as sounds*—something about their tone, or their steady pace. But isn't it possible that we are calmed by the sounds of rain or streams because they are the sounds *of water*? Rain has fallen on leaves for millennia, sending our ancestors—and now us—a timeless message: Water's here; you and your family will have enough to drink; food will be abundant; all is well.

3.1 The Importance of Water to Life

Human societies tend to come together precisely where water exists, of course. In places where it's plentiful, water seems less like a substance than an environment: People drink it, cook in it, bathe in it, wash wastes away in it, harness it for power, swim in it. Some 71 percent of Earth's surface is ocean water, and human bodies are about 66 percent water by weight, so that if we have, say, a 128-pound person, about 85 pounds of him or her will be water. If Earth amounts to a watery environment flecked by the landmasses we call continents, the human body amounts to a watery mass with significant proportions of other materials immersed in it.

Water Is a Major Player in Many of Life's Processes

This preponderance of water in living things gives rise to an important point. Recall that you learned, in Chapter 2, what a solution is. Well, a specific kind of solution—a so-called aqueous solution—is one in which water is the solvent, acting to break down the materials or *solutes* that have been put into it. If

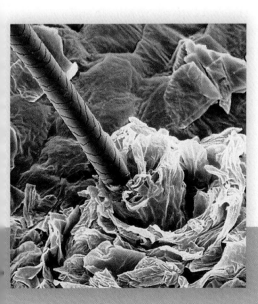

A protein found over almost all of your body.
(Section 3.7, page 58)

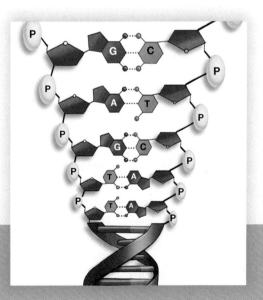

The building blocks of DNA.
(Section 3.8, page 61)

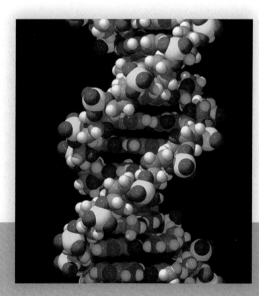

A portrait of the most famous molecule in all of biology.
(Section 3.8, page 61)

Figure 3.1
Water's Power as a Solvent

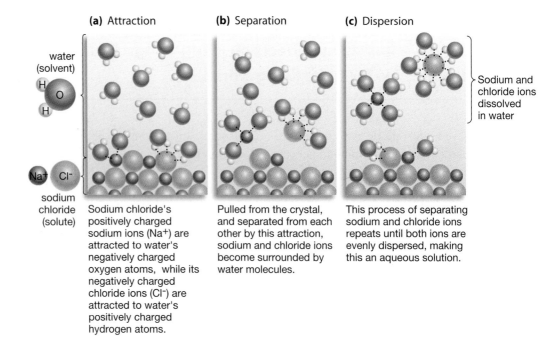

(a) Attraction   **(b)** Separation   **(c)** Dispersion

water
(solvent)

H   O   H

Na⁺   Cl⁻

sodium
chloride
(solute)

Sodium chloride's
positively charged
sodium ions (Na⁺) are
attracted to water's
negatively charged
oxygen atoms, while its
negatively charged
chloride ions (Cl⁻) are
attracted to water's
positively charged
hydrogen atoms.

Pulled from the crystal,
and separated from each
other by this attraction,
sodium and chloride ions
become surrounded by
water molecules.

This process of separating
sodium and chloride ions
repeats until both ions are
evenly dispersed, making
this an aqueous solution.

Sodium and
chloride ions
dissolved
in water

Web Tutorial 3.1
Chemistry and Water

you look at **Figure 3.1**, you can get a detailed view of water's solvent power in connection with sodium and chloride ions. Attracted by the polar nature of the water molecule, these ions separate from a crystal—and from each other. Each ion is then surrounded by several water molecules (Figure 3.1b). These units keep the sodium and chloride ions from getting back together. In other words, they keep the ions evenly dispersed throughout the water, which is what makes this a solution (Figure 3.1c). Water works as a solvent here because the ionic compound sodium chloride carries an electrical charge. What *generally* makes water work as a solvent, however, is its ability to form hydrogen bonds with other molecules.

Water's Structure Gives It Many Unusual Properties

When we note water's ability to act as a solvent, we're actually not giving water its due. Water is not just *a* solvent: Over the range of substances, nothing can match it as a solvent. It can dissolve more compounds in greater amounts than can any other liquid.

Ice Floats Because It Is Less Dense than Water

But solvency power is merely the beginning of water's abilities. It is a multitalented performer. And it achieves this status because it is . . . odd. Compared to other molecules, water is like some zany eccentric whose powers stem precisely from its eccentricity. Consider the fact that ice floats on water. This is so because the solid form of H₂O is less dense than the liquid form—a strange reversal of nature's normal pattern. Things work this way with H₂O because, when water molecules slow their motion in cooling,

they are able to form the maximum number of hydrogen bonds with each other. The result is that water molecules are spaced *farther apart* when frozen. Thus, ice is less dense than water. This may seem like some minor, quirky quality but it actually has the effect of making possible life as we know it. Ice on the surface of water acts to insulate the water beneath it from the freezing surface temperatures and wind above, creating a warmer environment for organisms such as fish (**see Figure 3.2**). If ice *sank*, on the other hand, the entire body of water would freeze solid at colder latitudes, creating an environment in which few living things could survive for long.

Water Has a Great Capacity to Absorb and Store Heat

Water serves as an insulator not only when it is frozen, but also when it is liquid or gas. This has to do with a quality called **specific heat**: the amount of energy required to raise the temperature of a substance by 1 degree Celsius. As it turns out, water has a high specific heat. Put a gram container of drinking alcohol (ethyl alcohol) side by side with one of water, heat them both, and it will take almost twice as much energy to raise the temperature of the water one degree as it will the alcohol. Having absorbed this much heat, however, water then has the capacity to *release* it when the environment around it is colder than the water itself. The result? Water acts as a great heat buffer for Earth. The oceans absorb tremendous amounts of radiant energy from the Sun, only to release this heat when the temperature of the air above the ocean gets colder. Without this buffering, temperature on Earth would be less stable. People who have spent a day and a night in a

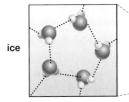

In ice, the maximum number
of hydrogen bonds form,
causing the molecules to be
spread far apart.

ice

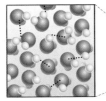

**liquid
water**

In liquid water, hydrogen
bonds constantly break and
reform, enabling a more
dense spacing than in ice.

**Figure 3.2
Life Under the Ice**
Water is unusual in that its solid
form (ice) is less dense than its liquid form. This means that ice floats,
and this in turn means that life can
flourish in cold-weather aquatic
environments. Pictured are some
shrimp-like krill, feeding on algae
that grow on the underside of ice
in the waters off Antarctica.

desert can attest to this effect. The searing heat of the desert day radiates off the desert floor; but at night, with little water vapor in the air to capture this heat, the desert cools off dramatically. In the same way, our *internal* temperature can remain much more stable be-cause the water that makes up so much of us is first able to absorb and then to release great amounts of heat. The sweat that we throw off in exercise has con-siderable cooling power because each drop of perspi-ration carries with it a great deal of heat.

Water derives these powers from its chemical structure, in particular from the hydrogen bonding described earlier. Weak and shifting though individ-ual hydrogen bonds may be, collectively they have a great strength. Heat is the motion of molecules. To get molecules moving, though, chemical bonds must first be broken. Because water has a formidable set of hydrogen bonds, it takes a lot of energy to break them and get its molecules moving. This very same set of bonds gives water molecules another notable characteristic: cohesion, meaning a tendency to stay together. With this quality, a chain of water mole-cules can be drawn (by evaporation) in one continu-ous column from a plant's roots all the way up to its leaves and out into the air as water vapor.

Water's Cohesion Gives It Surface Tension

Water's cohesion imparts a special quality, called *surface tension,* in places where water meets air. Water molecules below the surface are equally at-tracted in all directions to other water molecules. *At* the surface, however, water molecules have no such attraction to the air above them—they are pulled

down and to the side, but not up. This causes the "beading" that water droplets do on surfaces. More important for our purposes, surface water mole-cules pack together more closely than do interior molecules, allowing all kinds of small animals to move across the surface of water rather than sinking into it. Note the familiar water strider in **Figure 3.3**.

(a) Walking on water

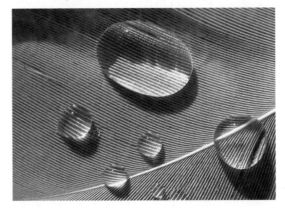

(b) Beading up

**Figure 3.3
Walking on Water**

(a) Water's high surface tension
enables small animals, such as this
pond skater, to walk on it.

(b) This same surface tension also
causes water to bead up on waxy
or oily surfaces, such as this goose
feather.

Having looked at the qualities of water, you may be tempted to think how *uncanny* it is that water has all these qualities that are conducive to life. But there actually is nothing surprising about this. Life went on solely in water for billions of years before living things ever came onto land. Thus, water has fundamentally conditioned life as we know it. Being surprised about water's life-enhancing qualities is like being surprised that a stage is a good place to put on a play.

What Water Cannot Do

With its great complexity, life requires molecules that *can't* be dissolved by water. Such is the case with nonpolar covalent molecules. Compounds made of hydrogen and carbon (**hydrocarbons**) are good examples of nonpolar molecules. Recall that Chapter 2 presented methane (CH_4) as an example of a nonpolar molecule. Petroleum products are more complex hydrocarbons than is methane, and you can see a vivid demonstration of their water-insolubility in oil spills (**see Figure 3.4**). Oil doesn't dissolve in water because the oil carries almost no electrical charge that water can bond with; thus water has no way to separate one oil molecule from another.

Two Important Terms: Hydrophobic and Hydrophilic

The ability of molecules to form bonds with water has a couple of important names attached to it. Compounds that will interact with water—such as the sodium chloride considered earlier—are known as **hydrophilic** ("water-loving"), while compounds that do not interact with water are known as **hydrophobic** ("water-fearing"). Both terms are mislead-

ing, in that no substance has any emotional relationship with water. *Hydrophobic* is particularly off the mark, because water does not repel hydrophobic molecules; instead, the strong bonding that water molecules do with each other causes them to form circles around concentrations of hydrophobic molecules, as if they had lassoed them.

The importance of hydrophobic molecules can be illustrated in part by the common milk carton. Why is the milk carton important? Because it can keep milk separate from everything else. We living organisms need some kind of "carton" that can separate the world outside of us from ourselves. Likewise, organisms have great use *within* themselves for compartments that can be sealed off to one degree or another. If water broke down every molecule of life it came in contact with, then it would break down all these divisions of living systems. Molecules do not have to be completely hydrophilic or hydrophobic. Indeed, as you'll see, a number of important molecules have both hydrophilic and hydrophobic portions.

3.2 Acids and Bases Are Important to Life

When considering the question of aqueous solutions, an important concept is that of acids, bases, and the pH scale used to measure their levels.

We've all had experience with acids and bases, whether we've called them by these names or not. Acidic substances tend to be a little more familiar: lemon juice, vinegar, tomatoes. Substances that are strongly acidic have a well-deserved reputation for being dangerous: The word *acid* is often used to mean something that can sear human flesh. It might seem to follow that bases are benign, but ammonia is a strong base, as are many oven cleaners. The safe zone for living tissue in general lies with substances that are neither strongly acidic nor strongly basic. Science has developed a way of measuring the degree to which something is acidic or basic—the pH scale. So widespread is pH usage that it pops up from time to time in television advertising ("It's pH-balanced!").

Acids Yield Hydrogen Ions in Solution; Bases Accept Them

The "H" in pH stands for hydrogen, while the "p" can be thought of as standing for power. Thus we get "hydrogen power," which describes what lies at the root of pH. An **acid** is any substance that *yields hydrogen ions* when put in aqueous solution. A **base** is any substance that *accepts* hydrogen ions in solution.

How might this yielding or accepting come about? Recall first that an ion is a charged atom, and that atoms become charged through the gain or loss

Figure 3.4
Oil and Water Do Not Mix
When there is an oil spill in the ocean, the oil stays concentrated even as it spreads out, because oil and water do not form chemical bonds with each other. Here trawlers are using a boom to clean up after an oil spill in Great Britain.

of one or more electrons. Because electrons carry a negative charge, the loss of an electron leaves an atom with a net *positive* charge. Also recall that the hydrogen atom amounts to one central proton and one electron that circles around it. A hydrogen ion, then, is a lone proton that has lost its electron—a positively charged ion whose symbol is H^+.

Now, suppose you put an acid—hydrochloric acid (HCl)—into some water. What happens is that HCl *dissociates* or breaks apart into its ionic components, H^+ and Cl^-. The HCl has therefore yielded a hydrogen ion (H^+). Now, with a greater concentration of hydrogen ions in it, the water is more acidic than it was (**see Figure 3.5**).

**Figure 3.5
Hydrogen Ions and pH**

(a) Starting with pure water

Pure water is a "neutral" substance in terms of its pH levels.

pure water
(H_2O)

HCl

(b) Making water more acidic

Hydrochloric acid (HCl), poured into the water, dissociates into H^+ and Cl^- ions. With a higher concentration of H^+ ions in it, the water moves towards the acidic end of the pH scale.

acid

NaOH

(c) Making water more basic

An equal concentration of sodium hydroxide, poured into water, dissociates into Na^+ and OH^- ions, moving the water toward the basic end of the scale.

base

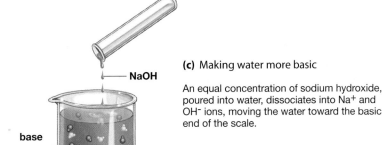

$H_2O \rightarrow H^+ + Cl^-$ $NaOH \rightarrow Na^+ + OH^-$

(d) Combining acidic and basic solutions

When the acid and base solutions are poured together, the OH^- ions from **(c)** accept the H^+ ions from **(b)**, forming water and keeping the solution at a neutral pH.

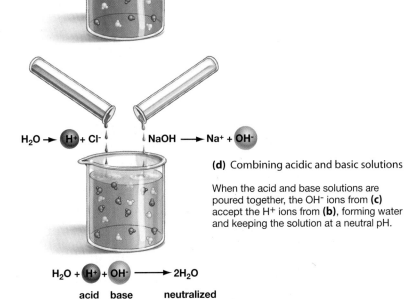

$$H_2O + H^+ + OH^- \longrightarrow 2H_2O$$

acid   base   neutralized
                solution

What about bases? There is a compound called sodium hydroxide (NaOH)—better known as lye—that, when poured into water, dissociates into Na^+ (sodium) and OH^- (hydroxide) ions. The place to look here is the OH^- ions. Negatively charged as they are, they would readily bond with positively charged H^+ ions. In other words, they would *accept* H^+ ions in solution, which is the definition of a base. This accepting of the H^+ ions makes the solution more basic—or, to look at it another way, *less acidic.* Thus, acids and bases are something like a teeter-totter: When one goes up, the other comes down. As you might have guessed, in the right proportions they can balance each other out perfectly. Look at Figure 3.5 to see how this would play out with the solutions you've looked at so far. Should H^+ and OH^- ions be poured together into water, for each pair of ions that interacts, the result is:

$$H^+ + OH^- \longrightarrow H_2O$$

Water, which is neutral on the pH scale. The acid and the base have perfectly balanced one another out. The OH^- ion, generally referred to as the **hydroxide ion**, is important because compounds that yield them in quantity are strongly basic and can be used to move solutions from the acidic toward the basic.

Many Common Substances Can Be Ranked According to How Acidic or Basic They Are

Look at **Figure 3.6** to get an idea of how acidic or basic some common substances are. Following from the notion of what pH amounts to, it's clear that battery acid, for example, is strongly acidic because, when it dissociates in solution, it yields a large number of hydrogen ions. As you move on to lemon juice and then tomatoes, however, you run into weaker acids, which is to say substances that yield fewer H^+ ions in solution. By the time you get to seawater, you've arrived at substances that *accept* hydrogen ions.

The pH Scale Allows Us to Quantify How Acidic or Basic Compounds Are

The net effect of all this yielding and accepting of hydrogen ions is the concentration of H^+ ions in solution. Through this concentration, the notion of pH can be *quantified*—can have numbers attached to it. What is employed here is the **pH scale**: a scale utilized in measuring the relative acidity of a substance. Look at Figure 3.6 again, this time in connection with the numbers that mark it off. You can see that zero on the scale is the most acidic while 14 is the most basic. It's important to note that the pH scale is *logarithmic:* A substance with a pH of 9 is 10 times as basic as a substance with a pH of 8 and 100 times as basic as a substance with a pH of 7.

Some Terms Used When Dealing with pH

Here are some notes on pH terminology:

- As a solution becomes more basic, its pH *rises.* Thus the higher the pH, the more basic the solution; and the lower the pH, the more acidic the solution. Oven cleaner is said to have a high pH, while lemon juice has a low pH.

- Given that a hydrogen ion amounts to a single proton, it is also correct to say that an acid is something that yields *protons* in solution, while a base is something that accepts protons. This is how you will often hear hydrogen ions talked about in biology.

- A solution that is basic is also referred to as an **alkaline** solution.

Why Does pH Matter?

So, why do we care about pH? The brief answer is: because living things are sensitive to its levels in many ways. There is, for example, a class of proteins called enzymes that you'll be reading about later in this chapter. Enzymes are chemical tools that must retain a very specific shape to function. However, if you put an enzyme in a solution whose pH is too acidic, the enzyme loses it shape. Why? Because the charged nature of the acidic solution starts breaking down the enzyme's chemical bonds. (Remember, lots of positively charged protons are floating around in an acidic solution.) Likewise, cell membranes—which serve the milk carton function noted earlier—can start to break down if pH levels start to go outside normal limits. Membranes and enzymes are so fundamental to life that when they are interfered with, death can result. It is not surprising, then, that many organisms have developed so-called acid-base **buffering systems**, meaning physiological systems that function to keep pH within normal limits. What are these limits? The usual range for living things is about 6–8, with the pH of the human cell being about 7 and that of the blood in our arteries being about 7.4. To break down food, our stomachs receive a digestive acid that gives the interior of the stomach a pH that lies below 2, and tiny digestive structures within cells called lysosomes have a low pH. But lysosomes and the stomach are sealed containers, in a sense, in that their contents don't mix with bodily fluids in general.

The molecules that function in buffering systems generally are weak acids or bases that work to neutralize any sudden infusion of acid or base. The challenge for human pH buffering systems usually is to keep our fluids from becoming too acidic, since the food we eat tends to pull our fluids toward the acidic side of the pH scale. Because the consequences of pH imbalance are so extreme, however, the body also has what might

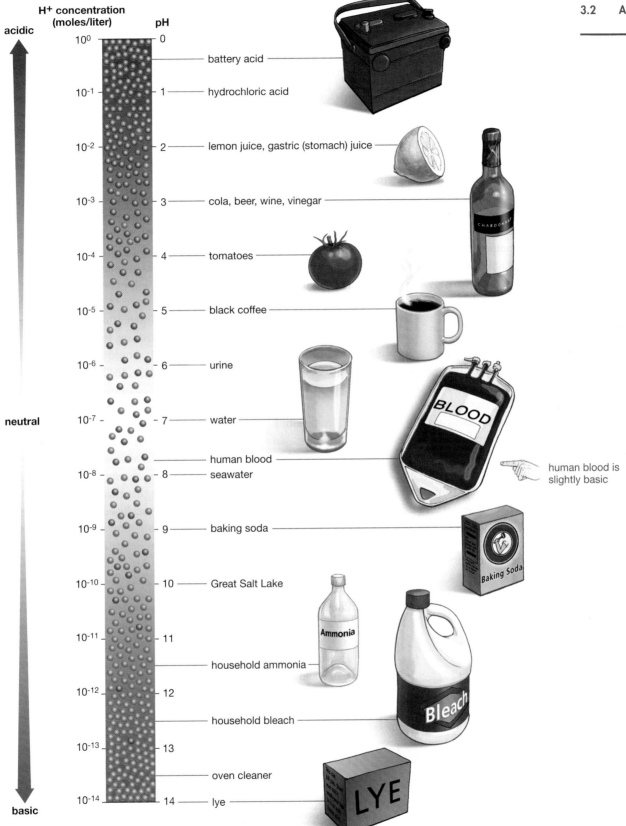

Figure 3.6
Common Substances and the pH Scale
Chemists use units called moles per liter to measure the concentration of substances in solution. The pH scale, derived from this framework, measures the concentration of hydrogen ions per liter of solution. The most acidic substances on the scale have the greatest concentration of hydrogen ions, while the most basic (or alkaline) substances have the least concentration of hydrogen ions. The scale is logarithmic, so that wine, for example, is 10 times as acidic as tomatoes, and 100 times as acidic as black coffee.

How does carbon link up with itself or other atoms? You've already had a look at a model of a very simple carbon compound, methane (CH_4). Look at it again:

```
      H
      |
  H — C — H
      |
      H
```

Note that the only elements present are hydrogen and carbon atoms. Now observe a slightly more complex hydrocarbon, the familiar gas propane (C_3H_8):

```
      H   H   H
      |   |   |
  H — C — C — C — H
      |   |   |
      H   H   H
```

You can see that instead of being surrounded by four hydrogen atoms, the carbons here link up with other carbons, as well as with the hydrogens. This process of extension keeps going with still more complex hydrocarbons. For obvious reasons, the preceding configuration is known as a *straight-chain* carbon molecule. But carbon has more tricks up its sleeve than simple straight-line extensions. The next hydrocarbon up the line is butane (C_4H_{10}), the fuel found in cigarette lighters. It can be just another straight-chain extension, looking like this:

```
      H   H   H   H
      |   |   |   |
  H — C — C — C — C — H
      |   |   |   |
      H   H   H   H
```

But it can also look like this:

```
      H   H   H
      |   |   |
  H — C — C — C — H
      |   |   |
      H   H   H
          |
      H — C — H
          |
          H
```

These two forms of butane are known as **isomers**—molecules that are the same in their chemical formulas, but differ in the spatial arrangement of their elements.

Carbon can also form rings. Here is the structure of benzene (C_6H_6), which is found in petroleum products:

```
           H
           |
           C
      H   ╱ ╲╲  H
       ╲ C    C ╱
        ‖      |
        C      C
      ╱  ╲╲  ╱  ╲
     H    C    H
          |
          H
```

Note the three sets of double lines in the molecule. Recall that this means there are double *bonds* between these atoms; the atoms involved are sharing *two* pairs of electrons.

It just so happens that all the carbon molecules introduced so far are made of nothing but carbon and hydrogen. But this is the exception rather than the rule. Here are two representations of a molecule mentioned already—glucose, better known as blood sugar:

```
           CH2OH
            |
            C ——————— O
       H   ╱ |           ╲   H
        ╲ C  H            C ╱
    HO ╱   | OH       H  | ╲  OH
            C ——————— C
            |           |
            H           OH
```

```
           CH2OH
            |
            C ——————— O
       H   ╱ |           ╲   H
      | ╲ C  H            C |
    HO    | OH       H    | OH
            C ═══════ C
            |           |
            H           OH
```

So here are some added oxygen atoms. Notice how the second model, below the first, with its heavy line down at the bottom, looks a little different than the benzene ring? Such a model is meant to give you a slightly more realistic picture of the actual arrangement of atoms when a carbon ring is present. Think of the ring as lying at an angle to the paper, with the heavy line closer to you than the line in the back; the H and OH groups then lie above and below the plane of the ring, as shown. You may also notice that when things get this complicated, space starts getting a little tight for writing in the letters that stand for the various atoms that are a part of the molecule. Because most rings are formed predominantly of carbon, it is common to dispense with the Cs when a structural formula is presented. Here is a third representation of the glucose model, written in this stripped-down form:

```
           CH2OH
       H   ╱ ————————— O ╲   H
        ╲ ╱    H            ╲ ╱
         ╱     OH        H    ╲
    HO ╱                       ╲ OH
         ═══════════════════
       H                     OH
```

You can see that carbon is *assumed* to exist at each bond juncture in the ring; if some other element occupies one of these points, it is explicitly noted. Often the solitary Hs that are attached to carbons are left out as well.

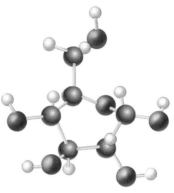

ball-and-stick model of glucose

space-filling model of glucose

Figure 3.8
Ball-and-Stick and Space-Filling Models of Glucose
The simple sugar glucose is the single most important energy source for our bodies. Because it has several OH groups, it is highly hydrophilic and thus readily breaks down in water. Note how the carbon atoms (in red) form the core of the molecule, with oxygen (in blue) and hydrogen (in white) at the periphery.

Such models as these, as you'll recall from the last chapter, don't show the actual three-dimensional form of a molecule. To do so, ball-and-stick or space-filling models will do. Shown in **Figure 3.8** are examples of these models.

3.4   Functional Groups

Having seen how carbon compounds start to become specialized through the addition of other kinds of atoms, it's worth noting that such atoms often are affixed to carbon compounds in *groups*. Each of these units is a **functional group**: a group of atoms that confers a special property on a carbon-based molecule. Just as an attachment on a wrench might make the difference between an Allen wrench and a socket wrench, so a functional group can make the difference between one molecule and another. Look here at the formula for the hydrocarbon ethane, which is a flammable gas:

$$H-\underset{\underset{H}{|}}{\overset{\overset{H}{|}}{C}}-\underset{\underset{H}{|}}{\overset{\overset{H}{|}}{C}}-H$$

Now let's substitute a functional group, called the hydroxyl or −OH group, for the rightmost hydrogen atom in ethane and thus get:

$$H-\underset{\underset{H}{|}}{\overset{\overset{H}{|}}{C}}-\underset{\underset{H}{|}}{\overset{\overset{H}{|}}{C}}-OH$$

With this, we no longer have a gas; we have a liquid—the drinkable liquid ethyl alcohol. Anytime an −OH group is added to a hydrocarbon chain, some sort of alcohol is formed, be it the isopropyl alcohol used to disinfect cuts, the methanol used to power turbine engines, or some other variety. Thus, the −OH group has a generalized function in connection with hydrocarbons and is therefore referred to as a functional group. Now note another effect of this group. The ethane we started with was nonpolar—there was no difference in charge at one end of the ethane molecule as opposed to the other. But the addition of the −OH group changed this; the oxygen atom it contains has such strong electronegativity that ethyl alcohol does have a polarity, meaning it can bond with other charged or polar molecules, including water. This exemplifies a general characteristic of functional groups, which is that they often impart an electrical charge, or at least a polarity, onto molecules, which makes a big difference in their bonding capacity. The −OH group serves this polarizing function not only when it is added to hydrocarbons, but when it is part of other molecules as well. If you look at **Table 3.1**, you can see some of the functional groups that are most important in living things.

Table 3.1 Functional groups

Group:	Structural formula:	Found in:
Carboxyl (−COOH)	$-C\overset{\displaystyle O}{\underset{\displaystyle OH}{\big<}}$	fatty acids, amino acids
Hydroxyl (−OH)	− OH	alcohols, carbohydrates
Amino (−NH$_2$)	$-N\overset{\displaystyle H}{\underset{\displaystyle H}{\big<}}$	amino acids
Phosphate (−PO$_4$)	$-O-\overset{\overset{\displaystyle O}{\|}}{P}-\underset{\underset{\displaystyle O^-}{\|}}{O^-}$	DNA, ATP

3.5 Carbohydrates

With this review of carbon structures under your belt, you're ready to begin looking at some of the classes of carbon-based molecules—the molecules of living things. You'll explore four groupings of these organic compounds: carbohydrates, lipids, proteins, and nucleic acids.

As you get into this section, it will be a great help to keep in mind that *complex* organic molecules often are made from *simpler* molecules. Many of the molecules you'll be reading about have a building-blocks quality to them: Take a simple sugar, or monosaccharide, such as glucose, put it together with another monosaccharide (fructose), and you have a larger *di*saccharide called sucrose (better known as table sugar). Put *many* monosaccharide units together and you have a polysaccharide, such as starch. The starch is an example of a **polymer**—a large molecule made up of many similar or identical subunits. Meanwhile, the glucose is an example of a **monomer**—a small molecule that can be combined with other similar or identical molecules to make a polymer. Look at **Table 3.2** for examples of both.

Web Tutorial 3.3
Monomers and Polymers

Carbohydrates: From Simple Sugars to Cellulose

Happily, for purposes of memory, the elements in the first molecules we'll look at, carbohydrates, are all hinted at in the name: **Carbohydrates** are organic molecules that always contain carbon, oxygen, and

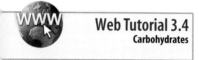

Web Tutorial 3.4
Carbohydrates

hydrogen and that in many instances contain nothing *but* carbon, oxygen, and hydrogen. Furthermore, they usually contain exactly twice as many hydrogen atoms as oxygen atoms. For example, the carbohydrate glucose ($C_6H_{12}O_6$) contains 12 atoms of hydrogen and 6 of oxygen. Most people think of carbohydrates purely in terms of foods such as breads and pasta (**see Figure 3.9**); but as you'll see, carbohydrates have more roles than this in nature.

The building blocks of the carbohydrates are the **monosaccharides** mentioned earlier, which also are known as **simple sugars**. You've already seen several views of one of these molecules—glucose. Glucose has a use in and of itself: Much of the food we eat is broken down into it, at which point it becomes our most important energy source. Glucose can also bond with other monosaccharides, however, to form more complex carbohydrates. Let's take a look at how this happens. If you look at **Figure 3.10**, you can see an example of two glucose molecules bonding to create the disaccharide called maltose.

Several things are worth noting here. In maltose, as you can see, the link that joins the two glucose monomers is a single oxygen atom linked to carbons of each of the glucose units. To get this, two atoms of hydrogen and one atom of oxygen are split off (on the left side of the equation) from the original glucose molecules. What becomes of these three atoms? Look at the far right-hand side of the reaction and see the $+H_2O$. The product of this reaction is a molecule of maltose and a molecule of water.

Now note the arrows in the middle of the reaction. You've been used to seeing a single, rightward-pointing arrow, but here the arrows go both ways. What this means is that this is a reversible reaction—it can proceed in either of the two directions. When maltose is placed in water, it can be split apart to yield two *glucose* molecules. Finally, note the existence of the $-OH$ functional group in both the glucose and maltose molecules.

Kinds of Simple Carbohydrates

You have thus far seen examples of one of the simplest carbohydrates, the monosaccharide glucose, and one slightly more complex carbohydrate, a disaccharide called maltose. There are many kinds of mono- and disaccharides, however. Among the monosaccharides are, for example, fructose and deoxyribose. Among disaccharides, there are sucrose and lactose. At the risk of pointing out the obvious, note that all these sugars have *-ose* at the end of their name: If it's an *-ose*, it's a sugar (**see Figure 3.11**).

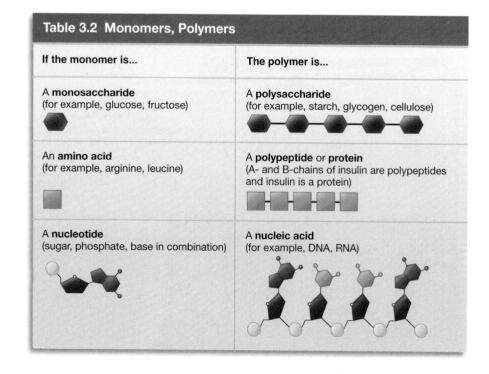

Table 3.2 Monomers, Polymers	
If the monomer is...	**The polymer is...**
A **monosaccharide** (for example, glucose, fructose)	A **polysaccharide** (for example, starch, glycogen, cellulose)
An **amino acid** (for example, arginine, leucine)	A **polypeptide** or **protein** (A- and B-chains of insulin are polypeptides and insulin is a protein)
A **nucleotide** (sugar, phosphate, base in combination)	A **nucleic acid** (for example, DNA, RNA)

Figure 3.9
Carbohydrates in Foods
Breads, cereals, and pasta make up a significant proportion of our diets. These foods are all rich in carbohydrates, one of the four main types of biological molecules.

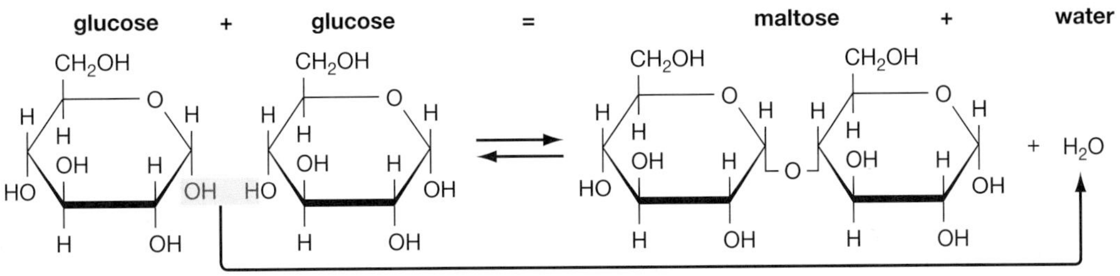

glucose   +   glucose   =   maltose   +   water

Figure 3.10
Carbohydrates Follow a Building-Blocks Model
In this example, two units of the monosaccharide (or simple sugar) glucose link to form the disaccharide maltose. In addition to maltose, the reaction yields water. The double arrows indicate that this reaction is reversible; a single maltose molecule can yield two glucose molecules.

Figure 3.11
Sugars Come in Many Forms
Sucrose or table sugar comes to us from sugarcane or sugar beets, and glucose is found in corn syrup. Fructose comes to us in sweet fruits and in high-fructose corn syrup, which often is used to sweeten soft drinks.

Complex Carbohydrates Are Made of Chains of Simple Carbohydrates

If you go another couple of bumps up from disaccharides, you get to the polymers of carbohydrates, the **polysaccharides**. The *poly* in polysaccharide means "many," while *saccharide* means "sugars," and the term is apt. In the polysaccharide molecule cellulose, for example, there may be 10,000 glucose units linked up with one another. The basic unit here is the six-carbon monosaccharide glucose, $C_6H_{12}O_6$, from which chains of glucose units are built up. The complexity of these molecules gives them their alternate name of complex carbohydrates. Four different types of complex carbohydrates interest us: starch, glycogen, cellulose, and chitin (**see Figure 3.12**).

Starch is a complex carbohydrate found in plants that exists in the form of such foods as potatoes, rice, carrots, and corn. In plants, these starches serve as the main form of carbohydrate *storage*, sometimes as seeds (rice and wheat grains), or sometimes as roots (carrots or beets); **see Figure 3.12a**.

Another complex carbohydrate, **glycogen**, serves as the primary form of carbohydrate storage in animals. Thus, glycogen does for animals what starch does for plants. For this reason, glycogen is sometimes called animal starch. The starches or sugars we eat are broken down, eventually into glucose, at which point some of the glucose may be used immediately. Some may not be needed right away, however, in which case it is moved into the muscle cells and liver to be stored as the more complex carbohydrate glycogen. The "carbohydrate loading" that swimmers and runners do before races (by eating lots of pasta, for example) could just as easily be called glycogen loading (**see Figure 3.12b**).

Cellulose is a rigid, complex carbohydrate contained in the cell walls of many organisms. Despite this innocuous-sounding function, cellulose is important because it makes up so much of the natural

Figure 3.12
Four Examples of Complex Carbohydrates
All complex carbohydrates are composed of chains of glucose, but they differ in the details of their chemical structure.

(a) Starch serves as a form of carbohydrate storage in many plants. Here starch granules can be seen within the cells of a slice of raw potato.

(b) Glycogen serves as a form of carbohydrate storage, as shown in this photo of glycogen globules in the liver.

(c) Cellulose, running as fibers through cell walls, provides structural support for plants. The photo is of sets of cellulose fibers running at right angles to one another in the cell wall of marine algae.

(d) Chitin provides structural support for some animals. The outer "skin" or cuticle of insects is composed mostly of chitin. The photo shows the exoskeleton of a tick.

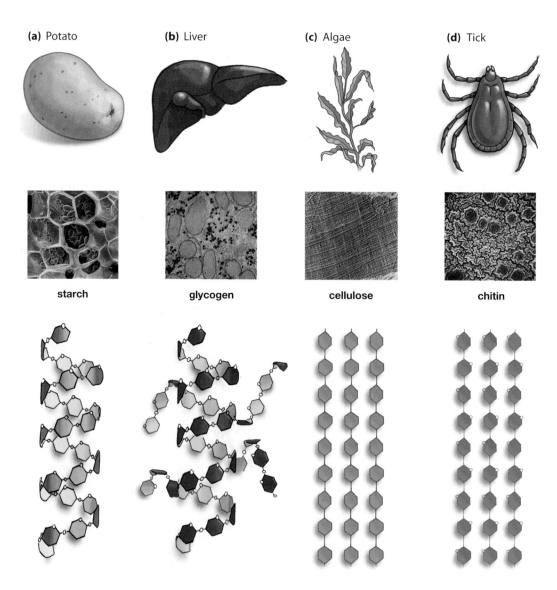

(a) Potato   **(b)** Liver   **(c)** Algae   **(d)** Tick

starch   glycogen   cellulose   chitin

world. It is easily the most abundant carbohydrate on Earth: Trees, cotton, leaves, and grasses are largely made of it. When cellulose is enmeshed with a hardening compound called *lignin*, the result is a set of cell walls that can hold up giant redwood trees. Because cellulose is so dense and rigid, it is not surprising that human beings and other mammals cannot digest it. This statement may make you wonder about grass-eating *cows*, but in fact cows have cellulose digested for them by special bacteria that reside in their digestive tract. Cellulose is important to humans in that it is our major source of insoluble fiber, which helps move foods through the digestive tract. Because cellulose exists in the cell walls of plants, we can get our fiber from such foods as whole grains and fresh fruits (**see Figure 3.12c**).

Chitin is a complex carbohydrate that forms the external skeleton of the arthropods—all insects, spiders, and "crustaceans," such as crabs. In all these animals, chitin plays a "structural" role similar to that of cellulose in plants: It gives shape and strength to the structure of the organism (**see Figure 3.12d**).

3.6 Lipids

We turn now to the **lipids**, a class of biological molecules whose defining characteristic is their relative insolubility in water. It turns out that lipids are made of the same elements as carbohydrates—carbon, hydrogen, and oxygen. But lipids have much more hydrogen, relative to oxygen, than do the carbohydrates. We're all familiar with some lipids; they exist as fats, as oils, as cholesterol, and as hormones such as testosterone and estrogen. Unlike the other biological molecules you'll be studying, a lipid is not a polymer composed of component-part monomers; no single structural unit is common to all lipids. Thus, the one characteristic shared by pure lipids is that they do not readily break down in water. Remember an earlier discussion about the need life has to create internal containers? Well, thanks to their insolubility, lipids are able to serve this function. In addition, lipids have considerable powers to store energy and to provide insulation.

One Class of Lipids Is the Glycerides

The glyceride, the most common kind of lipid, can be thought of as a molecule in two parts. The first is a "head" composed of a particular kind of alcohol, usually glycerol. Attached to the glycerol head is the second part of the glyceride, a series of fatty acids, each one consisting of a long chain of carbon and hydrogen atoms (see Figure 3.15 on page 50).

The structural formula for stearic acid, one of the fatty acids found in animal fat, is shown in

Figure 3.13. This monotonous chain of hydrogen and carbon is, as you might have guessed, the hydrocarbon portion of the stearic fatty-acid molecule. Fatty acids can have from 4 to 22 carbons linked up like this. On the left side, you can see that the chain terminates with one of the functional groups we talked about earlier, a COOH chemical group (indicated in red), properly known as a *carboxyl group*.

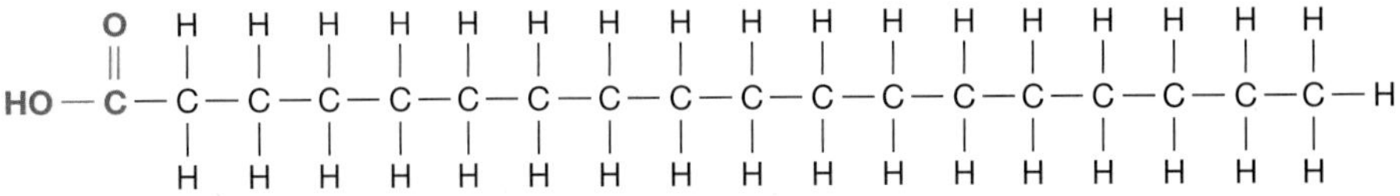

Figure 3.13
Structural Formula for Stearic Acid

Now look at the other part of the fat molecule we discussed, the alcohol known as glycerol:

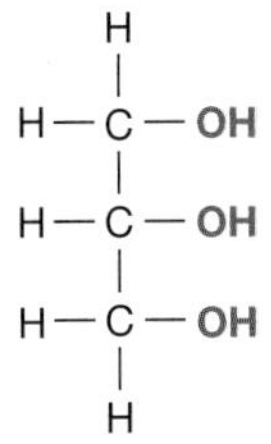

As you know, the hallmark of alcohols is that they have an −OH group, which you can see here. Bringing the −OH portion of glycerol together with the carboxyl part of the fatty-acid chain is what makes a glyceride. But you can see that the glycerol has *three* −OH groups on the right. Thus there is, you might say, docking space on glycerol for three fatty acids, and in the synthesis of many glycerides, this linkage takes place: Three fatty acids link up with glycerol to form a triglyceride, which is an important form of lipid. **Figure 3.14** shows how it works schematically.

Web Tutorial 3.5
Lipids

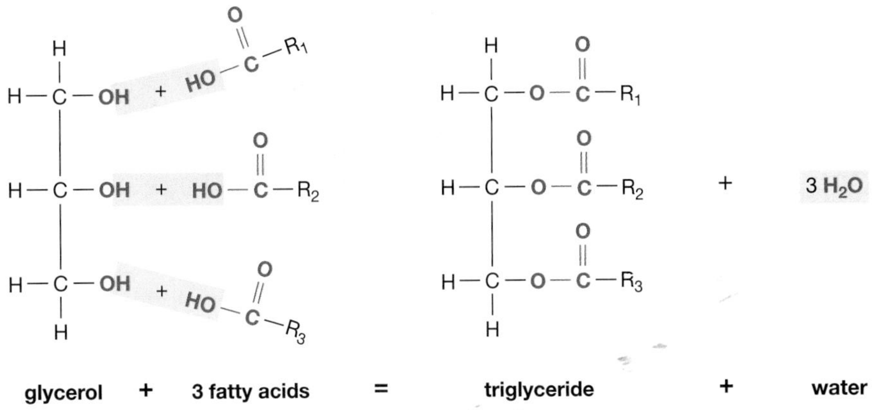

glycerol   +   3 fatty acids   =   triglyceride   +   water

Figure 3.14
Formation of a Triglyceride

The Rs on the right end of the COOH group stand for whatever the hydrocarbon chain is of that particular fatty acid. To get a better idea of the shape of a completed triglyceride, look at the space-filling model in **Figure 3.15**. This particular triglyceride, called tristearin, has three stearic fatty acids stemming like tines on a fork from the glycerol. But this is only one possibility among many, and is exceptional, rather than usual, in that all three fatty acids are the same. Among the dozens of fatty acids that exist, several *different* kinds of fatty acids generally will hook up in glycerol's three "slots" to form a triglyceride. Actual fat products, such as butter, are composed of different proportions of various fatty acids.

We've seen three fatty acids linking with glycerol to form *triglycerides*, but monoglycerides (one fatty acid joined to glycerol) and diglycerides (two fatty acids joined with glycerol) can be formed as well. Triglycerides are the most important of the glycerides, however, because they constitute about 90 percent of the lipid weight in foods. In short, the substances we call fats usually are composed mostly of triglycerides.

With all this under your belt, you're ready for a couple of definitions. A **triglyceride** is a lipid molecule formed from three fatty acids bonded to glycerol. A **fatty acid** is a molecule found in many lipids that is composed of a hydrocarbon chain bonded to a carboxyl group.

Saturated and Unsaturated Fatty Acids: A Linkage with Solids and Liquids and with Health

If you now look at **Figure 3.16**, you can see three different fatty acids: palmitic, oleic, and linoleic. There is an obvious difference among them, in that the oleic and linoleic hydrocarbon chains are "bent"

compared to the straight-line palmitic chain. Upon closer inspection, you can see that the palmitic acid has an unbroken line of single bonds linking its carbon atoms. Meanwhile, the oleic acid has one double bond between the carbons in its chain and the linoleic acid has two. Furthermore, note that these double bonds exist precisely where the "kinks" appear in these molecules.

Fatty Acids: From Solid to Liquid

What these variations describe are the differences between three *kinds* of fatty acids. First, there is a **saturated fatty acid**—a fatty acid with no double bonds between the carbon atoms of its hydrocarbon chain. Then there is a **monounsaturated fatty acid** (a fatty acid with one double bond between carbon atoms), and a **polyunsaturated fatty acid** (a fatty acid with two or more double bonds between carbon atoms). What a saturated fatty acid is saturated *with* is hydrogen atoms. A given monounsaturated fatty acid could theoretically have its lone carbon-carbon double bond replaced with a *single* bond between the carbons. This change would entail the addition of two more hydrogen atoms at the bond sites. Once this happened, this fatty acid would contain the maximum number of hydrogen atoms possible, meaning it would be a saturated fatty acid.

This may not sound like much of a difference, but it has several important consequences. First, at room temperature, as you move from saturated to unsaturated, you also move in general from fats in their solid form to fats in their liquid form, which we call **oils**. For decades, food producers have turned naturally occurring oils *into* fats by saturating them—by bubbling hydrogen through the oils in a process called hydrogenation. Why would this turn a liquid into a solid? The reason is a familiar one: molecular shape. Saturated fatty acids have a straight-line form that allows them to pack together tightly, like so many boards in a lumber yard. In contrast, unsaturated fatty acids stick out at varying angles to one another in a triglyceride, a disorder that generally makes them liquid at room temperature. The "kinks" in unsaturated fatty acids have consequences.

Saturated Fatty Acids and Health

The distinction between saturated and unsaturated fatty acids also has another important effect, this one related to human health. Think of it this way: To the degree that *fatty acids* are saturated, the *fats* they make up will be saturated—and consumption of saturated fats has been linked with heart disease.

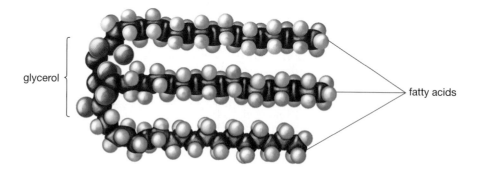

Figure 3.15
The Triglyceride Tristearin
This lipid molecule is composed of three stearic fatty acids, stemming rightward from the glycerol OH "heads." Tristearin is found both in beef fat and in the cocoa butter that helps make up chocolate.

glycerol

fatty acids

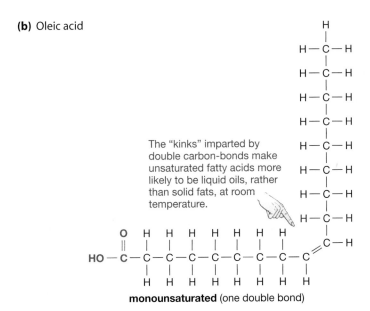

(a) Palmitic acid

saturated (no double bonds)

(b) Oleic acid

The "kinks" imparted by double carbon-bonds make unsaturated fatty acids more likely to be liquid oils, rather than solid fats, at room temperature.

monounsaturated (one double bond)

(c) Linoleic acid

polyunsaturated (more than one double bond)

Figure 3.16
Saturated and Unsaturated Fatty Acids
The degree to which fatty acid hydrocarbon chains are "saturated" with hydrogen atoms has consequences for both the form these lipids take and for human health.

(a) The hydrocarbon "tail" in palmitic acid is formed by an unbroken line of carbons, each with a single bond to the next.

(b) In oleic acid, a double bond exists at one point between two carbon atoms. An additional hydrogen atom could link to each of these carbon atoms instead, which would make this a saturated fatty acid—saturated with hydrogen atoms. As things stand, this is a monounsaturated fatty acid.

(c) The carbons in linoleic acid have double bonds in two locations, making this a polyunsaturated fatty acid.

The linkage here is that saturated fats increase the amount of cholesterol in circulation in the body. When this cholesterol lodges in the arteries of the heart in excessive amounts, heart disease can begin. The picture on fats in general and heart disease is a complicated one, however. The industrial hydrogenation process just noted plays a part in creating a particularly harmful sort of fat you may have heard of, called trans fat, but it turns out there are other fats that are actually good for your heart. You can read more about this in "From Trans Fats to Omega 3's" on page 54.

Energy Use and Energy Storage via Lipids and Carbohydrates

Lipids have something in common with carbohydrates in connection with energy *storage* on the one hand and energy *use* on the other. To be stored, lipids must be in their triglyceride form; to be used to provide energy, they must first be broken down into glycerol and fatty acids. Carbohydrates, meanwhile, are stored as the complex carbohydrate glycogen; to be used for energy expenditure, they must be broken down to simple carbohydrates, often glucose. This process of alternately building up molecules for energy storage, or breaking them down for energy expenditure, is a major task of living things. If you look at **Figure 3.17**, you can see how carbohydrates and lipids are alternately stored and used by plants and animals.

A Second Class of Lipids Is the Steroids

Steroids are a class of lipid molecules that have, as a central element in their structure, four carbon rings. What separates one steroid from another are the various side chains that can be attached to these rings (**see Figure 3.18**). When you see how different steroids are structurally from the triglycerides, you can understand why the monomers-to-polymers framework doesn't apply to lipids.

Among the most well known steroids are cholesterol and two of the steroid hormones, testos-terone and estrogen. Like fats in general, cholesterol has a bad reputation; but also like fats in general, it serves good purposes too. **Cholesterol** is a steroid molecule that forms part of the outer membrane of all animal cells, and that acts as a precursor for many other steroids. One of the steroids formed from it is the principal "male" hormone testosterone; another is the principal "female" hormone estrogen. (Both hormones actually are produced in both sexes, though in differing amounts.) The term *steroids* by itself undoubtedly rings a bell, because the phrase "on steroids" has come to mean artificially bulked up or supercharged. In this common usage, steroids refers to manufactured drugs that are close chemical cousins of the muscle-building "male" steroid hormones (**see Figure 3.19**).

Figure 3.17
Storage and Use of Carbohydrates and Lipids
Carbohydrates and lipids generally are stored in one form, but used in another.

(a) In plants, carbohydrates commonly are stored as starch, a complex carbohydrate that is a major component of such vegetables as carrots. Human beings can consume such a starch, after which it may be broken down into the smaller, simple carbohydrate glucose and used immediately—for exercise or other purposes—or stored in the liver and muscles as the larger, more complex carbohydrate glycogen.

(b) Cows use their fat for energy storage and insulation. The fat portion of the meat that humans consume may be used for energy, but only after being broken down into its glycerol and fatty-acid components. Conversely, humans may convert the components back into triglycerides to be stored in fat cells, providing energy reserves for later.

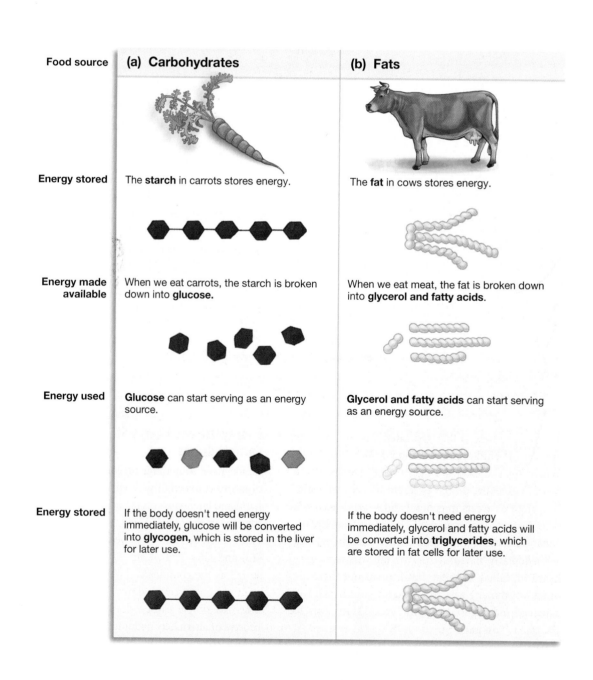

(a) Four-ring steroid structure

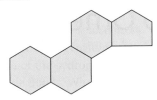

(b) Side chains make each steroid unique

testosterone

estrogen

cholesterol

Figure 3.18
Structure of Steroids

(a) The basic unit of steroids, four interlocked carbon rings.

(b) Types of steroids, each differentiated from the other by the side chains that extend from the four-ring skeleton. Testosterone is a principal "male" hormone, while estrogen is a principal "female" hormone. Both of these steroid hormones actually are found in both men and women, though in differing amounts. Cholesterol is also a prevalent and important steroid in both men and women. Although cholesterol has a bad reputation, it has several important functions—for example, breaking down fats.

(a) Natural steroids

(b) Pharmaceutical steroids

Figure 3.19
Steroids, in Uses Natural and Unnatural

(a) Some steroid hormones, such as estrogen and testosterone are important in natural processes, such as reproduction.

(b) Pharmaceutical steroids can be used in potentially harmful ways—for example, to build muscle mass. Pictured is British sprinter Linford Christie, who in 1999 tested positive for substances produced by a banned steroid called nandrolone.

3.7 Proteins

Living things must accomplish a great number of tasks just to get through a day, and the diverse biological molecules you've been looking at allow this to happen. You've seen carbohydrates do some things, and you've seen lipids do some more. But in the range of tasks that molecules accomplish, proteins reign supreme. Witness the fact that almost every chemical reaction that takes place in living things is hastened—or, in practical terms, *enabled*—by a particular kind of protein called an enzyme. These molecules function in nature like some vast group of tools, each one taking on a specific chemical task. Accordingly, an animal cell might contain up to 4,000 different types of enzymes.

We might marvel at proteins solely because of what enzymes can do, but the amazing thing is that enzymes are only *one class* of proteins. Proteins also form the scaffolding, or structure, of a good deal of tissue; they're active in transporting molecules from one site to another; they allow muscles to contract and cells to move; some hormones are made from them. If you factor out water, they account for about half the weight of the average cell. In short, it's hard to overestimate the importance of these molecules. **Table 3.3** lists some of the different kinds of proteins.

Web Tutorial 3.7
Proteins

Proteins Are Made from Chains of Amino Acids

Proteins are prime examples of the building-block type of molecule described earlier. The monomers in this case are called amino acids. String a minimum number of them together in a chain—some say 10, some say 30—and you have a **polypeptide**, defined as a series of amino acids linked in linear fashion. When the polypeptide chain *folds up* in a specific three-dimensional manner, you have a **protein**, defined as a large, folded chain of amino acids. As a practical matter, proteins are likely to be made of hundreds of amino acids strung together and folded up. As you'll see, it's not unusual for two or more polypeptide chains to be part of a single protein.

Figure 3.21a gives the fundamental structural unit for amino acids, followed by a couple of examples. You can see carbon at the center of this unit, an amino group off to its left, and a carboxyl group to its right. (The name *amino acid* comes from the fact that there is both an *amino* group and a carboxyl group, which is an *acid* group.) What differentiates one amino acid from another is the group of atoms that occupies the R or "side-chain" position. In **Figure 3.21b** you can see examples of actual amino acids, tyrosine and glutamine, with their different occupants of the R position.

Table 3.3 Types of Proteins

Type	Role	Examples
Enzymes	Quicken chemical reactions	Sucrase: Positions sucrose (table sugar) in such a way that it can be broken down into component parts of glucose and fructose
Hormones	Chemical messengers	Growth hormone: Stimulates growth of bones
Transport	Move other molecules	Hemoglobin: Transports oxygen through blood
Contractile	Movement	Myosin and actin: Allow muscles to contract
Protective	Healing; defense against invader	Fibrinogen: Stops bleeding Antibodies: Kill bacterial invaders
Structural	Mechanical support	Keratin: Hair Collagen: Cartilage
Storage	Stores nutrients	Ovalbumin: Egg white, used as nutrient for embryos
Toxins	Defense, predation	Bacterial diphtheria toxin
Communication	Cell signaling	Glycoprotein: Receptors on cell surface

A Group of Only 20 Amino Acids Is the Basis for All Proteins in Living Things

Although only two examples are shown here, it is a group of 20 amino acids that is the basis for all the proteins that occur in living organisms. The thousands of proteins that exist can be made from a mere 20 amino acids, because these amino acids can be strung together in different *order*. Substitute an alanine here for a glutamine there, and you've got a different protein. In this, amino acids commonly are compared to letters of the alphabet. In English, substituting one letter can take us from *bat* to *hat*. In the natural world, 20 amino acids can be put together in different order to create a multitude of proteins, each with a different function.

The stringing together of amino acids happens in a regular way: The carboxyl group of one amino acid joins to the amino group of another, with the loss of a water molecule resulting. Look at **Figure 3.22** to see how three amino acids come together.

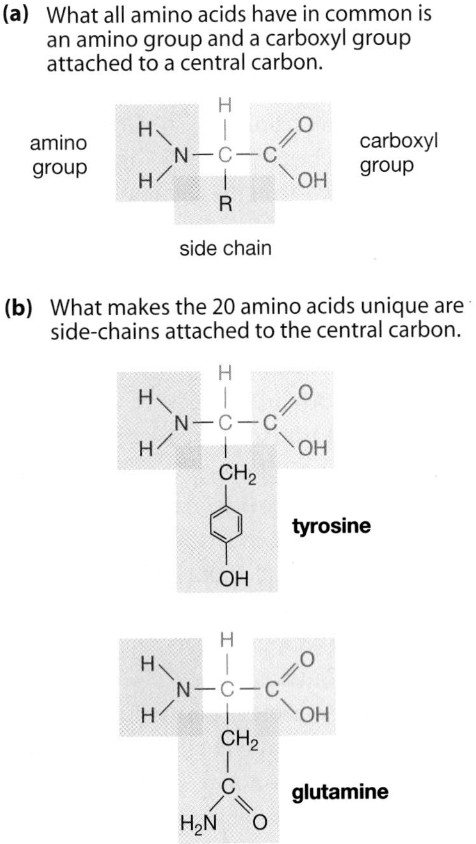

(a) What all amino acids have in common is an amino group and a carboxyl group attached to a central carbon.

(b) What makes the 20 amino acids unique are side-chains attached to the central carbon.

Figure 3.21
Structure of Amino Acids

(a) The structural element that is common to all amino acids is an amino group and a carboxyl group (on the left and right), linked by a central carbon with a hydrogen attached to it. What makes one amino acid different from another is the side-chain of atoms that occupies the R position.

(b) Two examples of actual amino acids; the differing occupants of the R position give us the amino acids tyrosine and glutamine.

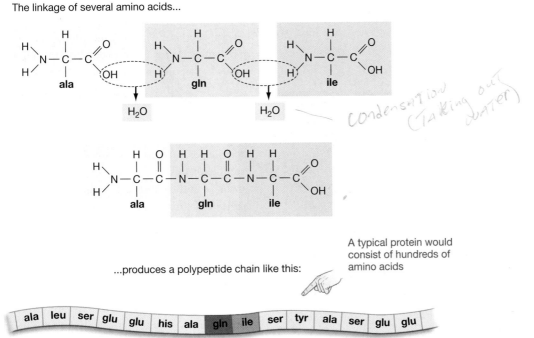

Figure 3.22
Beginnings of a Protein
Amino acids join together to form polypeptide chains, which fold up to become proteins. The linking of amino acids yields water as a by-product. In this figure, alanine (ala) first joins with glutamine (gln), which then is linked to isoleucine (ile). (A list of all 20 primary amino acids can be found on page 265.)

Shape Is Critical to the Functioning of All Proteins

Now, recall that a protein is a chain of amino acids that has become *folded up* in a specific way. As the amino acids are being strung together in sequence, all the kinds of chemical forces discussed in Chapter 2 begin to work on the chain; as a result, it begins to twist and turn and fold into a unique three-dimensional shape. And it turns out that in the functioning of proteins, this shape, or *protein conformation*, is utterly crucial. Here's an example of why. The hormone insulin, which is a protein, is released from a person's pancreas; it moves through the bloodstream and latches onto muscle cells; its presence then allows blood sugar to get into the muscles and provide energy. Now, how does insulin "latch onto" muscle cells? By *linking* itself with a molecule, called an insulin receptor, that lies on the surface of the muscle cell. But protein and receptor must be precisely shaped for such linkage or "binding" to occur.

This binding of a protein hormone to surface receptor sounds important enough; but recall that in addition to being hormones, proteins are the chemical enablers called enzymes, and they're transport molecules, and so forth. In *all* these functions, shape is critical. The American architect Frank Lloyd Wright had a famous dictum about his designs: "Form follows function." With proteins, we can turn this around and say that function follows form (**see Figure 3.23**).

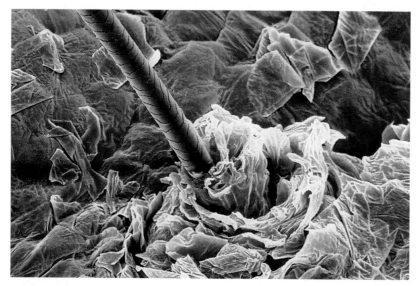

Figure 3.23
Made of Protein
A human hair, shown here emerging from a follicle in the skin, is composed mostly of the protein keratin. Each hair is composed of many groups of keratin strands that wrap around each other in alpha helices, giving hair its structure, which is flexible and much longer than it is wide.

There Are Four Levels of Protein Structure

So, what forms do proteins take? Well, the answer to this depends on what vantage point you adopt in looking at them. There are four levels of structure in proteins that determine their final shape. You can see all four levels in **Figure 3.24**. The first of these levels, the **primary structure** of a protein, is simply its sequence of amino acids. Everything about the final shape of a protein is dictated by this sequence. Electrochemical bonding and repulsion forces act on this structure, and the result is the folded-up protein.

As it turns out, when these forces begin to operate on the amino acid sequence, a couple of common shapes begin to emerge in the **secondary structure** of proteins, defined as structure proteins assume after having folded up. The **alpha helix**, a common secondary structure of proteins, has a shape much like a corkscrew. Another common secondary structure in proteins, the **beta pleated sheet**, takes a form like the folds of an accordion. Proteins can be made almost entirely of alpha helices. This is the case with hair, nails, horns, and the like. Likewise, proteins can be made entirely of beta pleated sheets. The most familiar example of this is silk, in which the beta sheets lie pancake-style on one another. Often, however, alpha helices and beta pleated sheets form what we might think of as design motifs within a larger protein structure; they periodically give way to the less regular segments called random coils. The larger-scale three-dimensional shape that a protein takes is its **tertiary structure**. The way in which *two or more* polypeptide chains come together to form a protein results in that protein's **quaternary structure**.

Proteins Can Come Undone

As noted, proteins fold up into a precise conformation in order to function. However, proteins can *lose* their shape, and thus their functionality. You saw an example of this earlier in connection with pH and enzymes. In the wrong pH environment, an enzyme can unfold, losing its tertiary structure and thus losing its ability to hasten a chemical process. Alcohol works as a disinfectant on skin because it *denatures* or alters the shape of the proteins of bacteria.

Lipoproteins and Glycoproteins

Some molecules in living things are hybrids, or combinations of the various types of molecules you've been looking at. **Lipoproteins**, as their name implies, are molecules that are a combination of lipids and proteins. Active in transporting fats throughout the body, lipoproteins are transport molecules that amount to a capsule of protein surrounding a globule of fat.

Two kinds of lipoproteins have managed to enter public consciousness despite the handicap of having long names. These are high-density lipoproteins and low-density lipoproteins, also known as HDLs and LDLs. What makes them more or less dense is the ratio of protein to lipid in them; lipid is less dense than protein, so a low-density lipoprotein has a relatively large amount of lipid, in comparison to protein. LDLs have acquired a reputation as villains, because they carry cholesterol *to* outlying tissues including the coronary arteries of the heart, where this cholesterol may come to reside, thickening eventually into "plaques" that can block coronary arteries and bring about a heart attack. HDLs, meanwhile, are regarded as the cavalry; they carry cholesterol *away* from outlying cells to the liver. A high proportion of HDLs in relation to cholesterol is predictive of keeping a healthy heart.

Glycoproteins are combinations of proteins and *carbohydrates*. Remember the discussion about insulin traveling through the bloodstream and latching onto a receptor on the surface of a muscle cell? Well, such receptors are usually glycoproteins, meaning mostly protein with a side-chain made of carbohydrate. A profusion of these receptors sits on cell surfaces like so many antennae, ready for a partner

Figure 3.24
Four Levels of Structure in Proteins

Four Levels of Structure In Proteins

(a) Primary structure

The primary structure of any protein is simply its sequence of amino acids. This sequence determines everything else about the protein's final shape.

(b) Secondary structure

Structural motifs, such as the corkscrew-like alpha helix, beta pleated sheets, and the less organized "random coils" are parts of many polypeptide chains, forming their secondary structure.

(c) Tertiary structure

These motifs may persist through a set of larger-scale turns that make up the tertiary structure of the molecule.

(d) Quaternary structure

Several polypeptide chains may be linked together in a given protein, in this case hemoglobin, with their configuration forming its quaternary structure.

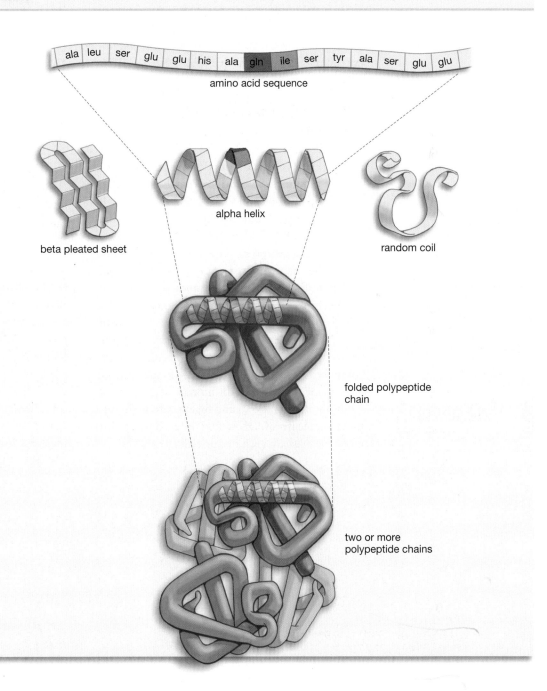

amino acid sequence

beta pleated sheet

alpha helix

random coil

folded polypeptide chain

two or more polypeptide chains

with just the right shape to come by and latch on. Some hormones themselves are glycoproteins, along with many other proteins released from cells.

3.8 Nucleotides and Nucleic Acids

Nucleotides are the last major type of biological molecule we'll study. They are important molecules in and of themselves and as building blocks for a couple of very important molecules.

As molecules in their own right, some nucleotides, called adenosine phosphates, serve as chemical energy carriers. When you study energy within cells, you'll find many references to a molecule called adenosine triphosphate, or ATP. As money is to shopping, so ATP is to getting things done in living organisms. In this text, the most detailed discussion of nucleotides, however, comes in connection with their role as monomers in building two very large and important polymers—the nucleic acids DNA and RNA, which are considered next.

DNA Provides Information for the Structure of Proteins

You've learned that proteins perform a large number of biological functions, and that one class of proteins, the enzymes, may be represented with up to 4,000 types in a single animal cell. If you had a factory that turned out four thousand different kinds of tools, you would obviously need some direction on how each of these tools was to be manufactured: This part of the tool goes here first, and then that goes there, and so on. There is a molecule that in essence provides this kind of information for the construction of proteins. It is **DNA**, or **deoxyribonucleic acid**: the primary information-bearing molecule of life, composed of two linked chains of nucleotides. How many nucleotides? In human beings, about 3 billion of them are strung together to form our main molecule of DNA, and one copy of this molecule exists in each of our cells.

Web Tutorial 3.8
Nucleic Acids

The information contained in the DNA molecule is much like the information contained in a cookbook, only what the DNA "recipes" are calling for are precisely ordered chains of amino acids, the building blocks of proteins. "Start with an alanine, then add a cysteine, then a tyrosine . . . " and so on for hundreds of steps, the DNA-encoded instructions will say, after which—through a series of steps—a protein becomes synthesized and then gets busy on some task. A player in this series of steps is another nucleic acid, **ribonucleic acid** or **RNA**, a molecule composed of nucleotides that is active in the synthesis of proteins. RNA's functions include ferrying the DNA-encoded instructions to the sites in the cell where proteins are put together. It also helps make up a protein-synthesizing workbench you'll look at later, called a ribosome.

The Structural Unit of DNA Is the Nucleotide

Look at **Figure 3.25a** and you'll see the structure of the building blocks that make up DNA. Each nucleic acid **nucleotide** is a molecule in three parts: A *phosphate* group, a *sugar* (deoxyribose), and a nitrogen-containing *base*. One nucleotide then attaches to another, forming a chain. Two of these chains then link together—as if a ladder, split down the middle, were coming together—forming the most famous molecule in all of biology, the DNA double helix.

On to Cells

Before getting to the story of this elegant DNA molecule, you need to learn a little bit about the territory in which it does its work. What you'll be looking at is a profusion of jostling, roiling, ceaselessly working chemical factories that make up all living things. These factories are called cells. Look at **Table 3.4** on page 62, and you'll find a summary of all the types of molecules that were reviewed in this chapter.

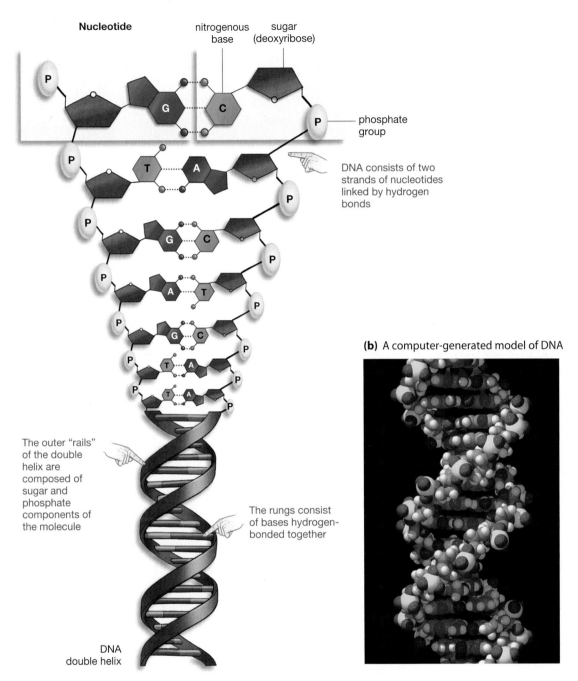

(a) Nucleotides are the building blocks of DNA

Nucleotide

nitrogenous base

sugar (deoxyribose)

phosphate group

DNA consists of two strands of nucleotides linked by hydrogen bonds

The outer "rails" of the double helix are composed of sugar and phosphate components of the molecule

The rungs consist of bases hydrogen-bonded together

DNA double helix

(b) A computer-generated model of DNA

Figure 3.25
Nucleotides Are the Building Blocks of DNA

(a) The basic unit of the DNA molecule is the nucleotide, a molecule in three parts: a sugar, (deoxyribose), a phosphate group, and a nitrogen-containing base. A given nucleotide might contain any of four bases: Adenine (A), Guanine (G), Cytosine (C), or Thymine (T). The sugar and phosphate components of the nucleotides link up to form the outer "rails" of the DNA molecule, while the bases point toward the molecule's interior. Two chains of nucleotides are linked, via hydrogen bonds, to form DNA's double helix. Note that the hydrogen bond is between one base and another.

(b) A computer-generated space-filling model of DNA.

Table 3.4 Summary Table of Biological Molecules

Type of Molecule	Subgroups	Examples and Roles
Carbohydrates	Monosaccharides	Glucose: Energy source
	Disaccharides	Sucrose: Energy source
	Polysaccharides	Glycogen: Storage form of glucose Starch: Carbohydrate storage in plants; used by animals in nutrition Cellulose: Plant cell walls, structure; fiber in animal digestion Chitin: External skeleton of arthropods
Lipids	Triglycerides 3 Fatty acids and glycerol	Fats, Oils (butter, corn oil): Food, energy, storage, insulation
	Fatty acids Components of Triglycerides	Stearic Acid: Food, energy sources
	Steroids Four-ring structure	Cholesterol: Fat digestion, hormone precursor, cell membrane component
	Phospholipids Polar head, nonpolar tails	Cell membrane structure
Proteins	Enzymes Chemically active	Sucrase: Breaks down sugar
	Structural	Keratin: Hair
	Lipoproteins Protein-lipid molecule	HDLs, LDLs: Transport of lipids
	Glycoproteins Protein-sugar molecule	Cell surface receptors
Nucleotides	Adenosine phosphates	Adenosine triphosphate (ATP): Energy transfer
	Nucleic acids Sugar, phosphate group, base	DNA, RNAs: Contain information for and facilitate synthesis of proteins

Chapter Review

Summary

3.1 The Importance of Water to Life

- Water has several qualities that have strongly affected life on Earth. It is a powerful solvent, being able to dissolve more compounds in greater amounts than any other liquid. Because water's solid form (ice) is less dense than its liquid form, bodies of water in colder climates do not freeze solid in winter, which allows life to flourish under the ice.

- Water has a great capacity to absorb and retain heat. Because of this, the oceans act as heat buffers for the Earth, thus stabilizing Earth's temperature. Water has a high degree of cohesion, which allows water to be drawn up through plants, via evaporation, in one continuous column, from roots through leaves.

- Some compounds do not interact with water. Hydrocarbons such as petroleum are examples of such hydrophobic compounds. Water cannot break down hydrophobic compounds, which is why oil and water don't mix. Compounds that do interact with water are polar or carry an electric charge and are called hydrophilic compounds.
Web Tutorial 3.1: Chemistry and Water

3.2 Acids and Bases Are Important to Life

- An acid is any substance that yields hydrogen ions when put in solution. A base is any substance that accepts hydrogen ions in solution. A base added to an acidic solution makes that solution less acidic, while an acid added to a basic solution makes that solution less basic.

- The concentration of hydrogen ions that a given solution has determines how basic or acidic that solution is, as measured on the pH scale. This scale runs from 0 to 14, with 0 being most acidic, 14 being most basic, and 7 being neutral. The pH scale is logarithmic; a substance with a pH of 9 is 10 times as basic as a substance with a pH of 8. Living things function best in a near-neutral pH, though some systems in living things have different pH requirements.

3.3 Carbon Is a Central Element in Life

- Carbon is a central element to life, because most biological molecules are built on a carbon framework. Carbon plays this central role because its outer shell has only four of the eight electrons necessary for maximum stability. Carbon atoms are thus able to form stable, covalent bonds with a wide variety of atoms, including other carbon atoms. The complexity of living things is facilitated by carbon's linkage capacity.
Web Tutorial 3.2: Chemistry of Carbon

3.4 Functional Groups

- Groups of atoms known as functional groups can confer special properties on carbon-based molecules. For example, the addition of an – OH group to a hydrocarbon molecule always results in the formation of an alcohol. Functional groups often impart an electrical charge or polarity onto molecules, thus affecting their bonding capacity.

3.5 The Molecules of Life: Carbohydrates

- Carbohydrates are formed from the building blocks or monomers of simple sugars, such as glucose. These can be linked to form the larger carbohydrate polymers such as starch, glycogen, cellulose, and chitin. Carbohydrates generally are stored in polymer form in living things (glycogen in animals, for example) but broken down into monomers to be used (glucose in animals).
Web Tutorial 3.3: Monomers and Polymers
Web Tutorial 3.4: Carbohydrates

3.6 Lipids

- There are several different varieties of lipids. Among the most important are the triglycerides, composed of a glyceride and three fatty acids. Most of the fats we consume are triglycerides. Another important variety of lipids is the steroids, which include cholesterol, and such hormones as testosterone and estrogen.
Web Tutorial 3.5: Lipids
Web Tutorial 3.6: Nutrition Labels

3.7 Proteins

- Proteins are a diverse group of biological molecules composed of the monomers called amino acids. Important groups of proteins include enzymes, which hasten chemical reactions, and structural proteins, which make up such structures as hair. The primary structure of a protein is its amino acid sequence; this sequence determines how a protein folds up. The activities of proteins are determined by their final, folded shapes.
Web Tutorial 3.7: Proteins

3.8 Nucleotides and Nucleic Acids

- Nucleic acids are polymers composed of nucleotides. The nucleic acid DNA is composed of nucleotides that contain a phosphate group, a sugar (deoxyribose), and one of four nitrogen-containing bases. DNA is a repository of genetic information. The sequence of its bases encodes the information for the production of the huge array of proteins produced by living things.
Web Tutorial 3.8: Nucleic Acids

Key Terms

acid 38	monounsaturated
alkaline 40	fatty acid 50
alpha helix 58	nucleotide 60
base 38	oil 50
beta pleated sheet 58	organic chemistry 43
buffering system 40	pH scale 40
carbohydrate 46	phosphate group 55
cellulose 48	phospholipid 55
chitin 49	polymer 46
cholesterol 52	polypeptide 56
deoxyribonucleic acid (DNA) 60	polysaccharide 48
fatty acid 50	polyunsaturated fatty acid 50
functional group 45	primary structure 58
glycogen 48	protein 56
glycoprotein 59	quaternary structure 58
hydrocarbon 38	ribonucleic acid (RNA) 60
hydrophilic 38	saturated fatty acid 50
hydrophobic 38	secondary structure 58
hydroxide ion 40	simple sugar 46
isomer 44	specific heat 36
lipid 49	starch 48
lipoprotein 58	steroid 52
monomer 46	tertiary structure 58
monosaccharide 46	triglyceride 50

Understanding the Basics

Multiple-Choice Questions (answers in the back of the book)

1. Near an ocean or other large body of water, air temperatures do not vary as much with the seasons as they do in the middle of a continent. This tendency of water to resist changes in temperature is the result of water's
   a. high density
   b. low density
   c. being a good solvent
   d. low specific heat
   e. high specific heat

2. A frog survives a freezing-cold winter on the bottom of a pond because
   a. Ice, which floats on water, insulates the water beneath it.
   b. The cells of a frog's body cannot freeze.
   c. Water has a low specific heat.
   d. The surface tension of water protects the frog's body.
   e. All of these are factors.

3. Janine has dry skin, so she uses body oil every morning. The oil seals in some of the water on her skin, so that it doesn't get as dry. This is possible because oils:
   a. are hydrophilic
   b. are rare in nature
   c. have a high specific heat
   d. are more dense than water
   e. are hydrophobic

4. Detergent added to water disrupts the network of hydrogen bonds between water molecules. One way entomologists (biologists who study insects) trap bees for population surveys is to place a tray of sugar water in the field to which they add a few drops of detergent. The bees are attracted to the sugar water and they land on its surface. Why do these investigators bother adding detergent when they know it's sugar that bees are attracted to?
   a. The detergent works with sugar to make the trays more attractive.
   b. They want to wash off any bee that's gathered.
   c. Water without hydrogen bonds is toxic to bees and therefore makes an effective trap.
   d. Water without hydrogen bonds is a much more effective bee attractant.
   e. Bees will sink into water with fewer hydrogen bonds, and so cannot escape.

5. When you eat starch such as spaghetti, an enzyme in your mouth breaks it down to maltose. Eventually, the maltose enters your small intestine, where it is broken down to glucose, which you can absorb into your bloodstream. The starch is a _____, the maltose is a _____, and the glucose is a(n) _____.
   a. protein   dipeptide   amino acid
   b. monosaccharide   disaccharide   polysaccharide
   c. triglyceride   fatty acid   glycerol
   d. amino acid   dipeptide   protein
   e. polysaccharide   disaccharide   monosaccharide

6. In some vintage science fiction movies space travelers find themselves on a planet orbiting a distant star in which there are curious forms of life based on silicon instead of carbon. Although the story clearly is sci-fi, there is an aura of plausibility in the choice of silicon, an atom with 14 protons, in place of carbon as this alien lifeform's central atom. This is because silicon:
   a. is lighter than carbon.
   b. is heavier than carbon.
   c. has one more proton than carbon, so is very similar to it.
   d. has 4 electrons in its outer-most shell.
   e. is an isotope of carbon.

7. The myoglobin protein, which carries oxygen in muscle cells, has only the first three levels of protein structure. In other words, it lacks a quaternary level. From this you can conclude that myoglobin
   a. is made of nucleic acids
   b. is made of only one polypeptide chain
   c. lacks hydrogen bonds
   d. is not helical or pleated
   e. is a fiber

8. You received your genetic information from your parents in the form of DNA. This DNA carried the instructions for making
   a. carbohydrates such as glycogen
   b. fatty acids

c. phospholipids for making the membranes of your cells

d. proteins

e. all of the above

9. John is lactose intolerant. The *–ose* ending indicates that John cannot digest a certain

a. sugar

b. polysaccharide

c. protein

d. steroid

e. enzyme

Brief Review

1. Describe two ways that water serves as a heat buffer.

2. Both low-density lipoproteins (LDLs) and high-density lipoproteins (HDLs) are involved with carrying fats through the bloodstream. If your LDL count is unusually high, should you be concerned? What if your HDL count is high? Why are they different?

3. List as many functions of proteins as possible. Why are proteins able to do so many different types of jobs? How does this affect the world we live in?

4. How does cohesion allow water to move through plants?

5. What effects does hydrogen bonding have on the properties of water?

Applying Your Knowledge

1. Would you buy a pH-balanced shampoo, or one that is not? Explain your answer.

2. Suppose that you are designing clothing to be worn at a space station on a very cold planet. Should you consider designing clothing that has a layer of water between two layers of plastic? Why or why not? What else might you need to consider?

3. Many species of beans produce a poisonous group of glycoproteins called lectins, which cause red blood cells to clump together (agglutinate) and cease to function. (Fortunately, cooking destroys most of them.) Using what you learned about the structure and function of glycoproteins, suggest how they might do this. What benefit do you think plants might get just by causing red blood cells to clump together?

4 *The Cell*

Are those muffins I smell?
(Section 4.1, page 70)

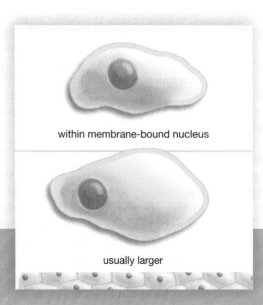

within membrane-bound nucleus

usually larger

Some cells need oxygen.
(Section 4.2, page 71)

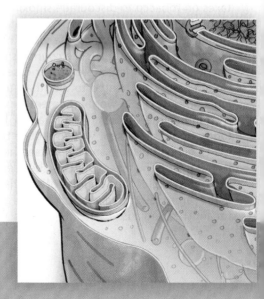

Inside the cell.
(Section 4.3, page 74)

All life exists within cells. These tiny entities can be compared to factories whose products maintain life.

The United States Senate has 100 members in it, and all of them do occasionally gather together to consider given issues. But if you want to know how the Senate actually gets something *done*, you must look elsewhere. The business of the nation is simply too complex for all Senate members to deal with all issues. Instead, issues are dealt with one at a time within the specialized working units of the Senate known as committees.

Something similar to this goes on in the natural world. If we look at birds or trees or people and ask how any of these living things actually gets something done, the answer once again is through working units—in this case the working units known as cells.

4.1 Cells Are the Working Units of Life

Muffins are in the oven and you are in the living room. Gas molecules from the baking muffins waft into the living room, and some of them happen to make their way to your nose, there to travel a short distance to your upper nasal cavity and land on a set

Nature's Fundamental Unit

(a) Human red and white blood cells inside a blood vessel. The large, dark ovals that can be seen in the background are flat cells that form the interior lining of the blood vessel.

(b) Cells of the fungus brewer's yeast, which is used to make wine, beer, and bread.

(c) Cells of a spinach leaf, with the leaf seen in cross-section. Two layers of outer or "epidermal" cells can be seen at top and bottom; many of the cells in between perform photosynthesis.

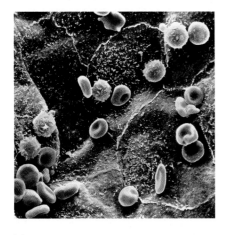

(a)

(b)

(c)

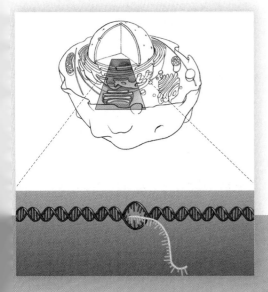

The protein production pathway.
(Section 4.4, page 75)

The center of attention.
(Section 4.4, page 76)

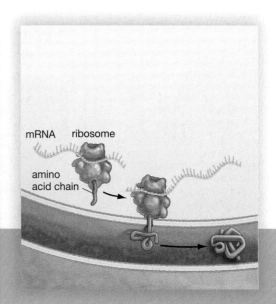

mRNA ribosome

amino acid chain

Where proteins take shape.
(Section 4.4, page 77)

of ceaselessly waving hair-like projections called cilia. These actually are the extensions of some specialized nerve cells (see **Figure 4.1**). If enough muffin molecules bind with enough of the cilia, an impulse is passed along (through other nerve cells) to trigger not only the *sensation* of smell, but the *association* of this smell with muffins you have smelled in the past. How do we know muffins are in the oven? Through cells. How do we move our hands or read this page? Through cells. Life's working units are cells, and in our amazingly complex natural world, there is a great specialization in them. The nerve cells in human beings are specialists in transmitting nerve signals, red blood cells are specialists in transporting oxygen, and muscle cells are specialists in contracting.

Running like a thread through this diversity, however, is the unity of cellular life. Every form of life either is a cell, or is composed of cells. The one possible exception to this is viruses, but even they must use the machinery of cells in order to reproduce. There is unity, too, in the way cells come about: Every cell comes *from* a cell. Human beings are incapable of producing cells from scratch in the laboratory, and so far as we can tell, nature has fashioned cells from simple molecules only once—back when life on Earth got started. The fact that all cells come from cells means that each cell in your body is a link in a cellular chain that stretches back more than 3.5 billion years.

Web Tutorial 4.1
Prokaryotic and
Eukaryotic Cells

4.2 All Cells Are Either Prokaryotic or Eukaryotic

So, what are these tiny, working units we call cells? It follows, from the variety of cells that exist, there is no such thing as a "typical" cell. There are, however, certain *categories* of cells that are important, and the two most important of these are prokaryotic and eukaryotic cells. Every cell that exists is one or the other, and this simple either-or quality extends to the organisms that fall into these camps. All prokaryotic cells either are bacteria or another microscopic form of life known as archaea. Setting bacteria and archaea aside, *all other cells* are eukaryotic. This means all the cells in plants, in animals, in fungi—all the cells in every living thing except in the bacteria and the archaea.

Prokaryotic and Eukaryotic Differences

The name *eukaryote* comes from the Greek *eu*, meaning "true," and *karyon*, meaning "nucleus," while *prokaryote* means "before nucleus." These terms describe the most critical distinction between the two cell types. **Eukaryotic cells** are cells whose primary complement of DNA is enclosed within a nucleus. **Prokaryotic cells** are cells whose DNA is not enclosed within a nucleus. To complete this circle, a **nucleus** is a membrane-lined compartment that encloses the primary complement of DNA in eukaryotic cells.

While having a nucleus is the most important difference between eukaryotic and prokaryotic cells, it is not the only difference. It may seem sensible to think of two single-celled creatures—one a prokaryote, the other a eukaryote—as being very similar, but the distance between them as life-forms is immense. Human beings and chimpanzees are nearly identical in comparison. Eukaryotic cells tend to be much larger than their prokaryotic counterparts; indeed, thousands of bacteria could easily fit into an average eukaryotic cell (see "The Size of Cells" on page 72). Eukaryotes are quite often multicelled organisms, while prokaryotes are always single-celled. And al-

Figure 4.1
Cells Can Specialize
In more complex organisms, different cells carry out different functions. In the picture at left, you can see one type of cell,—a nerve cell, in this case located in the lining of the nose and surrounded by gray accessory cells. A closer look at one of these nerve cells, in the picture at right, shows a number of hair-like extensions, called cilia, protruding from it. When we smell muffins in the oven, gas molecules that waft off the muffins bind with the cilia, which sets in motion a nerve impulse to the brain about the muffins.

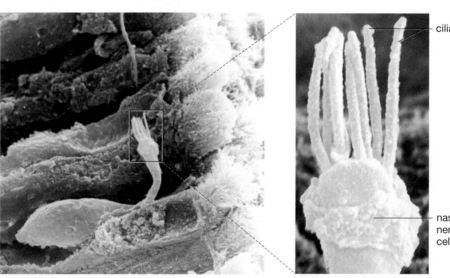

cilia

nasal
nerve
cell

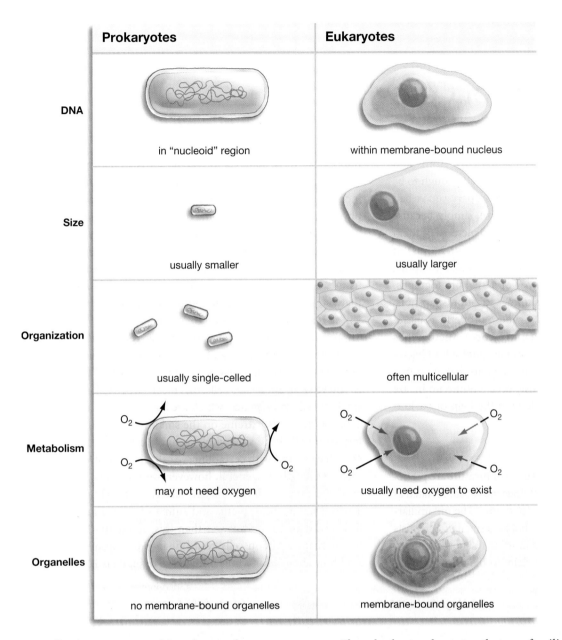

	Prokaryotes	Eukaryotes
DNA	in "nucleoid" region	within membrane-bound nucleus
Size	usually smaller	usually larger
Organization	usually single-celled	often multicellular
Metabolism	may not need oxygen	usually need oxygen to exist
Organelles	no membrane-bound organelles	membrane-bound organelles

**Figure 4.2
Prokaryotic and Eukaryotic
Cells Compared**
A prokaryotic cell is a self-contained organism, since the prokaryotes—bacteria and archaea—are all single-celled. Eukaryotic organisms, which may be single- or multicelled, include plants, animals, fungi, and another group called protists.

most all eukaryotes are aerobic—they need oxygen to exist—whereas many prokaryotes can get along with or without oxygen, while others actually are poisoned by it (**see Figure 4.2**).

Perhaps the most notable distinction between prokaryotic and eukaryotic cells, though, is that eukaryotic cells are *compartmentalized* to a far greater degree than is the case with prokaryotes. The nucleus in eukaryotic cells turns out to be only one variety of **organelle**: a highly organized structure, internal to a cell, that serves some specialized function. Eukaryotic cells contain several different kinds of these "tiny organs." There are organelles called mitochondria, for example, that transform energy from food, and organelles called lysosomes that recycle the raw materials of the cell. In prokaryotic cells, meanwhile, there is only a single type of organelle—a kind of workbench for producing proteins we'll look at later.

If we look at eukaryotes that are familiar to us, such as trees, and mushrooms, and horses, it's easy to see that there is a fantastic diversity in the *forms* of eukaryotes, compared to the strictly single-celled prokaryotes. It does not follow from this, however, that prokaryotes are uniform, nor that they are unsuccessful. On the contrary, prokaryotes differ greatly from one another in, say, the way they obtain nutrients, and they are extremely successful, if success is defined as living in a lot of places in huge numbers. As Lynn Margulis and Karlene Schwartz have observed, more bacteria are living in your mouth right now than the number of people who have ever existed. But for the range of biological structures and processes we want to look at, eukaryotic cells are more diverse, and hence are the initial focus of our study.

Components of eukaryotic cells

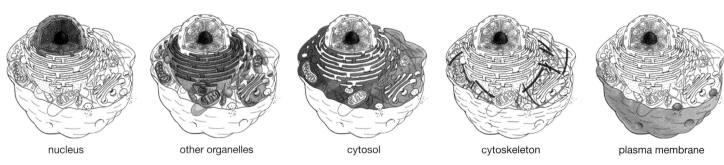

nucleus   other organelles   cytosol   cytoskeleton   plasma membrane

Figure 4.3
Eukaryotic Cell
This cutaway view of a eukaryotic cell displays the elements that nearly all such cells possess: a nucleus, other membrane-bound organelles, a jelly-like cytosol, a cytoskeleton, and an outer plasma membrane.

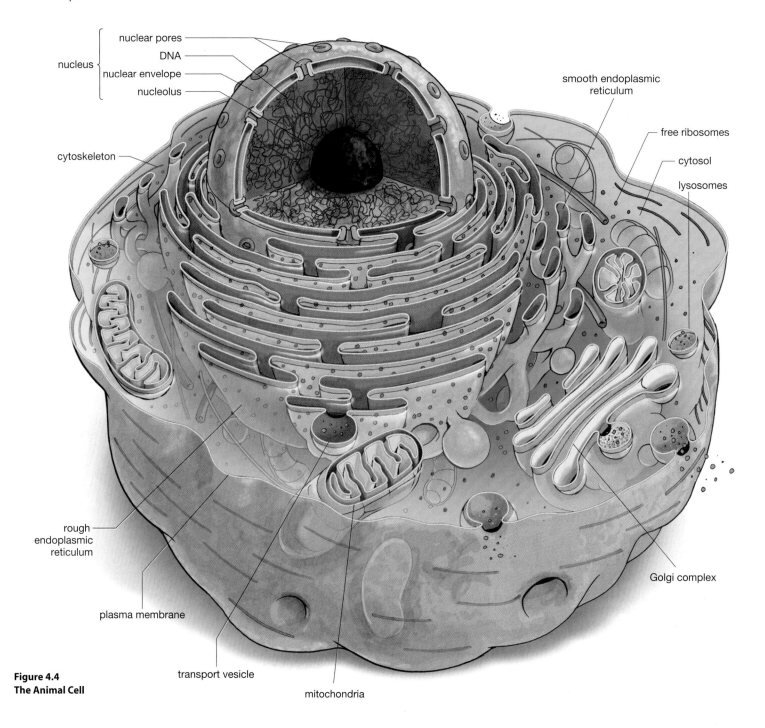

Figure 4.4
The Animal Cell

4.3   The Eukaryotic Cell

What are the constituent parts of eukaryotic cells? Here are five larger structures that in turn have smaller structures within them. The first two have been mentioned already (**see Figure 4.3**).

- The cell's nucleus

- Other organelles (which lie outside the nucleus)

- The **cytosol**, a protein-rich, jelly-like fluid in which the cell's organelles are immersed

- The *cytoskeleton*, a kind of internal scaffolding consisting of three sorts of protein fibers; some of these can be likened to tent poles, others to monorails

- The **plasma membrane**, the outer lining of the cell

In any discussion of a cell, you are likely to hear the term **cytoplasm**, which simply means the *region* of the cell inside the plasma membrane but outside the nucleus. The cytoplasm is different from the cyto*sol*. If you removed all the structures of the cytoplasm—meaning the organelles and the cytoskeleton—what would be left is the cytosol, which is mostly water. This does not mean that the cytosol is simply a passive medium for the other structures. But it is not an organized structure in the way the organelles are. Almost all the organelles you'll see are encased in their own membranes, just as the whole cell is encased in its plasma membrane. What are membranes? Until you get the formal definition next chapter, think of a membrane as the flexible, chemically active outer lining of a cell or of its compartments. In the balance of this chapter, we'll explore all of the constituent parts listed above except for the plasma membrane, which is so special it gets its own chapter.

The Animal Cell

Scientists sometimes make a convenient division of eukaryotic cells into two types: animal cells and plant cells. These cell types have more similarities than they do differences, but they are different *enough* that it will be helpful to look at them separately. We'll start by examining animal cells and then look at how plant cells differ from them.

Insofar as we can characterize that elusive creature, the "typical" animal cell (**see Figure 4.4**), it is roughly spherical, surrounded by, and linked to, cells of similar type, immersed in water; and about 25 micrometers (μm) in diameter, meaning that about 30 of them could fit side by side within the period at the end of this sentence.

4.4   A Tour of the Animal Cell: Along the Protein Production Path

You are now going to take an extended tour of the animal cell, which can be thought of as a living factory. Much of the first part of your trip will be spent tracing the way this factory puts together a product—a protein—for export outside itself. Just as a new employee might tour an assembly line as a means of learning about factory equipment, so you are going to follow the path of protein production as a means of learning about cell equipment.

Figure 4.5 shows the path you'll be taking, from nucleus to the outer edge of the cell. Don't be bothered by the unfamiliar terms in Figure 4.5, because they'll all be explained in the text. The important thing is that, before we begin our tour, you have some sense of the path that protein production takes.

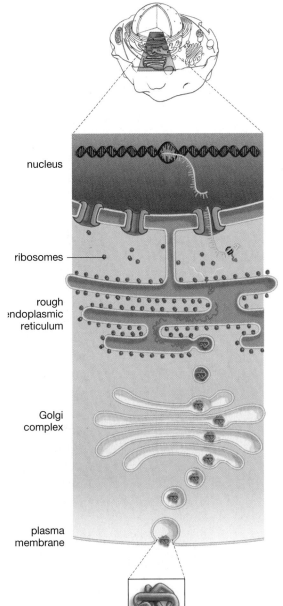

nucleus

ribosomes

rough
endoplasmic
reticulum

Golgi
complex

plasma
membrane

1. Instructions from DNA are copied onto mRNA.

2. mRNA moves to ribosome.

3. Ribosome moves to endoplasmic reticulum and "reads" mRNA instructions.

4. Amino acid chain growing from ribosome is dropped inside endoplasmic reticulum membrane. Chain folds into protein.

5. Protein moves to Golgi complex for additional processing and for sorting.

6. Protein moves to plasma membrane for export.

Figure 4.5
Path of Protein Production in Cells

Beginning in the Control Center: The Nucleus

As noted in Chapter 3, proteins are critical working molecules in living things, and DNA contains the information for producing these proteins. Our entire complement of DNA is like a cookbook that contains individual recipes, which we call genes. Each gene's chemical building blocks in effect say, "Now give me some of this, now some of this, then some of this," the final result being the specifications for a protein. (See Chapter 3, p. 60.) In the eukaryotic cell, as you've seen, DNA is largely confined within a nucleus bound by a membrane. If you look at **Figure 4.6**, you can see the nature of this membrane. It is the **nuclear envelope**: the double membrane that lines the nucleus in eukaryotic cells.

There comes a point in the life of most cells when they divide, one cell becoming two. Because (with a few exceptions) all cells must possess the set of instructions that are contained in DNA, it follows that when a cell divides, its original complement of DNA must *duplicate*, so that both cells that result from the cell division can have their own DNA. The nucleus, then, is not just the site where DNA exists; it is the site where new DNA is put together, or "synthesized," for this duplication.

Messenger RNA

At the end of Chapter 3, you saw that the process of protein synthesis requires that DNA's instructions first get copied onto another long-chain molecule, RNA.

Figure 4.6
The Cell's Nucleus
The DNA of eukaryotic cells is sequestered inside a compartment, the nucleus, which is lined by a double membrane known as the nuclear envelope. Compounds pass into and out of the nucleus through a series of microscopic channels called nuclear pores. The prominent spherical structure within the nucleus is the nucleolus, an area that specializes in the production of ribosomal RNA—the material that helps make up ribosomes. Protein production is dependent upon the information encoded in DNA's sequence of chemical building blocks. This information is copied onto a length of messenger RNA (mRNA), which then exits from the nucleus through a nuclear pore. (Micrograph: ×4400)

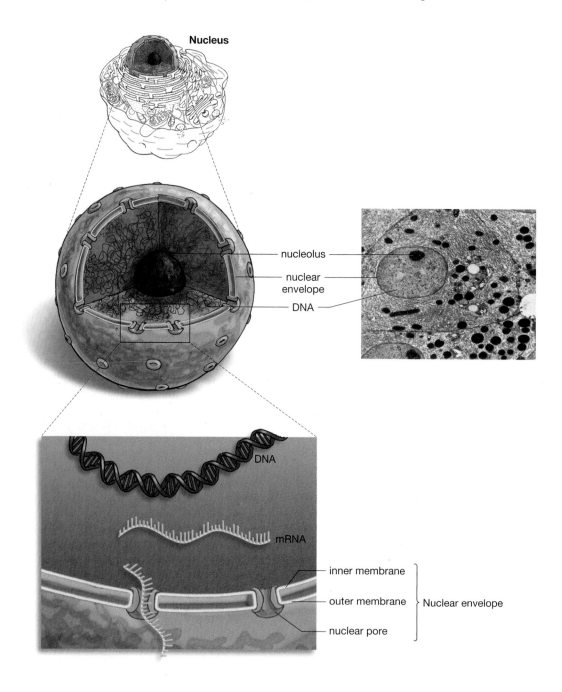

Nucleus

nucleolus

nuclear envelope

DNA

DNA

mRNA

inner membrane

outer membrane } Nuclear envelope

nuclear pore

This step is akin to having a cassette tape (of DNA) and then making a copy of it (onto RNA). Our RNA "tape" then moves out to the cell's cytoplasm to continue the process of protein synthesis. As it turns out, RNA comes in several forms. The one that the DNA instructions are copied onto is called *messenger RNA* (mRNA). Given that mRNA goes to the cytoplasm, it must, of course, have some way of getting out of the nucleus; its exit points turn out to be thousands of channels that stud the surface of the nuclear envelope—the *nuclear pores*. Materials can go the other way through the nuclear pores as well: Proteins and other materials pass from the cytoplasm into the nucleus by way of them.

mRNA Moves Out of the Nucleus

Imagine shrinking down in size so that the nucleus seems about as big as a house, with you standing outside it. What you would see in protein synthesis is lengths of mRNA, rapidly moving out through nuclear pores and dropping off into the cytoplasm. Several varieties of RNA actually come out like this, but for now, let's just follow the trail of the mRNA. You're about ready to leave the nucleus to continue your cell tour, but before you do, note two things. First, DNA contains information for making proteins, and the mRNA coming out of the nuclear pores amounts to a means of disseminating this information. Thus, it's not hard to conceptualize the nucleus as a control center for the cell. Beyond this, in looking at Figure 4.6, you've probably noticed that there is a rather imposing structure *within* the nucleus, called the nucleolus. For now, just *hold that thought*; the mRNA tapes have come out of the nuclear pores and are making a short trip to another part of the cell.

Ribosomes

Small structures called ribosomes are the destination for our mRNA tapes. For now, we can define the **ribosome** as an organelle that serves as the site of protein synthesis in the cell. In line with this, ribosomes are commonly described as the "workbenches" of protein synthesis, which is a fine metaphor. But following the notion of mRNA as a cassette tape, you might look at a ribosome as a kind of *playback head* on a cassette deck. What does such a head do in an actual deck? As a tape is run through it, it reads signals that have been laid down on it (as magnetic bits) and turns these into sound. Likewise, each ribosome acts as a site that an mRNA tape runs through, only the information on this tape results in the production of an *object* that grows from the ribosome: A chain of amino acids that folds up into the molecule we call a protein (**see Figure 4.7**).

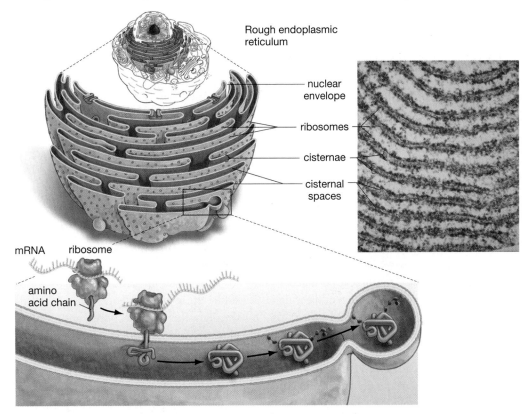

Rough endoplasmic reticulum

nuclear envelope

ribosomes

cisternae

cisternal spaces

mRNA   ribosome

amino acid chain

1. mRNA docks on ribosome. Amino acid chain production begins.

2. Ribosome docks on ER. Amino acid chain moves into cisternal space as it is completed.

3. Amino acid chain folds up making a protein.

4. Side chains added to protein.

5. Vesicle formed to house protein while in transport.

Figure 4.7
Proteins Taking Shape: Ribosomes and the Rough Endoplasmic Reticulum
The steps of the lower figure begin with a messenger RNA "tape," which has migrated from the nucleus of a cell. (**1**) The tape binds with a ribosome that is free-standing in the cytoplasm. The ribosome then starts to "read" the length of the mRNA, which bears instructions for producing a protein. The result is an amino acid (or polypeptide) chain that begins to grow from one part of the ribosome. The ribosome soon halts this activity, however, and moves toward the rough endoplasmic reticulum (rough ER). (**2**) The ribosome docks on the outside face of the rough ER and resumes reading the mRNA tape. When it is completed, the resulting amino acid chain drops into the cisternal space of the rough ER. (**3**) The amino acid chain folds up, thus becoming a protein. (**4**) The protein undergoes processing (in this example, by having a side-chain added). (**5**) The protein is then encased in a membrane-lined vesicle, which will bud off from the rough ER and be transported to the Golgi complex. (Micrograph: ×90,500)

You may remember that the kind of protein we are tracking will eventually be exported out of the cell altogether. The synthesis of this kind of protein actually stops when only a very short sequence of the amino acid (or "polypeptide") chain has grown from the ribosome. Why? "Export" polypeptide chains need to be processed within other structures in the cell before they can become fully functional proteins. The first step in this process is for the ribosome, and its associated cargo, to migrate a short distance in the cell and then attach to another cell structure.

The Rough Endoplasmic Reticulum

If you look again at Figure 4.4, you can see that, though the nucleus cuts a roughly spherical figure out of the cell, there is, in essence, a folded-up continuation of the nuclear envelope on one side. This mass of membrane has a name that is a mouthful: the **rough endoplasmic reticulum**. This structure can be defined as a network of membranes that aids in the processing of proteins in eukaryotic cells. The rough endoplasmic reticulum is rough because it is studded with ribosomes; it is endoplasmic because it lies within (endo) the cyto*plasm*; and it is a reticulum because it is a network, which is what *reticulum* means in Latin. Understandably, it is generally referred to as the rough ER or the RER.

Our ribosome, bound up with its mRNA and polypeptide chains, will migrate to the rough ER and dock on its outside face, thus joining a multitude of other ribosomes that have done the same thing. As noted, the polypeptide chain that is being output from the ribosome is, in essence, an unfinished protein that needs to go through more processing before it can be exported. The first step in this processing leads only to the other side of the ER wall the ribosome is embedded in. As the ribosome goes on with its work, the polypeptide chain it is producing drops into chambers inside the rough ER.

If you look at Figure 4.7, you can see that the whole of the rough ER takes the shape of a set of flattened sacs (called *cisternae*). The membrane that the rough ER is composed of forms the periphery of these sacs. Inside are the *cisternal spaces* of the rough ER (sometimes referred to as rough ER's *lumen*). As polypeptide chains enter the cisternal spaces, they first fold up into their protein shapes, as you saw in Chapter 3. Beyond this, most proteins that are exported from cells have sugar side-chains added to them here. Quality control of the production line is in operation in the RER as well. Polypeptide chains that have faulty sequences are detected in the RER cisternal space and ejected out of the protein production line altogether, after which they are degraded into their component parts. Other proteins will pass the quality control tests, however, and will then move out of the RER for more processing.

Several Locations for Ribosomes

All of the mRNA that comes out of the nuclear pores goes to ribosomes, but only some of these ribosomes end up migrating to the rough ER. A multitude of ribosomes will remain "free ribosomes," which is to say ribosomes that remain free-standing in the cytosol. What makes the difference? Remember how, in the ribosome we looked at, only a small stretch of the polypeptide chain it was producing emerged before the ribosome first halted its work and then migrated to the RER? That small stretch of the chain contained a chemical signal that said, in effect, "RER processing needed." The result was the ribosome's move to the RER. In general, these RER-linked ribosomes produce proteins that will reside in the cell's membranes or that will be exported out of the cell altogether (making them **secretory proteins**). Meanwhile, free ribosomes tend to produce proteins that will be used within the cell's cytoplasm or nucleus.

A Pause for the Nucleolus

Before you continue on the path of protein processing, think back a bit to the discussion about the nucleus, when you were asked to hold the thought about the large structure *inside* the nucleus, called the nucleolus. This is the point where its story can be told, because now you know what ribosomes are.

It turns out that *ribosomes* are mostly made of RNA. (They are made of a mixture of proteins and *ribo*nucleic acid.) But so great is the cell's need for ribosomes that a special section of the nucleus is devoted to the synthesis of RNA. This is the nucleolus. The *type* of RNA that's part of the ribosomes is, fittingly enough, called ribosomal RNA or rRNA, and it's one of the multiple varieties of RNA mentioned before. With this in mind, we can define the **nucleolus** as the area within the nucleus of a cell devoted to the production of ribosomal RNA.

Ribosomes are brought only to an unfinished state within the nucleolus, after which they pass through the nuclear pores and into the cytoplasm; when put together in final form, they begin receiving mRNA tapes. They are the one variety of organelle that is not lined by a membrane. (They are also the one variety of organelle that prokaryotic cells have.) The cell's traffic in ribosomes is considerable; with millions of them in existence, lots are going to be wearing out all the time. Perhaps a thousand need to be replaced every minute.

Elegant Transportation: Transport Vesicles

The proteins that have been processed within the rough ER need to move out of it and to their "downstream" destinations before being exported. But how do proteins move from one location to another within the cell? Recall that the rough ER and the nuclear membrane amount to one long, convoluted membrane. And, as just noted, all the organelles in the cell except ribosomes are "membrane-bound," meaning they have membranes at their periphery. Each of these membranes has its own chemical structure, but collectively they have an amazing ability to work together: A piece of one membrane can *bud off*, as the term goes, carrying inside it some of our proteins-in-process. Moving through the cytosol, this tiny sphere of membrane can then *fuse* with another membrane-bound organelle, releasing its protein cargo in the process. The membrane-lined spheres that move within this network, carrying proteins and other molecules, are called **transport vesicles**. The network itself is known as the **endomembrane system**: an interactive group of membrane-lined organelles and transport vesicles within eukaryotic cells.

This system gives cells a remarkable capability. One minute a piece of membrane may be an integral part of, say, the rough ER; the next it is separating off as a spheroid and moving through the cytosol, carrying proteins within. It is this system that makes it possible for our proteins-in-process to move out of the rough ER. Note, though, that many different *kinds* of proteins are being processed at any one time in the rough ER cisternae. It is as if the cellular factory has a lot of different assembly lines working at once. Most of the proteins under construction are, however, initially bound for the same place—the Golgi complex.

Downstream from the Rough ER: The Golgi Complex

Once a transport vesicle, bearing proteins, has budded off from the rough ER, it then moves through the cytosol to fuse with the membrane of another organelle, one first noticed by Italian biologist Camillo Golgi at the beginning of the twentieth century. The **Golgi complex** is a network of membranes that processes and distributes proteins that come to it from the rough endoplasmic reticulum. Some side-chains of sugar may be trimmed from proteins here, or phosphate groups may be added. But the Golgi complex does something else as well. Recall that some proteins in this production line are bound for export outside the cell, while others will end up being used within various membranes in the cell. It follows that proteins have to be *sorted and shipped* appropriately, and the Golgi does just this, acting as a kind of distribution center. Chemical "tags" that are part of the proteins often allow for this routing. Remember how, in the rough ER, carbohydrate side-chains might be attached to a newly formed protein? Oftentimes these side-chains serve as the routing tags; other times a section of the protein's amino acid sequence will serve this function.

The Golgi is similar to the ER in that it amounts to a series of cisternae, or connected membranous sacs with internal spaces (**see Figure 4.8**). Proteins

Figure 4.8
Processing and Routing: The Golgi Complex
Transport vesicles from the rough endoplasmic reticulum (RER) move to the Golgi complex, where they unload their protein contents by fusing with the Golgi membrane. The protein is then passed, within other vesicles, through the layers of disk-shaped Golgi cisternae, where editing of the protein may occur. At the part of the Golgi furthest from the rough ER, the proteins are sorted, packed in vesicles, and shipped to sites mostly in cell membranes or outside the cell altogether. The vesicles in the micrograph are the pink and purple spheres.

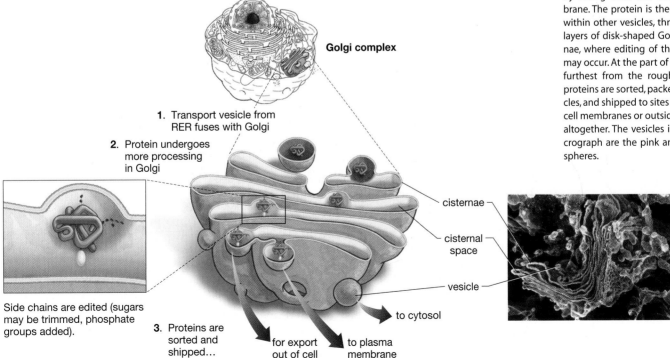

Golgi complex

1. Transport vesicle from RER fuses with Golgi

2. Protein undergoes more processing in Golgi

Side chains are edited (sugars may be trimmed, phosphate groups added).

3. Proteins are sorted and shipped...

for export out of cell

to plasma membrane

to cytosol

cisternae

cisternal space

vesicle

arrive at the Golgi housed in transport vesicles that fuse with the Golgi "face" nearest the RER, at which point the vesicles release their protein cargo into the Golgi cisternal sacs for processing. Once processed, proteins of the sort we are following eventually bud off from the outside face of the Golgi, now housed in their final transport vesicles.

From the Golgi to the Surface

For secretory proteins, the journey that began with the copying of DNA information onto messenger RNA is almost over. Once a vesicle buds off from the Golgi, all that remains is for it to make its way through the cytosol to the plasma membrane at the outer reaches of the cell. There, the vesicle fuses with the plasma membrane and the protein is ejected into the extracellular world. This last step, called exocytosis, is a process you'll be looking at next chapter. With it, one finished product of the cellular factory has rolled out the door.

4.5 Outside the Protein Production Path: Other Cell Structures

A functioning cell engages in more activities than the protein synthesis and shipment just reviewed, for the simple reason that cells do a lot more than produce proteins.

The Smooth Endoplasmic Reticulum

If you look back to Figure 4.4, you can see that there actually are *two* kinds of endoplasmic reticuli. The part of the ER membrane, farther out from the nucleus, that has no ribosomes is called the smooth endoplasmic reticulum or smooth ER. It's "smooth" because it is not peppered with ribosomes, and this very quality means it is not a site of protein synthesis. Instead, the **smooth endoplasmic reticulum** is a network of membranes that is the site of the synthesis of various lipids, and a site at which potentially harmful substances are detoxified within the cell. The tasks the smooth ER undertakes, however, will vary in accordance with cell type. The lipids we normally think of as "fats" are synthesized and stored in the smooth ER of liver and fat cells, while the "steroid" lipids reviewed last chapter—testosterone and estrogen—are synthesized in the smooth ER of the ovaries and testes. The detoxification of potentially harmful substances, such as alcohol, takes place largely in the smooth ER of liver cells.

Web Tutorial 4.2
The Structure of Cells

Tiny Acid Vats: Lysosomes and Cellular Recycling

Any factory must be able to get rid of some old materials, while recycling others. A factory also needs new materials, brought in from the outside, that probably will have to undergo some processing before being used. A single organelle in the animal cell aids in doing all these things. It is the **lysosome**: an organelle found in animal cells that digests worn-out cellular materials and foreign materials that enter the cell. Several hundred of these membrane-bound organelles may exist in any given cell. You could think of them as sealed-off acid vats that take in large molecules, break them down, and then return the resulting smaller molecules to the cytosol. What they cannot return, they retain inside themselves or expel outside the cell. They carry out this work not only on molecules entering the cell from the outside (say, invading bacteria) but on materials that exist inside the cell—on worn-out organelle parts, for example (**see Figure 4.9).**

A given lysosome may be filled with as many as 40 different enzymes that can break larger molecules into their component parts—an enzymatic array that allows each lysosome to break down most of what comes its way. A lysosome gets ahold of its macromolecule prey through the endomembrane system. A lysosome will fuse with the membrane surrounding a worn-out organelle part; proceeding to engulf it, the lysosome then goes to work breaking the organelle down. The small molecules that result then pass freely out of the lysosome and into the cytosol for reuse elsewhere. Thus, there is recycling at the cellular level. Cells carry out this kind of self-renewal at an amazing rate. Christian de Duve, who with his colleagues discovered lysosomes in the 1950s, has noted the effect of this activity on human brain cells. In an elderly person, he notes, such cells

> have been there for decades. Yet most of their mitochondria, ribosomes, membranes, and other organelles are less than a month old. Over the years, the cells have destroyed and remade most of their constituent molecules from hundreds to thousands of times, some even more than 100,000 times.

There are certain things, however, that even lysosomes cannot digest, and over time these substances can cause lysosomes to leak their acidic contents into the cell, causing great harm. You may have heard of an affliction called black lung disease that miners are prey to. One form of this disease, called silicosis, is caused by the inability of lysosomes to digest the silica fibers that miners inhale. Because these fibers have a needle-like shape, they can cause the lysosomes to leak. The result is an acid spill that can kill cells and produce scar tissue in the lungs, thus reducing the miners' ability to breathe.

Extracting Energy from Food: Mitochondria

Just as there is no such thing as a free lunch, there is no such thing as free lysosome activity, or ribosomal

action, or protein export. There is a price to be paid for all these things, and it is called energy expenditure. The fuel for this energy is contained in the food that cells ingest. But the energy in this food has to be converted into a molecular *form* that the cell can easily use, just as the energy in, say, coal needs to be converted by a power plant into a form that home appliances can easily use—electricity. The place to look for most of this conversion is inside

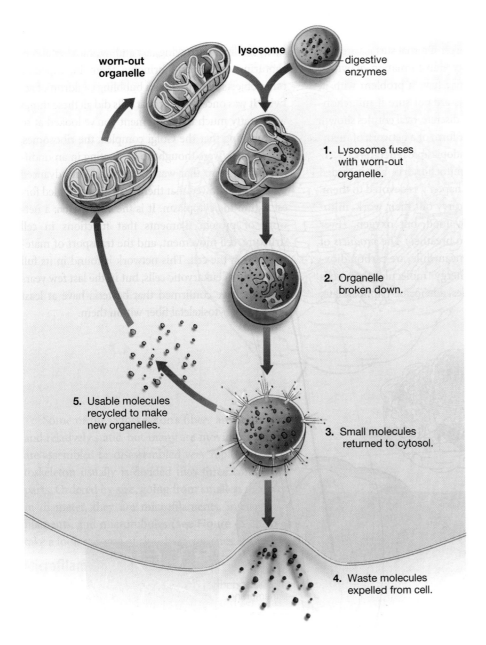

worn-out organelle

lysosome

digestive enzymes

1. Lysosome fuses with worn-out organelle.

2. Organelle broken down.

5. Usable molecules recycled to make new organelles.

3. Small molecules returned to cytosol.

4. Waste molecules expelled from cell.

Figure 4.9
Cellular Recycling: Lysosomes
Lysosomes are membrane-bound organelles that contain potent enzymes capable of digesting large molecules, such as worn-out organelles. The useful parts of such organelles will be returned to the cytosol and used elsewhere—a form of cellular recycling. If a lysosome cannot digest a given material, it may expel it outside the cell, though, in multicelled organisms, lysosomes generally will hold on to such materials, so as not to harm structures outside the cell. Lysosomes also digest small particles, such as food, that come from outside the cell, along with invading organisms, such as bacteria.

export, and so on). You've also seen that cells have other structures such as lysosomes for digestion and recycling, mitochondria for energy transformation, the smooth endoplasmic reticulum for detoxification and lipid synthesis, and the cytoskeleton for structure and movement. If you look at **Figure 4.14**, you can see, in metaphorical form, a "map" of these component parts within the cell. Table 4.1 (on page 88 in the discussion of plant cells) lists cellular elements found in plants and animals, as well as some elements found only in plant cells.

4.7 The Plant Cell

As noted earlier, plant and animal cells have more similarities than they do differences. This is understandable because plant cells do most of the things that animal cells do (produce proteins, transform energy, and so on). A quick look at **Figure 4.15** will confirm for you how structurally similar plant and animal cells are. As you see, a plant cell has a nucleus (with a nucleolus), the smooth and rough ERs, a cytoskeleton—most of the things you've just gone over in animal cells. Indeed, there is only one structure present in the animal cells you've looked at that

Figure 4.14
The Cell as a Factory
This comparison may help you to remember some roles of the different parts of the cell.

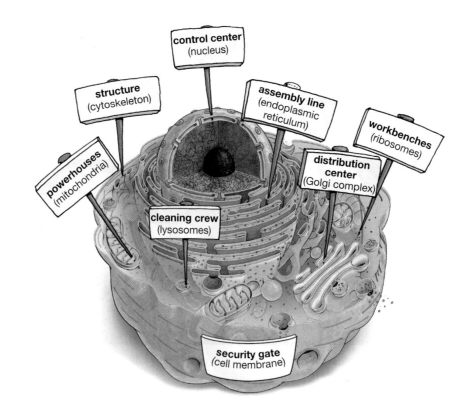

Figure 4.15
Common Structures in Animal and Plant Cells

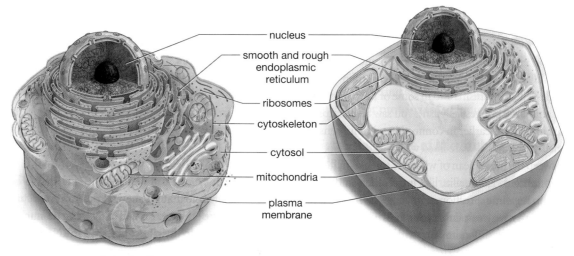

Animal cell

Plant cell

plant cells don't have: the lysosome. What jumps out at you when you look at plant cells is not what they lack compared to animal cells, but what they *have* that animal cells do not. As you can see in **Figure 4.16**, these additions are

- A thick cell wall
- A large structure called a central vacuole
- Structures called plastids, one important variety of which is the chloroplast

The Cell Wall

Plant cells have an outer protective lining, called a **cell wall**, that makes their plasma membrane, just inside it, look rather thin and frail by comparison. This is because it is thin and frail by comparison; the plasma membrane of a plant cell may be 0.01 μm thick, while the combined units of a cell wall may be 700 times this width—7 μm or more. Cell walls are nearly always present in plant cells. They also exist in many organisms that are neither plant nor animal—bacteria, protists, and fungi—though their cell walls differ in chemical composition from those of plants.

What do cell walls do for plant cells? They provide them with structural strength, put a limit on their absorption of water (as you'll see next chapter), and generally protect plants from harmful outside influences. So, if they're so useful, why don't *animal* cells have them? Cell walls make for a rather rigid, inflexible organism—like plants, which are stationary. Animals, meanwhile, are mobile and thus must remain flexible.

Cell walls in plants can come in several forms, but all such forms will be composed chiefly of a molecule you were introduced to last chapter:

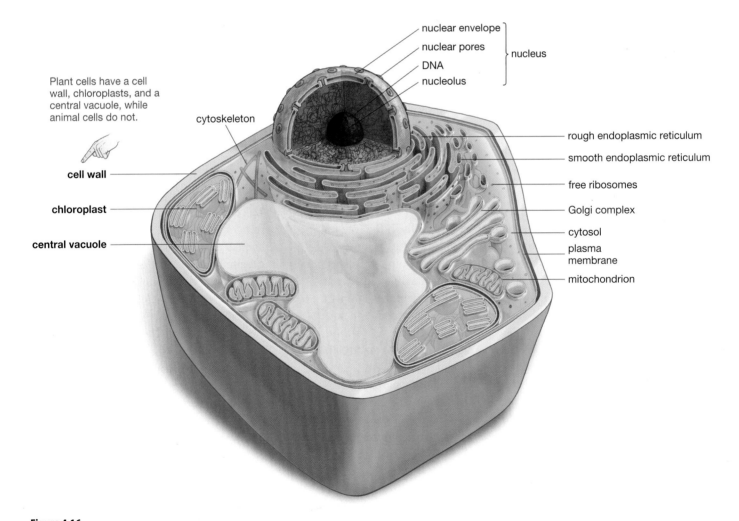

Plant cells have a cell wall, chloroplasts, and a central vacuole, while animal cells do not.

cytoskeleton

nuclear envelope
nuclear pores
DNA
nucleolus
nucleus

cell wall

chloroplast

central vacuole

rough endoplasmic reticulum
smooth endoplasmic reticulum
free ribosomes
Golgi complex
cytosol
plasma membrane
mitochondrion

Figure 4.16
The Plant Cell
The cell wall, central vacuole, and chloroplasts do not exist in animal cells, but all other components are common to both plant and animal cells.

Table 4.1 Structures in Plant and Animal Cells

Name	Location	Function
Cytoskeleton	Cytoplasm	Maintains cell shape, facilitates cell movement and movement of materials within cell
Cytosol	Cytoplasm	Protein-rich fluid in which organelles and cytoskeleton are immersed
Golgi complex	Cytoplasm	Processing, sorting of proteins
Lysosomes (in animal cells only)	Cytoplasm	Digestion of imported materials and cell's own used materials
Mitochondria	Cytoplasm	Transform energy from food
Nucleolus	Nucleus	Synthesis of ribosomal RNA
Nucleus	Inside nuclear envelope	Site of most of the cell's DNA
Ribosomes	Rough ER, Free-standing in cytoplasm	Sites of protein synthesis
Rough endoplasmic reticulum	Cytoplasm	Protein processing
Smooth endoplasmic reticulum	Cytoplasm	Lipid synthesis, storage; detoxification of harmful substances
Vesicles	Cytoplasm	Transport of proteins and other cellular materials
Cell walls (in plant cells only)	Outside plasma membrane	Limit water uptake; maintain cell membrane shape, protect from outside influences
Central vacuole (in plant cells only)	Cytoplasm	Cell metabolism, pH balance, digestion, water maintenance
Plastids (in plant cells only)	Cytoplasm	Nutrient storage, pigmentation, photosynthesis (chloroplasts)

cellulose, a complex sugar or "polysaccharide" that is embedded within cell walls in the way reinforcing bars run through concrete. In some cell walls, cellulose is joined by a compound called lignin, which imparts considerable structural strength. You can see a vivid demonstration of this in the material we know as wood, which is largely made of cell walls (**see Figure 4.17a**).

Cell walls can serve different functions over the life of an organism. Generally, they are the site of a good deal of metabolic activity, and thus should not be considered mere barriers. On the other hand, the outer portion of tree bark consists mostly of *dead* cell walls, which clearly are serving a barrier function (**see Figure 4.17b**).

The Central Vacuole

In looking at the diagram of the plant cell, you'll see one structure, the central vacuole, that is so prominent it appears to be a kind of organelle continent surrounded by a mere moat of cytosol. In a mature plant cell, one or two central vacuoles may comprise 90 percent of cell volume.

Although animal cells can have vacuoles, the imposing central vacuole in plants is different. For a start, it is composed mostly of water, which demonstrates just how watery plant cells are. A typical animal cell may be 70 to 85 percent water, but for plant cells the water proportion is likely to be 90 to 98 percent.

The watery milieu of the central vacuole contains hundreds of other substances. Many of these are nutrients, others are waste products. There are also hydrogen ions, pumped in to keep the cell's cytoplasm at a near-neutral pH. Given these diverse materials, the **central vacuole** can be defined as a large, watery plant organelle that has many functions, among them the storage of nutrients and the retention and degradation of waste products. In this latter role, the central vacuole uses digestive enzymes, much like the lysosomes in animal cells. There is even an aesthetic side to the central vacuole: Many red and blue flowers owe their colors to the pigments it contains.

Plastids

Plastids are a diverse group of organelles that are found only in plants and algae. Some gather and store nutrients for plant cells; others are pigment-containing organelles that give us, for example, the red color of the tomato skin. The best known of the plastids, however, are the chloroplasts.

Human beings are indebted, in a sense, to chloroplast-containing organisms, because most of the food we eat and the oxygen we breathe comes from them. This is so because these organisms carry out photosynthesis, the process by which organisms produce their own food (a sugar), using nothing but the energy contained in sunlight and the starting materials of carbon dioxide, water, and a few minerals. As it turns out, the **chloroplast** is the organelle that is the site of photosynthesis in plant and

(a) Wood is mostly cell walls

(b) A magnified view of bark

Figure 4.17
Great Strength from Small Things

(a) Redwoods such as this one can reach enormous heights because of the strength of their wood, which is largely made of cell walls.

(b) Cell walls play important roles in living cells, but they also are valuable as the strong, remaining components of dead cells; here they help make up this redwood bark. ($\times 240$)

algae cells (**see Figure 4.18**). This may not seem like such a big deal until you try to name a food you eat that is not a plant or does not itself eat plants. A by-product of photosynthesis is oxygen, whose significance to us can scarcely be overstated.

4.8 Cell-to-Cell Communication

Most of what you've seen so far has made the cell seem like an isolated entity, but this is not the case. Single-celled organisms can exist as separate entities, along with certain plant or animal cells (red blood cells, for example), but most plant and animal cells are linked together in organized collections referred to as tissues. Not surprisingly, these assemblages of cells—be they plant or animal—have the ability to communicate with one another.

Communication among Plant Cells

Having noted the thickness of something like the cell walls in plants, you might wonder how one plant cell could interact with another. Communication between plant cells takes place quite readily, however, through a series of tiny channels in the plant cell wall called **plasmodesmata** (singular, plasmodesma). The structure of these channels is such that the cytoplasm of one plant cell is continuous with that of another—so much so that the cytoplasm of an entire plant can be properly looked at as one continuous whole (**see Figure 4.19a**). The structure of plasmodesmata is more complex than that of a simple opening, but the basic idea is of a channel-like linkage between two plant cells.

Communication among Animal Cells

There are no plasmodesmata in animal cells, but there are three other kinds of *cell junctions*, or linkages, one of which serves to facilitate cell communication. It is called a gap junction, and it consists of clusters of protein structures that shoot through the plasma membrane of a cell from one side to the other, forming a kind of tube. When these tubes line up in adjacent cells, the result is a channel for passage of small molecules and electrical signals (see Figure 4.19b). Thus, we can define a **gap junction** as a protein assemblage that forms a communication channel between adjacent animal cells. Note that animal gap junctions and plant plasmodesmata are very different kinds of channels. Plasmodesmata can be thought of as permanent channels between plant cells, whereas gap junctions open only as necessary.

How Did We Learn?

The cellular world you have been reading about was completely unknown to human beings before the seventeenth century. It came as quite a shock back then to learn that the human body had, living within it, a multitude of creatures who were so small they could not be perceived with the naked eye. In "First Sightings: Anton van Leeuwenhoek" on page 92, you can read about the Dutchman who was one of the first visitors to the micro-world.

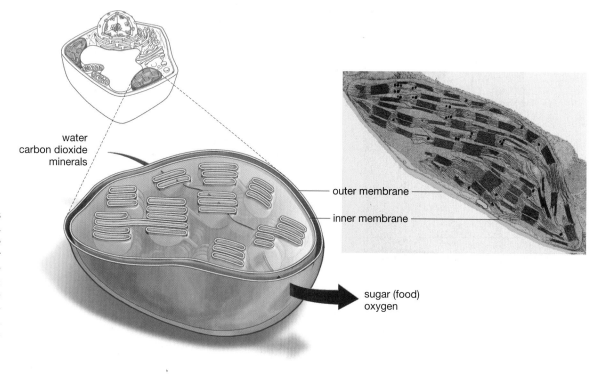

Figure 4.18
Food Source for the World
Chloroplasts, the tiny organelles that exist in plant and algae cells, are sites of photosynthesis—the process that provides food for most of the living world. Using the starting materials of water, carbon dioxide, and a few minerals, these organisms use energy from sunlight to produce their own food. A double membrane lines the chloroplasts.
(Micrograph: ×13,000)

water
carbon dioxide
minerals

outer membrane

inner membrane

sugar (food)
oxygen

(a) Plant tissues

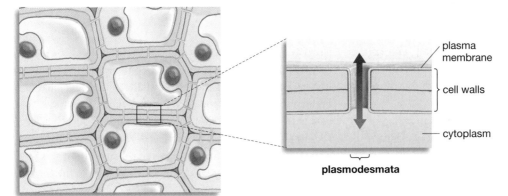

(b) Animal tissues

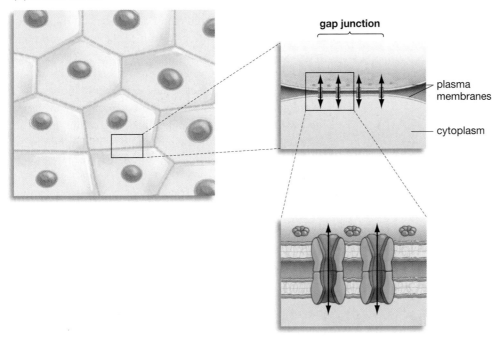

Figure 4.19
Cell Communication

(a) Plasmodesmata

In plants, a series of tiny pores between plant cells, the plasmodesmata, allow for the movement of materials among cells. Thanks to the plasmodesmata channels, the cytoplasm of one cell is continuous with the cytoplasm of the next; the plant as a whole can be thought of as having a single complement of continuous cytoplasm.

(b) Gap junctions

In animals, protein assemblies come into alignment with one another, forming communication channels between cells. A cluster of many such assemblies—perhaps several hundred—is called a gap junction.

On to the Periphery

Having looked at what is inside the cell, you've arrived at the cell's periphery, the plasma membrane, which is where you'll be staying for awhile. It may at first seem strange to devote a whole chapter to an outer boundary. How much attention would you pay, after all, to a factory's wall as opposed to its contents? The answer: A lot, if that wall could facilitate communication with the outer world, continually renew itself, and let some things in while keeping others out. Such is the case with the plasma membrane, a slender lining that manages to make one of the most fundamental distinctions on Earth: Inside, life goes on; outside it does not.

How Did We Learn?
First Sightings: Anton van Leeuwenhoek

To the list of explorers that includes Columbus and Balboa we could add the name Anton van Leeuwenhoek. This unassuming Dutchman was the great early voyager into another world, the micro-world.

It was not until the seventeenth century that human beings realized that things as small as cells existed. Leeuwenhoek began to report in the 1670s on what he saw with the aid of a device that had been invented at the end of the 1500s—the microscope (**Figure 1**). One of Leeuwenhoek's contemporaries, Englishman Robert Hooke, coined the term *cell* after viewing a slice of cork under a simple microscope, but Hooke's purpose was to reveal the detailed structure of familiar, small objects, such as the flea. Leeuwenhoek, by contrast, revealed the *existence* of creatures unimagined until his time. Moreover, he carried out this work in the most extraordinary fashion: Laboring alone in the small town of Delft with palm-sized magnifiers he himself had created, looking at anything that struck his fancy. (And many things struck his fancy; he once looked at exploding gunpowder under a microscope, nearly blinding himself in the process.) For 50 years, while working as shopkeeper and minor city official, this untrained amateur of boundless curiosity examined the micro-world and reported on it in letters he posted to the Royal Society in London.

Who could believe what he uncovered? How was it possible that there was a buzzing, blooming universe of "animalcules" (little animals) whose existence had been completely unsuspected? Prior to his work, no one thought that any creature smaller than a worm could exist within the human body. Yet, examining scrapings from his own mouth, Leeuwenhoek tells us:

> I saw, with as great a wonderment as ever before, an inconceivably great number of little animalcules, and in so unbelievably small a quantity of the foresaid stuff, that those who didn't see it with their own eyes could scarce credit it.

The animalcules that Leeuwenhoek beheld over his career were single-celled organisms that today are known as bacteria and protists. It would be two hundred years before Leeuwenhoek's findings were fully integrated into a modern theory of cells. Yet the man from Delft had shown that only a small portion of the world's living things are visible things.

(a) What Leeuwenhoek could see

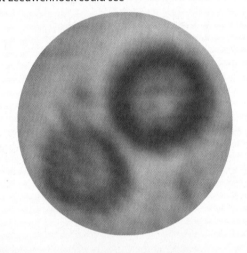

(b) Leeuwenhoek with two of his microscopes

Figure 1

(a) What Leeuwenhoek Could See
Biologist Brian Ford used an actual 300-year-old Leeuwenhoek microscope to capture this image of spiny spores from a truffle. ($\times$600)

(b) Leeuwenhoek's Microscopes Were Hand-Held and Paddle-Shaped
Leeuwenhoek revealing the micro-world to Queen Catherine of England, wife of King Charles II (1630–1685).

Chapter Review

Summary

4.1 Cells Are the Working Units of Life

- With the possible exception of viruses, every form of life on Earth either is a cell or is composed of cells. Cells come into existence only through the activity of other cells.

4.2 All Cells Are Either Prokaryotic or Eukaryotic

- All cells can be classified as prokaryotic or eukaryotic. Prokaryotic cells either are bacteria or another single-celled life-form called archaea. Setting bacteria and archaea aside, all other cells are eukaryotic.

- Eukaryotic cells have most of their DNA contained in a membrane-lined compartment, called the cell nucleus, whereas prokaryotic cells do not have a nucleus. Eukaryotic cells also tend to be much larger than prokaryotic cells, and they have more of the specialized internal structures called organelles than do prokaryotic cells. Many eukaryotes are multicelled organisms, whereas all prokaryotes are single-celled.
Web Tutorial 4.1: Prokaryotic and Eukaryotic Cells

4.3 The Eukaryotic Cell

- There are five principal components to the eukaryotic cell: the nucleus, other organelles, the cytosol, the cytoskeleton, and the plasma membrane. Organelles are "tiny organs" within the cell that carry out specialized functions, such as energy transfer and material recycling. The cytosol is the fluid outside the nucleus in which these organelles are immersed. (The cytosol should not be confused with the cytoplasm, which is the region of the cell inside the plasma membrane but outside the nucleus.) The cytoskeleton is a network of protein filaments that function in cell structure, cell movement, and the transport of materials within the cell. The plasma membrane is the outer lining of the cell. A membrane can be defined as the flexible, chemically active outer lining of a cell or of its compartments.

4.4 A Tour of the Animal Cell: Along the Protein Production Path

- Information for the construction of proteins is contained in the DNA located in the cell nucleus. This information is copied onto an informational "tape" of messenger RNA (mRNA) that departs the cell nucleus through its nuclear pores and goes to the sites of protein synthesis, structures called ribosomes, which lie in the cytoplasm. Many ribosomes that receive mRNA tapes process only a short stretch of them before migrating to, and then embedding in, one of a series of sacs in a membrane network called the rough endoplasmic reticulum (RER). The polypeptide chains produced by the ribosomal "reading" of the mRNA tapes are dropped from ribosomes into the internal spaces of the RER; there, the polypeptide chains fold up, thus becoming proteins, and undergo editing.

Some ribosomes are not embedded in rough endoplasmic reticulum but instead remain free-standing in the cytosol.

- Materials move from one structure to another in the cell via the endomembrane system, in which a piece of membrane, with proteins or other materials inside, can bud off from one organelle, move through the cell, and then fuse with another membrane-lined structure. Membrane-lined structures that carry cellular materials are called transport vesicles.

- Once protein processing is finished in the rough endoplasmic reticulum, proteins undergoing processing move, via transport vesicles, to the Golgi complex, where they are processed further and marked for shipment to appropriate cellular locations.

4.5 Outside the Protein Production Path: Other Cell Structures

- The smooth endoplasmic reticulum is a network of membranes that functions to synthesize lipids and to detoxify potentially harmful substances. Lysosomes are organelles that break down worn-out cellular structures or foreign material that comes into the cell. Once this digestion is completed, the lysosomes return the molecular components of these materials to the cytoplasm for further use. Mitochondria are organelles that function to extract energy from food and to transform this energy into a chemical form the cell can use.
Web Tutorial 4.2: The Structure of Cells

4.6 The Cytoskeleton: Internal Scaffolding

- Cells have within them a web of protein strands, called a cytoskeleton, that provide the cell with structure, facilitate the movement of materials inside the cell, and facilitate cell movement.

- There are three principal types of cytoskeleton elements. Ordered by size, going from smallest to largest in diameter, they are microfilaments, intermediate filaments, and microtubules. Microfilaments are made of the protein actin. They help the cell move and capture prey by forming rapidly in the direction of movement and decomposing rapidly at their other end. Intermediate filaments provide support and structure to the cell. Microtubules play a structural role in cells and facilitate the movement of materials inside the cell by serving as transport "rails."

- Cilia and flagella are extensions of cells composed of microtubules. Cilia extend from cells in great numbers, serving to move the cell or to move material around the cell. By contrast, one—or at most a few—flagella extend from cells that have them. The function of flagella is cell movement.

4.7 The Plant Cell

- Plant cells have almost all the structures found in animal cells. In addition, plant cells have a cell wall, a large central vacuole, and organelles called plastids, one variety of which is the chloroplasts that are the sites of photosynthesis. The cell wall gives the plant structural strength and helps regulate the intake and retention of water.

4.8 Cell-to-Cell Communication

- Cells are able to communicate with each other through special structures. Plant cells have channels, called plasmodesmata, that are always open and hence have the effect of making the cytoplasm of one plant cell continuous with that of another. Adjacent animal cells have channels, called gap junctions, that are composed of protein assemblages that open only as necessary, allowing the movement of small molecules and electrical signals between cells.

Key Terms

cell wall 87
central vacuole 88
chloroplast 89
cilia 84
cytoplasm 75
cytoskeleton 82
cytosol 75
endomembrane system 79
eukaryotic cell 70
flagella 85
gap junction 90
Golgi complex 79
intermediate filament 84
lysosome 80
microfilament 83
micrograph 72
micrometer 72

microtubule 84
mitochondria 82
nanometer 72
nuclear envelope 76
nucleolus 78
nucleus 70
organelle 71
plasma membrane 75
plasmodesmata 90
prokaryotic cell 70
ribosome 77
rough endoplasmic
  reticulum 78
secretory protein 78
smooth endoplasmic
  reticulum 80
transport vesicle 79

Understanding the Basics

Multiple-Choice Questions (answers in the back of the book)

1. Jerome has strep throat, a bacterial infection. The cause of the infection is
   a) the growth of a virus
   b) the presence of archaea
   c) eukaryotic cells dividing in his throat
   d) organelles that take control of his organs—in this case, his throat
   e) prokaryotic cells

2. Where would you expect to find a cytoskeleton?
   a) primarily inside the nucleus
   b) as the internal structure of a mitochondrion
   c) between bacterial cells
   d) throughout the cytosol
   e) as an outer coat on an insect

3. Suppose Dr. Hyde found a cell that had many mitochondria, a nucleus, a cell wall made of cellulose, and an endoplasmic reticulum, as well as many other parts. He might assume that he has found
   a) a plant cell
   b) an animal cell
   c) a bacterial cell
   d) one of these three types, but he will not know which type without further investigation
   e) either a plant cell or an animal cell, but not a bacterial cell

4. Cells in the pancreas manufacture large amounts of protein. Which of these would you expect to find a large amount of, or number of, in pancreatic cells?
   a) lysosomes
   b) rough endoplasmic reticulum
   c) smooth endoplasmic reticulum
   d) chloroplasts
   e) plasma membrane

5. Suppose that a cell could be seen to lack intermediate filaments. Which of these would be the most likely effect of this condition?
   a) an inability to get energy out of food
   b) an inability to manufacture proteins
   c) a tendency for the nucleus and organelles in the cell to drift around inside the cell
   d) an inability of the skeletal muscles to contract
   e) an inability to move proteins from one part of the cell to another

6. Heart muscle cells have a number of gap junctions connecting them to their adjoining cells. From this you can conclude that heart muscle cells
   a) exchange nuclei very frequently
   b) have plasmodesmata
   c) move vacuoles from cell to cell
   d) communicate frequently
   e) lack the ability to divide

7. Which is the correct ranking of these "small" things, from smallest to largest? (Use typical sizes.)
   a) animal cells < atoms < bacteria < proteins < amino acids
   b) atoms < amino acids < proteins < bacteria < animal cells
   c) atoms < proteins < amino acids < animal cells < bacteria
   d) bacteria < atoms < amino acids < proteins < animal cells
   e) bacteria < animal cells < atoms < amino acids < proteins

8. Cells need large amounts of ribosomal RNA to make proteins. The ribosomal RNA is made in a specialized structure known as _____, which is found in _____.
   a) a chloroplast … the cytosol
   b) a ribosome … the cytosol
   c) the endoplasmic reticulum … the nucleus
   d) the nuclear envelope … the nucleus
   e) the nucleolus … the nucleus

9. Many antibiotics work by blocking the function of ribosomes. Therefore, these antibiotics will:
   a) block DNA synthesis
   b) block RNA synthesis
   c) block protein synthesis
   d) prevent the movement of proteins through nuclear pores
   e) make the two nuclear membranes fuse into one

Brief Review

1. Suppose your entire body were just one gigantic cell with one central nucleus and lots of organelles outside it to perform the various functions. What problems can you envision with this system?

2. What are some of the differences between eukaryotic cells and prokaryotic cells?

3. Insulin, the hormone that controls sugar levels in the body, is a protein that is exported from special cells in the pancreas. Trace the path from a length of DNA in the nucleus that carries the code for insulin to the release of the hormone in the bloodstream.

4. What are cilia and flagella? What are some of the roles they play in the human body?

5. Consider what plants would be like if they had no plasmodesmata. Explain why this would be a problem.

6. What is a gap junction, and what is its function?

Applying Your Knowledge

1. The earliest organisms in the fossil record are all single-celled. Why do you think life had to "start small" like this?

2. Why do membranes figure so prominently in eukaryotic cells? What essential function do they serve?

5.1 The Nature of the Plasma Membrane

So, what is the nature of the plasma membrane? First, it's worth noting that it's very much like the membranes described in Chapter 4. Much of what follows about the plasma membrane could also be said of the various membranes that are internal to the cell—those that line lysosomes or that make up the Golgi apparatus, for example. There are some important differences between the plasma and other membranes, however, because the plasma membrane is constantly interacting with the world outside the cell, whereas internal membranes are not.

When we start to consider the qualities of the plasma membrane, we find that, even in the micro-world of cells, it is an extremely *thin* entity. If you stacked 10,000 of them on top of one another, their combined width would be about that of a sheet of paper. Next, the plasma membrane has a fluid and somewhat fatty makeup. If we looked for a counterpart to it in our everyday world, a soap bubble might come close, meaning this membrane is very flexible. Yet it is stable enough to stay together despite being constantly re-formed, due to the endless movement of materials in and out of it. If you imagine half the surface of your skin remaking itself every 30 minutes or so, you begin to get the picture about how dynamic this membrane is. Let's look now at the most important component parts of it. All of these components are illustrated in **Figure 5.1**. They are:

Web Tutorial 5.1
Plasma Membranes and Diffusion

- The phospholipid bilayer. Just as bread is made mostly from flour, the plasma membrane is made mostly from two layers of phospholipid molecules, whose "tails" mingle together, but whose "heads" point in opposite directions—to the interior of the cell in one direction, and to the world outside the cell in the other.

- Cholesterol. Like mortar between bricks, cholesterol acts as a "patching" material for the membrane; it also keeps the membrane at an optimal level of fluidity.

- Proteins. These molecules shoot through the membrane and lie on either side of it, serving as support structures, signaling antennas, identification markers, and cellular passageways.

- Glycocalyx. This is the collective term for a set of carbohydrate chains that layer the outside of the membrane.

Now let's look at each of these components in greater detail.

First Component: The Phospholipid Bilayer

You may recall the discussion in Chapter 3 of phospholipids—molecules that have two long fatty-acid chains linked to a phosphate-bearing group (**see Figure 5.2**). You may also recall that, because fatty-acid chains are lipids, they are hydrophobic or "water-fearing," meaning they will not bond with water. Conversely, phosphate groups are charged, hydro*philic* molecules, meaning they are "water-loving" and will *readily* bond with water. When such phosphate and lipid components are put together, the resulting phospholipid is a molecule with a dual nature. Its phosphate "head" will seek water, but its fatty-acid "tail" will avoid it. Drop a group of phospholipids in water, and they will arrange themselves into the form you see in Figure 5.2: two *layers* of phospholipids sandwiched together, the tails of each layer pointing inward (thus avoiding water) and the heads pointing outward (thus bonding with the watery environment that lies both inside and outside the cell). With this in mind, the **phospholipid bilayer** can be defined as a chief component of the plasma membrane, composed of two layers of phospholipids, arranged with their fatty-acid chains pointing toward each other.

This phospholipid structure has a couple of important consequences for the plasma membrane. First, it helps give the membrane its fluid nature: the bilayer's fatty-acid tails have a chemical makeup that is more like that of oil than shortening. Second, because these are two hydrophobic layers of tails, the only substances that can pass through them with ease are *other* hydrophobic substances or a few very small molecules. This stems from the rule of thumb about solvents noted in Chapter 3: Like dissolves like. Fats dissolve in fats, charged molecules dissolve in charged molecules. Thus, various steroid hormones (which, remember, are lipids) gain fairly easy entry into the cell, as do fatty acids, because these substances will dissolve in the lipid bilayer. Conversely, hydro*philic* substances—ions, amino acids, all kinds of polar molecules—will not dissolve in the membrane and thus cannot make it past the fatty-acid chains without help. What kind of help? Stay tuned.

Second Component: Cholesterol

Molecules of the lipid material cholesterol nestle between phospholipid molecules throughout the plasma membrane, performing two functions. First, they act as a kind of patching substance on the bilayer, keeping some small molecules from getting through. Second, they help keep the membrane at an optimum level of fluidity. Without cholesterol,

materials through. We'll review all the forms of protein-aided transport shortly. For now, just note the existence of **transport proteins** proteins that facilitate the movement of molecules or ions from one side of the plasma membrane to the other.

Fourth Component: The Glycocalyx

If you look at Figure 5.1 and focus on the extracellular face of the plasma membrane, you can see, protruding from proteins and phospholipids alike, a number of short, branched extensions. These are simple carbohydrate chains, meaning sugar chains. It turns out that these chains serve as the actual binding sites for many signaling molecules, including insulin. (Remember in Chapter 4 how many of the proteins that were constructed had sugar side-chains added to them? Well, now you can see what some of these chains do.) Sugar chains also serve to lubricate cells, and allow them to stick to other cells by acting as an adhesion layer. Collectively, all these chains form the **glycocalyx:** an outer layer of the plasma membrane, composed of short carbohydrate chains that attach to membrane proteins and phospholipid molecules.

The Fluid-Mosaic Membrane Model

With this review of component parts under your belt, you're ready for a definition of the **plasma membrane.** It is a membrane, forming the outer boundary of many cells, composed of a phospholipid bilayer that is interspersed with proteins and cholesterol and coated on its exterior face with carbohydrate chains.

Taking a step back, the general message here is that the plasma membrane is a loose, lipid structure peppered with proteins and coated with sugars. A better image, however, might be of a *sea* of lipids that has proteins floating on it, because as it turns out, the plasma membrane is fluid enough that most of its elements are able to move sideways or "laterally" through it fairly freely. Indeed, membrane proteins frequently are compared to icebergs drifting through an ocean. (The proteins are numerous enough, however, that it's tempting to liken the cell surface to the Arctic Ocean at about the time the ice pack starts to break up.) When we overlay this quality of movement on the basic membrane structure just reviewed, the result is the contemporary view of the plasma membrane, the **fluid-mosaic model:** A conceptualization of the plasma membrane as a fluid, phospholipid bilayer that has, moving laterally within it, a mosaic of proteins.

5.2   Diffusion, Gradients, and Osmosis

It may be obvious that one of the main roles of the plasma membrane is to let in whatever's needed and to keep out whatever's not. Having learned something about the membrane's makeup, you will now see how it carries out this task of passage and blockage.

It is necessary to begin this story, however, not with the cell, but with the more general notion of how substances go from places where they are more concentrated to places where they are less concentrated. Anyone who has put some food dye in water has a sense of how this works: A few drops of, say, red dye will tumble into the liquid, start dispersing, and eventually the result is a solution that is uniformly more reddish than plain water (**see Figure 5.4**). The question that will be addressed here is why this takes place.

(a) Dye is dropped in   **(b)** Diffusion begins   **(c)** Dye is evenly distributed

water molecules

dye molecules

Figure 5.4
From Concentrated to Dispersed
Diffusion is the movement of molecules or ions from areas of their greater concentration to areas of their lesser concentration. In this sequence of photos and diagrams, a few drops of red dye, added to a beaker of water, are at first heavily concentrated in one area but then begin to diffuse, eventually becoming evenly distributed throughout the solution.

Random Movement and Even Distribution

All molecules or ions are constantly in motion, and this motion is random. (The *degree* to which molecules are in motion defines their temperature; absolute zero is the point at which all molecular motion has ceased.) The laws of thermodynamics dictate that, because the motion of molecules is random, they will naturally move from any initial *ordered* state—in our example, their concentrated state when first dropped in the container—to their most disordered state, meaning evenly distributed throughout a given volume.

What is at work here is **diffusion**: the movement of molecules or ions from a region of their higher concentration to a region of their lower concentration. This notion carries with it the concept of a gradient. A **concentration gradient** is defined as the difference between the highest and lowest concentration of a solute within a given medium. In our example, the solute is the dye and the medium is water. As with bicycles coasting down a grade, the natural tendency for any solute is to move *down* its concentration gradient, from higher concentration to lower. Our dye did just that. Bikes and solutes can move up a grade or gradient; but there is a price to be paid, and that price is the expenditure of energy. (On its own, would red dye ever spontaneously come *back* to its concentrated state in the container? Not without some work being performed, which is to say not without some energy being expended.)

Diffusion through Membranes

So far, you have looked at molecules diffusing in an undivided container. But the subject here is divisions—those provided by the plasma membrane. Let us consider, therefore, what happens to a solution in a container that is divided by a membrane. If that membrane is *permeable* to both water and the solute—that is, if both water and the solute can freely pass through it—and the solute lies only on one side of the membrane, then the predictable happens: The solute moves down its concentration gradient, diffusing right through the membrane, eventually becoming evenly distributed on both sides of it.

Now let's imagine a *semi*permeable membrane, one that water can freely move through but that solutes cannot. **Figure 5.5** shows you what happens if more solute is put on the right side of the membrane than the left. Water flows through the membrane both ways, but *more* water flows into the right chamber, which has a greater concentration of solutes in it. The result is that the solution on the right side rises to a higher level than the one on the left.

This seems strange on first viewing, as if gravity were taking a vacation on the right side of the container. But what has been demonstrated here is **osmosis**: the net movement of water across a semipermeable membrane from an area of lower solute concentration to an area of higher solute concentration. Why should this occur? In the case of a solute like salt, water molecules will surround and bond with the sodium and chloride ions that salt separates into in solution. Because these solutes are not free to pass through the membrane, the water molecules bound to them will likewise remain confined to the right side. This means that more "free" water will exist on the left side, and the result is a net movement of water into the right side (see Figure 5.5).

The Plasma Membrane as a Semipermeable Membrane

So what does movement of water have to do with the cell's outer lining? The cell's phospholipid bilayer is itself a semipermeable membrane. It is somewhat permeable to water and lipid substances but not permeable to larger charged substances. Thus, osmosis can take place across the plasma membrane—indeed, it does all the time. It is the primary means by which plants get water, and it is a player in all sorts of routine metabolic processes in animals.

**Figure 5.5
Osmosis in Action**

(a) An aqueous solution divided by a semi-permeable membrane has a solute—in this case, salt—poured into its right chamber.

(b) As a result, though water continues to flow in both directions through the membrane, there is a net movement of water toward the side with the greater concentration of solutes in it.

(c) Why does this occur? Water molecules that are bonded to the sodium (Na^+) and chloride (Cl^-) ions that make up salt are not free to pass through the membrane to the left chamber of the container.

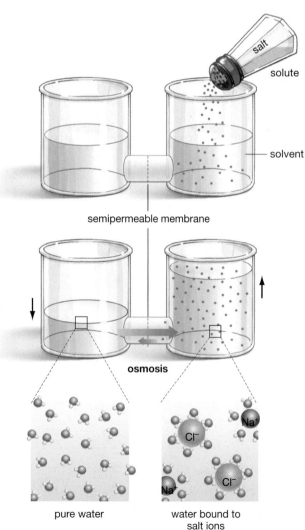

salt

solute

solvent

semipermeable membrane

osmosis

Na^+

Cl^-

Na^+

Cl^-

Na^+

Cl^-

pure water

water bound to salt ions

In humans, the fluid portion of blood is at one point driven out of the blood vessels called capillaries by the force of blood pressure. How does this fluid get back in? Osmosis. The proteins that remain in the capillaries are like the salt in the container: they bring about an osmosis that pulls the fluid back into the blood vessels.

It is possible, however, for cells to experience an "osmotic imbalance," meaning a situation in which osmotic pressure is either flooding cells or drying them out. In some of these situations, plant cells have a great advantage over animal cells—their cell walls. Remember from Chapter 4 that one function of the cell wall is to regulate water uptake. Well, now we can see why this is so valuable. Animal cells, which do not have a cell wall, can expand until they burst when water comes in (**see Figure 5.6**). Plant cells, conversely, will expand only until their membranes push up against the cell wall with some force, setting up a pressure, or turgor, that keeps more water from coming in. Such tight quarters actually are an optimal condition for plants. A nice, crisp celery stick is one that has achieved this kind of *turgid* state, while a droopy stick that has lost this quality has cells that are *flaccid*; flowers in the latter condition are *wilted*.

Osmosis and Cell Environments

Is a given cell likely to lose water to its surroundings, gain water from them, or have a balanced flow back and forth with them? Any of these things are possible, depending on what the solute concentration is outside the cell as opposed to inside. Three terms are helpful in describing the various conditions that can exist. A fluid that has a higher concentration of solutes than another is said to be a **hypertonic solution**. If a cell's surroundings are hypertonic to the cell's cytoplasm, water will flow out of the cell. Two solutions that have equal concentrations of solutes are said to be **isotonic**. If one of these solutions is the cell's cytoplasm and the other the fluid surrounding the cell, fluid flow will be balanced between cell and surroundings. Finally, a fluid that has a lower concentration of solutes than another is a **hypotonic solution**. If a cell's surroundings are hypotonic to the cell's cytoplasm, water will flow into the cell. Figure 5.6 gives examples of what can happen to cells in all three types of environments.

5.3 Moving Smaller Substances In and Out

This excursion into the land of diffusion and osmosis has prepared us to start looking at the ways materials actually move into and out of the cell,

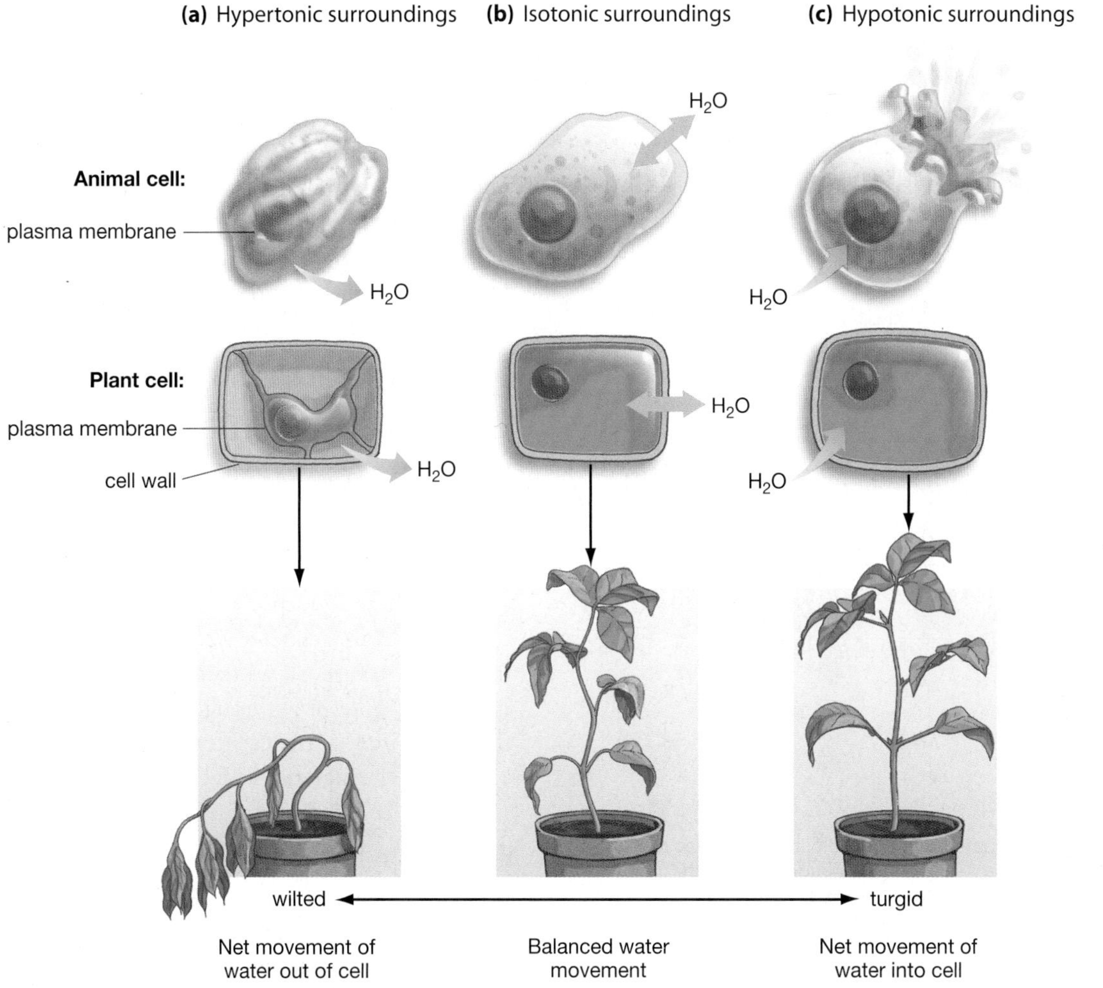

(a) Hypertonic surroundings **(b)** Isotonic surroundings **(c)** Hypotonic surroundings

Animal cell:

plasma membrane

H_2O

Plant cell:

plasma membrane

cell wall

H_2O

H_2O

H_2O

H_2O

wilted ◄————————————► turgid

Net movement of water out of cell

Balanced water movement

Net movement of water into cell

Figure 5.6
Osmosis in Cells

(a) When solutes (such as salt) exist in greater concentration outside the cell than inside, water moves out of the cell by osmosis and the cell shrinks. Here, the fluid surrounding the cell is hypertonic to the cell's cytoplasm.

(b) When solute concentrations inside and outside the cell are balanced, there is a balance of water movement into and out of the cell. Here, the fluid surrounding the cell is isotonic to the cell's cytoplasm.

(c) When solutes exist in greater concentration inside the cell than outside, water moves into the cell by osmosis. This influx may cause animal cells to burst, but plant cells are reinforced with cell walls and thus remain turgid—generally a healthy state for them. Here, the fluid surrounding the cell is hypotonic to the cell's cytoplasm.

across the plasma membrane. The big picture here is that some molecules are able to cross with no more assistance than is provided by diffusion or osmosis. Other molecules require these forces and the protein channels we talked about, and still others require channels and the expenditure of energy to get across. Different terms are applied to these different kinds of transport. **Active transport** is any movement of molecules or ions across a cell membrane that requires the expenditure of energy. **Passive transport** is any movement of molecules or ions across a cell membrane that does not require the expenditure of energy (**see Figure 5.7**). Our review of transport begins with two varieties of passive transport.

Passive Transport
Simple Diffusion

Gases such as oxygen and carbon dioxide are such small molecules that they need only move down their concentration gradients to pass into or out of the cell. Having been delivered by blood capillaries to an area just outside the cell, oxygen exists there in greater concentration than it does inside the cell. Moving down its concentration gradient, it diffuses through the plasma membrane and emerges on the other side. This is an example of **simple diffusion**, meaning diffusion through a cell membrane that does not require a special protein channel. Molecules that are larger than oxygen, but that are fat soluble, such as the steroid hormones

mentioned before, also move into the cell in this manner. Carbon dioxide, which is formed *in the* cell as a result of cellular respiration, has a net movement *out* of the cell through simple diffusion.

Facilitated Diffusion: Help from Proteins

Water can traverse the lipid bilayer through simple diffusion, moving through *despite* its hydrophilic nature primarily because it's such a small molecule. Yet with water itself—and certainly with polar molecules larger than water—protein channels begin to be required for passage in and out of the cell. With this, the method of passage is no longer *simple* diffusion; it is **facilitated diffusion**, meaning passage of materials through the plasma membrane that is aided by a concentration gradient and a transport protein.

Here you can begin to see the protein specificity referred to earlier. Each transport protein acts as a conduit for only one substance, or at most a small group of associated substances. The process of transport begins with the kind of binding explained earlier. A circulating molecule of glucose, for example, latches onto the binding site of a glucose transport protein. This binding causes the protein to change its shape, thus allowing the glucose molecule to pass through. In addition to glucose, other hydrophilic molecules, such as amino acids, move through the plasma membrane in this way. In some cases, a protein channel might not need to change shape. It simply remains open all the time, and is thus one of the passive passageways we talked about, though it will only let specific molecules pass through.

Note that this facilitated diffusion does not require the expenditure of energy, because it has a concentration gradient working in its favor. Glucose exists in greater concentration outside most cells than inside them. It is moving *down* its concentration gradient, then, by passing into the cell through a protein channel. So facilitated diffusion has something in common with simple diffusion. *Both* processes are driven by concentration gradients, meaning that neither requires metabolic energy. Thus, both are specific examples of passive transport.

Though the glucose in our example moved only one way in facilitated diffusion (into the cell), in general transport proteins are a channel for movement *either* way through the membrane. All that's required is a concentration gradient and a binding of the material with the transport protein.

Active Transport

If passive transport—either in simple or facilitated diffusion—were the only means of membrane passage available, cells would be totally dependent

Web Tutorial 5.2
Passive and Active Transport

Figure 5.7
Transport through the Plasma Membrane

Passive transport

simple diffusion

facilitated diffusion

Active transport

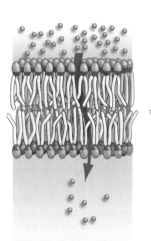

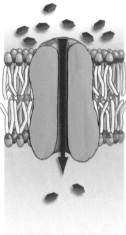

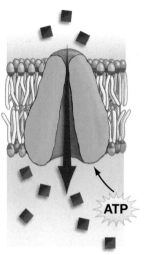

ATP

Materials move down their concentration gradient through the phospholipid bilayer.

The passage of materials is aided both by a concentration gradient and by a transport protein.

Molecules again move through a transport protein, but now energy must be expended to move them against their concentration gradient.

on concentration gradients. If, for example, molecules of a given amino acid come to exist in the same concentration on both sides of the plasma membrane, these amino acid travelers will continue to be transported, but they will be flowing out of the cell at the same rate they are flowing in. For cells, the problem is that some solutes are *needed* in greater concentration, say, inside the cell as opposed to outside. For example, the liver needs to take in large amounts of glucose because it is a prime storage site for glucose. Left alone, however, the glucose molecules that are so abundant in the liver would follow their concentration gradient and diffuse *out* of liver cells and back into the bloodstream. How do cells deal with such a situation? The cell's solution is pumps: it expends energy and pumps molecules against their concentration gradients. This is active transport at work. Many kinds of pumps are in operation in a cell, each of them being specific to one or perhaps two substances. The energy source for this transport often is ATP (adenosine triphosphate), though the cell also harnesses the power of oppositely charged ions to move substances across the membrane. One of the most important and best studied of the active transport mechanisms is known as the sodium-potassium pump.

The Sodium-Potassium Pump

The sodium-potassium pump allows the cell to maintain an environment of high potassium (K^+) *inside* the cell, and high sodium (Na^+) *outside* the cell. Imagine that we could take a snapshot of a cell at a given moment in time that would allow us to see these ion concentrations. With lots of Na^+ ions outside the cell, the result is predictable: A slow "leakage" of them into the cell as they follow their concentration gradient down and move in through transport proteins. K^+ ions, meanwhile, are moving out.

But our fantasy snapshot would also reveal something else. An abundance of protruding proteins that are busy pumping these molecules the *other way*, acting like so many bilge pumps moving seawater that has leaked into a ship back out into the ocean. If you look at **Figure 5.8**, you can see how the sodium-potassium pump works. The point here is not to count the movement of so many molecules in and out, but just to observe the process. One critical aspect of this is the shape-changing noted before with transport proteins.

You may wonder, though: Why would a cell spend so much energy to maintain this imbalance in sodium and potassium ions? Well, to take one example, nerve cells undertake this task for the same reason you undertake the task of charging your cell

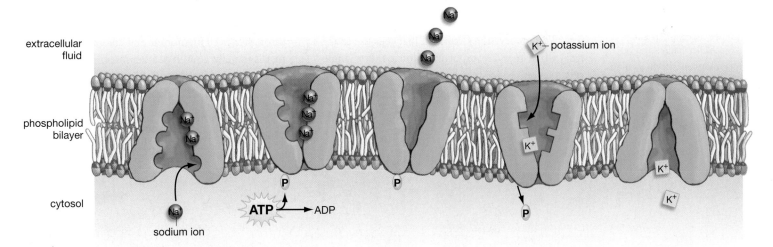

extracellular
fluid

phospholipid
bilayer

cytosol

K^+ potassium ion

ATP → ADP

sodium ion

1. Three sodium ions (Na^+) located within the cell's cytoplasm bind with a transport protein.

2. The energy molecule ATP gives up an energetic phosphate group to the transport protein.

3. This binding causes the protein to open its channel to the extra-cellular fluid; to lose its Na^+ binding sites, thus releasing the ions into the fluid; and to create binding sites for potassium ions (K^+).

4. Two K^+ move into the protein's K^+ binding sites, which brings about the release of the phosphate group.

5. This loss returns the protein to its original shape, releasing the K^+ into the cytoplasm and readying the protein for binding with another set of Na^+.

Figure 5.8
Active Transport in the Sodium-Potassium Pump

phone batteries: to have ready access to a capability. In your case, the capability is talking on the phone. In the nerve cell's case, the capability is being able to instantly take in large numbers of Na^+ ions. This capability is crucial in getting a nerve signal under way—something you can read about in Chapter 26.

5.4 Getting the Big Stuff In and Out

Pumps and channels, diffusion and osmosis—these mechanisms move substances in and out of the cell, but as it turns out they only move relatively *small* substances. Remember an earlier discussion about the immune system cells that check to see if a given cell is friend or foe? Well, if an immune sentry finds a bacterial foe, it may have to ingest this *whole cell*. As you might guess, a cell cannot do this by employing little channels or pumps. You learned a bit in Chapter 4 about what does happen in these cases; now you'll be looking at it in somewhat greater detail.

The methods in question here are endocytosis, which brings materials into the cell; and exocytosis, which sends them out. What these mechanisms have in common is their use of vesicles, the membrane-lined enclosures that alternately bud off from membranes or fuse with them.

Movement Out: Exocytosis

Exocytosis is defined as the movement of materials out of the cell through a fusion of a vesicle with the plasma membrane. As **Figure 5.9** shows, exocytosis involves a transport vesicle making its way to the plasma membrane and fusing with it, whereupon the vesicle's contents are released into the extracellular fluid. You observed in Chapter 4 that cells use exocytosis when they are exporting proteins. For single-celled creatures, waste products may be released into the extracellular fluid through this process.

Movement In: Endocytosis

Endocytosis is the movement of relatively large materials into the cell by infolding of the plasma membrane. It can take any of three forms: pinocytosis, receptor-mediated endocytosis, and phagocytosis.

Pinocytosis

Pinocytosis means "cell drinking," and if you look at **Figure 5.10**, you can see why this term is fairly accurate. The cell invaginates, creating a kind of harbor on its exterior. Whatever material happens to be enclosed in the harbor when it pinches off to become a vesicle is brought into the cell. What is brought in, of course, is mostly water, with some solutes in it. **Pinocytosis** can be defined as a form of endocytosis that brings into the cell a small volume of extracellular fluid and the materials suspended in it. Remember how it was noted earlier that the plasma membrane is constantly being "remade," due to the movement of materials in and out of it? With pinocytosis you can get some sense of this. Through this form of endocytosis, whole sections of the plasma membrane are continually budding off and moving to the interior of the cell. These sections have to be replaced, of course, with membrane brought up from the cell's interior. The result is a membrane in a constant state of flux.

Receptor-Mediated Endocytosis (RME)

As its name implies, the second form of endocytosis, **receptor-mediated endocytosis** (RME), depends on receptors, whose role is to bind to specific molecules and then hold onto them. Groups of receptors bearing these molecules then make lateral migration through the cell membrane and congregate in a place that is literally a pit; it is a depression in the cell, referred to as a *coated pit*. (What it is coated with is a layer of a protein, called clatharin, on the cytoplasmic

Figure 5.9
Movement Out of the Cell

(a) In exocytosis, a transport vesicle—perhaps loaded with proteins or waste products—moves to the plasma membrane and fuses with it. This section of the membrane then opens, and the contents of the former vesicle are released to the extracellular fluid.

(b) Micrograph of material being expelled from the cell through exocytosis.

(a) Exocytosis

(b) Micrograph of exocytosis

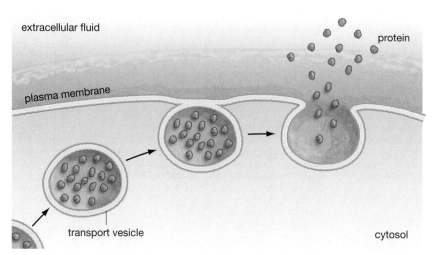

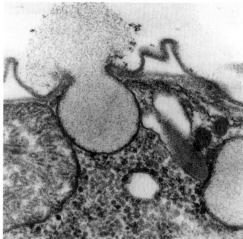

(a) Pinocytosis

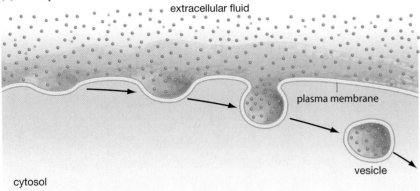

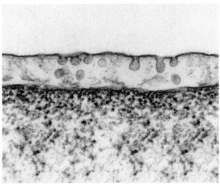

In pinocytosis, the plasma membrane invaginates to create a kind of harbor. The harbor then encloses completely, pinches off as a vesicle, and moves into the cell's cytoplasm, carrying with it whatever material was enclosed.

Micrograph of pinocytosis in a capillary cell.

(b) Receptor-mediated endocytosis

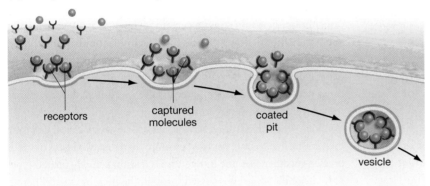

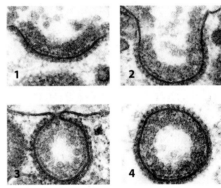

In receptor-mediated endocytosis, many receptors bind to molecules. Then, while holding on to the molecules, the receptors migrate laterally through the cell membrane, arriving at a depression called a coated pit. The coated pit pinches off, delivering its receptor-held molecules into the cytoplasm.

Formation of an RME vesicle.

(c) Phagocytosis

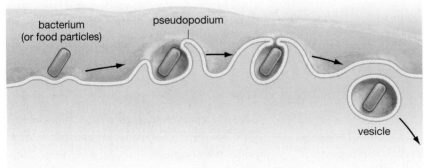

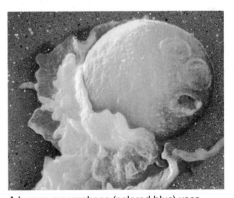

In phagocytosis, food particles—or perhaps whole organisms—are taken in by means of "false feet" or pseudopodia that surround the material. Pseudopodia then fuse together, forming a vesicle that moves into the cell's interior with its catch enclosed.

A human macrophage (colored blue) uses phagocytosis to ingest an invading yeast cell.

Figure 5.10
Three Ways to Get Relatively Large Materials into the Cell

side.) Eventually, the pit deepens and pinches off, creating the familiar vesicle moving into the cell. RME is very important in getting nutrients and other substances into cells. (It is the way cholesterol gets into your cells, for example.) The scale of the RME operation can be judged from the fact that, in mammalian cells, coated pits might take up 20 percent of a plasma membrane's surface area. RME is also a clear example of the kind of icebergs-on-the-ocean migration that proteins are capable of within the plasma membrane.

How Did We Learn?
The Fluid-Mosaic Model of the Plasma Membrane

The richly detailed illustrations found in science textbooks do a great job of presenting information, but they also can be a little misleading. Given their clarity, such illustrations might leave students with the impression that many of the discoveries that scientists make would be apparent to anyone who simply looked in the right place with the right microscope.

Figure 1a presents an example of the kind of pictures (or "micrographs") we have of the plasma membrane, while **Figure 1b** presents an example of the kind of artist's rendering you've been seeing of it.

Our view of the cell membrane was constructed piece by piece in an intellectual chain that stretched out for decades.

You'll probably agree that no one could have discerned the detail in 1b just by looking at 1a. How, then, do we know that the plasma membrane actually is structured as you see it in 1a? How do we know that it is largely lipid, and that proteins move about on the surface of it like so many icebergs? As it turns out, our view of the cell membrane was constructed piece by piece in an intellectual chain that stretched out for decades.

The idea that the surface of the cell is largely made of lipid material was set forth as early as 1855 by investigator Carl Nägeli. In 1899, Charles E. Overton came to the same conclusion by working with some cellular extensions of plant roots. What Overton found was that lipid-soluble substances moved quite easily into these "root hairs," while water-soluble substances did not. Remember now the adage about like dissolving like: Lipid substances would dissolve within *lipid materials.* Overton's conclusion was that the outside lining of the root cells was made of just such materials. From our vantage point—knowing about the phospholipid bilayer that makes up much of the membrane—we can see that he was right.

In 1925, two Dutch scientists, E. G. Gorter and F. Grendel, decided to take red blood cells and measure their lipid content. They knew from the work of another scientist that a given volume of phospholipids would cover a certain *area* when arrayed in a layer. What they found, however, was that the lipids they extracted from the blood cells spread out into an area *twice* as large as would be expected from this benchmark. This doubling, they concluded, came about because the cell membrane was doubled; it existed as the *bi*layer you've been reading about.

Following this advance, however, it was apparent that the plasma membrane had to be composed of more than just two layers of lipids pressed together. For one thing, it was clear that closely related types of hydrophilic molecules passed through the membrane at different rates. This was difficult to reconcile with a uniform phospholipid membrane, which would have let all such substances pass through at nearly the same rate. Given such evidence, James Danielli, writing with Hugh Davson, proposed a view of the plasma membrane that ended up having a long run in science. With modifications over time, the Davson-Danielli model, as it was called, wore the mantle of being "generally accepted" by scientists from the time it was set forth in 1935 until the early 1970s. It proposed that a *layer of proteins* coated each side of the phospholipid bilayer.

Phagocytosis

The third form of endocytosis is a means of bringing even larger materials into the cell. It is phagocytosis (literally, "cell eating"). This is the mechanism mentioned earlier, by which a human immune-system cell might ingest a whole bacterium (see Figure 5.10c). This is also the way that many one-celled creatures eat, so it's probably apparent that relatively large materials are being ingested here. How large? Vesicles of perhaps 1–2 µm are created in this process, as opposed to pinocytosis and RME, where the vesicles might be a tenth this size. Not all the materials brought into the cell by phagocytosis are whole cells; parts of cells and large nutrients are fair game as well.

As you can see in Figure 5.10, phagocytosis begins when the cell sends out extensions of its plasma membrane called pseudopodia ("false feet"). These surround the food and fuse their ends together. What was once outside is now inside, encased in a vesicle and moving toward the cell's interior. Once this happens, lysosomes in the cell move to the vesicle, fuse with it, and begin breaking down the material in it. **Phagocytosis** can be defined as a process of bringing relatively large materials into a cell by means of wrapping extensions of the plasma membrane around the materials and fusing the extensions together.

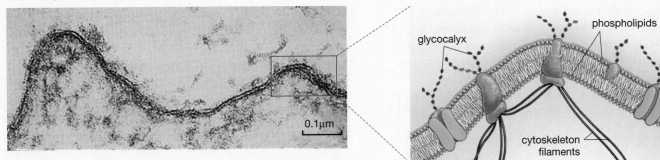

(a) Micrograph of the plasma membrane

(b) Artist's rendering of it

0.1μm

glycocalyx

phospholipids

proteins

cytoskeleton filaments

Figure 1
How Much Can a Picture Tell Us?

By the 1950s, electron microscopes had become refined enough that they yielded a "railroad track" picture of the plasma membrane: two dark bands separated by a space in the middle (the view you see in **Figure 2** on the next page). Seduced, perhaps, by the comforts of what was "known" to be true, scientists took such pictures to be visual proof of Davson-Danielli. The two dark lines were protein bands, it was thought, and the space in the middle was the phospholipid bilayer. We now know, however, that the "tracks" actually were electrons that bound with both membrane proteins and the charged portions of the phospholipid bilayer, thus creating the impression of a continuous track. In reality, there were no protein bands.

Through these years, it was also becoming apparent that there wasn't just one membrane in the cell, there were many: membranes encasing the mitochondria, membranes making up the Golgi complex, and so forth. In 1960, J. David Robertson set forth a proposal that saw *all* the cell's membranes as existing within the framework of Davson-Danielli. By that time the venerable model had been modified to include the notion of some proteins *spanning* the bilayer, from cytoplasm to outside face. This, then, was the generally accepted view of cell membranes in the early 1960s: Davson-Danielli-Robertson which posited a phospholipid bilayer coated on each side with a protein layer, and shot through here and there with other proteins.

Into this situation came a chemist, then at Yale, named Jonathan Singer. In 1962, Singer got to thinking about cell membranes and realized that Davson-Danielli-Robertson had a real problem: It would require hydro*phobic* parts of the supposed protein coat to be in contact with *water*. This was like assuming you could set one marble on top of another and expect it to stay there. By 1964, Singer was convinced not only that Davson-Danielli-Robertson had some shortcomings, but that he had a model that improved on it. It's one thing, however, to believe you're right, based on general chemical principles, and quite another to prove it.

"What happened was that I wasn't going to propose a model without some kind of experiments to go with it," Singer said in an interview. Working with a postdoctoral student of his, John Lenard, Singer published a scientific paper in 1966 that made the basic break with Davson-Danielli-Robertson. The view that emerged from this paper was that the membrane had a series of *individual* proteins that either were embedded in the bilayer or lay on it. By 1971, Singer was able to set forth a case for his model in great detail in a book chapter. The problem was that nobody paid much attention to it. So he set out to write a review of the same subject in *Science* magazine. In the interval between the book chapter and the magazine article, however, one last piece of the membrane puzzle fell into place.

(continued)

How Did We Learn?

How do we know that the plasma membrane has the structure you've just been reading about—the phospholipid bilayer, the proteins moving about like icebergs on the ocean, and so forth? This knowledge was gained through a series of investigations that lasted from the nineteenth century to the 1970s. You can read about this chain of discovery in "The Fluid-Mosaic Model of the Plasma Membrane" above.

On to Energy

Your tour of the cell and its plasma membrane is now complete. What's coming up next is something called bioenergetics. In our story of biology so far, there have been continuing vague references to proteins that are called enzymes and to a molecule called ATP and to energy expenditures. This has been rather like a description of an elephant that avoids the subject of its trunk. We will now embark on the subject of bioenergetics, however, which describes forces that are so basic they condition all of biology.

(continued)

Recall that the modern view of the cell membrane is the fluid-mosaic model. It is *mosaic* because it posits discrete proteins within the phospholipid bilayer; but it is *fluid* because these proteins (and the phospholipids themselves) can *move* laterally through the membrane. In 1971, Singer had the mosaic part; by 1972, he had the fluid as well.

"I came to Dr. Singer's lab in 1967," said Garth Nicolson, who was a graduate student with Singer in those days and is now president of the Institute for Molecular Medicine in Huntington Beach, California. "I became very interested in the polarity of membranes ... what I found when doing this work was that the distribution of components [on the surface of the cell] seemed to change dynamically." What Nicolson saw, in other words, was that the constituent parts of the membrane were shifting around on it.

Singer and Nicolson are agreed that a paper by L. D. Frye and Michael Edidin made Singer a believer in the rapid movement of materials across the cell surface. The upshot was a paper called "The Fluid Mosaic Model of the Structure of Cell Membranes," which appeared in *Science* magazine in February of 1972, with Singer and Nicolson as authors. It turned out to have great impact. Read any biology textbook today, and the account of the cell's plasma membrane is essentially the view set forth in this paper. The illustrations contained in textbooks likewise all look like a drawing in the *Science* article. Such was its effect that the fluid-mosaic model is just as likely to be referred to as the "Singer-Nicolson" model, even though, as Garth Nicolson is the first to say, "this is really [Singer's] model ... he should get the credit for the membrane as far as I'm concerned."

Does this mean that the fluid-mosaic model has at last given us an accurate picture of the membrane? We might wonder about this because, after all, Davson-Danielli was generally accepted for more than 30 years. Singer believes, however, that fluid-mosaic does describe reality. "It's not going to change," he says. "Things rattle around for awhile, but then you get settled in on the right picture."

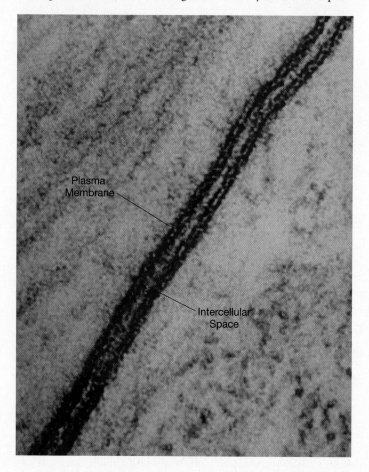

Plasma Membrane

Intercellular Space

Figure 2
A Close-up of the Plasma Membrane
Electron micrograph of two adjoining cells, each of which has a plasma membrane with a "railroad track" appearance. The intercellular space is the space between the two cells. (×470,000)

Chapter Review

Summary

5.1 The Nature of the Plasma Membrane

- The plasma membrane is a thin, fluid, lipid entity that manages to be very flexible and yet stable enough to stay together despite being continually remade, due to the constant movement of materials in and out of it. The plasma membrane has four principal components: (1) A phospholipid bilayer, (2) molecules of cholesterol interspersed within the bilayer, (3) proteins that are embedded in or that lie on the bilayer, and (4) short carbohydrate chains on the cell surface, collectively called the glycocalyx, that function in cell adhesion and as binding sites on proteins.

- Phospholipids are molecules composed of two fatty-acid chains linked to a charged phosphate group. The fatty-acid chains are hydrophobic, meaning they avoid water, while the phosphate group is hydrophilic, meaning it readily bonds with water. Such phospholipids arrange themselves into "bilayers"—two layers of phospholipids in which the fatty-acid "tails" of each layer point inward (avoiding water) while the phosphate "heads" point outward (bonding with it). Phospholipids take on this configuration in the plasma membrane because a watery environment lies on either side of the membrane.

- The cholesterol molecules that are interspersed between phospholipid molecules in the plasma perform two functions: they act as a patching material that helps keep some small molecules from moving through the membrane; and they keep the membrane at an optimal level of fluidity.

- Some plasma membrane proteins are integral, meaning they are bound to the hydrophobic interior of the phospholipid bilayer. Others are peripheral, meaning they lie on either side of the membrane but are not bound to its hydrophobic interior.

- The functions that membrane proteins serve include (1) structural support; (2) cell identification, by serving as external recognition proteins that interact with immune system cells; (3) communication, by serving as external receptors for signaling molecules; and (4) transport, by providing channels and associated structures for the movement of compounds into and out of the cell.

- The plasma membrane today is described by a conceptualization called the fluid-mosaic model, which views the membrane as a fluid, phospholipid bilayer that has moving laterally within it a mosaic of proteins.
Web Tutorial 5.1: Plasma Membranes and Diffusion

5.2   Diffusion, Gradients, and Osmosis

- Diffusion is the movement of molecules or ions from a region of their higher concentration to a region of their lower concentration. A concentration gradient defines the difference between the highest and lowest concentrations of a solute within a given medium. Through diffusion, compounds naturally move from higher to lower concentrations—meaning down their concentration gradients. Energy must be expended, however, to move compounds against their concentration gradients.

- A semipermeable membrane is one that allows some compounds to pass through freely while blocking the passage of other compounds. Osmosis is the net movement of water across a semipermeable membrane from an area of lower solute concentration to an area of higher solute concentration. Because the plasma membrane is a semipermeable membrane, osmosis operates in connection with it. Osmosis is a major force in living things; it is responsible for much of the movement of fluids into and out of cells. Osmotic imbalances can cause cells either to dry out from losing too much water or (in the case of animal cells) to break from taking too much in. Plant cells generally do not have this latter problem, because their cell walls limit their uptake of water.

- Cells will gain or lose water relative to their surroundings in accordance with what the solute concentration is inside the cell as opposed to outside. A cell will lose water to a surrounding solution that is hypertonic—a solution that has a greater concentration of solutes in it than does the cell's cytoplasm. A cell will gain water when the surrounding solution is hypotonic to the cytoplasmic fluid. Water flow will be balanced between the cell and its surroundings when the surrounding fluid and the cytoplasmic fluid are isotonic to each other—when they have the same concentration of solutes.

5.3   Moving Smaller Substances In and Out

- Some compounds are able to cross the plasma membrane strictly through diffusion and osmosis; others require these forces and special protein channels; still others require protein channels and the expenditure of cellular energy. Active transport is any movement of molecules or ions across a cell membrane that requires the expenditure of energy. Passive transport is any movement of molecules or ions across a cell membrane that does not require the expenditure of energy.

- Two forms of passive transport are simple diffusion and facilitated diffusion. Simple diffusion is diffusion that does not require a protein channel. Facilitated diffusion is diffusion that requires a protein channel. In facilitated diffusion, transport proteins function as channels for larger hydrophilic substances—substances that, because of their size and electrical charge, cannot diffuse through the hydrophobic portion of the plasma membrane.

- Cells cannot rely solely on passive transport to move substances across the plasma membrane. It may be that a given cell needs to have, on one side of its membrane a greater concentration of a given substance than exists on the other side. Yet passive transport serves to equalize concentrations of substances on both sides of the plasma membrane. To deal with such needs, cells employ chemical pumps to move compounds across the plasma membrane against their concentration gradients. One prime example of such a transport mechanism is the sodium-potassium pump.
Web Tutorial 5.2: Passive and Active Transport

5.4   Getting the Big Stuff In and Out

- Larger materials are brought into the cell through endocytosis and moved out through exocytosis. Both mechanisms employ vesicles, the membrane-lined enclosures that alternately bud off from membranes or fuse with them.

- There are three principal forms of endocytosis: (1) pinocytosis, in which the plasma membrane invaginates, creating an enclosure that pinches off to become a vesicle that moves into the cell, carrying with it extracellular fluid and the materials suspended in it; (2) receptor-mediated endocytosis, in which cell-surface receptors bind with materials to be brought into the cell and then migrate laterally through the cell membrane, congregating in a "coated pit," where vesicle-budding will bring them into the cell; and (3) phagocytosis, in which certain cells engulf whole cells, fragments of them, or other large organic materials, and bring these materials into the cell inside vesicles that bud off from the plasma membrane.
Web Tutorial 5.3: Exocytosis and Endocytosis

Key Terms

active transport 106
concentration gradient 104
diffusion 104
endocytosis 108
exocytosis 108
facilitated diffusion 106
fluid-mosaic model 103
glycocalyx 103
hypertonic solution 105
hypotonic solution 105
integral protein 101
isotonic solution 105

osmosis 104
passive transport 106
peripheral protein 101
phagocytosis 110
phospholipid bilayer 100
pinocytosis 108
plasma membrane 103
receptor-mediated
  endocytosis (RME) 108
receptor protein 102
simple diffusion 106
transport protein 103

Understanding the Basics

Multiple-Choice Questions (answers in the back of the book)

1. If the plasma membrane lacked a glycocalyx, the cell would have difficulty
   a. keeping its fluid state
   b. transporting materials across the membrane
   c. adhering to other cells
   d. binding to the cytoskeleton on the inside of the membrane
   e. keeping small particles from passing easily across the membrane

2. You eat spaghetti for dinner and digest the starch in it to glucose. The glucose is now in high concentration in the cells of your small intestine, relative to its concentration in your bloodstream. Predictably, the glucose diffuses across the plasma membrane of your intestinal cells, into your bloodstream.
   a. True.
   b. False, because molecules cannot move across membranes by diffusion.
   c. False, because diffusion moves materials from low concentration to high concentration.
   d. False, because molecules are too large to move across membranes.
   e. False, because energy is required for diffusion.

3. When a human cell is placed into water, it swells. This is said to be due to osmosis. What's happening to make the cell swell?
   a. Solutes are growing larger inside the cell
   b. Solutes are being drawn into the cell across the plasma membrane.
   c. Water is diffusing across the plasma membrane from a region of lower solute concentration (outside the cell) to a region of higher solute concentration (inside the cell).
   d. Water is diffusing across the plasma membrane from a region of higher solute concentration (outside the cell) to a region of lower solute concentration (inside the cell).
   e. The cell synthesizes new biological macromolecules as a protective response against a watery environment.

4. Tyrone gave Lakeisha a rose for the anniversary of their first date. Now the rose has wilted, because
   a. The cells have taken too much water in and burst.
   b. The cell walls have broken down.
   c. The plasma membrane has lost the proteins that keep water in.
   d. The phospholipid bilayer has broken down.
   e. The plant's cells have insufficient water in them to push against their cell walls.

5. The protein that functions in the sodium-potassium pump is an *integral* protein. The word *integral* tells you that this protein
   a. functions on only one side of the membrane
   b. is connected to a cholesterol molecule
   c. is essential to the cell
   d. is connected to the hydrophobic interior of the plasma membrane
   e. is connected to the glycocalyx

6. You get a cut on your finger and some bacteria enter. Your immune-system cells kill off the invaders by ingesting them. This is an example of
   a. pinocytosis
   b. phagocytosis
   c. receptor-mediated endocytosis
   d. exocytosis
   e. active transport

7. According to the fluid-mosaic model of membrane structure, the membrane is made up of a
   a. continuous base of proteins with phospholipids in it that have fairly free lateral movement
   b. bilayer of phospholipids with proteins interspersed in it that have fairly free lateral movement
   c. bilayer of protein with layers of phospholipids on the outside and inside
   d. bilayer of phospholipid with layers of proteins on the outside and inside
   e. bilayer of phospholipids sandwiched between two bilayers of proteins

8. Calcium ions (Ca^{++}) are present in the endoplasmic reticulum at 1000-fold higher concentrations than in the cytosol (the gel-like material of the cell outside of organelles). One protein on the endoplasmic reticulum membrane is devoted to allowing Ca^{++} to move from the endoplasmic reticulum to the cytosol and another protein moves Ca^{++} from the cytosol to the endoplasmic reticulum. These processes are not trivial, as Ca^{++} released into the cytosol allows such things as muscle contraction and communication between nerve cells. In this pair of Ca^{++} transport proteins, you would predict that the protein moving Ca^{++} from the endoplasmic reticulum to the cytosol _____ and the protein that moves Ca^{++} from cytosol to the endoplasmic reticulum _____.
   a. is large; is small
   b. is small; is large
   c. requires ATP; works without ATP
   d. works without ATP; requires ATP
   e. transports calcium slowly; transports calcium rapidly

9. Endocytosis and exocytosis are similar in that both involve the use of
   a. vesicles that carry materials into and out of the cell
   b. the Golgi apparatus to package materials
   c. diffusion to carry food and wastes across the plasma membrane
   d. starch to supply large amounts of energy
   e. the rough ER to manufacture proteins

10. Osmosis refers to the net movement of _____ across a _____
    a. energy . . . medium
    b. solutes . . . semipermeable membrane
    c. solutes . . . permeable membrane
    d. glucose . . . protein channel
    e. water . . . semipermeable membrane

Brief Review

1. Imagine a receptor on the surface of one of your muscle cells that responds only to glucose. Plenty of insulin molecules are passing by this receptor, but it does not respond to them. Why not?

2. You often hear how dangerous it is for athletes to take steroids to "bulk up" and how many problems steroids cause in various parts of the body. Why do these drugs affect all parts of the body—not just the muscles?

3. Explain why the phospholipid "heads" of the plasma membrane are always pointed toward the cytosol and extracellular fluid, whereas the "tails" are always oriented toward the middle of the membrane.

4. Describe the process of receptor-mediated endocytosis.

5. As you saw in this chapter, all cellular membranes are similar in structure. However, what differences are likely to exist between the plasma membrane and membranes in the interior of the cell?

Applying Your Knowledge

1. All life-forms on Earth exist either as single cells or as collections of cells that have a plasma membrane at their periphery (though some cells have cell walls outside their plasma membrane). Why do you think the plasma membrane exists in all living things? Why aren't there living things that don't have a membrane at their periphery?

2. Because materials move from high concentration to low concentration by simple diffusion, and they move the same way by facilitated diffusion, why do cells make proteins to carry out facilitated diffusion? Isn't this just a waste of energy?

3. Place a number of marbles on one end of a cafeteria tray. Shake the tray gently a few times, keeping it level. What happens to the marbles? Start over and shake faster the next time. What happens now? Explain how this is like diffusion.

6 An Introduction to Energy

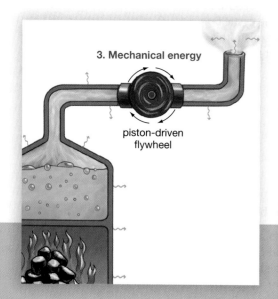

Freeing energy from chemical bonds.
(Section 6.2, page 121)

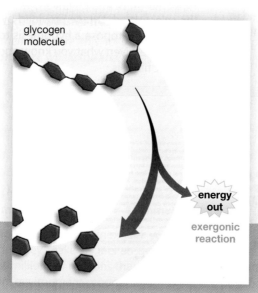

A downhill reaction releases energy.
(Section 6.3, page 122)

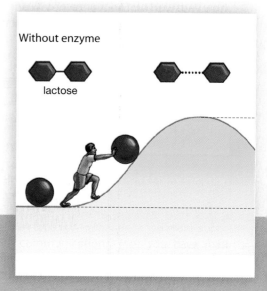

Activation energy.
(Section 6.6, page 126)

When my daughter Tessa was young, she had a teddy bear with a music box in it that let us listen to "Brahms's Lullaby." All I had to do was wind up the key that attached to the music box. After a few turns of the wrist, the melody would cycle through several times, then slow for a last few bars, then stop. This process could be repeated as many times as I wound the music box. However—no surprise here—the music box never seemed to wind itself. As we listened, a mechanism in the music box kept the melody coming out at an even tempo. Even the slowing at the end seemed just right for a lullaby, since the audience was a sleepy child.

6.1 Energy Is Central to Life

The forces that determined how Tessa's teddy bear worked also turn out to condition life, since, in order to function, both the music box and life must be supplied with the same thing: energy. Just as the music from the box depended on energy from an "outside hand" (mine) to continue, so almost all life on Earth depends on energy from an outside source: the sun, whose daily showering of energetic rays lets the green world of plants and algae flourish. These organisms then pass their bounty along to animals and other living things in the form of food to eat and oxygen to breathe.

Beyond this, the sun and the music box have something else in common: their energy always runs downhill, from more concentrated to less. The spring in the music box spontaneously unwinds, but it does not rewind itself. The sun sends out light and heat, but this energy does not spontaneously come together to form another sun. The music from the music box eventually must come to a stop, and in a similar way, even the sun will not shine forever. With all its grandeur, its energy is winding down, its brilliance dispersing. The difference is that once the sun's showering of energy comes to a stop, nothing will restart it.

While the sun's gift lasts, however, living things can transform its energy and use it to do things like sprout leaves and swim through oceans. Such complex activities make it imperative, though, that living things be able to *control* the energy they capture. The spring within the music box unwinds at a measured rate, spinning out the melody at just the right pace; similarly, living things have elaborate mechanisms in place that allow them to make the most out of the energy they receive.

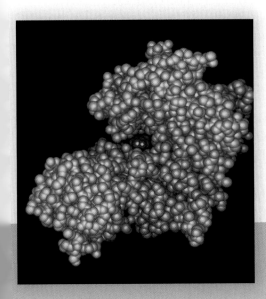

The site of action in an enzyme.
(Section 6.6, page 127)

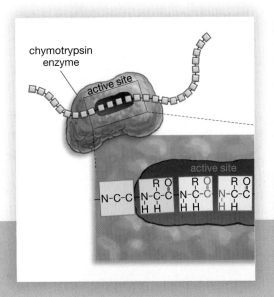

How an enzyme does its job.
(Section 6.6, page 127)

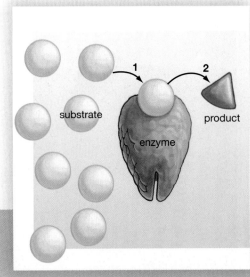

Regulating enzymatic activity.
(Section 6.7, page 128)

Principles such as these are important enough to biology that this chapter and the two that follow are devoted to the subject of energy and life. In this chapter you'll learn some basics about energy and some energy-saving molecules called enzymes. Then, in Chapter 7, you'll see how living things "harvest" energy, meaning how they extract energy from food. Finally, in Chapter 8, you'll look at the ways that plants, algae, and some bacteria capture the sun's energy in the process called photosynthesis. We'll start now by looking at the energy basics.

6.2 What Is Energy?

Energy is commonly defined as the capacity to do work, but this definition merely passes the question mark along, to the word *work*. Given this, here's one possible definition of **energy:** the capacity to bring about movement against an opposing force. The thing that makes energy so tricky as a concept is that although we can measure it with great precision, and we can experience its effects, we cannot grasp it or see it. We can see the water that drives a waterwheel, but who has seen the energy the water contains? (**See Figure 6.1.**)

The Forms of Energy

Energy can be thought of as coming in different forms, many of them familiar. Mechanical energy is captured in the wound-up spring of the musical teddy bear. Chemical energy is the energy held in the chemical bonds of a lump of coal. One other way to conceptualize energy is as potential and ki-

netic. **Potential energy** is stored energy: the rock perched precariously at the top of the hill; the charged ions kept on one side of a cell membrane. Conversely, **kinetic energy** is energy in motion, as with the rock tumbling down the hill or water driving a waterwheel. If we define energy as the capacity to bring about movement against an opposing force, we can see how kinetic energy fits into this scheme. A rock that tumbles off a cliff and into a lake would cause the water in the lake to splash upward—against the force of gravity.

The Study of Energy: Thermodynamics

Given how important energy is to life, it makes sense that there is a branch of biology, called *bioenergetics*, that links biology with a scientific discipline known as **thermodynamics**, the study of energy. Though thermodynamics is a mathematical and rather abstract discipline, research into it was prompted originally by a very down-to-earth goal: the desire to build a better steam engine. British inventors had harnessed steam power in the eighteenth century, and this discovery had the awesome impact of sparking the Industrial Revolution. The word *revolution* is fitting here because, before the advent of steam power, people used the stored energy in, say, wood or coal solely to produce *heat*. The steam engine was a device that allowed them to use heat to perform *work* on a scale that was previously impossible. Heat could be channeled, it was discovered, to drive an engine, which in turn could power any number of industrial processes.

With this development, new questions confronted scientists: How exactly did heat drive a steam engine? Was there a finite quantity of energy, such that it could all be used up? Research carried out in England, Germany, and France during the nineteenth century yielded the answers to these questions, and the result was the development of some insights known as the laws of thermodynamics.

The First Law of Thermodynamics: The Transformation of Energy

The **first law of thermodynamics** states that energy is never created or destroyed, but is only transformed. The sun's energy is not used up by green plants; rather, some of it is *converted* by the plants into chemical form. Plants use the sun's energy to put together carbohydrates, whose chemical bonds contain some of the energy that previously existed in the sun's rays.

A critical point here is the qualification that plants turn *some* of the sun's energy into chemical form. As it turns out, a plant cannot convert *all* of

Figure 6.1
Where's the Energy?
Part of the water moving over these falls is channeled to a water wheel, which turns as a result. Though the fall of the water is plainly visible, the energy released is not.

the solar energy it receives into carbohydrates. Likewise, a steam engine actually converts only a fraction of the energy contained in coal into the movement of a piston. What happens to the energy that is not stored in carbohydrates or transformed into motion? It can't be destroyed; that's clear from the first law. What happens is that it is converted into the form of energy known as *heat*. For every energy "transaction" that takes place, at least some of the original energy is converted into heat. This is true for every transaction in every energy system: cars burning gasoline, electric utilities burning coal, and your brain decoding this sentence.

The way this concept is usually phrased, however, is that some of the original energy is *lost* to heat. At first glance, this may seem like an unfair knock on heat, which after all has intrinsic value. Why, then, is energy "lost" to heat?

Well, consider what might be called the before-and-after of this energy transaction. In the "before," a lump of coal is a very ordered object, containing carbon atoms that exist in a precise spatial relationship to one another. Thus bonded, these atoms are not free to move. However, bringing a flame to the lump of coal causes the coal's carbon atoms to react with oxygen. Chemical bonds are broken through this process, and the energy stored in these bonds is released as heat. But heat is the random motion of molecules, and these molecules have no tendency to stay concentrated. Thus, it's clear that we have gone from the *ordered, concentrated* chemical bonds of coal to the *disordered, dispersed* energy of heat.

The Second Law of Thermodynamics: The Natural Tendency toward Disorder

And this gets us to the critical thing. Energy transformations will spontaneously run *only* from greater order to lesser order. Some of the heat produced in a steam engine is useful; it's driving a piston. Some of it is not, however—it's simply dissipated. But *all* of it amounts to a form of energy that is more dispersed than coal and that will quickly disperse further. Once this happens, there is no chance that it will spontaneously convert into anything *but* heat. We expect to see a lump of coal burn to ashes, and to feel the hot air that results disperse. But we do not expect to see the ashes re-form into a lump of coal, nor to feel the hot air concentrate itself again.

In yielding energy, matter goes from a more-ordered state to a less-ordered state. The scientific principle that speaks to this is the **second law of thermodynamics**: Energy transfer always results in a greater amount of disorder in the universe. In con-

nection with this law there is a term you will hear—*entropy*, which is a measure of the amount of disorder in a system. The greater the entropy, the greater the disorder. **Figure 6.2** summarizes the thermodynamic principles of the first and second laws of thermodynamics. **Figure 6.3** shows you how much energy is lost to heat in several energy systems.

The Consequences of Thermodynamics

All of this may seem like just so much … *hot air*. But it turns out that few things so profoundly condition the universe we live in as these thermodynamic rules. Indeed, the relentless increase in entropy that the second law entails may dictate nothing less than the fate of the universe.

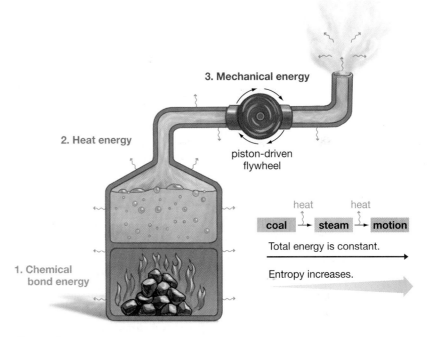

Figure 6.2
The Transformations of Energy
In a steam engine, energy locked up in the chemical bonds of coal is transformed into heat energy and mechanical energy. There is no loss of energy in this process, but energy is transformed from a more-ordered, concentrated form (the chemical bonds of coal) to a less-ordered, more dispersed form (heat). Thus, the amount of disorder—or entropy—increases in the transaction.

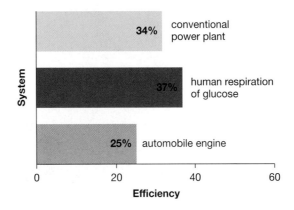

Figure 6.3
Energy Efficiency
The efficiency of several energy systems, as measured by the proportion of energy they receive relative to what they then make available to perform work. In measuring the efficiency of the car engine, the question is, how much of the energy contained in the chemical bonds of gasoline is converted by the car into the kinetic energy of wheel movement? In each system, most of the energy not available for work is lost to heat.

Allow yourself, for a second, to think of the universe as consisting solely of the sun, the Earth, and the space between them. The sun is a concentration of hydrogen that is undergoing a nuclear reaction, thus releasing light and heat. Living things on Earth manage to capture buckets of this energy, through the activities of plants and the other photosynthesizers. But every time one of these buckets is poured into another—every time light energy from the sun is transformed into energy stored in plants, for example—there is some spillage into heat. And once heat is generated, there is no spontaneous way it can make its way back into the orderly form of sunlight or carbohydrates. With every firing of a car's cylinder or contraction of a muscle, then, some amount of energy is transformed into heat.

Against this, life as we know it is made possible by ordered concentrations of energy. The sun can sustain life, but dispersed heat will not. One day, our sun will have wound down like the spring in a music box, bringing to an end the energy source for life on Earth. One possible scenario for the universe as a whole is that *all* its stars eventually will darken, leaving a cold universe with little potential for life—indeed, with little activity of any sort. The laws of thermodynamics are powerful indeed.

6.3 How Is Energy Used by Living Things?

A sprouting plant is constantly building itself up; it is making larger, more complex molecules (proteins, starches) from smaller, simpler ones (amino acids, simple sugars). This increasing organization stands in contrast to the spontaneous course of things—breakdown and disorder—and thus may seem to be a violation of the second law. But it is

not. In the context of the universe as a whole, living things are contributors to its entropy. Any given biological activity, such as building a leaf, generates heat and thus increases the disorder in the universe as a whole. But living things can bring about *local* increases in order (in themselves) by using the energy that comes ultimately from the sun.

Up and Down the Great Energy Hill

The building of complex molecules from simpler ones is an example of so-called synthetic work, and in it, we begin to see how living things use energy. A starchy carbohydrate is a more complex thing than the simple sugars it is made of. It is a more *ordered* arrangement of atoms than is a group of individual simple sugars. As such, the second law tells us, it should *take* energy to make a starch from simple sugars, and this is indeed the case. Going in the opposite direction, the breakdown of a starchy carbohydrate into simple sugars is an action favored by the second law, because it brings about *less* order. Such a process therefore should not take energy—no more than it takes energy for a boulder to roll down a hill. Indeed, this process should *release* energy, and this too is the case.

Breakdown operations, such as starches breaking down into sugars, are examples of **exergonic reactions**: reactions in which the starting set of molecules (the reactants) contains more energy than the final set of molecules (the products). Meanwhile, buildup operations are examples of **endergonic reactions**: reactions in which the products contain more energy than the reactants. This is what happens when, for example, simple glucose molecules are brought together to form glycogen, which is a storage form of carbohydrates (**see Figure 6.4**).

Web Tutorial 6.1
Energy and Biology

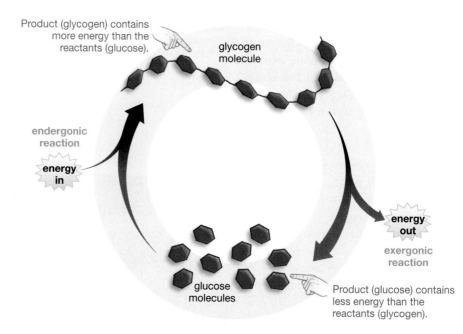

Product (glycogen) contains more energy than the reactants (glucose).

glycogen molecule

endergonic reaction

energy in

glucose molecules

energy out

exergonic reaction

Product (glucose) contains less energy than the reactants (glycogen).

Figure 6.4
Energy Stored and Released
It takes energy to build up a more complex molecule (in this case glycogen) from simpler molecules (in this case glucose). Such a buildup is thus an endergonic or uphill reaction. Conversely, energy is released in reactions in which more complex molecules are broken down into simpler ones. Such reactions are exergonic or downhill reactions.

Coupled Reactions

With this as background, we arrive at a critical insight: endergonic and exergonic reactions are constantly linked in living things. Why should this be so? If you want to get something uphill—into a higher energy state—you have to power that process. And that means *using* some energy. Just as it takes energy to get a bicycle uphill, so it takes energy to power the endergonic reaction that results in glycogen. Now, where is this energy going to come from? It must come from an energy-releasing *exergonic* reaction. The result is a **coupled reaction**: a chemical reaction in which an exergonic reaction powers an endergonic reaction. You actually observed some coupled reactions in Chapter 5, though they weren't labeled as such then. Remember how the sodium-potassium pump moved sodium ions out of the cell against their concentration gradient? (See page 107.) For the sodium ions, this was an uphill reaction: They finished the reaction in a higher energy state than they started, because they were moved up their electrical and concentration gradients.

And how was this movement powered? Through a downhill reaction. All you observed in Chapter 5 was that a molecule in the cell furnished the energy that powered the sodium pump. What this really meant was that a complex, energy-rich molecule was broken down to yield the energy for the pump. And what was this molecule? ATP.

6.4   The Energy Currency Molecule: ATP

Adenosine triphosphate. If you look at **Figure 6.5,** you can see that ATP is one of the nucleotides discussed in Chapter 3—a three-part molecule containing a sugar (ribose in this case), a nitrogen-containing base (adenine) and one or more phosphate groups. Actually, it is ATP's *three* phosphate groups that are of most interest to us. Notice that they all are negatively charged. This is the key to understanding why ATP serves as such an effective energy transfer molecule. Remember that like charges repel each other, and in this case three phosphate groups are doing just that. Energy was required to put these phosphate groups together as they are, in this relatively unstable state. This linkage represents a move *up* the energy hill. With ATP, it is as though someone squeezed a jack-in-the-box back into the box and shut the lid. It should be no surprise that when the lid is opened, a good deal of energy will be released.

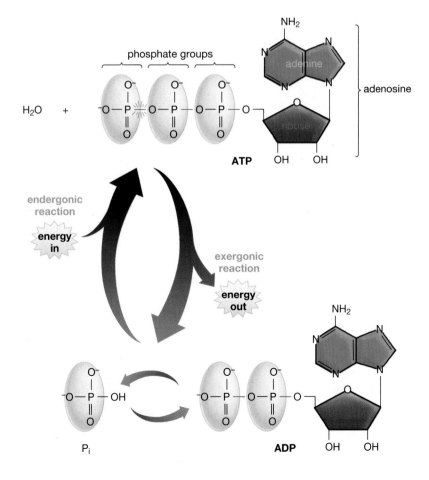

Figure 6.5
Energy Release from Breakdown of ATP
ATP (adenosine triphosphate) is life's most important energy transfer molecule. It stores energy in the form of chemical bonds between its phosphate groups. When the bond between the second and outermost phosphate group is broken, the outermost phosphate separates from ATP and energy is released. This separation transforms ATP into adenosine diphosphate or ADP, which then goes on to pick up another phosphate group, becoming ATP again.

How ATP Functions

Now, how is the energy from ATP used? As a starting point, consider an action that ATP molecules carry out within muscle cells (**see Figure 6.6**). Under the right conditions, ATP will bind to a type of protein, called an enzyme, that spans the membrane of this cell. The ATP molecule is then split, with the outermost of its phosphate groups breaking off from the rest of the molecule. This is a downhill reaction: It took energy to get the third of ATP's phosphate groups onto it (in an uphill push), and this stored energy is now being released. The subsequent attachment of this phosphate group to the protein causes the protein to change its shape. This shape change happens to drive the transport of calcium ions across the protein. For our purposes,

however, what's important is that something that would not have happened on its own took place because of the energy provided by the ATP molecule.

The ATP/ADP Cycle

There is another side to this ATP-driven reaction, however. Once the outermost phosphate group has split off, ATP has become a molecule containing only *two* phosphate groups, adenosine *di*phosphate or ADP. It is now free to have a third phosphate group added to it again (in an uphill reaction). This is just what happens: ADP returns to being ATP, which is capable of providing energy for yet another reaction. This shuttling from ATP to ADP and back takes place constantly in cells.

ATP as Money

It is often said that ATP is the cell's "energy currency" molecule. Now you're in a position to see why: It really does function like money. Imagine working at a typical job, in which you're called upon to do dozens of different things. You do this and you do that, but in the end what you get for all this activity is *money*. You can then take this cash and pay the rent, buy some CDs, or go bowling. Now think of animals and their breakdown of food. They have to obtain food in the first place and then digest the resulting fats and carbohydrates and proteins; but what they get in the end from all this work is ATP. It is the final outcome of the energy-harvesting process. And, like money, ATP then can be *used* to do any number of things, such as transport calcium ions across a cell membrane. Summing up, **ATP** can be defined as a nucleotide that functions as the most important energy transfer molecule in living things.

Between Food and ATP

The essence of our story so far is that, in animals, energy harvesting starts with the ingestion of food and ends with the synthesis of ATP. This is, however, such a bare-bones description that it is rather like a synopsis of *Cinderella* that goes: There was a young girl who was mistreated by her stepmother; then she married a prince. What's missing is everything that happened in between to make things come out this way. In the case of energy in ourselves, the end of the story is getting that third phosphate molecule attached to ADP, making it ATP. The beginning is that we have energy contained in food. Now, how do we *channel* that energy from food, and thus drive the attachment of the phosphate group onto ADP? It is almost time to answer this question—it will be in Chapter 7—but first, you need to know something about a group of proteins that living things use to control energy.

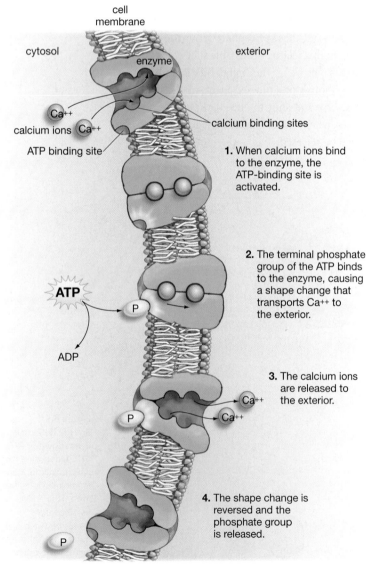

1. When calcium ions bind to the enzyme, the ATP-binding site is activated.

2. The terminal phosphate group of the ATP binds to the enzyme, causing a shape change that transports Ca++ to the exterior.

3. The calcium ions are released to the exterior.

4. The shape change is reversed and the phosphate group is released.

Figure 6.6
How ATP Functions
By transferring a phosphate group to the enzyme shown in this figure, ATP causes the enzyme to change shape in a manner that transports calcium ions across the cell membrane.

6.5 Efficient Energy Use in Living Things: Enzymes

Lactose—better known as milk sugar—is composed of two simple sugars, glucose and galactose, that are linked together by a single atom of oxygen. It follows from what you've seen so far that lactose will *split* into its glucose and galactose parts without any energy being required. Indeed, a little energy will be released in this process, because chemical bonds have been broken and two smaller molecules have been produced. Such a splitting is therefore a downhill reaction.

If we actually took a small amount of lactose, however, and put it in some plain water, it would certainly be hours—it might be days—before *any* of the lactose molecules would break down into glucose and galactose. How can this be? Most of us drink milk and it doesn't seem to pile up in us.

Accelerating Reactions

The secret is that when lactose is metabolized in living things, it isn't being split up in plain water. Something else is present that speeds up this process immensely, accelerating it perhaps a billionfold. That something is an **enzyme**, a type of protein that accelerates a chemical reaction. This particular enzyme is called lactase, but it is merely one of thousands of enzymes known to exist. Each of these compounds is working on some chemical process, but not all are involved in splitting molecules. Some enzymes combine molecules, some rearrange them.

Given their numbers, it may be apparent that enzymes are involved in a lot of different activities; but this doesn't begin to give them the recognition they deserve. Enzymes facilitate nearly every chemical process that takes place in living things. No organism could survive without them. Technically, these compounds are only *accelerating* chemical reactions that would happen anyway, as with lactase splitting lactose. But in practical terms, they are *enabling* these reactions, because no living thing could wait days or months for the milk sugar it ingests to be broken down, or for hormones to be put together, or for bleeding to stop.

To get an idea of the importance of enzymes, consider the millions of Americans who are known as lactose intolerant because, after childhood, their bodies reduced their production of lactase. When these people consume milk, the lactose in it *does* simply stay in their intestines. It is eventually digested not by them, but by the bacteria that live in the human digestive tract (something that causes bloat and gas).

Specific Tasks and Metabolic Pathways

From the sheer number of enzymes that exist, you can probably get a sense of how specifically they are matched to given tasks. Although some enzymes can work on *groups* of similar substances, a specific enzyme usually facilitates a specific reaction: It works on one or perhaps two molecules and no others. Thus, lactase breaks down lactose—and *only* lactose—and the products of its activity are always glucose and galactose. The substance that is being worked on by an enzyme is known as its **substrate**. Thus, lactose is the substrate for lactase. The *-ase* ending you see in *lactase* is common to most enzymes. As you can see, the substrate (lactose) is identified in the enzyme's name (lactase).

Certain activities in living things, such as blood clotting, are much more complex than the breakdown of lactose. They are multistep processes, *each step of which* requires its own enzyme. Most large-scale activities in living things work this way: leaf growth, digestion, hormonal balance. The result is a **metabolic pathway**: a set of enzymatically controlled steps that results in the completion of a product or process in an organism. In such a sequence, each enzyme does a particular job and then leaves the succeeding task to the next enzyme, with the product of one reaction becoming the substrate for the next (**see Figure 6.7**). The sum of all the chemical reactions that a cell or larger organism carries out is known as its **metabolism**. To put things simply, enzymes are active in all facets of the metabolism of all living things.

Web Tutorial 6.2
Enzymes

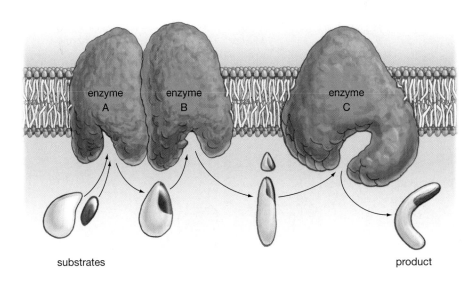

substrates

product

Figure 6.7
Metabolic Pathway: Sequence of Enzyme Action
Most processes in living organisms are carried out through a metabolic pathway—a sequential set of enzymatically controlled reactions in which the product of one reaction serves as the substrate for the next. Enzymes perform specific tasks on specific substrates. In this example, enzyme A combines two substrates, enzyme B removes part of its substrate, and enzyme C changes the shape of its substrate.

6.6 Lowering the Activation Barrier through Enzymes

All this information about enzymes is well and good, you may be saying at this point, but what does it have to do with energy? The answer is that in carrying out their tasks, enzymes are in the business of lowering the amount of energy needed to get chemical reactions going. And this means that the reactions can get going faster.

An analogy may be helpful here. Consider the now-familiar rock perched at the top of a hill. Situated in this way, it has a good deal of potential energy. If it rolled down the hill, it would release this potential energy as kinetic energy. But here is the critical question: What would be required to get this rock going? It is perched at the top of the hill because it is *stable* at the top of the hill. To get it going would require *additional* energy in the form of, say, a push.

Now consider the lactose molecule. It is "perched" like the rock, in a sense, because it lies "uphill" from the glucose and galactose it can break into. But just as the rock is stable because of its position in the ground, so the lactose is stable because it has a strong set of chemical bonds holding its atoms in a fixed position.

The question thus becomes: Is there anything that can lower the amount of energy required to break these bonds? The answer is yes—an enzyme. If you look at **Figure 6.8**, you can see this in schematic terms. Imagine that you had to push a boulder up the hill in (a) to get it to roll down the other side. Now imagine that you only had to push a boulder up the hill in (b) to get the same result. In which situation would you be able to get your desired result *faster*? What is at work here is a lowering of **activation energy**: the energy required to initiate a chemical reaction.

How Do Enzymes Work?

How do enzymes carry out this task of lowering activation energy? As you'll see shortly, they bind to their substrates, and in so doing make these substances more vulnerable to chemical alteration. The amazing thing is that they do this without being permanently altered themselves. Enzymes are **catalysts**—substances that retain their original chemical composition while bringing about a change in a substrate. At the end of its cleaving of the lactose molecule, lactase has exactly the same chemical structure as it did before. It is thus free to pick up another lactose molecule and split *it*. All this takes place in a flash: The fastest-working enzymes can carry out 100,000 chemical transformations per second.

Enzymes generally take the form of globular or ball-like proteins whose shape includes a kind of pocket into which the substrate fits. If you look at **Figure 6.9**, you can see a space-filling model of one enzyme, called hexokinase. You can also see, buried there in the middle, its substrate—glucose, which is better known as blood sugar. As with all proteins, enzymes are made up of amino acids, but only a few of the hundreds of amino acids in an enzyme are typically involved in actually binding with the substrate. Five or six would be common. These amino acids help form the substrate pocket, known as the **active site**: the portion of an enzyme that binds with and transforms a substrate.

In some cases, the participants at the active site include, along with amino acids, one or more accessory molecules. One variety of these molecules is the **coenzymes**: molecules other than amino acids that facilitate the work of enzymes by binding with them. If you've ever wondered what vitamins do, participation in enzyme binding is a big part of it. Once we have ingested them, many vitamins are transformed into coenzymes that sit in the active site of an enzyme and provide an added chemical attraction or repulsion that allows the enzyme to do its job.

Figure 6.8
Enzymes Accelerate Chemical Reactions
How is lactose split into glucose and galactose? Without an enzyme, the amount of energy necessary to activate this reaction is high. In the presence of the enzyme lactase, however, a low activation energy is sufficient to get the process started. The energy released from the splitting of lactose is the same in both cases.

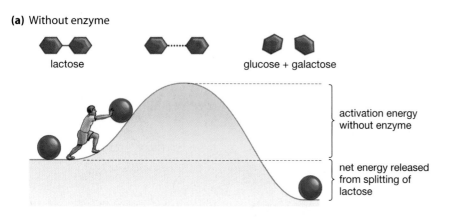

(a) Without enzyme

lactose   glucose + galactose

activation energy without enzyme

net energy released from splitting of lactose

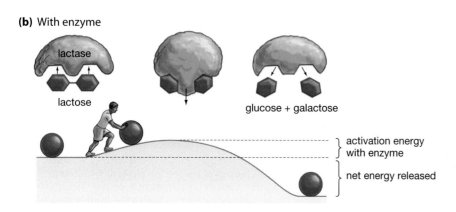

(b) With enzyme

lactase

lactose   glucose + galactose

activation energy with enzyme

net energy released

An Enzyme in Action: Chymotrypsin

The details of how enzymes carry out their work are complex, and they vary according to enzyme. Let's consider in a general way, however, the activity of a much-studied enzyme called chymotrypsin.

Chymotrypsin is delivered from the human pancreas to the small intestine, where it works with water to break down proteins we have ingested. Its function is to clip protein chains in between their building-block amino acids. It does this by breaking the single bond that binds one amino acid to the next (**see Figure 6.10**). How does it work? After having bound to part of a protein chain, chymotrypsin then interacts with it to create a transition-state molecule. In effect, it distorts the shape of the protein—and holds it in this new shape briefly—in such a way that the protein becomes vulnerable to bonding with ionized water molecules. This allows carbon and nitrogen atoms to latch onto new partners, and the protein chain is clipped. Chymotrypsin then returns to its original form and proceeds to a new reaction.

6.7   Regulating Enzymatic Activity

The activity of enzymes is regulated in several ways. The question here is: What factors influence the amount of "product" an enzyme turns out? One of these factors, quite logically, is the amount of substrate in the enzyme's vicinity. If there's no substrate to work on, no product can be turned out.

The work of enzymes can, however, be reduced in several other ways. For example, some molecules that are not the normal substrate for an enzyme can still bind to the enzyme at its active site. An enzyme that is "occupied" in this way is not free to bind with its normal substrate and thus is not able to carry out its usual catalytic activity.

This happens all the time in minor ways in our bodies, but to sharpen the point, consider an extreme example of it. There is a nerve gas, called sarin, that was developed in the 1930s to kill people. In 1995 it did just that, when it was maliciously released in a Tokyo subway. Sarin works by binding very strongly to the active site of an enzyme called acetylcholinesterase. Under normal circumstances, acetylcholinesterase's substrate is acetylcholine—a compound that travels, ferryboat-like, out of one nerve cell and binds to the next. The effect of this binding is to cause the second nerve cell to fire. Each binding is, however, supposed to bring about *one* nerve-signal firing, after which acetylcholine is broken down by acetylcholinesterase. But if acetylcholinesterase is occupied by sarin, it cannot do its job. The result is that there is no end to the nerve-signal firing. This means a breakdown of

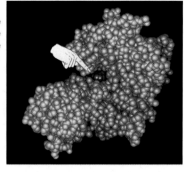

Glucose molecule binding to active site of hexokinase enzyme.

Figure 6.9
Shape Is Important in Enzymes
Substrates generally fit into small grooves or pockets in enzymes. The location at which the enzyme binds the substrate is known as the enzyme's active site. Pictured is a computer-generated model of a glucose molecule (in red) binding to the active site of an enzyme called hexokinase (in blue).

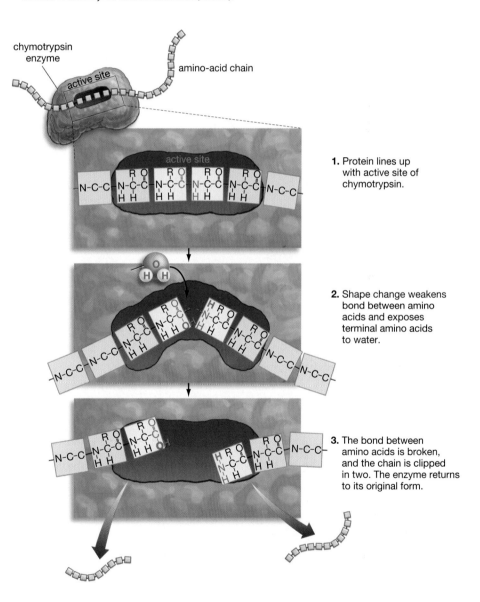

chymotrypsin enzyme

amino-acid chain

active site

1. Protein lines up with active site of chymotrypsin.

2. Shape change weakens bond between amino acids and exposes terminal amino acids to water.

3. The bond between amino acids is broken, and the chain is clipped in two. The enzyme returns to its original form.

Figure 6.10
How the Enzyme Chymotrypsin Works
In this example, the enzyme chymotrypsin is facilitating the breakdown of a protein by changing the protein's shape. In the absence of chymotrypsin, this process would take a billion times longer.

**Figure 6.11
Allosteric Regulation of
Enzymes**
The frequency of chemical reactions can be controlled by the binding of a molecule to an enzyme at a site other than its active site. In this instance, the enzyme's own product binds with the enzyme in this way, thereby reducing the enzyme's activity.

the nervous system and, in some cases, death from a paralysis of the breathing muscles. A number of toxins work by carrying out this kind of enzyme occupation, which is known as *competitive inhibition*.

Allosteric Regulation of Enzymes

Living things routinely regulate enzyme activity by another means. Earlier, you saw a model of an enzyme called hexokinase, which works on the substrate glucose. The product of this reaction is something called glucose-6-phosphate. Though this is an enzymatic product, it can bind to hexokinase at a site other than its active site, and in so doing change hexokinase's shape. This change renders hexokinase less able to bind with its substrate, glucose, meaning hexokinase's work is temporarily reduced. If you look at **Figure 6.11**, you can see how this process works schematically.

This is but one example of the process known as **allosteric regulation**—the regulation of an enzyme's activity by means of a molecule binding to a site on the enzyme other than its active site. Many variations on this theme are possible. In the case of hexokinase, its own product (glucose-6-phosphate) bound with it and cut down on its activity, but molecules other than the enzymatic product can bind with an enzyme at one of its additional binding sites. Furthermore, the effect of such binding need not be negative; allosteric binding can increase the activity of enzymes as well as reducing it.

The importance of such regulation is that enzymes are not simply fated to turn out product as long as there is substrate to work on. If a cell has too much of a product, allosteric control can cut down on it; if there is too little, its concentration can be increased.

On to Energy Harvesting

So, to review for a second, you've seen that, for animals, the energy that comes bound in food can be extracted from it. The result of this energy "harvesting" is the production of the energy currency molecule, ATP, which is used to power various processes. The question we will tackle next is: How does the body actually do its energy harvesting—what are the steps by which it extracts energy from food and then channels this energy into pushing phosphate groups onto ADP? That is the subject of Chapter 7.

(a) Substrate becomes product

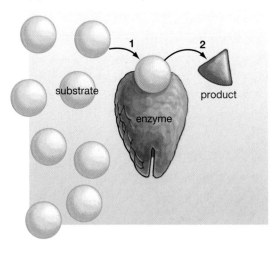

substrate

product

enzyme

1. Substrate binds to enzyme.

2. Enzyme transforms substrate to product.

(b) Product feeds back on enzyme

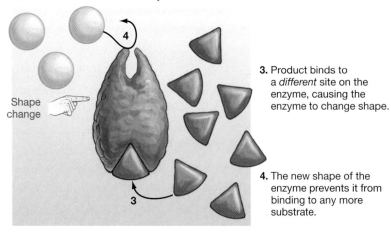

Shape change

3. Product binds to a *different* site on the enzyme, causing the enzyme to change shape.

4. The new shape of the enzyme prevents it from binding to any more substrate.

Chapter Review

Summary

6.1 Energy Is Central to Life

- All living things require a source of energy. The sun is the ultimate source of energy for most living things. The sun's energy is captured on Earth by photosynthesizing organisms (such as plants), which then pass on this energy to other life-forms.

6.2 What Is Energy?

- Energy can be defined as the capacity to bring about movement against an opposing force. Energy can be conceptualized as being either potential energy, meaning stored energy; or kinetic energy, meaning energy in motion. The study of energy is known as thermodynamics.

- Two fundamental principles of energy are the first law of thermodynamics, which states that energy is never created or destroyed, but is only transformed; and the second law of thermodynamics, which states that energy transfer will always result in a greater amount of disorder in the universe.

- In line with the second law of thermodynamics, in every energy transaction, some energy will be lost to the disordered form of heat. Entropy is a measure of the amount of disorder in a system; the greater the entropy, the greater the disorder.

6.3 How Is Energy Used by Living Things?

- Living things can bring about local increases in order (in themselves), through their metabolic processes. They can, for example, build up more-ordered molecules (starches, proteins) from less-ordered molecules (simple sugars, amino acids). However, it takes energy to do this.

- Energy in living things is stored away in endergonic (uphill) reactions, in which the products of the reaction contain more energy than the starting substances (or "reactants"). Conversely, energy is released in exergonic (downhill) reactions, in which the reactants contain more energy than the products. The linkage of simple sugars to form a complex carbohydrate is an endergonic reaction. Such reactions will not occur without an input of energy. Conversely, the breakdown of a complex carbohydrate into simple sugars is an exergonic reaction. Such reactions release energy. Endergonic and exergonic reactions are linked in coupled reactions—reactions in which an energy-yielding exergonic reaction powers an energy-using endergonic reaction. The molecule most often used in living things to power coupled reactions is ATP.
Web Tutorial 6.1: Energy and Biology

6.4 The Energy Currency Molecule: ATP

- Adenosine triphosphate (ATP) is the most important energy transfer molecule in living things. It is the final product of the energy-harvesting process, in which energy is extracted from food. ATP's energy transfer powers stem from the fact that it contains three phosphate groups, each of which is negatively charged, meaning these groups repel each other. ATP drives chemical reactions by donating its third phosphate group to them. In the process, it becomes the two-phosphate molecule adenosine diphosphate (ADP). To once again become ATP, it must have a third phosphate group attached to it. In animals, the energy supplied by food powers the process by which third phosphate groups are attached to ADP.

6.5 Efficient Energy Use in Living Things: Enzymes

- An enzyme is a type of protein that increases the rate at which a chemical reaction takes place in an organism. Nearly every chemical process that takes place in living things is facilitated by an enzyme. For example, the enzyme lactase facilitates the splitting of the sugar lactose into its component sugars, glucose and galactose.

- Many activities in living things are controlled by metabolic pathways, in which a series of interrelated reactions is undertaken in sequence, each one of them facilitated by an enzyme. The substance that an enzyme helps transform through chemical reaction is called its substrate. Lactose is the substrate of the enzyme lactase. The sum of all the chemical reactions that a cell or larger living thing carries out is its metabolism.
Web Tutorial 6.2: Enzymes

6.6 Lowering the Activation Barrier through Enzymes

- Enzymes work by lowering activation energy, meaning the energy required to initiate a chemical reaction. Enzymes are catalysts: They bring about a change in their substrates without being chemically altered themselves.

- Enzymes generally take the form of globular or ball-like proteins whose shape includes a pocket into which the enzyme's substrate fits. This pocket is the active site—that portion of an enzyme that binds with and transforms a substrate.

6.7 Regulating Enzymatic Activity

- Enzyme activity can be controlled in several ways. One of these is allosteric regulation: A molecule binds with the enzyme at a site other than its active site, thus changing the enzyme's shape, and thereby decreases or increases the enzyme's ability to bind with its substrate.

Key Terms

Understanding the Basics

Multiple-Choice Questions (answers in the back of the book)

1. According to the first law of thermodynamics, energy
   a. is never lost or gained, but is only transformed
   b. always requires an ultimate source such as the sun
   c. can never be gained, but can be lost
   d. can never really be harnessed
   e. can never be transformed

2. Each time there is a chemical reaction, some energy is exchanged. According to the second law of thermodynamics, with each exchange
   a. Some energy is lost, but other energy is created.
   b. Some energy must come from the sun.
   c. Some energy is transformed into heat.
   d. Energy is gained for future use.
   e. Some energy is permanently and completely destroyed.

3. Your car uses gasoline to fuel the engine. Can you explain why you need a coolant system for your car?
   a. it helps the car move through the air easier
   b. it is needed to absorb the heat energy that results from fuel combustion
   c. it is required to help move the heat to the drive shaft
   d. it is required to move the fuel to the correct location
   e. it is required to assist in making the fuel the correct temperature

4. ATP stores energy in the form of
   a. mechanical energy
   b. heat
   c. complex carbohydrates
   d. chemical bond energy
   e. amino acids

5. People who are lactose intolerant lack a compound called lactase in their digestive tract. You know that this compound, lactase, is probably a(n)

   a. hormone, because it causes a change in the person
   b. coenzyme, because it helps the enzyme, lactose, do its job
   c. vitamin, because it is essential to health
   d. substrate, because it is involved with food
   e. enzyme, because its name ends in *-ase*

6. Glycolysis is a metabolic pathway that helps living things extract energy from food. From this we know that glycolysis
   a. consists of a series of chemical reactions
   b. uses a number of enzymes
   c. involves the modification of a series of substrates
   d. proceeds by means of each enzyme leaving a succeeding reaction to a different enzyme
   e. all of the above

7. Each molecule of ATP stores a good deal of energy because
   a. ATP is such a large protein
   b. ATP captures energy directly from the sun
   c. ATP's third phosphate group exists in a highly energetic state
   d. a long metabolic pathway is required for the production of ATP

8. Which of the following will lower the activation energy of a reaction in a cell?
   a. lowering the temperature
   b. lowering the pressure
   c. using an enzyme
   d. changing the amount of the reactants
   e. supplying ATP

9. The active site of enzyme A is occupied by a molecule other than its substrate. Enzyme B has had its activity reduced by means of a molecule binding to it at a site other than its active site. What is taking place in connection with enzyme A and then enzyme B?
   a. a reaction acceleration, a coupled reaction
   b. a lowering of activation energy, ATP use
   c. allosteric regulation, competitive inhibition
   d. allosteric regulation, coenzyme participation
   e. competitive inhibition, allosteric regulation

10. Enzymes
    a. accelerate specific chemical reactions
    b. are not permanently altered by binding with a substrate
    c. lower the activation energy of specific chemical reactions
    d. all of the above
    e. a and c only

Brief Review

1. A plant leaf is more ordered and complex than are carbon dioxide and water. In photosynthesis, plants take small, relatively disordered carbon dioxide and water molecules and make complex, ordered sugars out of them. How is it possible for this to happen without violating the second law of thermodynamics?

2. The text notes that in the sodium-potassium pump, sodium ions are moved out of the cell against their concentration gradient and that ATP plays a part in this. Explain how this is a coupled reaction.

3. Where does the energy come from that is stored and released by ATP? In what way is ATP an energy currency molecule?

4. The average person thinks of enzymes only as substances in their digestive tract that break down food. How would you explain the role of enzymes to your little sister, who has heard about digestion in grade school biology?

5. What is meant by allosteric regulation of an enzyme?

Applying Your Knowledge

1. In light of the laws of thermodynamics, explain why you get so much hotter when you run than when you are sitting still.

2. Which is easier to have: a messy, disordered room or a neat, ordered room? Is the second law of thermodynamics at work in this situation? What is the price of bringing order to a room?

3. Physicists say that it is impossible to create a perpetual motion machine—a machine whose own activity keeps it running perpetually. What energy principle precludes the possibility of a perpetual motion machine?

7 Deriving Energy from Food

One effect of oxidation: rust.
(Section 7.2, page 135)

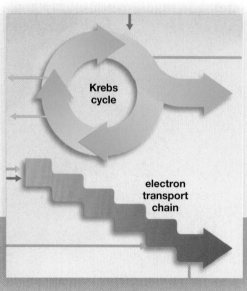

The stages of energy harvesting.
(Section 7.3, page 137)

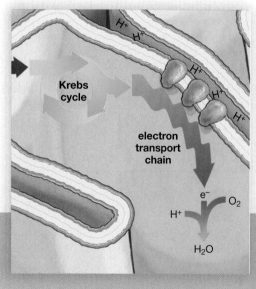

A transition step in respiration.
(Section 7.5, page 141)

In May of 1996, the world's attention was riveted on news coming out of Nepal, in Southeast Asia. Disaster had struck several groups of mountain climbers who had been attempting to scale the world's highest peak, Mt. Everest. A vicious storm had taken the members of two Everest expeditions by surprise as they were moving down from Everest's summit, toward the safety of lower altitudes. Within 36 hours, five climbers had lost their lives and another was so badly frostbitten that one of his hands had to be amputated.

One of the expeditions was led by a highly respected guide from New Zealand, Rob Hall. Courageously remaining near the summit in an effort to bring down one of his clients (who had collapsed), Hall ended up spending a night in a howling storm on Everest without a tent. When the sun rose the next day, the factors working against him included cold and wind, which he might have encountered in lots of inhospitable places on Earth. Also bedeviling him, however, was something that human beings rarely encounter: a lack of oxygen. At his altitude—sitting in the snow at 8,700 meters (or about 28,500 feet)—the oxygen he took in with each breath was little more than a third of the amount he would have inhaled with each breath at sea level. Technology might have come to his aid, as he had two oxygen bottles with him, but his intake valve for them had become clogged with ice.

Down at Everest's lower altitudes, other climbers—some of them Hall's friends—had to endure the agony of talking to him by two-way radio while not being able to reach him physically. All of them knew that a lack of oxygen was draining not only his muscles but his mind. Here is Hall in one of his radio transmissions asking about another guide on the ill-fated expedition: "Harold was with me last night, but he doesn't seem to be with me now. Was Harold with me? Can you tell me that?" In the end, the mountain without mercy claimed Rob Hall. He never got up from the place where he spent his night on Everest.

Exhausted Near the Everest Summit
Two members of the ill-fated expeditions that climbed Mount Everest in May 1996 approach the mountain's summit. At this altitude, each breath brings with it only about one-third the amount of oxygen available at sea level.

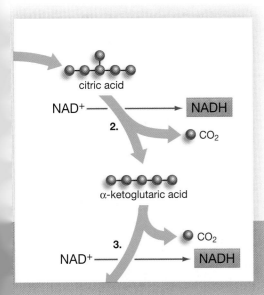

A closer look at energy harvesting.
(Section 7.5, page 142)

Fruit of the vine.
(Essay, page 140)

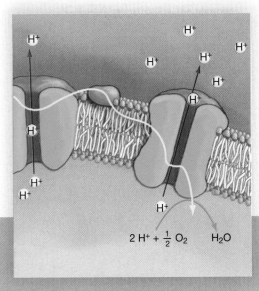

Big ATP harvest.
(Section 7.6, page 143)

The tragedy of the 1996 Everest expeditions drives home a point and raises a paradox: No one needs to be reminded that we need to breathe in order to live, and most people are aware that oxygen is the most important thing that comes in with breath. That said, of the next 100 people you meet, how many could tell you what oxygen is *doing* to sustain life? Put another way, why do we need to breathe?

The short answer is that breathing and oxygen are in the energy transfer business. They are part of a system that allows us to extract, from food, energy that is then used to put together the "energy currency" molecule, ATP. As you'll recall from Chapter 6, the body uses ATP (adenosine triphosphate) to power activities that range from muscle contraction to thinking to cell repair. Living things need large amounts of ATP to live, and organisms such as ourselves use oxygen to produce much of our ATP. If we don't get enough oxygen to keep making ATP, our bodies and minds start failing.

Oxygen is not required for all the "harvesting" we do of the energy contained in food. When you lift a box, the short burst of energy that is required comes largely from a different sort of energy harvesting, which you'll learn more about shortly. But even here, the essential *product* of energy harvesting is ATP. Keep that in mind and you won't get lost in the byways of energy transfer. How does the body produce enough ATP to allow us to read a book or climb a mountain? In this chapter, you'll find out.

7.1 Energizing ATP

Given the importance of ATP to this story, let us begin with a brief review of how this molecule works. Recall from Chapter 6 that each ATP molecule has three phosphate groups attached to it, and that ATP powers a given reaction by losing the outermost of these phosphate groups. In this process ATP is transformed into a molecule with *two* phosphate groups, adenosine diphosphate or ADP. To return to its more "energized" ATP state, it must have a third phosphate group attached again. This is, however, a trip *up* the energy hill, because the product (ATP) contains more energy than the reactant (ADP). Thus, getting that third phosphate group onto ADP requires energy. It is like preparing a spring-loaded mousetrap for action. It takes energy to pull a mousetrap bar back, and the same is true of putting a third phosphate group onto ATP (**see Figure 7.1**). Where does the energy come from? For animals such as ourselves, it comes from food; energy that is extracted from food powers the phosphate group up the energy hill—and literally onto ADP.

7.2 Electrons Fall Down the Energy Hill to Drive the Uphill Production of ATP

In tracking the extraction of energy from food, we will use as an example one particular food molecule, glucose, to see how energy is harvested from it. Though the details here are complex, the essential

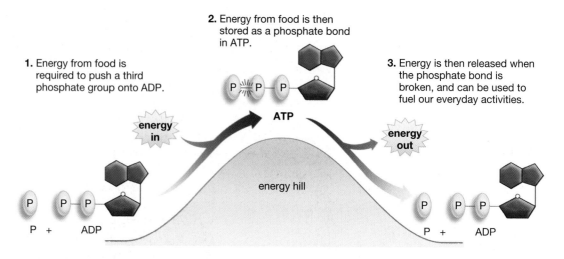

1. Energy from food is required to push a third phosphate group onto ADP.

2. Energy from food is then stored as a phosphate bond in ATP.

3. Energy is then released when the phosphate bond is broken, and can be used to fuel our everyday activities.

energy in

ATP

energy out

energy hill

P + ADP

P + ADP

Figure 7.1
Storing and Releasing Energy
Adenosine triphosphate (ATP) is the most important energy-releasing molecule in our bodies. The energy it contains is used to power everything from muscle contraction to thinking.

story is simple. *Electrons* derived from glucose, which is high in energy, will be running downhill; they will be channeled off, a few at a time, and their downhill drop will power the *uphill* push needed to attach a phosphate group onto ADP. The glucose-derived electrons will be transferred to several intermediate molecules in their downhill journey, but the *final* molecule that will receive them at the bottom of the energy hill is oxygen. We need to breathe because we need oxygen to serve as this final electron acceptor.

As you are following this process, you may well wonder why so many steps are required to transfer energy from food to ATP. Why isn't there just *one* step in energy transfer—food to ATP? It turns out that the gradual transfer of energy, through many steps, allows the body to make the most of the energy it receives. Think of it this way. You could have water drop 1,000 feet straight off a cliff onto a waterwheel below. In this case, the wheel would simply be pulverized by the water crashing onto it. In the process, the potential energy that had been contained in the water at the top of the cliff would be transformed almost entirely into useless heat. Conversely, you could have a stream that ran briskly downhill for 1,000 vertical feet with waterwheels spaced periodically to channel the energy off a little at a time, thereby conserving the energy for work. The steps you are going to read about amount to this kind of controlled channeling off of energy—the energy that's contained in food.

The Great Energy Conveyors: Redox Reactions

The basis for electron transfers down the energy hill is straightforward: Some substances more strongly attract electrons than do others. A substance that loses one or more electrons to another is said to have undergone **oxidation**. We hear the related word "oxidized" all the time—when paint on an outdoor surface has become dulled, for example, or when metal rusts (**see Figure 7.2**). Meanwhile, the substance that *gains* electrons in this reaction is said to have undergone **reduction**. (This seems about as logical as saying a country lost a war by winning. One way to think about it is that, because electrons carry a negative charge, any substance that gains electrons has had a reduction in its *positive charge*.)

In cells, oxidation and reduction never occur independently. If one substance is oxidized, another must be reduced. It's like a teeter-totter; if one side goes down, the other must come up. The combined operation is known as a reduction-oxidation reaction, or simply a **redox reaction**: the process by which electrons are transferred from one molecule

Figure 7.2
An Effect of Oxidation
Rust has formed on a hook atop a corroded steel sheet.

to another. Critically, the substance being oxidized in a redox reaction has its electrons traveling energetically *downhill*.

Many Molecules Can Oxidize Other Molecules

The term *oxidation* might give you the idea that oxygen must be involved in any redox reaction, but this is not the case. Any compound that serves to pull electrons from another is a so-called oxidizing agent. In living things, a large number of molecules are involved in energy transfer, and each has a certain tendency to gain or lose electrons relative to the others.

This is how electrons can be passed down the energy hill: The starting "energetic" molecule of glucose is oxidized by another molecule, which in turn is oxidized by the *next* molecule down the hill. The whole thing might be thought of as a kind of downhill electron bucket brigade. Molecules that serve to transfer electrons from one molecule to another in ATP formation are known as **electron carriers**. The thing that makes their role a little complicated is that many of the electrons they accept are bound up originally in hydrogen atoms. You may remember from Chapter 3 that a hydrogen atom amounts to one proton and one electron. In transferring a hydrogen atom, then, a molecule is transferring a single electron (bound to a proton), which means a redox reaction has taken place.

Redox through Intermediates: NAD

The most important electron carrier in energy transfer is a molecule known as **nicotinamide adenine dinucleotide**, or **NAD**, which can be thought of as a city cab. It can exist in two states: loaded with passengers or empty. And like a cab, it can switch very easily between those two states. The passengers that NAD picks up and drops off are electrons (**see Figure 7.3**).

The "empty" state that NAD comes in is ionic: NAD^+. Remembering back to the discussion about ions, you can see that NAD^+ is positively charged, meaning that it has fewer electrons than protons. In a redox reaction, what NAD^+ does is pick up, in effect, one hydrogen atom (an electron and a proton) and one solo electron (from a second hydrogen atom). The isolated electron that NAD^+ picks up turns it from positively charged to neutral $(NAD^+ \rightarrow NAD)$; the whole hydrogen atom takes it from NAD to NADH. Keeping an eye on redox reactions here, NAD^+ has become NADH by oxidizing a substance—by accepting electrons from it.

So much for half of NAD's role: picking up passengers. *Now*, as NADH, it is loaded with these passengers. It can proceed down the energy hill to donate them to molecules that have a greater potential to *accept* electrons than it does. Having dropped its passengers off with such a molecule, it returns to being the empty NAD^+ and is ready for another pickup. Through this process, NADH transfers energy from one molecule to another.

How Does NAD Do Its Job?

The molecule that NAD^+ is oxidizing is the starting glucose molecule (or actually derivatives of it). To carry out its oxidizing role, NAD^+ obviously needs to be brought together with the glucose derivatives. What have you been looking at that brings substances together in this way? Enzymes! One of the things you'll be seeing, therefore, is NAD^+ and glucose derivatives being brought together by enzymes, at which point NAD^+ accepts electrons from these derivatives, later to hand them off to another molecule.

Electron transfer through intermediate molecules such as NAD^+ provides the energy for most of the ATP produced. Thus, when looking at the diagrams in this chapter that outline respiration, you'll see:

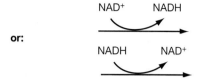

What these drawings indicate is the electron carrier shifting between its empty state (NAD^+) and its loaded state (NADH), or vice versa.

7.3 The Three Stages of Cellular Respiration: Glycolysis, the Krebs Cycle, and the Electron Transport Chain

All the energy players are finally in place: glucose, redox reactions, electron carriers, enzymes, and the rest. In terms of a molecular formula, the big picture on respiration looks like this:

$$C_6H_{12}O_6 + 6O_2 + 36ADP + 36P \rightarrow$$
$$6CO_2 + 6H_2O + 36ATP$$

The starting molecule on the first row $(C_6H_{12}O_6)$ is glucose, which is food, storing chemical energy. You can also see, next to it, the oxygen $(6O_2)$ that is needed as the final electron acceptor. Then there is ADP, the two-phosphate molecule; and inorganic phosphate molecules (the 36P), which are pushed on ADP to make it ATP. The *products* of this reaction, on the second row, are carbon dioxide $(6CO_2)$ and water $(6H_2O)$ as by-products, and energy in the form of ATP. How *much* ATP does the breakdown of one molecule of glucose yield? The answer is contained in the formula: a maximum of 36 molecules.

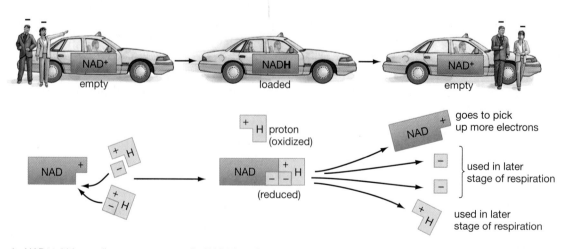

Figure 7.3
The Electron Carrier NAD$^+$
In its unloaded form (NAD^+) and its loaded form (NADH), this molecule is a critical player in energy transfer, picking up energetic electrons from food and transferring them to later stages of respiration.

1. NAD$^+$ within a cell, along with two hydrogen atoms that are part of the food that is supplying energy for the body.

2. NAD$^+$ is reduced to NAD by accepting an electron from a hydrogen atom. It also picks up another hydrogen atom to become NADH.

3. NADH carries the electrons to a later stage of respiration then drops them off, becoming oxidized to its original form, NAD$^+$.

Though many separate steps are involved in actually getting energy from glucose to ATP, respiration can be divided into three main phases. These are known as *glycolysis*, the *Krebs cycle,* and the *electron transport chain*.

In overview, energy harvesting through these three phases goes like this: We get some ATP yield in glycolysis and the Krebs cycle, and glycolysis is useful in providing small amounts of ATP in quick bursts. But for most organisms, the main function of glycolysis and the Krebs cycle is the transfer of electrons to the electron carriers such as NAD^+. Bearing their electron cargo, these carriers then move to the third phase of energy harvesting, the electron transport chain (ETC). Here the electron carriers are themselves oxidized and the resulting movement of their electrons through ETC provides the main harvest of ATP. Indeed, about 32 of the approximately 36 molecules of ATP that are netted per glucose molecule are obtained in the ETC. If you look at **Figure 7.4**, you can see two representations of the three stages of energy extraction.

Web Tutorial 7.2
Stages of Cellular
Respiration

Figure 7.4
Overview of Energy Harvesting

(a) In metaphorical terms

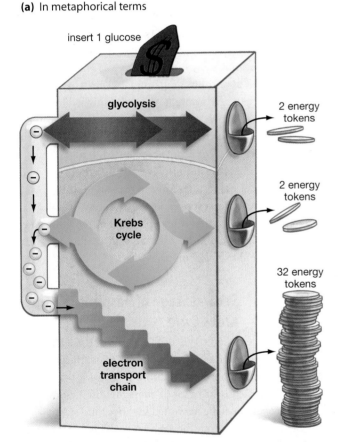

(b) In schematic terms

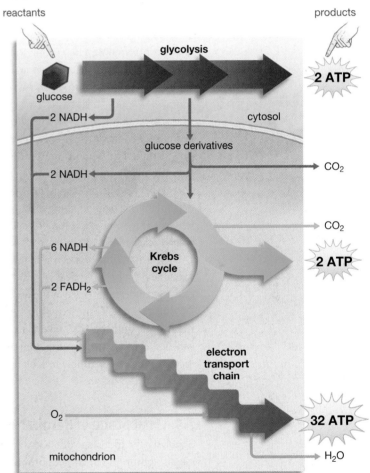

36 ATP maximum per glucose molecule

Just as the video games in some arcades can use only tokens (rather than money) to make them function, so our bodies can use only ATP (rather than food) as a direct source of energy. The energy contained in food—glucose in the example—is transferred to ATP in three major steps: glycolysis, the Krebs cycle, and the electron transport chain. Though glycolysis and the Krebs cycle contribute only small amounts of ATP directly, they also contribute electrons (on the left of the token machine) that help bring about the large yield of ATP in the electron transport chain. Our energy-transfer mechanisms are not quite as efficient as the arcade machine makes them appear. At each stage of the conversion process, some of the original energy contained in the glucose is lost to heat.

As with the arcade machine, the starting point in this example is a single molecule of glucose, which again yields ATP in three major sets of steps: glycolysis, the Krebs cycle, and the electron transport chain (ETC). These steps can yield a maximum of about 36 molecules of ATP: 2 in glycolysis, 2 in the Krebs cycle, and 32 in the ETC. As noted, however, glycolysis and the Krebs cycle also yield electrons that move to the ETC, aiding in its ATP production. These electrons get to the ETC via the electron carriers NADH and $FADH_2$, shown on the left. Oxygen is consumed in energy harvesting, while water and carbon dioxide are produced in it. Glycolysis takes place in the cytosol of the cell, but the Krebs cycle and the ETC take place in cellular organelles, called mitochondria, that lie within the cytosol.

Glycolysis: First to Evolve, Less Efficient

The division among the three energy-harvesting phases is, in one sense, a division of evolutionary history. For most organisms, energy harvesting's first phase, glycolysis, could be likened to a nineteenth-century steam engine working side by side with a state-of-the art electric turbine. By itself, glycolysis produces only two ATP molecules per glucose molecule, which is pretty small in comparison to the 36 ATP molecules likely to come from all three phases. And glycolysis doesn't even yield many electrons compared to energy harvesting's second phase, the Krebs cycle. Yet glycolysis takes place in all living things—at the very least, as the first phase in energy harvesting—and for certain primitive one-celled organisms it can be the *only* phase in energy harvesting.

All of this points to glycolysis being a more ancient form of energy harvesting than the other two phases. It existed first, and then at some time in the evolutionary past, most organisms added to it the processes we now call the Krebs cycle and the ETC. The critical factor in this change was the use of oxygen in energy harvesting—specifically the use of oxygen in the ETC. Because the products of glycolysis feed ultimately into the ETC, the entire three-stage energy-harvesting process is referred to as oxygen-dependent or *aerobic* energy transfer. Taken as a whole, the aerobic harvesting of energy is known as **cellular respiration**.

The separation between glycolysis and the other two phases is also a physical separation. In eukaryotic cells, glycolysis takes place in the watery material that lies outside the cell's nucleus—the cytosol—while the Krebs cycle and the ETC take place within tiny organelles, called mitochondria, that are immersed in the cytosol. Let's start now to look at all three phases of cellular respiration.

7.4   First Stage of Respiration: Glycolysis

Glycolysis, the first stage of energy harvesting, means "sugar splitting," which is appropriate because our starting molecule is the simple sugar glucose. Briefly, here is what happens during glycolysis. First, this one molecule of glucose has to be prepared, in a sense, for energy release. ATP actually has to be *used*, rather than synthesized, in two of the first steps of glycolysis, so that the relatively stable glucose can be put into the form of a less-stable sugar. This sugar is then split in half. The two molecules formed have three carbons each (whereas the starting glucose has six). Once this split is accomplished, the steps of glycolysis take place in *duplicate*: It is like taking one long

piece of cloth, dividing it in two, and then proceeding to do the same things to both of the resulting pieces. If you look at **Figure 7.5**, you can see the individual steps of glycolysis. (If your class is studying glycolysis in detail, you will need to look carefully at Figure 7.5, as it is the only place in the text where the individual steps of glycolysis are reviewed.)

Now, what are the results of these steps? Glycolysis accomplishes three valuable things in energy harvesting: It yields two ATP molecules, it yields two energized molecules of NADH, and it results in two molecules of pyruvic acid. These pyruvic acid molecules are the derivatives of the original glucose molecule. Now *they* are the molecules that will be oxidized—that will have electrons removed from them—in the next stage of energy harvesting. Keeping our eye on the ball here, we have some net ATP through glycolysis, some traveling electrons that will help make more ATP in the electron transport chain, and a derivative of the original glucose molecule (pyruvic acid) that will now move to the next stage of energy harvesting, where more energy will be extracted from it.

Before continuing on to this next stage, you may wish to read more about what might be called the consequences of glycolysis. Even oxygen-using organisms such as ourselves rely on glycolysis when short bursts of energy are needed. And, as noted, for some organisms, glycolysis is the end of the line in energy harvesting. You can read about this in "When Energy Harvesting Ends at Glycolysis" on page 140. Glycolysis fits into a special place in providing energy for human exercise; you can read about this in "Energy and Exercise" on page 144.

7.5   Second Stage of Respiration: The Krebs Cycle

Since glycolysis yielded only two of the 36 molecules of ATP that eventually are harvested, the glucose derivative that resulted from glycolysis, pyruvic acid, still clearly has a lot of energy left in it. Some of this energy will be harvested in the second stage of cellular respiration, the **Krebs cycle**. This stage is named for the German and English biochemist Hans Krebs, who in the 1930s used the flight muscles of pigeons to find out how aerobic respiration works. Because the first product of the Krebs cycle is citric acid, the cycle is sometimes referred to as the **citric acid cycle**. The ATP yield in this cycle is once again a paltry two molecules. But the Krebs cycle serves to set the stage for the big harvest of ATP in the electron transport chain by supplying energy-bearing electron carriers, as you will see.

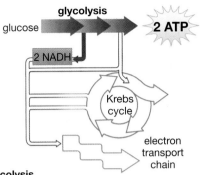

↗ 6 carbon compound

Steps in glycolysis

1. Delivered by the bloodstream, glucose enters a cell and immediately has a phosphate group from ATP attached to it. Because this process, called phosphorylation, attaches the phosphate to the sixth carbon of glucose, it now goes under the name glucose-6-phosphate. Note that one molecule of ATP has been *used* in this step.

   ATP ledger now reads: -1 ATP

2. Glucose 6-phosphate is rearranged to become a molecule called fructose-6-phosphate.

3. Another molecule of ATP is used to add a second phosphate to fructose-6-phosphate, which now becomes fructose-1,6-diphosphate.

   ATP ledger now reads: -2 ATP

4. The single, six-carbon sugar fructose-1,6-diphosphate now becomes two molecules of a 3-carbon sugar, glyceraldehyde-3-phosphate, each with a phosphate group attached. From here on out, glycolysis happens in duplicate: What happens to one of the glyceraldehyde molecules happens to the other.

5. An enzyme brings together glyceraldehyde-3-phosphate, the electron carrier NAD$^+$, and a phosphate group. The glyceraldehyde-3-phosphate molecule is oxidized by NAD$^+$, which in its new form, NADH, moves to the electron transport chain bearing its electron cargo. The oxidation of NAD$^+$ is energetic enough that it allows the phosphate group to become attached to the main molecule, now called 1,3-diphosphoglyceric acid. Because everything is happening in duplicate, two NADH molecules are produced.

6. 1,3-diphosphoglyceric acid loses one of its phosphate groups, thus becoming 3-phosphoglyceric acid. The reaction is energetic enough to push this phosphate group onto an ADP molecule, yielding ATP. Because of duplication, two ATP are produced.

   ATP ledger now reads: 0 ATP

7. In two reactions, 3-phosphoglyceric acid becomes phosphoenolpyruvic acid, which generates more ATP as it transfers its phosphate group to ADP. Two more ATP molecules are produced. The phosphate transfer turns phosphoenolpyruvic acid into pyruvic acid—the derivative of the original glucose that now will enter the Krebs cycle.

   ATP ledger now reads: +2 ATP

molecules in **molecules out**

glucose

ATP ———————————————→ ADP

1. glucose-6-phosphate

   Red balls are carbons and gold ovals are phosphate groups

2. fructose-6-phosphate

ATP ———————————————→ ADP

3. fructose-1,6-diphosphate

4. glyceraldehyde-3-phosphate

2 NAD$^+$ + 2 P ———————————————→ **2 NADH** + 2 H$^+$

5. 1,3-diphosphoglyceric acid

2 ADP ———————————————→ **2 ATP**

6. 3-phosphoglyceric acid

2 ADP ———————————————→ **2 ATP**

7. pyruvic acid

↘ 3 carbon compounds

Figure 7.5
Glycolysis
In glycolysis, the single glucose molecule is transformed in a series of steps into two molecules of a substance called pyruvic acid. These two molecules then move on to the next stage of cellular respiration (the Krebs cycle). Glycolysis also produces two molecules of electron-carrying NADH, which move to the electron transport chain. Although two molecules of ATP are used in the early stages of glycolysis, four are produced in the later stages, for a net production of two ATP.

Into the Krebs Cycle: Why Is It a Cycle?

Acetyl CoA now enters the Krebs cycle. The reason this is a *cycle* becomes clear when you look at **Figure 7.8**. You can see that, in the first step of the cycle, acetyl CoA combines with a substance called oxaloacetic acid to produce citric acid. If you look over at about 11 o'-clock on the circle, however, you can see that the *last* step in the cycle is the *synthesis* of oxaloacetic acid. Thus, a substance that's necessary for this chain of events to take place is itself a product of the chain.

Stroll around the circle now to see how the Krebs cycle functions in a little more detail. In essence, what's happening is that, as the entering acetyl CoA molecule is transformed into these various molecules, it is being oxidized by electron carriers, with the resulting electrons moving on to the ETC. ATP is also derived, and CO_2 is a product. The only major player that's unfamiliar here is the $FAD–FADH_2$ that can be seen taking part in a redox reaction at about 8 o'clock. It is simply another electron carrier, similar to $NAD^+/NADH$.

In counting up the Krebs cycle's yield of both ATP and electron carriers (NADHs and so on), recall that *two turns* around this cycle result from the original glucose molecule (because it provided two molecules of the starting acetyl CoA). This means a total yield of 6 NADH, 2 $FADH_2$, and 2 ATP from Krebs for each glucose molecule. Figure 7.8 reviews the Krebs cycle step by step. (If your class is studying the Krebs cycle in detail, you should look carefully at Figure 7.8, as it is the only place in the text where the individual steps of the cycle are reviewed.)

Figure 7.8
The Krebs Cycle
The Krebs cycle is the major source of electrons that are transported to the electron transport chain by the carriers NADH and $FADH_2$. For each molecule of acetyl coenzyme A that enters the cycle, 3 NADH molecules, 1 $FADH_2$, and 1 ATP are produced. Because two acetyl CoA molecules enter per molecule of glucose, the yield per glucose is 6 NADH, 2 $FADH_2$, and 2 ATP.

Steps in the Krebs cycle

1. Acetyl CoA combines with the four-carbon oxaloacetic acid and the CoA fragment separates from this compound. The result is the energetic six-carbon molecule citric acid, which will now be oxidized.

2. A citric acid derivative is oxidized by NAD^+; the resulting NADH carries electrons to the ETC. An intermediate molecule then loses a CO_2 molecule. Citric acid is now alpha-ketoglutaric acid.

3. Alpha-ketoglutaric acid loses a CO_2 molecule and the resulting four-carbon molecule is oxidized by NAD^+.

4. An alpha-ketoglutaric acid derivative is split, releasing enough energy to attach a phosphate to ADP, making it ATP. Alpha-ketoglutaric acid has become succinic acid.

5. Succinic acid is oxidized by FAD, losing two complete hydrogen atoms to it. The resulting $FADH_2$ then moves to the ETC. In a series of steps, succinic acid is transformed into malic acid.

6. Malic acid is oxidized by NAD^+. This step transforms malic acid to oxaloacetic acid—the molecule that enters the first step of the Krebs cycle.

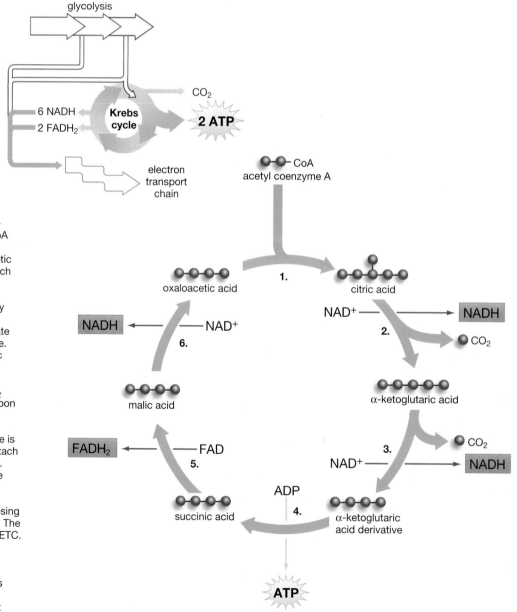

7.6 Third Stage of Respiration: The Electron Transport Chain

Having moved through the Krebs cycle, you've reached the main event in cellular respiration: the production of 32 more ATP molecules through the work of the **electron transport chain**, the third stage of aerobic energy harvesting. Glycolysis took place in the cytosol, and the Krebs cycle took place in the inner compartment of the mitochondria. Now, however, the action shifts to the mitochondrial inner membrane, the site of the ETC (**see Figure 7.9**). The "links" in this chain are a series of molecules. You've been looking for some time now at how NADH and $FADH_2$ carry off the electrons derived from glycolysis and the Krebs cycle. This is the destination of these electron carriers and their cargo.

Upon reaching the mitochondrial inner membrane, the electron carriers donate electrons and hydrogen ions to the ETC. Once again, this is a trip down the energy hill, only *NADH* is now the molecule whose electrons exist at a higher energy level. Thus, when NADH runs into the appropriate enzyme in the ETC, NADH is oxidized, donating its electrons and proton to the ETC. This process then is simply repeated down the whole ETC, each carrier donating electrons to the next electron carrier in line. A carrier is reduced by receiving these electrons; donating them to the next carrier, it then returns to an oxidized state. Each carrier thus alternates between oxidized and reduced states.

Looking at Figure 7.9, you can see the carriers in the ETC: there are three large enzyme complexes with

Figure 7.9
The Electron Transport Chain (ETC)
The movement of electrons through the ETC powers the process that provides the bulk of the ATP yield in respiration. The electrons carried by NADH and $FADH_2$ are released into the ETC and transported along its chain of molecules. The movement of electrons along the chain releases enough energy to power the pumping of hydrogen ions (H^+) across the membrane into the outer compartment of the mitochondrion. It is the subsequent energetic "fall" of the H^+ ions back into the inner compartment that drives the synthesis of ATP molecules by the enzyme ATP synthase.

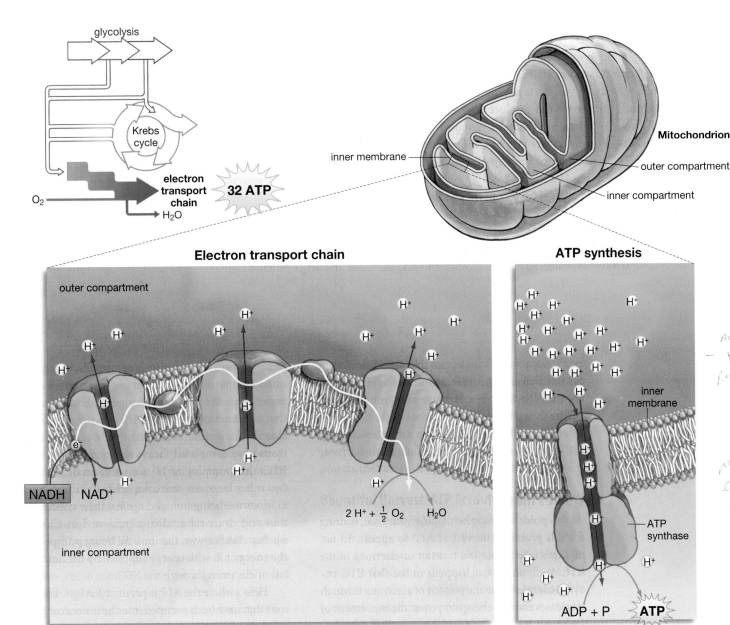

from the Krebs cycle joins the ETC a little later than does NADH and as such produces only about two ATP molecules, compared to NADH's three. The upshot of all this is summarized in **Table 7.1** which presents the final count on our energy harvest.

What's apparent from looking at this table is that the ETC is a mighty energizer. Think about how efficient this process is compared to glycolysis (34 ATP compared to 2). Imagine one car that got 34 miles to the gallon compared to another that got 2. The 36 total ATP yield actually is the amount that can be derived under optimal conditions. (The "loss, due to active transport" in the table refers to the expenditure of ATP required to move the NADH produced in glycolysis into the mitochondria.)

Table 7.1 Energy Harvesting Accounting

Stage	NADH	FADH$_2$	ATP yield
Glycolysis	2		2
Pyruvic acid conversion	2		
Krebs cycle	6	2	2
Electron transport chain	10		30
		2	4
Loss, due to active transport:			−2
Total ATP:			**36**

Finally, Oxygen Is Reduced, Producing Water

The story of the breakdown of glucose—the saga of *one molecule's* transformation—is nearly complete. The only unfinished business lies at the end of the ETC. As noted earlier, oxygen is the final acceptor of the working electrons. In the mitochondrial inner compartment, a single oxygen atom $\left(\frac{1}{2}O_2\right)$ accepts two electrons from the ETC and two H^+ ions. The result is H_2O: water. That's why the formula for respiration is written as:

$$C_6H_{12}O_6 + 6O_2 + 36ADP + 36P \rightarrow$$
$$6CO_2 + 36\,ATP + 6H_2O$$

Glucose, plus oxygen, plus ADP and phosphate yield carbon dioxide, plus ATP, plus water: That's the short story of aerobic respiration.

7.7 Other Foods, Other Respiratory Pathways

By looking at aerobic respiration as it applies to glucose, which was used as the starting molecule, you've been able to go through the whole respiratory chain. But it goes without saying that there are many kinds of nutrients besides glucose—for example, fats, proteins, and other sugars. All these can provide energy by being oxidized within the chain of reactions you've just gone over. However, none of them proceeds through the chain in exactly the same way as glucose. In addition, the body has more uses for foods than breaking them down to produce ATP. At any given moment, an organism may need to go the other way: toward the *buildup* of proteins from amino acids or the synthesis of the fats we call triglycerides. Organisms are able to handle this variability because nutrients and their derivatives can be *channeled* through pathways in different directions in the respiratory chain in accordance with cell needs. The alternation of the breakdown of food molecules and the buildup of food storage molecules is one of the central functions that living things carry out.

Alternate Respiratory Pathways: Fats as an Example

To give one example of how this channeling of nutrients works, consider the possible fates of triglycerides. As you learned in Chapter 4, triglycerides are fats made of a 3-carbon glycerol "head" and three long hydrocarbon chains that look like tines on a fork. Imagine that a given cell needs energy from fat. The question is, how do *triglycerides* go through aerobic respiration, as opposed to glucose?

The first thing that happens is that a triglyceride molecule is split by enzymes into its constituent parts of fatty acids and glycerol. Glycerol is then converted into glyceraldehyde phosphate. This formidable name may not ring a bell, but if you look back to step 4 in Figure 7.5, you'll see that it is one of the *downstream* derivatives of glucose in glycolysis. To put this another way, glycerol does not first get converted to glucose and then march through *all* of the steps of glycolysis. Instead, it joins the glycolytic pathway several steps down from glucose and is then converted to pyruvic acid. Then, as pyruvic acid, it goes through the Krebs cycle and ETC, producing ATP. And the fatty acids? They are converted to acetyl CoA, which you may remember is the substrate that enters the Krebs cycle, yielding energy by becoming oxidized there.

What Happens When Less Energy Is Needed?

So much for what happens when an organism needs a good deal of energy from fats. But what happens to the triglyceride components when this need is reduced? Glycerol, which had been converted to glyceraldehyde phosphate and begun marching through glycolysis, can be converted to *glucose* in a process that, in its main steps, is glycolysis in reverse. The resulting glucose molecules can then be used to do any number of things. They might, for example, be put together to create the storage form of glucose, glycogen.

You may be getting a sense that the metabolism is rather like a traffic circle: Molecules can enter at different points and leave at different points for varying destinations, depending on the needs of the organism. In **Figure 7.10**, you can see a summary of the way foods are broken down through various respiratory pathways.

On to Photosynthesis

This long walk through cellular respiration has illustrated how living things harvest energy from food. Recall, however, that almost all living things have one source to thank for this food: the sun's energy, which is trapped by plants in the chemical bonds of carbohydrates. This capture of energy takes place through a process that has a beautiful symmetry with the respiration you have just looked at. That process is called photosynthesis.

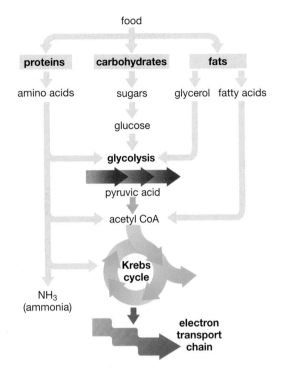

Figure 7.10
Many Respiratory Pathways
Glucose is not the only starting material for cellular respiration. Other carbohydrates, proteins, and fats can also be used as fuel for cellular respiration. These reactants enter the process at different stages.

Chapter Review

Summary

7.1   Energizing ATP: Adding a Phosphate Group to ADP

- The molecule adenosine triphosphate (ATP) supplies the energy for most of the activities of living things. For ATP to be produced, a third phosphate group must be added to adenosine diphosphate (ADP), a process that requires energy.

7.2   Electrons Fall Down the Energy Hill to Drive the Uphill Production of ATP

- In animals, the energetic fall of electrons derived from food powers the process by which the third phosphate group is attached to ADP, making it ATP.

- Electron transfer in the production of ATP works through redox reactions, meaning reactions in which one substance loses electrons to another substance. The substance that loses electrons in a redox reaction is said to have been oxidized, while the substance that gains electrons is said to have been reduced. In energy transfer in living things, starting food molecules are oxidized, and the electrons they lose ultimately power the process by which ATP is produced.

- Electrons are carried between one part of the energy-harvesting process and another by electron carriers, the most important of

which is nicotinamide adenine dinucleotide or NAD. In its "empty" state, this molecule exists as NAD^+. Through a redox reaction, it picks up one hydrogen atom and another single electron from food, thus becoming NADH, a form it will retain until it drops off its energetic electrons (and a proton) in a later stage of the energy-harvesting process.
Web Tutorial 7.1: Oxidation and Reduction

7.3   The Three Stages of Cellular Respiration: Glycolysis, the Krebs Cycle, and the Electron Transport Chain

- In most organisms, the harvesting of energy from food takes place in three principal stages, glycolysis, the Krebs cycle, and the electron transport chain (ETC). Taken as a whole, this oxygen-dependent three-stage harvesting of energy is known as cellular respiration.

- Some organisms rely solely on glycolysis for energy harvesting. For most organisms, glycolysis is a primary process of energy extraction only in certain situations—when quick bursts of energy are required, for example—but it is a necessary first stage to the Krebs cycle and the ETC.

- Glycolysis takes place in the cell's cytosol, while the Krebs cycle and the ETC take place in cellular organelles, called mitochondria, that lie within the cytosol. Glycolysis yields two net molecules of ATP per molecule of glucose, as does the Krebs cycle. The net yield in the electron transport chain is a maximum of about 32 ATP molecules per molecule of glucose. Glycolysis and the Krebs cycle are critical, however, in that they yield electrons that are carried to the ETC (via electron carriers such as NAD^+) for the final high-yield stage of energy harvesting.

Web Tutorial 7.2: Stages of Cellular Respiration

7.4 First Stage of Respiration: Glycolysis

- When glycolysis begins with a single molecule of glucose, the ultimate products are two molecules of NADH (which move to the ETC, bearing their energetic electrons) and two molecules of ATP (which are ready to be used). Glycolysis also produces two molecules of pyruvic acid—the derivatives of the original glucose molecule—which move on to the Krebs cycle.

7.5 Second Stage of Respiration: The Krebs Cycle

- There is a transition step in respiration between glycolysis and the Krebs cycle. In it, each pyruvic acid molecule that was produced in glycolysis combines with coenzyme A, thus forming acetyl coenzyme A (acetyl CoA), which enters the Krebs cycle. There are also two other products of this reaction. One is a molecule of carbon dioxide, which diffuses to the bloodstream; the other is one more molecule of NADH, which moves to the ETC. For each starting molecule of glucose, two molecules of pyruvic acid go through this step; thus its product per molecule of glucose is two molecules each of carbon dioxide, NADH, and acetyl CoA.

- In the Krebs (or citric acid) cycle, the derivatives of the original glucose molecule are oxidized, with the result that more energetic electrons are transported by the electron carriers NADH and $FADH_2$ to the ETC. The net energy yield of the Krebs cycle is six molecules of NADH, two molecules of $FADH_2$, and two molecules of ATP per molecule of glucose.

7.6 Third Stage of Respiration: The Electron Transport Chain

- The ETC is a series of molecules that are located within the mitochondrial inner membrane. Upon reaching the ETC, the electron carriers NADH and $FADH_2$ are oxidized by molecules in the chain. Each carrier in the chain is then reduced by accepting electrons from the carrier that came before it. The last electron acceptor in the ETC is oxygen.

- The movement of electrons through the ETC releases enough energy to power the movement of hydrogen ions (H^+ ions) through the three ETC protein complexes, moving them from the mitochondrion's inner compartment to its outer compartment. The movement of these ions down their concentration and charge gradients, back into the inner compartment through an enzyme called ATP synthase, drives the synthesis of ATP from ADP and inorganic phosphate. In the inner compartment, oxygen accepts the electrons from the ETC and hydrogen ions, thus forming water.

7.7 Other Foods, Other Respiratory Pathways

- Nutrients and their derivatives can be channeled through different pathways in cellular metabolism, in accordance with the needs of the organism. At any given moment, a cell may need to work more on synthesizing organic molecules than on breaking them down. Proteins and lipids enter the metabolic pathway for ATP production at different points than does the glucose reviewed in detail in the chapter.

Key Terms

alcoholic fermentation 140
ATP synthase 145
cellular respiration 138
citric acid cycle 138
electron carrier 135
electron transport chain (ETC) 143
glycolysis 138

Krebs cycle 138
lactate fermentation 140
lactic acid 140
nicotinamide adenine dinucleotide (NAD) 135
oxidation 135
redox reaction 135
reduction 135

Understanding the Basics

Multiple-Choice Questions (answers in the back of the book)

1. You need energy to think, to keep your heart beating, to play a sport, and to study this book. This energy is directly supplied by _____, which is (are) produced in the process of cellular respiration.
   a. enzymes
   b. ATP
   c. NAD^+
   d. vitamins
   e. proteins

2. Energy transfer in living things works through redox reactions, in which one substance is _____ by another substance, thereby _____.
   a. transported … becoming more energetic
   b. digested … becoming more energetic
   c. reduced … losing electrons to it
   d. oxidized … losing electrons to it
   e. oxidized … gaining electrons from it

3. _____ and _____ are important not so much for the ATP produced in them, but for their _____.
   a. Glycolysis … the Krebs cycle … yield of electrons transported to the ETC
   b. Glycolysis … the ETC … yield of electrons transported to the cytosol
   c. Redox reactions … fatty acid breakdown … yield of calories
   d. The Krebs cycle … the ETC … numerous redox reactions
   e. The Krebs cycle … the ETC … fatty-acid breakdown

4. What happens when a cell in your body momentarily has an abundance of ATP?
   a. It stops oxidizing food and starts building up storage food molecules from smaller food molecules.
   b. It continues to produce ATP at the same rate so that it is ready for the next emergency.
   c. It starts harvesting energy solely through glycolysis.
   d. It forms glucose out of extra ATP.
   e. It dies.

5. At most, how many molecules of ATP can be produced per glucose molecule in cellular respiration?
   a. 2
   b. 8
   c. 24
   d. 36
   e. 75

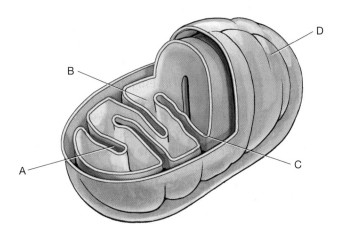

6. Where does the Krebs cycle occur in this model of a mitochondrion?
   a. A
   b. B
   c. C
   d. D
   e. none of these

7. When NADH passes its electrons to the ETC, it is
   a. electrolyzed
   b. polarized
   c. reduced
   d. oxidized
   e. deformed

8. We need to breathe because we need
   a. both atmospheric nitrogen and the oxygen for energy transformation
   b. oxygen to donate electrons to NAD^+
   c. nitrogen to donate phosphate groups to oxygen
   d. oxygen to act as the final acceptor of electrons in the ETC
   e. oxygen to donate phosphate groups to ADP, making it ATP

9. Why is the Kreb's Cycle so important for cells?
   a. it is where most ATP molecules are made
   b. it serves to assist in protein synthesis
   c. it is where many of the NADH molecules are made that will be used later
   d. it is where glucose is first broken down
   e. it is not that important.

Brief Review

1. Living things need ATP to power most of the processes that go on within them. In essence, how does cellular respiration yield this ATP?

2. Where do you expect people would have more mitochondria: their skin cells or their muscle cells? Why?

3. In the ETC, what is physically moved across the inner membrane—and energetically moved to a higher state—in order to drive the attachment of phosphate groups to ADP?

4. Where does the energy come from that powers this movement across the inner membrane in the ETC?

5. Most of the ATP we use is produced in the ETC. So when we have to run across the street to avoid traffic, does the ATP we are using come primarily from the ETC?

Applying Your Knowledge

1. In Madeleine L'Engle's children's novel *A Wrinkle in Time*, the mitochondria in one of the characters start to die. Describe what would happen to people who lost their mitochondria, and explain why it would happen.

2. Most of the heat in the human body is generated within mitochondria. Why should this be?

3. The food we eat is broken down in our digestive tract, but where are most of the calories in this food actually "burned"?

cells and releases the CO_2, which will be used in the Calvin cycle. These special cells are wrapped around leaf veins and are hence named the *bundle-sheath cells* (**see Figure 8.10**). What the C_4 system provides—even at times when little CO_2 may be coming in through the stomata—is a relatively high concentration of CO_2 where it's needed: in the cells where the Calvin cycle takes place. We can thus define **C_4 photosynthesis** as a form of photosynthesis in which carbon dioxide is first fixed to a four-carbon molecule and then transferred to special cells in which the Calvin cycle is undertaken, bundle sheath cells.

The C_4 Pathway Is Not Always Advantageous

The C_4 pathway is known to be employed in about 1,000 species of plants, most notably in some grasses, and in corn, sugarcane, and sorghum. But there are vastly more C_3 than C_4 plants. You might wonder why evolution hasn't, you might say, weeded out C_3 plants in favor of the C_4 variety, since the latter seem to be so much more efficient at photosynthesis. The answer is that C_4 fixation does not confer an across-the-board advantage. It has a cost, which is the expenditure of ATP in shuttling carbon dioxide to the bundle-sheath cells. What this means is that C_4 fixation is advantageous only where the weather is warm enough to bring about a significant increase in photorespiration. Where the weather is cooler, C_4 plants will be out-competed by C_3 species.

8.7 Another Photosynthetic Variation: CAM Plants

When plants live in climates that are not just warm but *dry*, a large part of their survival comes down to retaining water. Photosynthesis, however, works against water retention, because as you have seen, when CO_2 can pass in, water vapor can pass out.

The plants we call succulents—a group that includes most cacti—have a solution for this. It is to close the stomata during the day and open them at night. The plants then carry out C_4 metabolism at night, but only up to a point. They fix carbon dioxide into an initial four-carbon molecule and then stand pat. The CO_2 stays "banked" in them, awaiting the energy of the next day's sun, which will power the production of the ATP they need to carry out the Calvin cycle (**see Figure 8.11**). This is **CAM photosynthesis**, defined as a form of photosynthesis, undertaken by plants in hot, dry climates, in which carbon fixation takes place at night, and the Calvin cycle during the day. This process gets the job done for cactus, pineapple, orchid, and some mint family plants, among others. As you've no doubt guessed, CAM is an acronym; it stands for *crassulacean acid metabolism*. The succulent plant family *Crassulaceae* (which includes jade plants) was the first group of plants in which CAM metabolism was discovered. The three methods of photosynthesis—C_3, C_4, and CAM—are summarized in **Table 8.1**.

How Did We Learn?

It was not until after World War II that scientists pieced together the story of how plants join carbon dioxide to a sugar to create food. As you've seen, this happens through a set of steps that today are called the Calvin cycle. You can read about the man the cycle was named for, Melvin Calvin, and his process of discovery in "Plants Make Their Own Food, But How?" on page 164.

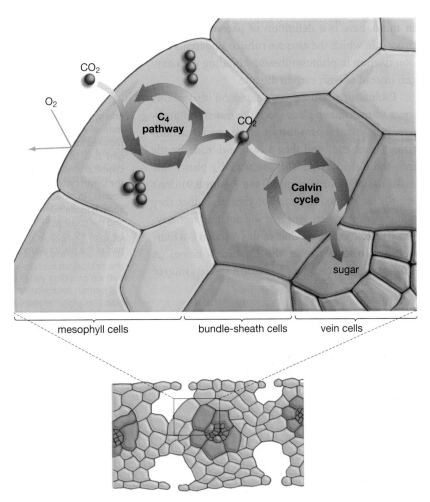

Figure 8.10
C_4 Photosynthesis
C_4 photosynthesis is an adaptation of some warm-climate plants to the problem of photorespiration. Photorespiration occurs when the C_3 enzyme rubisco binds to oxygen rather than carbon dioxide (thereby undercutting the production of sugar). C_4 plants contain a different enzyme in their mesophyll cells that binds to carbon dioxide, but not to oxygen. Then the carbon dioxide is escorted into special bundle-sheath cells, where rubisco can bind to it, thus initiating the Calvin cycle reviewed earlier, meaning the production of sugar can proceed.

Closing Thoughts on Photosynthesis and Energy

Readers who have stayed the course on this account of photosynthesis now know at least one thing very well about it: How complicated it is, with its many components and long metabolic pathways. It's also probably clear by this point why scientists continue to study photosynthesis in such detail. It is the foundation of plant growth, and upon plant growth hinges nothing less than the survival of all animals—including human beings. Without an understanding of this linkage, it's easy to see plants as a set of mute fixtures whose main contribution to human life is aesthetic. But with this knowledge, you can begin to see the central position that plants occupy in the interconnected web of life.

Over the last three chapters, you have learned about the endless back-and-forth of oxygen, carbon dioxide, and energy in the living world. *Cycle* has been a recurring word in this long discussion, because the only one-way trip you've encountered has been the relentless "spillage" of energy from the sun down into heat. Looked at in a cynical way, Earth and its inhabitants constitute a kind of leaky holding tank for energy that comes from the sun. Looked at another way, however, the living world has been able, through pho-

Figure 8.11
Dry-Weather Photosynthesis
Plants such as this saguaro cactus in Arizona utilize CAM photosynthesis, thereby preserving precious water.

tosynthesis, to take the sun's energy and build a remarkable edifice with it. Think of the forms and sheer mass of living material on Earth. One of the most amazing things about this structure is that we humans get to be both a part of it and witnesses to it.

Web Tutorial 8.4
**Different Kinds of
Photosynthesis**

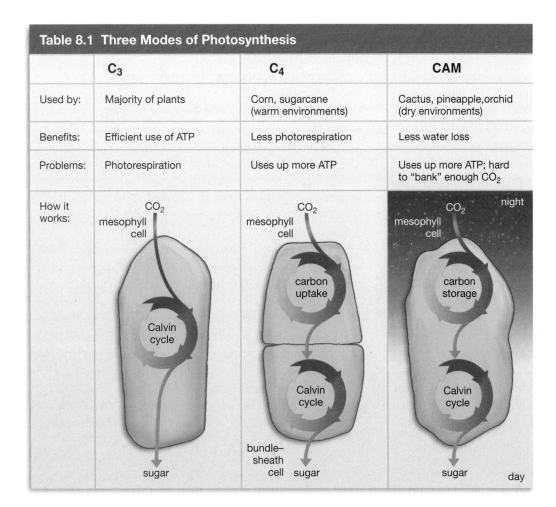

Table 8.1 Three Modes of Photosynthesis

	C_3	C_4	CAM
Used by:	Majority of plants	Corn, sugarcane (warm environments)	Cactus, pineapple, orchid (dry environments)
Benefits:	Efficient use of ATP	Less photorespiration	Less water loss
Problems:	Photorespiration	Uses up more ATP	Uses up more ATP; hard to "bank" enough CO_2
How it works:			

How Did We Learn?
Plants Make Their Own Food, But How?

Science constantly is confronted with what are known as "black box" problems. In these problems, researchers know what goes into a process and they know what comes out of it, but what happens *in between* is a mystery. At the conclusion of World War II, plant science had a doozy of a black-box problem. It was clear that plants were starting with carbon dioxide from the air and ending with high-energy carbohydrates, but no one knew what came in between. What were the *intermediate* substances that carbon dioxide was being made part of on the route to food? Getting inside this black box turned out to require a three-part combination: insightful researchers, new scientific techniques, and old-fashioned hard work.

One of the peacetime benefits of World War II's atomic weapons research was that scientists could produce quantities of substances known as *radioactive isotopes*. As you saw in Chapter 2, isotopes result from variations in the number of neutrons an element has in its nucleus. Radioactivity, meanwhile, is the release of radiant energy from atoms undergoing rapid decay. The element carbon is everywhere in nature, and as it turns out, one of its isotopes, carbon-14, is radioactive. Carbon-14 occurs naturally, but the ability to *manufacture* this substance opened a host of possibilities for research in the mid-1940s.

Stepping up to the plate in 1946 to take advantage of these possibilities was a University of California, Berkeley, chemist named Melvin Calvin. Then just 35 years old, Calvin was encouraged by the physicist Ernest Lawrence of Berkeley to apply radiocarbon techniques to organic chemistry research. With carbon-14, Calvin realized, he could *tag* carbon dioxide to find out what it becomes part of during photosynthesis, much as a zoologist might tag a bear with a signaling device to find out where it goes during winter. The radiation that carbon-14 emits might be thought of as a signal whose constant message is: "Here I am."

The steps that Calvin and his colleagues followed were simple in conception but very difficult in execution. They first took some photosynthetic algae and put them in a flask that had light shining on it. Like any photosynthetic organism, these algae would, of course, be taking in carbon dioxide to carry out their work. In this case, however, the carbon dioxide in question contained carbon-14. The trick was to allow the algae to perform a little photosynthesis using this $^{14}CO_2$. Then the algae were plunged into boiling alcohol, which killed them. By the time the plunge was taken, the carbon dioxide would have made it to *some point* in its transformation to carbohydrate; it would

have been incorporated into one molecule in the series of intermediate molecules involved. Stopping photosynthesis at differing points—two seconds after it started, five seconds, and so on—meant stopping the progress of carbon dioxide as it marched through its changes.

All of this would have been for naught, however, without a way to figure out where the carbon was when the process was stopped. This is where the carbon-14 could do its job, but it needed to be joined to another technique that was also newly developed at that time: two-dimensional paper chromatography. If you look at **Figure 1a**, you can see how this technique works. Put a spot of ink on some filter paper and then stick one end of the paper in a solvent—water, for example. The water moves through the paper, carrying different components of the ink along with it at different rates. You can tell something about what the ink is made of by measuring how far its components move through the paper. You can tell still more by doing this in two dimensions: Let the solvent move through the spot one way, then set the paper down at a right angle in a *second* solvent and let it move through in another direction (**Figure 1b**). This is what Calvin and his colleagues did, using as their "ink" the algae that had been killed following photosynthesis.

The paper chromatography left the algae cells drawn out in two directions by the solvents. But where was the original carbon in this material? Because it was radioactively *tagged* carbon, it was sending out its signal saying, "I'm here now." But only a special "receiver" could pick up this signal: photographic film (**Figure 1c**). Setting X-ray film tightly on top of the chromatography paper for long periods—usually about two weeks—caused the radiation from the carbon-14 to expose the film. Spots would show up saying, in effect, "carbon-14 is now here, in this concentration."

There is a wide gap, however, between seeing a black blotch on a piece of film and *identifying* that blotch as a specific substance. The difficulty the researchers encountered can be measured in time—10 years in this case. That's how long it took these workers to complete their investigation. During that time, Calvin and his principal colleagues—postdoctoral researcher Andrew Benson and graduate student James Bassham—worked extraordinarily long hours in deciphering what the downstream products of photosynthesis were and the order in which they were produced.

This long operation was a great success, however. Calvin's team dismantled the black box of carbon fixation, and the information they uncovered can be *used* to aid in such things as increased food production. The set of steps these workers elucidated has variously been referred to as the C₃ cycle, Calvin-Benson cycle, or the Calvin cycle. In 1961, Melvin Calvin received the ultimate in congratulations for a piece of scientific work when he was awarded the Nobel Prize in chemistry for shining a light on photosynthesis.

Figure 1

(a) Paper chromatography in first dimension

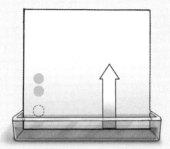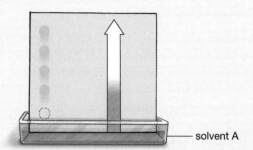

solvent A

Material of interest (algae for Calvin) is put on paper in solution form.

Solvent A moves through paper, separating out components of original material. (Components, shown here as green dots, would probably not be visible in experiments.)

(b) Paper chromatography in second dimension

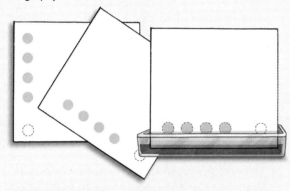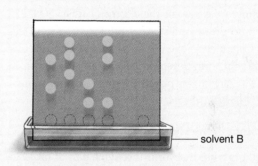

solvent B

Paper is rotated 90 degrees and put in second solvent.

Component compounds are separated in second dimension. Some compounds contain radioactively labeled carbon.

(c) Autoradiography

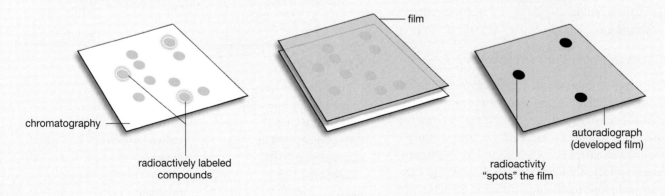

film

chromatography

radioactively labeled compounds

radioactivity "spots" the film

autoradiograph (developed film)

Chapter Review

Summary

8.1 Photosynthesis and Energy

- Photosynthesis has made possible life as we know it on Earth, because the organic material produced in photosynthesis (a sugar) is the source of food for most of Earth's living things. Photosynthesis also is responsible for the atmospheric oxygen used by many living things in cellular respiration.

8.2 The Components of Photosynthesis

- In plants and algae, photosynthesis takes place in the organelles called chloroplasts, which can exist in great abundance in the mesophyll cells of leaves. The energy for photosynthesis comes mostly from various blue and red wavelengths of visible sunlight that are absorbed by pigments in the chloroplasts. Plant leaves contain microscopic pores called stomata that can open and close, letting carbon dioxide in and water vapor out.
  Web Tutorial 8.1: Properties of Light

- There are two primary stages to photosynthesis. In the first stage, called the light reactions, electrons derived from water are energetically boosted by the power of sunlight. These electrons physically move in this process; they are passed along through a series of electron carriers, ending up as part of the electron carrier NADPH, which carries them to the second stage of photosynthesis. In this second stage, the Calvin cycle, the electrons are brought together with carbon dioxide and a sugar to produce a high-energy sugar, in a process powered by ATP that is produced in the light reactions.
  Web Tutorial 8.2: Photosynthesis

8.3 Stage 1: The Steps of the Light Reactions

- In its first stage, photosynthesis works through a pair of molecular complexes, photosystems II and I. These photosystems are composed partly of antennae molecules—chlorophyll and some accessory molecules—that absorb and transmit solar energy. Other photosystem components include reaction center molecules, which accept both this energy and electrons derived from water; and primary electron acceptors (a part of the reaction centers), to which the electrons move after being energetically boosted.

8.4 What Makes the Light Reactions So Important?

- Two actions of great consequence take place in the light reactions. The first is that water is split, yielding both the electrons that move through the light reactions and the oxygen that organisms such as ourselves breathe in. The second is that the electrons that are derived from the water—and then given an energy boost by the sun's rays—are transferred to a different molecule: the initial electron acceptor. The energetic fall of electrons through the electron transport chain between photosystems II and I also yields energy that produces ATP, which is used to power the second stage of photosynthesis.

8.5 Stage 2: The Calvin Cycle

- In the second stage of photosynthesis, carbon dioxide from the atmosphere is brought together with a sugar, RuBP, by the enzyme rubisco. The resulting compound is energized with the addition of phosphate groups and electrons supplied by the first stage of photosynthesis. The result is the high-energy sugar G3P, which is the product of photosynthesis. All these steps are carried out within a process known as the Calvin or C_3 cycle.
  Web Tutorial 8.3: The Calvin Experiments

- G3P can be used for energy or for plant growth. Everything in the plant ultimately is derived from this sugar, in association with minerals that the plant absorbs through its roots.

8.6 Photorespiration and the C_4 Pathway

- In C_3 plants, the enzyme rubisco frequently binds with oxygen, rather than carbon dioxide. When such photorespiration takes place, no new carbon dioxide is fixed and CO_2 that has been fixed in a plant is lost, thus undercutting productivity in photosynthesis. This problem increases as the temperature rises, because as plants close their stomata to keep in moisture, they also keep out CO_2, thus increasing the likelihood that rubisco will bind with oxygen.

- Some warm-climate plants have evolved a means of dealing with photorespiration. Called C_4 photosynthesis, it employs an enzyme that binds with carbon dioxide but not with oxygen. The carbon dioxide is then shuttled to special bundle-sheath cells in the plant and released, after which it moves into the Calvin cycle. With high levels of CO_2 in the bundle-sheath cells, rubisco binds with CO_2 (and not oxygen), thus greatly reducing photorespiration.

8.7 Another Photosynthetic Variation: CAM Plants

- Dry-weather plants such as cacti employ another form of photosynthesis, CAM photosynthesis. In it, the plant's stomata open only at night, letting in and fixing carbon dioxide that is "banked" until sunrise, when the sun's rays will supply the energy needed to power the Calvin cycle.
  Web Tutorial 8.4: Different Kinds of Photosynthesis

Key Terms

C_4 photosynthesis 161	**photosystem** 156
Calvin cycle 159	**reaction center** 156
CAM photosynthesis 162	**rubisco** 160
chlorophyll *a* 156	**stomata** 155
chloroplast 155	**stroma** 156
fixation 160	**thylakoid** 156
photorespiration 161	**thylakoid compartment** 156
photosynthesis 154	

riety of them forms a football-shaped cage around the nuclear material, while a second variety attaches to the chromosomes themselves. Taken together, the microtubules active in cell division are known as the **mitotic spindle**.

Metaphase

By the time metaphase begins, the nuclear envelope has disappeared completely, and the microtubules that were growing toward the chromosomes now *attach* to them. Through a lively back-and-forth movement, the microtubules align the chromosomes at the equator. With this, each chromatid now faces the pole *opposite* that of its sister chromatid, and each chromatid is attached to its respective pole by perhaps 30 microtubules.

Anaphase

At last the genetic material divides. As you may have guessed, this is a parting of sisters. The sister chromatids are pulled apart, each now becoming a full-fledged chromosome. All 46 chromatid pairs divide at the same time, and each member of a chromatid pair moves toward its respective pole, pulled by a shortening of the microtubules to which it is attached.

Telophase

Telophase represents a return to things as they were before mitosis started. The newly independent chromosomes, having arrived at their respective poles, now unwind and lose their clearly defined shape. New nuclear membranes are forming. When this work is complete, there are two finished daughter nuclei lying in one elongating cell. Even as this is going on, though, something else is taking place that will result in this one cell becoming two.

Cytokinesis

Cytokinesis actually began back in anaphase and is well under way by the time of telophase. It works through the tightening of a cellular waistband that is composed of two sets of protein filaments working

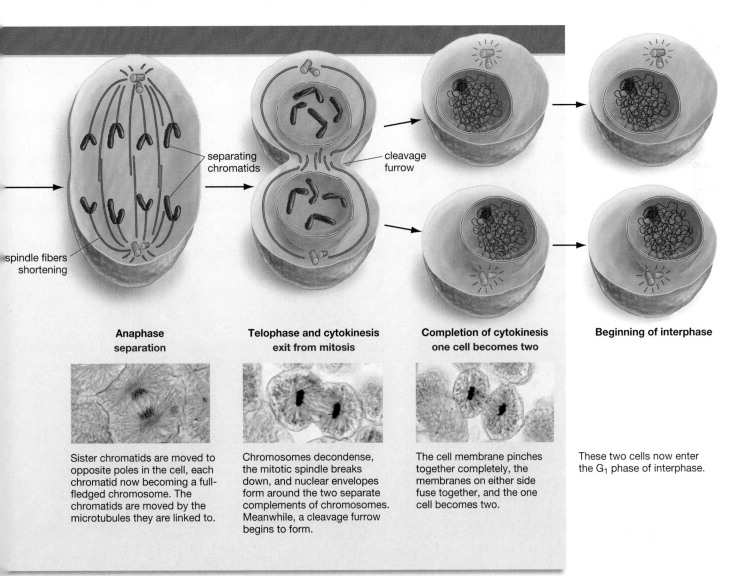

separating chromatids

cleavage furrow

spindle fibers shortening

Anaphase separation

Telophase and cytokinesis exit from mitosis

Completion of cytokinesis one cell becomes two

Beginning of interphase

Sister chromatids are moved to opposite poles in the cell, each chromatid now becoming a full-fledged chromosome. The chromatids are moved by the microtubules they are linked to.

Chromosomes decondense, the mitotic spindle breaks down, and nuclear envelopes form around the two separate complements of chromosomes. Meanwhile, a cleavage furrow begins to form.

The cell membrane pinches together completely, the membranes on either side fuse together, and the one cell becomes two.

These two cells now enter the G₁ phase of interphase.

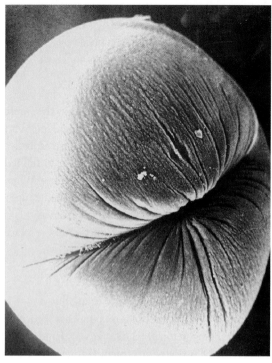

Figure 9.11
Cytokinesis in Animals
Cytokinesis in animal cells begins with an indentation of the cell surface, a cleavage furrow, shown here in a dividing frog egg. (×85)

together. These filaments—the same type that allow your muscles to contract—form a *contractile ring* that narrows along the cellular equator (**see Figure 9.11**). An indentation of the cell's surface (called a cleavage furrow), results from the ring's contraction; consequently, the fibers in the mitotic spindle are pushed closer and closer together, eventually forming one thick pole that is destined to break. The dividing cell now assumes an hourglass shape; as the contractile ring continues to pinch in, one cell becomes two by means of something you looked at in Chapter 4:

membrane fusion. The membranes on each half of the hourglass circle toward each other and then fuse. With this, the two cells become separate. Mitosis and cytokinesis are over, and the two daughter cells slip back into the relative quiet of interphase.

9.5 Variations in Cell Division

Now, we'll look at variations in cell division in certain different types of cells, including plant cells and prokaryotes.

Plant Cells

For the most part, plant cells carry out the steps of mitosis just as animal cells do. In the splitting of cytokinesis, however, plant cells must deal with something that animal cells don't have: the cell wall (see page 86 in Chapter 4). The way animal cells carry out cytokinesis—pinching the plasma membrane inward via the contractile ring—wouldn't work with plant cells, because their thick cell wall lies *outside* the plasma membrane.

The plant cell's solution to cytokinesis is to grow a new cell wall and plasma membrane that run roughly down the middle of the parent cell. If you look at **Figure 9.12**, you can see how this works. A series of membrane-lined vesicles begins to accumulate near the metaphase plate; inside these vesicles is a complex sugar. The vesicles then begin to fuse together; eventually they will form a flat *cell plate* that runs from one side of the parent cell to the other. The membrane portion of the cell plate then fuses with

Figure 9.12
Cytokinesis in Plants
The cell walls of plants are too rigid to form cleavage furrows, as animal cells do when dividing. Therefore, plant cells use a different strategy in cytokinesis. They build a new plasma membrane and cell wall down the middle of the parent cell to separate the two daughter cells from the inside out.

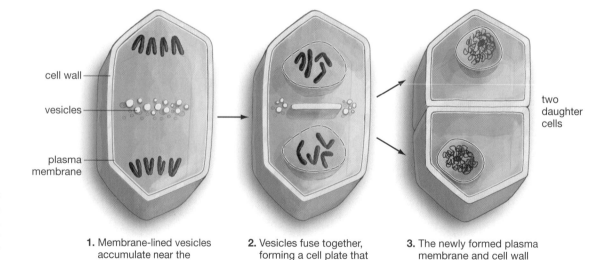

cell wall

vesicles

plasma
membrane

two
daughter
cells

1. Membrane-lined vesicles accumulate near the metaphase plate. The vesicles contain precursors to the cell wall.

2. Vesicles fuse together, forming a cell plate that grows toward the parent cell wall.

3. The newly formed plasma membrane and cell wall fuse with the parent plasma membrane and cell wall, forming two distinct daughter cells.

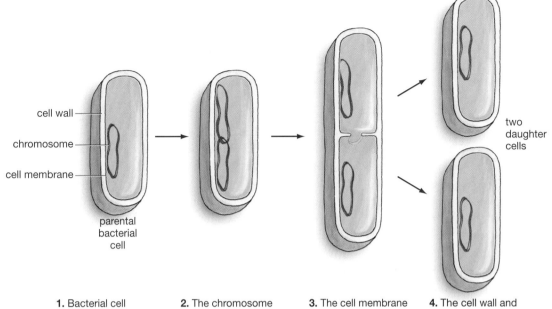

1. Bacterial cell
starts with a single,
circular chromosome
attached to its plasma
membrane.

2. The chromosome
replicates and the
daughter chromosomes
attach to different
sites on the plasma
membrane.

3. The cell membrane
and wall grow an
extension between
the attachment
points of the two
chromosomes.

4. The cell wall and
membrane join
together in the
middle, resulting
in two new cells.

labels: cell wall, chromosome, cell membrane, parental bacterial cell, two daughter cells

Figure 9.13
Binary Fission in Bacteria
The key to bacterial cell division is that the bacterial chromosome attaches to the cell's plasma membrane. Both daughter chromosomes attach to the plasma membrane, but in different locations. The cell then produces an outgrowth of both plasma membrane and cell wall—called a septum—that grows in from each side of the cell, between the two daughter chromosomes. When the septum runs completely across the cell, it divides the cell in two, leaving one chromosome in each daughter cell.

Web Tutorial 9.3
Cell Division for Bacteria

the original plasma membrane of the parent cell, and the result is two new adjacent plasma membranes that split the parent cell in two. Meanwhile, the material enclosed *inside* the cell-plate membrane is the foundation for the new cell-wall segments of the daughter cells.

Prokaryotes

It's also important to recognize that some cells have fundamentally different ways of replicating. The cells of prokaryotes—the bacteria and archaea you looked at in Chapter 4—are a case in point.

If you look at **Figure 9.13**, you can see how cell division works among the prokaryotes. Bacteria are single-celled and have a single, circular chromosome. Bacterial cells do not, however, have their chromosome sequestered inside a nucleus, as with eukaryotic cells, for the simple reason that they *have* no nucleus. Instead, their lone chromosome is attached to their plasma membrane.

A bacterial cell begins its division by duplicating its chromosome, utilizing the same DNA unwinding and replication described earlier. As the daughter chromosomes are completed, however, they attach to *different sites* on the plasma membrane. The cell then begins developing an outgrowth of its plasma membrane and cell wall in between these attachment sites. This outgrowth, called a septum, begins growing from opposite sides of the cell. When the

two septum extensions join in the middle, they divide the one cell into two (**see Figure 9.14**). This process is obviously very different from the eukaryotic mitosis and cytokinesis you looked at. As such, prokaryotic cell division goes under an entirely different name: **binary fission**. The simplicity of binary fission makes for a short cell cycle. Remember how the animal cells we looked at took about 24 hours to go through a cycle? Bacteria can complete their cycle in as little as 20 minutes.

Figure 9.14
Cell Division for a Resident of the Human Gut
E. coli bacteria such as this one exist in great quantity in the human digestive tract, generally harmlessly. It is certain rare varieties of *E. coli*, generally found in food, that have caused outbreaks of sickness and even death in recent years. (×30,000)

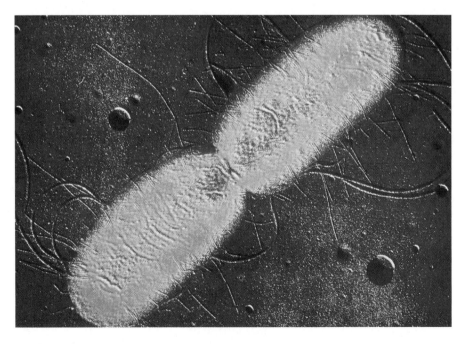

When the Cell Cycle Runs Amok: Cancer

The cell cycle reviewed in this chapter is, in one sense, a common natural process: Cells grow; they duplicate their chromosomes; these chromosomes separate; one cell divides into two. This goes on like clockwork, millions of times a second in each one of us. But then one day we learn that an aunt or a grandfather or a friend is experiencing an *unrestrained* division of cells. Things aren't explained to us in this way, of course. We are simply told that someone we know has cancer.

At root, all cancers are failures of the cell cycle. Put another way, all cancers represent a failure of cells to limit their multiplication in the cell cycle. What is liver cancer, for example? It is a damaging multiplication of liver cells. First one, then two, then four, then eight liver cells move repeatedly through the cell cycle, and as their numbers increase, they destroy the liver's working tissues. Given that cancer manifests in this way, it's not surprising that a large portion of modern cancer research is *cell-cycle* research. The logic here is simple: To the extent that uncontrolled cell division can be stopped, cancer can be stopped.

Research done to date has revealed the general outline of the way cells become cancerous. To use a common but apt metaphor, two things are required to bring cells to this state: their accelerators must get stuck and their brakes must fail. The control mechanisms that *induce* cell division must become hyperactive, and the mechanisms that *suppress* cell division must fail to perform. You may have heard a couple of terms used to describe the genetic components of this process. There are normal genes that induce cell division, but that when mutated can cause cancer; these are the stuck-accelerator genes, called oncogenes. Then there are genes that normally suppress cell division, but that can cause cancer by acting like failed brakes. These are tumor suppressor genes. Note that *both* kinds of genes must malfunction for cancer to get going; indeed, it usually takes a long succession of genetic failures to induce cancer. This is why cancer is most often a disease of the middle-aged and elderly: It can take decades for the required series of mutations to fall into line in a single cell, such that it becomes cancerous.

How do oncogenes interact with tumor suppressor genes? In the normal case, cells will not begin division until prompted to do so by a signal from outside themselves. A protein (called a growth factor) will bind to a cell, setting off a cascade of chemical reactions inside it that triggers division. One of the links in this chemical cascade is a protein called Ras that could be thought of as an old-time railway switch. When Ras is chemically pointed one way (toward "on"), the cell moves through the cell cycle. When it is pointed the other (toward "off"), the cell stays in G_1. The gene that codes for Ras can become mutated, however, and when this happens, the Ras protein changes shape and points in the "on" direction all the time—no matter what signals it is getting from the outside. Thus, *ras* is an example of an oncogene. It normally prompts the cell to divide intermittently; when mutated it prompts the cell to divide continuously.

A cell with a mutated *ras* gene is not doomed to become cancerous, however. What can save it is a good set of brakes, in the form of tumor suppressor genes. The most important of

Variations in the Frequency of Cell Division

There are also variations in cell division among different types of eukaryotic cells. As noted, not all cells divide throughout their existence. Most human brain cells are formed in the first three months of embryonic existence and then live for decades, with relatively few of them ever dividing again. Much the same is true of the leaf cells in plants; they divide only when the leaf is very small, then grow for a time, and then function at this mature size for as long as the leaf lives. At the other extreme, cells located in human bone marrow never *stop* dividing as they produce red blood cells (which have a short life span). The output here is staggering: about 180 million new red blood cells are produced inside us each minute.

On to Meiosis

The cell division reviewed in this chapter concerns "somatic" cells—the kind that form bone, muscle, nerve, and many other sorts of tissue. Given a distribution this wide, you might well ask: What kind of cells are *not* somatic? The answer is only one kind—the kind that forms the basis for each succeeding generation of sexually reproducing living things.

these genes—one known as *p53*—is so vital to human health that it is sometimes referred to as the "guardian of the genome." In the presence of certain kinds of mutations, *p53* protein levels rise in the cell, and these levels start turning selected genes on and off. The result is the cell's first line of defense against cancer—the cell cycle is shut down until the mutation has been repaired. This shutdown doesn't happen at just any point in the cycle, however. As cancer researchers Leland Hartwell and Ted Weinert discovered in the 1980s, cells have specific *checkpoints* in their cycle. Just as NASA mission control will stop at a defined point in a countdown to see if "all systems are go" for a launch, so a cell has specific points at which it makes sure that all its systems are healthy enough for cell division to continue. The first of its checkpoints comes in G_1, as it is about to enter S phase (during which it doubles its DNA). The second point comes in G_2, as it is about to enter into mitosis and cytokinesis. Thus, if a dividing skin cell, for example, has acquired some mutations in G_1, it will not enter S phase until its DNA repair enzymes have fixed the problem.

But what happens if the DNA damage spotted in G_1 can't be fixed? Then, prompted by *p53*, the cell goes to level-two of its emergency responses: it commits suicide. Through an orderly process called *apoptosis*, the cell shuts down its activities, breaks up, and dies. If you have ever been sunburned, you have probably seen the effects of apoptosis up close. The ultraviolet light in the sun damages the DNA in skin cells. When this damage cannot be fixed with repair enzymes, these cells undergo apoptosis; their remains are the peeling skin that comes with a bad sunburn.

In sum, when genes such as *ras* become mutated, they can lead to an out-of-control cell cycle, but this process can be halted in several ways by tumor suppressor genes such as *p53*. With this, you can probably see what comes next. What will stave off cancer if a cell has *both* a mutated *ras* gene and a mutated *p53* gene? Perhaps nothing, because now the accelerator is stuck and the brakes have failed. As noted, there is generally more to cancer than two mutated genes, but when both *ras* and *p53* malfunction, a cell is well on its way to cancer.

As it turns out, *ras* is probably the most important human oncogene among the hundred or so that have been identified. And *p53* certainly is the most important tumor suppressor gene among the two dozen that have been identified. A mutated *ras* gene is found in about 30 percent of all human cancers, while a mutated *p53* is found in half of all human cancers. As you can imagine, there is intense research interest in both these genes. From 1989 to 2000, more than 17,000 scientific publications were written on *p53* alone. The war on cancer may not have been won, but it certainly is being waged.

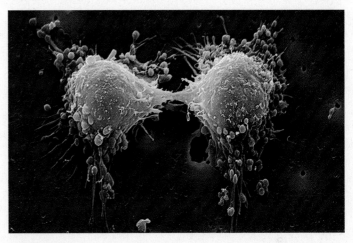

Harmful Division
Pictured are two prostate cancer cells in the final stages of cell division.

Chapter Review

Summary

9.1   An Introduction to Genetics

- DNA is an information-bearing molecule that plays a critical role in the reproduction, development, and everyday functioning of living things. DNA contains the information for the production of proteins, which carry out an array of tasks in living things.

- The information in DNA is encoded in chemical substances called "bases," which are laid out along the DNA double helix in four varieties: adenine (A), thymine (T), guanine (G), and cytosine (C). One series of bases contains information for the production of one protein, while a different series of bases specifies a different protein. Each series of protein-specifying bases is known as a gene.

- Protein synthesis begins with the information in a sequence of DNA bases being copied onto a molecule called messenger RNA (mRNA). This molecule moves out of the cell's nucleus to a structure in its cytoplasm called a ribosome. There, the mRNA "tape" is brought together with the building blocks of proteins, amino acids. As the ribosome "reads" the mRNA tape, it strings together a sequence of amino acids called for by the tape. The result is a chain of amino acids that folds into a protein.

- Most of the cells in an organism contain a complete copy of that organism's genome, meaning its collection of genetic information. Before cells divide, their genome must first be copied and the resulting copies apportioned evenly into what will become two daughter cells.

9.2 An Introduction to Cell Division

- Cell division takes place because old cells die and because there are many instances in which an organism needs quantities of new cells above "replacement" level. Cell division includes the duplication of DNA (replication); the apportioning of the copied DNA into two quantities in a parent cell (mitosis); and the physical splitting of this parent cell into two daughter cells (cytokinesis). In DNA replication, the two strands of the double helix unwind, after which each single strand serves as a template for construction of a second, complementary strand of DNA. The result is a doubling of the original quantity of DNA.
  Web Tutorial 9.1: The Cell Cycle

9.3 DNA Is Packaged in Chromosomes

- DNA comes packaged in units called chromosomes. These chromosomes are composed of DNA and its associated proteins—a combined chemical complex called chromatin. Chromosomes exist in an unduplicated state until such time as DNA replicates, prior to cell division. DNA replication results in chromosomes that are in duplicated state, meaning one chromosome composed of two identical sister chromatids.

- Chromosomes in human beings (and many other species) come in matched pairs, with one member of each pair inherited from the mother, and the other member of each pair inherited from the father. Such homologous chromosomes have closely matched sets of genes on them, though many of these genes are not identical. A given paternal chromosome may have genes that code, for example, for different hair or skin color than the counterpart genes on the homologous maternal chromosome. Human beings have 46 chromosomes—22 matched pairs and either a matched pair of X chromosomes (in females) or an X and a Y chromosome (in males).

- Cell division fits into the larger framework of the cell cycle, meaning a repeating pattern of growth, genetic replication, and cell division. The cell cycle has two main phases. The first is interphase, in which the cell carries out its work, grows, and duplicates its chromosomes in preparation for division. The second is mitotic phase, in which the duplicated chromosomes separate and the cell splits in two.

9.4 Mitosis and Cytokinesis

- There are four stages in mitosis: prophase, metaphase, anaphase, and telophase. The essence of the process is that duplicated chromosomes line up along an equatorial plane of the parent cell, called the metaphase plate, with the sister chromatids that make up each duplicated chromosome lying on opposite sides of the plate. Attached to fibers called microtubules, the sister chromatids are then pulled apart, to opposite poles of the parent cell. Once cell division is complete, sister chromatids that once formed a single chromosome will reside in separate daughter cells, with each sister chromatid now functioning as a full-fledged chromosome.
  Web Tutorial 9.2: Mitosis

- Cytokinesis in animal cells works through a ring of protein filaments that tightens at the middle of a dividing cell. Membranes on the portions of the cell being pinched together then fuse, resulting in two daughter cells.

9.5 Variations in Cell Division

- Because of their cell walls, plant cells must carry out cytokinesis differently from animal cells. The plant's solution is to grow new cell walls and plasma membranes near the metaphase plate, thus dividing the parent cell into two daughter cells. Prokaryotes such as bacteria employ a process called binary fission: They double their single, circular chromosome, with the two resulting chromosomes attaching to different sites on the plasma membrane. Then an outgrowth of plasma membrane and cell wall, called a septum, begins growing from opposite sides of the cell, in between the two chromosomes. When the two septum extensions join in the middle, they divide the one cell into two.
  Web Tutorial 9.3: Cell Division for Bacteria

Key Terms

binary fission 183	**genome** 171
cell cycle 178	**homologous chromosomes** 177
centrosome 180	**interphase** 178
chromatid 177	**karyotype** 177
chromatin 176	**metaphase plate** 180
chromosome 175	**microtubule** 180
cytokinesis 178	**mitosis** 178
enzyme 172	**mitotic phase** 178
genetics 173	**mitotic spindle** 181

Understanding the Basics

Multiple-Choice Questions (answers in the back of the book)

1. Proteins serve in living things as:
   a. hormones
   b. enzymes
   c. cell receptors
   d. structural building blocks
   e. all of the above

2. The four bases found along the DNA double helix are
   a. adenine, thymine, guanine, and cytosol
   b. adenine, thyroxine, glucose, and cytosine
   c. adrenaline, thymine, glucosamine, and uracil
   d. adenine, thymine, guanine, and cytosine
   e. none of the above

3. A gene can be described as
   a. a protein that contains enough information to carry out a task
   b. a hormone that prompts some action inside a cell
   c. a series of DNA bases that transfers energy within a cell

   d. a series of DNA bases that contains information for production of a protein

   e. a protein that delivers information to a ribosome

4. Cell division in bacteria operates very differently from cell division in eukaryotes. Which statement is *incorrect* regarding the cell-division process in bacteria?

   a. bacteria are incapable of rapid cell division

   b. bacteria have a single circular chromosome

   c. the bacterial chromosome is sequestered inside a nucleus

   d. the bacterial chromosome is attached to the plasma membrane during replication

   e. the bacterial chromosome is duplicated before cell division

5. DNA is primarily

   a. a molecule that protects cells from invaders

   b. a molecule that contains information

   c. a molecule that helps transfer energy

   d. a "workbench" on which proteins are put together

   e. a form of protein

6. Prophase has begun when

   a. we can see well-defined chromosomes.

   b. a cleavage furrow has formed.

   c. chromosomes have aligned along the metaphase plate.

   d. chromosomes have duplicated.

   e. sister chromatids have separated.

7. A _____ is composed of two sets of protein filaments that help bring about cytokinesis.

   a. dividing fiber

   b. metaphase plate

   c. spindle

   d. contractile ring

   e. centrosome

8. A root-tip cell in a plant has 22 chromosomes. How many chromatids can be seen during prophase when this cell is undergoing mitosis?

   a. 22

   b. 44

   c. 66

   d. 11

   e. 88

9. The cell cycle is linked to cancer because cancer always entails

   a. too much time spent in S phase

   b. a slow-down in mitosis

   c. unrestrained cell division

   d. a misalignment of chromosomes

   e. not enough time spent in M phase

10. If a cell is not dividing, and has not yet begun replicating its DNA (duplicated its chromosomes), what phase of the cell cycle would it be in?

   a. gap-two (G_2) phase

   b. synthesis (S) phase

   c. gap-one (G_1) phase

   d. prophase

   e. mitotic phase

Brief Review

1. In what way does DNA help form and then maintain our bodies?

2. The drawing shows a cell in a phase of mitosis. What phase is shown? How many chromatids are present in this stage? What is the total number of chromosomes each daughter cell will have?

3. How do mitosis and cytokinesis differ?

4. Name the four phases of mitosis and describe the major events occurring in each phase.

5. What additional cell structure must plant cells deal with when they are undergoing cytokinesis (as compared to animal cells)? Given this difference, how do plant cells carry out cytokinesis?

6. In what way are the 23 pairs of human chromosomes "matched" chromosomes?

7. The protein synthesis pathway—DNA → mRNA → protein—is a very important concept in biology.

   a. What are the information-bearing "building blocks" of DNA?

   b. Where is DNA located in the cell?

   c. What does the "m" in mRNA stand for?

   d. Using this knowledge, what is the function of mRNA, where is it made, and where is it used?"

   e. What are the building blocks of proteins?

   f. Where does protein formation occur?

Applying Your Knowledge

1. Why is it accurate to think of each human being as the owner of a library of ancient information?

2. People who are born with the condition called Down syndrome generally have three copies of a certain chromosome (chromosome 21), rather than the usual two copies. What can you deduce from this regarding the proteins produced by a person with Down syndrome?

3. Given the conceptualization of genetics as information management, what would you predict about the relative size of the human genome as opposed to the genome of a bacterium? What about the human genome compared to that of a flowering plant?

10 *Meiosis*

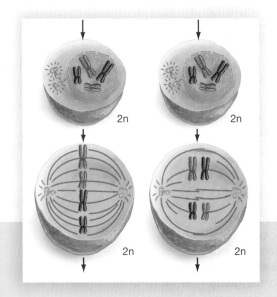

Meiosis and mitosis compared.
(Section 10.2, page 193)

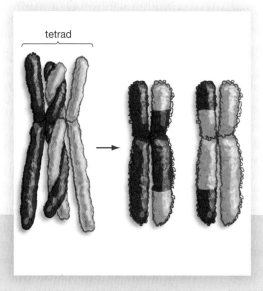

Swapping sections.
(Section 10.2, page 194)

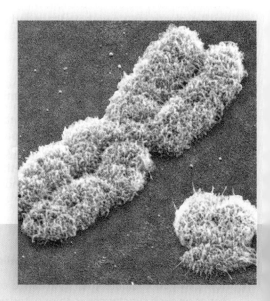

The large X and the small Y.
(Section 10.4, page 199)

A real-life couple—let's call them Jack Fennington and Jill Kent—combine their last names when they get married and thus become Jack and Jill Fennington-Kent. Now, what would happen if the Fennington-Kent's children were to continue in this tradition? Their daughter Susie might marry, say, Ralph Reeson-Dodd, which would make her Susie Fennington-Kent-Reeson-Dodd. If *Susie's* son, John, were to keep this up, he might be fated to become John Fennington-Kent-Reeson-Dodd-Garcia-Lee-Minderbinder-Green, and so on.

This growing chain of names illustrates a kind of problem that living things have managed to address in carrying out reproduction. *Reproduction* in this context does not mean the mitotic cell division you looked at in Chapter 9. Reproduction here means *sexual* reproduction, in which specialized reproductive cells come together to pro-

duce offspring. The cells described in Chapter 9—the ones that undergo mitosis—are known as *somatic* cells, which in animals means all the cells in the organism except for one type. What type is this? Sex cells; cells called *gametes*. These are reproductive cells, known more commonly as eggs and sperm.

Eggs and sperm are not commonly thought of as cells, but that's just what they are, complete with plasma membranes, nuclei, chromosomes, and all the rest (though sperm eventually lose much of this material). Eggs and sperm are *special* cells, however, in that they are not destined to divide like somatic cells and thus make more eggs and sperm. Rather, they will *fuse*—sperm fertilizing egg—to make a zygote, which grows into a whole organism. It is sperm and egg that take part in sexual reproduction and that thus bear some

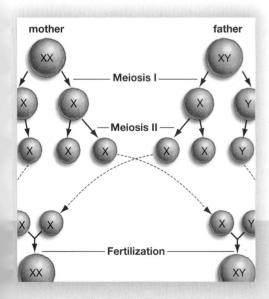

On the way to a boy or a girl.
(Section 10.4, page 199)

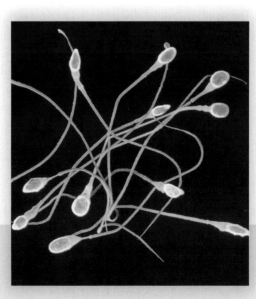

About 250 million of these are made every day.
(Section 10.5, page 201)

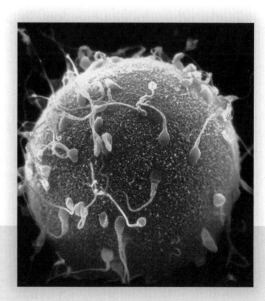

End of two journeys.
(Section 10.5, page 201)

relation to the Fennington-Kent problem. Here's how it plays out in humans. Human somatic cells, as you have seen, have 23 pairs of chromosomes or 46 chromosomes in all. If an egg and sperm each brought 46 chromosomes to their *union*, the result would be a zygote with 92 chromosomes. The *next* generation down would presumably have 184 chromosomes, the next after that 368, and so on. This would be about as functional as having Fennington-Kent-Reeson-Dodd-Garcia-Lee-Miderbinder-Green as a last name.

10.1 An Overview of Meiosis

How do chromosomes avoid the problem of doubling with each generation? In sexual reproduction, chromosome union is preceded by chromosome *reduction*. The reduction comes in the cells that give rise to sperm and egg. When these cells divide, the result is sperm or egg cells that have only *half* the usual, somatic number of chromosomes. Human sperm or egg, in other words, have only 23 chromosomes in them. (Not the 23 *pairs* of chromosomes that somatic cells have, mind you; 23 chromosomes.) Each 23-chromosome sperm can then unite with a 23-chromosome egg to produce a 46-chromosome zygote that develops into a new human being. In each generation, then, there is first a halving of chromosome number (when egg and sperm cells are produced), followed by a coming together of these two halves (when sperm and egg unite). This basic pattern holds true for sexual reproduction in all eukaryotes.

Some Helpful Terms

Now here are a few terms that will be helpful. The kind of cell division that results in the halving of chromosome number is called *meiosis*, in contrast with *mitosis*, which you looked at in Chapter 9. In mitosis, there is no halving of chromosome number; just a duplication of chromosomes and then a division of these chromosomes into each of two daughter cells.

In animals, cells that are produced in meiosis—thus getting the half-number of chromosomes—are always reproductive cells, the term for which is **gametes**, as we've seen. Such reduced-chromosome cells are said to be in the haploid state, the term *haploid* literally meaning "single vessel." When egg and sperm unite, however, it marks a return to the *diploid*, or "double vessel" state of cellular existence, meaning 46 chromo-

somes in human beings. (More formally, **haploid** means possessing a single set of chromosomes while **diploid** means possessing two sets of chromosomes.) **Meiosis** can thus be defined as a process in which a single diploid cell divides to produce haploid reproductive cells. Diploid cells are also sometimes referred to as **2n** cells (the "2" here standing for a doubled number of chromosomes), while haploid cells are said to be **1n**. Thus, eggs and sperm are 1n, while the diploid cells that give rise to them are 2n.

In what follows, we will be looking at meiosis as it occurs in human beings; toward the end of the chapter we'll go over some variations on reproduction in other kinds of organisms.

10.2 The Steps in Meiosis

How does the chromosomal halving take place? Let's go over the process of meiosis and see. **Figure 10.1** shows meiosis, in a stripped-down form, as compared to the mitosis described in Chapter 9. As you can see, meiosis includes one chromosome duplication followed by two cellular divisions. This is the key to how the chromosome numbers work out. In mitosis, there was one chromosome duplication and one subsequent division of the doubled chromosomes (into two cells). In meiosis there is one duplication followed by *two* divisions (into four cells). The first of the meiotic divisions is a separation not of chroma*tids*, as in mitosis, but of homologous chromo*somes*. The second meiotic division is then the separation of the chromatids that make up these homologous chromosomes.

But what is this term *homologous*? You may remember from Chapter 9 that homologous chromosomes are chromosomes that are the same in size and function. These chromosomes do similar things, which is to say they have genes on them that code for similar proteins. We have a chromosome 1 that we inherit from our mother, and it is homologous to the chromosome 1 we inherit from our father. Likewise we inherit a chromosome 2 from our mother that is homologous with chromosome 2 from our father, and so forth for 23 pairs. The lone exception to this pairs rule is that the X and Y chromosomes males have are not homologous with one another. Let's walk through the steps of meiosis now. These details will reveal a process that not only divides up genetic material, but provides for a great deal of the diversity

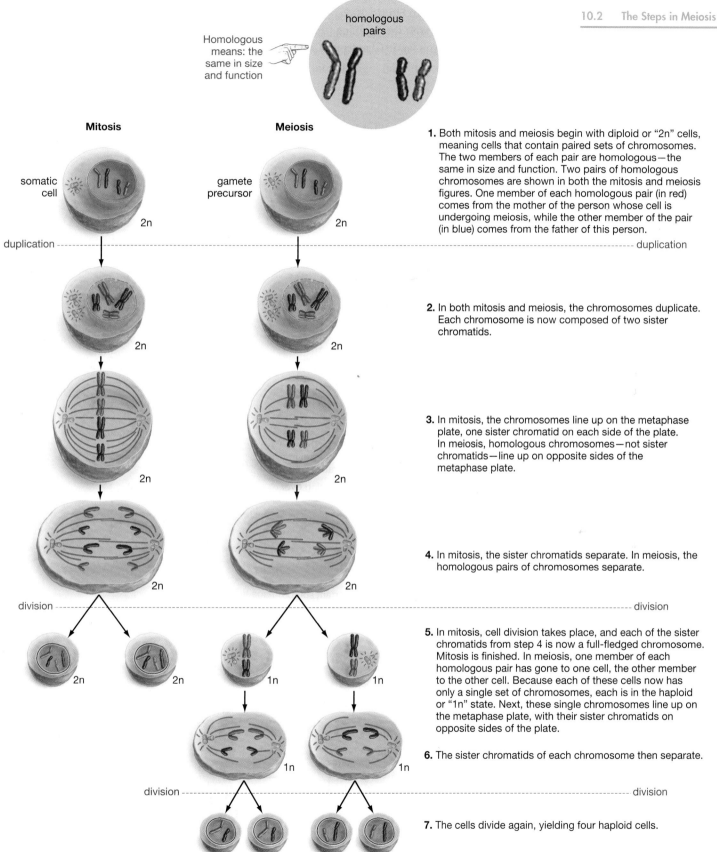

homologous pairs

Homologous means: the same in size and function

Mitosis

Meiosis

somatic cell

2n

gamete precursor

2n

duplication - duplication

2n

2n

2n

2n

2n

2n

division - division

2n

2n

1n

1n

1n

1n

division - division

1n

1n

1n

1n

1. Both mitosis and meiosis begin with diploid or "2n" cells, meaning cells that contain paired sets of chromosomes. The two members of each pair are homologous—the same in size and function. Two pairs of homologous chromosomes are shown in both the mitosis and meiosis figures. One member of each homologous pair (in red) comes from the mother of the person whose cell is undergoing meiosis, while the other member of the pair (in blue) comes from the father of this person.

2. In both mitosis and meiosis, the chromosomes duplicate. Each chromosome is now composed of two sister chromatids.

3. In mitosis, the chromosomes line up on the metaphase plate, one sister chromatid on each side of the plate. In meiosis, homologous chromosomes—not sister chromatids—line up on opposite sides of the metaphase plate.

4. In mitosis, the sister chromatids separate. In meiosis, the homologous pairs of chromosomes separate.

5. In mitosis, cell division takes place, and each of the sister chromatids from step 4 is now a full-fledged chromosome. Mitosis is finished. In meiosis, one member of each homologous pair has gone to one cell, the other member to the other cell. Because each of these cells now has only a single set of chromosomes, each is in the haploid or "1n" state. Next, these single chromosomes line up on the metaphase plate, with their sister chromatids on opposite sides of the plate.

6. The sister chromatids of each chromosome then separate.

7. The cells divide again, yielding four haploid cells.

Figure 10.1
Meiosis Compared to Mitosis

Prophase I

Meiosis I begins, in prophase I, with the same appearance of 46 identifiable chromosomes that you observed in mitosis. The first big difference between meiosis and mitosis also appears in prophase I. It is that, in meiosis, *homologous chromosomes pair up*. In mitosis, 46 individual chromosomes condensed, but did not pair up in any way. In meiosis, however, each pair of homologous chromosomes link up as they condense. Maternal chromosome 5, for example, intertwines with *paternal* chromosome 5, maternal chromosome 6 with paternal 6, and so on. When two homologous chromosomes have paired up in this way, they are called a **tetrad** (see the "Crossing over" section of Figure 10.2).

A critical bit of part-swapping then takes place between the non-sister chromatids of the paired chromosomes. This process is called *crossing over* (or *recombination*), and you'll be looking at it in detail later. Once this crossing over has finished, the homologous chromosomes begin to unwind from one another, though they remain overlapped.

Metaphase I

Still existing as tetrads, the homologous chromosomes, attached to microtubules, are moved to the metaphase plate; in this step of meiosis, the maternal member of a given pair lies on one side of the plate and the paternal member on the other. A critical point about this, however, is that the alignment adopted by any one pair of chromosomes bears no relation to the alignment adopted by any other pair. It may be, for example, that paternal chromosome 5 will line up on what we might call side A of the metaphase plate; if so, then maternal chromosome 5 ends up on side B. Shift to chromosome 6, however, and things could just as easily be reversed: The *maternal* chromosome might end up on side A, and the paternal on side B. It is thus a throw of the dice as to which side of the plate a given chromosome lines up on. More important, it is this chance event that determines which *cell* a chromosome joins, because each side of the plate becomes a separate cell once cell division is finished. As you will see, this randomness is critical in the shuffling of genetic material.

Anaphase I

In anaphase I the paired, homologous chromosomes now begin to move away from each other, toward their respective poles, pulled through the disassembly of the microtubule spindles they are attached to. Though homologous chromosomes have now separated, the sister chromatids that make them up are still together.

Telophase I

With chromosome movement toward the poles completed, the original cell now undergoes cytokinesis, dividing into two completely separate daughter cells. With this, there are two haploid cells, whereas in the beginning there was one diploid cell. (Why are the cells now haploid? Meiosis I moved one set of chromosomes to one cell and the second set to another cell. Therefore, each cell now has only one set of chromosomes.)

Meiosis II

There is little pause for interphase between meiosis I and meiosis II; little time for the daughter cells to regroup and carry on regular metabolic activities. What happens instead is another division—or set of divisions, because there are now two cells. Each of these cells now has 23 duplicated chromosomes in it. After a very brief prophase, these chromosomes line up at a new metaphase plate, with sister chromatids on opposite sides of each plate. What happens next is the same thing that happened in mitosis. The *sister chromatids* now separate, moving toward opposite poles; with this, they assume the role of full-fledged chromosomes. Once this separation is completed, cytokinesis occurs. Where once there were two cells, there are now four. The difference between this process and mitosis is that each of these cells has 23 chromosomes in it instead of 46.

10.3 What Is the Significance of Meiosis?

You have now examined the mechanics of meiosis. But what are the effects of this process? First of all, the "Fennington-Kent" problem noted at the beginning of the chapter has been solved. There will be no 92-chromosome zygotes, because egg and sperm will each bring only 23 chromosomes to their meeting, thus yielding a combined 46 chromosomes in human somatic cells. Just as notable, however, are two kinds of diversity that meiosis brings about. The first of these these is genetic diversity in offspring; the second is a large-scale diversity in the natural world.

Genetic Diversity through Crossing Over

What is genetic diversity? Recall that, back in prophase I of meiosis, the homologous chromosome pairs intertwined for a time and then engaged in something called crossing over (or recombination). In this, there is an exchange of *parts* of chromosomes prior to their lining up at the metaphase plate.

Look at the "Crossing over" section of Figure 10.2 to see how this important process works. In the first stage of meiosis, there is a physical breaking of non-sister chromatids and then a "reunion" of these recip-

Web Tutorial 10.1
Meiosis

rocal chromosomal sections onto new chromatid partners. With this, what had been a "maternal" chromosome now has a portion of the paternal chromosome within it, and vice versa. Formally, then, **crossing over** can be defined as a process, occurring during meiosis, in which homologous chromosomes exchange reciprocal portions of themselves.

The linkage between crossing over and genetic diversity probably is apparent; the ability of reciprocal lengths of DNA to be exchanged between chromosomes provides a means by which the genetic deck can be reshuffled prior to the formation of gametes. This is so because of what you saw in Chapter 9 about homologous chromosomes: While the genetic information on any two will be similar, it will not be identical. A gene on one chromosome may code for red hair color, but the gene on its homologous chromosome may code for blonde hair color. In crossing over, such genetic variants are being swapped between chromosomes.

Genetic Diversity through Independent Assortment

Another way that meiosis ensures genetic diversity is a crucial event that took place in metaphase I (right after crossing over). It is the random alignment or *independent assortment* of chromosomes at the metaphase plate. (See the "Independent assortment" section of Figure 10.2.) Recall that the alignment of any one pair of homologous chromosomes along the plate bears no relation to the alignment of any other homologous pair. If you looked at, say, chromosome 5, the paternal member of the pair might line up on "side A" of the metaphase plate, meaning maternal chromosome 5 will be on side B. Shift to chromosome 6, however, and things could just as easily be reversed. This random alignment of chromosomes then leads to a random *separation* of them, because a chromosome that lines up on *side* A of the plate will end up in *cell* A, while a chromosome that lines up on side B will end up in cell B. **Independent assortment** can thus be defined as the random distribution of homologous chromosome pairs during meiosis. If we wonder how traits from a mother and father can get "mixed up" in their children, independent assortment joins crossing over in providing an answer. Indeed, for complex creatures such as ourselves, independent assortment assures that, with the exception of identical twins, offspring produced through sexual reproduction will not just be diverse. Each offspring will be *unique*—no other offspring will be exactly like it in genetic terms. Let's see why.

If human beings had only two chromosomes in each cell that went through meiosis—one from the father, one from the mother—there would be only two possible "states" of line-up at the metaphase plate: "maternal on side A" of the plate or "maternal on side B" (with the paternal homologue on the opposite side). The actual number of possibilities, however, is this number of states (2) *raised to the power* of the number of chromosome pairs. Thus, if three chromosome pairs existed, the number of possible outcomes would go to 2^3 or 8 ($2 \times 2 \times 2$). Now get your calculator ready. With the 23 pairs of chromosomes human beings have, the possibilities go to 2^{23}, or about 8 million. General Motors may be able to mix and match engines and upholstery to come up with scores of different models of a given car; but in any given human meiosis, chromosome pairs can line up along the metaphase plate in 8 million different ways. This means any given egg or sperm will be one of about 8 million possible "models." To get a basic idea of how this works out, look at the "Independent assortment" section of Figure 10.2.

Why does this matter? Remember that there are *genes* on these chromosomes, and that the genes on any given paternal chromosome may differ somewhat from those on its matched maternal chromosome. Which chromosome the offspring gets thus makes a difference—in eye color, height, or any of thousands of other characteristics. Given this, each of us is very much affected by the way paternal and maternal chromosomes happened to line up in two very special meioses: those that produced the egg cell and the sperm cell that gave rise to us.

When we add independent assortment to crossing over, it's easy to see why separate children from the same parents can come out looking so different—from each other as well as from the parents themselves (**see Figure 10.3**). Overall, meiosis generates

Figure 10.3
Ensuring Variety
The shuffling of genetic material that occurs during meiosis is the primary reason that children look different from their parents, and from each other.

genetic *diversity*, while its counterpart, mitosis, does not. Mitosis makes genetically exact copies of cells; it *retains* the qualities that cells have from one generation to the next. Meiosis mixes genetic elements each time it produces reproductive cells and thus brings about genetic variation in succeeding generations.

From Genetic Diversity, a Visible Diversity in the Living World

This genetic variation has an importance way beyond that of making siblings look different from each other. It turns out that this diversity is responsible, in significant part, for the fantastically diverse natural world around us. Think of it: two-foot ferns, 100-yard redwood trees, bats, bees, whales, mushrooms. How did the natural world come to be such a remarkably diverse place? The short answer is that evolution is spurred on by the *differences* in offspring that meiosis brings about. Let's see why.

Three frog siblings survive to reproductive age from a clutch of eggs laid by a female and fertilized by a male. Thanks to crossing over and independent assortment, *one* of these frogs may end up having, say, a darker coloration than all the other frogs of its species in its area. In the formation of the eggs and sperm that led to this frog, chromsome parts were swapped (in crossing over) and chromosomes lined up (in independent assortment) in such a way that this frog was the darkest "model" possible, among all the frogs in its population. Further, it happens that, because of shifting rain patterns in this frog's habitat, this dark coloration is starting to be useful for evading predators. This frog will thus survive to have more offspring of its own than do other frogs around it. By being selected for survival in this way—by undergoing what is called natural selection—this frog may have started its population down the road to having a darker coloration.

Now, it's easy to see that coloration is only one of a countless number of traits that could be selected for like this. Taller height in a tree? Diving stamina in a dolphin? Meiosis and sexual reproduction would lead the way to these traits by producing, in one generation after another, *this* tree that grew a little taller or *that* dolphin that dove a little deeper. Just as General Motors provides an array of models for consumers to choose from each year, meiosis and sexual reproduction provide an array of models for *nature* to choose from each generation.

To put a finer point on this, consider the alternative to meiosis and sexual reproduction. Bacteria are single-celled organisms that possess a single, circular chromosome. As you saw in Chapter 9, bacteria divide through the process of binary fission, one bacterium becoming two in the course of cell division. As it turns out, that's it for bacteria as far as reproduction goes. There is no chromosomal part-swapping or independent assortment of chromosomes in reproduction. And there is no fusion of two gametes from separate organisms, as occurs with eggs and sperm. Instead, there is just an exact duplication of the bacterial chromosome and a splitting of one bacterial cell into two. In general, then, every "daughter" bacterial cell is a genetic replica of its parent cell—a clone of it. What this means is that differences among bacteria are not a built-in feature of their reproduction. Instead, such differences as come to pass are dependent on their DNA mutating (which is rare) or on a form of gene transfer they are capable of (which is intermittent). Natural selection simply does not have as much to choose from in asexually reproducing organisms such as bacteria. Where is the equivalent of our darker-colored frog in a group of bacterial cells descended in a line? How could an equivalent difference come about, given that each bacterium is a clone of its predecessor?

What evidence do we have that sexual reproduction makes a practical difference? Well, consider that bacteria and their fellow single-celled organisms, the archaea, came into existence perhaps 3.8 billion years ago. Then, for 1.8 billion years after that, the living world consisted of nothing *but* bacteria and archaea—it took that long for anything else to evolve from them. We are not sure when sexual reproduction got going, but we have some evidence that it was taking place by 1.2 billion years ago. So, in the last 1.2 billion years, all the fish and birds and trees and fungi and flowers that we see before us developed through sexual reproduction, whereas in the 1.8 billion years after bacteria and archaea got going, they remained pretty much as they were when they began. Though other factors contributed to the ramping up of diversity on Earth, meiosis and sexual reproduction were key players in it. The British geneticist J. B. S. Haldane is said to have remarked that "Evolution could roll on fairly efficiently without sex but such a world would be a dull one to live in." In his statement, we can see the large-scale power of the small-scale processes we've been reviewing.

10.4 Meiosis and Sex Outcome

Let us think now about what meiosis means in relation to passing on sex to offspring. How do humans, in particular, come to be male or female? In humans there is one exception to the rule that chromosomes come in homologous pairs. Human females do indeed have 23 pairs of matched chromosomes, including 22 pairs of "autosomes" and one matched pair of **sex chromosomes**, meaning the chromosomes that determine what sex an individual will be. The sex chromosomes in females are called X chromosomes, and each female possesses two of them. In males, conversely, there are 22 autosomes, one X sex chromosome, and then one Y sex chromosome. It is this Y chromosome that confers the male sex. (See **Figure 10.4** for the differences in the X and Y sizes.)

In meiosis I in a female, the female's two X chromosomes, being homologous, line up together at the metaphase plate. Then these chromosomes separate, each of them going to different cells (**see Figure 10.5**). In males, the non-homologous X and Y chromosomes line up as if they were homologues. Then one resulting cell gets an X chromosome while the other gets the Y chromosome.

Sex determination is then simple. Each of the eggs produced by a female bears a single X chromosome. If this egg is fertilized by a *sperm* bearing an X chromosome, the resulting child will be a female. If, however, the egg is fertilized by a sperm bearing a Y chromosome, the child will be a male.

Because females don't have Y chromosomes to pass on, it follows that the Y chromosome that any male has had to come from his father. Meanwhile, the single X chromosome that any male has had to come from his mother. Females, conversely, carry one X chromosome from their mother and one from their father.

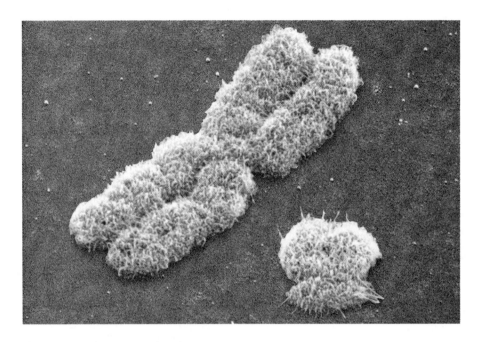

10.5 Gamete Formation in Humans

You've seen that, through meiosis, a single diploid cell gives rise to four haploid gametes. But what are the diploid cells that go through meiosis? As you might expect, there is a whole sequence of actions involved in the process by which egg and sperm are produced—a process known as *gamete formation*. In humans, the starting female cells in gamete formation are known as **oogonia**, while the starting male cells are **spermatogonia**. These are diploid cells

Figure 10.4
The X and the Y
The human X chromosome can be seen on the left, while the human Y chromosome is on the right. The difference in size between them is in part a reflection of the difference in the number of genes each of them contains. The Y chromosome only has about 50, while the X has about 1,500. (Magnified × 10,000)

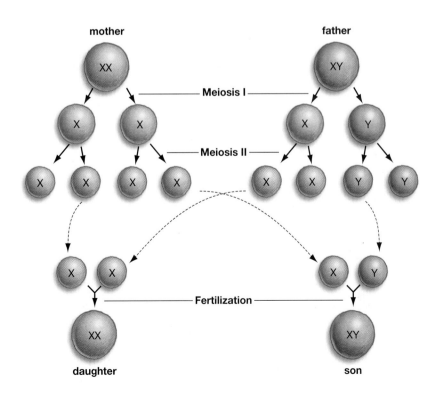

Figure 10.5
Sex Outcome in Human Reproduction

1. Early in meiosis I, the mother's two X chromosomes line up at the metaphase plate (XX). Meanwhile, in the father's meiosis, his X and Y chromosomes line up (XY).

2. The X-X and X-Y pairs then separate into different cells.

3. The chromatids that made up the duplicated chromosomes separate, yielding individual eggs and sperm.

4. Should an X-bearing sperm from the male reach the egg first, the child will be a girl; should the male's Y-bearing sperm reach the egg first, the child will be a boy.

that give rise to two other sets of diploid cells, the **primary oocytes** and **primary spermatocytes**. These then are the cells that undergo meiosis, yielding haploid sperm and egg cells.

Sperm Formation

The starting male spermatogonia exist in the male testes and are capable of giving rise to primary spermatocytes and of generating more spermatogonia. It is this latter, self-generating ability that qualifies spermatogonia as "stem" cells and that allows males to *keep* producing sperm throughout their lives, beginning at puberty. These sperm-producing "factories," in other words, generate not only sperm but more factories as well (**see Figure 10.6**).

The primary spermatocytes that come from spermatogonia go through meiosis I and thus produce the haploid *secondary spermatocytes*. Meiosis II then ensues, and the result is four haploid *spermatids*. These are not yet the familiar, tadpole-like sperm;

those come only after spermatids develop further, a process that takes about 3 weeks. There is thus an assembly-line quality to sperm production. Once a male reaches puberty, he starts producing sperm in this manner, with each sperm moving right through the entire process in a matter of weeks and the process continuing until the male dies. As you'll see, female egg production takes a much different route.

What is the nature of the sperm produced? A given sperm cell must fully develop its best-known feature, its tail-like flagellum. As it turns out, these "tails"—the only such structure in the human body—are composed of none other than the microtubules you have been seeing so much of in cell division. Moving the flagellum in a whip-like motion, microtubules serve to propel the sperm on their journey (**see Figure 10.7**). Sperm also need to "travel light," to which end developing sperm get rid of nearly all the cellular organelles reviewed back in Chapter 4: lysosomes, the Golgi complex, and so forth. A mature

Web Tutorial 10.2
Sex and Human Reproduction

Figure 10.6
Sperm and Egg Formation in Humans

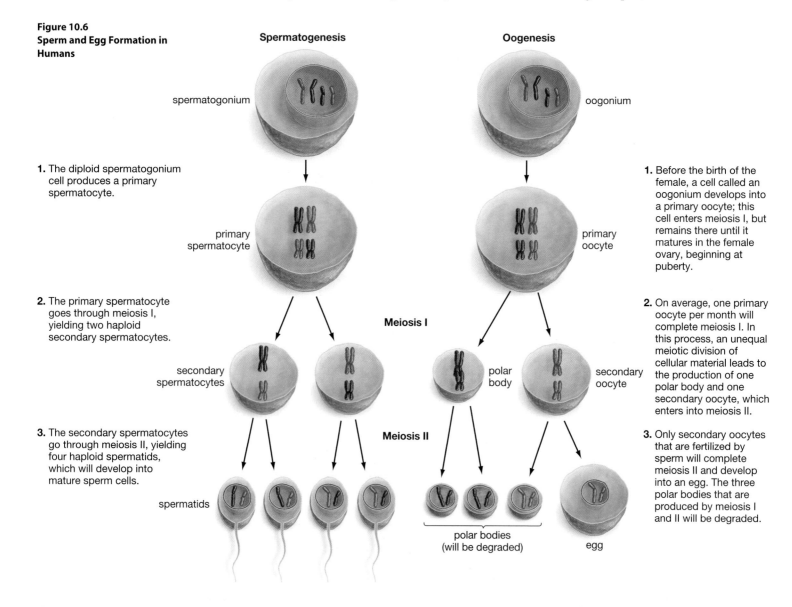

Spermatogenesis

Oogenesis

spermatogonium

oogonium

primary spermatocyte

primary oocyte

Meiosis I

secondary spermatocytes

polar body

secondary oocyte

Meiosis II

spermatids

polar bodies (will be degraded)

egg

1. The diploid spermatogonium cell produces a primary spermatocyte.

2. The primary spermatocyte goes through meiosis I, yielding two haploid secondary spermatocytes.

3. The secondary spermatocytes go through meiosis II, yielding four haploid spermatids, which will develop into mature sperm cells.

1. Before the birth of the female, a cell called an oogonium develops into a primary oocyte; this cell enters meiosis I, but remains there until it matures in the female ovary, beginning at puberty.

2. On average, one primary oocyte per month will complete meiosis I. In this process, an unequal meiotic division of cellular material leads to the production of one polar body and one secondary oocyte, which enters into meiosis II.

3. Only secondary oocytes that are fertilized by sperm will complete meiosis II and develop into an egg. The three polar bodies that are produced by meiosis I and II will be degraded.

sperm thus amounts to little more than a haploid set of DNA-containing chromosomes in front, a collection of "engines" (mitochondria) in the middle, and a propeller (the flagellum) in the back. About 250 million sperm are made each day, and about that number will be released with each ejaculation. Sperm that are not released will age and eventually are destroyed by the body's immune system.

Egg Formation

The need of male sperm to become stripped-down DNA packages stands in contrast to the needs of female ova or eggs, which must have rich resources of nutrients and building materials in order to sustain life once they have been fertilized. This difference in function is reflected in size: The volume of an egg is 200,000 times the volume of a sperm (**see Figure 10.8** to get an idea of this difference). Accordingly, eggs are created in a process that maximizes cell content, at least for the fraction of eggs that become viable.

Egg formation begins with the cells called the *oogonia* of the female. These may sound like the counterpart to the male spermatogonia, but actually they are very different. Whereas spermatogonia are self-generating stem cells, oogonia are produced in large numbers only up to the seventh month of embryonic life, and none are created after birth. Of the millions produced, most will die while the female is still an embryo; but some survive to become primary oocytes and thus, as diploid cells, enter meiosis I (**see Figure 10.6**).

Amazingly enough, meiosis I is where these oocytes *stay*—for years. Imagine a collection of oocytes in the female ovaries (which are walnut-sized structures just right and left of center in the female pelvic area). Once puberty begins, each oocyte is theoretically capable of maturing into an egg, but an average of only *one* oocyte per month actually starts this maturation process. Which oocyte is this? The one released from the ovary each month in the process called ovulation. All the other oocytes in the ovaries halt their meiosis in prophase of metaphase I. Their chromosomes are duplicated and they have completed crossing over, but they have not yet lined up at the metaphase plate. And this is where things stay for most of the oocytes a female possesses. Only the one oocyte per month that is expelled from the ovary will complete meiosis I and thus enter into meiosis II. Even this meiosis is "arrested," however, until this oocyte is fertilized by a sperm. It is the union with a sperm that prompts the completion of meiosis II. Thus, an oocyte that came into being when a female was an embryo might remain sus-

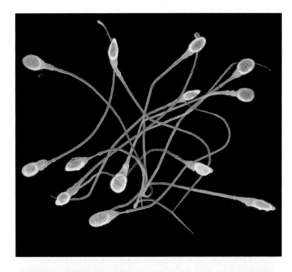

Figure 10.7
Mobile Cells, Bearing Chromosomes
Human sperm, colored to show detail.

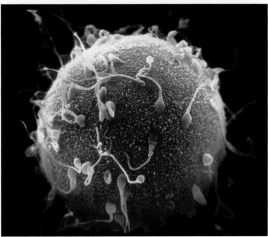

Figure 10.8
Big Difference in Size
A human egg surrounded by much smaller human sperm.

pended in the cell cycle for years or even decades. Of the millions of oocytes originally created, only a few hundred will mature even through metaphase I in a woman's lifetime.

How do certain oocytes get "selected" to mature through metaphase I—how do they get selected to be ovulated, in other words? This occurs in connection with an intricate interplay that involves their surrounding tissue, hormones, and the happenstance of timing. You'll read more about this process in Chapter 30.

One Egg, Several Polar Bodies

For the few oocytes selected, the process of going through meiosis I and II means proceeding from one diploid cell to four haploid cells. In this process, however, almost all the cytoplasm will be shunted to *one* of the four cells, the better to build up its stores of nutrients and other cytoplasmic material. This happens in two rounds: meiosis I and meiosis II. When cytokinesis comes in meiosis I, one daughter cell gets almost all the cytoplasm while the second daughter cell gets almost none. This same thing happens in meiosis II. The result? One richly endowed haploid ovum and two, or perhaps three, very small haploid **polar bodies**: nonfunctional cells produced during meiosis in females (Figure 10.6). (Only two polar bodies and the ovum will be the end result if the polar body produced in meiosis I does not continue through meiosis II, which often happens.) The ovum has a chance at taking part in producing offspring, but the polar bodies are bound for oblivion rather than a shot at immortality. They will eventually degrade into their constituent substances.

10.6 Life Cycles: Humans and Other Organisms

The process of the formation of the human gametes called egg and sperm is followed, as you have seen, by the union of these gametes. In humans, the oocyte that is expelled from the ovary makes a slow trip down the female's uterine (or Fallopian) tube, during which time it may encounter a swimming sperm. If it does, the result will be a fertilized egg (or zygote), which then develops into a whole human being. One of the things that will eventually occur in this new human being is the development of another group of eggs or sperm through the process of meiosis. Thus do we see a **life cycle**: the repeating series of steps that occur in the reproduction of an organism. You have looked at the human life cycle, but it's important to recognize that it amounts to only one among many.

Figure 10.9
All-Female Species
Pictured is a Sonoran spotted whiptail lizard, *Cnemidophorus sonorae*, which reproduces through parthenogenesis. Because the eggs these lizards produce are not fertilized by males, they contain only one set of chromosomes, passed on by the mother. Hatchlings develop from these haploid eggs, however, and as a consequence the hatchlings themselves are haploid and genetically identical to the mother. The end result is a species that is entirely female. This lizard was photographed in Arizona.

Not All Reproduction Is Sexual

As noted earlier, human reproduction is **sexual reproduction**: the union of two reproductive cells to create a new organism. But in many other species there is **asexual reproduction** or reproduction that does not involve sex. Asexual reproduction actually is seen in a vast array of organisms. In fact, the only two groupings of living things that *never* carry out asexual reproduction are birds and mammals. Admittedly, asexual reproduction is rare in creatures as complex as fish or reptiles, but it does happen. In the American southwestern desert, for example, there are a few species of whiptail lizard that are entirely female (**see Figure 10.9**). The female whiptail lays haploid eggs, but they are never fertilized. Instead, they develop into haploid clones of her through mitotic cell division. This method of reproduction (called *parthenogenesis*) also results in "drone" male honeybees. Those eggs that a queen bee chooses not to fertilize (with stored-up sperm) develop into drones through parthenogenesis.

Other forms of asexual reproduction also are possible. In fact, you've already been introduced to one of them. Recall that a bacterial cell engages in a form of asexual reproduction, called binary fission, in which it duplicates its single chromosome and then divides in two. Beyond this, anyone who has ever seen a "cutting" taken from a plant knows that a snipped-off stem or branch will, when properly planted, sprout its own roots and grow into an independent plant, complete with the ability to take part in sexual reproduction. This process is called *vegetative reproduction*, and it has a counterpart in the animal world. Cut the arm and part of a central disk from a sea star, for example, and, through *regeneration*, this severed limb will grow into a whole, multi-armed sea star (**see Figure 10.10**).

As we go through the branching tree of life, from more complex creatures to less complex ones, asexual reproduction becomes more common. By the time we reach bacteria, it is the only form of reproduction used. Note that all forms of asexual reproduction produce offspring that are genetically identical to their parents. There is no mixing and matching of chromosomes from different parents in these asexual processes; just a cloning of one genetically identical individual from its parent.

Variations in Sexual Reproduction

Within the framework of sexual reproduction, as you can imagine, there are many variations. For example, separate gametes need not come from separate organisms. Some animals, such as tapeworms, are hermaphrodites, meaning they have both male and female reproductive parts. In some cases, hermaphrodites fertilize themselves, but generally they are fertilized by another member of their species. Most plants contain both male and female parts, and some plants "self-fertilize"—their sperm can fertilize their eggs—as with a pea plant you will be looking at in Chapter 11.

On to Patterns of Inheritance

The tour of cell division over the last couple of chapters has yielded not only a look at how one cell becomes two, but at how chromosomes operate within this process. Given the paired nature of chromosomes and their precisely ordered activity during meiosis, it stands to reason that there would be some predictability or pattern to the passing on of traits. This is the case, as it turns out. And the person who first recognized this pattern was a monk who worked in nineteenth-century Europe in an obscurity he did not deserve.

Figure 10.10
Regenerating the Whole from a Part
A sea star off the coast of Papua New Guinea is regenerating a body and other arms from a piece of one arm and part of its central disk.

Chapter Review

Summary

10.1 An Overview of Meiosis

- In human beings, nearly all cells have paired sets of chromosomes, meaning these cells are diploid. Meiosis is the process by which a single diploid cell divides to produce four haploid cells—cells that contain a single set of chromosomes.

- The haploid cells produced through meiosis are called gametes. Female gametes are eggs; male gametes are sperm. They are the reproductive cells of human beings and many other organisms.

- When the haploid sperm and haploid egg fuse, a diploid fertilized egg (or zygote) is produced, setting into development a new generation of organism.

10.2 The Steps in Meiosis

- In meiosis, there is one round of chromosome replication, followed by two rounds of cell division. There are two primary stages to meiosis, meiosis I and meiosis II. In meiosis I, chromosome duplication is followed by a pairing of homologous chromosomes with one another, during which time they exchange reciprocal sections of themselves. Homologous chromosome pairs then line up at the metaphase plate—one member of each pair on one side of the plate, the other member on the other side. These homologous pairs are then separated, in the first round of cell division, into separate daughter cells. In meiosis II, the chromatids of the duplicated chromosomes are separated into separate daughter cells.
  Web Tutorial 10.1: Meiosis

10.3 What Is the Significance of Meiosis?

- Meiosis generates genetic diversity by ensuring that the gametes it gives rise to will differ from one another. In this, it is unlike regular cell division, or mitosis, which produces daughter cells that are exact genetic copies of parent cells.

- Meiosis generates genetic diversity in two ways. First, in prophase I of meiosis, homologous chromosomes pair with each other and, in the process called crossing over or recombination, exchange reciprocal chromosomal segments with one another. Second, in metaphase I of meiosis, there is a random alignment or independent assortment of maternal and paternal chromosomes on either side of the metaphase plate. This chance alignment determines which daughter cell each chromosome will end up in.

- The genetic diversity brought about by meiosis and sexual reproduction is responsible, to a significant extent, for the great diversity of life-forms seen in the living world today. Evolution is spurred on by differences among offspring, and meiosis and sexual reproduction ensure such differences. By contrast, asexual reproduction, as is seen in bacteria and other organisms, produces organisms that are exact genetic copies, or clones, of the parental organism.

10.4 Meiosis and Sex Outcome

- Human females have 23 matched pairs of chromosomes—22 autosomes and two X chromosomes. Human males have 22 autosomes, one X chromosome, and one Y chromosome. Each egg that a female produces has a single X chromosome in it. Each sperm that a male produces has either an X or a Y chromosome within it. If a sperm with a Y chromosome fertilizes an egg, the offspring will be male. If a sperm with an X chromosome fertilizes the egg, the offspring will be female.

10.5 Gamete Formation in Humans

- In humans, the starting cells in male gamete formation, spermatogonia, are diploid cells that give rise to other diploid cells, called primary spermatocytes, that go through meiosis. The result of this meiosis is cells called spermatids that develop into mature sperm. Female gamete formation begins with cells called oogonia, all of which are produced prior to the birth of the female. These give rise to primary oocytes—the cells that will go through meiosis I and II, producing the haploid female ovum or egg. The vast majority of primary oocytes never complete meiosis, however. It is only the single primary oocyte released each month, in the process of ovulation, that completes meiosis I. Only those ovulated oocytes that are fertilized by sperm complete meiosis II.

  Web Tutorial 10.2: Sex and Human Reproduction

10.6 Life Cycles: Humans and Other Organisms

- Not all reproduction is sexual reproduction, meaning reproduction that works through a fusion of two reproductive cells. Most types of organisms are capable of asexual reproduction, though such reproduction is rare among more complex organisms and is never carried out by mammals or birds. Asexual reproduction can take several forms. Bacteria reproduce through one form of asexual reproduction, called binary fission, in which a given bacterial cell replicates its single chromosome and then divides in two. Plants can engage in vegetative reproduction; other organisms, such as worms and sea stars, can carry out regeneration in which a new, complete organism can be formed from a portion of an existing one.

Key Terms

1n 192	**meiosis** 192
2n 192	**oogonia** 199
asexual reproduction 202	**polar body** 202
crossing over 197	**primary oocyte** 200
diploid 192	**primary spermatocyte** 200
gamete 192	**sex chromosome** 199
haploid 192	**sexual reproduction** 202
independent assortment 197	**spermatogonia** 199
life cycle 202	**tetrad** 196

Understanding the Basics

Multiple-Choice Questions (answers in the back of the book)

1. In reference to chromosomes, meiosis involves
   a. no duplications, 1 reduction
   b. 1 duplication, 2 reductions
   c. 2 duplications, 1 reduction
   d. 2 duplications, 2 reductions
   e. none of the above

2. Imagine, at prophase of meiosis I, that a diploid (2n) cell that is precursor to the gamete has eight chromosomes. How many chromatids are in the cell?
   a. 16
   b. 8
   c. 4
   d. 2
   e. twice that of mitosis prophase

3. Independent assortment occurs during
   a. prophase I
   b. anaphase II
   c. metaphase I
   d. metaphase II
   e. both c and d

4. Which of the following statements is *not* true regarding egg formation in humans?
   a. Oogonia are not produced until the eleventh or twelfth year.
   b. At the completion of division, one oocyte and two or three polar bodies exist.
   c. In both meiosis I and II, the metaphase plate is shifted to one side of the cell.
   d. An oocyte may take decades to go through the cell cycle.
   e. Each egg contains 23 chromosomes.

5. What are the two sources of genetic diversity in meiosis?
   a. binary fission and regeneration
   b. crossing over and independent assortment
   c. oogenesis and crossing over
   d. spermatogenesis and oogenesis
   e. vegetative reproduction and regeneration

6. Meiosis and sexual reproduction have fostered diversity in the natural world because they
   a. are so accurate in copying genetic information
   b. are nearly certain to produce differences among offspring
   c. are such ancient processes
   d. produce genetic replicas, one generation after another
   e. are used by such a wide range of organisms

7. Eggs are relatively large because they _____; sperm are relatively small because they _____.
   a. must contain spare sets of chromosomes … function through mobility

b. must contain room for the growing zygote … have only one set of chromosomes

c. must contain room for the sperm … have a short life span

d. must contain nutrients for the zygote … function through mobility

e. exist in a harsh environment … exist in a nurturing environment

8. In prophase I of meiosis,
   a. maternal chromosome 1 joins with maternal chromosome 2.
   b. maternal chromosomes exchange genetic material with other maternal chromosomes.
   c. paternal chromosomes exchange genetic material with other paternal chromosomes.
   d. homologous chromosomes exchange sections.
   e. both c and d are correct.

9. At the completion of meiosis I, each human cell contains
   a. 23 chromosomes, 46 chromatids
   b. 46 chromosomes, 46 chromatids
   c. 46 chromosomes, 92 chromatids
   d. 23 chromosomes, 23 chromatids
   e. none of the above

10. Asexual reproduction is _____ and always results in _____.
    a. found in only a few types of organisms … offspring that are different from parents
    b. found in all types of organisms except mammals and reptiles … offspring that are sometimes genetically identical to parents
    c. reproduction that results from a fusion of two reproductive cells … twice as many offspring as parents
    d. found more commonly in simple organisms than in complex ones … offspring that are genetically identical to parents
    e. found in only a few mammals and birds … offspring that are different from parents

Brief Review

1. Does chromosome reduction take place during meiosis I or II?

2. Is asexual reproduction confined to the plant world?

3. Why are men able to keep producing sperm throughout their lifetime, from puberty forward?

4. How does meiotic cytokinesis in the production of eggs that females produce differ from mitotic cytokinesis?

5. Define and distinguish between somatic cells and gametes.

6. A diploid organism has four chromosomes. Name the phases of mitosis or meiosis shown in the following figures.

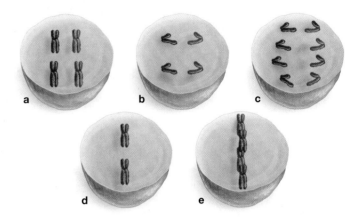

7. In what ways does meiosis ensure genetic diversity in offspring?

8. When human egg and sperm come together, how many chromosomes does each kind of gamete have? What is the genetic makeup of the resulting zygote?

Applying Your Knowledge

1. Ultimately, is it the paternal or maternal gamete that delivers the sex-determining chromosome? Describe the process.

2. Mammals have now been cloned by scientists (think of Dolly the sheep). Suppose that half the world's human births were achieved through cloning. How would our world be different?

3. All the bananas we eat come from trees that are produced through "cuttings"—a stem from an existing tree is planted in the ground, resulting in a new tree. Thus, each tree is a clone of another. Why would growers find it advantageous to produce an enormous series of identical clones, rather than using trees that reproduce sexually?

11 *Mendel and His Discoveries*

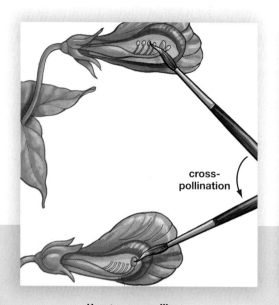

How to cross-pollinate.
(Section 11.2, page 209)

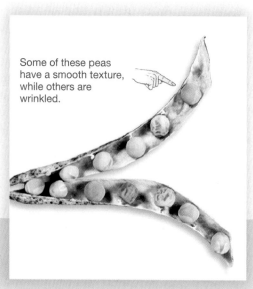

Same pod, different peas.
(Section 11.3, page 211)

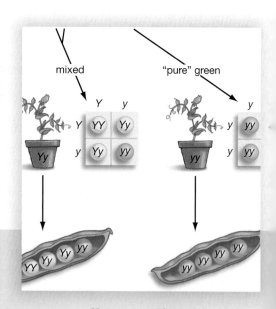

More pea experiments.
(Section 11.4, page 213)

In the Czech Republic, in the oldest part of the city of Brno, stands the former Monastery of St. Thomas. Now a research center, St. Thomas has a small, fenced-in garden that sits just outside its historic living quarters. At one end of the garden stands an ivy-covered tablet, inscribed with a few simple words: "Prelate Gregor Mendel made his experiments for his law here." Visitors to the garden can cast their eyes upward and see the rooms that Gregor Johann Mendel lived in while carrying out his work in the years between 1856 and 1863 (**Figure 11.1**). Mendel labored during this time on a species of common peas, *Pisum sativum*, using tweezers and an artist's paintbrush as his tools. Like many scientific discoverers, Mendel looked at ordinary objects; and as with many scientists, his work was tedious and exacting. Nevertheless, the distance from what Mendel did to what Mendel learned is the distance from the ordinary to the historic.

Coming to adulthood in the mid-nineteenth century, Mendel knew nothing of the elements of genetics we have reviewed so far. Chromosomes and the genes that lie along them had not even been discovered when he was working with his peas; little knowledge existed of meiosis or mitosis, to say nothing of DNA. And as it happened, Mendel himself played no direct role in dis-

covering any of these things. Yet this unassuming monk, the son of eastern European peasant farmers, generally is accorded the title of the father of genetics. What contribution earned him this honor?

Figure 11.1
Austrian Monk and Naturalist Gregor Mendel

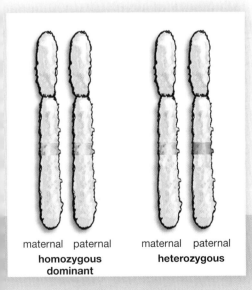

maternal paternal   maternal paternal
homozygous   heterozygous
dominant

Dominant and recessive.
(Section 11.4, page 215)

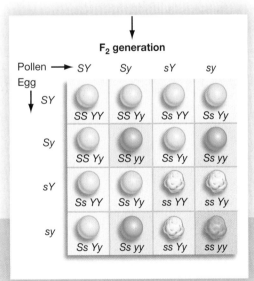

F₂ generation

Pollen → SY   Sy   sY   sy
Egg
SY   SS YY   SS Yy   Ss YY   Ss Yy
Sy   SS Yy   SS yy   Ss Yy   Ss yy
sY   Ss YY   Ss Yy   ss YY   ss Yy
sy   Ss Yy   Ss yy   ss Yy   ss yy

Breeding for two characters.
(Section 11.5, page 216)

Genes and environment.
(Section 11.10, page 222)

11.1 Mendel and the Black Box

Mendel's achievement was to comprehend what was going on inside what might be called the "black box" of genetics without ever being able to look inside that box himself. Science is filled with so-called black-box problems, in which researchers know what goes *into* a given process and what comes *out*. It is what is going on in between—in the black box—that is a mystery. In the case of genetics, what lies inside the black box is DNA and chromosomes and meiosis, and so forth—all the component parts of genetics, in other words, and the way they work together. Because of the timing of his birth, Mendel had no knowledge of any of these things. Until Mendel, however, nobody had looked carefully at even the starting and ending points of this black-box problem: at what went in and what came out. Mendel's original pea plants represented the "input" side to the black box of genetics, while the offspring he got from breeding these plants represented the output. By looking

carefully at generations of parents and offspring—at both sides of the box, in a sense—Mendel was able to infer something about what had to be going on within. As you will see, his main inferences were correct: (1) that the basic units of genetics are material elements; (2) that these elements come in pairs; (3) that these elements (today called genes) can retain their character through many generations; and (4) that gene pairs *separate* during the formation of gametes.

These insights may sound familiar, because all of them were approached from another direction in Chapters 9 and 10. Here are the lessons you went over then:

1. Genes are material elements—lengths of DNA.

2. In human beings, genes come in pairs, residing in pairs of homologous chromosomes.

3. Chromosomes make copies of themselves, thus giving the genes that lie along them the ability to be passed on intact through generations.

4. In meiosis, homologous chromosomes line up next to each other and then *separate*, with each member of a pair ending up in a different egg or sperm cell.

By observing generations of pea plants and applying mathematics to his observations, Mendel inferred that something like this had to be happening in reproduction. Because Mendel was the first to perceive a set of principles that govern inheritance, we date our knowledge of genetics from him.

11.2 The Experimental Subjects: *Pisum sativum*

Beginning work in his monastery garden after a period of study in Vienna, Mendel managed to pick, in *Pisum sativum*, a nearly perfect species on which to carry out his experiments. If you look at **Figure 11.2**, you can see something of the life cycle of this garden pea. Note that what we think of as peas in a pod are seeds in this plant's ovary. Each of these seeds begins as an unfertilized egg, just as a human baby begins as a maternal egg that is unfertilized. Sperm-bearing pollen, landing on the plant's stigma, then set in motion the fertilization of the eggs.

Importantly, *Pisum* plants can *self*-pollinate. The anthers of a given flower release pollen grains that land on that flower's stigma. Each seed that develops from the resulting fertilizations can then be planted in the ground and give rise to a new generation of plant. The way to think of the seeds in a pod is as

Figure 11.2
Life Cycle of the Pea Plant

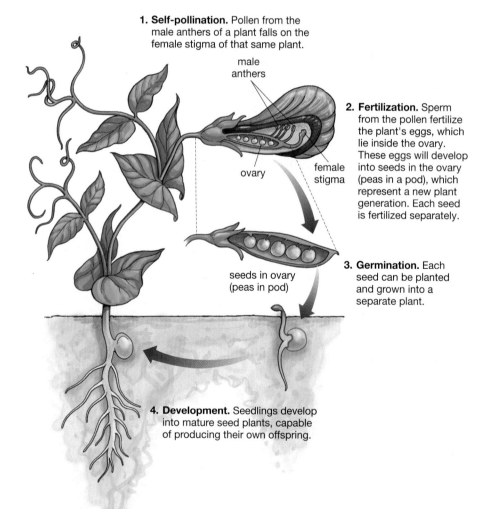

1. **Self-pollination.** Pollen from the male anthers of a plant falls on the female stigma of that same plant.

male anthers

2. **Fertilization.** Sperm from the pollen fertilize the plant's eggs, which lie inside the ovary. These eggs will develop into seeds in the ovary (peas in a pod), which represent a new plant generation. Each seed is fertilized separately.

ovary

female stigma

seeds in ovary (peas in pod)

3. **Germination.** Each seed can be planted and grown into a separate plant.

4. **Development.** Seedlings develop into mature seed plants, capable of producing their own offspring.

multiple offspring from separate fertilizations—one pollen grain fertilizes this seed, another pollen grain fertilizes another. This is why, as you'll see, seeds that are in the same pod can have different characteristics.

Though the pea plants can self-pollinate, Mendel could also **cross-pollinate** the plants at will—he could have one plant pollinate another—by going to work with his tweezers and paintbrushes, as shown in **Figure 11.3**.

In directing pollination, Mendel could control for certain attributes in his pea plants—qualities now referred to as *characters*. If you look at **Table 11.1** on page 210, you can see that there were seven characters in all—such things as stem length, seed color, and seed shape. Note that each of these characters comes in two varieties. There are yellow or green seeds, for example, and purple or white flowers. Such character variations are

**Figure 11.3
How to Cross-Pollinate Pea
Plants**

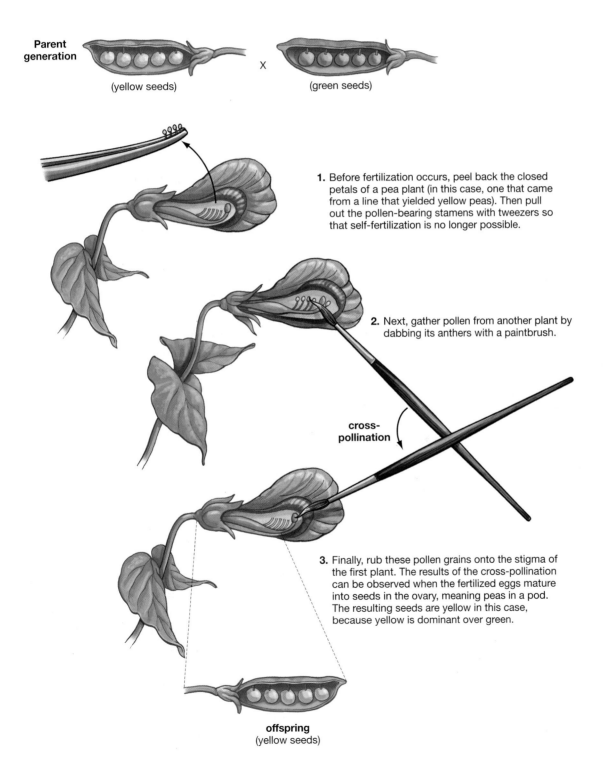

**Parent
generation**

(yellow seeds) X (green seeds)

1. Before fertilization occurs, peel back the closed petals of a pea plant (in this case, one that came from a line that yielded yellow peas). Then pull out the pollen-bearing stamens with tweezers so that self-fertilization is no longer possible.

2. Next, gather pollen from another plant by dabbing its anthers with a paintbrush.

**cross-
pollination**

3. Finally, rub these pollen grains onto the stigma of the first plant. The results of the cross-pollination can be observed when the fertilized eggs mature into seeds in the ovary, meaning peas in a pod. The resulting seeds are yellow in this case, because yellow is dominant over green.

offspring
(yellow seeds)

Figure 11.5
(a) Mendel's P Generation Crosses

(b) Examples of Punnett Squares

element, which got expressed only in the F₂ generation. Finally, because his phenotypes came in pairs, it was reasonable for Mendel to hypothesize that the elements likewise came in pairs.

If you think Mendel's *pairs of elements* sound suspiciously like the "pairs of genes" on homologous chromosomes we've seen before, you are right. However, it's time to start referring to these "matched pairs of genes" in scientific terminology. It is more accurate to think of matched pairs of genes as alternative forms of a single gene. The proper name for an alternative form of a gene is an **allele**. Thus the pea plant had a single gene for seed color that came in two alleles—one of which coded for yellow seeds, the other of which coded for green seeds. These alleles resided on separate homologous chromosomes.

11.4 Another Generation for Mendel

Having seen recessive phenotypes reappear in F₂, Mendel decided to extend his breeding. In his next experiment, he took a set of F₂ generation seeds—all of them yellow—planted them and let the plants they grew into self-pollinate. Then he looked at the color of the F₃ seeds contained in the pods of these plants. What he found was that, of 519 plants grown, 166 yielded only yellow seeds, while the remaining 353 plants had a mixture of yellow and green seeds in the pods, in the familiar 3:1 ratio. But when he similarly took F₂ green seeds, planted them, and let them self-pollinate, these plants produced nothing but green seeds.

What could have caused these results? The answer was the underlying elements Mendel had hit upon, meaning the alleles for the color gene. When he planted his yellow seeds, some of them yielded all-yellow offspring, but others yielded mixed (green and yellow) offspring. He reasoned that there must be *two kinds* of yellow plants: Those that are "pure" yellow (the 166 that yielded solely yellow seeds) and those that were "mixed" yellow (the 353 that produced both yellow and green seeds). Pure yellow had *nothing but* yellow alleles within, and would produce nothing but yellow seeds when self-pollinated. Meanwhile, the mixed yellow would have one yellow and one *green* allele within, and thus would produce pods that contained seeds of both colors. Finally, the green seeds had nothing but green alleles and thus produced nothing but green seeds.

Mendel's Generations in Pictures

Let's make these points clearer by reviewing all three generations of experiments schematically.

The F₁ Generation

It will be helpful to introduce a convention here. Let us refer to dominant and recessive alleles by uppercase and lowercase letters, with uppercase used for dominant types and lowercase used for recessive types. Thus a "pure" yellow-seeded plant, having *two* yellow alleles, would be symbolized as *YY*. Meanwhile, pure green-seeded plants are *yy*, while mixed

(a) P generation crosses

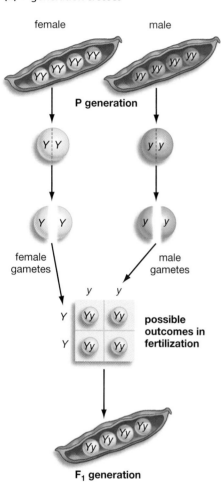

P generation

female gametes   male gametes

possible outcomes in fertilization

F₁ generation

1. Female gametes are being provided by a plant that has the dominant, yellow alleles (*YY*); male gametes are being provided by a plant that has the recessive, green alleles (*yy*).

2. The cells of the pea plants that give rise to gametes start to go through meiosis.

3. The two alleles for pea color, which lie on separate homologous chromosomes, separate in meiosis, yielding gametes that each bear a single allele for seed color. In the female, each gamete bears a *Y* allele; in the male, each bears a *y* allele.

4. The Punnett square shows the possible combinations that can result when the male and female gametes come together in the moment of fertilization. (If you have trouble reading the Punnett square, see Figure 11.5b). The single possible outcome in this fertilization is a mixed genotype, *Yy*.

5. Because *Y* (yellow) is dominant over *y* (green), the result is that all the offspring in the F₁ generation are yellow, because they all contain a *Y* allele.

(b) How to read a Punnett square

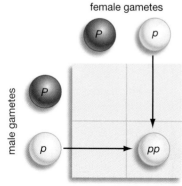

1. A *p* gamete from the male combines with a *p* gamete from the female to produce an offspring of *pp* genotype (and white color).

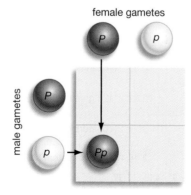

2. A *p* gamete from the male combines with a *P* gamete from the female to produce an offspring of *Pp* genotype (and purple color).

seeds are *Yy*. Let us say, just as an example, that female gametes are being supplied by a plant that is *YY*, while male gametes are coming from a plant that is *yy*. In meiosis, as you've seen, homologous chromosomes separate. This happens in the pea plants, meaning that each of these plants will contribute *one member* of its gene pair to its respective gametes—the *YY* female contributing a *Y* gamete, and the *yy* male contributing a *y* gamete. When these gametes fuse in the moment of fertilization, the result is a *Yy* hybrid in the F₁ generation. If you look at **Figure 11.5**, you will be introduced to a time-honored way to represent such outcomes, the Punnett square, and see Mendel's F₁ results symbolized as well.

Note that, because *Y* is dominant, all the seeds in the F₁ offspring pods will have a yellow *phenotype*, even though every one of these seeds contains a mixed *genotype*. (Every one is *Yy*.) It takes only one *Y*

allele for a seed to be yellow; for a seed to be green, then, it must have two *y* alleles.

The F₂ Generation

Next come the F₁ crosses that yielded the F₂ generation. The starting point here is the F₁ seeds, which are all of the mixed *Yy* type, as you can see in **Figure 11.6**. Meiosis then occurs in the gamete precursors, which results in a separation of these alleles: Half the gametes now contain *Y* alleles and the other half *y*. You can see from the figure why this cross can now give us back the green-seed phenotype, on average in a 1:3 proportion with yellow seeds. Note that there are twice as many mixed-genotype seeds (*Yy*) as either variety of pure seed (*YY* or *yy*). Important mathematical principles underlie this outcome. You can read more about them in "Proportions and Their Causes." on the next page.

**Figure 11.6
From the F₁ to the F₃
Generation**

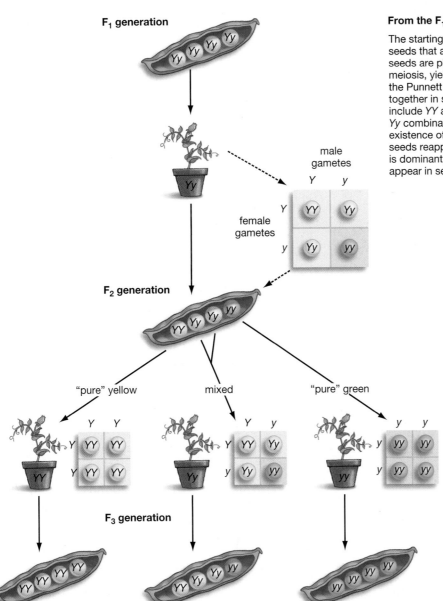

From the F₁ to the F₂ generation

The starting point is the F₁ generation, a set of seeds that all have the *Yy* genotype. These seeds are planted and the plants go through meiosis, yielding the gametes shown around the Punnett square. When these gametes come together in self-fertilization, the possibilities include *YY* and *yy* combinations, as well as the *Yy* combination seen in the F₁ generation. The existence of *yy* individuals is the reason green seeds reappear in the F₂ generation. Because *Y* is dominant, the green phenotype could not appear in seeds that had even a single *Y* allele.

From the F₂ to the F₃ generation

With three starting genotypes (*YY*, *Yy*, *yy*) the F₂ generation yields plants that have these three genotypes, though there are more F₂ plants of "mixed" genotype than of either "pure" genotype.

Proportions and Their Causes: The Rules of Multiplication and Addition

To be a good songwriter, a person needs to be both a good musical composer and a good lyricist. Of all the people who decide to take up songwriting, say that 1 in 10 is a good composer and that 1 in 10 is a good lyricist. Now, what is the probability that a person who takes up songwriting will be *both* a good composer and a good lyricist? The answer to this question turns out to have relevance to all kinds of questions in our world, including the results that Gregor Mendel got.

When Mendel examined the F_2 pods that contained the F_3 seeds, he found that there were many more pods that had mixed seed colors within (yellow and green) than pods whose seeds were either all-yellow or all-green. It was the F_1 meiosis that set the stage for this outcome, as you can see here:

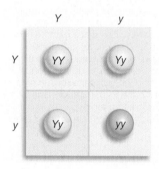

The Punnett square provides an intuitive visual sense of why things turned out the way they did. Meiosis in the F_1 plants produced twice as many mixed F_2 seeds (Yy) as either type of "pure" seeds (YY or yy). It's even more helpful, however, to understand a principle that underlies this outcome. A simple coin-tossing example can illustrate what is at work. Suppose that you are going to throw two coins in the air—a nickel and a dime. What is the likelihood that *both* coins will come up heads? There is only one way to get this result:

1. Nickel = heads; dime = heads

Now consider tossing the two coins and specifying a *mixed* result. There are *two* ways that this condition can be satisfied:

1. Nickel = heads; dime = tails
2. Dime = heads; nickel = tails

The seeds are in the same situation. What is the likelihood of getting a "pure" yellow seed? As you can see by looking again at the

Punnett square, there is only one way to get this: female *Y*, male *Y*. Now consider specifying a mixed result. There are two ways to get this: female *y*, male *Y*; and female *Y*, male *y*.

The Rule of Multiplication

In order to state a principle that can generalize to any situation, let's look at this concept in a more formal way. What are the odds of tossing up one coin and getting heads? Fifty-fifty, right?

What are the odds of tossing up two coins and having both come up heads?

This means a 50 percent chance, which equals 0.5. Now what are the odds of tossing up two coins and having both come up heads? A principle called the **rule of multiplication** comes into play here. It states that the probability of any *two* events happening is the product of their respective probabilities. In this case, the probability for each head coming up is 0.5, so the equation is $0.5 \times 0.5 = 0.25$, or a 25 percent probability of two heads.

The Rule of Addition

Now what is the probability of getting one coin that comes up heads and one that comes up tails—when either of them can be heads or tails? When an outcome can occur in two or more *different* ways, as is the case here, the probability of this happening is the *sum* of the respective probabilities. This principle is known as the **rule of addition**. The probability of nickel = heads, dime = tails is the same 0.25 probability you saw in the rule of multiplication. But the probability of nickel = tails, dime = heads is also 0.25. The probability that *either* of these outcomes will take place is thus $0.25 + 0.25 = 0.50$, or 50 percent.

All this gives us a way to figure the songwriter probability—or any set of independent probabilities. Remember the stipulation that 1 in 10 aspiring songwriters is a good composer and 1 in 10 is a good lyricist. Thus, under the rule of multiplication, $0.10 \times 0.10 = 0.01$, meaning 1 in 100 will be good at both composing and lyric writing.

The F₃ Generation

Finally, in Figure 11.6, you can see why Mendel got the results he did when he went from F_2 to F_3. Recall that when Mendel planted and self-fertilized his *yellow* F_2 seeds, he got 519 plants, of which 166 produced nothing but yellow seeds. Now you can see why: These were not only phenotypically yellow; they were "pure" yellow in terms of genotype (*YY*). Mendel also got 353 plants that produced both yellow and green seeds. This is because the seeds these plants grew from were of mixed genotype (*Yy*); in reproduction, their offspring would be of both pure (*YY*, *yy*) and mixed (*Yy*) genotype. Finally, when Mendel planted and self-fertilized *green* F_2 seeds, all the plants that resulted from these seeds bred true for green seeds, because all of them began as *yy*. **Figure 11.7** illustrates how the three genotypes discussed here yield only the two phenotypes of yellow or green.

The Law of Segregation

For inheritance to work this way, Mendel saw something we noted earlier: That, though plant cells may contain *two* copies (alleles) of a gene relating to a given character, these copies must *separate* in gamete formation. How else could two *Yy* parents ever give rise to *yy* (or *YY*) progeny, unless the *Yy* elements could first separate (in the parents) and then recombine in different ways (in the offspring)? Thus did Mendel derive his insight, sometimes called Mendel's First Law or the **Law of Segregation**: Differing characters in organisms result from two genetic elements (alleles) that separate in gamete formation, such that each gamete gets only one of the two alleles. As noted, the physical basis for this law, which Mendel knew nothing of, is the separation of homologous chromosomes during meiosis.

Homozygous and Heterozygous Conditions

It's time to add a couple more terms to the concepts you have been considering. There is scientific terminology for the genotypically "pure" and "mixed" organisms noted earlier. An organism that has two identical alleles of a gene for a given character is said to be **homozygous** for that character (as with *YY* or *yy*). An organism that has differing alleles for a character is said to be **heterozygous** for that character (as with *Yy*). You often see homozygous used in combination with the terms dominant and recessive. For example, a *yy* plant is a *homozygous recessive*, while a *YY* is a *homozygous dominant* (**see Figure 11.8**). Knowledge of these terms puts you in a position to understand formal definitions of dominant and recessive. **Dominant** is a term used to desig-

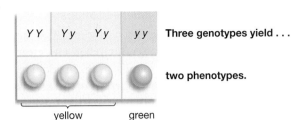

Three genotypes yield . . .

two phenotypes.

yellow   green

Figure 11.7
Three Genotypes Yield Two Phenotypes
The two alleles for seed color (*Y* = yellow and *y* = green) can result in three genotypes (*YY*, *Yy*, *yy*), but these can yield only two phenotypes (yellow and green).

nate an allele that is expressed in the heterozygous condition. In a heterozygous pea plant (*Yy*), the yellow allele (*Y*) is expressed, meaning it is dominant over the green allele (*y*). **Recessive** is a term used to designate an allele that is not expressed in the heterozygous condition. The green allele (*y*) is recessive because it is not expressed when it exists heterozygously with the yellow allele (*Yy*).

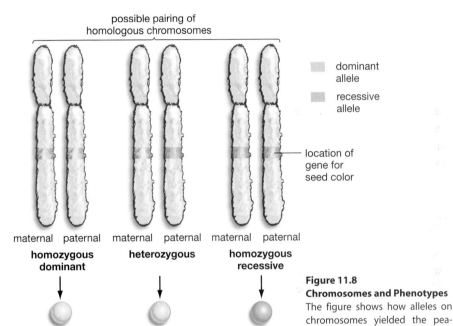

possible pairing of homologous chromosomes

dominant allele

recessive allele

location of gene for seed color

maternal   paternal   maternal   paternal   maternal   paternal

homozygous dominant   **heterozygous**   **homozygous recessive**

yellow seeds   yellow seeds   green seeds

Figure 11.8
Chromosomes and Phenotypes
The figure shows how alleles on chromosomes yielded the pea-color phenotypes that Mendel observed.

11.5 Crosses Involving Two Characters

Thus far, the subject has been how pea plants come to differ in *one* of their characters, seed color. When people breed organisms for a single difference such as this, looking to see how the offspring will come out, the procedure is known as a **monohybrid cross**. Mendel, however, went on to ask: What happens if you breed organisms for *two* characters? What happens, in other words, if you undertake what is known as **dihybrid cross**? The question Mendel was seeking to answer in carrying out his dihybrid crosses was: Do given characters travel together in heredity, or do they travel separately? If you get one character passed on to a plant, do you necessarily get the other?

Crosses for Seed Color *and* Seed Shape

One dihybrid cross that Mendel performed involved, for its first character, the yellow and green seeds you've become so familiar with. In addition, this cross involved a second character of these seeds: their shape, which can be smooth or wrinkled (as you saw in Figure 11.4). It's clear that yellow color is dominant to green and, as Table 11.1 shows, smooth seed shape is dominant to wrinkled. We have been denoting the alleles for seed color as Y for yellow and y for green. In the same fashion, let us now denote seed-shape alleles as S for smooth and s for wrinkled.

You can see, in **Figure 11.9a**, what the phenotypic outcomes were for these crosses from the P to F_1 gen-

erations. In the F_1 generation, all the seeds were smooth and yellow. When the F_1s were self-pollinated, however, the F_2 phenotypes that resulted came out as 315 smooth yellow seeds; 108 smooth green seeds; 101 wrinkled yellow seeds; and 32 wrinkled green seeds. These numbers work out to a 9:3:3:1 ratio, meaning 9 parts smooth yellow; 3 parts smooth green; 3 parts wrinkled yellow; and 1 part wrinkled green.

A Hidden, Underlying Ratio

For Mendel, these imposing figures represented another opportunity to perceive something about how the black box of genetics operated. He realized early on that this ratio might be a composite, hiding an underlying reality for each of the *single* characters

**Figure 11.9
Phenotype Ratios in a Dihybrid Cross**

(a) Results of Mendel's dihybrid cross

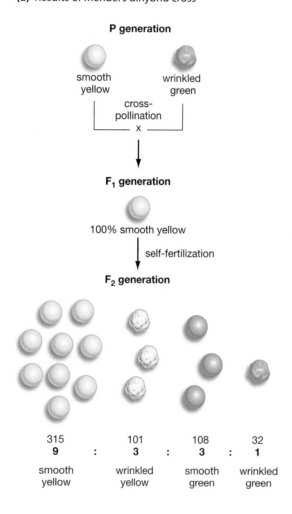

In one of his dihybrid crosses, Mendel cross-bred plants that had smooth yellow seeds with those that had green wrinkled seeds. The result was a generation of plants that all had smooth yellow seeds. When these plants self-fertilized, the result was an F_2 generation that had the phenotypes shown in a 9:3:3:1 ratio.

(b) Why Mendel got these results

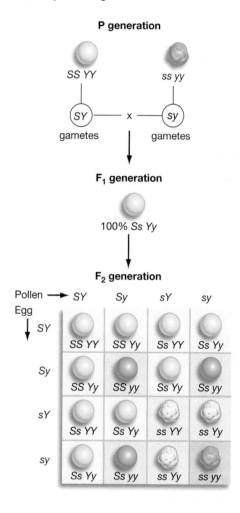

The Punnett square demonstrates why Mendel got the 9:3:3:1 phenotypic ratio in his dihybrid cross. Nine combinations yield smooth yellow seeds, 3 yield smooth green seeds, 3 yield wrinkled yellow seeds, while only 1 results in a wrinkled green seed.

he was studying. Think for a second, as Mendel did, of only one of these characters, the color of the seeds. Here, the result is:

315 (smooth) yellow seeds   108 (smooth) green seeds
101 (wrinkled) yellow   32 (wrinkled) green
416 yellow seeds   140 green seeds

This equals a yellow:green ratio of about 3:1. In other words, the familiar 3:1 ratio in the F_2 generation still holds. It held as well when Mendel looked only at seed *shape*. This suggested something to Mendel that turned out to be another of his major insights: that characters—in this case, seed shape and seed color—are transmitted *independently* of one another. When these two characters were being crossed at the same time, the results for each were the same as when they were crossed as single characters in the earlier experiments. Thus, one character's transmission did not appear to affect the other's.

Understanding the 9:3:3:1 Ratio

Still, what's the basis of the formidable 9:3:3:1 ratio that Mendel got in his dihybrid cross, with its mixture of wrinkled yellows, smooth greens, and so forth? This is simply a more complex example of the kind of outcomes you saw earlier with the aid of the Punnett square. If you look at the square in **Figure 11.9b** and start adding up internal squares *by phenotype*, Mendel's 9:3:3:1 ratio starts making sense. You can see, for example, where the 1 in the ratio comes from: There is only 1 part wrinkled green seeds, because it is only the *sy* (pollen) and *sy* (egg) combination that yields this phenotype. Likewise, you will get three parts wrinkled yellow seeds by adding up the three possible male and female gamete combinations that could bring this about.

The Law of Independent Assortment

This outcome could only come about, however, if Mendel's fundamental insight about dihybrid crosses was correct: that characters were being transmitted independently of one another. If one character had affected another's transmission, these ratios would have been very different. This insight of Mendel's is now known as Mendel's Second Law, or the **Law of Independent Assortment**. It states that during gamete formation, gene pairs assort independently of one another.

Independent Assortment and Chromosomes

Now recall that the underlying physical basis for this law was set forth in Chapter 10: In meiosis, pairs of homologous chromosomes *assort independently* from one another at the metaphase plate. It may be, for example, that paternal chromosome 5 will line up on side A of the metaphase plate; if so, then maternal chromosome 5 ends up on side B. For chromosome 6, however, things could just as easily be reversed: The *maternal* chromosome might end up on side A, and the paternal on side B (see Figure 10.2, page 194.) The genes for Mendel's seed color exist on the plant's chromosome 1, while the genes for seed shape exist on its chromosome 7. Because these are separate chromosomes, they assort independently at the metaphase plate, meaning they are passed on independently to future generations.

11.6 Reception of Mendel's Ideas

When Mendel finished his experiments, he delivered two lectures on them to a local scientific society in 1865. It would be nice to report that the society members' jaws dropped open once Mendel let them in on how inheritance worked in the living world, but no such thing happened. His findings were ultimately published in a scientific journal that was distributed in Germany, Austria, the United States, and England. Mendel even took it upon himself to get 40 reprints of his paper, thereafter sending some out to various scientists in an effort to spark some exchange on his findings. All to no avail. His work sank nearly without a trace, finally to be rediscovered in 1900, 16 years after his death. Today, the consensus within the scientific community is that nobody cared about Mendel's findings in his own time simply because nobody grasped their significance (see "Why So Unrecognized?" on the next page).

This poor early reception notwithstanding, Mendel's insights have stood the test of time. For certain kinds of phenotypes, his rules of inheritance are extremely reliable. To this day, biologists will begin an observation by noting that "Mendelian rules tell us that …" or "Mendelian inheritance operates here." Even allowing that he had intellectual predecessors, the fact is that Mendel did not just *add* to the discipline of genetics; he founded it.

11.7 Incomplete Dominance

Mendel knew as well as anyone that the rules he discovered did not apply in all instances of inheritance. In his peas, crossing white- and purple-flowered plants resulted in an F_1 generation in which all flowers were purple. In some other

Applying Your Knowledge

1. Stands of aspen trees often are a series of genetically identical individuals, with each succeeding tree growing from the severed shoot of another tree. Using what you've learned of genetics in this chapter, would you expect one aspen tree in a stand to differ greatly from another in its phenotype? Would you expect each to look exactly like the next in terms of phenotype?

2. The ABO blood-typing system mentioned in the text has great practical significance because the human immune system will attack blood-cell surface proteins that it recognizes as "foreign"— that is, proteins that a person does not have on his or her own blood cells. Given this, what is the "universal donor" blood type? What is the "universal recipient" type?

3. If, as the text states, the effects of genes can be expected to vary in accordance with the environment in which these genes work, would you expect this phenomenon to be applicable to human genes, such as those that help produce such traits as height, weight, or introversion and extroversion?

Genetics Problems

One of Mendel's many monohybrid (single character) crosses was between a true-breeding (homozygous) parent bearing purple flowers and a true-breeding (homozygous) parent bearing white flowers. All the offspring had purple flowers. When these offspring were allowed to self-fertilize, Mendel observed that their offspring occurred in the ratio of 3 purple-flowered:1 white-flowered plants. The allele determining purple flowers is symbolized P and the allele determining white flowers is symbolized p. Use this information to answer the following questions.

1. What are the genotypes of the parents?
   a. PP; pp
   b. PP; Pp
   c. Pp; Pp
   d. pp; pp

2. Which generation has only purple-flowered plants?
   a. P
   b. F_1
   c. F_2
   d. F_3

3. Which generation tells you which of the traits is dominant?
   a. P
   b. F_1
   c. F_2
   d. F_3

4. How many phenotypes are present in the F_2 generation?
   a. 1
   b. 2
   c. 3

5. How many different genotypes are present in the F_2 generation?
   a. 1
   b. 2
   c. 3

6. Mendel allowed the white-flowered F_2 plants to self-fertilize. What proportion of these F_2 plants produced only white-flowered offspring?
   a. none
   b. $1/4$
   c. $1/3$
   d. $3/4$
   e. all

7. Mendel allowed the purple-flowered F_2 plants to self-fertilize. What proportion of these F_2 plants produced only purple-flowered offspring?
   a. none
   b. $1/4$
   c. $1/3$
   d. $3/4$
   e. all

8. If a heterozygous purple-flowered pea plant is crossed with a homozygous white-flowered plant, what proportion of offspring will be white-flowered?
   a. none
   b. $1/4$
   c. $1/2$
   d. $3/4$
   e. all

9. Given what you know about the alleles that determine purple and white color in pea flowers, why is it possible to have a purple-flowered heterozygote, but not a white-flowered heterozygote?

10. In considering flower color in pea plants, we are dealing with _____ gene(s) and _____ allele(s).
    a. 1; 1
    b. 1; 2
    c. 1; 4
    d. 2; 1
    e. 2; 2

11. The next three questions require you to apply rules of probability.
    a. In a toss of two coins, what is the chance of obtaining two heads?
       (1) 0
       (2) 12.5 percent
       (3) 25 percent
       (4) 50 percent
       (5) 75 percent
    b. What rule of probability was used to determine the chance of obtaining two heads?
    c. Now, use this rule to solve the following problem. In a cross between two purple-flowered heterozygotes 75 percent of their offspring are expected to be purple-flowered. In a cross between two heterozygotes with yellow-colored seeds, 25 percent of their offspring are expected to be green. If two purple-flowered, yellow-seeded individuals, each heterozygous for both genes, are crossed, what proportion of their offspring are expected to be purple-flowered with green seeds?

12. By observing flower color in snapdragons, is it possible to unambiguously determine the genotype? Is the same true for flower color in peas? Why or why not?

13. If the alleles for snapdragon flower color were codominant, instead of incompletely dominant, a possible phenotype for a heterozygous individual would be:
    a. pure red flowers
    b. pure pink flowers
    c. pure white flowers
    d. a checkerboard pattern of red and white regions on flowers

14. If two individuals with AB blood type marry, what proportion of their children are expected to have AB blood type?
    a. none
    b. 0.25
    c. 0.50
    d. 0.75
    e. all

MediaLab

Where Did I Get This Nose? Understanding Mendelian Genetics

Can you roll your tongue? Is there a cleft in your chin? What is the color of your eyes? These and countless other characteristics that make you a unique individual were determined by the traits you inherited from your parents. In this *MediaLab*, you'll firmly establish your understanding of the rules governing the inheritance of simple genetic traits. In the *Web Tutorial*, you will work through these rules using some of Gregor Mendel's original genetic studies of peas. In the *Web Investigation*, you can practice these rules of probability with your own family, and make predictions about the inheritance of these traits in the *Communicate Your Results* section.

The online version of this *MediaLab* can be found in Chapter 11 on your Companion Website (www.prenhall.com/krogh3).

WEB TUTORIAL

It is obvious to most people that physical traits are inherited in families. But the rules that govern them were a mystery until a nineteenth-century Augustinian friar, Gregor Mendel, discovered the pattern behind inheritance of traits, by conducting breeding experiments on garden peas. With this *Web Tutorial*, you can review Mendel's experiments in order to lay the foundation for applying these rules to human genetics.

Activity

1. *First, you will examine how different versions of a trait (like green versus yellow seed color) are passed from parents to offspring, and how rules of probability can explain the outcome.*

2. *Then you will learn to use a Punnett square, a chart that shows all possible combinations of alleles in the offspring of any parents, allowing you to predict the chances of inheriting one or more different traits.*

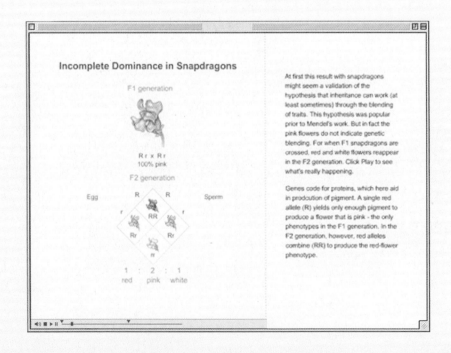

Incomplete Dominance in Snapdragons

F1 generation

R r x R r
100% pink

F2 generation

Egg   R   R   Sperm

r   RR
         Rr   Rr
                  rr

1   :   2   :   1
red   pink   white

At first this result with snapdragons might seem a validation of the hypothesis that inheritance can work (at least sometimes) through the blending of traits. This hypothesis was popular prior to Mendel's work. But in fact the pink flowers do not indicate genetic blending. For when F1 snapdragons are crossed, red and white flowers reappear in the F2 generation. Click Play to see what's really happening.

Genes code for proteins, which here aid in prodcution of pigment. A single red allele (R) yields only enough pigment to produce a flower that is pink - the only phenotypes in the F1 generation. In the F2 generation, however, red alleles combine (RR) to produce the red-flower phenotype.

Investigation 1
Estimated time for completion = 10 minutes

Most complex human genetic traits are not as simple to understand as the pea traits that Mendel studied. For example, researchers have found that children of two obese parents have an 80 percent chance of being obese. Can we conclude that it is due to their genes, or is it due to their environment? Maybe people are more likely to overeat and not exercise if their parents exhibit these behaviors. Select the keyword **OBESITY** on the Website for this *MediaLab*, to read an article on the linkage between genes and obesity.

Investigation 2
Estimated time for completion = 20 minutes

Before the introduction of DNA testing, ABO blood type was commonly used as evidence in paternity tests. Select the keywords **BLOOD TYPE** on the Website for this *MediaLab* and read a tutorial about the inheritance of blood type. Then, try your hand at answering some questions on blood type by clicking next on the keywords **ABO QUESTIONS**, and then select the blood-type questions on the genetics quiz.

Investigation 3
Estimated time for completion = 20 minutes

Understanding the inheritance of Mendelian traits is just the beginning of understanding how genes help shape our identity. Our appearance, health, and even behavior is strongly influenced by our genes. Currently under way is a multinational project, called the Human Genome Project, which is intended to uncover all of the genes located on all the human chromosomes. Select the keyword **GENOME** on the Website for this *MediaLab* to view a current map of all the human genes that have been located thus far. Select a chromosome and read as much information as you can about a human disease-causing gene that has been located there. What is the name of the gene, and what disease does it cause? Is there a genetic test available for the gene? Has identifying the gene helped create any new therapies?

Now that you have practiced some of Mendel's laws in the preceding investigations, apply and extend your abilities by addressing some genetics problems.

Exercise 1
Estimated time for completion = 15 minutes

How much of our behavior, personality, even physical appearance is due to our genes, and how much is due to our environment? Having read the evidence given in the article in *Web Investigation 1* for a genetic basis for obesity, what is your conclusion about how our weight is determined? Be prepared to discuss this with several of your peers by answering the following questions: In your mind, what was the most convincing point in the article? How did the researchers control for environmental differences versus genetic differences? Can you really assume that two individuals, even siblings, are ever raised with the same environment? How many participants would need to be included in this type of study to make it statistically significant?

Exercise 2
Estimated time for completion = 20 minutes

You just learned how to use ABO blood type to determine parentage. Now find out your own blood type and the blood type of your parents, and try to determine your family's genotypes. If you don't know your blood type, you can find it out easily when you donate blood at your local blood drive. Create a problem like the ones from the **ABO QUESTIONS** in *Web Investigation 2*, and then generate a Punnett square showing all possible blood-type genotypes for the children of your family. What is your phenotype? Genotype? If your parents had more children, what is the probability that they could donate blood for you?

Exercise 3
Estimated time for completion = 15 minutes

Prepare a 50-word description of the gene you investigated in *Web Investigation 3* to inform your classmates about the current state of human genetics research.

12 *Chromosomes and Inheritance*

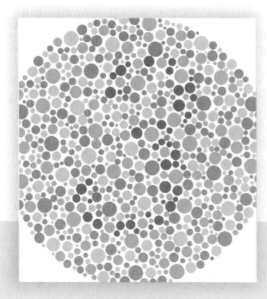

What number do you see?
(Section 12.1, page 232)

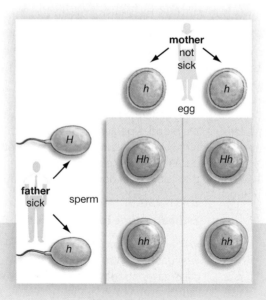

A dominant disorder.
(Section 12.2, page 234)

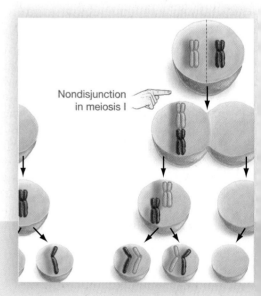

A mistake in meiosis.
(Section 12.5, page 237)

Chromosomes are critical players in the process by which living things pass on traits. Some of our worst diseases result from chromosomes that have failed to function properly.

The common condition known as Down syndrome is almost always caused by the inheritance of an extra chromosome. But how can a parent who does not have an extra chromosome pass one on to a son or daughter? The gene that causes sickle-cell anemia seems also to offer some protection against malaria. How can it have both effects? The chromosomal makeup of women offers them a kind of protection against being color blind. Why should this be so?

From these questions, it may be apparent that chromosomes and the genes on them are centrally involved in issues of human health. You saw last chapter that chromosomal activity lies at the root of several of the genetic principles that Gregor Mendel uncovered. Mendel's "pairs of elements," which separate in reproduction, had as their physical basis the pairs of homologous chromosomes—one inherited from the male, one from the female—that first pair up and then separate from one another in meiosis. The differing traits that Mendel investigated so thoroughly, such as yellow or green peas in a pod, had as their physical basis the pairs of *alleles* or differing forms of a gene that lie on homologous chromosomes.

But there are aspects of chromosomal functioning that Mendel's work scarcely touched on. For one, there is the issue of sex chromosomes. As noted in Chapter 10, the sex of human beings is determined by their chromosomal makeup. Human females have two X chromosomes, while human males have one X and one Y chromosome. But apart from conferring sex, what part do these special chromosomes play in heredity?

Next, how do the terms *dominant* and *recessive* that you looked at last chapter play out with respect to human disease? And how can genetically based diseases be traced through generations of a family?

Finally, you saw in Chapter 10 that a critical step in meiosis is the *crossing over* (or recombination) that occurs early in it. In this process, homologous chromosomes first intertwine and then swap pieces of themselves, after which they separate in meiosis. But what happens when chromosomes fail to properly carry out this part swapping or separation? The short answer is that such mistakes are responsible for a number of the afflictions that affect human beings, including Down syndrome.

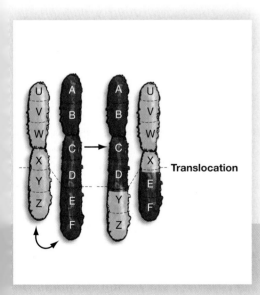

Translocation.
(Section 12.6, page 241)

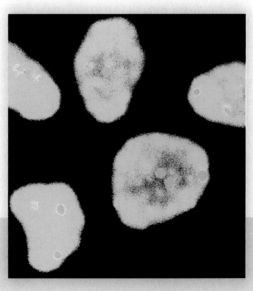

Down syndrome's signature.
(Essay, page 241)

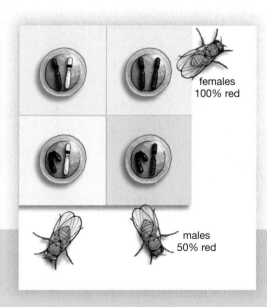

females
100% red

males
50% red

Only the males had white eyes.
(Essay, page 243)

So, in this chapter you will look at sex chromosomes, dominant and recessive disease genes, tracing diseases through generations, and diseases caused by chromosomal mistakes. All the disorders that will be reviewed are transmitted through so-called "Mendelian" inheritance, meaning they are all disorders in which a single gene or chromosome means the difference between sickness and health. In contrast, many widespread diseases, such as heart disease and cancer, are polygenic—they are based on many genes working in concert, with a given set of faulty genes only predisposing a person toward sickness, rather than ensuring it. Let's start now with sex and heredity.

12.1 X-Linked Inheritance in Humans

Minor cuts are simply an irritation to most of us, but for people suffering from the condition known as hemophilia, such cuts can be life threatening. Hemophilia is a failure of blood to clot properly. A group of proteins interact to make blood clot, but about 80 percent of hemophiliacs lack a functioning version of just one of these proteins (called Factor VIII). Genes contain the information for the production of proteins, of course, so at root hemophilia is a genetic disease—a faulty gene is causing the faulty Factor VIII protein. This same process is at work in a couple of other well-known afflictions: the disease called Duchenne muscular dystrophy, and a far less serious condition, red-green color blindness. These disorders have more in common than being genetically based, however. All of them are known as *X-linked* disorders, because the genes that cause them lie on the X chromosome. As it turns out, X-linked disorders claim more male victims than female. Why should this be so? A look at color blindness can provide the answer.

Web Tutorial 12.1
X-Linked Recessive Traits

The Genetics of Color Vision

Despite its name, "color blindness" almost never means a *total* inability to perceive color. Rather, it most often means an inability to distinguish among certain shades of red and green (**Figure 12.1**). To understand how color vision can malfunction in this way, it's important to understand how it commonly works. When light enters our eyes, it travels to the back of them and lands on a group of cells (called cones) that have within them molecules called pigments. A pigment is simply a substance that strongly absorbs a given color of light. The pigments in cone cells turn out to be molecules composed of two main parts, one of which is a protein called opsin. Because genes code for proteins, it's easy to see how color blindness is genetically based: We have genes that are coding for the protein portions of our light-absorbing pigments. More specifically, we have a gene that codes for blue-absorbing pigment, and it lies on chromosome 7; and then we have genes that code for both red- and green-absorbing pigments, and they lie very close to one another on the X chromosome. A red-green color-blind person, then, is one in whom these X-located genes fail to code for the proper pigment proteins.

Alleles and Recessive Disorders

In the normal case, as you've seen, a given gene comes in two variant forms or *alleles* in each person. One allele for a gene might reside on, say, the chromosome 1 we inherit from our mother; if so, the second allele for this gene will reside on the chromosome 1 we inherit from our *father*. There is therefore a kind of backup system in human genetics: genes come in pairs of alleles that lie on separate, homologous chromosomes. If one allele for a given trait is defective, there will usually exist a second, functional allele for that same trait on a homologous chromosome.

This genetic structure usually works fine when a condition is a so-called **recessive disorder**, meaning a genetic disorder that will not exist in the presence of a functional allele. In such a condition, a person does not have to have *two* functional alleles; a single "good" allele will do. And red-green color blindness turns out to be a recessive disorder: a person who has even one functional allele for red pigment and one functional allele for green pigment will not be color blind. So why do men suffer more from this condition than women? Because men have no backup: they have only one allele for red pigment and one allele for green pigment because they have only one *X chromosome* (and one Y). Meanwhile, women have two X chromosomes. If the pigment alleles are defective on one of their chromosomes, the alleles on their other X chromosome will still provide them with full color vision.

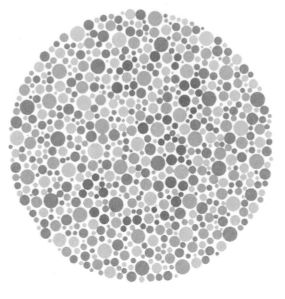

Figure 12.1
Typical Test for Red-Green Color Blindness
If you do not see a number inside the large circle, you may have this recessive trait.

Given this, think about the interesting way that color blindness is passed along. Say there is a mother who is not color blind herself, but who is heterozygous for the trait. That is, she has functional color alleles on one of her X chromosomes but dysfunctional alleles on her other X chromosome. Should her son happen to inherit this second X chromosome, he will be color blind, because the X chromosome he got from his mother is his *only* X chromosome (**Figure 12.2**). Meanwhile, a daughter who inherited this chromosome would likely be protected by her second X chromosome—the one she got from her father. Not surprisingly, then, more males are color blind than females. About 8 percent of the male population has some degree of color blindness, while for females the figure is about half a percent. Women enjoy a similar protection in connection with other X-linked disorders, such as hemophilia and muscular dystrophy.

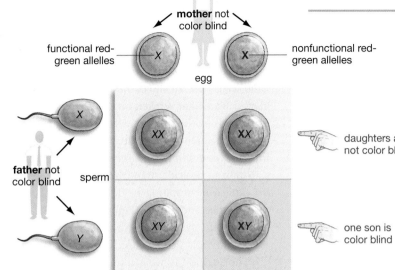

mother not color blind

functional red-green alleles

nonfunctional red-green alleles

egg

father not color blind

sperm

daughters are not color blind

one son is color blind

Figure 12.2
The Value of Having Two X Chromosomes
The mother in the figure is not herself color blind but has one dysfunctional set of red-green alleles (represented by the red X), which she passes on to one son and one daughter. The daughter is not color blind because she has inherited, from her father, a second X chromosome—one that has functional red-green alleles. Meanwhile the son is color blind because his only X chromosome is the flawed one he inherited from his mother. (The other son, shown in the square's lower-left cell, is not color blind because he has inherited his mother's functional alleles.)

12.2 Autosomal Genetic Disorders

The X and Y chromosomes are, of course, only two of the chromosomes in the human collection, and any chromosome can have a malfunctioning gene on it. The chromosomes other than the X and Y chromosomes are called *autosomes*, as you've seen. A recessive dysfunction related to an autosome is thus known as an **autosomal recessive disorder**. (For a list of all the disorders you'll be looking at and more, see **Table 12.1**.)

Web Tutorial 12.2
Some Human Genetic Disorders

Table 12.1 Selected Examples of Human Genetic Disorders

Type	Name of condition	Effects
X-linked recessive disorders	Hemophilia	Faulty blood clotting
	Duchenne muscular dystrophy	Wasting of muscles
	Red-green color blindness	Inability to distinguish shades of red from green
Autosomal recessive disorders	Albinism	No pigmentation in skin
	Sickle-cell anemia	Decreased oxygen to brain and muscles
	Cystic fibrosis	Impaired lung function, lung infections
	Phenylketonuria	Mental retardation
	Tay-Sachs disease	Nervous system degeneration in infants
	Werner syndrome	Premature aging
Autosomal dominant disorders	Polydactyly	Extra fingers or toes
	Campodactyly	Inability to straighten little finger
	Huntington disease	Brain tissue degeneration
Aberrations in chromosome number	Down syndrome	Mental retardation, shortened life span
	Turner syndrome	Sterility, short stature
	Klinefelter syndrome	Dysfunctional testicles, feminized features
Aberrations in chromosome structure	Cri-du-chat syndrome	Mental retardation, malformed larynx
	Fragile-X syndrome	Mental retardation, facial deformities

Sickle-Cell Anemia

A well-known example of an autosomal recessive disorder is sickle-cell anemia, which affects populations derived from several areas on the globe, including Africa. In the United States it is, of course, most widely known as a disease affecting African Americans. The "sickle" in the name comes from the curved shape that is taken on by the red blood cells of its victims. Red blood cells carry oxygen to all parts of the body; in their normal shape they look a little like doughnuts with incomplete holes. When they take on a sickle shape, however, red cells clog up capillaries, resulting in decreased oxygen supplies to brain and muscle (**Figure 12.3**). In the United States, the average life expectancy for men with the condition is 42 years; for women, it is 48 years.

The question is, what causes a red blood cell to take on this lethal, sickled shape? There is a protein, called hemoglobin, that carries the oxygen within red blood cells. The vast majority of people in the world have one form of hemoglobin, called hemoglobin A; but sickle-cell anemia sufferers have another form of this protein, hemoglobin S, which coalesces into crystals that distort the cell.

Now, how is sickle-cell anemia passed from parents to offspring? Because it is a recessive condition, any child who gets it must inherit *two* alleles for it—one allele from the mother, one from the father, each of them coding for the hemoglobin S protein. This means, of course, that both parents must themselves have at least one hemoglobin S allele; both parents

must be at least heterozygous for the trait. In this situation, the laws governing inheritance of this disorder are simply those of Mendel's monohybrid cross: a child of these two parents has a 25 percent chance of inheriting the disorder. You can look at the Punnett square in **Figure 12.4a** to see how this works out.

It's important to note the status of the parents in this example. They are heterozygous for a recessive debilitation, meaning they do not suffer from the condition themselves because they are protected by their single "good" allele. Each of them is, however, a **carrier** for the condition: a person who does not suffer from a recessive genetic debilitation, but who carries an allele for it that can be passed along to offspring. In this respect, they are just like the mother in the color-blindness example.

Malaria Protection: Hemoglobin S as a Useful Protein

Sickle-cell anemia offers another lesson in connection with the notion of hemoglobin S as a "faulty" protein. It is in fact quite a functional protein in certain circumstances; namely, when it appears heterozygously (with hemoglobin A) in people who live in regions where malaria is common. The most severe form of malaria is caused by a single-celled parasite that is transmitted into human beings by the *Anopheles* mosquito. Traveling through the bloodstream, these parasites invade and ultimately destroy red blood cells. For reasons we still don't understand, having hemoglobin S in one's system makes red blood cells resistant to invasion. Thus, it is likely that hemoglobin S became as widespread as it is because of its value in resisting malaria. The price of this resistance, however, is that some offspring are not heterozygous for hemoglobin S. They are homozygous for it, meaning they have sickle-cell anemia.

Dominant Disorders

Though sickle-cell anemia is an autosomal disorder and red-green color blindness an X-linked disorder, both are recessive disorders. That is, a person with even a single properly functioning allele will not suffer from them. However, there are also **dominant disorders**: genetic conditions in which a single faulty allele can cause damage, even when a second, functional allele exists. This leads to the concept of an **autosomal dominant disorder**, simply meaning a dominant genetic disorder caused by a faulty allele that lies on an autosomal chromosome. There is, for example, an autosomal dominant disorder called Huntington disease. Affecting about 30,000 Americans, it results in both mental impairment and uncontrollable spastic

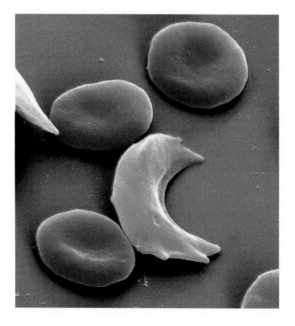

Figure 12.3
Healthy Cells, Sickled Cell
Three normal red blood cells flank a single sickled cell. ($\times$ 7400)

(a) Sickle-cell anemia: transmission of a recessive disorder.

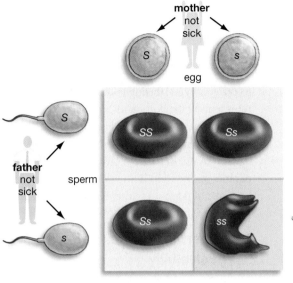

Sickle-cell anemia is a recessive autosomal disorder; both the mother and father must carry at least one allele for the trait in order for a son or a daughter to be a sickle-cell victim. When both parents have one sickle-cell allele, there is a 25 percent chance that any given offspring will inherit the condition.

☞ 25% probability of inheriting the disorder

**Figure 12.4
Transmission of Recessive and
Dominant Disorders**

(b) Huntington disease: transmission of a dominant disorder.

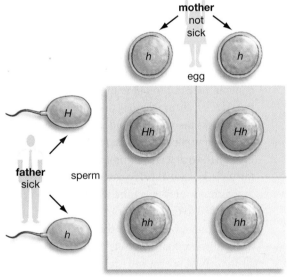

☞ 50% probability of inheriting the disorder

In Huntington disease, if only a single parent has a Huntington allele there is a 50 percent chance that a son or daughter will inherit the condition.

movements called *chorea*. Perhaps its most famous victim was the American folksinger Woody Guthrie. Like most Huntington sufferers, Guthrie did not begin to show symptoms of the disease until well into adulthood—*after* he had children, who then may have had the disease passed along to them. You can look at the Punnett square in **Figure 12.4b** to see how the inheritance pattern of an autosomal dominant disease, such as Huntington, differs from that of an autosomal recessive illness. Because a parent need only pass on a single Huntington allele for a son or

daughter to suffer from the condition, the chances of any given offspring getting the disease from a single affected parent are one out of two.

12.3 Tracking Traits with Pedigrees

Confronted with a medical condition, such as Huntington, that is running through a family, scientists find it helpful to construct a medical **pedigree**, defined as a familial history. Normally set forth as diagrams, medical pedigrees do more than give a

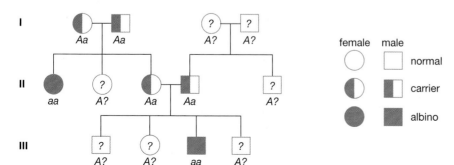

family history of a disease. They can be used to ascertain whether a condition is dominant or recessive, and X-linked or autosomal. This can help establish probabilities for *future* inheritance of the condition—something that can be very helpful for couples thinking of having a child.

If you look at **Figure 12.5**, you can see a simple pedigree for albinism: a lack of skin pigmentation, which is known to be an autosomal recessive condition. In the figure, you can see some of the standard symbols used in pedigrees. A circle is used for a female and a square for a male. Parents are indicated by a horizontal line connecting a male and a female, while a vertical line between the parents leads to a lower row that denotes the parents' offspring. This second row—a horizontal line of siblings—has an order to it: oldest child on the left, youngest on the right. A circle or square that is filled in indicates a family member who has the condition (in this case albinism), while a symbol that is half filled in indicates a person known to be heterozygous for a recessive condition (a carrier). One of the strengths of a pedigree is that it can sometimes tell researchers which persons are heterozygous carriers of a recessive condition. Remember that a carrier does not display *symptoms* of the condition. So how can a pedigree reveal this?

Take a look at Figure 12.5. The only thing that would be apparent from actually seeing any of the people in the pedigree is that two of them—a female in generation II and a male in generation III—have the condition of albinism. But a little knowledge of Mendelian genetics also allows some other deductions. If the condition had been dominant, then it would have manifested itself in at least one of the parents on the left in generation I. Because this was not the case, it is fair to deduce that this is a recessive condition. The fact that it is recessive, however, means that *both* parents in generation I had to be carriers for the allele—both have to be heterozygous for the condition, which is why their symbols can be half shaded in. Things are less clear with the parents on the right in generation I. One of their sons did not manifest the condition himself, but went on to have a son who did.

From this, we know that the son in generation II had to have been heterozygous for the condition. But from which parent did this son get his albinism allele? Either, or both, of his parents in generation I could have been heterozygous for the condition and yet between them, only have passed along a single albinism allele.

This mixture of certainty and uncertainty leads to the genotype labeling you see in the figure, with *A* representing the dominant "normal" allele and *a* representing the recessive albinism allele. We know, for example, that both parents on the left in generation I had to be *Aa*, but all we can say for sure about the parents on the right in the same generation is that at least one of them was a carrier.

12.4 Aberrations in Chromosomal Sets: Polyploidy

All of the maladies you've looked at so far have resulted from dysfunctional genes that exist among a standard array of chromosomes. In humans this array is 22 pairs of autosomes and either an XX combination (in females) or an XY (in males). Meanwhile, Mendel's peas had 7 pairs of chromosomes, and the common fruit fly has 4 pairs. Whatever the *number* of chromosomes, note the similarity in all these species: Their chromosomes come in pairs; or to put it another way, they all have two *sets* of chromosomes, meaning the organisms that posses them are diploid.

In many situations, however, organisms don't end up with the two sets of chromosomes that are standard for their species. This is called **polyploidy**: a condition in which one or more entire sets of chromosomes has been added to the genome of a diploid organism.

You may wonder how a human being could end up with more than two sets of chromosomes. After all, in meiosis, sperm and egg end up with just one set of chromosomes in them. When they later fuse, in the moment of conception, the result is then two sets of chromosomes in the fertilized egg. But what if *two* sperm manage to fuse with one egg? The result is a fertilized egg with three sets of chromosomes, and the outcome is polyploidy. Normally an egg lets in only one sperm, and it blocks the others hovering about with a kind of chemical gate-slamming mechanism. But occasionally more than one sperm manages to make it in. Human polyploidy can also get started before fertilization, when meiosis malfunctions and gives a single egg or sperm two sets of chromosomes, meaning there will be three sets in total when egg and sperm fuse.

Polyploidy actually is tolerated well in some organisms. Indeed, it has come about countless times in

**Figure 12.5
A Hypothetical Pedigree for Albinism through Three Generations**

plant species and generally gives rise to perfectly robust plants. In human beings, however, polyploidy is a disaster. Perhaps only 1 percent of human embryos with the condition will survive to birth, and none of these babies lives long. Concerns about chromosome number in living persons, then, center not on addition or deletion of whole sets of chromosomes, but rather on the gain or loss of *individual* chromosomes.

12.5 Incorrect Chromosome Number: Aneuploidy

The condition called **aneuploidy** is one in which an organism has either more or fewer chromosomes than normally exist in its species' full set. Put another way, aneuploidy is a condition in which an organism has one chromosome too many, or one too few. Among human genetic malfunctions, aneuploidy is unusual in that it occurs quite commonly and yet goes largely unrecognized. This is so because it most often occurs not in fully formed human beings, but in embryos. A would-be mother may know only that she is having a hard time getting pregnant. What she may not know is that she actually *has* been pregnant—perhaps several times—but that aneuploidy has doomed the embryo in each case. Doomed it in what way? The most common outcome of aneuploidy is a pregnancy that ends in miscarriage. And many of these miscarriages happen so early in a pregnancy that a woman never realizes she was pregnant; the microscopic embryo may simply be part of material that is discharged, perhaps with

the woman's next menstrual period. Even when a miscarriage is recognized as such, tests can seldom be run that can identify aneuploidy as the cause. But by undertaking careful examinations of thousands of pregnancies, scientists have concluded that aneuploidy is responsible for about 30 percent of miscarriages. And how common is miscarriage? One 1988 study found that 31 percent of embryos miscarry after they have been implanted in the woman's uterus. A small proportion of embryos do manage to survive aneuploidy, but even in these instances there are consequences. Down syndrome is one of the outcomes of aneuploidy.

Aneuploidy's Main Cause: Nondisjunction

What brings about the harmful condition of aneuploidy? The cause usually is a phenomenon known as **nondisjunction**, which simply means a failure of homologous chromosomes or sister chromatids to separate during meiosis. If you look at **Figure 12.6**, you can see how this works. Note that it can occur either in meiosis I or in meiosis II. In meiosis I, two homologous chromosomes, on the opposite side of the metaphase plate, can be pulled to the *same* side of a dividing cell, producing daughter cells with imbalanced numbers of chromosomes. Conversely, nondisjunction can take place in meiosis II, by means of sister chromatids going to the same daughter cell after failing to "disjoin" (hence the cumbersome term *nondisjunction*). Such actions then produce an egg or sperm that has 24 or 22 chromosomes instead of the standard haploid number of 23; after union with a normal egg or sperm, this re-

Figure 12.6
A Mistake in Meiosis Brings about an Abnormal Chromosome Count
Nondisjunction can occur either in meiosis I or meiosis II, when either chromosomes or chromatids fail to separate properly. When it occurs in meiosis I, 100 percent of the resulting gametes will be abnormal; when it takes place in meiosis II, only 50 percent will be abnormal.

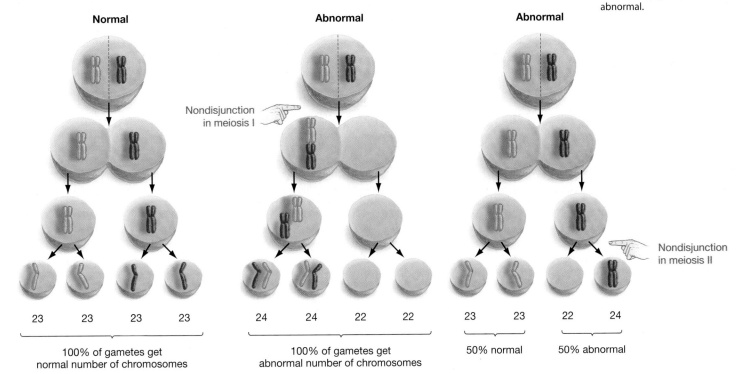

| Normal | Abnormal | Abnormal |

Nondisjunction in meiosis I

Nondisjunction in meiosis II

| 23 | 23 | 23 | 23 | 24 | 24 | 22 | 22 | 23 | 23 | 22 | 24 |

100% of gametes get normal number of chromosomes

100% of gametes get abnormal number of chromosomes

50% normal   50% abnormal

sults in an embryo that has either 47 or 45 chromosomes, instead of the standard 46.

The effect of this is that the embryo has a genetic imbalance—one that perhaps can be made clear by an analogy. Imagine that while baking chocolate chip cookies, you decided to increase the amount of flour called for in the recipe. This change might be of little consequence if you increased all the other ingredients in the same proportion. If, however, you increased nothing but the flour, the final product would be hard and flavorless because the ingredients in it would be unbalanced. An added chromosome has this same effect: it unbalances the proportions of *biological* ingredients (the proteins) that help produce and maintain a living thing.

The Consequences of Aneuploidy

As noted, embryos with the wrong number of chromosomes are not likely to survive. Setting aside aneuploidies that affect the X and Y chromosomes, *all* aneuploidies result in miscarriage except for those that give an embryo an additional chromosome 13, 18, or 21. It is the gain of an additional chromosome 21 that causes the most well known outcome of aneuploidy in human beings. **Down syndrome** is, in some 95 percent of cases, a condition in which a person has three copies of chromosome 21, rather than the standard two. Ninety percent of these cases come about because of nondisjunction in egg formation, with the remaining 10 percent caused by nondisjunction in sperm formation. Seen in about 0.1 percent of all live births, Down syndrome results in an array of effects: smallish, oval heads, IQs that are well below normal, infertility in males, short stature and reduced life span in both sexes.

It is well known that when women pass the age of about 35, their risk of giving birth to a Down syndrome

child increases dramatically. It seems less well known that even at maternal age 40, the odds of conceiving such a child are less than 1 in 100 (**Figure 12.7**). Scientists are not certain why the mother's age should figure so prominently in Down syndrome, though several hypotheses have been put forward to account for this. Most scientists agree that one piece of the puzzle is something you saw in Chapter 10: that in egg development, the process of meiosis generally stretches on for decades. Any given egg that develops in a woman starts meiosis I before the woman is born, but it may not complete this meiosis until she is, say, in her thirties. In the intervening decades, this egg's cellular machinery has aged, perhaps to the point that it can no longer separate its chromosomes or chromatids properly.

Abnormal Numbers of Sex Chromosomes

Aneuploidy can affect not only autosomes but sex chromosomes as well, meaning the two X chromosomes that females have and the X and Y chromosomes that males have. Embryos often survive sex chromosome aneuploidies, but the result, once again, is usually debilitations. An example is Turner syndrome, which produces people who are phenotypically female, but who have only one X chromosome. Such females, then, have only 45 chromosomes, rather than the usual 46. Their state is sometimes referred to as XO, the "O" signifying the missing X chromosome. The absence of a second sex chromosome causes a range of afflictions. Females with Turner syndrome have ovaries that don't develop properly (which causes sterility), they are generally short, and they often have brown spots (called nevi) over their bodies.

Turner syndrome results from a loss of a chromosome, but all manner of sex chromosome *additions* can also take place. There are, for example, XXY men, who while phenotypically male in most respects, tend

(a) Living with Down syndrome

(b) Maternal age and Down syndrome risk

Mother's age	Chances of giving birth to a child with Down syndrome
20	1 in 1925
25	1 in 1205
30	1 in 885
35	1 in 365
40	1 in 110
45	1 in 32

Figure 12.7
Down Syndrome: Increasing Risk with Age
(a) An adolescent girl with Down syndrome works with her teacher.
(b) The risk of giving birth to a Down syndrome child increases dramatically past maternal age 35.

to have a number of feminine features: some breast development, a more feminine figure, and lack of facial hair. When coupled with other characteristics (such as tall stature and dysfunctional testicles), the result is a condition called Klinefelter syndrome.

Beginning in the late 1960s, it became possible to test developing fetuses for genetic abnormalities. In recent years, however, reproductive technology has moved beyond testing. Prospective parents can now choose, from a group of embryos they have created, embryos that do not have any recognizable genetic defects. You can read more about this in "PGD: Screening for a Healthy Child" on the next page.

12.6 Structural Aberrations in Chromosomes

Aberrations can occur *within* a given chromosome, sometimes resulting from interactions *between* chromosomes. A frequent cause of such change is that pieces of chromosomes can break off from the main chromosomal body. Such a chromosomal fragment may then be lost to further genetic activity, or it may rejoin a chromosome—either the one it came from or another—often to harmful effect. These structural aberrations may come about spontaneously, but they may also be caused by exposure to such agents as radiation, viruses, and chemicals.

Deletions

A chromosomal **deletion** occurs when a chromosome fragment breaks off and then does not rejoin any chromosome. Such an event might take place during meiosis in a parent whose own somatic cell chromosomes were perfectly normal. In such an instance, a healthy parent could pass along a complement of 23 chromosomes, one chromosome of which would be missing a segment. If the deleted piece were large enough, the zygote would likely not survive. You can see how critical even a portion of a chromosome is by looking at children who suffer from a rare condition, called *cri-du-chat* ("cry of the cat") syndrome, which results from the deletion of the far end of the "short arm" of chromosome 5. If you look at **Figure 12.8**, you can see a **karyotype**—a visual image of a chromosome set—that shows you this deletion. Children born with the cri-du-chat condition exhibit a host of maladies, among them mental retardation and an improperly constructed larynx that early in life produces sounds akin to those of a cat.

Inversions and Translocations

When a chromosome fragment rejoins the chromosome it came from, it may do so with its orientation "flipped," so that the fragment's chemical sequence is out of order. This is an inversion: a chromosomal abnormality that comes about when a chromosomal fragment that rejoins a chromosome does so with an inverted orientation (see **Figure 12.9** on page 241). A translocation is a chromosomal abormality that occurs when two chromosomes that are not homologous exchange pieces, leaving both with improper gene sequences, which can have phenotypic effects.

Duplications

There is, of course, a perfectly normal process of chromosomal part swapping—crossing over. But consider what might happen if two homologous chromosomes exchanged *unequal* pieces of themselves in crossing over. One would lose genetic material while the other would gain it. Because these are homologous (meaning paired) chromosomes, the fragment added to the latter chromosome would duplicate some of the material it already has. This is but one of the ways in which

Figure 12.8
Chromosomal Deletion
(a) A five-year-old boy affected by cri-du-chat syndrome, which results from a chromosomal deletion. Note the small head and low-set ears.
(b) Karyotype of a person with cri-du-chat syndrome. Note, in chromosome 5, the abnormality in the homologous chromosome on the right. It lacks a section that is present in the left chromosome.

(a) Cri-du-chat syndrome

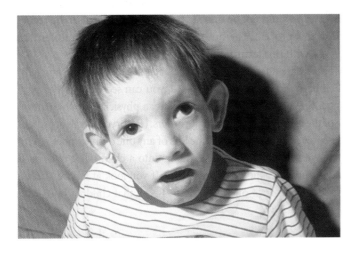

(b) Cri-du-chat karyotype

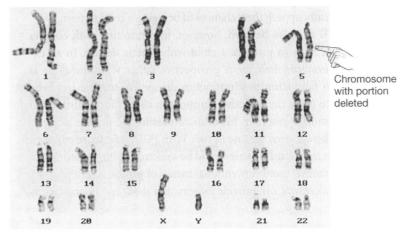

Chromosome with portion deleted

How Did We Learn?
Thomas Hunt Morgan: Using Fruit Flies to Look More Deeply into Genetics

Gregor Mendel used pea plants in his work, but generations of researchers after him have employed a common fruit fly, *Drosophila melanogaster*, in trying to unravel the secrets of genetics. One of the earliest *Drosophila* workers was a Kentuckian named Thomas Hunt Morgan, who greatly deepened our understanding of heredity with work he began in 1908 in his lab at Columbia University.

The *Drosophila* fly has a lot going for it as an experimental subject: It is small, it has unusually large chromosomes, and it has what is known as a short generation time. Whereas Mendel got one generation of pea plants a year, Morgan got one generation of *Drosophila* every 12 days or so. In practical terms, this meant that Morgan's group had to wait at most a month to see the results of their experiments.

Lessons from a Mutation: White-Eyed Flies

Morgan's path to discovery began with a chance event. One day he looked into one of the empty milk bottles in which he kept his flies and saw something strange: Among numerous normal or "wild-type" *Drosophila* with red eyes was a single male with white eyes (**see Figure 1**). Morgan correctly assumed that the variation he saw had come about because of a spontaneous genetic change or "mutation" in the fly. He then crossed his white-eyed male with females that he knew bred true for red eyes, and got all red-eyed flies in the F₁ generation. No surprise there; it simply looked as though red eyes were the same type of dominant trait that Mendel had seen in, for example, his yellow peas. Indeed, when Morgan went on to breed the F₁ generation—those that had mixed red and white alleles—he got a 3:1 red-to-white ratio in the F₂ generation. But the strange thing was that every white-eyed fly was a *male*. Why should that be?

(a) Eye of red-eyed *Drosophila*   **(b)** Eye of white-eyed *Drosophila*

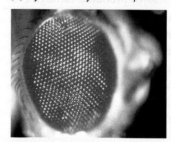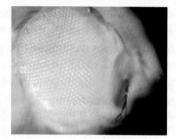

Figure 1
The Mutation Morgan Saw

The importance of Morgan's insight was that he had linked a particular trait to a particular chromosome, which was a first.

As it turns out, sex chromosomes exist in *Drosophila* as they do in human beings: Females have two X chromosomes, while males have one X and one Y. Doing further experiments on these flies, Morgan eventually decided that the gene for eye color had to lie on a *particular* chromosome: the X chromosome. Why? Any fly that was white-eyed had to have only white-eyed alleles (because white was recessive). Normally, of course, this would mean *two* white-eyed alleles, but there was an exception: Males could be white-eyed if they had but a single white allele—because they have only a *single X chromosome*. You can look at **Figure 2** to see how this plays out schematically. The importance of Morgan's insight was not that he had grasped something about eye color in the fly. It was that he had linked a particular trait to a particular chromosome, which was a first.

On to DNA

For several chapters, you've been looking at genetics from the viewpoint of the genetic packages called chromosomes. Important as chromosomes are, it's probably been apparent that the real genetic control lies with the units that help

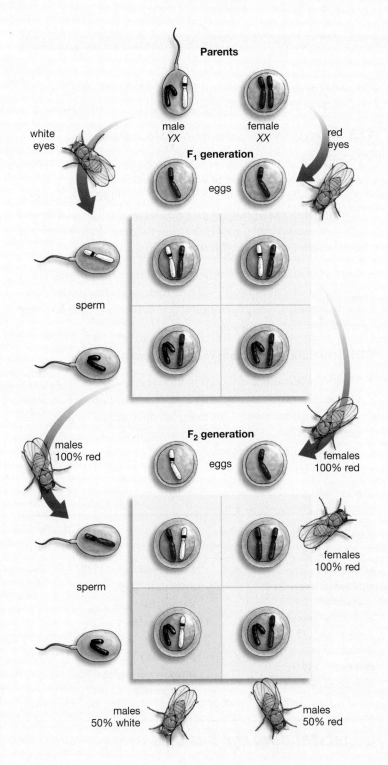

When Traits Separate: Recombination

Continuing with *Drosophila* breeding work, Morgan and his colleagues soon were finding all kinds of mutations in the flies. There was, for example, a mutation for "miniature" wings, as opposed to full ones. This trait followed the rules described earlier for the white-eyed mutation and accordingly was deemed to be an X-linked characteristic. Not surprisingly, the white-eyed and miniature-winged mutations tended to be transmitted *together* when both of these mutations were bred for, because the genes for both of them lay on the same chromosome. The question was, though, did genes on the same chromosome *always* travel together? Surprisingly, the answer was no. For example, miniature wings were usually passed on together with white eyes, but not always. Sometimes one trait would appear in an offspring but the other would not. Since genes for these traits were on one chromosome, how did they get separated from one another?

A paper published in Europe in 1909 gave Morgan a clue. Viewing cells under a microscope, F. A. Janssens had observed chromosomes entwining with each other during an early phase of meiosis. Given this and his own data, Morgan made a creative leap. He suggested that chromosomes can swap parts with each other during their meiotic intertwining. Sound at all familiar? Remember, in the Chapter 10 review of meiosis, the discussion of the phenomenon known as crossing over? Here, with Morgan, is the insight that led to this now-established tenet of science. Appropriately enough, insights such as these won Morgan a Nobel Prize in 1934.

Web Tutorial 12.3
Morgan's Flies

Figure 2
Why All Morgan's White-Eyed F₂'s Were Males
The F₂ female flies that inherited one white-eyed X-chromosome allele also inherited a second red-eyed X allele, making them red-eyed. However, the males inheriting the white-eyed allele had no second X chromosome. Instead, they inherited a Y chromosome—which has no genes on it for eye color—leaving them white-eyed.

make them up: the DNA sequences we call genes. As you've seen, it is genes that make pea-plant flowers purple or white—and that give a person Huntington disease or not. It's time to explore just what these genes are and how they operate.

Chapter Review

Summary

12.1 X-Linked Inheritance in Humans

- Certain human conditions, such as red-green color blindness and hemophilia, are called X-linked conditions, because they stem from dysfunctional genes located on the X chromosome. Men are more likely than women to suffer from these conditions because men have only a single X chromosome. A woman with, for example, a dysfunctional blood-clotting allele on one of her X chromosomes usually will be protected from hemophilia by a functional allele that lies on her second X chromosome.

- Hemophilia and red-green color blindness are examples of recessive genetic conditions, meaning conditions that will not exist in the presence of functional alleles.

- Given the nature of recessive genetic conditions, persons who do not suffer from such conditions themselves may yet possess an allele for it, which they can pass on to their offspring. Such persons, referred to as carriers of the condition, are heterozygous for it—the alleles they have for the trait differ, one of them being functional, the other being nonfunctional.
Web Tutorial 12.1: X-Linked Recessive Traits

12.2 Autosomal Genetic Disorders

- Sickle-cell anemia is an example of a recessive autosomal disorder. It is autosomal because the genetic defect that brings it about involves neither the X nor Y chromosome. It is recessive because persons must be homozygous for the sickle-cell allele in order to suffer from the condition—they must have two alleles that code for the same sickle-cell hemoglobin protein.

- Some genetic disorders are referred to as dominant disorders, meaning those in which a single allele can bring about the condition, regardless of whether a person also has a normal allele.
Web Tutorial 12.2: Some Human Genetic Disorders

12.3 Tracking Traits with Pedigrees

- In tracking inherited diseases, scientists often find it helpful to construct medical pedigrees, meaning familial histories that normally take the form of diagrams. Pedigrees allow experts to make deductions about the genetic makeup of several generations of family members.

12.4 Aberrations in Chromosomal Sets: Polyploidy

- Human beings and many other species have diploid or paired sets of chromosomes. In human beings, this means 46 chromosomes in all: 22 pairs of autosomes and either an XX chromosome pair (for females) or an XY pair (for males). The state of having more than two sets of chromosomes is called polyploidy. Many plants are polyploid, but the condition is inevitably fatal for human beings.

12.5 Incorrect Chromosome Number: Aneuploidy

- Aneuploidy is a condition in which an organism has either more or fewer chromosomes than normally exist in its species' full set. Aneuploidy is responsible for a large proportion of the miscarriages that occur in human pregnancies. A small proportion of embryos survive aneuploidy, but the children that result from these embryos are born with such conditions as Down syndrome.

- The cause of aneuploidy usually is nondisjunction, in which homologous chromosomes or sister chromatids fail to separate correctly in meiosis, leading to eggs or sperm that have one too many or one too few chromosomes.

12.6 Structural Aberrations in Chromosomes

- Harmful aberrations can occur within chromosomes, with many of these aberrations coming about because of mistakes in chromosomal interactions. Chromosomal aberrations include deletions, inversions, translocations, and duplications.
Web Tutorial 12.3: Morgan's Flies

Key Terms

aneuploidy 237	**Down syndrome** 238
autosomal dominant disorder 234	**inversion** 239
	karyotype 239
autosomal recessive disorder 233	**nondisjunction** 237
	pedigree 235
carrier 234	**polyploidy** 236
deletion 239	**recessive disorder** 232
dominant disorder 234	**translocation** 239

Understanding the Basics

Multiple-Choice Questions (answers in the back of the book)

1. Much genetic information can be derived from pedigrees (family genetic histories). Such information is especially helpful in connection with humans because
   a. humans cannot be crossed experimentally
   b. other animals have long life spans
   c. we know relatively little about human genetics
   d. humans have so many chromosomes
   e. humans are too complicated

2. Sickle-cell anemia patients have a different form of _____, which leads to _____.
   a. hemoglobin, increased muscle activity
   b. a sex chromosome, sterility
   c. pigment, color blindness
   d. skin cell, nevi
   e. hemoglobin, decreased oxygen transport

3. Individuals most protected against malaria are those who
   a. are heterozygous for sickle-cell anemia (one hemoglobin A, one hemoglobin S)
   b. are homozygous for sickle-cell anemia (two hemoglobin S)
   c. are hemophiliacs
   d. are homozygous for hemoglobin A
   e. both a and c

4. Inheritance of a dominant autosomal disorder differs from inheritance of an autosomal recessive disorder in that:
   a. A dominant disorder may be passed on only if both parents are affected.
   b. A dominant disorder is evident only if the offspring is homozygous for the allele.
   c. A dominant disorder is more often seen in females.
   d. A dominant disorder may be passed on even if only one parent is affected.
   e. both b and d

5. An individual having 44 autosomes and one X chromosome would be classified as:
   a. polyploid
   b. aneuploid
   c. having Klinefelter syndrome
   d. having Turner syndrome
   e. both b and d

6. A human embryo with 69 chromosomes would
   a. die in the womb or shortly after birth
   b. be considered polyploid
   c. have fewer problems than a plant with the same condition
   d. both a and b
   e. all of the above

7. Which of the following statements is *not* true regarding Down syndrome?
   a. Affected individuals usually have three copies of chromosome 21.
   b. The condition is usually caused by nondisjunction in egg formation.
   c. A woman over 35 increases her chance of having a Down syndrome child to 1 out of every 40 births.
   d. Approximately 0.1 percent of all live births are children with Down syndrome.
   e. The physical characteristics associated with the disorder include mental retardation, male infertility, short stature, and reduced life span.

8. Nondisjunction can occur in
   a. meiosis I or mitosis
   b. mitosis or meiosis II
   c. only meiosis II
   d. only meiosis I
   e. either meiosis I or II

9. A carrier is a person who
   a. suffers from a recessive disorder
   b. is heterozygous for a recessive disorder
   c. is homozygous for a recessive disorder
   d. does not suffer from a recessive disorder but possesses an allele for it
   e. b and d

10. What do all human males inherit from their mother?
    a. an X chromosome
    b. genes for red-green color vision
    c. two X chromosomes
    d. all of the above
    e. a and b only
    f. Turner syndrome
    g. none of the above

Brief Review

1. How can the serious genetic defect known as aneuploidy be both widespread and relatively unrecognized?

2. In sex-linked diseases, which gender is more frequently affected—male or female? Why?

3. Why can people with red-green color blindness still see other colors, such as blue?

4. Compare and contrast the four chromosome structural aberrations discussed in this chapter: deletions, inversions, translocations, and duplications.

5. Given the following situations, answer accordingly:
   a. A human cell with 47 chromosomes has probably undergone _____.
   b. A cell duplicates DNA, but fails to undergo cytokinesis. This would lead to _____.

6. Why would a person with Klinefelter syndrome exhibit both male and female features?

7. Certain genotypes may be advantageous in specific situations. Give an example of a heterozygous genotype that is advantageous over either homozygous genotype. (Hint: especially in mosquito-infested areas.) Why is this the case?

Applying Your Knowledge

1. The text notes that an alternate allele of the hemoglobin gene can cause sickle-cell anemia when a person is homozygous for this allele, but that a person who is heterozygous for the allele actually can derive a benefit from it—protection from malaria. In the United States, 8 percent of African Americans are carriers for the sickle-cell allele, while in central Africa the figure is 20 percent. What could account for this difference?

2. The text notes that the technique known as pre-implantation genetic diagnosis (PGD) allows prospective parents to screen embryos not only for serious genetic defects such as Down syndrome, but for traits that may be regarded as more or less desirable, such as male or female gender. What limits, if any, should be placed on parents' ability to choose from among embryos for traits such as this?

3. Thinking back to what you learned in Chapter 10 about how sperm are produced, why is it that the cells that give rise to sperm undergo nondisjunction less frequently than the cells that give rise to eggs?

Genetics Problems

1. A man who is a carrier for sickle-cell anemia, a recessive genetic disease, marries a normal, noncarrier woman. What proportion of their children are expected to be afflicted with sickle-cell anemia?

2. Hemophilia is an X-linked recessive disease that prevents blood clotting. If a woman who is a carrier for hemophilia marries a normal man, what proportion of their sons are expected to have hemophilia? What is the chance that their first child will have hemophilia? (Remember, the sex of this child is unknown until birth.)

3. Neurofibromatosis is a genetic disorder associated with an allele of one autosomal gene. Individuals with neurofibromatosis have uneven skin pigmentation and skin tumors. A man with neurofi-

bromatosis marries a normal woman who does not carry the allele for this disorder. The couple has 5 children, 3 of whom have neurofibromatosis. The most likely explanation for this outcome is that neurofibromatosis is a _____ trait and the man is _____.
   a. recessive; homozygous recessive
   b. recessive; heterozygous
   c. dominant; homozygous dominant
   d. dominant; heterozygous
   e. a and b are both possible explanations

Consider the following pedigree for a human autosomal trait. (Note that in pedigrees, generations are indicated by Roman numerals and individuals within a generation are numbered from left to right with Arabic numerals.)

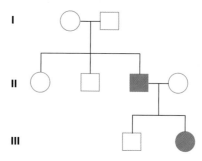

4. Is the allele that determines this trait dominant or recessive? Explain your reasoning.

5. Using symbols A and a for the dominant and recessive alleles, respectively, what is the genotype of the affected male in the second generation (individual II-3)?

6. What is the genotype of this individual's mate?

7. If female III-2 marries a heterozygous man, what proportion of their children are expected to have this trait?

8. Consider the following pedigree, which posits a fantasy gene for ear shape, symbolized by *P* for the regular allele or *p* for a pointy-shaped allele if this is an autosomal trait; or by X^+ for the regular allele or X^p for the pointy allele if this is an X-linked trait.

   This pedigree indicates the trait is most likely inherited as an:
   a. autosomal dominant
   b. autosomal recessive
   c. X-linked dominant
   d. X-linked recessive

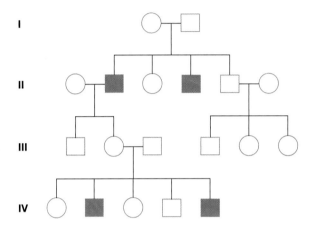

9. What is the genotype of individual I-1?

10. If a woman who is a carrier for this trait married individual III-1, what proportion of their children are expected to show this trait? What proportion of their sons are expected to show this trait?

11. Construct a pedigree to answer the following question. Red-green color blindness is an X-linked recessive trait. A woman who has a color blind mother and a father with normal color vision marries a man with normal vision. This couple has a son. What is the chance that the son is color blind?

12. Hyperphosphatemia, a disease that causes a form of rickets (abnormal bone growth and development), is inherited as an X-linked dominant trait. A woman who is heterozygous for the disease allele marries a normal man. What proportion of their sons will have hyperphosphatemia? What proportion of their daughters will have hyperphosphatemia? If hyperphosphatemia were an X-linked recessive trait, how would your answer differ?

13 DNA Structure and Replication

James Watson.
(Section 13.2, page 252)

Francis Crick.
(Section 13.2, page 252)

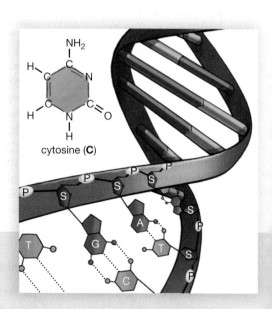

cytosine (**C**)

The double helix.
(Section 13.3, page 254)

"In research the front line is almost always in a fog," wrote Francis Crick in his 1988 memoir, *What Mad Pursuit*. Crick has first-hand knowledge of this. In 1953, with James Watson, he discovered the structure of DNA, an effort that brought him to the front line of research, which is where he has remained ever since. Crick's book makes clear what a tremendous difference there is between learning something about nature and discovering something about it. In learning, dozens of voices stand ready to instruct; there are books, lectures, videos, CDs—a world of well-organized information. The person who seeks to discover something about nature, meanwhile, is confronted with silence. Nature goes on: Cells divide, chromosomes condense, birds migrate. But *why* do these things operate the way they do? All the researcher has as a guide are the things themselves, working away. The trick is to devise some test (called an experiment) that can make nature's routine operations yield up information. In this process of teasing out truth, knowledge is gained in very small increments. Scientists are like the first people down a darkened maze of tunnels; they must make their way exceedingly slowly, feeling each square inch of wallspace as they go, after which they can only leave lights on *behind* them (in the form of their scientific papers, books, and so forth).

13.1 What Do Genes Do, and What Are They Made of?

To get a feel for the process of discovery, consider the state of genetics early in the twentieth century. By 1920, it had been demonstrated beyond any doubt that genetic information resided on chromosomes. Within a decade or so, by looking at some abnormally large fly chromosomes, scientists could observe in great detail such chromosomal processes

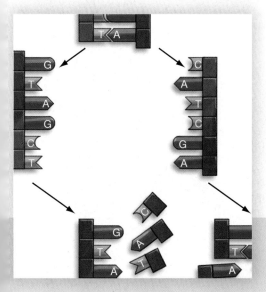

DNA replication begins.
(Section 13.3, page 255)

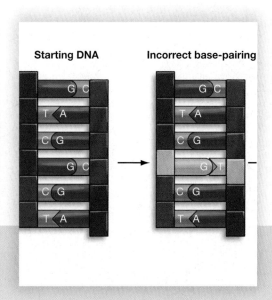

One kind of mutation.
(Section 13.4, page 257)

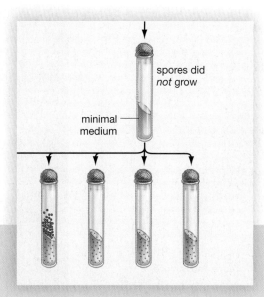

A famous experiment.
(Essay, page 259)

251

as crossing over. By viewing the banding patterns on these chromosomes, they could even identify the rough location of some genes.

Yet what was a gene? What was the physical nature of this unit of heredity, and how did it work? No one knew. Because genes are much smaller than the chromosomes on which they reside, there was no hope of simply viewing one under a microscope. In observing chromosomes, scientists were only "looking" at genes in the way that any of us is "looking" at lunar rocks by glancing up at the moon. Through the 1920s and 1930s, then, genes could be described only in vague, functional terms, as in: A gene is an entity that lies along a chromosome and brings about a phenotypic trait in an organism (for example, the green or yellow peas of Mendel's experiments).

Things were to clear up somewhat in the ensuing years. The seminal achievement in genetics in the late 1930s and early 1940s was a more concrete description of what genes do: They bring about the production of proteins. The central achievement in genetics from the mid-1940s to the early 1950s was strong evidence indicating what genes are composed of: deoxyribonucleic acid, or DNA for short.

DNA Structure and the Rise of Molecular Biology

By the early 1950s, a key question became *how* DNA carried out its genetic function. To find out, scientists had to piece together its exact chemical structure. This investigation turned out to be a watershed event in biology. Just as opening a mechanical watch and observing how its parts fit together would allow a person to understand how the watch works, so deciphering the structure of DNA allowed biologists to understand how genetics works at its most fundamental level. How could a section of DNA specify a protein? How could genetic information be copied? The structure of the DNA molecule suggested answers to these questions, as you'll see.

Apart from what this investigation uncovered, the inquiry itself stood as a symbol of a new era in biology. When Gregor Mendel did his experiments with peas, he was looking at whole organisms. When T. H. Morgan was doing his work with *Drosophila* flies, he was interested in whole chromosomes, which he knew to be composed of several types of molecules. In the early 1950s, however, the search turned to a single molecule. How was DNA structured, and how did this structure allow it to carry out its various func-

Figure 13.1
Young and Famous
James Watson, on the left, and Francis Crick, with a model of the DNA double helix, shortly after they published their paper on the molecule's structure.

tions? Biological research of this sort has grown ever more important in the decades since the 1950s. It is today known as **molecular biology**: the investigation of life at the level of its individual molecules.

13.2 Watson and Crick: The Double Helix

James Watson and Francis Crick may not be instantly recognized scientific names in the way that, say, Albert Einstein or Louis Pasteur are, but there's a certain public awareness that these two researchers did something important in connection with DNA (**Figure 13.1**). What they did was present to the world, in 1953, the structure of DNA: Atom-by-atom, bond-by-bond, this is how DNA fits together, they said in unveiling DNA's now-famous configuration, the double helix.

The two were a seemingly unlikely pair to make an epochal scientific discovery. Watson, an American, was a 23-year-old who was scarcely more than a year out of graduate school, and Crick, an Englishman, was a 35-year-old just then working on his doctorate when the two met in the fall of 1951 at Cambridge University in England. Ending up together by coincidence at the same laboratory, they realized in short order their mutual interest in the structure of DNA and, to the neglect of projects they were supposed to be working on, they began several rounds of model building and brainstorming that resulted in their breakthrough.

Watson and Crick were greatly aided in their investigation by the work of others. Though the DNA molecule was too small to be seen by even the most powerful microscopes of the time, something about its structure could be inferred from a technique called *X-ray diffraction*. In this process, a purified form of a molecule is bombarded with X rays. The way these rays scatter upon impact then reveals something about the structure of the molecule.

If you look at **Figure 13.2**, you can see the results of some X-ray diffraction. As you can imagine, it takes a highly trained observer to be able to deduce anything about the structure of a molecule from such an image. Fortunately for Watson and Crick, such a person was working just up the road from them. She was Rosalind Franklin, a researcher at King's College in London and one of the handful of individuals then skilled in performing X-ray diffraction on DNA. She and her colleague, Maurice Wilkins, were themselves working on the structure of DNA at the time, as

were other researchers in America. Thus did Watson, at least, regard the search for DNA structure to be a race between several teams, a fact that concentrated his efforts wonderfully. In 1962 Watson, Crick, and Wilkins were awarded the Nobel Prize in Medicine or Physiology for their work on DNA. Rosalind Franklin died of cancer in 1958 at the age of 37. Nobel Prizes are not awarded posthumously; it's unknown what would have happened had she lived.

Let's start now to look at DNA's structure, which will put you in a position to appreciate the achievement of Watson, Crick, and their fellow researchers.

Web Tutorial 13.1
DNA

(a) Rosalind Franklin

(b) X-ray diffraction image of DNA

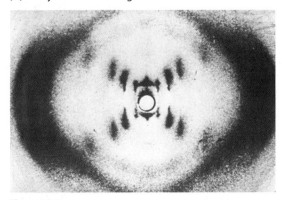

Figure 13.2
(a) DNA Investigator
Rosalind Franklin, whose work in X-ray diffraction was important in revealing the structure of the DNA molecule.

(b) Imaging DNA
One of Franklin's X-ray diffraction images of DNA. The "cross" formed of dark spots indicated the molecule had a helical structure.

13.3 The Components of DNA and Their Arrangement

If you look at **Figure 13.3**, you can see the component parts of DNA. First, there is a phosphate group, and second, a sugar called *deoxyribose*. Third, there are four possible DNA "bases": adenine guanine, thymine, and cytosine—A, G, T, and C, for short. When you link together a phosphate group, a deoxyribose molecule and one of the four bases, you get the basic building block of DNA, the *nucleotide*, one of which is boxed in the larger DNA molecule in the figure.

When you look at this larger figure as a whole, you can see that the double-helix looks something like a spiral staircase. By focusing on its exterior "handrails,"

you can see how sugar and phosphate fit together to form a kind of chain, with its links going: sugar-phosphate-sugar-phosphate (symbolized in the figure by S and P). The "steps" lying between these handrails are composed of DNA's bases. As the figure shows, each of the bases amounts to a *half*-stair-step, if you will. Each extends inward from one handrail of the double helix and is then joined to base extending inward from the *other* handrail of the DNA molecule. (The bases are linked via hydrogen bonds, symbolized by the dotted lines in the figure.)

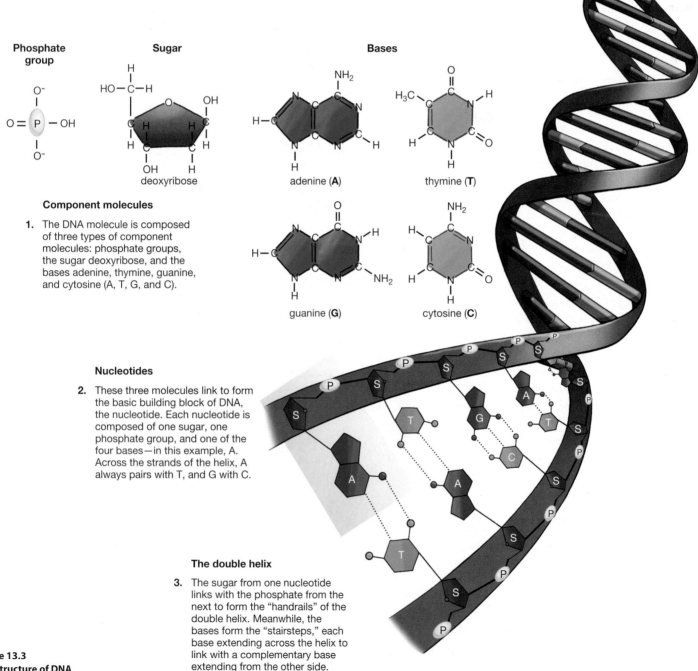

Phosphate group

Sugar

deoxyribose

Bases

adenine (**A**)

thymine (**T**)

guanine (**G**)

cytosine (**C**)

Component molecules

1. The DNA molecule is composed of three types of component molecules: phosphate groups, the sugar deoxyribose, and the bases adenine, thymine, guanine, and cytosine (A, T, G, and C).

Nucleotides

2. These three molecules link to form the basic building block of DNA, the nucleotide. Each nucleotide is composed of one sugar, one phosphate group, and one of the four bases—in this example, A. Across the strands of the helix, A always pairs with T, and G with C.

The double helix

3. The sugar from one nucleotide links with the phosphate from the next to form the "handrails" of the double helix. Meanwhile, the bases form the "stairsteps," each base extending across the helix to link with a complementary base extending from the other side.

Figure 13.3
The Structure of DNA

The figure shows some examples of one base being paired with another. At the bottom, for example, an A base on the left is paired with a T base on the right. Go two sets of nucleotides up and a G on the left is being paired with a C on the right. This turns out to be one of the fundamental rules about DNA structure. If you viewed a billion DNA *base pairs*, as they're called, you would find the same thing: A always pairing with T, and G always pairing with C, across the helix. (Any two bases that can pair together in this way are said to be *complementary*; thus A is complementary to T, and G is complementary to C. Likewise, you can have complementary chains of DNA.)

The Structure of DNA Gives Away the Secret of Replication

It was DNA's very structure that suggested the answer to one of the great questions of genetics: How is genetic information passed on? To put this another way, how does a cell make a copy of its own DNA? As you've seen, a full complement of our DNA is contained in nearly every cell in our body. Yet cells divide, and the daughter cells contain exactly the same DNA complement as the parent cell. This means that genetic information is passed on by means of DNA being copied, with one copy of this molecule ending up in each daughter cell. Indeed, such copying extends to the formation of egg and sperm cells, meaning this is the way genetic information is passed on from one *generation* to the next. But how could this work?

The structure Watson and Crick discovered suggested a way. We've observed that A must always pair with T, and G with C, across the two strands of the double helix. This rule, Watson and Crick saw, meant that each single strand of DNA could serve as a *template* for the synthesis of a new single strand (**see Figure 13.4**). Each A on an old strand would specify the place for a T on the new, each G on the old a place for C on the new, and so forth. All that was required was for the two old strands to separate—splitting the stair steps right down the middle—and for new strands to be synthesized that were complementary to the old. As it turned out, this was exactly how things worked, as you'll see shortly.

The Structure of DNA Gives Away the Secret of Protein Production

The second thing the structure suggested was a partial answer to another great question of genetics: How the molecule of heredity could be versatile enough to specify the dazzling array of proteins that all living things produce. The handrails of DNA's double helix are monotonous—phosphate, sugar, phosphate, sugar. But the bases can be laid out along

these handrails in an extremely varied manner. A has to pair with T and G with C *across* the helix; but *along* the handrails, the bases can come in any order. Look at the top drawing in Figure 13.4 and see the order this hypothetical group of bases comes in; starting from the bottom right, A-G-C-T-A-C. Strung together by the thousands, these bases can specify a particular protein, just as in Morse code a series of short and long clicks can specify a particular word.

The Building Blocks of DNA Replication

You have already looked briefly at the basic process by which DNA is copied or "replicated," but we now need to review some of the details of this process. The basic steps here are straightforward. The essential building block of DNA is the unit called the **nucleotide**, which can be seen in the box in Figure 13.3. Every nucleotide contains one sugar and one phosphate group. What differentiates one nucleotide from another is the base

Figure 13.4
DNA Replication
The result of DNA replication is two identical molecules of DNA, whereas the process began with one.

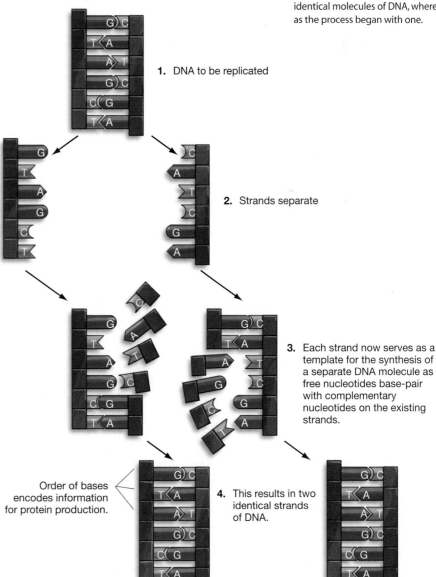

1. DNA to be replicated

2. Strands separate

3. Each strand now serves as a template for the synthesis of a separate DNA molecule as free nucleotides base-pair with complementary nucleotides on the existing strands.

Order of bases encodes information for protein production.

4. This results in two identical strands of DNA.

that is attached to it; in this case the base is A, but as you can see, others have T, C, or G.

Figure 13.4 shows the process of DNA replication in overview. Note that the first thing that happens is that the joined strands of the double helix unwind, separating from one another. The nucleotides on each of the single strands are then paired with free-floating nucleotides that line up in new, complementary strands. Because *both* strands of the original double helix are being paired with new strands, the end product is two double helices, where before there was one.

You can see in Figure 13.4 that the addition of nucleotides moves in differing directions on the two strands. This is so because of something you can see in Figure 13.3: the two strands of the double helix have opposite orientations. Note that the sugars of the right-hand strand, for example, are upside down relative to the sugars of the left-hand strand. (If the strands were people, we would think of them as lying head to *foot* relative to each other.) Because new nucleotides can be added to only one end of a DNA strand—what's known as the three-prime end—the nucleotides are added in different directions.

DNA Replication: Something Old, Something New

It's worth drawing a little finer point on one aspect of this process: Each resulting double helix is a combination of the old and the new. Each has one "parental" strand of DNA and one newly synthesized complementary strand. This combination is conceptually important because it is how life builds upon itself. You can see this illustrated in **Figure 13.5**.

Eventually, the two double helices are fully formed. A little later they part company, during mitosis, with each newly formed double helix moving into a separate cell and thereafter serving as an independently functioning segment of DNA. (At the chromosomal level, the newly replicated helices form the core of something you've seen a lot of in the last few chapters: sister chromatids, which eventually part and become independent chromosomes.)

However simple this process may sound in overview, its details are enormously complicated. As you might imagine, such a process could not proceed without enzymes to catalyze it. To name just two groups of them, there are enzymes called *helicases* that unwind the double helix, separating its two strands to make the bases on them available for base pairing. There is also another group of enzymes, collectively known as **DNA polymerases**, that move along each strand of the double helix, joining together nucleotides as they are added—one by one—to form the new, complementary strands of DNA.

Editing Out Mistakes

The base pairing that goes on in replication happens hundreds, then thousands, then millions of times in a given stretch of DNA: Free-standing A's are aligned with complementary T's, while C's are aligned with G's. Given the numbers of base pairs involved, the amazing thing is how few mismatches there are by the time the process is completed. The error rate in DNA replication—the rate at which the *wrong* bases have been brought together—might be only one in every billion bases by the end of replication. Yet *during* replication, such a mistake might be made once in every 100,000 bases. Obviously, to start with one number of mistakes but to end up with far fewer, the cell's genetic machinery has to be capable of correcting its errors.

This happens partly through the services of the versatile DNA polymerases, which are able to perform a kind of DNA editing: They remove a mismatched nucleotide and replace it with a proper one. Interestingly enough, this happens through a kind of backspacing. Normally the DNA polymerases are moving along a DNA chain, linking together recently arrived nucleotides. When an error is detected, however, they stop, move "backward," remove the incorrect nucleotide, put in the correct one, and then move forward again.

13.4 Mutations: Another Name for a Permanent Change in DNA Sequence

With this consideration of alterations in DNA structure, you have arrived at an important concept: that of a **mutation**, which can be defined as a permanent alteration of a DNA base sequence. Such alterations come about because the cell's various DNA error-correcting mechanisms are not foolproof; they do not correct all the mistakes that occur.

DNA's makeup can be altered in many ways. A slight change in the chemical form of a base might,

Figure 13.5
How Life Builds on Itself
Each newly synthesized DNA molecule is a combination of the old and the new. An existing DNA molecule unwinds, and each of the resulting single strands (the old) serves as a template for a complementary strand that will be formed through base pairing (the new).

old

new

for example, cause a G to link up across the helix with a T (instead of with its normal partner, C), as you can see in **Figure 13.6**. Then, in a subsequent round of DNA replication, the cell might "repair" this error in such a way that a permanent mistake is introduced: An A-T pair now exists, whereas the original sequence had a G-C pair. Permanent mistakes like these are called **point mutations**, meaning a mutation of a single base pair in the genome. Such mutations stand in contrast to the kind of whole-chromosome aberrations presented in Chapter 12, though these also qualify as mutations under some definitions.

In what way are mutations "permanent"? Think of how DNA replication works. Before any cell can divide, it must first make a copy of its complement of DNA. Should this DNA contain an uncorrected mistake, it too will be copied again and again with each succeeding cell division. Most mutations have no noticeable effect on an organism; and, as you'll see, mutations are vital to the process of evolution. But the concept of mutation is quite rightly fearful to us, because in relatively rare instances, mutations can have disastrous effects.

Examples of Mutations: Cancer and Huntington

A cancerous growth is a line of cells that has undergone a special kind of mutation—one that causes the affected cells to proliferate wildly. As an example, the skin cancer known as melanoma causes skin cells called melanocytes to start dividing very rapidly. For our purposes, the question is: How does this process get going? In 2002, British scientists looked at numerous lines of melanoma cells that had been taken from cancer patients. They found that 59 percent of these cells contained a mutation in a gene called *BRAF*. This gene, like all others, amounts to a sequence of DNA nucleotides—a sequence of C's, T's, A's, and G's. When the British scientists looked at how the *BRAF* gene had actually mutated, they found that in 80 percent of cases, there had been a single change: an A had been substituted for a T at *BRAF*'s 1,796th nucleotide. What they found, in other words, was a point mutation, nearly two thousand bases within the sequence of nucleotides that makes up the *BRAF* gene. And the effect of this mutation? It produces a protein that is a slightly altered version of the normal BRAF protein. This protein is altered enough, however, that it keeps melanocytes moving through the cell-division cycle, causing them to multiply wildly.

Cell proliferation is not the only kind of trouble mutations can bring about. Those of you who read Chapter 12 are familiar with the Huntington disease that causes spastic movements, severe de-

mentia, and usually death among its victims. In all people there is a gene, called *IT15*, that has within it a number of repeats of a particular "triplet" of bases, CAG. People with 37 or fewer CAG repeats in their *IT15* gene are fine; they will suffer no illness at all from the gene. People with 39 CAG repeats, however, probably will develop Huntington, though their first symptoms are not likely to appear until they are 66; people who have 50 or more repeats will descend into dementia at an average age of 27. Unlike the case with cancer, however, the Huntington mutation does not cause cells to multiply wildly. Instead, the mutated *IT15* gene creates a faulty protein called huntingtin, which cannot be broken down by nerve cells and ends up building up inside them, eventually killing them. Because Huntington is caused by a repeating group of three nucleotides, it is referred to as a "trinucleotide repeat" disease. At least eight other diseases of the nervous system fall into this category.

Heritable and Non-Heritable Mutations

Though Huntington disease and melanoma are both caused by mutations, there is an important distinction to be made between them. Most mutations come about in the body's **somatic cells**, which is to say cells that do not become eggs or sperm. This is the case with the mutations that bring about melanoma. Conversely, some mutations arise in **germ-line cells**, meaning the cells that do become eggs or sperm. This is the case with the Huntington mutation. The important point here is that germ-line cell mutations are *heritable*—they can be passed on from one generation to the next, in the ways reviewed in Chapter 12. In contrast, though a line of melanoma cells may be quite harmful, it is separate from the line of cells that

Web Tutorial 13.2
Mutations

Figure 13.6
One Route to a Mutation

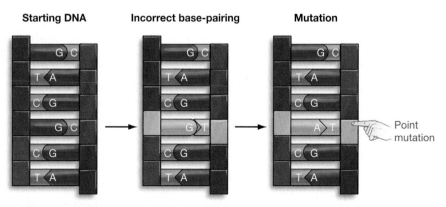

Starting DNA   **Incorrect base-pairing**   **Mutation**

Point
mutation

1. In replicating a cell's DNA, mistakes are sometimes made, such that one base can be paired with another base that is not complementary to it (G with T in this case).

2. The next time a cell replicates its DNA, the replication repair mechanism may "fix" this error in such a way that a permanent alteration in the DNA sequence results. The original G will be replaced, instead of the wrongly added T. The result is an A-T base pair, whereas the cell started with a G-C base pair.

gives rise to eggs or sperm. For this reason, melanoma cannot be passed on from one generation to the next.

What Causes Mutations?

A critical question, of course, is what causes DNA to mutate? One answer is so-called environmental insults. The chemicals in cigarette smoke are powerful *mutagens*, meaning substances that can mutate DNA. So is the ultraviolet light that comes from the sun. To look at the latter example in a little more detail, ultraviolet light is a form of radiation that can link adjacent T's together in a single strand of DNA; sometimes it even causes both strands of the DNA helix to break. When this latter damage takes place in a cell, one of three things can happen. First, enzymes can successfully repair the damage. Second, the cell will recognize that this damage *cannot* be repaired, in which case the cell will commit suicide in the process known as apoptosis. Third, the enzymes will not repair the damage, but the cell will fail to initiate apoptosis. In this instance, the *failed* repair action may mean permanent change in base sequence—a mutation, in short.

Not all mutations are caused by environmental influences. Mutations happen simply as random, spontaneous events. Molecules collide in a cell, for example, causing DNA damage. The very process of eating and breathing produces so-called free radicals that can damage DNA. And the DNA replication machinery itself may introduce errors, irrespective of any outside influences.

Once such errors occur, it is understandable that some of them will go uncorrected, simply because of the size of the replication operation. When a human cell divides, billions of DNA base pairs have to be copied. And some 25 million cells are dividing each second in human beings. The surprise, therefore, is not that mistakes happen, but that so *few* of them happen, and that fewer still become permanent. (Recall the one-in-a-billion error rate in DNA replication that was observed earlier.) After learning about the kinds of effects mutations can have on organisms, however, you can understand why this error rate is so low: Life as we know it could not exist with a *high* error rate, at least in the portions of genome that code for proteins. Remember that within 3 billion A's, T's, G's, and C's, the substitution of a single A for a single T can help bring about melanoma. How could life exist if such mistakes were common?

The Value of Mistakes: Evolutionary Adaptation

While it's true that, for individuals, mutations can have a negative effect, for the living world as a whole mutations have been vitally important because of a role they play in evolution. It turns out that germ-line muta-

tions are the only means by which completely new genetic information can be added to a species' genome, in the form of new alleles (meaning variant forms of a gene). Organisms can combine *existing* alleles, in myriad ways. Think of meiosis, with its chromosomal part swapping (crossing over) and reshuffling of chromosomes (independent assortment). Valuable as these processes are, no amount of genetic recombination could have produced, for example, the eyes that some living things possess. To go from no eyes to eyes, there had to have been some mutations along the line—some accidental reorderings of DNA sequences such that entirely new proteins were produced. Such adaptations are vital to living things, given their struggle to get along in environments that are constantly changing. If environments change, species need to change too, in order to survive. And the only way major changes can come about is through mutations. You'll be looking at this topic again in the evolution unit of this book. For now, however, isn't it interesting to ponder the fact that the living world adapts partly through its mistakes?

How Did We Learn?

It may be clear to us today that genes are information-bearing units and that the information they contain is used to produce proteins. But 60 years ago, things were not so clear. Indeed, scientists had to work hard to attain this insight. To find out more about their efforts, see "Getting Clear about What Genes Do" on the next page.

On to How Genetic Information Is Put to Use

Back in Chapter 9, you saw that since antiquity, human beings have speculated about how one generation of living things can give rise to another—about *how* it is that life goes on. The questions posed by the Greeks, by scientists in the Renaissance, by Mendel and Morgan and others were finally answered in the 1950s and 1960s, when scientists came to understand DNA replication at the molecular level. The something-old, something-new quality to this replication—parental DNA strands serving as templates for new strands—was the detailed answer to the ancient question of how the qualities of living things can be passed down through generations.

Splendid as this function is, it is only one of the two great tasks carried out by our genetic machinery. You have just reviewed the process by which genetic information is replicated; now let's look at the process by which this information is used. How can a stretch of DNA bring about the production of a protein? You'll see in the chapter coming up.

Web Tutorial 13.3
One Gene, One Enzyme
Hypothesis

How Did We Learn?
Getting Clear about What Genes Do: Beadle and Tatum

In the 1930s, biologists were confused not only about what genes were made of (DNA) but also about what it was that genes did. We now know that a large part of what they do is contain the information for the production of proteins, including the reaction-hastening proteins called *enzymes*. But it took work over a period of years by Stanford University researchers George Beadle and Edward Tatum to make this clear. Their experimental subject was a red fungus, the bread mold *Neurospora*.

Following the Growth of a Fungus

Research scientists are always looking for ways to control the confusing conditions of nature in order to tease out the rules it plays by. Beadle and Tatum's control mechanism was, first, to *induce* genetic mutations in *Neurospora* spores by bombarding them with X-rays. These mutated spores were then transferred one at a time into test tubes, where the second element of control came in. Natural or "wild type" *Neurospora* can manufacture or "synthesize" most of what they need to live, given just a few basic biological molecules in their surroundings. But could the descendants of the *mutated* group of *Neurospora* do this? Could they be put in a so-called minimal medium—a liquid environment that forces the fungus to manufacture almost all its necessary biological molecules—and still multiply into a thriving colony? The answer for many of the mutant strains was no. The mutations these organisms had undergone had rendered them unable to synthesize some substance that they needed to thrive. By being selective about *adding* substances to the minimal media, Beadle and Tatum could tell what substance the mutant cells were unable to synthesize. If they added the right thing, the colony would grow, as you can see in **Figure 1**. Thus, they reasoned, whatever substance they added was the substance a mutated strain could no longer synthesize on its own. For some of the mutant strains, the missing substance turned out to be the amino acid arginine.

We now know that arginine is produced in one of the metabolic pathways reviewed in Chapter 6. In such a pathway, a precursor compound is modified several times in a chain, with a different enzyme facilitating each step of the operation. By selectively supplying different strains of mutant *Neurospora* with arginine's *precursor* substances, Beadle and Tatum could discern the step at which the arginine metabolic pathway had been blocked. One strain required only precursor A to produce arginine, for example, while another strain required precursors A and B. In the tradition of T. H. Morgan, the researchers linked these blockages to genes on three separate *Neurospora* chromosomes. The summary result then was clear. A given *Neurospora* strain had become a mutant by having its genes altered through radiation. As a result, it no longer carried out the transformation of, say, precursor A to precursor B. It took an *enzyme* to facilitate this transformation. What genes were doing, then, was bringing about the production of enzymes.

The One-Gene, One-Enzyme Hypothesis

Eventually, a famous label was applied to these experimental results: The *one-gene, one-enzyme hypothesis*. What each gene does is call up production of one enzyme, which undertakes a particular metabolic task. Though the notion of what a gene is has now been broadened, the essential insight was not only correct, it was critical. With it, scientists at last understood what genes were doing. For this work, which they completed in 1941, Beadle and Tatum shared in the Nobel Prize in 1958.

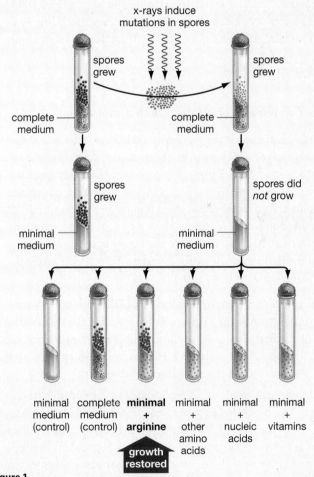

Figure 1
Beadle and Tatum Experiment with Mutated Spores
Using spores of the fungus *Neurospora*, Beadle and Tatum found that some mutated versions of the spores could not grow in a medium that required them to synthesize all their biological molecules. By selectively adding molecules to this "minimal medium," they demonstrated that the mutation was preventing the spores from producing certain compounds, among them the amino acid arginine. By tracing the metabolic pathway leading to arginine, Beadle and Tatum realized that the *Neurospora* genes that had been made dysfunctional were coding for enzymes that facilitated the steps leading from the precursors of arginine to arginine itself.

Chapter Review

Summary

13.1 What Do Genes Do, and What Are They Made of?

- James Watson and Francis Crick discovered the chemical structure of DNA in 1953. This event ushered in a new era in biology because it allowed researchers to understand some of the most fundamental processes in genetics.

- In trying to decipher the structure of DNA, Watson and Crick were performing work in molecular biology. This is the investigation of life at the level of its individual molecules. Molecular biology has grown greatly in importance since the 1950s.

13.2 Watson and Crick: The Double Helix

- Watson and Crick met in the early 1950s at Cambridge University in England and set about to decipher the structure of DNA. Their research was aided by the work of others, including Rosalind Franklin, who was using X-ray diffraction to learn about DNA's structure.
  Web Tutorial 13.1: DNA

13.3 The Components of DNA and Their Arrangement

- The DNA molecule is composed of building blocks called nucleotides, each of which consists of one sugar (deoxyribose), one phosphate group, and one of four bases: adenine, guanine, thymine, or cytosine (A, G, T, or C). The sugar and phosphate groups are linked together in a chain that forms the "handrails" of the DNA double helix. Bases then extend inward from the handrails, with base pairs joined to each other in the middle by hydrogen bonds. In this base pairing, A always pairs with T across the helix, while G always pairs with C.

- DNA is copied by means of each strand of DNA serving as a template for the synthesis of a new, complementary strand. The DNA double helix first divides down the middle. Each A on an original strand then specifies a place for a T in a new strand, each G specifies a place for a C on the new strand, and so forth.

- Each double helix produced in replication is a combination of one parental strand of DNA and one newly synthesized complementary strand. This is how life builds upon itself. A group of enzymes known as DNA polymerases are central to DNA replication; they move along the double helix, bonding together new nucleotides in complementary DNA strands.

- DNA can encode the information for the huge number of proteins utilized by living things because the sequence of bases along DNA's handrails can be laid out in an extremely varied manner. A collection of bases in one order encodes the information for one protein, while a different sequence of bases encodes the information for a different protein.

13.4 Mutations: Another Name for a Permanent Change in DNA Sequence

- The error rate in DNA replication is very low, partly because repair enzymes are able to correct mistakes. When such mistakes are made and then not corrected, the result is a mutation: a permanent alteration in a cell's DNA base sequence.

- Most mutations have no effect on an organism, but when they do have an effect, it is generally negative. Cancers result from a line of cells that have undergone types of mutations that cause them to proliferate wildly.

- Some mutations come about in the body's germ-line cells, meaning cells that become eggs or sperm. Such mutations are heritable: They can be passed on from one generation to another. The gene for Huntington disease, which is expressed in nerve cells, is a heritable, mutated form of a normal gene.
  Web Tutorial 13.2: Mutations
  Web Tutorial 13.3: One Gene, One Enzyme Hypothesis

Key Terms

DNA polymerase 256	nucleotide 255
germ-line cell 257	point mutation 257
molecular biology 253	somatic cell 257
mutation 256	

Understanding the Basics

Multiple-Choice Questions (answers in the back of the book)

1. James Watson and Francis Crick
   a. proved that DNA contains information for the production of proteins
   b. elucidated the nature of genetic mutations
   c. discovered DNA
   d. proved that chromosomes were made partly of DNA
   e. discovered the structure of DNA

2. The type of biological study that was exemplified through the discovery of DNA structure is called
   a. detailed biology
   b. component biology
   c. microscopic biology
   d. elemental biology
   e. molecular biology

3. The components of the DNA handrails are _____ and _____.
   a. nucleotides, phosphates
   b. sugars, nucleotides

c. sugars, bases

d. sugars, phosphates

e. none of the above

4. Which of the following combinations is an example of a nucleotide?

a. sugar + phosphate + adenine

b. sugar + phosphate + guanine

c. sugar + phosphate + thymine

d. sugar + phosphate + cytosine

e. all of the above

5. Which of the following DNA sequences contains a mismatched base pair?

a. TCAA
AGTT

b. CAGC
GTCG

c. GATA
CTCT

d. AATT
TTAA

e. GACG
CTGC

6. If cytosine makes up 26 percent of the nucleotides in a sample of DNA from an organism, then adenine would make up what percent of the bases?

a. 26

b. 52

c. 28

d. 24

e. It cannot be determined from the information provided

7. If DNA's structure is compared to a spiral staircase, then its stair-steps would be

a. phosphate-sugar-phosphate chains

b. free-floating DNA bases bonded together randomly

c. chromatin

d. paired DNA bases, always linking A and T, or G and C

e. none of the answers above is correct

8. DNA polymerases work to

a. unwind the double helix during replication

b. cause mutations in DNA

c. join together nucleotides in a growing, complementary DNA strand

d. correct errors in DNA base sequences during replication

e. both c and d

9. Watson and Crick made their momentous discovery about DNA in _____.

a. 1943

b. 1953

c. 1921

d. 1973

e. 1962

10. How are the base pairs of the DNA double helix linked together?

a. phosphate bonds

b. hydrogen bonds

c. carbon–carbon bonds

d. sugar bonds

e. basic bonds

Brief Review

1. How does the phrase "something old, something new" describe the method of DNA replication employed by the cell?

2. What did Watson and Crick's discovery reveal that was so important?

3. True or False: DNA polymerases work only in a forward direction.

4. What is a mutation? Are all mutations harmful?

5. What is the nature of the Huntington disease mutation, and what is its effect at a cellular level?

Applying Your Knowledge

1. Given the following DNA sequence, determine the complementary strand that would be added in replication:

ATTGCATGATAGCC

2. Why does a germ-line mutation carry greater potential significance than a somatic mutation?

3. A gene's nucleotide bases are known to be composed of 15 percent guanine. What percentages of each of the other three bases are contained in the same gene?

4. Why is it important that DNA replication have a low error rate?

5. Would you expect cancer to arise more often in types of cells that divide frequently (such as skin cells) or in types of cells that divide rarely or not at all (such as nerve cells)? Explain your reasoning.

14 *Genetic Transcription, Translation, and Regulation*

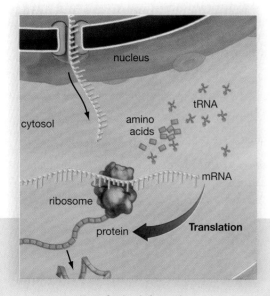

Transferring information.
(Section 14.2, page 265)

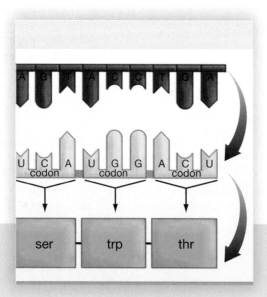

The triplet code.
(Section 14.4, page 269)

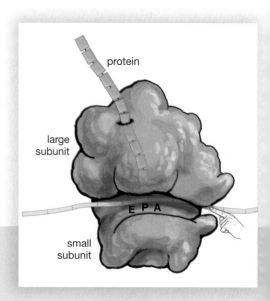

A ribosome.
(Section 14.4, page 271)

Like workers using a set of blueprints to construct a house, cells use the information contained in DNA to construct proteins. Each cell can finely tune its production of proteins in accordance with its needs.

The scientific advance that was reviewed in Chapter 13, the discovery of the structure of DNA in 1953, bears some comparison to the birth of Napoleon: Both events had momentous consequences, but not for a while. Only slowly did the discoveries come that, in time, were seen as hinging on James Watson and Francis Crick's work. So significant were these advances, however, that the years between 1953 and 1966 are now regarded by some scholars as a kind of golden age of genetics; a time when basic findings about this field came fast and furious, such that by the mid-1960s scientists had at last grasped the essentials of how the genetic machinery works.

These discoveries were made in connection with both of the two great tasks of genetics: on the one hand, the copying or "replication" of DNA that you reviewed in Chapter 13; on the other, the means by which DNA's genetic infor-

mation brings about the production of proteins. The first task has to do with how genetic information is preserved; the second with how genetic information is utilized. This second task is the subject of this chapter.

14.1 The Structure of Proteins

Because you'll be dealing extensively with proteins here, now may be a good time to review some basic information about them. As you saw in Chapter 3, proteins fit into the "building blocks" model of biological molecules. The blocks in this case are amino acids. String a number of these together and you have a **polypeptide** chain, which then folds up in a specific three-dimensional manner, resulting in a protein. Proteins are likely to be made of hundreds of amino acids strung together, often in several linked polypeptide chains.

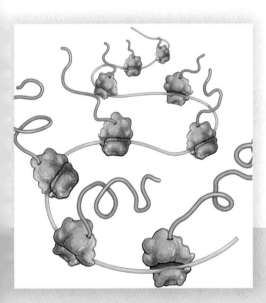

Mass production.
(Section 14.4, page 273)

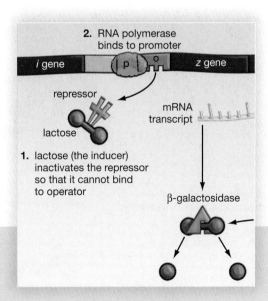

Genetic regulation in action.
(Section 14.5, page 275)

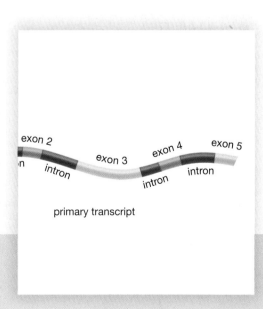

Ready for editing.
(Section 14.5, page 277)

Synthesizing Many Proteins from 20 Amino Acids

Though there are hundreds of thousands of different proteins, each one of them is put together from a starting set of a mere 20 amino acids. If you wonder how such diversity can proceed from such simplicity, think of the English language, which has thousands of words, but only 26 letters in its alphabet. It is the *order* in which letters occur that determines whether, for example, "cat" or "act" is spelled out; just so, it is the order of amino acids that determines what protein is synthesized.

If you look at **Figure 14.1a**, you can see the chemical structure of two free-standing amino acids, glycine (gly) and isoleucine (ile). If you look at **Figure 14.1b**, you can see that these two amino acids are the first two that occur in one of the two polypeptide chains that make up the unusually small

protein we call insulin. **Figure 14.1c** then shows a three-dimensional representation of insulin—the folded-up form this protein assumes before taking on its function of moving blood sugar into cells. A list of all 20 primary amino acids and their three-letter abbreviations can be found in **Table 14.1**.

For our purposes, the question is: How do the chains of amino acids that make up a protein come into being? How do gly, ile, val, and so forth come to be strung together in a specific order to create this protein? You know from what you've studied so far that genes can be thought of as "recipes" for proteins, and that these recipes are set forth as a series of chemical "bases"— the A's, T's, C's, and G's that lie along a DNA strand. How, then, do we get from *this* series of DNA bases to *that* series of amino acids—to a protein? Let's find out.

Figure 14.1
The Structure of Proteins

glycine **(gly)**

isoleucine **(ile)**

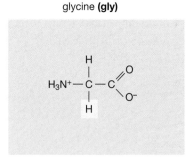

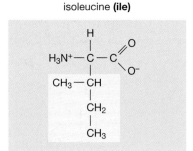

(a) Amino acids

The building blocks of proteins are amino acids such as glycine and isoleucine, which differ only in their side-chain composition (light colored squares).

(b) Polypeptide chain

These amino acids are strung together to form polypeptide chains. Pictured is one of the two polypeptide chains that make up the unusually small protein insulin.

(c) Protein

Polypeptide chains function as proteins only when folded into their proper three-dimensional shape, as shown here for insulin. Note the position of the glycine and isoleucine amino acids in one of the insulin polypeptide chains (colored light green).

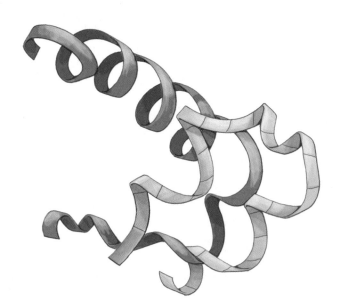

Table 14.1 Amino Acids

Amino Acid	Abbreviation
Alanine	ala
Arginine	arg
Asparagine	asn
Aspartic acid	asp
Cysteine	cys
Glutamine	gln
Glutamic acid	glu
Glycine	gly
Histidine	his
Isoleucine	ile
Leucine	leu
Lysine	lys
Methionine	met
Phenylalanine	phe
Proline	pro
Serine	ser
Threonine	thr
Tryptophan	trp
Tyrosine	tyr
Valine	val

The destination of this mRNA is a molecular workbench in the cell's cytoplasm, a structure called a ribosome. More formally, a **ribosome** is an organelle, located in the cell's cytoplasm, that is the site of protein synthesis. At the ribosome, both the recipe (the mRNA tape) and the raw materials (amino acids) come together to make the product (a protein). As the mRNA tape is "read" within the ribosome, something grows from it: a chain of amino acids that have been linked together in the ribosome in the order specified by the mRNA. When the chain is finished and folded up, a protein has come into existence. And how do amino acids get to the ribosomes? They are brought there by a second type of RNA, transfer RNA (tRNA).

As may be apparent from this account, protein synthesis divides neatly into two sets of steps. The first set is called **transcription**: the process by which the genetic information encoded in DNA is copied onto messenger RNA. The second set is called **translation**: the process by which information encoded in messenger RNA is used to assemble a protein at a ribosome. Let's look in more detail now at both of these processes, starting with transcription.

Web Tutorial 14.2
The Genetic Code

Figure 14.2
The Two Major Stages of Protein Synthesis

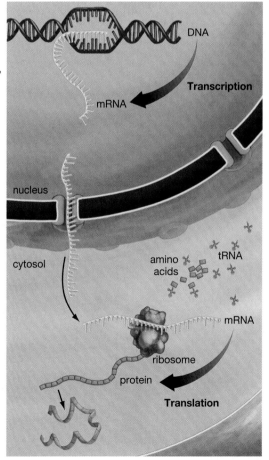

Transcription

1. In transcription, a section of DNA unwinds and nucleotides on it form base pairs with nucleotides of messenger RNA, creating an mRNA "tape."

2. This segment of mRNA then leaves the cell nucleus, headed for a ribosome in the cell's cytoplasm, where translation takes place.

Translation

3. Joining the mRNA tape at the ribosome are amino acids, brought there by transfer RNA molecules. The length of messenger RNA is then "read" within the ribosome. The result? A chain of amino acids is linked together in the order specified by the mRNA tape.

4. When the chain is finished and folded up, a protein has come into existence.

14.2 Protein Synthesis in Overview

In overview, the process of protein synthesis can be described fairly simply. In eukaryotic organisms such as ourselves, the DNA just referred to is contained in the nucleus of the cell. The first step is that a stretch of it unwinds there, and its message—the order of a string of A's, T's, C's and G's—is copied onto a molecule called messenger RNA (mRNA). This length of mRNA, which could be thought of as an information tape being copied off a "master" DNA tape, then exits from the cell nucleus (**see Figure 14.2**).

14.3 A Closer Look at Transcription

From what's been reviewed so far, you can see that a key player in transcription (and translation) is RNA, whose full name is ribonucleic acid. If you look at **Figure 14.3**, you can see just how similar RNA is to DNA. For one thing, RNA has the sugar and phosphate "handrail" components you saw in DNA last chapter. Then, in both molecules, this two-part structure is joined to a third element, a base.

There are, however, differences between DNA and RNA. For one, RNA is usually single-stranded, whereas DNA is structured in two strands that form its famous double helix. Beyond this, recall that any given DNA building block or "nucleotide" has one of four bases: adenine, guanine, cytosine, or thymine (A, G, C, or T). RNA utilizes the first three of these, but then substitutes uracil (U) for the thymine (T) found in DNA.

Passing on the Message: Base Pairing Again

Given the chemical similarity between DNA and RNA, it's not hard to see how DNA's genetic message can be passed on to messenger RNA: Base pairing is at work again. Recall from Chapter 13 how DNA is replicated. The double helix is unwound, after which bases along the now-single DNA strands are paired up with complementary DNA bases. Every T on a single DNA strand is paired with a free-floating A, and every C is linked with a G, thus yielding a complementary DNA *chain*. Because of RNA's similarity to DNA, the bases RNA has can also form base pairs with DNA. The twist to this RNA-DNA base pairing is that each *A* on a DNA strand links up with a *U* on the RNA strand, instead of the T that would be A's partner in DNA-to-DNA base pairing. The length of RNA that results from the process of transcription is called a *primary transcript*. With all this in mind, here's a more formal definition of **messenger RNA**

(a) Comparison of RNA and DNA nucleotides

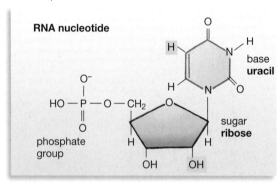
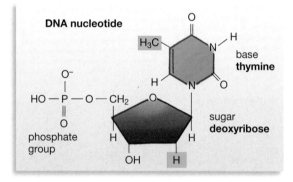

(b) Comparison of RNA and DNA three-dimensional structure

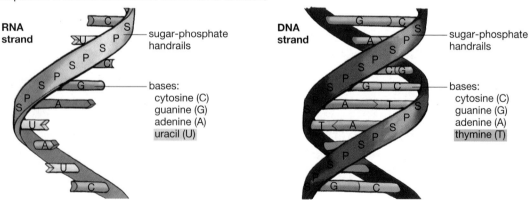

Figure 14.3
RNA and DNA Compared
(a) Each building-block nucleotide of both DNA and RNA is composed of a phosphate group, a sugar—ribose in RNA, deoxyribose in DNA—and one of four bases. RNA and DNA both utilize the bases adenine, guanine, and cytosine (A, G, and C), but RNA utilizes the base uracil (U) instead of the thymine (T) that DNA utilizes.
(b) Both RNA and DNA amount to linked chains of these nucleotides. The "handrails" of both RNA and DNA are composed of the sugar molecule of one nucleotide linked to the phosphate molecule of the next nucleotide (thus the S-P-S-P labeling). The "stair steps" stemming from the handrails are formed by the bases of the nucleotides, as with the G and C bases extending inward at the top of the DNA strand. While DNA is double-stranded, RNA generally is single-stranded.

(mRNA): a type of RNA that encodes, and carries to ribosomes, information for the synthesis of proteins. You can see how transcription works by looking at **Figure 14.4**. It will come as no surprise to you that an enzyme is critically involved in this process. **RNA polymerase**, as this complex of enzymes is known, actually undertakes two critical tasks: It unwinds the DNA sequence and then strings together the chain of RNA nucleotides that is complementary to it, thus producing the primary transcript.

Messenger RNA Processing

In prokaryotes such as bacteria, transcription works just as you've seen so far: messenger RNA is synthe-sized directly from a DNA template. It turns out, however, that in eukaryotes such as ourselves, there is an intervening step between getting an initial RNA nucleotide chain and getting a finished messenger RNA molecule. Just as a writer's first draft must undergo some editing before it is a finished work, so the original RNA chain must undergo some editing before it is referred to as messenger RNA. This is so because the starting DNA sequences in eukaryotes are not made up of a continuous series of bases that code for one amino acid after another. Instead, coding sequences of DNA are interspersed with *non*-coding sequences. The initial RNA chain synthesized from the DNA is a copy of

1. RNA polymerase unwinds a region of the DNA double helix.

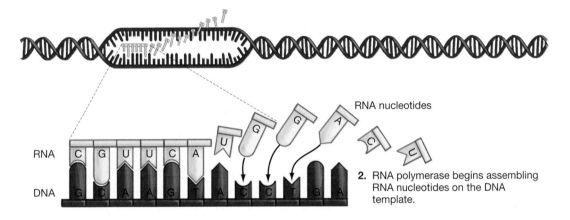

RNA nucleotides

2. RNA polymerase begins assembling RNA nucleotides on the DNA template.

3. The completed portion of the RNA transcript separates from the DNA. Meanwhile, RNA polymerase unwinds more of the untranscribed region of the DNA.

4. The RNA transcript is released from the DNA, and the DNA is rewound into its original form. Transcription is completed.

Figure 14.4
Transcription Works through Base Pairing
Thanks to their chemical similarity, DNA and RNA can engage in base pairing, and this base pairing is how RNA transcripts are synthesized. The enzyme complex RNA polymerase undertakes two tasks in transcription: It unwinds the DNA sequence to be transcribed, and it brings together RNA nucleotides with their complementary DNA nucleotides, thus producing an RNA chain.

both kinds of sequences, but this soon changes. The original RNA chain—the primary transcript—undergoes a kind of molecular editing in which non-coding sequences are snipped out, and the remaining coding sequences are spliced back together. You can see this pictured in **Figure 14.5**. The product of this editing is the messenger RNA molecule. The portions of the primary transcript that are edited out in this process are called *introns* (because they are *in*tervening sequences). Meanwhile, the portions of the primary transcript that are left in during editing are called *exons* (because most of these sequences are *ex*pressed as proteins). Of course, these two kinds of sequences have counterparts in the original DNA. Thus, we can define an **intron** as that part of a primary transcript (or the DNA encoding it) that is removed in mRNA editing. And we can define an **exon** as that part of a primary transcript (or the DNA encoding it) that is retained during mRNA editing.

Taking a step back from this, the message is that eukaryotic DNA has sequences—the introns—that do not code for protein, and that don't even make it into the final mRNA chain. This, of course, raises the question: What are introns doing in DNA in the first place? As you can see in "Making Sense of 'Junk' DNA" on page 280, clues are steadily emerging about the function of these sequences.

A Triplet Code

Thus far, you have been looking at what might be called the flow of genetic information (from DNA to mRNA). A separate issue in protein synthesis, however, has to do with the linkage between DNA and mRNA on the one hand and amino acids on the other. Think of this as scientists did in the 1950s. Allowing that DNA bases are coding for amino acids in a protein, a primary question that arose was: How *many* DNA bases does it take to code for each amino acid? It seemed clear that the answer was not one. Because there are only four DNA bases (A, T, C, and G), if each coded for its own amino acid, only four amino acids could have been incorporated into proteins, as opposed to the 20 that actually are. Nor could the answer be two, because the number of possible amino acids this number could yield up was only 16. Thinking that nature probably would work as economically as possible, the South African–born biochemist Sydney Brenner suggested that DNA worked in a *triplet code*, meaning that each *three* DNA bases specified a single amino acid. Brenner turned out to be right. As **Figure 14.6** shows, each three bases in a DNA sequence pairs with three mRNA bases, but each group of three mRNA bases then codes for a *single* amino acid. Each coding triplet of mRNA bases is known, appropriately enough, as a **codon**.

Once scientists knew that protein production worked by means of a triplet code, they were confronted with a different question: If you have a given three bases, *which* amino acid do they specify? In Morse code, the sounds .—. (short-long-short) code for the letter *P*. But if we know that a given mRNA codon has the base sequence UCC, what *amino acid* does this code for? Today we know that the answer is serine (ser), but at one time this was not clear at all. Indeed, it took years for scientists to figure out all the linkages between codons and amino acids. When this work was completed, however, the result was the **genetic code**, which can be defined as the inventory of linkages between nucleotide triplets and the amino acids they code for. If you look at "Learning to Read Nature's Rule Book" on page 278, you can learn more about the nature and importance of this code.

14.4 A Closer Look at Translation

With a length of RNA having first been transcribed from a length of DNA, having been edited, and having then moved to a ribosome as mRNA, many of the players are in place for translation—the second stage

Figure 14.5
Editing Out Non-Coding Sequences

In eukaryotes, lengths of DNA that are transcribed contain sequences that code for amino acids as well as sequences that do not code for them. Both kinds of sequences are copied onto primary transcripts of RNA. Once this is done, however, the non-coding sequences (called introns) are removed by enzymes through the editing process shown, while the coding sequences (called exons) are spliced back together. The result is a messenger RNA molecule.

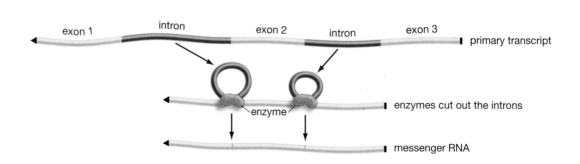

exon 1   intron   exon 2   intron   exon 3   primary transcript

enzyme   enzymes cut out the introns

messenger RNA

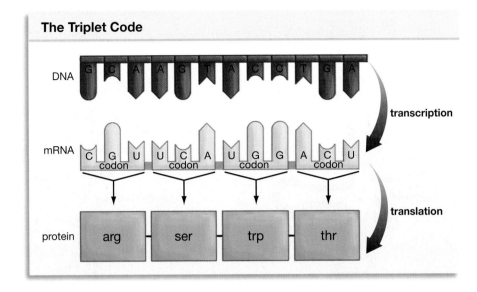

The Triplet Code

DNA

mRNA

transcription

protein   arg   ser   trp   thr

translation

Figure 14.6
Triplet Code
Each triplet of DNA bases codes for a triplet of mRNA bases (a codon), but it takes a complete codon to code for a single amino acid.

of protein production. What's also needed, however, are the building blocks of proteins, amino acids. As noted, they are brought to ribosomes by a second form of RNA, called transfer RNA (or tRNA).

The Nature of tRNA

If you look at **Figure 14.7**, you can see why tRNA is aptly placed within the translation phase of protein synthesis. When, in common discourse,

we think of a *translator*, we think of someone who can communicate in *two* languages. Transfer RNA effectively does this. One end of each tRNA molecule links with a specific *amino* acid, which it finds floating free in the cytoplasm. Then, transferring this amino acid to the ribosome, this tRNA molecule bonds with a *nucleic* acid—a triplet of bases on the mRNA that is moving through the ribosome. Thus, tRNA is bonding

Web Tutorial 14.3
Manufacturing Proteins

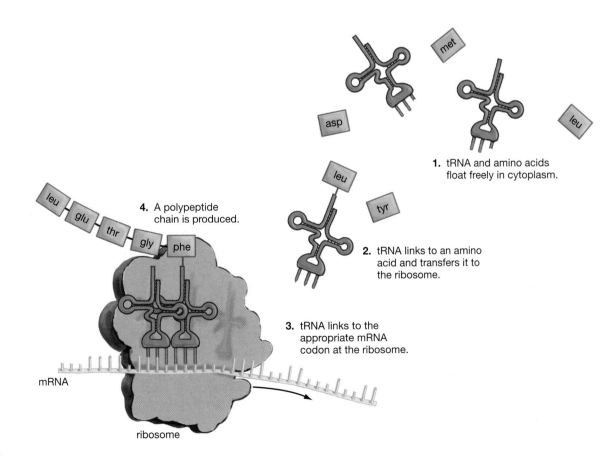

4. A polypeptide chain is produced.

1. tRNA and amino acids float freely in cytoplasm.

2. tRNA links to an amino acid and transfers it to the ribosome.

3. tRNA links to the appropriate mRNA codon at the ribosome.

mRNA

ribosome

Figure 14.7
Bridging Molecule
Transfer RNA (tRNA) molecules link up with amino acids on the one hand and mRNA codons on the other, thus forming a chemical bridge between the two kinds of molecules in protein synthesis. They also transfer amino acids to ribosomes, as shown in the steps of the figure.

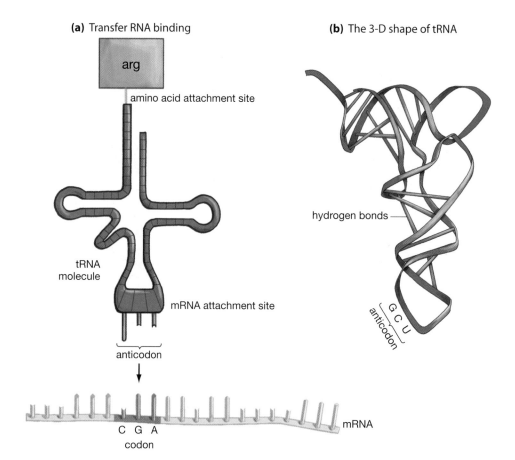

(a) Transfer RNA binding

arg

amino acid attachment site

tRNA
molecule

mRNA attachment site

anticodon

C G A
codon

mRNA

(b) The 3-D shape of tRNA

hydrogen bonds

G C U
anticodon

Figure 14.8
The Structure of Transfer RNA
(a) In this two-dimensional model of transfer RNA, one end of the tRNA molecule can be seen bonding to a specific amino acid (arg), while the other end bonds to its counterpart mRNA codon (CGA).

(b) This three-dimensional model of tRNA shows the molecule in the "folded up" form it takes when carrying out its dual-bonding function.

Table 14.2 Types of RNA		
Type of RNA	**Functions in**	**Function**
Messenger RNA (mRNA)	Nucleus, migrates to ribosomes in cytoplasm	Carries DNA sequence information to ribosomes
Transfer RNA (tRNA)	Cytoplasm	Provides linkage between mRNA and amino acids; transfers amino acids to ribosomes
Ribosomal RNA (rRNA)	Cytoplasm	Structural component of ribosomes

with two very different kinds of molecules—amino acids on the one hand, nucleic acids on the other—thereby serving as a translator between them. **Transfer RNA (tRNA)** can be defined as a form of RNA that, in protein synthesis, bonds with amino acids, transfers them to ribosomes, and then bonds with messenger RNA.

Figure 14.8a provides a more detailed look at how transfer RNA carries out this function. You can see that it bonds with an mRNA codon by means of three bases it possesses, called an **anticodon**. At its other end (the 12 o'clock position in the figure), there is an attachment site for the amino acid. Each tRNA molecule is specific for a particular amino acid, meaning the correct amino acid sequence will result from a given mRNA sequence. **Figure 14.8b** gives you a more realistic idea of what tRNA looks like in its folded-up form.

The Structure of Ribosomes

You've now been introduced to almost all the players that will be active in protein synthesis at the ribosome. But what is the structure of the ribosome itself? If you look at **Figure 14.9**, you can begin to get an idea. Note

(a) Large and small ribosomal units

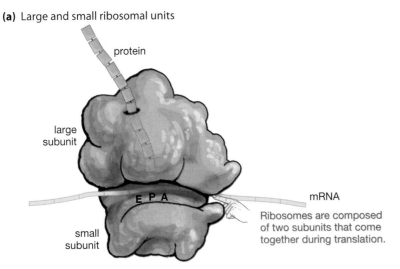

protein

large
subunit

mRNA

E P A

small
subunit

Ribosomes are composed
of two subunits that come
together during translation.

(b) Binding sites in the ribosome

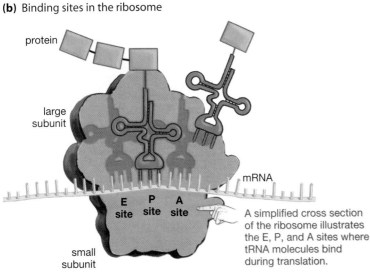

protein

large
subunit

mRNA

E   P   A
site site site

small
subunit

A simplified cross section
of the ribosome illustrates
the E, P, and A sites where
tRNA molecules bind
during translation.

Figure 14.9
The Structure of Ribosomes

that ribosomes are composed of two "subunits"—one larger than the other—both made of a mixture of proteins and yet another type of RNA, **ribosomal RNA (or rRNA)**. The two subunits may float apart from another in the cytoplasm until prompted to come together by the process of translation. You can see that when the subunits have been joined, there exist three binding sites (which it is convenient to think of as slots), one of them an "E" site, the next a "P" site, and the third an "A" site. You'll be looking at their roles

shortly. **Table 14.2** sets forth the three types of RNA reviewed so far that are active in protein synthesis.

The Steps of Translation

With all the players introduced, let's see how translation works. To keep things simple, we'll follow the process as it occurs in prokaryotes. Our starting point is an mRNA transcript that is ready to begin binding with a ribosome. Meanwhile, nearby tRNA molecules have linked to their appropriate amino acids.

transcribed for rRNA. Likewise, there are sections of DNA that code for transfer RNA, which also migrates to the cytoplasm to be involved in translation. Note that in the case of both transfer RNA and ribosomal RNA, DNA is being transcribed. But the ultimate *product* of this transcription is not a protein; it's RNA in two forms. With all this in mind, here is a more accurate definition of a gene: A **gene** is a segment of DNA that brings about the transcription of a segment of RNA.

14.5 Genetic Regulation

The basic understanding you now have of how genes code for proteins leaves a critical question unanswered: Why does a given gene turn on or off? Why does transcription of this gene begin, or not begin? Most genes don't simply stay "on." The cells in our pancreas, for example, turn out the protein insulin *intermittently*—in response to what we've eaten—rather than ceaselessly transcribing the gene for insulin. So, what are the start and stop signals for this protein-synthesizing machinery? For that matter, why is it that only *pancreatic* cells turn out insulin when almost all other cells in the body possess copies of the insulin gene? To come to an understanding of these things, you need to learn something about how protein synthesis is regulated.

DNA Is the Cookbook, Not the Cook

We have thus far thought of the "genome" or genetic inventory of a living thing as a kind of cookbook, containing recipes (DNA sequences) that lead to products (proteins) that undertake a wide variety of tasks (hastening chemical reactions, serving as receptors). It's tempting to go from these ideas to thinking that DNA is the boss, at least for the metabolic processes lying outside our conscious control. After all, it's giving "orders" for the body to put together these all-important proteins, isn't it?

Well, sticking with the metaphor, a cookbook does not give orders; it merely contains information, which an *agent* (called a cook) then carries out. It becomes clear that DNA is not the cook when you consider that by itself, DNA can't synthesize anything. Think back to transcription. The DNA double helix does not unwind itself, nor does it bring RNA nucleotides into alignment with its own bases, nor does it snip out introns in the primary transcript, nor does it affect any of the other steps you looked at. Instead, all these tasks are carried out by enzymes, which are *proteins*. (Recall that the protein complex known as RNA poly-

merase is central in carrying out the steps of transcription.) DNA is indeed the cookbook, but it is powerless to prepare its own recipes. A critical insight is that, through chemical bonding, information comes *to* the double helix as well as from it, in a process known as genetic regulation.

A Model System in Genetic Regulation: The Operon

Genetic regulation actually can occur at any step in the protein synthesis process. But our main focus here will be on regulation as it occurs in transcription—the copying of the DNA information onto a length of RNA. The essential question is: What factors stand to facilitate or impede this process? What factors stand to make it more or less likely that RNA polymerase will latch onto a length of DNA and then go down the line, pairing DNA bases up with complementary RNA bases? To get an idea, you'll look at what's known as a "model" regulatory system in a model organism. The system is called the *operon*, and the organism is the *Escherichia coli* bacterium.

Jacques Monod and François Jacob's Experiments

Jacques Monod and François Jacob were two French researchers, who in the 1950s wanted to learn how genes turn on and off in living creatures. By then it was clear that in most cases, genes didn't simply *stay* on. The activity of at least some genes had to be *inducible*, meaning triggered by conditions in the organism's environment. One means of inducement seemed clear; scientists had known for many years that bacteria would synthesize certain enzyme proteins only if the substance these enzymes worked on (their "substrate") was present in the bacterial cell. Accordingly, Monod and Jacob focused on a group of enzymes that allow *E. coli* first to obtain the sugar lactose and then to break it down into two simple sugars. This latter step is a clipping operation in which lactose (commonly known as milk sugar) is broken into its component simple sugars, galactose and glucose.

In the simplified version of this operation that we'll look at, two genes are involved in *E. coli* lactose metabolism. As you can see in **Figure 14.12a**, there is a *y* gene that codes for a so-called *permease* enzyme, which transports lactose into the cell. Then there is a *z* gene, which codes for the enzyme beta-galactosidase (or β-galactosidase), which does the lactose clipping mentioned before. (A third, *a* gene seems to be part of the system, but no one has yet discovered what it does.)

Web Tutorial 14.4
Genetic Regulation

(a) Large and small ribosomal units

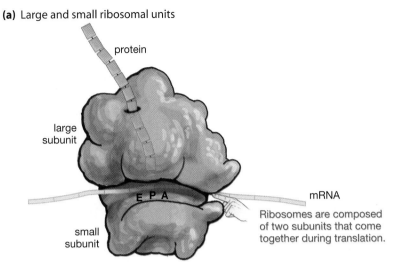

protein

large
subunit

E P A

mRNA

Ribosomes are composed
of two subunits that come
together during translation.

small
subunit

(b) Binding sites in the ribosome

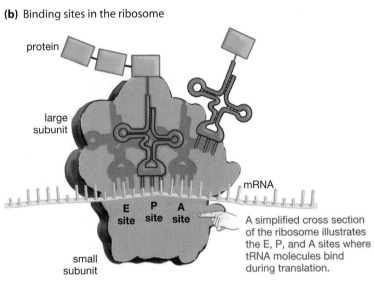

protein

large
subunit

mRNA

E   P   A
site   site   site

A simplified cross section
of the ribosome illustrates
the E, P, and A sites where
tRNA molecules bind
during translation.

small
subunit

Figure 14.9
The Structure of Ribosomes

that ribosomes are composed of two "subunits"—
one larger than the other—both made of a mixture of
proteins and yet another type of RNA, **ribosomal RNA
(or rRNA)**. The two subunits may float apart from another in the cytoplasm until prompted to come together by the process of translation. You can see that
when the subunits have been joined, there exist three
binding sites (which it is convenient to think of as
slots), one of them an "E" site, the next a "P" site, and
the third an "A" site. You'll be looking at their roles

shortly. **Table 14.2** sets forth the three types of RNA
reviewed so far that are active in protein synthesis.

The Steps of Translation

With all the players introduced, let's see how translation works. To keep things simple, we'll follow the
process as it occurs in prokaryotes. Our starting point
is an mRNA transcript that is ready to begin binding
with a ribosome. Meanwhile, nearby tRNA molecules
have linked to their appropriate amino acids.

mRNA Binds to Ribosome, First tRNA Arrives

In this first step, the mRNA chain arrives at the ribosome and binds to the ribosome's small subunit (see Figure 14.10). The mRNA codon AUG is the usual "start" codon for a polypeptide chain. Next, a tRNA molecule with the appropriate anticodon se-

The steps of translation

1. A messenger RNA transcript binds to the small subunit of a ribosome as the first transfer RNA is arriving. The mRNA codon AUG is the "start" sequence for most polypeptide chains. The tRNA, with its methionine (met) amino acid attached, then binds this AUG codon.

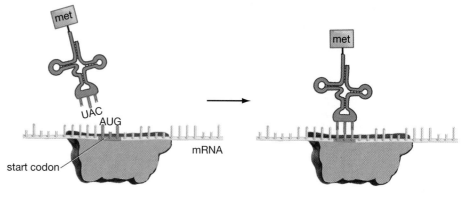

2. The large ribosomal subunit joins the ribosome, as a second tRNA arrives, bearing a leucine (leu) amino acid. The second tRNA binds to the mRNA chain, within the ribosome's A site.

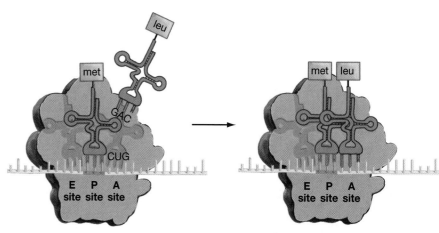

3. A bond is formed between the newly arrived leu amino acid and the met amino acid, thus forming a polypeptide chain. The ribosome now effectively shifts one codon to the right, relocating the original P site tRNA to the E site, the A site tRNA to the P site, and moving a new mRNA codon into the A site.

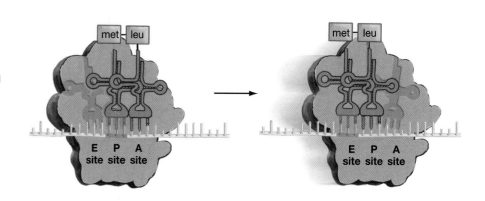

4. The E site tRNA leaves the ribosome, even as a new tRNA bonds with the A site mRNA codon, and the process of elongation continues.

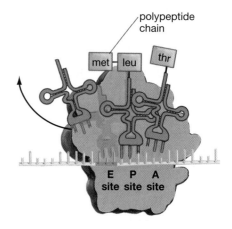

Figure 14.10
The Steps of Translation

quence (UAC) binds to this AUG codon. This tRNA arrives bearing its appropriate amino acid, which is methionine (met). Following this, the large ribosomal subunit becomes part of the ribosome, providing the ribosome's A, P, and E binding sites.

Polypeptide Chain Is Elongated

Now amino acids will start to be joined together in a chain. As you can see, this elongation process begins with a second incoming tRNA molecule binding to an mRNA codon in the A site. Because it happens to be a CUG codon, a tRNA with a GAC anticodon binds to it. This tRNA comes bearing the amino acid leucine (leu). Next, the met amino acid attached to the tRNA in the P site bonds with the leu amino acid attached to the tRNA in the A site. In this process, the bond is broken between met and its original tRNA.

Once this occurs, a kind of molecular musical chairs ensues: The ribosome effectively shifts one codon to the right. With this, the tRNA that had been in the P site is relocated to the E site. What does the E stand for? Exit. The tRNA in this site bears no amino acid now; soon it will be ejected from the ribosome altogether. Meanwhile, the tRNA that had been in the A site moves to the P site, and a new codon shifts into the now-vacated A site. You can see what comes next: A tRNA bonds with this new codon in the A site. Soon, the growing polypeptide chain will bond with *this* tRNA's amino acid, and the process will continue.

Termination of the Growing Chain

There are three separate codons that don't code for any amino acid, but that instead act as stop signals for polypeptide synthesis. Any time one of these "termination" codons moves into the ribosome's A site, it doesn't bind with an incoming tRNA, but instead brings about a severing of the linkage between the P-site tRNA and the polypeptide chain. Indeed, the whole translation apparatus comes apart at this point, with the polypeptide chain being released to fold up and be processed as a protein. Translation has been completed.

Speed of the Process; Movement through Several Ribosomes

How fast does this process go? Very fast. An *E. coli* bacterium can string together up to 40 amino acids per second, meaning that an average-sized protein of about 400 amino acids could be put together in 10 seconds. Note, however, that mRNA tapes often are read not by one ribosome, but by many. As you can see in **Figure 14.11**, several ribosomes—perhaps scores of them—might move over a given mRNA transcript, with the result that identical polypeptide chains grow out of each ribosome. This greatly increases the number of proteins that can be put together in a given period of time.

What Is a Gene?

This trip through transcription and translation has made it possible for you to get a more accurate definition of the most basic unit in genetics, the gene. Thus far, a gene has been viewed as a length of DNA that codes for a protein. But recall that ribosomes are made partly of RNA. Where does this ribosomal RNA come from? Just like messenger RNA, ribosomal RNA must be coded for by DNA. A segment of DNA unwinds and then forms base pairs with RNA nucleotides to produce an RNA sequence. Only this is a *ribosomal* RNA sequence that doesn't code for anything; it simply migrates to the cytoplasm to become part of a ribosome. So great is the cell's need for ribosomes that there is an entire section of the nucleus (called the nucleolus) whose DNA is constantly being

(a)

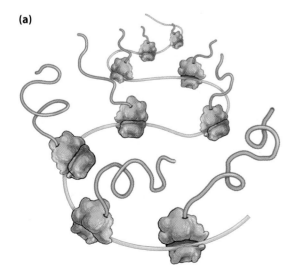

(b)

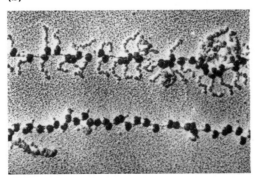

Figure 14.11
Mass Production
(a) An mRNA transcript can be translated by many ribosomes at once, resulting in the production of many copies of the same protein.

(b) A micrograph of this process in operation. The figure shows two mRNA strands with ribosomes spaced along their length. In the upper strand, translation is under way and polypeptides can be seen emerging from the ribosomes.

transcribed for rRNA. Likewise, there are sections of DNA that code for transfer RNA, which also migrates to the cytoplasm to be involved in translation. Note that in the case of both transfer RNA and ribosomal RNA, DNA is being transcribed. But the ultimate *product* of this transcription is not a protein; it's RNA in two forms. With all this in mind, here is a more accurate definition of a gene: A **gene** is a segment of DNA that brings about the transcription of a segment of RNA.

14.5 Genetic Regulation

The basic understanding you now have of how genes code for proteins leaves a critical question unanswered: Why does a given gene turn on or off? Why does transcription of this gene begin, or not begin? Most genes don't simply stay "on." The cells in our pancreas, for example, turn out the protein insulin *intermittently*—in response to what we've eaten—rather than ceaselessly transcribing the gene for insulin. So, what are the start and stop signals for this protein-synthesizing machinery? For that matter, why is it that only *pancreatic* cells turn out insulin when almost all other cells in the body possess copies of the insulin gene? To come to an understanding of these things, you need to learn something about how protein synthesis is regulated.

DNA Is the Cookbook, Not the Cook

We have thus far thought of the "genome" or genetic inventory of a living thing as a kind of cookbook, containing recipes (DNA sequences) that lead to products (proteins) that undertake a wide variety of tasks (hastening chemical reactions, serving as receptors). It's tempting to go from these ideas to thinking that DNA is the boss, at least for the metabolic processes lying outside our conscious control. After all, it's giving "orders" for the body to put together these all-important proteins, isn't it?

Web Tutorial 14.4
Genetic Regulation

Well, sticking with the metaphor, a cookbook does not give orders; it merely contains information, which an *agent* (called a cook) then carries out. It becomes clear that DNA is not the cook when you consider that by itself, DNA can't synthesize anything. Think back to transcription. The DNA double helix does not unwind itself, nor does it bring RNA nucleotides into alignment with its own bases, nor does it snip out introns in the primary transcript, nor does it affect any of the other steps you looked at. Instead, all these tasks are carried out by enzymes, which are *proteins*. (Recall that the protein complex known as RNA poly-

merase is central in carrying out the steps of transcription.) DNA is indeed the cookbook, but it is powerless to prepare its own recipes. A critical insight is that, through chemical bonding, information comes *to* the double helix as well as from it, in a process known as genetic regulation.

A Model System in Genetic Regulation: The Operon

Genetic regulation actually can occur at any step in the protein synthesis process. But our main focus here will be on regulation as it occurs in transcription—the copying of the DNA information onto a length of RNA. The essential question is: What factors stand to facilitate or impede this process? What factors stand to make it more or less likely that RNA polymerase will latch onto a length of DNA and then go down the line, pairing DNA bases up with complementary RNA bases? To get an idea, you'll look at what's known as a "model" regulatory system in a model organism. The system is called the *operon*, and the organism is the *Escherichia coli* bacterium.

Jacques Monod and François Jacob's Experiments

Jacques Monod and François Jacob were two French researchers, who in the 1950s wanted to learn how genes turn on and off in living creatures. By then it was clear that in most cases, genes didn't simply *stay* on. The activity of at least some genes had to be *inducible*, meaning triggered by conditions in the organism's environment. One means of inducement seemed clear; scientists had known for many years that bacteria would synthesize certain enzyme proteins only if the substance these enzymes worked on (their "substrate") was present in the bacterial cell. Accordingly, Monod and Jacob focused on a group of enzymes that allow *E. coli* first to obtain the sugar lactose and then to break it down into two simple sugars. This latter step is a clipping operation in which lactose (commonly known as milk sugar) is broken into its component simple sugars, galactose and glucose.

In the simplified version of this operation that we'll look at, two genes are involved in *E. coli* lactose metabolism. As you can see in **Figure 14.12a**, there is a *y* gene that codes for a so-called *permease* enzyme, which transports lactose into the cell. Then there is a *z* gene, which codes for the enzyme beta-galactosidase (or β-galactosidase), which does the lactose clipping mentioned before. (A third, *a* gene seems to be part of the system, but no one has yet discovered what it does.)

Activity Induced by Lactose

The genes involved in lactose metabolism were known to be inducible, and the *inducer*—the substance that prompted their activity—was known to be lactose. Put simply, in the presence of lactose, these genes turned on. Yet what was the control mechanism here? The experiments of Jacob and Monod eventually led them to propose that *E. coli*'s lactose metabolism is governed by an elegant multipart genetic system that they dubbed the *operon*.

(a) Structure of the lac operon system

A simplified structure of the lac operon system. The system consists of (left to right) an *i* gene that codes for a repressor protein (the two green crosses); RNA polymerase, shown bound to the DNA strand's promoter region (P); a region of the DNA strand called the operator (O); two genes and their protein products: The *z* gene that codes for the lactose-clipping beta-galactosidase enzyme; and the *y* gene, which codes for a permease enzyme, which transports lactose into the cell. (The *a* gene, while part of the system, codes for a protein whose function is unknown.)

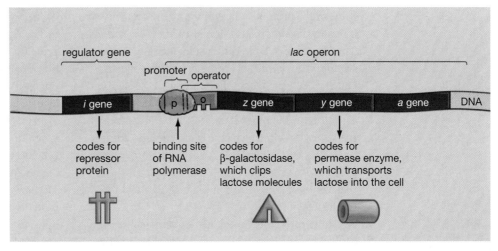

(b) When lactose is absent …

The repressor protein binds to the operator and prevents RNA polymerase from ever beginning transcription of the lactose-processing enzymes.

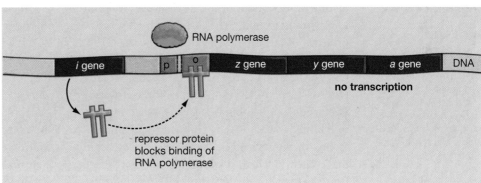

(c) When lactose is present …

The repressor is inhibited from binding with the operator, thus allowing transcription and hence production of the lactose-processing enzymes.

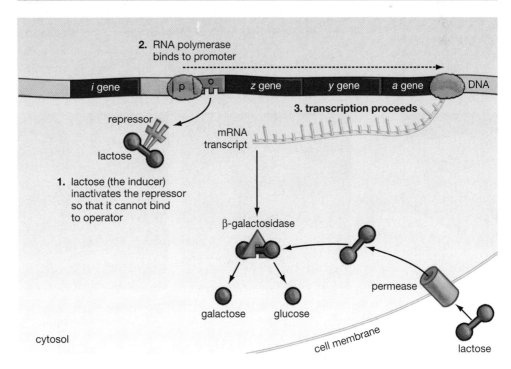

Figure 14.12
Genetic Regulation in Action

In the *lac* operon, the *z*, *y*, and *a* genes are transcribed as a unit onto a single messenger RNA transcript. "Upstream" from the *z* gene is a segment of DNA known as the promoter sequence. Any **promoter sequence** is the site on a segment of DNA to which RNA polymerase attaches prior to beginning transcription. Note, in the figure, however, that *in between* the promoter site and the *z* gene sequence is a region that Jacob and Monod called the *operator*. This sequence of DNA can exist in one of two states. In one, it is bound to a protein called a *repressor*, which effectively *blocks* the binding of RNA polymerase to the promoter, thus shutting down transcription. In the other state, it is unbound—free of the repressor—in which case transcription can proceed. Because the repressor is a protein, it is itself the product of a gene (the *i* gene) that lies just a little farther yet upstream. (You can see it in Figure 14.12a, on the left.)

Now, the critical question is: What causes the repressor either to bind with the operator (thus halting transcription) or to stay clear of it (thus allowing transcription)? The answer is the absence or presence of lactose. If you look at **Figure 14.12b**, you can see how this plays out. The critical factor is that the repressor protein has *two* binding sites. One of them binds to the operator, as already noted, but the other binds to *lactose*. When this lactose binding takes place, a repressor that is bound to an operator changes shape; like a clothespin that's been pinched, it opens up and is released from the operator site. The result? RNA polymerase is no longer blocked, meaning it can bind to the promoter site and begin transcription of the *z*, *y*, and *a* genes. They produce their enzymes, and lactose metabolism proceeds. When all the lactose has been broken down, however, the repressor molecules return to their original shape—meaning the shape that can bind to the operator—thereby shutting down gene transcription.

With their insights about this system, Jacob and Monod elucidated a great deal about gene regulation. First, they confirmed the existence of genetic sequences whose sole functions are *regulatory*. Think of it: The *i* gene sequence codes for a protein, but it is a protein whose function is to allow DNA transcription to go forward, or to turn it off. Meanwhile, the operator DNA segment doesn't code for a protein at all; instead it exists as a binding site for this repressor protein. Such regulatory sequences stand in distinction to the DNA segments that code for end-product proteins—in this series, the *z*, *y*, and *a* genes. Next, the operon concept confirmed the nature of *feedback* mechanisms that operate in genetic regulation. The proteins that are produced through DNA's instructions feed back on the DNA molecule itself, thus helping to control the production of other proteins.

The *lac* operon concept has been deepened since this original understanding was arrived at. In particular, it was found that a completely different protein (called CAP) had to bind with DNA upstream from the *lac* promoter for RNA polymerase to readily bind with the *lac* promoter site. Looked at one way, the CAP protein facilitates the *lac* operation, while the repressor protein blocks it. And with this added layer of control, you can see the basic elements that operate in almost all transcription regulation, no matter how elaborate it may get (and it gets very elaborate). Some molecules are going to facilitate the binding of RNA polymerase to the promoter, while other molecules are going to impede this binding. Transcription is contolled by the sum of all these influences.

Figure 14.13
Not as Much Difference as We Thought
At one time, scientists assumed that the human genome contained about 100,000 genes. Genome sequencing revealed, however, that the human genome probably contains about 30,000 genes. This is only about 11,000 more genes than the tiny roundworm *C. elegans*, and about 17,000 more than the *Drosophila* fruit fly. Scientists can make comparisons among genomes because, though the human genome has gotten most of the attention, the genomes of several other organisms have now been sequenced as well, including those shown here.

Saccharomyces cerevisiae
(baker's yeast)

Drosophila melanogaster
(fruit fly)

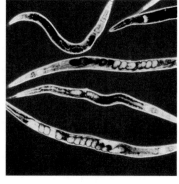

Caenorhabditis elegans
(roundworm)

Arabidopsis thaliana
(mustard plant)

Estimated number of genes:	6,034	13,061	19,099	25,000

Another Layer of Control: Alternate Splicing

If we ask how the genetic machinery is able to turn out its fabulous array of proteins, an additional layer of control can be found in a step leading to protein synthesis that comes after the transcription we've just gone over. When biologists completed the first stage of the Human Genome Project in 2001, one of the surprises was that the human genome was shown to contain far fewer genes than the 100,000 that had been the standard estimate for years. The teams that sequenced the human genome came up with consensus estimates of about 30,000 genes (though some scientists still believe the number could be twice that). The 30,000 figure actually held a double surprise because it meant that, contrary to our previous assumptions, human beings do not have five times or ten times as many genes as, say, fruit flies or roundworms. If you look at **Figure 14.13**, you can see that we humans—with our vision, speech, and 100 trillion cells—probably have only 11,000 more genes than the microscopic roundworm *C. elegans*, which is eyeless, speechless, and made up of 959 cells.

Once this was recognized, the challenge to science was to understand how human beings do so much more with their 30,000 genes than *C. elegans* does with its 19,000. One part of the answer seems clear: Human beings have a relatively high proportion of multifunctional genes, meaning genes capable of coding for more than one protein. If you look at **Figure 14.14**, you can see the "alternative splicing" process by which organisms such as ourselves can get different proteins from a single gene. Note that the introns you looked at earlier aid in this process by dividing genes into "modules" that can be spliced together in different ways. This is one more indication of the valuable role these non-coding sequences play in human genetics. By some estimates, 60 per-

cent of human genes may undergo alternate splicing. The upshot of this is that, though we probably have only 30,000 genes, we may have hundreds of thousands of proteins functioning within us.

14.6 The Magnitude of the Genetic Operation

Our 30,000 genes are contained in a genome that is about 3.2 billion base pairs long. To make this large number a little less abstract, consider the following exercise. If we took the base sequence of the human genome and simply arrayed the single-letter symbols for the bases on a printed page, going like this:

> AATCCGTTTGGAGAAACGGCCCTATTG
> GCAGCAAGGCTCTCGGGTCGTCAACG
> CGTATTAAACATATTTCAAGGCTCTA …

it would take about 1,000 telephone books, each of them 1,000 pages long, merely to record it all. We're talking, then, about an unbroken series of these base symbols that goes on for a million pages. That is one measure of the size of the human genome. Simply maintaining this genome—making sure that it is passed on from one cell and generation to another—requires that all of these 3.2 billion bases be faithfully copied each time a cell divides. (They have to be copied twice, actually, because we have two copies of each chromosome.) When we start thinking about *using* this genome—about producing proteins from it—all the layers of genetic regulation sketched out earlier (and more) come into play. Genes help bring about proteins; some of these proteins feed back on other genes and control their production of other proteins; RNA sequences get

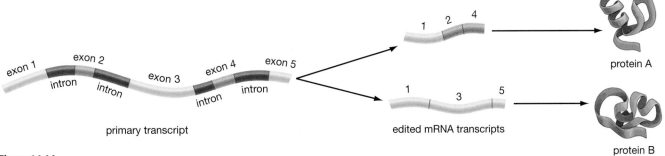

Figure 14.14
Alternative RNA Splicing
The human genome may contain only 30,000 genes, but that does not mean these genes code for only 30,000 proteins. Through the process of alternative RNA splicing, it is possible for a single gene to code for several proteins. In the figure, a single primary transcript, containing five exons or protein-coding portions, has been transcribed from a single gene. The human genetic machinery can then splice these exons together in alternative ways to yield different proteins—A or B in the figure.

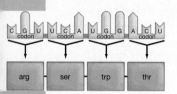

Learning to Read Nature's Rule Book: The Genetic Code

Like scholars trying to extract meaning from an ancient written language, biologists in the early 1960s were trying to extract meaning from an ancient molecular language. They knew that triplets of mRNA bases were coding for amino acids, but what was the linkage? If you had a given triplet, which amino acid did it code for? Though it took them the better part of a decade to find out, eventually they succeeded. If you look at **Figure 1**, you can see a summation of their work—the genetic code in its entirety. Let's look a little closer at this code to understand its nature and significance.

One of the notable things about the code is that not every mRNA triplet in it codes for an amino acid. Note, in Figure 1, that three mRNA codons specify "stop" codes that bring an end to the synthesis of a polypeptide chain. One triplet (AUG) specifies a special amino acid, methionine (met), which serves as the starting amino acid for most polypeptide chains. Note also that the genetic code is redundant. The amino acid phenylalanine (phe), for example, is coded for not only by the UUU triplet, but by UUC as well. Indeed, almost all the amino acids are coded for by more than one mRNA codon. Leucine (leu) and serine (ser) are coded for by no fewer than six.

With the deciphering of these linkages, scientists were handed a powerful tool: If they knew one "end" of a genetic linkage, they instantly were given insight into the other. If they saw, with-in a protein suspected of causing disease, an abnormally high number of "repeats" of the amino acid glutamine (gln), they knew just who the culprit was: an mRNA sequence that had within *it* an abnormally high number of repeats of the triplet CAG (or CAA). And this, in turn, led directly to the faulty DNA sequence—they could start looking for a DNA sequence whose telltale signature was the faulty series of repeats. With the cracking of the code, knowing the protein meant being able to make predictions about DNA sequences and vice versa. It was as if scientists had been given a critical page from a book called *Important Rules of Nature*.

A second lesson from the code was that it revealed how unified life is at the genetic level. As biologists looked among various species at the linkages between DNA sequences and amino acids, they found that with only a few exceptions, the genetic code is universal in all living things. This means that the base triplet CAC codes for the amino acid histidine, whether this coding is going on in a bacterium or in a human being.

This insight turned out to be important in two ways. First, it is evidence that all life on Earth is derived from a single ancestor. How else can we explain all of life's diverse organisms sharing this very specific code? Such a complex molecular linkage—this triplet equals that amino acid—is extremely unlikely to have evolved *more* than once. What seems likely is that

spliced in different ways, yielding different proteins from the same original transcript, and so forth.

The obvious message here is that the human genetic operation is unimaginably complex. But it turns out that humans are not special in this regard. If we were to walk through the genetic operation of even a humble bacterium, you would see that its complexity is only slightly less stunning than ours. What's important to recognize is that this complexity is necessary because it *enables life*. Living things must obtain energy, they must grow, they must respond constantly to their environ-ment, they must reproduce, and they must coordinate all these activities. Only an incredibly sophisticated system of information storage and use could allow a self-contained entity to do all these things. And genetics is exactly such a system. With each twitch of a muscle or swig of a soft drink, information repositories called genes are opened or closed within us, setting in motion the delivery of the tiny working entities called proteins—or stop-ping delivery of them, if that's what is called for. Not surprisingly, we are nowhere close to having a complete understanding of how any genetic sys-

it came to be employed in an ancient common ancestor, after which it was passed on to all of this organism's descendants—every creature that has subsequently lived on Earth. Note, then, that this code has been passed on from one generation to the next, and from one evolving species to the next, for billions of years. We humans share in an informational linkage that stretches back billions of years and runs through all the contemporary living world.

Apart from this, the universality of the genetic code has a very practical consequence. It means that genes from one organism can function in another. This has both good and bad consequences for human beings. First the bad: Viruses that cause diseases ranging from colds to AIDS can "hijack" the human cellular machinery for their own purposes, precisely because their genes function in human cells. (The human cellular machinery will put together proteins whether these proteins are called for by human or viral DNA.) Now the good news: Using "biotechnology" processes you'll be learning about in Chapter 15, human beings can today use viruses and bacteria to manufacture all kinds of products, including medicines such as human insulin and human growth hormone. In these cases, human genes are being put to work inside microorganisms. This transferability of genes from one species to another has, as its basis, the universality of the genetic code.

first base	second base: U	second base: C	second base: A	second base: G	third base
U	UUU, UUC: phe UUA, UUG: leu	UCU, UCC, UCA, UCG: ser	UAU, UAC: tyr UAA **Stop** UAG **Stop**	UGU, UGC: cys UGA **Stop** UGG trp	U C A G
C	CUU, CUC, CUA, CUG: leu	CCU, CCC, CCA, CCG: pro	CAU, CAC: his CAA, CAG: gln	CGU, CGC, CGA, CGG: arg	U C A G
A	AUU, AUC, AUA: ile AUG met (start)	ACU, ACC, ACA, ACG: thr	AAU, AAC: asn AAA, AAG: lys	AGU, AGC: ser AGA, AGG: arg	U C A G
G	GUU, GUC, GUA, GUG: val	GCU, GCC, GCA, GCG: ala	GAU, GAC: asp GAA, GAG: glu	GGU, GGC, GGA, GGG: gly	U C A G

Figure 1
The Genetic Code Dictionary
If we know what a given mRNA codon is, how can we find out what amino acid it codes for? This dictionary of the genetic code offers a way. Say you want to confirm that the codon CGU codes for the amino acid arginine (arg). Looking that up here, C is the first base (go to the C row along the "first base" line), G is the second base (go to the G column under the "second base" line) and U is the third (go to the codon parallel with the U in the "third base" line).

tem works. But, as one scientist has said, we are at "the end of the beginning" of our understanding—we understand the process in outline; now it's on to the details.

Biotechnology Is Next

You have seen in this chapter that DNA is only one of the molecules governing protein synthesis. But saying that something is simply one part of a chain is not saying it is not powerful. DNA may "only" be the cookbook, but think of the profound effect this single function stands to have. Following the metaphor, think how different a pie would be if an editing error in a recipe resulted in the insertion of the word *cinnamon* where *sugar* should be (and if our cook had to insert the cinnamon). Now imagine the change for the better if you could *change the recipe* to include sugar instead of cinnamon. The ability to manipulate DNA's instructions lies at the heart of what is commonly called biotechnology. This exciting, promising area of biology is the subject of Chapter 15.

to bond with DNA sequences, thereby controlling the production of other proteins.

- The human genome contains an estimated 30,000 genes—only about 11,000 more genes than exist in the primitive roundworm *C. elegans*. Human complexity is partly attributable to the fact that most human genes appear to be able to code for more than one protein, through the process known as alternative splicing of RNA transcripts.

  Web Tutorial 14.4: Genetic Regulation

14.6 The Magnitude of the Genetic Operation

- Life is made possible by the fantastic ability of genetic systems to store and use information.

Key Terms

anticodon	270	promoter sequence	276
codon	268	ribosomal RNA (rRNA)	271
exon	268	ribosome	265
gene	274	RNA polymerase	267
genetic code	268	transcription	265
intron	268	transfer RNA (tRNA)	270
messenger RNA (mRNA)	266	translation	265
polypeptide	263		

Understanding the Basics

Multiple-Choice Questions (answers in the back of the book)

1. A chain of _____ results in a polypeptide chain that folds up into a _____.
   a. RNA bases … transcript
   b. amino acids . … gene
   c. proteins . … ribosome
   d. DNA bases … gene
   e. amino acids … protein

2. How does DNA differ from RNA?
   a. DNA uses the bases A, T, C, G; RNA uses the bases A, U, C, G.
   b. DNA is a double-stranded molecule; RNA is usually single-stranded.
   c. DNA is a nucleic acid; RNA is a protein.
   d. a and b

3. The press reported recently that, within a large liver cancer tumor, doctors found small bone fragments, fingernails, and skeletal muscle tissue. What can you infer from this about the tumor cells?
   a. that they are functioning normally
   b. that genetic regulation in them has broken down
   c. that cancerous cells need many different kinds of tissue
   d. that the cells came originally from bone-, fingernail-, and muscle-generating tissue
   e. that liver cells can give rise to many kinds of tissue

4. What mRNA sequence signals the start of a sequence to be translated?
   a. ATG
   b. UGA
   c. AUG
   d. UAG
   e. UAA

5. An intron is:
   a. a portion of a primary transcript or DNA that is transcribed that does not contribute to mRNA
   b. the structure in the cytoplasm that is the site of protein synthesis
   c. the coding portion of a length of DNA or a primary transcript of RNA
   d. always a non-coding sequence of bases, as found either in DNA or a primary transcript
   e. both a and d

6. Which of the following statements is true regarding ribosomes?
   a. They are composed of two separate units, each made of rRNA combined with protein.
   b. They are used to translate the mRNA sequence into a protein composed of amino acids.
   c. They have three different sites for tRNA binding.
   d. a and b only
   e. all of the above

7. Why can transfer RNA be referred to as a "bridging" molecule?
   a. It forms a chemical bridge between DNA and messenger RNA.
   b. It creates the linkage between the base pairs of the double helix.
   c. It links species through evolution.
   d. It links ribosomes together.
   e. It forms chemical bonds with both amino acids and messenger RNA.

8. A typical protein might be _____ amino acids long and could be produced in as little as _____ in translation.
   a. 50 … five minutes
   b. 50,000 . … five hours
   c. 5 … five seconds
   d. 400 … 10 seconds
   e. 4,000 … four minutes

9. The *lac* operon provided an example of feedback in genetic regulation. What was feeding back onto what?
   a. A protein produced through transcription and translation of one gene went on to regulate transcription of a different gene.
   b. Ribosomes produced through the information encoded in DNA went on to serve as the sites of protein synthesis.
   c. Messenger RNA strands synthesized as a complement to DNA went on to reduce DNA transcription.

d. Transfer RNA molecules limited translation at the ribosomes.

e. Transfer RNA molecules sped up translation in mitochondria.

10. The genetic code is an inventory of which _____ specify(ies) which _____.

   a. three mRNA bases … amino acid
   b. mRNA base . … amino acid
   c. amino acids … proteins
   d. polypeptide chain … protein
   e. three DNA bases … three mRNA bases

Brief Review

1. What two great tasks are carried out by our genetic machinery?

2. What are the three types of RNA discussed in this chapter? What do their abbreviations stand for? What are their functions?

3. What is the difference between transcription and translation?

4. During transcription, DNA neither unwinds itself nor pairs its own bases with complementary RNA bases. How are these tasks accomplished?

5. What is the difference between introns and exons?

6. Why is it inaccurate to conceptualize DNA as the sole controller of a cell's production of proteins?

7. How many DNA bases does it take to code for an RNA codon? How many amino acids does an RNA codon code for?

Applying Your Knowledge

1. Why does the activity of at least some genes need to be inducible?

2. Given the following sequence of mRNA, what would the resulting amino acid sequence be?

   AUGAAACGGGGACCAAUGGAUAACUAA

3. In a hypothetical situation, you have identified a new species that uses six bases, instead of three, in its genetic code. It has also been discovered that the proteins this species forms are made up of 220 amino acids. To ensure that each amino acid has a separate and distinct encoding sequence, how many bases need to be included in each codon?

The Future Isn't What It Used to Be

15 *Biotechnology*

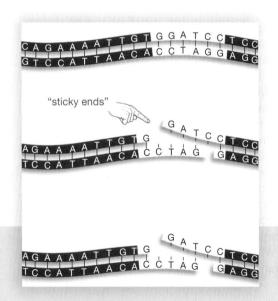

Snipped by restriction enzymes.
(Section 15.2, page 290)

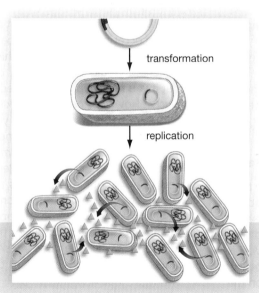

Bacterial protein factories.
(Section 15.2, page 291)

Hello, Dolly!
(Section 15.3, page 293)

During summertime in America, many a suburban and country resident takes pride in growing a lush, green lawn. The bane of weekend life for many of these same people, however, is that these lawns need to be mowed so often. For millions of Americans, one of the givens in life—right up there with paying taxes and dying—is the necessity of regularly cutting the grass.

Now, however, this given may be up for grabs. A giant American lawn products company currently is field testing a strain of grass that has been genetically engineered to make it grow slowly—so slowly that it might need to be cut only twice from spring through fall. And this "low-mow" trait may be just one of several changes in store for grass. By manipulating grass genes, it should be possible for grass to come in different *colors* (reds, blues, various greens) or to stay green even in a drought.

When you hear the word "biotechnology," designer grass may not immediately come to mind. But in this one product, it's possible to see most of the elements that make biotechnology seem both so promising and yet so threatening. Note that designer grass:

- Shakes up our notions of what is possible in the world.
- Has a big commercial potential. The market for low-mow grass has been estimated at a whopping $10 billion per year.
- Could have a personal impact. If brought to market, low-mow grass would affect not only businesses (think of golf courses) but personal lives as well. It would mean a change in how our lawns look and how we spend our time.
- Has critics who would like to see it go away entirely. The American Society of Landscape Architects petitioned the U.S. Department of Agriculture to suspend low-mow field tests until the department determined that pollen from the grass won't mix or "hybridize" with strains of natural grasses. The fear is that low-mow traits could spread, with unknown ecological consequences.

How it was done.
(Section 15.3, page 294)

Can extinction be reversed?
(Section 15.3, page 296)

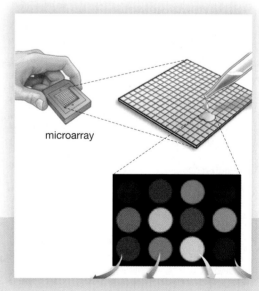

An array of possibilities.
(Section 15.5, page 300)

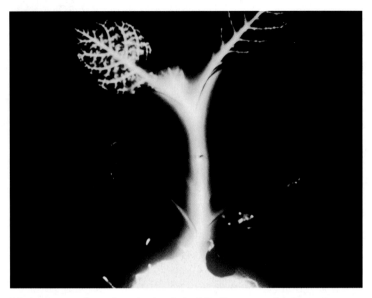

(a) A tobacco plant glows in the dark. A firefly gene spliced into the plant has allowed it to produce the firefly's fluorescent enzyme, luciferase.

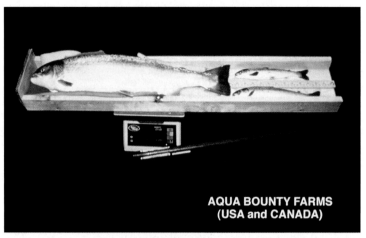

**AQUA BOUNTY FARMS
(USA and CANADA)**

(b) Bigger salmon faster. All the Atlantic salmon in the picture are about 14 months old, but the salmon on the left grew to 3 kilograms in this time-frame, while the other salmon won't reach this "market weight" for another year. The difference? Scientists from the biotech firm Aqua Bounty Farms spliced a DNA sequence from an ocean pout into the salmon's genome . At maturity, however, the genetically engineered salmon will weigh no more than the wild variety.

(c) Mother and clone. The Italian foal in the foreground, named Prometea, is a clone of the mare standing by her. DNA was extracted from skin cells of the mare and then put into an egg whose DNA had been removed. In a twist on the cloning procedure, the resulting embryo was then implanted back into this very mare, who went on to give birth to Prometea. In genetic terms, the two horses could be thought of as identical twins.

(d) Human medicine from goats. A firm called GTC Biotherapeutics has specialized in splicing DNA into the genomes of goats. The result is goat's milk that contains therapeutic human proteins, among them a protein called AT3 that prevents the formation of blood clots.

**Figure 15.1
Biotechnology in Many Forms**

If the word "revolution" can be defined as "overturning an established order," then surely biotechnology is ushering in a revolution. To judge for yourself, look at the pictures in **Figure 15.1**. Our assumption may be that human beings and goats are fundamentally separate varieties of living things. But note that the goats in Figure 15.1 have a *human* gene in them. As a result, the milk they give contains a medically useful human protein. Our assumption may be that fireflies glow and that plants don't. But note that the tobacco plant in Figure 15.1 is all lit up, thanks to a firefly gene that has been spliced into it. If biotechnology were about nothing more than this kind of genetic "mixing" be-tween species, it would be important enough. But as you'll see, its reach is much broader yet. It affects the criminal justice system, the diagnosis of illness, and even the preservation of endangered species.

15.1 What Is Biotechnology?

The term biotechnology is familiar to most of us, but what does it really mean? In the broadest sense, **biotechnology** can be defined as the applica-tion of technology to natural biological process-es. One such process involves the way a cell copies its DNA every time it divides. In the 1980s, a Cal-ifornia scientist realized there was a way to use technology to exploit this natural copying, such that a tiny DNA sample could rapidly be copied over and over, thus yielding a relatively large DNA sample. One result of this "PCR" process is that police can now recover, from the tiniest specks of blood or tissue, enough DNA to get a "match" in a criminal case. (Even a licked postage stamp can yield enough DNA to get a convic-tion.) Like other biotech processes we'll be look-ing at, PCR was a marriage of biological and technological insight. By the time you finish this chapter, you'll see why "biotechnology" is such an appropriate term.

Because biotechnology is so varied, any account of it has to be limited to a few examples chosen from among thousands that exist. In this chapter, we will consider four aspects of biotechnology—aspects that overlap one another somewhat, but that are nevertheless distinct. These are:

- Transgenic biotechnology—the splicing of DNA from one species into another
- Reproductive cloning—the production of mam-mals through cloning
- Forensic biotechnology—the use of biotechnolo-gy to establish identities (of criminals, of crime victims, and so forth)
- Personalized medicine—the development of med-ical treatments that are tailored to individuals in accordance with their unique genetic makeup

Let's start now by reviewing an area of biotechnology in which genes are moved from one organism to another.

15.2 Transgenic Biotechnology

Human beings grow to their full height under the influence of human growth hormone (HGH), which normally is secreted by the human pituitary

gland. HGH's role in promoting growth is, of course, most important during childhood and ado-lescence. A faulty pituitary gland can greatly reduce the amount of HGH young people have in their system, leaving them abnormally short. For years, the only way to get HGH was to laboriously extract it from the pituitaries of dead human beings, a practice that not only yielded too little HGH to go around but that also turned out to be unsafe.

Enter biotechnology, which in the mid-1980s produced synthetic HGH in the following way. Using collections of human cells, the gene for HGH was snipped out of the human genome and inserted into the *E. coli* bacterium. Each bacterium that took on the HGH gene began transcribing and then translating this gene, which is to say turning out a small quantity of HGH. These bacteria were then grown in vats by the billions. The result? Collectible quantities of HGH, clinically indistinguishable from that produced in human pituitary glands, manufactured by a biotech firm, and shipped to pharmacies worldwide.

Note that in this process, a gene was taken from one species (a human being) and spliced into an-other species (a bacterium). With this, the bacteri-um became a **transgenic organism**: an organism whose genome has stably incorporated one or more genes from another species. This same phenome-non is repeated time and time again in biotechnolo-gy. The goats in Figure 15.1 are transgenic—they incorporated a human gene into their genome. The tobacco plant in Figure 15.1 is transgenic—it incor-porated a firefly gene into its genome. But how is this possible? How do you cut DNA out of one genome and paste it into another? To get an answer, we'll take a walking tour of some basic biotech processes. As a starting point, imagine that as with human growth hormone, the goal is to produce quantities of a hypothetical human protein through the use of a living organism.

A Biotech Tool: Restriction Enzymes

In the early 1970s it became possible for scientists to cut genomes at particular places, with the dis-covery of **restriction enzymes**. These are enzymes, occurring naturally in bacteria, that are used in biotechnology to cut DNA into desired fragments. (In nature, bacteria use them to cut up the DNA of invading viruses.) In isolating restriction enzymes, scientists found that many of them had a wonder-ful property: They didn't just cut DNA randomly; they cut it at very specific places. Here is how it works with an actual restriction enzyme called *Bam*HI. The two strands of DNA's double helix are

Web Tutorial 15.1
Restriction Enzymes and Plasmids

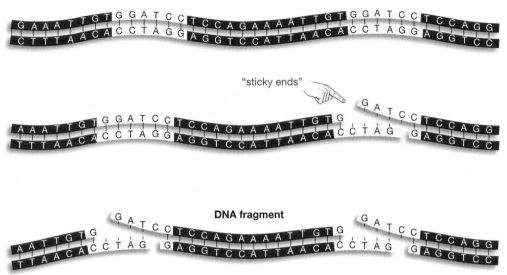

1. A portion of a DNA strand has the highlighted recognition sequence GGATCC.

2. A restriction enzyme moves along the DNA strand until it reaches the recognition sequence and makes a cut between adjacent G nucleotides.

3. A second restriction enzyme makes another cut in the strand at the same recognition sequence, resulting in a DNA fragment.

Figure 15.2
The Work of Restriction Enzymes

complementary, as we've seen, and they also happen to run in opposite directions. Thus, the sequence GGATCC would look like **Figure 15.2** if we were viewing both strands of the helix.

Now, the *Bam*HI restriction enzyme will move along the double helix, leaving the DNA alone *until* it comes to this series of six bases, known as its *recognition sequence*, and here it will make identical cuts on both strands of the DNA molecule, always between adjacent G nucleotides. When another *Bam*HI molecule encounters another GGATCC sequence, it too makes cuts, which effectively is like making a second cut in a piece of rope, giving us rope *fragments*.

*Bam*HI's recognition sequence may be GGATCC, but another restriction enzyme will have a *different* recognition sequence, and will make its cuts between a different pair of bases. Indeed, nearly 1,000 different restriction enzymes have been identified so far, which cut in hundreds of different places. The fact that they make cuts at so many specific locations has given scientists a terrific ability to cut up DNA in myriad ways.

Note that with *Bam*HI, each of the resulting DNA fragments has one strand that protrudes. Restriction enzymes that make this kind of cut are particularly valuable, for they produce "sticky ends" of DNA, so named because they have the potential to *stick to* other complementary DNA sequences. In the fragment on the lower-left in Figure 15.2, for example, the protruding sequence CTAG could now easily form a base pair with *any* piece of DNA whose sequence is the complemen-

tary GATC. So great is the need for restriction enzymes that they can now be ordered from biochemical suppliers, much as a person might order a set of socket wrenches from a hardware store.

Another Tool of Biotech: Plasmids

From the overview of manufacturing human growth hormone, recall that the human gene for HGH was inserted into *E. coli* bacteria, which then started turning out quantities of this protein. The question is, How did this human gene get into a bacterium? Several methods of transfer are available, but for now let's focus on a specific kind of DNA delivery vehicle. As it turns out, bacteria have small DNA-bearing units that lie *outside* their single chromosome. These are the **plasmids**, extrachromosomal rings of bacterial DNA that can be as little as 1,000 base pairs in length (**see Figure 15.3**). Plasmids can replicate independently of the bacterial chromosome; but just as important for biotech's purposes, they can *move into* bacterial cells.

How do plasmids do this? Bacteria are capable of taking up DNA from their surroundings, after which this DNA will function—that is, code for proteins—inside the bacterial cells. Appropriately enough, this process is known as **transformation**: a cell's incorporation of genetic material from outside its boundary. Some bacterial cells are naturally adept at transformation, while others, such as *E. coli*, can be induced to perform it by means of chemical treatment. Critically, plasmid DNA can be taken in via transformation and continue to function, as plasmid DNA, inside the bacteria.

Using Biotech's Tools: Getting Human Genes into Plasmids

At this point, you know about a couple of tools in the biotech tool kit: restriction enzymes and the transformation process involving plasmids. Let's now see how they work together. As you can see in **Figure 15.4**, the process starts with a gene of interest in the human genome. The first step is to use restriction enzymes on this human DNA. Knowing, say, the starting and ending sequence of the gene of interest, a restriction enzyme is selected that allows part of the genome to be cut into a manageable

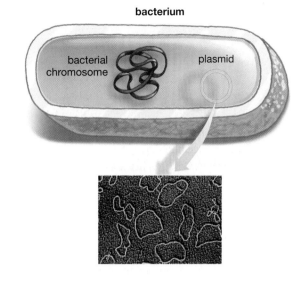

Figure 15.3
Transfer Agent
Plasmids are small rings of bacterial DNA that are not a part of the bacterial chromosome. They can exist outside bacterial cells and then be taken up by these cells. This artificially colored micrograph shows a type of plasmid, from *E. coli* bacteria, that is commonly used in genetic engineering.

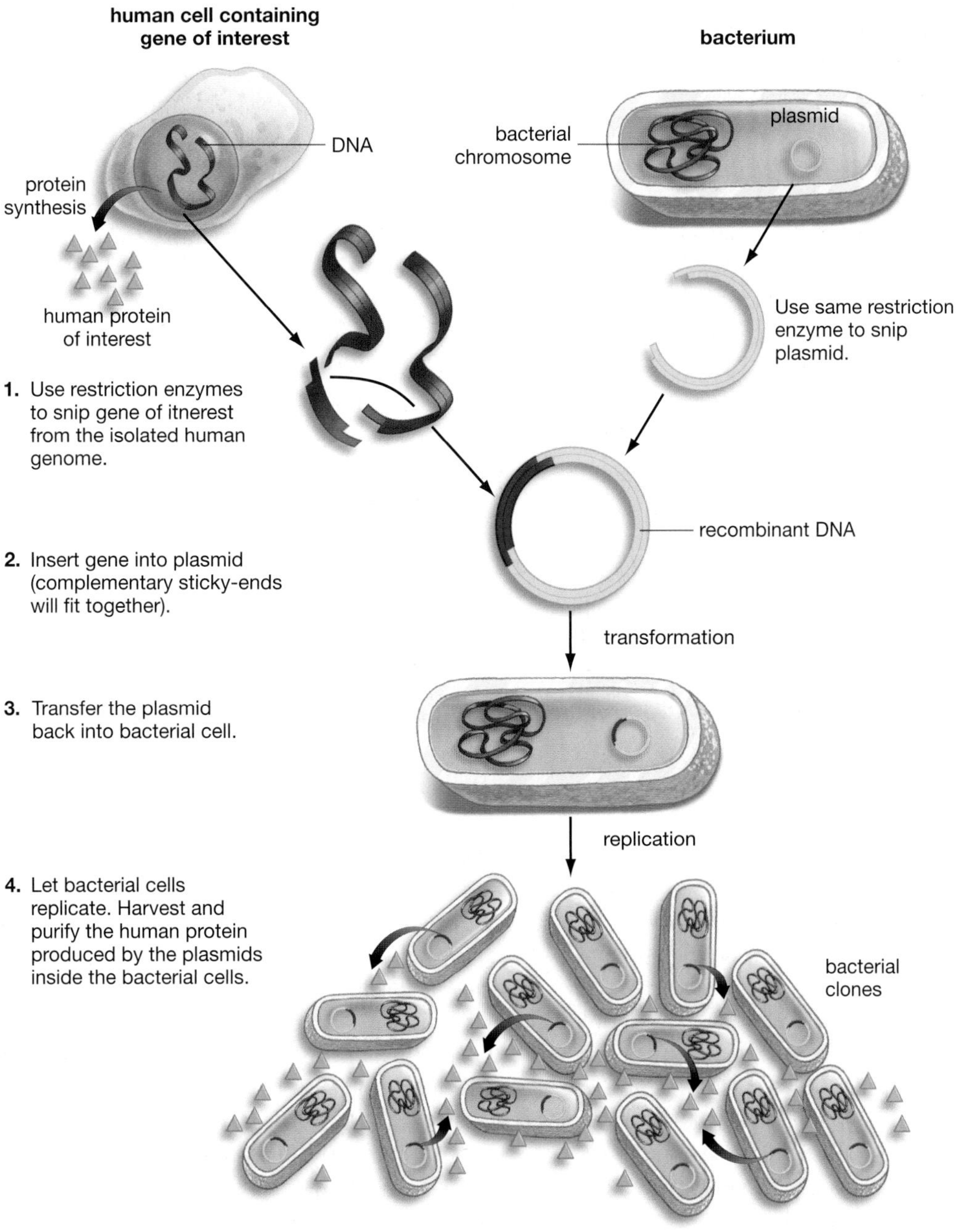

Figure 15.4
How to Use Bacteria to Produce a Needed Human Protein

human cell containing gene of interest

protein synthesis

human protein of interest

DNA

1. Use restriction enzymes to snip gene of itnerest from the isolated human genome.

2. Insert gene into plasmid (complementary sticky-ends will fit together).

3. Transfer the plasmid back into bacterial cell.

4. Let bacterial cells replicate. Harvest and purify the human protein produced by the plasmids inside the bacterial cells.

bacterium

bacterial chromosome

plasmid

Use same restriction enzyme to snip plasmid.

recombinant DNA

transformation

replication

bacterial clones

fragment including this sequence of interest, preferably in a sticky-ended form.

Here's where the beauty of restriction enzymes really comes into play. If the same restriction enzyme is now used on the DNA of isolated *plasmids*, the result is *complementary* sticky ends of plasmid and human DNA. In other words, these segments of human and plasmid DNA, through sticky-ended base pairing, fit together like puzzle pieces.

When the DNA fragments are mixed with the "cut" plasmids, that's just what happens: Human and plasmid DNA form base pairs, and the human DNA is incorporated into the plasmid circle. With this, a segment of DNA that was once part of the human genome has now been *re-combined* with a different stretch of DNA (the plasmid sequence). This process gives us the term **recombinant DNA**, defined as two or more segments of DNA that have been combined by humans into a sequence that does not exist in nature.

Getting the Plasmids Back inside Cells, Turning out Protein

To this point, what has been done is to produce a collection of independent plasmids having a human gene as part of their makeup. Remember, though, the goal is to turn out quantities of whatever protein the human gene is coding for. To do this requires a vast quantity of plasmids working away, which means working away *back inside* bacterial cells, and it's here that transformation comes into play. If the plasmids are put into a medium containing specially treated bacterial cells, a few of these cells will take up plasmids through transformation (at which point these cells have become transgenic, as noted earlier). Once this happens, the plasmids start *replicating* along with the bacterial cells themselves. As one bacterial cell becomes two, two become four, and so on, the plasmids are replicating away as well. And as the cell count reaches into the billions, collectible quantities of the protein begin to be turned out via instructions from the human gene inserted into the plasmid DNA.

A Plasmid Is One Kind of Cloning Vector

In this example, plasmids were the vehicle that served to take on some foreign DNA and then to ferry it into working bacterial cells. Note, however, that plasmids are only one vehicle among several that can be used. Collectively, such vehicles are known as **cloning vectors**, meaning self-replicating agents that serve to transfer and replicate genetic material. Next to plasmids, the most common cloning vector is a type of virus that infects bacteria—a bacteriophage.

Real-World Transgenic Biotechnology

The example we just went over concerned a hypothetical medicine, but lots of actual medicines are being produced through biotechnology today. By one industry count, more than 150 biotechnology drugs and vaccines have been approved by the U.S. Food and Drug Administration. We've already touched on one biotech medicine: the human growth hormone that is produced inside the *E. coli* bacterium. As it turns out, *E. coli* is something of a workhorse in medical biotech. Human insulin and several kinds of cancer-fighting compounds are also produced through it.

E. coli and its fellow bacteria are not the only "bioreactors" used in medical biotechnology, however. Yeast, hamster cells, and even mammals are utilized as well. Goats of the type pictured in Figure 15.1 produce several kinds of proteins in their milk, including one called AT3 that prevents the formation of blood clots. You might wonder why large, expensive goats would be used to turn out proteins when bacteria are available. The answer is that lactating goats naturally produce a tremendous amount of protein, in the form of the proteins that help make up their milk. Through genetic manipulation, scientists can tie the production of *human* proteins to the goat's natural production of its milk proteins, and the result is a bonanza: A single goat can produce over a kilogram (2.2 pounds) of needed human protein per year. You also might wonder how a human gene can be spliced into a goat genome. Human DNA is injected into goat embryos when they are at the one-cell stage.

Genetically Modified Food Crops

Transgenic organisms do more than produce medicines. You may have heard of genetically modified or "GM" food crops. It won't surprise you to hear that most of these crops are transgenic. If you look at **Figure 15.5**, you can see a transgenic rice that has been dubbed "golden rice" because of its ability to produce its own beta carotene, which the human body coverts to vitamin A. Conventional rice does not contain beta carotene, and this fact has health consequences. The World Health Organization estimates that half a million children go blind each year from vitamin A deficiency and that one to two million children die from it. No one claims that golden rice can completely solve this problem, but the hope is that it can help alleviate it. Golden rice is transgenic in more ways than one: Genes from both a bacterium and a daffodil have been spliced into it, thus enabling it to produce the beta carotene.

Field tests are just beginning on golden rice, but several other transgenic food crops are in full commercial production right now. Indeed, in the United States 80 percent of all soybeans, 70 percent of all cotton, and 38 percent of all corn crops currently are genetically modified in one way or another. In general, the purpose of transgenic crops is to get greater crop yields with less use of chemical pesticides and herbicides. No other biotech product has sparked as much controversy as GM crops, however. We'll review some of the issues surrounding GM crops in the chapter's concluding section.

15.3 Reproductive Cloning

You may have noticed that the word *cloning* popped up in the discussion of transgenic biotechnology. Thanks to movies and recent real-life events, this word has taken on some sinister implications, as if biological cloning were inherently a Frankenstein-like procedure. But it need be no more threatening than the process of making copies of a single gene. To **clone** simply means "to make an exact genetic copy of." An individual clone is one of these exact genetic copies. Thus, in the biotech world, the gene for human growth hormone is cloned by means of being snipped out of the human genome and then copied within bacterial cells. A collection of such cells can be thought of as a single clone, because all of them are genetically identical. (One bacterium splits into two, and both "daughter" cells are genetic replicas of the parent cell.)

Cloning can involve not just bacteria, however, but larger organisms as well. Human beings actually have been making clones for centuries, though by low-tech rather than high-tech methods. A "cutting" taken from a plant and put into soil can sometimes grow into a whole new plant. When this happens, there is no mixing of egg and sperm. The new plant comes entirely from one individual—the original plant—and thus meets the definition of a clone, in that it is an exact genetic copy of another entity.

All this said, in recent years biotechnology has greatly expanded the range of what can be cloned. The potential for such cloning appears to be very broad indeed—something that has inspired both great optimism and great concern. Biotechnology has now joined with various reproductive technologies to produce clones not just of genes or cells or plants, but of mammals, with the starting cells for these mammals coming from *adult* mammals. The most famous example is Dolly the sheep, cloned by researcher Ian Wilmut and his colleagues in Scot-

Figure 15.5
A More Nutritious Rice
Grains of the genetically engineered "golden rice" stand out next to grains of ordinary rice. The rice has a golden color because it is able to produce its own beta carotene, which the human body converts into Vitamin A.

land in 1997 (**see Figure 15.6**). This is **reproductive cloning**: cloning intended to produce genetically identical adult animals. Powerful in its own right, it gains added potential when combined with the basic biotech processes you've been reviewing.

Reproductive Cloning: How Dolly Was Cloned

Dolly the sheep, who died in 2003, was a clone as that term is defined: She was, to a first approximation, an exact genetic replica—in this case, of an-

Web Tutorial 15.2
Recombinant DNA

Figure 15.6
Revolutionary Sheep
Dolly, the first mammal ever cloned from an adult mammal, is shown here as a young sheep with her surrogate mother. Note that Dolly is white-faced, like the sheep she was cloned from, while her surrogate mother is black-faced.

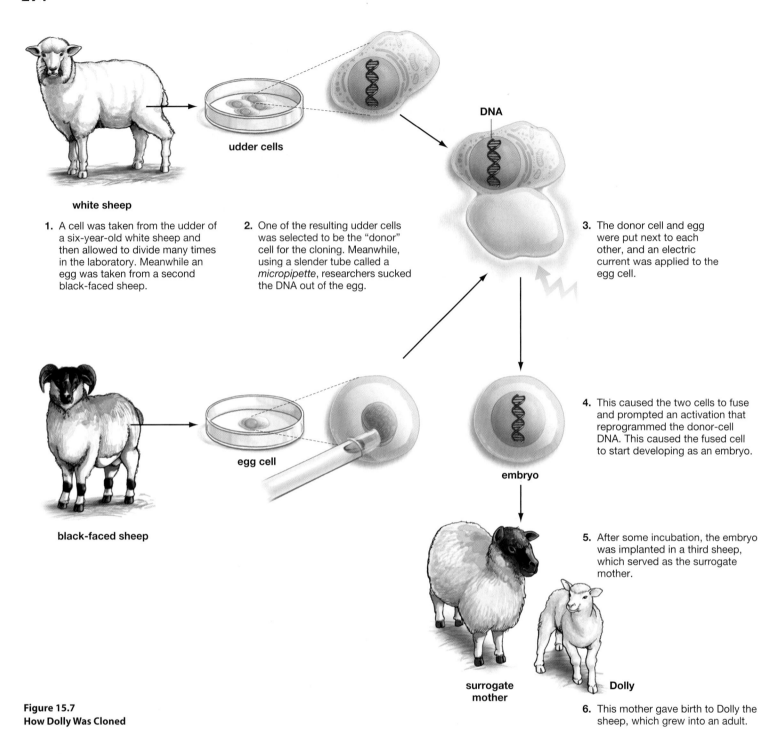

white sheep

1. A cell was taken from the udder of a six-year-old white sheep and then allowed to divide many times in the laboratory. Meanwhile an egg was taken from a second black-faced sheep.

2. One of the resulting udder cells was selected to be the "donor" cell for the cloning. Meanwhile, using a slender tube called a *micropipette*, researchers sucked the DNA out of the egg.

udder cells

DNA

3. The donor cell and egg were put next to each other, and an electric current was applied to the egg cell.

4. This caused the two cells to fuse and prompted an activation that reprogrammed the donor-cell DNA. This caused the fused cell to start developing as an embryo.

black-faced sheep

egg cell

embryo

5. After some incubation, the embryo was implanted in a third sheep, which served as the surrogate mother.

surrogate mother

Dolly

6. This mother gave birth to Dolly the sheep, which grew into an adult.

Figure 15.7
How Dolly Was Cloned

other sheep. Here's how she was produced (**see Figure 15.7**). A cell was taken from the udder of a six-year-old adult sheep and then grown in culture in the laboratory, meaning that the original cell divided into many "daughter" cells. While this was going on, researchers took an *egg* from a second sheep and removed its nucleus, meaning they removed all its nuclear DNA. Then they placed the udder cell (which had DNA) next to the egg cell (which did not have DNA) and applied a small electric current to the egg. This had two effects: It caused the two cells to fuse into one, and it mimicked the stimulation normally provided when a *sperm* cell fuses with an egg. With this, the udder cell DNA began to be reprogrammed. As a result of this reprogramming, the fused cell started to develop as an embryo. (Though the egg cell had its DNA removed, it still contained all kinds of egg-cell proteins whose normal function is to trigger development of an embryo. It was these factors still in the egg that reprogrammed the donor-cell DNA.) After the embryo had developed to a certain point, the researchers implanted it in a third sheep, which served as a surrogate mother. The result, 21 weeks later, was the lamb Dolly, who went on to give birth to two sets of her own lambs.

Every cell in Dolly had DNA in its nucleus that was an exact copy of the DNA in the six-year-old donor sheep. Thus Dolly was a clone of that sheep. There was no second parent contributing chromosomes to Dolly, no mixing of genetic material at the moment of conception—just a copying of one individual from another.

Cloning and Recombinant DNA

Once Dolly was born, science was off to the races in terms of producing clones of mammals. To date, horses, mules, cows, pigs, cats, and mice have been cloned—all of them essentially through the same process that was used with Dolly. But to what end? That is, why are scientists cloning these animals? Well, to use Dolly as an example, Ian Wilmut and his colleagues were not so much interested in cloning a sheep as they were in coupling the power of reproductive cloning to the techniques of recombinant DNA that you just looked at. The workers at Wilmut's research institute went on to produce Polly the sheep, who was not only a clone, but a clone who had had a human gene inserted into her genome and was thus able to produce a blood-clotting protein useful to hemophiliacs. A *group* of such clones would be valuable because they all would have the same starting genetic set—the same genome—and thus would all be able to produce the blood-clotting protein. The power of cloning is the power to produce one organism after another that has the same traits. Cloning makes it possible to sidestep the *mixing* of traits that comes with sexual reproduction.

Human Cloning

Reproductive cloning may interest scientists because of its potential to yield products such as medicines. It has captured the attention of the average person, however, because it raises the possibility of *human* cloning. For reasons noted below, human cloning seems less likely now than it did a couple of years ago. Nevertheless, it is still possible that in the near future, a person will be produced not through sex, but through cloning. How could a person be cloned? Pretty much through the same method used with Dolly—take a cell from an adult human being, fuse this cell with an enucleated human egg cell, and implant the fused cell in the uterus of a woman willing to bring the resulting embryo to term.

The prospect of a human clone is so dizzying that it's worthwhile to think about what such a person would represent in biological terms. He or she would be a genetic replica of the person who provided the donor cell with the DNA in it. One helpful way to think of this person's biological status is in terms of a more familiar concept: that of an identical twin. As it happens, identical twins also are genetic replicas of one another. Early in a human pregnancy, separate cells from an embryo start forming as two separate human beings, and the result is identical twins. Cloning is a *different* means of producing genetically identical individuals, but the critical point is that a donor and a clone would be no more alike than two identical twins. Indeed, we would expect DNA donors and their clones to be *less* alike than identical twins. This is so because identical twins usually share similar environments—in the womb, in the home, and so on—while donor and clone would likely grow up in very different environments.

Be this as it may, the prospect of a human clone still has the power to stun us—generally to repulse us, to judge by news reports. It now appears, however, that nature's rules may get in the way of human cloning. For several years after Dolly was produced, scientists puzzled over the fact that they could not clone a *primate* and have it live past the early embryonic stage. In 2003, scientists discovered what the problem was. In primate cells, proteins that help control cell division are concentrated in the very part of the egg cell that is removed in the cloning procedure. The result is that, when cells divide in a primate clone, the chromosomes are misapportioned—some cells get too many chromosomes while others get too few. An embryo cannot survive such a genetic imbalance. As a result, we have no cloned primates so far, and no clear idea of whether we will get one.

Saving Endangered Species

Could *Jurassic Park* someday become a reality through cloning? That is, could scientists recover some dinosaur DNA and use it to produce a living dinosaur via the cloning process we've just gone over? The chances are slim, for the simple reason that DNA degrades over time, and dinosaurs have been gone a *long* time—the last of them died 65 million years ago. The DNA of animals who died only a few thousand years ago is so fragmentary that nothing could be cloned from it. If you think about it, however, even if intact dinosaur DNA could be found, there would still be another problem. Recall that in Dolly's case, donor DNA from a sheep had to be placed in a sheep egg that had its nucleus removed. Where are the dinosaur eggs that could be used for dinosaur cloning? There aren't any, of course. But scientists working today have dealt with this very problem in trying to bring back species who have *recently* gone extinct, or who are threatened with extinction.

If you look at **Figure 15.8**, you can see a picture of a Spanish mountain goat called a bucardo. There are no longer any living bucardos, as the last one was killed in January of 2000 when a tree fell on it. Fortunately, just months before this goat's death, scientists took tissue samples from it and preserved them. One problem in bringing back the bucardo through cloning involves the *Jurassic Park* question: Where do you get a bucardo egg, given that there are no more bucardos? Scientists from a U.S. company called Advanced Cell Technology (ACT) believe they have an answer. Donor DNA cells can fuse with eggs from *closely related* species, thus bringing about a clone. Indeed, ACT has already pulled this trick off with a threatened ox-like species, called a gaur, that is found in India and Southeast Asia. In 2000, a cell taken from a male gaur was fused with an enucleated egg from a common cow. In January 2001, a seemingly healthy gaur was born to a surrogate mother (once again, a cow). The gaur died shortly after birth, but the cause apparently was unrelated to the cloning. It seems possible, then, that severely threatened species could have their numbers increased through cloning, while other species might be brought back to the world *following* extinction. Is biotechnology exciting or what?

15.4 Forensic Biotechnology

After New York's World Trade Center (WTC) was destroyed on September 11, 2001, public officials had to undertake the grim task of trying to identify all the victims of the attack, and to do so quickly, so as to provide some sense of closure to the victims' loved ones. In the weeks following the attack, therefore, officials collected samples of DNA not only from WTC victims, but from the toothbrushes and razors of persons who were missing since the tragedy and hence *presumed* to be victims. They also collected DNA from relatives of these presumed victims. With this DNA database in hand, workers then made an effort to match one kind of DNA with another—to match the DNA found on a toothbrush, for example, with the DNA from a WTC victim.

In 1998, then-President Bill Clinton publicly denied allegations that he had been sexually involved with a young intern at the White House, Monica Lewinsky. The matter might have stayed right there—Ms. Lewinsky saying one thing, the president saying another—except that Ms. Lewinsky had preserved a dress with a stain on it that she said came from the president's semen. The FBI subsequently extracted DNA from this stain and soon thereafter took a sample of blood from the president. Eventually, the FBI reported on the odds that anyone *but* the president was the source of the semen. These turned out to be roughly 7.8 trillion to one. With this, the president ceased to make claims about not having had sex with Ms. Lewinsky.

High-profile cases such as these are the most visible manifestation of a huge change that has come about in the way society *identifies* persons—criminals, biological fathers, and victims, most notably. In the criminal arena, the most highly valued type of physical evidence today is not fingerprints; it is tiny bits of human tissue from which DNA can be extracted. But how do police use DNA to get a "match" in a criminal case? Let's find out.

The essence of "forensic DNA typing," as it is known, is to get two sets of physical samples—first, a sample left by the perpetrator in such forms as blood or semen; and second, a blood sample from a suspect. By comparing the DNA patterns of both samples, forensic scientists can establish odds as to whether the suspect *is* the perpetrator. (Conversely, it would be possible to use a victim's blood sample and test it against, for example, blood found on the suspect's clothing.)

The Use of PCR

When forensic DNA typing first came into use in the 1980s, a significant limitation on it was that relatively *large* samples of human tissue had to be obtained in order to get enough DNA to work with. Eventually, however, this problem was all but eliminated by something called the **polymerase**

Figure 15.8
Is Extinction Forever?
This Spanish mountain goat, the bucardo, is now extinct, but scientists hope to bring the species back through reproductive cloning.

chain reaction (PCR): a technique for quickly producing many copies of a specific segment of DNA. Thanks to PCR, if a criminal so much as licks an envelope and leaves it behind, police can retrieve enough DNA to get a match.

The PCR process is very simple in outline. Four kinds of materials are mixed together to carry it out. The first of these is a quantity of DNA itself. Then there is a collection of DNA nucleotides and DNA polymerase (which, remember, goes down the line on single-stranded DNA, affixing nucleotides to available bases). Finally, there are two DNA "primer" sequences—short sequences of single-stranded DNA that act as signals to DNA polymerase, saying "start adding nucleotides here."

The first step in the process is that the starting quantity of DNA is heated until the two strands of DNA's double helix separate, resulting in two single strands of DNA (**see Figure 15.9**). As the heated DNA mixture cools, primers will attach to each of the now-separate strands of DNA. Then the DNA polymerases go down the line, starting from the primers, linking nucleotides to the template DNA and thus producing strands that are complementary to the original strands. The result is *two* double-stranded lengths of DNA, both identical to the original double strand. In short, the DNA sample has been doubled in one copying "cycle." Then the entire process of heating and cooling is repeated. Since each copying cycle takes only 1–3 minutes, by the time 90 minutes have passed, millions of copies can be created. Moreover, thanks to the specificity of primers, the sequence that is copied can be limited to a small sequence of interest within the larger genome. So useful is PCR that it is employed not just in forensic biotechnology, but in countless facets of biological research. We are focusing here on one specific use for PCR, but it has so many applications that in 1993, its inventor, Kary Mullis, was awarded the Nobel Prize in Chemistry for his achievement.

Finding Individual Patterns

Once PCR has done its job, the police have enough DNA to begin looking for the individual patterns within it—one set of patterns found in the crime-scene DNA, one set of patterns seen in the suspect's DNA. But what are these patterns that each of us carries? Human genomes are filled with short sequences of DNA that are repeated over and over, from 3 to 50 times, like this:

TCAT TCAT TCAT TCAT TCAT

This TCAT example is not hypothetical. It is an actual *short tandem repeat* (or STR) that police de-

partments use in today's DNA typing. The usefulness of STRs is that, at a given location in the genome, one person will have one number of tandem repeats, while another person is likely to have a *different* number of repeats, like so:

Individual 1: TCAT TCAT TCAT TCAT

Individual 2: TCAT TCAT TCAT TCAT TCAT TCAT

Now, the chances are small that two unrelated individuals will have an identical number of repeats at even one location in the genome. But modern forensic DNA typing looks at STR repeats at 13 different locations. Imagine yourself, then, as a forensic investigator who is reviewing both DNA from a crime scene and DNA from a suspect in the crime. At genomic location 1, the STR patterns are:

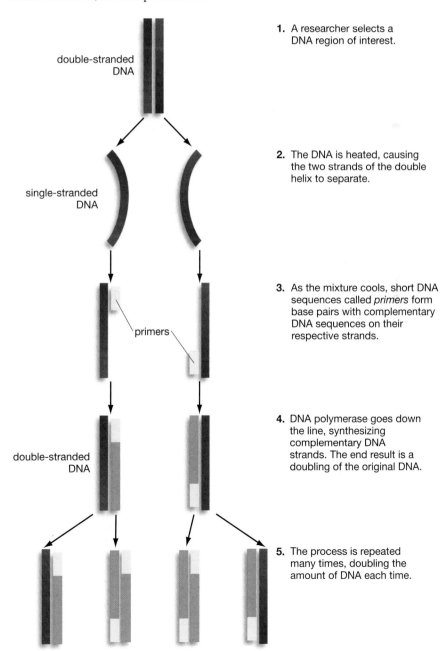

Figure 15.9
DNA Copying Machine
The polymerase chain reaction (PCR) makes copies of a given length of DNA.

1. A researcher selects a DNA region of interest.

2. The DNA is heated, causing the two strands of the double helix to separate.

3. As the mixture cools, short DNA sequences called *primers* form base pairs with complementary DNA sequences on their respective strands.

4. DNA polymerase goes down the line, synthesizing complementary DNA strands. The end result is a doubling of the original DNA.

5. The process is repeated many times, doubling the amount of DNA each time.

drug, while patients who have an SNP-pattern 2 will do better on another drug. Why? Because these genetic differences could lead to differences in the way the patients metabolize the drugs they are taking.

It's easy to see that such SNP analysis could get more personalized over time. Conceivably, each of us could have an "SNP profile" that physicians could match up against any number of disease treatments and drugs. The questions the physicians would ask would be: Is this drug safe for a person with this SNP profile? Is this treatment likely to be the most effective treatment for a person with this SNP profile? Note, however, that a personalized genetic profile might also reveal genetic *susceptibility* to diseases. We'll return to this subject later, but for now just ask yourself: Who should have access to genetic profiles?

Microarrays: Catching Genes in the Act

As you know, the outcome of genetic activity is the production of proteins. Given this, another way to personalize medicine would be to look not at a person's genome, but at what that person's genome is *doing*. The question here is what genes within the person's genome are being "expressed" at a given point in time? Which genes are actively coding for proteins? A technological advance called DNA microarrays now allows researchers to take "snapshots" of the genetic activity going on within individual cells.

Recall that, for any protein to be produced, a gene will first code for a messenger RNA (mRNA) that ferries the gene's information to a ribosome, where the protein is actually put together. Thus, if you know which mRNAs are in the cytoplasm of a cell, you know which *genes* are active in that cell. Of course, a cell is likely to have thousands of different mRNAs in its cytoplasm at any one time. So how could you learn what all of them are? You can put a microarray to work, as shown in **Figure 15.11a** on page 301.

Each microarray "chip" is a glass slide smaller than a postage stamp. Through use of the same photolithography technique that is used to make computer chips, thousands of tiny "spots" of single-stranded DNA are applied to a microarray chip, with each spot in effect containing multiple copies of a single gene. In the example shown, two different cells are being compared, one of them a cancerous cell from a breast-cancer patient, the other a normal breast cell. In the microarray procedure, mRNAs are isolated from each of these cells. Then these mRNAs are converted into a more stable nucleic acid, complementary DNA (cDNA). Critically, the cDNAs derived from the cancerous cell are fluorescently labeled red, while the cDNAs derived from the normal, "control" cell are labeled green.

Given your knowledge of complementary base pairing, you may be able to guess what comes next. When the cDNAs are poured onto the microarray, they will hybridize or bond with complementary DNA segments that are on the chip. But what are these complementary segments? Individual genes. And each cDNA molecule is complementary to *one* of these genes (because each cDNA represents a gene that was expressed back in the cell). Therefore, when a cDNA bonds with an individual gene on the chip, this is in effect a message saying, "This gene is active in this cell right now."

Recall, however, that we are comparing gene activity in two cells. Every time a red-labeled cDNA from the cancerous cell bonds with its complementary DNA, it adds to the red light that will be seen on a DNA spot when a laser is shined on it. Every time a green-labeled cDNA from the normal cell bonds to its complementary DNA, a small amount of green light will be added to a spot. A computer can calculate the *ratio* of red to green light at any one spot, and the result is what you see at the bottom of the figure. Genes in the cancerous cell that are strongly active, relative to genes in the control cell, show up as red dots. Genes in the cancerous cell that have weak activity relative to the control show up as green dots. Meanwhile, yellow dots indicate genes in the two cells that are equally active, and black dots indicate genes that are inactive in both cells.

You probably can see one use for this right away. At given points in time, what genes are active in a cancer cell, as opposed to a normal cell? Researchers would like to know about this in the same way that an army officer would like to know how an enemy's forces are being utilized over time.

But there's a more individualized aspect to microarrays as well. No two cancer patients will have exactly the same kind of tumor; different tumors will express different genes. Under a microscope, these tumors may look the same, but they are not the same in terms of gene expression. And such expression can affect how aggressive a tumor is, and how it responds to a given treatment. Imagine, then, using microarrays to develop "expression profiles" for individual tumors. This is now being done all the time. It has been done in Europe in recent years for breast-cancer tumors, and the results have been encouraging. Women whose tumors had one kind of expression profile were found to be more likely to have a recurring cancer than women whose tumors had another profile. As a result, trials are now underway in which women are being assigned to get one kind of *treatment* or another, based on what kind of expression profile their tumor had. As you can imagine, individualized genetic snap-

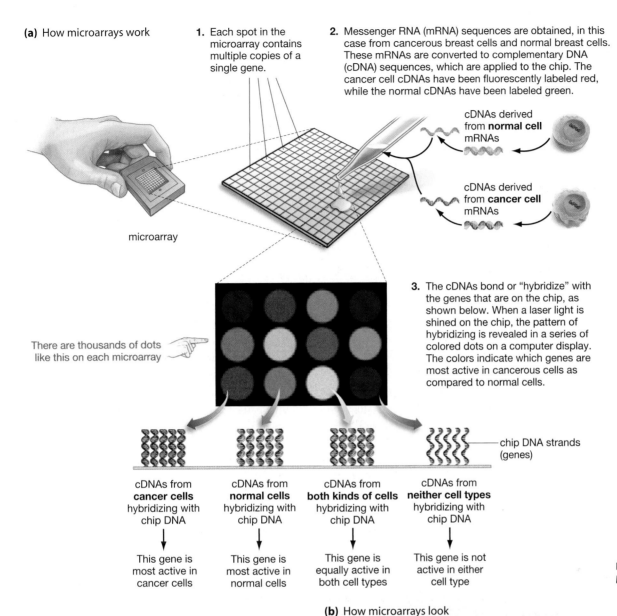

(a) How microarrays work

1. Each spot in the microarray contains multiple copies of a single gene.

2. Messenger RNA (mRNA) sequences are obtained, in this case from cancerous breast cells and normal breast cells. These mRNAs are converted to complementary DNA (cDNA) sequences, which are applied to the chip. The cancer cell cDNAs have been fluorescently labeled red, while the normal cDNAs have been labeled green.

cDNAs derived from **normal cell** mRNAs

cDNAs derived from **cancer cell** mRNAs

microarray

There are thousands of dots like this on each microarray

3. The cDNAs bond or "hybridize" with the genes that are on the chip, as shown below. When a laser light is shined on the chip, the pattern of hybridizing is revealed in a series of colored dots on a computer display. The colors indicate which genes are most active in cancerous cells as compared to normal cells.

chip DNA strands (genes)

| cDNAs from **cancer cells** hybridizing with chip DNA | cDNAs from **normal cells** hybridizing with chip DNA | cDNAs from **both kinds of cells** hybridizing with chip DNA | cDNAs from **neither cell types** hybridizing with chip DNA |
| This gene is most active in cancer cells | This gene is most active in normal cells | This gene is equally active in both cell types | This gene is not active in either cell type |

Figure 15.11
Microarray Technology

A small portion of a microarray display, with individual DNA spots visible. This microarray was used to study gene expression in baker's yeast.

(b) How microarrays look

shots are not limited to cancer; all kinds of human diseases can be analyzed through them. And this technology is still young. In the future, physicians will be able to take account of the genetic activity within us in ways that we can't yet imagine.

15.6 Controversies in Biotechnology

Having learned something about biotechnology, consider the following question. Let us suppose that through biotechnology, you could fashion an egg-laying chicken that had no eyes and no legs. If you could do such a thing, would you? On a less theoretical level, a safer, cheaper human growth hormone is now available thanks to biotechnology. But who should get human growth hormone? Should it be reserved for children who are truly short, or should it be made available to children who are simply shorter

than they'd like to be? And if, in the future, it becomes possible to manipulate qualities other than height—say hair or eye color—what limits should be placed on the use of these techniques? Do we face the prospect of "designer human beings," and if so, do cosmetic traits get apportioned out strictly in accordance with who can pay for them? It's hard to tell whether society will have to grapple with such issues in the future, simply because it's impossible to know where biotechnology might lead us. But there are plenty of current controversies in biotechnology. Here's a brief look at a few of them.

Genetically Modified Foods

No biotechnology issue has raised more furor than that of genetically modified foods. Consider that the "golden rice" reviewed earlier had to be grown in a greenhouse in Switzerland that was made grenade-proof because of fear it might be attacked by opponents of "GM" foods. What is this controversy about? In a nutshell, proponents of GM foods see in them the potential to feed a hungry world and to lessen the environmental damage caused by such human practices as pesticide application. Opponents of them see the potential to harm human health and wreak havoc on Earth's ecosystems.

If you look at **Figure 15.12**, you can see a case in point in this battle. Pictured are two sets of cotton plants, one set genetically modified, the other not. The modified plants are transgenic. They contain genes from a bacterium, called *Bacillus thuringiensis*, that is found naturally in the soil and that produces proteins that are toxic to a number of insects. Collectively, these proteins constitute a natural insecticide known as Bt, which has been sprayed on crops for years, mostly by "organic" farmers, meaning those who do not use human-made pesticides or herbicides. Enter biotechnology firms, which in the 1990s spliced Bt genes into crop plants, with the result that these plants—corn, cotton, and potatoes—now produce their *own* insecticide. The results have in some ways been an environmentalist's dream. In one survey conducted in the American Southeast, farmers who planted Bt cotton reduced the amount of chemical insecticides they applied to their fields by 72 percent. They did this, moreover, while increasing cotton yields by more than 11 percent.

Despite such benefits, we can also see, in the use of Bt seeds, the qualities that make environmentalists uneasy about GM crops in general. Those few insects that survive in Bt-enhanced fields are likely to give rise to populations resistant to the natural toxin. This

raises the prospect of a valued natural insecticide losing its effectiveness against insects over the long run. So alarming is this possibility that the U.S. Environmental Protection Agency requires that at least 20 percent of any given farmer's crops must be non-Bt plants. The idea is to create non-Bt "refuges" near the Bt fields that will harbor bugs that have not built up a resistance to Bt. These bugs will then mate with Bt-resistant bugs, thus helping to ensure that Bt resistance does not spread.

Meanwhile, another fear is that some GM crops will take on the role of "super-plants" that will spread by means of their pollen fertilizing nearby plants. (This is the same threat landscape architects see in low-mow grass, for example.) Then there is the idea that the foods made from GM crops could set off allergic reactions in consumers or have other adverse health effects on them. There is no evidence that GM foods have caused either health or environmental problems so far. But European countries have been concerned enough about these issues that, since 1998, they have enforced a moratorium on the planting of any new commercial GM crops.

Cloning

The ethical issues connected to cloning center almost entirely around human cloning. There is near-universal agreement that it would be unethical to attempt to clone a human being with the intention of having it develop from embryo to child to adult. This is so in part because the reproductive cloning of any mammal is currently a hit-or-miss process. Hundreds of attempts may be required to produce a viable embryo that can be successfully implanted in a uterus and brought to term. And the clones that are produced often have physical defects or are abnormally large at birth.

Even if each attempt at cloning a horse or a cat resulted in a healthy individual, however, science would still be confronted with the fundamental question of whether it is ethical to clone a human being. While the answer from almost all scientists seems to be no, a few scientists from several countries have expressed an interest in human cloning. As such, the possibility of its taking place cannot be ruled out.

Genetic Profiles

As noted, in the future it might be possible for people to get genetic profiles of themselves—inventories of the SNPs that make their genome unique. Such a profile might well reveal the susceptibilities that individuals have to genetically linked diseases, however, and this is where the controversy starts. The worst-case scenario is that individuals could be branded as disease-prone by reason of their genetic profiles, with

Figure 15.12
Built-in Resistance to Pests
The cotton plants on the left have had genes spliced into them from the bacterium *Bacillus thuringiensis*. These genes code for proteins that function as a pesticide. Meanwhile, the cotton plants on the right are not transgenic in this way—they have not had genes from another organism spliced into them. Both stands of cotton were under equal attack from insects, but the Bt-enhanced plants fared much better.

the result that employers would not want to hire them and insurers would not want to insure them. (An extreme version of this kind of genetic discrimination is portrayed in the film *Gattaca*.)

Because a small number of genetically based diseases can be tested for currently, this issue has some relevance right now. Certainly governments are acting as if genetic discrimination is a current threat. Most states have enacted laws that prohibit the use of genetic information in the issuance of medical insurance policies; most have passed laws prohibiting employment discrimination based on genetics; and President Clinton signed an executive order prohibiting the federal government from making any decision to hire, promote, or dismiss a worker based on genetic information. Thus far, however, there is little evidence that any institutions have had an interest in carrying out genetic discrimination in connection with employment or health insurance.

The most contentious area in this debate seems to be life insurance, which is special in that it is seen less as a "right" than a consumer option. As a result, life insurance has not been subject to the same kind of government regulation as has health insurance or employment practice. Imagine what would happen, though, if individuals could know, through genetic testing, that they were at significant risk of an early death, but that insurance companies were forbidden to know the genetic background of any of their life-insurance applicants. The result would be droves of disease-susceptible people signing up for life insurance, dying young (on average), and thus driving up life-insurance rates for everyone. Such are the economic issues that come with the new world of science and technology.

On to Evolution

In touring the strands of DNA's double helix, you've had many occasions to see how this microscopic molecule can profoundly affect our macroscopic world. DNA replicates, it mutates, it is passed on from one generation to the next. And some organisms will have more *success* in passing on their DNA—in reproducing, in other words—than will others. This latter fact has been critical in bringing about the huge variety of life-forms that Earth houses, from bacteria to bats to trees. Over billions of years, genetics has interacted with Earth's myriad environments to produce the grand story of life. That story is called evolution, and it is the subject of the next four chapters.

Chapter Review

Summary

15.1 What Is Biotechnology?

- Biotechnology is the application of technology to natural biological processes. Innovations in biotechnology result from insights that are both biological and technological in nature.

15.2 Transgenic Biotechnology

- A transgenic organism is an organism whose genome has stably incorporated one or more genes from another species. Many biotechnology products are produced within transgenic organisms. Human growth hormone, for example is produced within a bacterium that has been made transgenic by means of incorporating a human gene.

- Restriction enzymes are proteins derived from bacteria that can cut DNA in specific places. Plasmids are small, extrachromosomal rings of bacterial DNA that can exist outside of bacterial cells and that can move into these cells through the process of transformation.
  Web Tutorial 15.1: Restriction Enzymes and Plasmids

- Human DNA can be inserted into plasmid rings. Scientists use the same restriction enzyme on both the human DNA of interest and the plasmids. Complementary "sticky ends" of the fragmented human and plasmid DNA will then bond together, thus splicing the human DNA into the plasmid. This produces recombinant

DNA: two or more segments of DNA that have been combined by humans into a sequence that does not exist in nature.

- Once plasmids have had human DNA spliced into them, the plasmids can then be taken up into bacterial cells through transformation. As these cells replicate, producing many cells, the plasmid DNA inside them replicates as well. These plasmids are producing the protein coded for by the human DNA that has been spliced into them. The result is a quantity of the human protein of interest.

- Plasmids are one type of cloning vector, meaning a self-replicating agent that, in cloning, functions in the transfer of genetic material. Viruses known as bacteriophages are another common cloning vector.

- A large number of medicines and vaccines are produced today through transgenic biotechnology. Transgenic organisms that are used for this purpose include not only bacteria but also yeast, hamster cells, and mammals such as goats. Transgenic food crops are planted in abundance today in the United States. These genetically modified or GM crops are a subject of great controversy.

15.3 Reproductive Cloning

- A clone is a genetically identical copy of a biological entity. Genes can be cloned, as can cells and plants. Reproductive cloning is the process of making adult clones of animals. Dolly the sheep was a reproductive clone. Today, reproductive cloning of mammals is carried out through variants of the process that was used with Dolly: An egg cell has its nucleus removed and is fused with an adult cell containing a nucleus and, therefore, DNA. The fused cell then starts to develop as an embryo and is implanted in a surrogate mother.

- Reproductive cloning can work in tandem with various recombinant DNA processes. Reproductive clones may be made transgenic and produce proteins of interest to humans.

- It is possible that a human being might be cloned in the near future though, for technical reasons, such cloning seems less likely today than it did a few years ago.

- It appears possible that severely threatened animal species could have their numbers increased through reproductive cloning and that species recently made extinct might be brought back from extinction through cloning.
  Web Tutorial 15.2: Recombinant DNA

15.4 Forensic Biotechnology

- Identities of criminals, biological fathers, and disaster victims often are established today through the use of DNA. Forensic DNA typing is the use of DNA to establish identities in connection with legal matters, such as crimes.

- The polymerase chain reaction (PCR) is a technique for quickly producing many copies of a segment of DNA. It is useful in situations, such as crime investigations, in which a large amount of DNA is needed for analysis, yet the starting quantity of DNA is small.
  Web Tutorial 15.3: PCR: Rapid Copying of DNA

- Forensic DNA typing usually works through comparisons of short tandem repeat (STR) patterns that are found in all human genomes. Police will, for example, compare the STR pattern in a suspect's DNA with the STR pattern in DNA extracted from a crime scene.
  Web Tutorial 15.4: DNA Size and Sequencing

15.5 Personalized Medicine

- The genomes of any two unrelated human beings differ, on average, by one base pair per thousand. Each of these individual variations is called a single-nucleotide polymorphism or SNP. Researchers today are trying to find out if individuals might react differently to drugs or other medical treatments based on the SNP patterns in their genome.

- DNA microarrays are research devices that allow scientists to take a "snapshot" of the genetic activity going on at a given moment in a given type of cell. Through microarrays, scientists can learn which genes are being expressed in one type of cell, relative to another. Microarrays are being used today to develop "expression profiles" of cells—profiles that show the pattern of gene expression that is taking place in specific cells, such as cancer cells.

15.6 Controversies in Biotechnology

- Society is increasingly having to grapple with issues that biotechnology is raising. For example, what limits should there be on

modifying organisms other than humans? What limits should be placed on the cosmetic modification of humans?

- No aspect of biotechnology has been as controversial as the production of genetically modified food (GM food). Critics charge that GM foods are potentially dangerous to human beings and to the environment. Proponents of GM foods say these foods are of enormous potential to benefit humanity and the environment.

- Two of the fears most commonly expressed about GM foods are that they stand to cause allergic or other adverse health reactions in some people, and that as crops they may spread uncontrollably, through hybridization with wild plants. There is no evidence thus far that GM foods are having either effect.

- Genetic testing may be used in the near future to diagnose genetic susceptibility to a broad range of diseases. There is concern that individuals diagnosed as genetically susceptible will be subject to genetic discrimination, either by insurance companies or by employers.

Key Terms

biotechnology	289	**recombinant DNA**	292
clone 293		**reproductive cloning**	293
cloning vector	292	**restriction enzyme**	289
plasmid 290		**transformation** 290	
polymerase chain reaction		**transgenic**	
(PCR) 296		organism 289	

Understanding the Basics

Multiple-Choice Questions (answers in the back of the book)

1. How is biotechnology defined?
   a. the technical advances made by researchers in biology over the last decade
   b. the study of technical biology
   c. the application of technology to natural biological processes
   d. the process of cloning genes
   e. the use of commercialization in biological research

2. In transgenic biotechnology involving bacteria, which sequence of events would be followed in the process of harvesting a protein of interest?
   a. Replicate cloning vectors in host cells; isolate gene that codes for protein; insert gene into vector; harvest protein of interest; transform the vector into bacteria.
   b. Transform vector into bacteria; isolate gene that codes for protein; replicate vectors in host cells; insert gene into vector; harvest protein of interest.

c. Harvest protein of interest; isolate gene that codes for protein; insert gene into vector; replicate vectors in host cells; transform the vector into bacteria.

d. Isolate gene that codes for protein; insert gene into vector; replicate vectors in host cells; transform the vector into bacteria; harvest protein of interest.

e. Isolate gene that codes for protein; insert gene into vector; transform the vector into bacteria; replicate vectors in host cells; harvest protein of interest.

3. The most common way to visualize DNA molecules is to employ gel electrophoresis. The reason the DNA molecule migrates from one end of a gel electrophoresis tray to another is that:

a. DNA is a very small molecule

b. DNA is naturally active

c. DNA molecules are negatively charged; the opposite ends of the trays are positively charged

d. DNA molecules are positively charged; the opposite ends of the trays are negatively charged

e. None of the above is correct

4. Criminal investigators who want to make copies of a DNA sequence from a crime do so through

a. the polymerase chain reaction

b. transformation

c. gel electrophoresis

d. DNA sequencing

e. microarrays

5. Which of the following has biotechnology helped accomplish?

a. genetically engineered medicines

b. identification of criminals

c. cloning of mammals

d. genetically engineered crops

e. all of the above

6. Which of the following is an example of a transgenic organism?

a. *E. coli* bacterium with a human gene inserted

b. Dolly the sheep

c. rice with bacterium and daffodil genes inserted

d. all of the above

e. a and c only

7. Dolly the sheep was cloned by means of

a. applying an electric current to an egg and sperm in a laboratory

b. taking an embryo from one sheep and implanting it in another

c. taking a cell from an adult sheep and fusing it with a sheep egg that had its DNA removed

d. inserting new genes into an existing mammary cell

e. getting sperm to fuse with egg outside a living sheep

8. The basis of personalized medicine is that

a. genes can be moved from one organism to another

b. individual human genomes differ slightly from one another

c. cloning genes is so easy

d. cloning mammals has become routine

e. individual human genomes are exactly the same

9. Microarrays provide

a. specific lengths of DNA

b. knowledge of which genes are being expressed in a given cell

c. a way to reliably sequence entire genomes

d. a means for inserting genes into genomes

e. a way to compare crime-scene and suspect DNA

10. What is a clone?

a. an exact genetic copy of a gene or organism

b. a cell that can take up plasmids

c. a segment of DNA of a specified length

d. an offspring produced by test-tube fertilization

e. a segment of DNA that can prime the polymerase chain reaction

Brief Review

1. Why is it useful to have organisms such as bacteria turning out human proteins?

2. Wolf-like marsupials called Tasmanian tigers became extinct in the 1930s. What would be the key to bringing the Tasmanian tiger back through cloning?

3. Why is the polymerase chain reaction (PCR) so valuable?

4. Why are restriction enzymes that cut to produce sticky ends useful? If the following piece of DNA were cut with *Bam*HI, what would the fragment sequences be?

ATCGGATCCTCCG
TAGCCTAGGAGGC

5. Why did Dolly the sheep end up looking like the sheep that "donated" the udder cell, rather than the sheep that donated the egg?

Applying Your Knowledge

1. One of the motives put forth for human cloning is that people want to replace children or other loved ones who have died. To what extent could a clone of a loved one be a replacement for that person? If the technique had then been available, should doctors in the nineteenth century have preserved the DNA of Abraham Lincoln for cloning?

2. One of the concerns mentioned in the chapter regarding the use of genetic technology centers around privacy issues such as "genetic discrimination." Once we improve our ability to screen the human genome for disease susceptibility, is there any level of testing that should be required by society as a means of reducing health care costs or reducing transmission of genetic diseases across generations?

3. Should society demand that there be no risks to genetically modified foods before it allows them to be developed? Does any significant technology have no risks associated with it?

16 Charles Darwin, Evolutionary Thought, and the Evidence for Evolution

Charles Darwin.
(Section 16.2, page 311)

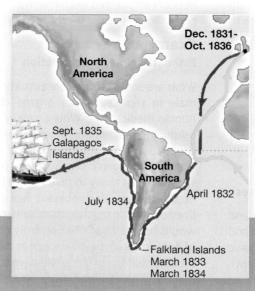

Fantastic voyage.
(Section 16.2, page 312)

Island of discovery.
(Section 16.4, page 314)

Charles Darwin is the person most responsible for deepening our understanding of how life evolved on Earth. Evidence uncovered from his time to ours has confirmed his insights.

Standing in a coastal redwood grove in California, we see trees stretching hundreds of feet into the air and wonder: Why are they so tall? Watching a nature program on television, we observe the exquisite plumage of the male peacock, but are puzzled that such a *cumbersome* display could exist on a creature that has to be wary of predators. Leafing through a book, we discover that the color vision of honeybees is most sensitive to the colors that exist in the very flowering plants they pollinate. We wonder: Can this be an accident?

For as long as humans have existed, people have asked questions like these. Yet for most of human history, even the most informed individuals had no way to make *sense* of the variations that nature presented. Why does the redwood have its height or the peacock its plumage? Through the middle of the nineteenth century, the answer was likely to be either a shrug of the shoulders or a statement such as: "The creator gave unique forms to living things for reasons we can't understand."

Jumping ahead to our own time, modern-day biologists would give different answers to these questions. Their replies would be something like this:

- Redwoods (and trees in general) are so tall because, over millions of years, they have been in competition with one another for the resource of sunlight. Individual trees that had the genetic capacity to grow tall got the sunlight, flourished, and thus left relatively many offspring, which made for taller trees in successive generations.

- Extravagant plumage exists on male peacocks because it is attractive to *female* peacocks when they are choosing mating partners. Such finery is indeed cumbersome and carries a price (in lack of mobility), but this price is outweighed by the value of the feathers. Over time, the peacock ancestors with more attractive plumage mated more often and thus sired more offspring than peacocks who had less impressive plumage. This made for successive generations of peacocks with more elaborate plumage.

- It is not an accident that honeybee eyesight is most sensitive to the colors that exist in the flowering plants they pollinate. Both bees and flowering plants flourish because of their interactions with one another—bees get food and flowers get pollinated. It is likely that flower color, or bee vision, or both, were *modified* over time to maximize this interaction.

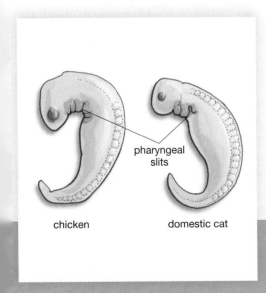

A clue to common descent.
(Section 16.6, page 316)

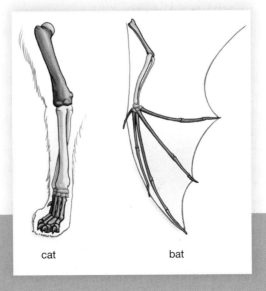

How is a cat like a bat?
(Section 16.9, page 319)

Tracking evolution.
(Section 16.9, page 322)

Comparing contemporary answers such as these to that single nineteenth-century answer, you might well ask: What is it that came *in between,* so that we no longer must simply stand in puzzled wonderment before the diversity of nature?

16.1 Evolution and Its Core Principles

What intervened was a set of intellectual breakthroughs in the mid-nineteenth century that today go under the heading of the *theory of evolution.* Two principles lie at the core of this theory.

Common Descent with Modification

One of these principles has been labeled **common descent with modification.** It holds that particular groups, or species, of living things can undergo modification in successive generations, with such change sometimes resulting in the formation of new, separate species—one species separating into two, these two then separating further. If you had a videotape of this "branching" of species and then ran it *backward,* what you'd eventually see is a reduction of all these branches into one, meaning that all living things on Earth ultimately are descended from a single, ancient ancestor. Evolution itself can have several definitions. In this chapter, we will think of it as synonymous with common descent with modification. In Chapter 17, you'll get a more technical definition of it.

Natural Selection

Descent with modification is joined to a second evolutionary principle, **natural selection:** A process in which the differential adaptation of organisms to their environment selects those traits that will be passed on with greater frequency from one generation to the next. In the redwood example, trees that grew taller were better adapted to their environment than shorter trees, in that the taller trees got more sunlight. Accordingly, the taller trees left more offspring than the shorter trees. These offspring would likewise tend to inherit the genetic capacity to grow tall, meaning they too would have more offspring than other trees of *their* generation. In this way, the trait of tallness is being *selected* for transmission to ever-greater proportions of the tree population. As a result, the population evolves in the direction of greater tallness. As you will see, there are other processes that shape evolution, but none is as important as natural selection.

The Importance of Evolution as a Concept

Evolution is not just important to biology; it is central to it. If biology were a house, evolution's principles would be the mortar binding its stones together. Why? Because all life on Earth has been shaped by evolution's key principles, natural selection and common descent with modification. This is easy to see once you think about any given organism. Take your pick: a fish, a tree, a human being? *All* of them evolved from ancestors, all of them ultimately evolved from a common ancestor, and natural selection was the primary process that shaped their evolution.

Given this, the theory of evolution provides a means for us to understand nature in all its buzzing, blooming complexity. It allows us to understand not just things that are familiar to us, such as the forms of the redwoods and peacocks, but all manner of *new* natural phenomena that come our way. When physicians observe that strains of infectious bacteria are becoming resistant to antibiotics, they don't have to ask: How could this happen? General evolutionary principles tell them that organisms evolve; that bacteria are capable of evolving very quickly (because they can produce many generations a day); and that antibiotic resistance is simply an evolutionary adaptation.

Evolution Affects Human Perspectives Regarding Life

So far-reaching is the theory of evolution that its importance stretches beyond the domain of biology and into the realm of basic human assumptions about the world. Once persuasive evidence for the theory had been amassed, it meant that human beings could no longer view themselves as something *separate* from all other living things. Common descent with modification tells us that we are descended from ancestors, just like other organisms. Thus we sit not on a pedestal, viewing the rest of nature beneath us, but rather on one tiny branch of an immense evolutionary tree.

Beyond this, evolution meant the end of the idea of a *fixed* living world, in which, for example, birds have always been birds and whales have always been whales. Birds are the descendants of dinosaurs, while whales are the descendants of medium-sized mammals that walked on land. Life-forms only *appear* to be fixed to us, because human life is so short relative to the time frames in which species undergo change.

Finally, in evolution human beings are confronted with the fact that, through natural selection, life has been shaped by a process that has no mind and no goals. Like a river that digs a canyon, natural selec-

Web Tutorial 16.1
Principles of Evolution

tion shapes, but it does not design; it has no more "intentions" than does the wind. Trees that received more sunlight left more offspring, and thus trees grew taller over time. This is not a "decision" on the part of nature; it is an outcome of impersonal process. And if evolution has no consciousness, it follows that it can have no morals. This force that has so powerfully shaped the natural world is neither cruel nor kind, but simply indifferent.

16.2 Charles Darwin and the Theory of Evolution

How did such a far-reaching theory come into existence? Those who have read through the unit on genetics in this book know that the basic principles of that field were developed over the course of about a century by a large number of scientists. In contrast, a single person—nineteenth-century British naturalist Charles Darwin—is credited with bringing together the essentials of the theory of evolution (**see Figure 16.1**).

Darwin's Contribution

Darwin's contribution was twofold. First, he developed existing ideas about descent with modification while providing a large body of evidence in support of them. Second, he was the first to perceive that descent with modification is primarily driven by the process of natural selection.

To be sure, the study of evolution has been refined and expanded since Darwin. And it should be noted that before Darwin had ever published anything on his theory, a contemporary of his, fellow Englishman Alfred Russel Wallace, independently arrived at his insight about the importance of natural selection to evolution. (This makes Wallace the codiscoverer of this principle.) Yet, there is universal agreement that it is Charles Darwin who deserves primary credit for providing us with the core ideas that exist in evolutionary biology even today.

Darwin's Journey of Discovery

Darwin was born on February 12, 1809, in the country town of Shrewsbury, England. He was the son of a prosperous physician, Robert Darwin, and his wife, Susannah Wedgwood Darwin, who died when Charles was eight. Young Charles seemed destined to follow in his father's footsteps as a doctor, being sent away to the University of Edinburgh at age 16 for medical training. But he found medical school boring, and his medical career came to a halt when, in the days before anesthesia, he found it unbearable to watch surgery being performed on children. His fa-

ther then decided that he should study for the ministry. At the age of 20, Darwin set off for Cambridge University to spend three years that he later recalled as "the most joyous in my happy life." Darwin's happiness came in part from the fact that theology at Cambridge took a backseat to what had been his true passion since childhood: the study of nature. From his early years, he had collected rock, animal, and plant specimens and was an avid reader of nature books. His studies at Cambridge did yield a divinity degree, but they also gave Darwin a solid background in what we would today call life science and Earth science.

An Offer to Join the Voyage of the *Beagle*

Darwin's training, and the contacts he made at school, came together in one of the most fateful first-job offers ever extended to a recent college graduate. One of Darwin's Cambridge professors arranged to have him be the resident naturalist aboard the HMS *Beagle*, a ship that was to undertake a survey of coastal areas

Figure 16.1
Scientist of Great Insight
Charles Darwin, late in his life. (Painting by John Collier, 1883. London, National Portrait Gallery. ©Archiv/Photo Researchers.)

given areas, after which the creator had carried out *new* acts of creation, bringing more complex life-forms into being with each act.

If you were to survey a broad swath of scientific thought leading up to Darwin, you'd repeatedly find what you've seen with Lamarck and Cuvier: a given scientist getting things *partly* right and partly wrong (which is almost always the way in science). The result was a rich mix of scientific findings and fanciful speculation that existed in Darwin's world as he boarded the *Beagle.* He himself believed at the time that species were fixed entities; that they did not change over time.

16.4 Darwin's Insights Following the *Beagle's* Voyage

Darwin observed and collected wherever he went, but the most important stop on his journey took place nearly four years into it, when the *Beagle* stopped for a scant five weeks at the remote series of volcanic outcroppings called the Galapagos Islands (**see Figure 16.6**). There, in the dry landscape, amidst broken pieces of black lava, he saw strange iguanas and tortoises and mockingbirds that varied from one island to another. He shot and preserved several of these mockingbird "varieties" for transport back home, along with a number of small birds that he took to be blackbirds, wrens, warblers, and finches.

Figure 16.6
The Galapagos Islands
Galapagos is Spanish for *tortoise.* A part of Ecuador, the islands are still an important site of evolutionary research. Here a giant tortoise moves through the Alcedo Volcano area on one of the islands, Isabela.

Perceiving Common Descent with Modification

Arriving back in England in October of 1836, Darwin soon donated a good deal of his *Beagle* collection to the Zoological Society of London. If there was a single most important flash of insight for him regarding descent with modification, it came when one of the Society's bird experts gave him an initial report on the birds he had collected on the Galapagos. Three of the mockingbirds were not just the "varieties" that Darwin had thought; they were separate species, as judged by the standards of the day. Beyond that, the small birds he had believed to be blackbirds, finches, wrens, and warblers were all finches—separate species of finches, each of them found nowhere but the Galapagos (**see Figure 16.7**).

With this, Darwin began to see a pattern: The Galapagos finches were related to an ancestral species that could be found on the mainland of South America, hundreds of miles to the east. Members of that ancestral species had come by air to the Galapagos and then, fanning out to separate islands, had diverged over time into separate species. Thus it was with Galapagos tortoises and iguanas and cactus plants as well; they had common mainland ancestors, but on these islands, had diverged into separate species. Darwin began to perceive the infinite branching that exists in evolution.

Perceiving Natural Selection

But what drives this branching? In England and set on a career as gentleman naturalist, Darwin married and settled in London for a time before moving to a small village south of London, where he would live out his days (**see Figure 16.8**). Two years after his homecoming, inspiration on evolution's key process came to him not from looking at nature, but from reading a book on human population and food supply—*An Essay on the Principle of Population,* by T. R. Malthus. As Darwin later wrote in his autobiography:

> I happened to read for amusement Malthus on Population, and being well prepared to appreciate the struggle for existence which everywhere goes on ... it at once struck me that under these circumstances favourable variations would tend to be preserved and unfavourable ones to be destroyed.

Thus did natural selection occur to Darwin as the driving force behind evolution. *Organisms* that had "favorable variations" were preserved (they lived and left many offspring) while those that had unfavorable variations were destroyed (they left fewer offspring or none at all). As a result, over generations, organisms evolved in the direction of the favorable variations.

You might think that, with two major insights in place—common descent with modification and natural selection—Darwin would have rushed to inform the world of them. In fact, more than 20 years were to elapse between his reading of Malthus and the 1859 publication of his great book, *On the Origin of Species by Means of Natural Selection.* In between, though incapacitated much of the time with a mysterious illness, Darwin published on geology, bred pigeons, and spent eight years studying the variations in barnacles—activities that each had relevance to his theory, and that yet constituted a kind of holding pattern for him as he contemplated informing the world of a theory that he knew would be controversial. Darwin finally got to work on what he thought would be his "big species book" in 1856.

16.5 Alfred Russel Wallace

Two years after Darwin began this labor, half a world away and unbeknownst to him, a fellow English naturalist lay in a malaria-induced delirium on an Indonesian island. Alfred Russel Wallace made a living collecting bird and butterfly specimens from the then-exotic lands of South America and Southeast Asia, selling his finds to museums and collectors. He had thought long and hard (and even published) about how species originate. Now, shivering with malarial fever in a hut in the tropical heat, with the question on his mind again, Wallace came upon the very insight that had come to Darwin 20 years earlier: Natural selection is the process that shapes evolution. Recovering from his illness, Wallace wrote out his ideas over the next few days

and sent them to a scientific hero of his in England—Charles Darwin! He asked Darwin to read his manuscript and submit it to a journal if Darwin thought it worthy.

Darwin was stunned as he faced the prospect of another man being the first to bring to the world an insight he thought was his alone. Nevertheless, he would not allow himself to be underhanded with the younger scientist. He informed some of his scientist friends about his plight, and (without informing Wallace) they arranged for both Wallace's paper and some of Darwin's letters, sketching out his ideas, to be presented at a meeting of a scientific society in London on July 1, 1858. The readings before the scientific society turned out to have little immediate effect, but they did prod Darwin into working on what would become *On the Origin of Species,* published some 16 months later. This event had a great effect. Indeed, it set off a thunderclap whose reverberations can still be heard today.

16.6 Descent with Modification Is Accepted

Sparking both scorn and praise, Darwin's ideas were fiercely debated in the years after 1859. At least within scientific circles, however, it did not take long for common descent with modification to be accepted. Fifteen years or so after *On the Origin of Species* was published, almost all naturalists had become convinced of it, and it's not hard to see why. With evolution as a framework, so many things that had previously seemed curious, or even bizarre, now

Figure 16.7
One of Darwin's Finches
Geospiza fuliginosa feeds on sedge seeds in the Galapagos Islands.

Figure 16.8
Where *On the Origin of Species* Was Written
In this study in his house in rural England, Charles Darwin conducted scientific research and wrote his ground-breaking work, *On the Origin of Species by Means of Natural Selection.*

made sense. For example, years before *Origin* was published, scientists had known that, at a certain point in their *embryonic* development, species as diverse as fish, chickens, and humans all have structures known as pharyngeal slits (**see Figure 16.9**). In fish, these structures go on to be *gill* slits; in humans they develop into the eustachian tubes, among other things. The question was: Why would such different animals share this common structure in embryonic life? With evolution, the answer became clear. All of them shared a common vertebrate *ancestor*, who had the slits. The vertebrate ancestral line had branched out into various species, yet elements of the common ancestor persisted in these different species in their embryonic stage.

In bringing together countless "loose ends" such as this, evolution became the mortar that unified the study of the living world. (It's interesting to think of biologists from this period as readers who start going over a detective novel a *second* time, saying, "Of course!" and "Why didn't I see that?") Given this, scientists had little trouble accepting the *fact* of evolution—that is, the occurrence of descent with modification. But the primary process that Darwin said was driving evolution, natural selection, was not embraced so quickly. Indeed, the twentieth century would be almost halfway over before it was generally accepted.

16.7 Darwin Doubted: The Controversy over Natural Selection

The essential stumbling block for the theory of evolution by natural selection was this. It asserted that traits can vary in ways that confer reproductive advantages on given individuals, and that these variations can be passed on from one generation to the next. With the trees considered earlier, the trait was height, and the variation that conferred a reproductive advantage was greater height. In a given environment, therefore, added height represents a *difference* that allows some members of the species to leave more offspring than others, leading eventually to a taller species over all. But in the middle of the nineteenth century, no one could imagine how such differences could reliably be passed down over many generations. Even scientists of the time could not understand this, because they had no grasp of the field that we today call genetics.

Coming to an Understanding of Genetics

Those of you who have read the genetics section of this book may remember, from Chapter 11, that a trait such as tree height is likely to be governed by many genes working in tandem, with each of these genes coding for a protein that has some small, additive effect on height. Indeed, such multigene or "polygenic" inheritance is at work with most of the traits in living things. (See "Multiple Alleles and Polygenic Inheritance" in Chapter 11, page 221.) A tree's "height genes" will be shuffled in countless ways when tree egg and sperm are formed, and when they fuse in fertilization. As a result, gene combinations that yield particularly tall trees will come together in certain instances. In addition, outright mutations—alterations in genetic information—can take place, showing up as new physical traits, one of them being a taller tree. Most important for our purposes, whatever the genetic information is, it *persists*; it is passed on largely intact from one generation to another in the small informational packets we call genes. The result is small physical differences that can be reliably passed on from one generation to the next.

Pharyngeal slits exist in these five vertebrate animals ...

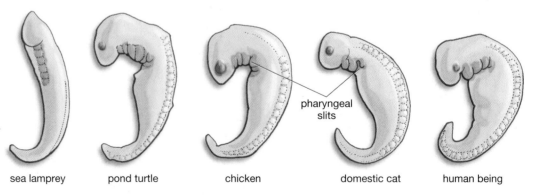

pharyngeal
slits

sea lamprey   pond turtle   chicken   domestic cat   human being

... evidence that all five evolved from a common ancestor.

**Figure 16.9
Different Embryos, Same
Structure**
(Adapted from
M. K. Richardson, 1997.)

Darwin could not have imagined a better mechanism for carrying out natural selection as he had envisioned it. The problem was that neither he nor anyone else in his time knew that things worked like this. Most nineteenth-century scientists believed that inheritance worked through a "blending" process akin to mixing different-colored paints. If you blend red and blue paint, for example, you get something that lies halfway between them, purple. Under such a system, differences in traits would not be preserved over time; they would be averaged into oblivion. Blending would work to make groups of organisms the *same*, not to allow differences among them. How could natural selection work if there were no persistent differences to select from?

Vindicating Natural Selection's Role in Evolution

The vindication of natural selection required, first, a demonstration (by Gregor Mendel) that inheritance doesn't work by blending. Rather, it works through the transmission of the stable genetic units we now call genes. Then, in the early part of the twentieth century, came the demonstration that, through polygenic inheritance, genes could account for very small physical differences that arise within a given population. Trees come in a *range* of heights, with the differences in this range governed by the combinations of individual genes described earlier.

Darwin Triumphant: The Modern Synthesis

Advances in genetics such as this were later joined to other types of evolutionary research, such that in the period roughly from 1937 to 1950 there took place what is known as the **modern synthesis**: The convergence of several lines of biological research into a unified evolutionary theory. Taxonomists reported on how species are distributed throughout the world. Mathematical geneticists provided guidance on how evolution *could* work, given the shifting around of genes among organisms. Paleontologists studied evolution in the fossil record. Critically, these different types of scientists were involved in a process of "mutual education" with one another; the findings of geneticists, for example, were now known to taxonomists, and vice versa. The upshot was a greatly deepened understanding of how evolution works—an understanding that provided conclusive evidence for Darwin's assertions about the importance of natural selection. The architect of evolutionary theory had been vindicated nearly 100 years after publishing his great work. (For another example of evolution's explanatory power, see "The Evolution of Skin Color" on page 320.)

16.8 Opposition to the Theory of Evolution

So, why are Darwin's essential insights so widely believed to be correct? We'll now review some of the evidence for these insights, but first a word about why this is desirable. When you studied cells in this book, you did not review the "evidence for cells." Nor did you go over the "evidence for energy transfer" when studying cellular respiration, nor the "evidence for lipids" when studying biological molecules. What makes evolution different from these topics?

The False Notion of a Scientific Controversy

Evolution has an unusual status among major biological subdisciplines. Almost alone among them, its findings are regularly challenged as being unproven or simply wrong. For the average person, who can't be bothered with the details of such a controversy, these attacks make it appear that the theory of evolution is not a body of knowledge, solidly grounded in evidence, but rather a kind of scientific guess about the history of life on Earth.

What Is a Theory?

Several factors are critical in allowing the continued appearance of a "scientific debate" on the validity of the theory of evolution, when in fact none exists. The first of these has to do with a simple misunderstanding regarding terminology. You saw in Chapter 1 that, to the average person, the word *theory* implies an idea that certainly is unproven and that may be pure speculation. In science, however, a theory is a general set of principles, supported by evidence, that explains some aspect of the natural world. Accordingly, we have Isaac Newton's theory of gravitation, Albert Einstein's general theory of relativity, and many others. Lots of principles go into making up the theory of evolution, some of them on surer footing than others. Over time, however, some of these principles have achieved the status of established fact—we are as sure of them as we are sure that the Earth is round. Among these principles are the fact that evolution has indeed taken place and the fact that this has occurred over billions of years.

The Nature of Historical Evidence

A second factor that provides an opening for the opponents of evolution is the nature of the evidence for it. If you were called upon to provide the "evidence for cells" referred to earlier, a look through a microscope at some actual cells might be enough to stop this "debate" in its tracks. Conversely, one of evolution's most important manifestations—radical transformations of life-forms—can never be observed in this way, because these transformations take place over vast expanses of time. Asking to "see" the equivalent of a dinosaur evolving into a bird before you'll believe it is like asking to see the European colonization of North America before you'll acknowledge it.

Evolution is taking place right now, all around us, but the evidence for evolution is *historical* evidence to a degree that is not true of, say, the study of genetics or of photosynthesis. Any ancient historical record is fragmentary, and evolution's historical record is ancient indeed. The fossils that scientists analyze are what remain from hundreds of millions of years of weathering and decay. Such an incomplete record leaves room for a great deal of interpretation among scientists—far more than is the case in purely experimental science. This interpretation has to do, however, with the *details* of evolution, not with its core principles. It has to do with what group descended from what other; with the rate at which evolution has proceeded, or with the role that pure chance has played in evolution. It does not have to do with *whether* evolution has occurred.

16.9 The Evidence for Evolution

So, what makes scientists so sure about the core principles of evolution? In any search for truth, we are more comfortable when lines of evidence are internally consistent and then go on to agree with *other* lines of evidence. The evidence for evolution satisfies both criteria. If we were to find even a single glaring inconsistency within or between lines of evidence, the whole body of evolutionary theory would be called into question. But no such inconsistencies have turned up yet, and scientists have been looking for them for about 145 years.

Radiometric Dating

One claim of evolutionary theory, as you have seen, is that evolution proceeds at a leisurely pace, with billions of years having elapsed between the appear-

ance of life and the present. Yet how do we know that Earth isn't, say, 46,000 years old, as opposed to the 4.6 billion that scientists believe it to be? The conceptual bedrock upon which we have determined the age of the Earth and its organisms is **radiometric dating**: a technique for determining the age of objects by measuring the decay of the radioactive elements they contain. As volcanic rocks are formed, they incorporate into themselves various elements that are in their surroundings. Some of these elements are radioactive, meaning they emit energetic rays or particles and "decay" in the process.

With such decay, one element can be transformed into another, the most famous example being uranium-238, which becomes lead-206 through a long series of transformations. The critical thing is that this transformation proceeds at a fixed *rate*; it is as steady as the most accurate clock imaginable. It takes 4.5 billion years for half a given amount of uranium-238 to decay into lead-206 (hence the term *half-life*). When such a transformation takes place in a cooled rock, the original or "parent" element is trapped within the rock, as is the "daughter" element, and the atoms of both elements can be counted. Therefore, comparing the *proportion* of a parent element to the daughter element in a rock sample provides a date for the rock with a fair amount of precision. This is so because of the fixed rate, noted earlier: It takes a set amount of time for a given number of parent atoms to yield the number of daughter atoms found in a sample. There are now more than 40 different radiometric dating techniques used, each of them employing a different radioactive isotope. (See Chapter 2, page 22, for information on isotopes.) Some of these isotopes have very long half-lives (as with uranium-238), but some have relatively short half-lives. Given the range of such "radiometric clocks," scientists have been able to date objects from nearly the formation of the Earth to the present, though there are some gaps in the picture.

Fossils

Today, one line of evidence for evolution is the similarity of fossil types by sedimentary layers. Looking at the same geologic layers of sediment worldwide, scientists find similar types of fossils, with a general movement toward more complex organisms as they go up through the newer strata. In Chapter 1, it was noted that scientific claims must be *falsifiable*; they must be open to being proved false upon the discovery of new evidence. The fossil record presents a falsifiable claim. For example, creatures called trilobites (**see Figure 16.10**) had a long run on Earth, existing in the ancient oceans from about 500

**Figure 16.10
Ancient Organism, Now Extinct**
Shown is a fossilized trilobite, a kind of arthropod that became extinct long before the animals known as primates came to exist. This fossilized trilobite dates from about 370 million years ago and was found in present-day Morocco.

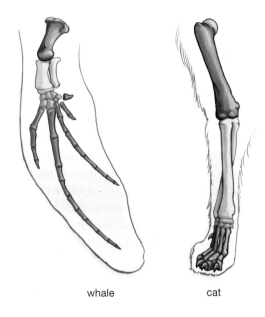
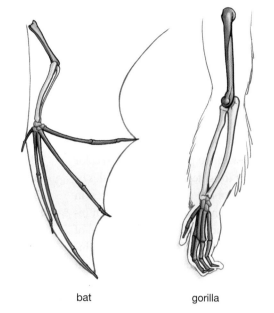

whale   cat   bat   gorilla

Figure 16.11
Four Animals, One Forelimb Structure
Whales, cats, bats, and gorillas are all descendants of a common ancestor. As a result, the bones in the forelimbs of these diverse organisms are very similar despite wide differences in function. Four sets of homologous bones are color-coded for comparison. Note that in each case there is one upper bone, joined to two intermediate bones, joined to five digits.

million years ago until some 245 million years ago, when they became extinct. By contrast, the evolutionary lineage that humans are part of, the primates, *began* about 60 million years ago. If scientists were to find, in a single fossil bed, fossils of trilobites existing side by side with those of early primates, our whole notion of evolutionary sequences would be called into question. No such incompatible pairing has happened, however, and the strong betting is it won't. Given this, the line of fossil evidence is internally consistent, but there's more. When we compare fossil placement with the dates we get from radiometric dating, we get excellent agreement *between* these two lines of evidence. We don't find trilobites embedded in sediments that turn out to be 60 million years old.

Comparative Morphology and Embryology

Morphology is the study of the physical forms that organisms take, while **embryology** is the study of how animals develop, from fertilization to birth. As it turns out, the evidence provided for evolution by comparative embryology was touched on earlier, in noting the pharyngeal slits that exist in the embryos of creatures as diverse as fish and humans.

In comparative morphology, some classic evidence for evolution is seen in the similar forelimb structures found in a very diverse group of mammals—in a whale, a cat, a bat, and a gorilla, as seen in **Figure 16.11**. Such features are said to be **homologous**, meaning the same in structure owing to inheritance from a common ancestor. Look at what exists in each case: one upper bone, joined to two intermediate bones, joined to five digits. Evolutionary biologists postulate that the four mammals evolved from a com-

mon ancestor, adapting this 1-2-5 structure over time in accordance with their varying needs.

Evidence from Gene Modification

In the genetics unit you saw that every living thing on Earth employs DNA and utilizes an almost identical "genetic code" (*this* triplet of DNA bases specifies *that* amino acid). At the very least, this means there is a unity running through all earthly life; it is also consistent with the idea that all life on Earth ultimately had a single starting point—the single common ancestor mentioned earlier.

In recent years, molecular biology has also provided another check on what scientists have long believed to be true about evolutionary relationships—about which species are more closely related and about how long it's been since they've shared a common ancestor. All genes amount to a sequence of A, T, G, and C bases along DNA's double helix. And there are some genes that do very similar things in different organisms. There is, for example, a gene called *hedgehog* that helps regulate embryonic development in the *Drosophila* fruit fly, and a gene called *sonic hedgehog* that helps regulate embryonic development in mice.

Though the *hedgehog* and *sonic hedgehog* genes are similar, we would not expect them to be identical. This is so because gene base sequences *change* over time through the process of mutation. If the *rate* of this change is constant, then mutations are like a molecular clock that's ticking, with the passage of a given amount of time, you get a set number of mutations. Given this, the longer it has been since two organisms shared a common ancestor, the greater the number of base differences we should see in the sequence of genes like *hedgehog* and *sonic hedgehog*.

Figure 16.12
Using Molecules to Track Evolution
Diverse organisms—such as yeast, moths, and pigs—all have genes that code for an enzyme called cytochrome *c* oxidase. These organisms inherited cytochrome *c* oxidase genes from a common ancestor many millions of years ago. Over time, however, the cytochrome *c* oxidase genes have undergone mutations that have altered the sequence of their DNA "building blocks," called bases. The longer it has been since any two species shared a common ancestor, the more differences there should be in their cytochrome *c* oxidase bases. There are 13 differences between the bases found in human cytochrome *c* oxidase genes and those found in pigs, but there are 66 differences between the human and yeast genes. (Data from Whitfield, Philip. 1993. From *So Simple a Beginning: The Book of Evolution*. New York: Macmillan: Maxwell Macmillan International.)

To better understand this last point, it may be helpful to think in terms of something more familiar: natural language. The European settlers of Australia were predominately English. Once the Australians and English became fundamentally *separate*, however, their manner of speaking began to evolve independently. For example, we could think of the word *outback* as a "mutation" in Australian speech. Meanwhile, England would be developing its own "mutations," meaning words not used in Australia. And the longer the Australians went on being separate from the English, the greater the number of such differences we would see between the two languages. Just so, the longer it has been since two species went down separate evolutionary lines, the greater the number of base differences we would see in similar *genes* they have, as measured by the different mutations each has acquired.

This gene-modification hypothesis has been put to the test in connection with a gene that codes for an enzyme called cytochrome *c* oxidase, which exists in organisms as different as humans, moths, and yeasts. Going into this experiment, evolutionary theory predicted that there should be fewer DNA base-pair differences between the cytochrome *c* oxidase genes of, say, a human and a duck than between a human and a moth, because all the *other* evidence we had told us that humans and ducks share a more recent common

ancestor than do humans and moths. If you look at **Figure 16.12**, you can see that evolutionary theory was fully borne out by the DNA sequencing. There are 17 sequence differences between a human and a duck, but 36 between a human and a moth. Indeed, all the differences between the species pictured fall into line with evolutionary theory. Thus we have another confirmation for evolution between lines of evidence—between DNA sequencing *and* the fossil record *and* radiometric dating *and* comparative morphology.

Experimental Evidence

Finally, much evidence for evolution has been provided in recent years by experiment and observation. At first glance, this may seem rather improbable because as you've seen, evolutionary change can often be perceived only over long stretches of time. Scientists have devised clever ways to catch evolution in the act, however.

Consider the experiments of John Endler, who believed that the male guppies he was studying were being pulled in two directions by natural selection. On the one hand, those who were larger and had brighter coloration were chosen more often by females for mating. On the other, these very characteristics made the males more vulnerable to predators. Endler saw that he could test natural selection by putting some guppies in a predator-free environment. In only a few generations, the males evolved brighter coloration and larger tails. When Endler then reintroduced predators into this population, things went the other way; over several generations, the males evolved smaller tails and less brilliant colors. Both of these outcomes are precisely what evolutionary theory would predict. In both instances the guppies evolved in a direction that would maximize their reproductive success—more mates with the brighter colors in the predator-free environment, and a longer life (and hence more reproduction) with the drab coloration they evolved in the predator-laden environment.

On to How Evolution Works

Given the abundance of evidence for evolution, biologists long ago stopped asking whether it occurred. The really interesting questions for decades have been: Through what means has evolution proceeded? At what pace? In what direction, if any? These are the questions this text will address in the next three chapters.

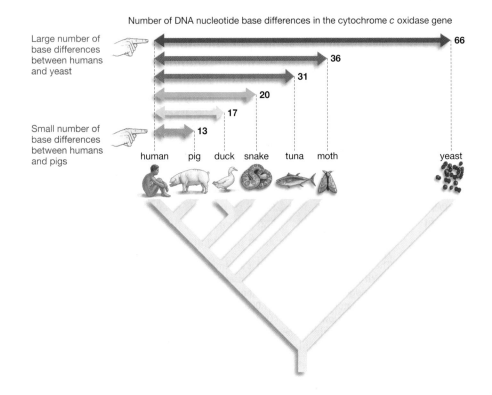

Number of DNA nucleotide base differences in the cytochrome *c* oxidase gene

Large number of base differences between humans and yeast

66
36
31
20
17
13

Small number of base differences between humans and pigs

human   pig   duck   snake   tuna   moth   yeast

Chapter Review

Summary

16.1 Evolution and Its Core Principles

- Within the theory of evolution, a key principle is that of descent with modification. This principle describes the process by which species of living things can undergo modification over time, with such change sometimes resulting in the formation of new, separate species. All species on Earth have descended from other species, and a single, common ancestor lies at the base of the evolutionary tree.

- A second key principle in the theory of evolution concerns natural selection: a process in which the differential adaptation of organisms to their environment selects those traits that will be passed on with greater frequency from one generation to the next.

- The theory of evolution has an importance beyond the domain of biology. Through it, human beings have become aware that (1) they are descended from other varieties of living things, and (2) the organisms that populate the living world are not fixed entities, but instead are constantly undergoing change.
Web Tutorial 16.1: Principles of Evolution

16.2 Charles Darwin and the Theory of Evolution

- Charles Darwin deserves primary credit for the theory of evolution. He developed existing ideas about descent with modification while providing a large body of evidence in support of them. And he was the first to perceive that natural selection is the primary process that drives evolution.

- Darwin's insights were inspired by the research he carried out during a five-year voyage he took around the world on the ship *HMS Beagle*, beginning in 1831.

16.3 Evolutionary Thinking before Darwin

- Some of Darwin's ideas can be traced to the work of Charles Lyell, Jean-Baptiste de Lamarck, and Georges Cuvier, who respectively noted the dynamic geological nature of the Earth, the possibility of descent with modification, and the extinction of species on Earth.

16.4 Darwin's Insights Following the *Beagle*'s Voyage

- Darwin understood descent with modification for several years before he comprehended that natural selection was the most important process driving it. It was his reading of a work by Malthus that sparked his realization about natural selection.

16.5 Alfred Russel Wallace

- English naturalist Alfred Russel Wallace is the co-discoverer of the principle that evolution is shaped by natural selection.

16.6 Descent with Modification Is Accepted

- Descent with modification was accepted by most scientists not long after publication of Darwin's *On the Origin of Species by Means of Natural Selection* in 1859. Scientists accepted it because it explained so many facets of the living world.

16.7 Darwin Doubted: The Controversy over Natural Selection

- The hypothesis that natural selection is the most important process underlying evolution was not generally accepted until the middle of the twentieth century. Its acceptance hinged on a modern synthesis in the theory of evolution that brought together lines of evidence from genetics, the fossil record, and the distribution of organisms throughout the world.

16.8 Opposition to the Theory of Evolution

- Even today the theory of evolution is regularly challenged as being unproven or simply wrong. One factor leading to the appearance of a "scientific debate" over evolution is confusion about the meaning of the word theory. Though the average person may equate "theory" with speculation, in science, a theory is a general set of principles, supported by evidence, that explains some aspect of the natural world.

16.9 The Evidence for Evolution

- Five principal lines of evidence are consistent with the theory of evolution. First, radiometric dating indicates that the Earth is greater than 4 billion years old. Second, the placement of fossils is consistent with the theory of evolution and with radiometric dating. Third, the theory of evolution explains the common occurrence of homologous physical structures in different organisms. Fourth, the theory of evolution is consistent with variations found in the DNA sequences of various organisms. Fifth, experimental demonstrations of evolution have been carried out in the laboratory and in nature.

Key Terms

common descent with modification 310	**modern synthesis** 317
	morphology 319
embryology 319	**natural selection** 310
homologous 319	**radiometric dating** 318

Understanding the Basics

Multiple-Choice Questions (answers in the back of the book)

1. In science, a theory is
   a. an untested hypothesis
   b. a general set of principles, supported by evidence, that explains some aspect of the natural world
   c. speculation about the natural world, based on general knowledge of a field
   d. an observation
   e. the first idea proposed to explain some aspect of the natural world

2. Which of the following are central ideas in the theory of evolution by natural selection? (Select all that apply.)
   a. Organisms vary in traits that affect their reproduction.
   b. Descent with modification occurs over generations.
   c. Evolution has a goal.
   d. Traits that aid in reproduction will become more widespread in succeeding generations.
   e. Traits that are acquired during the life of an individual contribute to evolution.

3. Which of the following observations provide evidence for evolution? (Select all that apply.)
   a. Monkey and trilobite fossils are never found in the same fossil beds.
   b. Athletic training can produce an increase in muscle mass.
   c. DNA sequences of genes shared by various species vary in accordance with predictions about how related those species are.
   d. Almost all organisms use a common genetic code.
   e. Species whose adult forms look very different may have similar features in embryonic life.

4. During the formulation of his theory of evolution by natural selection, Darwin brought together ideas and results from several disciplines. Match the person with the phrase that describes his work.
   a. Lyell — catastrophic extinction and new creations explain fossil record
   b. Wallace — inheritance of acquired characters
   c. Lamarck — natural selection is differential survival or reproduction
   d. Cuvier — geological forces observable today caused changes in Earth

5. Some of the following structures are homologous with each other. Which ones are they, and which two do not belong to this group?
   a. whale flipper
   b. insect leg
   c. bat wing
   d. cat forelimb
   e. octopus tentacles

6. The discovery of _____ provided critical evidence for natural selection to be widely accepted as the chief mechanism driving evolution (select one):
   a. methods for radiometric dating
   b. extinctions in the fossil record
   c. genes as the cause of disease
   d. genes as the units of inheritance
   e. fossil bacteria

7. Important implications of the theory of evolution by natural selection include (select all that apply):
   a. All organisms, including humans, are descended from a common ancestor.
   b. The biological world is constantly evolving.
   c. Nature designs things intentionally.
   d. The characteristics of organisms are molded by an impersonal process without a goal.
   e. Humans are just a small part of the tree of life.

8. John Endler performed an experiment involving guppies in which he demonstrated evolution driven by natural selection that worked through a predator-prey relationship. Which of the following are true statements about the results of his experiments? (Select all that are correct.)
   a. Males with longer tails and brighter coloration were more frequently eaten by predator fish.
   b. Females preferred to mate with dull-colored males.
   c. When predator fish were added to a population that had not previously had predators, males with dull coloration were favored by natural selection.
   d. When predators were removed, the average brightness of male tails in the population increased.
   e. When predators were removed, the average brightness of females increased.

Brief Review

1. What is homology, and why does it provide evidence for evolution?

2. How does the evidence presented in Figure 16.12 support the conclusion that humans are more closely related to pigs than to yeast?

3. Describe two examples in which agreement among different lines of evidence provides evidence for evolution.

4. What is one observation that Darwin made in the Galapagos that influenced his thinking about evolution, and why was it important? Would he have been as likely to formulate his theory had he not had the opportunity to sail on the *Beagle*? Why or why not?

Applying Your Knowledge

1. Using evolutionary principles, explain why large ears might be expected to evolve in a terrestrial plant-eater such as a rabbit or deer, but not in an aquatic mammal such as a seal.

2. Explain the evolutionary steps by which bacteria may become resistant to antibiotics, using the core requirements for evolution by natural selection.

3. Critics of the theory of evolution say it leaves no room for human purpose, since the theory asserts that human beings evolved through a process—natural selection—that has no goals or intentions. Does the theory undercut the idea of purpose in human life?

17 *Microevolution*

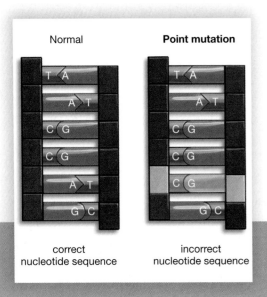

Route to new genetic information.
(Section 17.3, page 330)

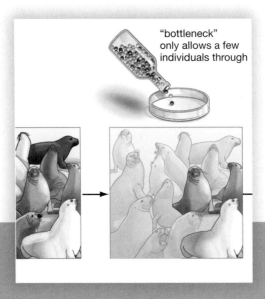

The bottleneck effect.
(Section 17.3, page 332)

Ladies, take your pick!
(Section 17.3, page 333)

When people think of the word *evolution*, what generally comes to mind is the grand sweep of evolution: The story of how microorganisms were the first life-forms; of how dinosaurs came and went; of how human beings arose as newcomers on the planet. It is quite a story, to be sure, but to fully appreciate it we need to look first not at these *outcomes* of evolution, but rather at evolution's *processes*. How do generations of organisms become modified over time? That's the subject of this chapter. When you understand it, you'll be in a position to comprehend evolution on its grander scale.

17.1 What Is It That Evolves?

In approaching the topic of how organisms become modified over generations, the first question is: What is it that evolves? If you think about it, it's pretty clear that it is not individual organisms. A tree may inherit a mix of genes that makes it slightly taller than other trees, but if this tree is considered in isolation, it is simply one slightly taller tree. If it were to die without leaving any offspring, nothing could be said to have evolved, because no persistent quality (added height) has been passed on to any group of organisms.

In Chapter 16 you were introduced to the idea of a **species**, which can briefly be defined as a group of organisms who can successfully interbreed with one another in nature, but who don't successfully interbreed with members of other such groups. You might think that it is species—such as horses or American elm trees—that evolve. Indeed, scientists often speak this way when they refer, for example, to an amphibious mammal having evolved into today's whales. But this is a kind of shorthand whose inaccuracy becomes apparent when species are considered as they live in the real world.

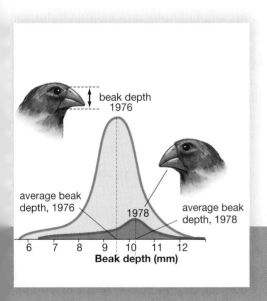

Surviving a drought.
(Section 17.4, page 336)

Evolving towards smoothness.
(Section 17.5, page 339)

No medium beaks.
(Section 17.5, page 339)

www

Web Tutorial 17.1
Evolution and Genetics

Figure 17.1
Evolution within Populations

(a) A hypothetical species of frog lives and interbreeds in one expanse of tropical forest.

(b) After several years of drought, the forest has been divided by an expanse of barren terrain. The single frog population has thus been divided into two populations, separated by the barren terrain. The two frog populations now have different environmental pressures, such as the kinds of predators each faces, and they no longer interbreed; after many generations their coloration diverges as they adapt to the different pressures.

Populations Are the Essential Units That Evolve

Think of a hypothetical species of frogs living in, say, equatorial Africa in a single expanse of tropical forest. Suppose that a drought persists for years, drying up the forest such that this single lush range is now broken up into two ranges separated by an expanse of barren terrain (**see Figure 17.1**). When the separation occurs, there is still only a single species of frog; but that species is now divided into two *populations* that are geographically isolated and hence no longer interbreeding. Each population now faces the natural selection pressures of its *own* environment—environments that may differ greatly even though they are nearly adjacent. One area may get more sun, while the other has a larger population of frog predators, for example. Thus, each population stands to be modified individually over time—to evolve. What is it that evolves? The essential unit that does so is a **population**, which can be defined as all the members of a species that live in a defined geographic region at a given time.

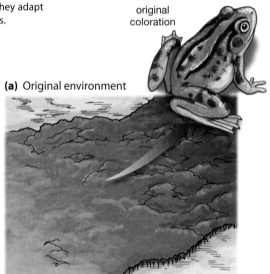

original
coloration

(a) Original environment

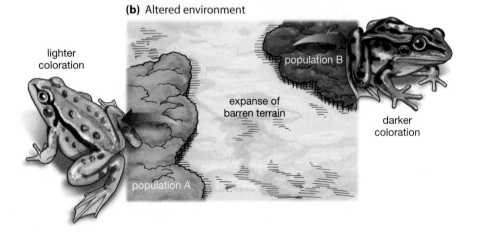

(b) Altered environment

lighter
coloration

population B

expanse of
barren terrain

darker
coloration

population A

The question then becomes, *how* do populations evolve? The population of frogs that has more predators may, over many generations, evolve a coloration that makes them less visible to predators. But how does this slightly different coloration come about? Through genes. A given frog may be unable to change its spots, but *generations* of frogs can have theirs changed through the shuffling, addition, and deletion of genetic material.

Genes Are the Raw Material of Evolution

Recall from Chapter 11 that the genetic makeup of any organism is its **genotype**—and that a genotype provides an underlying basis for an organism's **phenotype**, meaning any observable traits that an organism has, including its physical characteristics and behavior. Many genes are likely to be involved in producing a phenotype such as coloration. In sexually reproducing organisms, *each* of these coloration genes will likely come in two variant forms, called **alleles**, with offspring inheriting one allele from their father and one from their mother (**see Figure 17.2**). Though both alleles help code for coloration, one may result in slightly lighter or darker coloration than the other.

Genes don't just come in two allelic variants, however; they can come in many. In most species, no one organism can possess more than two alleles of a given gene, but a *population* of organisms can possess many allelic variants of the same gene (which is one reason human beings don't just come in one or two heights, but in a continuous *range* of heights.)

When the concept of a population is put together with that of genes and their alleles, the result is the concept of the **gene pool**: all the alleles that exist in a population. This gene pool is the raw material that evolution works with. If evolution were a card game, the gene pool would be its deck of cards. Individual cards (alleles) are endlessly shuffled and dealt into different "hands" (the genotypes that individuals inherit), with the strength of any given hand dependent upon what game is being played. (Survival in a drought? Survival against a new set of predators?)

17.2 Evolution as a Change in the Frequency of Alleles

Thinking of genes in terms of a gene pool gives us a new perspective on what evolution is at root: a change in the frequency of alleles in a population. This may sound a little abstract until you consider the frog coloration example. A frog inherits, from its mother and father, a coloration that allows it to evade predators slightly more successfully than other frogs of the same generation. It thus lives longer and leaves more offspring than the other frogs. It is successful in this way because of the advantageous set of alleles it inherited. Because this frog is more successful at breeding, its alleles are being passed to the *next* generation of frogs in relatively greater numbers than the alternative alleles carried by less successful frogs. Thus, this frog's alleles are increasing in frequency in the frog population. In looking at any example of evolution in a population over time, you would find this kind of change in allele frequency as its basis, assuming a stable environment.

With this perspective in mind, you're ready for a definition of **microevolution**: A change of allele frequencies in a population over a relatively short period of time. Why the *micro* in microevolution? Because evolution *within* a population is evolution at its smallest scale. This conception of microevolution allows you to understand the formal definition of evolution promised to you last chapter. **Evolution** can be defined as any genetically based phenotypic change in a population of organisms over successive generations. Taking a step back, the large-scale *patterns* produced by microevolution eventually become visible, as with the evolution of, say, mammals from reptiles. This is **macroevolution**, defined as evolution that results in the formation of new species or other groupings of living things.

17.3 Five Agents of Microevolution

So, what is it that causes microevolution? Put another way, what causes the frequency of alleles to change in a population? In the frog-coloration example, a familiar process was at work: natural selection. The frog's coloration allowed the frog to evade predators, thus helping the frog to produce more offspring than did other frogs. This particular coloration was thus *selected* for greater transmission to future generations. With this selection, allele frequencies began to be altered in the frog population. Over generations, the alleles for protective coloration increased relative to other sets of alleles.

Natural selection is not the only process that can change allele frequencies, however. There are five "agents" of microevolution that can alter allele frequencies in populations. These are mutation, gene flow, genetic drift, nonrandom mating, and natural selection. You can see these processes summarized in **Table 17.1**. (For a review of how scientists tell whether one of these processes actually is at work changing allele frequencies, see "Detecting Evolution: The Hardy-Weinberg Principle," on p. 340.) We'll now look at each of these agents of change in turn.

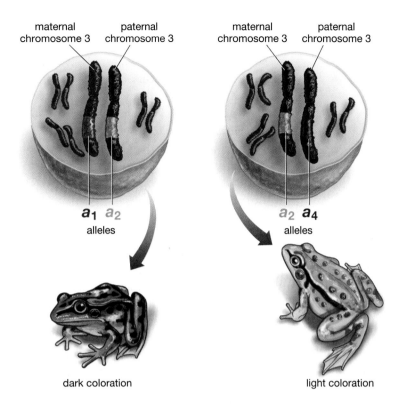

Figure 17.2
Genetic Basis of Evolution
Many genes can produce a trait such as body coloration, and each gene often has many alleles or variants. Each individual in a population, however, can possess only two alleles for each gene, one allele inherited from its mother and one inherited from its father. The two frogs in the figure both have maternal and paternal copies of chromosome 3 that house genes for coloration. The chromosomes of the two frogs may differ, however, in the allelic variants they have of these genes. The frog with dark coloration may possess alleles a_1 and a_2 while the light-colored frog may possess alleles a_2 and a_4 of this same gene.

Table 17.1 Agents of Change: Five Forces That Can Bring about Change in Allele Frequencies in a Population

Agent	Description
Mutation	Alteration in an organism's DNA; generally has no effect or a harmful effect. But beneficial or "adaptive" mutations are indispensable to evolution.
Gene flow	The movement of alleles from one population to another. Occurs when individuals move between populations or when one population of a species joins another, assuming the second population has different allele frequencies than the first.
Genetic drift	Chance alteration of gene frequencies in a population. Most strongly affects small populations. Can occur when populations are reduced to small numbers (the bottleneck effect) or when a few individuals from a population migrate to a new, isolated location and start a new population (the founder effect).
Nonrandom mating	Occurs when one member of a population is not equally likely to mate with any other member. Includes sexual selection, in which members of a population choose mates based on the traits the mates exhibit.
Natural selection	Some individuals will be more successful than others in surviving and hence reproducing, owing to traits that give them a better "fit" with their environment. The alleles of those who reproduce more will increase in frequency in a population.

Mutations: Alterations in the Makeup of DNA

As you saw in the genetics unit, a mutation is any permanent alteration in an organism's DNA. Some of these alterations are heritable, meaning they can be passed on to future generations. Mutations can be as small as a change in a single base pair in the DNA chain (a point mutation) or as large as the addition or deletion of a whole chromosome or parts of it. Whatever the case, a mutation is a change in the informational set an organism possesses (**see Figure 17.3**). Looked at one way, it is a change in one or more alleles.

The rate of mutation is very low in most organisms; during cell division in humans, it might be just one DNA base pair per billion. And of the mutations that do arise, very few are beneficial or "adaptive." Most do nothing, and many are harmful to organisms. Thus mutations usually are not working to *further* survival and reproduction, as the frog-coloration alleles did. Given this, they generally are not likely to appear with greater frequency in successive generations. The upshot is that mutations are not likely to account for much of the change in allele frequency that is observed in any population.

But a few mutations occur that are adaptive. These genetic alterations are something like creative thinkers in a society: They are rare but very important. Such mutations are the only means by which *new* genetic information comes into being—by which new proteins are produced that can modify the form or capabilities of an organism. The evolution of eyes or wings had to involve mutations. No amount of shuffling of *existing* genes could get the living world from no eyes to eyes. Of course, no mutation can bring about a feature such as eyes in a single step; such changes are the result of many mutations, followed by rounds of genetic shuffling and natural selection, generally over millions of years.

Web Tutorial 17.2
Agents of Change

Gene Flow: When One Population Joins Another

Allele frequencies in a population can also change with the mating that can occur after the arrival of members from a *different* population. This is the second microevolutionary agent, **gene flow**, meaning the movement of genes from one population to another. Such movement takes place through **migration**, which is the movement of individuals from one population into the territory of another. Some populations of a species may truly be isolated, such as those on remote islands, but migration and the gene flow that goes with it are the rule rather than the exception in nature. It may seem at first glance that migration would be limited to animal species, but this isn't so. Mature plants may not move, but plant seeds and pollen do; they are carried to often-distant locations by wind and animals (**see Figure 17.4**). Of course, for a migrating population to alter allele frequencies of another population, its gene pool must be different from that of the population it is joining.

Genetic Drift: The Instability of Small Populations

To an extent that may surprise you, evolution turns out to be a matter of chance. You can almost see the dice rolling in the third microevolutionary agent, genetic drift. Imagine a hypothetical population of 10,000 individuals. An allele in this gene pool is carried by one out of ten of them, meaning that 1,000 individuals carry it. Now imagine that some disease sweeps over the population, killing half of it. Say that this allele had nothing to do with the disease, so the illness might be expected to decimate the allele carriers in rough accordance with their proportion in the population. If this were the case, 5,000 individuals in the population would survive, and 1 in

(a)

Normal   **Point mutation**

correct nucleotide sequence   incorrect nucleotide sequence

Figure 17.3
Basis of New Genetic Information
A mutation is any permanent alteration in an organism's DNA. Examples of mutations include **(a)** point mutations, in which the nucleotide sequence is incorrect and **(b)** deletions, in which part of a chromosome is missing.

(b)

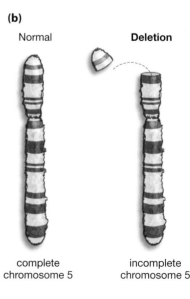

Normal   **Deletion**

complete chromosome 5   incomplete chromosome 5

(a) Hawaiian silversword

(b) Tarweeds in California

North
America

gene flow

Hawaiian
Islands

Pacific
Ocean

Figure 17.4
Migration in Plants
Brought about by volcanic erup-tions, the Hawaiian Islands have al-ways been surrounded by the Pacific Ocean. Therefore all the plant species existing on the islands today are descended from species that were introduced to the islands through one means or another—human activity, wind currents, water currents, or animal dispersal of seeds and pollen. **(a)** Hawaiian sil-verswords are derived from **(b)** a lineage of California plants com-monly known as tarweeds.

10—or about 500 of them—could be expected to be carriers of this allele. Let us say, however, that just by chance, *550* of the allele carriers were killed, thus leaving the surviving population of 5,000 with only 450 allele carriers. In that scenario, the frequency of the allele in this population would drop from 10 percent to 9 percent (**see Figure 17.5a**).

Now for the critical step. Imagine the same allele, with the same 1-in-10 frequency, only now in a pop-ulation of *10*. There is now but a single carrier of the allele, and that individual may not be one of the 5 members of the population to survive the disease. In this case, the frequency of this allele drops from 10 percent to zero (**see Figure 17.5b**). It can be replaced

(a) Large population = 10,000
(allele carriers in red)

allele frequency = $\frac{1,000}{10,000}$ = 10%

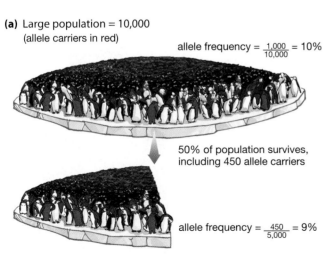

50% of population survives,
including 450 allele carriers

allele frequency = $\frac{450}{5,000}$ = 9%

**little change in allele frequency
(no alleles lost)**

(b) Small population = 10
(allele carriers in red)

allele frequency = $\frac{1}{10}$ = 10%

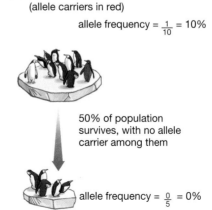

50% of population
survives, with no allele
carrier among them

allele frequency = $\frac{0}{5}$ = 0%

**dramatic change in allele frequency
(potential to lose one allele)**

Figure 17.5
Genetic Drift

(a) In a hypothetical population of 10,000 individuals, 1 in 10 carries a given allele. The population loses half its members to a disease, in-cluding 550 individuals who car-ried the allele. The frequency of the allele in the population thus drops from 10 percent to 9 percent.

(b) A population of 10 with the same allele frequency likewise loses half its members to a disease. Because the one member of the population who carried the allele is not a survivor, the frequency of the allele in the population drops from 10 percent to zero.

only by a mutation (which is extremely unlikely) or by migration from another population. Assuming that neither happens, no matter how this population grows in the future, in genetic terms it will be a different population than the original in that it lost this allele. The allele might be helpful or harmful, but its adaptive value doesn't matter. It has been eliminated strictly through chance. This is an example of **genetic drift**: the chance alteration of allele frequencies in a population, with such alterations having the greatest impact on small populations. It is true that some genetic drift has taken place in the larger population, but consider how small the effect is. The allele simply went from a 10 percent frequency to a 9 percent, with no loss of allele. Chance events can have much greater effects on small populations than on large ones.

Two scenarios, common in evolutionary history, produce the small populations that are most strongly affected by genetic drift. Populations can be greatly reduced through disease or natural catastrophe; or a small subset of a population can migrate elsewhere and start a new population. The first of these scenarios is called the *bottleneck effect*; the second is called the *founder effect*. Let's have a look at both of them in turn.

The Bottleneck Effect and Genetic Drift
The **bottleneck effect** can be defined as a change in allele frequencies in a population due to chance following a sharp reduction in the population's size. Real populations, or even species, can go through dramatic reductions in numbers. For example,

northern elephant seals, which can be found off the Pacific coast of North America, were prized for the oil their blubber yielded and thus were hunted so heavily that by the 1890s, fewer than 50 animals remained. Thanks to species protection measures, the seals' numbers have rebounded somewhat in recent decades. But genetically, all the members of this species are very similar today, because they all descended from the few seals who made it through the nineteenth-century bottleneck. What occurs in these reductions is a "sampling" of the original population—the "sample" being those who survived the devastation (**see Figure 17.6**).

The reason allele frequencies change in such an event has to do with the nature of probability and sample size. Imagine a box filled with M&M's candies, with equal numbers of red, green, and yellow M&M's inside. You close your eyes and grab a handful of M&M's and pull out 12. With such a small sample, you might get, say, six reds, four greens, and two yellows, rather than the four-of-each-kind that would be expected from probability. If you pulled out *120* M&M's, conversely, your reds, greens, and yellows would be much more likely to approach the 40-of-each-kind that would be expected. In just such a way, a small sample of *alleles* is likely to yield a gene pool that's different from the distribution found in the larger population.

The Founder Effect and Genetic Drift
Genetic drift can also result from the **founder effect**, which is simply a way of stating that when a small subpopulation *migrates* to a new area to start a new

Figure 17.6
The Bottleneck Effect
Northern elephant seals were hunted so heavily by humans that in the 1890s, fewer than 50 animals remained. The population has grown from the few survivors (represented here by the three seals in the second frame), but the resulting genetic diversity of this population is very low. (Seal coloration for illustrative purposes only.)

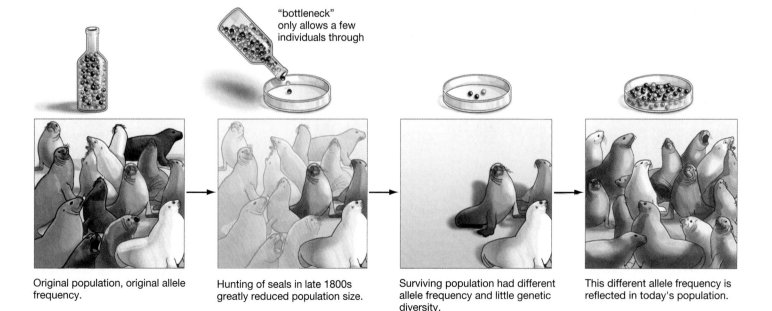

"bottleneck" only allows a few individuals through

Original population, original allele frequency.

Hunting of seals in late 1800s greatly reduced population size.

Surviving population had different allele frequency and little genetic diversity.

This different allele frequency is reflected in today's population.

population, it is likely to bring with it only a portion of the original population's gene pool. This is another kind of sampling of the gene pool, in other words; but in this case it's caused by the migration of a few individuals rather than the survival of only a few. This sample of the gene pool now becomes the founding gene pool of a new population. As such, it can have a great effect; whatever genes exist in it become the genetic set that is passed on to all future generations, as long as this population stays isolated.

The power of the founder effect can be seen most clearly when a founding population brings with it the alleles for rare genetic diseases. There is, for example, a very rare genetic affliction of the eyes called *cornea plana* that results in a misshapen cornea—the first structure of the eye through which light passes. The result can be impaired close-range vision and a general clouding of eyesight. Cornea plana is known to affect only 113 people worldwide. The strange thing is that 78 of these people are in Finland, most of them in an area in northern Finland. Current research indicates that about 400 years ago, a small population arrived in this isolated area, with at least one member of this population carrying the recessive allele that causes this affliction. Since then, this allele has continued to profoundly affect subsequent generations in the area. This will always be the case if the descendants of a founder population stay relatively isolated over time—that is, if the descendents breed mostly among themselves.

From this and the earlier examples, you may have perceived by now that there is an inherent value in genetic diversity; you can read about this in "Lessons from the Cocker Spaniel," on page 334.

Nonrandom Mating: When Mating Is Uneven across a Population

You've looked at mutation, gene flow, and genetic drift as agents of change with respect to allele frequencies in a population. Now let's look at a fourth agent, **nonrandom mating**, which is simply mating in which a given member of a population is not equally likely to mate with any other given member.

Some forms of nonrandom mating do not directly affect allele frequencies. Such is the case with **assortative mating**, which occurs when males and females that share a particular characteristic tend to mate with one another. Humans practice assortative mating, in that short individuals tend to pick short mates, while tall individuals tend to pick tall mates. Such mating tends to bring similar alleles together ("short" with short, etc.) but it does not directly alter the frequency of alleles in a population.

Nonrandom mating does affect allele frequency, however, when some members of a population mate *more* than others. In practice, this is mating based on *phenotype*, which you may recall is any observable trait in an organism, including differences in appearance and behavior. It is the appearance of particularly nice plumage on a male peacock that makes female peacocks choose it for mating, rather than another male, while it is the behavior of strutting and chest-swelling in a male sage grouse that causes a succession of female grouse to mate with it rather than nearby competitors (**see Figure 17.7**). These are examples of **sexual selection**, defined as a form of nonrandom mating that produces differential reproductive success, based on differential success in obtaining mating partners. It's easy to see that if one male mates four times as much as the average male of his generation, his alleles stand to increase proportionately in the next generation.

Differential mating success among members of one sex in a species often is based on choices made by members of the opposite sex in that species. In general, it is females who are doing the choosing in these situations. Differential mating success can also, however, be based on differences in combative abilities that give individuals of one sex (generally males) greater access to members of the opposite sex. Sexual selection has something in common with natural selection, in that both processes result in some individuals passing along more of their genes to future generations than others. But while natural selection has to do with differential survival and reproductive capacity (and hence reproduction), sexual selection has to do with differential *mating* (and hence reproduction). In sum, sexual selection is a form of nonrandom mating that can affect allele frequencies in populations.

Figure 17.7
Sexual Selection
Individuals in some species choose their mates based on appearance or behavior. Female sage grouses prefer to mate with males who put on superior "displays," which include sounds, a kind of strutting, and a puffing up of their chests. Here, two males display before females on a sage grouse breeding ground.

Figure 17.8
A Product of Evolution on the Galapagos
A male of the finch species *Geospiza fortis*, which is native to the Galapagos Islands.

Figure 17.9
Who Survives in a Drought?
A large percentage of the population of *Geospiza fortis* died on a Galapagos Island, Daphne Major, during a drought in 1977. Peter and Rosemary Grant observed in 1978 that individuals who survived the drought had a greater average beak depth than average individuals surveyed before the drought, in 1976. Individuals with larger beaks were better able to crack open the large, tough seeds that were available during the drought. The offspring of the survivors likewise had larger average beak size than did the population before the drought. Thus, evolution through natural selection was observed in just a few years on the island.

but rather "survival of those who fit—for now." Let's look at a real-life example of natural selection to see why this is true.

Galapagos Finches: The Studies of Peter and Rosemary Grant

When Charles Darwin stopped at the Galapagos Islands in 1835, some of the animal varieties he collected were various species of finches. Over the years, biologists kept coming back to "Darwin's finches" because of the very qualities Darwin found in them: They seemed to present a textbook case of evolution, with their 13 species having evolved from a single ancestral species on the South American mainland. Yet for more than 100 years after Darwin, it was a puzzle for scientists to figure out how Darwin's posited mechanism of evolu-

tion, natural selection, could have been at work with the finches. This changed beginning in the 1970s, when the husband-and-wife team of Peter and Rosemary Grant began a painstakingly detailed study of the birds.

Natural selection in the finches came into sharp focus in 1977, when a tiny Galapagos island, Daphne Major, suffered a severe drought. Rain that normally begins in January and lasts through July scarcely came at all that year. This was a disaster for the island's two species of finches; in January 1977 there were 1,300 of them, but by December the number had plunged to fewer than 300. Daphne's medium-sized ground finch, *Geospiza fortis*, lost 85 percent of its population in this calamity. The staple of this bird's diet is plant seeds. When times get tough, as in the drought, the size and shape of *G. fortis* beaks—their beak "morphology"—begins to define what one bird can eat as opposed to another (**see Figure 17.8**). In *G. fortis*, larger body size and deeper beaks turned out to make all the difference between life and death in the drought of 1977. Measuring the beaks of *G. fortis* who survived the drought, the Grant team found they were larger than the beaks of the population before the drought by an average of some 6 percent. This was a difference of about half a millimeter, or roughly two-hundredths of an inch; by such a difference were the survivors able to get into large, tough seeds and make it through the catastrophe, eventually to reproduce (**see Figure 17.9**).

This is natural selection in action, but there is more to be learned from the Grant study, which is to say *evolution* made visible. The Grant team knew that beak depth had a high "heritability" in the finches, meaning that beak depth is largely under genetic control. As it turned out, the *offspring* of the drought survivors had beaks that were 4 to 5 percent deeper than the average of the population before the drought. In other words, the drought had preferentially preserved those alleles from the starting population that brought about deeper beaks, and the result was a population that evolved in this direction.

But the Grant study yielded one more lesson. In 1984–1985 there was pressure in the opposite direction: Few large seeds and an abundance of small seeds provided an advantage to *smaller* birds, and it was they who survived this event in disproportionate numbers.

Lessons from *G. fortis*

So, where is the "fittest" bird in all this? There isn't any. Evolution among the finches was not marching toward some generally superior bird. Different traits were simply favored under different environmental conditions. Secondly, there is no evolutionary movement toward combativeness or general intelligence here. Survival had to do with size—and not necessarily *larger* size at that. Looking around in nature, it's

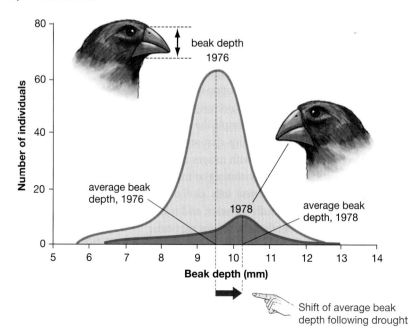

Shift of average beak depth following drought

true that some showcase species, such as lions and mountain gorillas, gain success in reproduction by being aggressive. And it's true that our own species owes such success as it has had to intelligence. But in most instances, it is not brawn or brain that make the difference; it is something as seemingly benign as beak depth and its fit with the environment.

Finally, consider how imperfect natural selection is at the genetic level. Suppose the smaller *G. fortis* that disproportionately died off in 1977 also disproportionately carried an allele that would have aided just slightly in, say, long-distance flying. In the long run this might have been an adaptive trait, but it wouldn't matter. The flight allele would have been *reduced* in frequency in the population (as the smaller birds died off) because flight distance didn't matter in 1977, beak depth did. Evolution operates on the phenotypes of whole organisms, not individual genes. As such, it does not work to spread *all* adaptive traits more broadly. Instead, the destiny of each trait is tied to the constellation of traits the organism possesses. Genes are "team players," in other words, that can only do as well as the team they came in on.

17.5 Three Modes of Natural Selection

In what directions can natural selection push evolution? As noted in Chapter 11, a character such as human height is under the control of many different genes and is thus **polygenic**. Such polygenic characters tend to be "continuously variable." There are not one or two or three human heights, but an innumerable number of them in a range. (See "Multiple Alleles and Polygenic Inheritance" on p. 221.) When natural selection operates on characters that are polygenic and continuously variable, it can proceed in any of three ways. The essential question here is: Does natural selection favor what is average in a given character, or what is extreme? (**See Figure 17.10.**)

Web Tutorial 17.3
Natural Selection

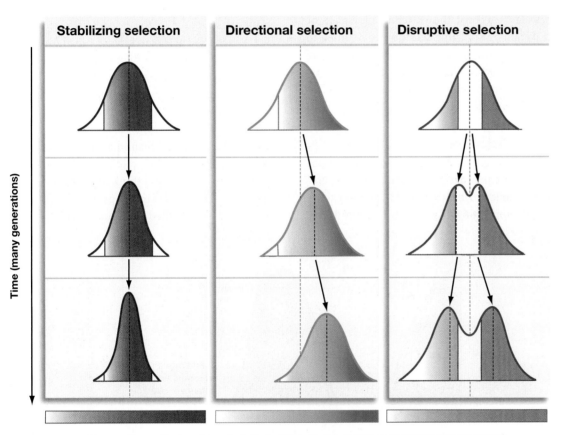

Range of a particular characteristic (in this instance, lightness or darkness of coloration)

In stabilizing selection, individuals who possess extreme values of a characteristic–here, both the lightest and the darkest colors–are selected against and die or fail to reproduce. Over succeeding generations, an increasing proportion of the population becomes average in coloration.

In directional selection, one of the extremes of a characteristic is better suited to the environment, meaning that individuals at the other extreme are selected against. Over succeeding generations, the coloration of the population moves in a directon–in this case toward darker coloration.

In disruptive selection, individuals with average coloration are selected against and die. Over succeeding generations, part of the population becomes lighter, while part becomes darker, meaning the range of color variation in the population has increased.

Figure 17.10
Three Modes of Natural Selection

Stabilizing Selection

In **stabilizing selection**, intermediate forms of a given character are favored over extreme forms. A clear example of this is human birth weights. If you look at **Figure 17.11**, you can see, first, the weights that human babies tend to be. Notice that there are not only relatively few 3-pound babies, but relatively few 9-pound babies as well. A great proportion of birth weights fall at the average or "mean" of a little less than 7 pounds. Now look at the infant-mortality curve. Infant deaths are highest at both extremes of birth weight; low-birth-weight babies *and* high-birth-weight babies are more at risk than are average-birth-weight babies (though low birth weight poses the greater risk). Put another way, the children most likely to survive (and reproduce) are those carrying alleles for intermediate birth weights. Thus, natural selection is working to make intermediate weights even more common. It is not working to move birth weights toward the extremes of higher or lower weights. Stabilizing selection is assumed to be the most common type of selection operating in the natural world. This should not be surprising, because most organisms are well adapted to their environments.

Directional Selection

When natural selection moves a character toward *one* of its extremes, **directional selection** is in operation—the mode in which we most commonly think of evolution operating. If you look at **Figure 17.12**, you can see an example of directional selection that took place over a very long period of time—about 10 million years—involving evolution toward smoothness of shells in certain species of brachiopods.

Disruptive Selection

When natural selection moves a character toward *both* of its extremes, the result is **disruptive selection**, which appears to occur much less frequently in nature than the other two modes of natural selection. This mode of selection is visible in the beaks of yet another kind of finch. A species of these birds found in West Africa (*Pyrenestes ostrinus*) has a beak that comes in only two sizes. Thomas Bates Smith, who has studied the birds, has observed that if human height followed this pattern, there would be some Americans who are 4 to 5 feet tall, and some who are 6 to 7 feet tall, but no one who is 5 to 6 feet tall (**see Figure 17.13**).

The environmental condition that leads to this mode of selection in the finches is, once again, diet. When food gets scarce, large-billed birds specialize in cracking a very hard seed, while small-billed birds begin feeding on several soft varieties of seed, with each type of bird being able to outcompete the other for its special variety. Given how these birds have evolved, it's probably safe to assume that a bird with an intermediate-sized bill would get less food than one with a bill of either extreme type. You might think that what really exists here are two separate species of bird, but large- and small-billed birds mate with one an-

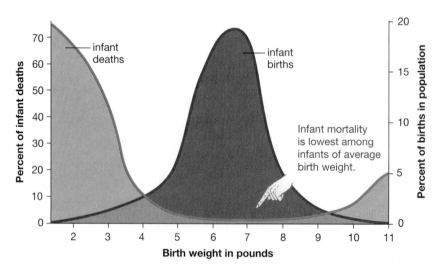

Figure 17.11
Stabililzing Selection: Human Birth Weights and Infant Mortality
Note that infant deaths are more prevalent at the upper and lower extremes of infant birth weights.

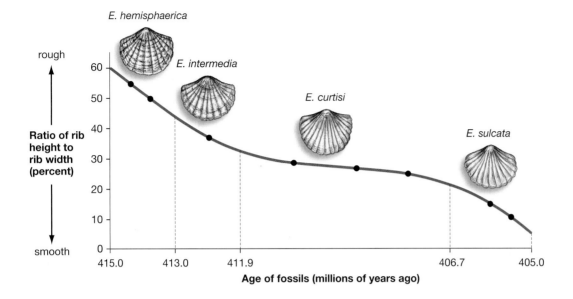

Figure 17.12
Directional Selection: Evolving toward Smoothness
Over a period of 10 million years, beginning about 415 million years ago, directional selection operated on these brachiopods, such that their shells became progressively smoother over time. (Adapted from Ziegler, A.A. 1966. The Silurian Brachiopod *Eocoelia hemisphaerica* (J. de C. Sowerby) *Palaeontology* 9:523-524. Reproduced by permission of The Palaeontological Association.)

Figure 17.13
Disruptive Selection: Evolution That Favors Extremes in Two Directions
Finches of the species *Pyrenestes ostrinus*, found in West Africa. Birds of this species have beaks that come in two distinct sizes—large and small.

other. Bill size seems to be under the control of a single genetic factor, so that bills are able to come out either large or small. This genotype was presumably shaped by natural selection over generations, such that any alleles for intermediate-sized bills were weeded out.

On to the Origin of Species

Stabilizing selection does what it says: It stabilizes given traits of a population, thereby keeping it a single entity. However, both disruptive and directional selection can serve as the basis for speciation—for bringing about the transformation of a single species into one or more *different* species. How is it, though, that such speciation works? And how do we classify the huge number of species that are the outcome of evolution's myriad branchings? These are the subjects of the next chapter.

(continued from previous page)

So, we can see from this example that 60.84 percent of the population is likely to be *RR*, 34.32 percent *Rr*, and 4.84 percent *rr*.

Now we have only one more step to go, which is to calculate expected allele frequencies in the *second* generation, based on the first-generation genotype frequencies that we got. This is just a matter of unpacking, we might say, the genotype frequencies. To calculate what the *R* allele frequencies will be, we start by dividing the 0.3432 *Rr* genotype in half—yielding 0.1716—because half the alleles in this genotype are, by definition, *R* (and the other half *r*). Then we add to this number all of the .6084 alleles in the *RR* genotype, because all these alleles are *R*. To get the frequency of *r* alleles, we take the 0.1716 figure and add to it the 0.0484 figure for the *rr* genotype. Thus:

$$R = 0.6084 + 0.1716 = 0.78$$
$$r = 0.0484 + 0.1716 = 0.22$$

Hopefully, these last two figures look familiar because they are, in fact, the allele frequencies we started with in the first generation. Thus Hardy-Weinberg has demonstrated that allele frequencies do not change from one generation to the next simply as a matter of reproduction.

To grasp the importance of this lack of change, imagine that you are a geneticist who has sampled a human population for the frequency of an allele suspected of contributing to disease in young adults. With your allele frequencies in hand, you run your Hardy-Weinberg equation to get the genotype frequencies for this population. Now the critical step: You sample the population for its *actual* genotype frequencies and find that they don't match the Hardy-Weinberg prediction. You discover that there are fewer people in the population with the homozygous recessive (*qq*) genotype frequency than predicted by Hardy-Weinberg. This means that at least one of the agents noted earlier in the chapter has to be at work. One or more of them is disturbing Hardy-Weinberg equilibrium. Among the possibilities is that the agent at work is natural selection. The *qq* genotype may in fact contribute to people to dying younger, and hence having fewer children. This could be why there are fewer people with this genotype in the population than predicted by the formula.

Hardy-Weinberg, then, can tell scientists not only about evolution, but about human health. Indeed, it can reveal information about human health in several ways. Recall that, in our flowering plant example, the equation showed that 4.84 percent of the population should have the *rr* genotype. This turns out to be another way of saying that 4.84 percent of the flowers should be *white*. All the rest should be red, because all the rest have at least one dominant red allele (*R*), which is all it takes to be red. In a simple genetic scenario such as this, then, Hardy-Weinberg allowed us to derive not only genotype frequencies, but phenotype frequencies—what percentage of plants will have what characteristics. Phenotypes can, however, just as easily have to with human health as flower color. What percentage of the U.S. caucasian population is likely to have cystic fibrosis, as a result of being homozygous for its recessive allele? The answer, 1 in every 2,000 individuals, is provided by Hardy-Weinberg.

Chapter Review

Summary

17.1 What Is It That Evolves?

- A population is defined as all the members of a species living in a defined geographical area at a given time. Natural selection acts on individuals, but it is populations that evolve.

17.2 Evolution as a Change in the Frequency of Alleles

- Genes may be found in variant forms, called alleles. In most species, no individual will possess more than two alleles for a given gene, but a population may possess many such allelic variants. The sum total of alleles in a population is referred to as that population's gene pool.

- The basis of evolution is a change in the frequency of alleles within a population (a phenomenon that includes the appearance of new alleles through mutation). Evolution at this level is called microevolution.
  Web Tutorial 17.1: Evolution and Genetics

17.3 Five Agents of Microevolution

- Five evolutionary forces can result in changes in allele frequencies within a population. These agents of microevolution are mutation, gene flow, genetic drift, nonrandom mating, and natural selection.

- Mutation happens fairly infrequently, and most mutations either have no effect or are harmful; yet rare adaptive mutations are vital to evolution in that they are the only means by which entirely new genetic information comes into being.

- Gene flow, the movement of genes from one population to another, takes place through migration, meaning the movement of individuals from one population into the territory of another.

- Genetic drift, the chance alteration of allele frequencies in a population, has its greatest effects on small populations. Genetic drift works on small populations in two ways. The first of these is the bottleneck effect, defined as a change in allele frequencies due to chance during a sharp reduction in a population's size. The second is the founder effect: the fact that when a small subpopulation migrates to a new area to start a new population, it is likely to bring with it only a portion of the original population's gene pool.

- Nonrandom mating is mating in which a given member of a population is not equally likely to mate with any other member. Sexual selection is a form of nonrandom mating that can directly affect the frequency of alleles in a gene pool. It occurs when differences in reproductive success arise because of differential success in mating.

- Some individuals will be more successful than others in surviving and hence reproducing, owing to traits that better adapt them to their environment. This phenomenon is known as natural selection. Natural selection is the only agent of microevolution that consistently acts to adapt organisms to their environments. As such, it is generally regarded as the most powerful force underlying evolution.
Web Tutorial 17.2: Agents of Change

17.4 What Is Evolutionary Fitness?

- The phrase "survival of the fittest" is misleading, because it implies that evolution works to produce generally superior beings who would be successful competitors in any environment. Evolutionary fitness, however, has to do only with the relative reproductive success of individuals in a given environment at a given time. One individual is said to be more fit than another to the extent that it has more offspring than another. But individuals are not born with invariable levels of fitness; instead, fitness can change in accordance with changes in the surrounding environment.

17.5 Three Modes of Natural Selection

- Natural selection has three modes: stabilizing selection, directional selection, and disruptive selection. Stabilizing selection moves a given character in a population toward intermediate forms and hence tends to preserve the status quo; directional se-

lection moves a given character toward one of its extreme forms; and disruptive selection moves a given character toward two extreme forms.
Web Tutorial 17.3: Natural Selection

Key Terms

adaptation	334	**genotype**	328
allele	328	**macroevolution**	329
assortative mating	333	**microevolution**	329
bottleneck effect	332	**migration**	330
directional selection	338	**natural selection**	334
disruptive selection	338	**nonrandom mating**	333
evolution	329	**phenotype**	328
fitness	335	**polygenic**	337
founder effect	332	**population**	328
gene flow	330	**sexual selection**	333
gene pool	328	**species**	327
genetic drift	332	**stabilizing selection**	338

Understanding the Basics

Multiple-Choice Questions (answers in the back of the book)

1. In most populations of living things, each individual has ___ copies of each gene, which may be the same or different allelic variants. A population may have _____ allelic variants for a given gene.
   a. 1, 2
   b. 2, 4
   c. 2, many
   d. 1, many
   e. 2, 2

2. Match the terms with their meanings:
   a. gene pool   a variant form of a gene
   b. allele   the set of all alleles in a population
   c. allele frequency the genetic makeup of an organism
   d. genotype   exchange of genes between populations
   e. gene flow   the relative representation of a given form of a gene in a population

3. Which of the following statements is true of microevolution? (Select all that apply.)

a. It is a change in allele frequency within populations.
b. It is caused by inheritance of characters acquired during the life of an organism.
c. It is seen when some allelic variants cause individuals to have increased reproductive success.
d. It always leads to adaptation.
e. It is caused only by natural selection.

4. Agents of change in allele frequency in populations include (select all that apply):
   a. genetic drift
   b. meiosis
   c. mutation
   d. segregation
   e. natural selection

5. Agents that consistently produce adaptive evolution (adaptation) include (select all that apply):
   a. genetic drift
   b. mutation
   c. meiosis
   d. natural selection
   e. gene flow

6. Mutations are (select all that apply):
   a. generally beneficial
   b. changes in the genetic material that may be inherited
   c. always caused by changes in single base pairs of DNA
   d. the ultimate source of new genetic variation
   e. prone to occur with great frequency

7. In stabilizing selection, _____ individuals have the highest fitness:
   a. large
   b. intermediate
   c. bright
   d. extreme
   e. small

8. Match the agent of evolution to its description:
   a. mutation — chance alterations of allele frequencies in a small population
   b. gene flow — a process in which the fit of an organism with its environment selects those traits that will be passed on with greater frequency from one generation to the next
   c. natural selection — movement of alleles between populations by migration
   d. genetic drift — genetic drift due to a few colonizing drift genotypes
   e. founder effect — changes in the genetic material

9. When an insect population becomes resistant to insecticides, this is an example of a response to _____ selection.
   a. disruptive
   b. stabilizing
   c. sexual
   d. directional
   e. random

10. Finches in the Galapagos experienced _____ selection on beak depth following a drought, while the African finch *P. ostrinus* experiences _____ selection on beak size because it eats either very small, soft seeds or much larger, hard seeds.
    a. stabilizing; directional
    b. sexual; disruptive
    c. directional; disruptive
    d. disruptive; directional
    e. stabilizing; disruptive

11. A large lizard population (1,000 individuals) on the coast contains mostly individuals that are plain brown, but a few have white spots. Coloration is genetically determined. One day during a storm, two spotted lizards hitch a ride on a piece of driftwood to a nearby island, where they join a population of 100 lizards, some of which are spotted and some of which are plain brown. The island's lizard population allele frequency has thus changed due to
    a. natural selection
    b. mutation
    c. gene flow
    d. nonrandom mating
    e. inbreeding

Brief Review

1. Explain the statement, "Individuals are selected, populations evolve."

2. What is a gene pool, and why do changes in gene pools lie at the root of evolution?

3. What are the five causes of allele frequency changes (microevolution), and how does each work?

4. Why is evolutionary fitness always a measure of relative fitness in a population?

5. Why is natural selection the only agent of evolution that consistently produces adaptation?

6. Is the evolutionary fitness of an individual expected to be the same or different in different environments? How did the Grant's study of beak depth of finches after the drought and then again in 1984–1985 provide a real example of this?

Applying Your Knowledge

1. Cheetahs have long legs relative to other large cats. However, leg length in a cheetah population is more likely to be under stabilizing rather than directional selection. Why?

2. The text notes that natural selection is the process that pushes populations to adapt to environmental change. Is natural selection still going on in human populations? Why or why not?

3. Explain why genetic drift may be important when captive populations of animals or plants are started with just a few individuals.

4. Two moth populations of the same species utilize different host plants. One rests on leaves and has evolved a green color that allows it to escape predation by blending in with the leaves. The other population rests on tree trunks, and is brown in color. The colors of these moths are genetically determined by different alleles of the same gene. What would be some consequences of migration of moths leading to gene flow between these two populations?

Sympatric Speciation in a Fruit Fly

To see how sympatric speciation works, consider one of its best-studied examples, a species of fruit fly named *Rhagoletis pomonella*. Prior to the European colonization of North America, *R. pomonella* existed solely on the small, red fruit of hawthorn trees (**see Figure 18.5**). The Europeans brought *apple* trees with them, however, and by 1862 some *R. pomonella* had moved over to them. It seemed to at least one mid-nineteenth-century observer, however, that with the introduction of apples there had arisen separate varieties of the flies—what are sometimes called apple flies and haw flies—with each variety courting, mating, and laying eggs almost exclusively on its own type of tree. A modern researcher named Guy Bush led the way, beginning in the early 1960s, in investigating whether this was the case, and the answer turned out to be yes. The two varieties of flies are separated from each other in all these ways. In one study undertaken on them, only 6 percent of the apple and haw flies interbred with one another. The apple and haw *R. pomonella* are not separate species yet, but they certainly give indication of being in transition to that status.

For our purposes, the first thing to note is that this separation has not come about because of *geographic* division. The apple and hawthorn trees that the two varieties of flies live on may scarcely be separated in space at all. Given this, how did this single species move toward becoming two separate species? Bush offers a likely scenario, based on a critical difference between the hosts the two species live on. Apples tend to ripen in August and September, while hawthorn fruit ripens in September and October. In the sum-

mer, all fruit flies emerge as adults after wintering underground as larvae, after which they fly to their host tree to mate and lay eggs in the fruit.

Bush believes that about 150 years ago (when hawthorns were hosts to all the flies), some individuals in a population of *R. pomonella* experienced either a mutation or perhaps a chance combination of rare existing alleles. In either event, this change did two things: It caused these flies to emerge slightly *earlier* from their underground state than did most flies; and it drew these flies to the smell of *apples* as well as hawthorns. Because apples mature slightly earlier than the hawthorn fruit, these flies had a suitable host waiting for them. More important, these flies would have bred *among themselves* to a high degree. Recall that the other flies were emerging later, and the adult fly only lives for about a month. Thus, the variant alleles were passed on to a selected population in the next generation. Today, the apple *R. pomonella* flies indisputably do emerge earlier than the hawthorn flies. This is one of the things that ensures reproductive isolation between the two groups; their periods of mating don't fully overlap.

Such a lack of overlapping mating periods may sound familiar, because it is one of the intrinsic reproductive isolating mechanisms you looked at earlier. (It is temporal isolation.) You also can see ecological isolation in operation with these flies. Though the two types of flies live in the same area, they meet up with relatively little frequency, because they have different habitats (hawthorn vs. apple trees). In short, these populations have developed intrinsic reproductive isolating mechanisms without ever having been separated from one another geographically. They are headed toward speciation, but in this case it is sympatric speciation.

Speciation through Hybridization

Sympatric speciation can also take place, at least in plants, through the hybridization mentioned earlier, wherein egg and sperm from different species come together to produce offspring. As you can imagine, if plant species A occupies one habitat, and a closely related plant species B occupies another nearby habitat, there will be a so-called "hybrid zone" running between these two habitats where both plants exist and fertilize each other by means of wind-borne pollen. In some instances, healthy hybrid offspring can be produced from this mixing of different species' gametes. However, most hybrid offspring are infertile; they grow, but like mules, they will be incapable of giving rise to a *succeeding* generation of plants.

In rare instances, however, hybrids will be able to have fertile offspring of their own, not by mating

Figure 18.5
A Species Undergoing Sympatric Speciation?
The fruit-fly species *Rhagoletis pomonella*, pictured here on the skin of a green apple, may be undergoing speciation.

Applying Your Knowledge

1. Cheetahs have long legs relative to other large cats. However, leg length in a cheetah population is more likely to be under stabilizing rather than directional selection. Why?

2. The text notes that natural selection is the process that pushes populations to adapt to environmental change. Is natural selection still going on in human populations? Why or why not?

3. Explain why genetic drift may be important when captive populations of animals or plants are started with just a few individuals.

4. Two moth populations of the same species utilize different host plants. One rests on leaves and has evolved a green color that allows it to escape predation by blending in with the leaves. The other population rests on tree trunks, and is brown in color. The colors of these moths are genetically determined by different alleles of the same gene. What would be some consequences of migration of moths leading to gene flow between these two populations?

MediaLab

Are Bacteria Winning the War?
Natural Selection in Action

The last time your doctor prescribed an antibiotic for you, was it for a common cold or a real bacterial infection? If you had a bacterial infection, the first antibiotic might not have worked. Why not? You may have been infected with a strain of bacteria resistant to that antibiotic. The rise and spread of drug-resistant bacteria are consequences of natural selection—evolution in action. The *Web Tutorial* will help you review how the environmental conditions can mold populations by natural selection. The following *Web Investigation* introduces the serious threat of evolving antibiotic-resistant bacteria. In the *Communicate Your Results* section, you can look for causes and solutions to this new problem.

This *MediaLab* can be found in Chapter 17 on your Companion Website (www.prenhall.com/krogh3).

WEB TUTORIAL

In any population, organisms vary. In a group of bacteria, for example, there may be some rare antibiotic-resistant members. When the environment changes, let's say you take a prescribed antibiotic, and natural selection acts. All the sensitive bacteria die, leaving the antibiotic-resistant bacteria to thrive, thus changing the makeup of the population. This is just one of three types of natural selection that are animated in this activity.

Activity

1. First, you will see how evolution, the modification of a trait in a population over time, is based in the change in frequencies of alleles for that trait.

2. Next, you will compare the three types of natural selection—directional, stabilizing, and disruptive selection—and see how they might act to change allele frequencies.

   Use the Internet in the following *Web Investigations* to view modern examples of natural selection, including antibiotic resistance in bacteria, and see how natural selection affects us all.

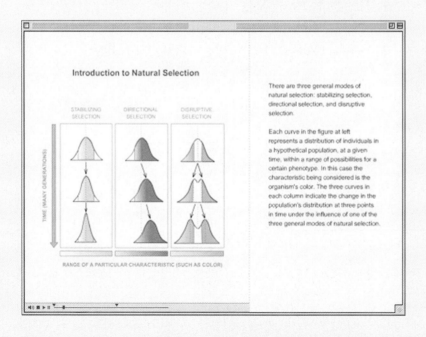

WEB INVESTIGATION

Investigation 1
Estimated time for completion = 20 minutes

Why should people care if antibiotic resistance occurs in bacteria? Read an online article by selecting the keyword **ANTIBIOTIC RESISTANCE** on your Website for this *MediaLab*. Note the biological causes of antibiotic (or antimicrobial) resistance, and the problems predicted. Two factors seem to be operating in this situation: antibiotic resistance by the bacteria in the context of natural selection, and the overuse (and misuse) of antibiotics by people.

Investigation 2
Estimated time for completion = 5 minutes

It's not just bacteria; many microbes have evolved to overcome the medical arsenals hurled against them. Americans were even beginning to see AIDS as a chronic but manageable disease after the introduction of the combination drug therapies that decimated the virus to nearly unde-

tectable levels. Now, however, many are finding these treatments no longer effective, and drug companies must develop new therapies. You can view the current state of several drug-resistant microbes by selecting the keyword **CDC** or **HIV** on your Website for this *MediaLab*.

Investigation 3
Estimated time for completion = 10 minutes

The curse of most infections isn't usually the direct damage inflicted by the virus, but rather the body's defensive response and the accompanying symptoms like runny nose, sneezing, fever, pain, even diarrhea. From an evolutionary standpoint these maybe very beneficial. Select the keyword **EVOLVED DEFENSES** on your Website for this *MediaLab* to read about several examples ranging from morning sickness to coughing.

Now that you have seen some real-life examples, do some of the following exercises to gain an understanding of the factors influencing natural selection and human diseases.

COMMUNICATE YOUR RESULTS

Exercise 1
Estimated time for completion = 15 minutes

What will happen when antibiotics are rendered useless against more and more antibiotic-resistant bacteria? What should we be doing to stop this trend? Using information from *Web Investigation 1*, compile a list of bacterial species that have developed antibiotic resistance, and indicate when. How has natural selection been a factor in the development and spread of antibiotic resistance? Write a list of common guidelines that you could use in explaining this problem to your friends and family. Include suggestions for stopping the spread of antibiotic resistance—and the consequences if we do not.

Exercise 2
Estimated time for completion = 15 minutes

In *Web Investigation 2*, you read about the emerging threat of drug-resistant microbes. Prepare a short presentation of one example to add to a list created by other students. Include the type of natural selection involved, the organism, and what science is doing about solving the problem.

Exercise 3
Estimated time for completion = 10 minutes

So you took some over-the-counter remedy for your common cold; but is it really best to completely obliterate those miserable symptoms? Using the article from *Web Investigation 3*, explain why or why not. Why might evolved defenses like runny noses, fevers, and diarrhea actually be beneficial for survival of our species? How did they evolve? Are the costs of these defenses actually worth it?

18 *Macroevolution*

Endangered or not?
(Section 18.1, page 350)

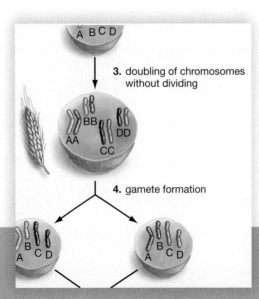

New species in a generation.
(Essay, page 355)

Derived from three species.
(Section 18.2, page 357)

New species come about when populations of existing species cease to breed with one another. Each new species that arises constitutes another branch on life's family tree. Scientists are trying hard to figure out what this tree looks like.

How many types of living things are there on Earth? How many varieties of life-forms are there that we can recognize as being fundamentally different from one another? It may surprise you to learn that we haven't the foggiest idea. The lowest estimate is about four million species, but higher-end estimates often come in at 10 to 15 million and the highest of them all is 100 million. Scientists simply have been unable to catalogue the vast diversity that exists on our planet. In 250 years of digging, netting, bagging, and observing they have identified about 1.8 million species—a large number, to be sure, but only a fraction of the species that actually exist, even if our lowest estimates are correct.

Since no one knows how many species there are, let's suppose that there are "only" 4 million on Earth. Isn't this number staggering? This seems particularly true given its starting point. A single type of organism arose perhaps 3.8 billion years ago, branched into two types of organisms, and then the process continued—branches forming on branches—until at least 4 million different types of living things came to exist on Earth. Now here's the real surprise: These are just the *survivors*. The fossil record indicates that more than 99 percent of all species that have ever lived on Earth are now extinct. Branching indeed.

The questions to be answered in this chapter are: How does this branching work? How do we go from the microevolutionary mechanisms explored in Chapter 17 to the actual divergence of one species into two? And how do we classify the creatures that result from this branching?

18.1 What Is a Species?

The category of life that's been mentioned so far, the species, has a great importance not only to biology, but to society in general. If we allow that some bacteria are harmful to human beings while others are helpful, that some varieties of rice carry disease-resistance genes while others do not, that some species of birds are endangered while others are not, then it follows that there must be some means of distinguishing one of these groupings from another (**see Figure 18.1** on next page). The whole notion of knowing something about the natural world begins to break down if we can't say *which* grouping it is that is endangered or harmful.

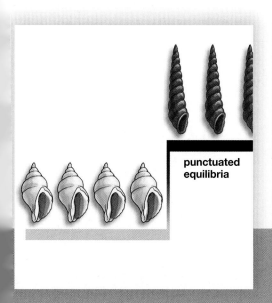

Jerky and smooth speciation.
(Section 18.3, page 359)

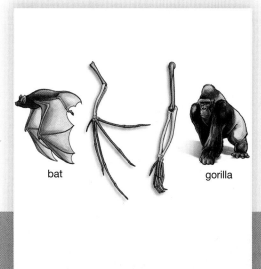

Same structure, different creatures.
(Section 18.4, page 362)

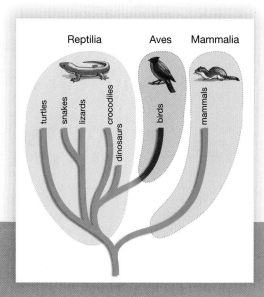

Are birds reptiles?
(Section 18.5, page 365)

Over the centuries, human beings have devised various ways of defining the groups now called species. But in science today, the most commonly accepted definition stems from what is known as the **biological species concept**, which uses breeding behavior to make classifications. We actually looked at a version of the biological species concept in Chapter 17. Here it is again, as formulated by the evolutionary biologist Ernst Mayr:

> Species are groups of actually or potentially interbreeding natural populations which are reproductively isolated from other such groups.

Note that the breeding behavior Mayr talks about can be real or potential. Two populations of finch may be separated from one another by geography, but if, upon being reunited in the wild, they began breeding again, they are a single species. Note also that Mayr stipulates that species are groups of *natural* populations. This is important because breeding may take place in captivity that would not in nature. No one doubts that lions and tigers are separate species; yet they will mate in zoos, producing little tiglons or ligers (depending on whether the father was a tiger or a lion). In natural surroundings, however, they apparently have never interbred, even when their ranges overlapped centuries ago.

You might think that, with the biological species concept in hand, scientists would be able to study any organism and, by discerning its breeding behavior, pronounce it to be a member of this or that species. Nature is so vast and varied, however, that this doesn't always work. We can't look at the breeding activity of the single-celled bacteria or archaea, for example, because they don't *have* any breeding behavior; they multiply instead by simple cell division. (Microbiologists define bacterial and archaeal groupings by sequencing their DNA or RNA and looking for the degree to which these sequences are the same.) Then there are separate species that sometimes interbreed in nature, producing so-called hybrid offspring as a result. Such mixing between species happens more in the plant world than the animal, but it does take place in both. If species are supposed to be "reproductively isolated" from one another, what are we to make of these crossings? (Mayr points out that it is *populations* in his definition that are reproductively isolated, not individuals, and that whole populations do tend to stay within their species confines.)

Despite these difficulties, the biological species concept provides a useful way of defining the basic category of Earth's living things. Bacteria and archaea notwithstanding, most multicellular species carry out sexual reproduction at least part of the time, and relatively few species outside the plant kingdom regularly

(a) Endangered species

(b) Not endangered

Figure 18.1
Separate Entities
The concept of a species can have very practical effects.

(a) This bird is a northern spotted owl (*Strix occidentalis*), which is an endangered species, protected by federal law.

(b) This bird is a barred owl (*Strix varia*), which is not endangered.

produce hybrids. Moreover, this species concept is rooted in a critical behavior of organisms themselves—mating, which controls the flow of genes. And as you saw in Chapter 17, it is the change in the genetic makeup of a population (a change in its allele frequencies) that lies at the root of evolution.

18.2 How Do New Species Arise?

Having defined a species, let's now see how one variety of them can be transformed into another in the process called **speciation**: the development of new species through evolution. As you've observed, a single species can diverge into two species, the "parent" species continuing while another branches off from it. A key question for scientists is: What brings this branching evolution about? The answer has to do with the flow of genes reviewed in Chapter 17. As you saw then, evolution within a population means a change in that population's allele frequencies. Now imagine *two* populations of a single species of bird, with one being separated from the other—say, by one of them having flown to a nearby location. To the extent that they continue to breed with one another, with individuals moving between the locations, each population will *share* in whatever allele frequency changes are going on with the other population. Hence the two populations will evolve together, remaining a single species. Now imagine that the migration stops between the two populations. Each population continues to undergo allele frequency changes, but it no longer shares these changes with the other population. Alterations in form and behavior may accompany allele changes, and these alterations may pile up over time. Coloration or bill lengths may change; feeding habits may be transformed. After enough time, should the two populations find themselves geographically reunited, they may no longer freely interbreed. At that point, they are separate species. Speciation has occurred.

The critical change here came when the two populations quit interbreeding. For scientists, the key question thus becomes: Why would this happen? What could drastically reduce interbreeding, and hence gene flow, between two populations of the same species?

The Role of Geographic Separation: Allopatric Speciation

In the preceding example, a very clear factor reduced gene flow between the bird populations: They were separated geographically. And geo-

graphical separation turns out to be the most important starting point in speciation.

Populations can become separated in lots of ways. On a large scale, glaciers can move into new territory, cutting a previously undivided population into two. Rivers can change course, with the result that what was a single population on one side of the river may now be two populations on different sides of it. On a smaller scale, a pond may partially dry up, leaving a strip of exposed land between what are now two ponds with separate populations. Such environmental changes are not the only ways that populations can be separated, however. Part of a population might *migrate* to a remote area, as did the bird population, and in time be cut off from its larger population. For a real-life example of how migration can bring about speciation, **see Figure 18.2**.

When geographical barriers divide a population and the resulting populations then go on to become separate species, what has occurred is **allopatric speciation**. *Allopatric* literally means "of other countries"

Figure 18.2
Speciation in Action
Millions of years ago, the salamander *Ensatina eschscholtzii* began migrating southward from the Pacific Northwest. When it reached the Central Valley of California—an uninhabitable territory for it—populations branched west (to the coastal range) and east (to the foothills of the Sierra Nevada mountains). Over time, the populations took on different colorations as they moved southward. By the time the two populations were united in Southern California, they differed enough, genetically and physically, that they either did not interbreed or produced infertile hybrid offspring when they did. (Salamanders are falsely colored in drawing for illustrative purposes.)

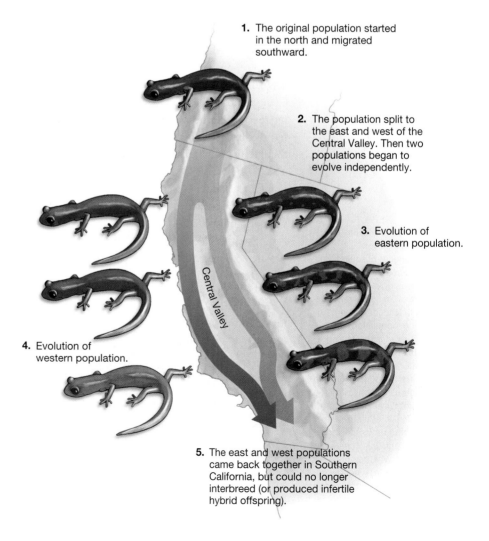

1. The original population started in the north and migrated southward.

2. The population split to the east and west of the Central Valley. Then two populations began to evolve independently.

3. Evolution of eastern population.

4. Evolution of western population.

5. The east and west populations came back together in Southern California, but could no longer interbreed (or produced infertile hybrid offspring).

(**see Figure 18.3**). Look along the banks on either side of the Rio Juruá in western Brazil, and you will find small monkeys, called tamarins, that differ genetically from one another in accordance with how wide the river is at any given point. Where it is widest, the members of this species do not interbreed, while at the narrow headwaters they do. Are the nonbreeding populations on their way to becoming separate species? Perhaps, but only if gene flow is drastically reduced between them, probably for a very long time.

Reproductive Isolating Mechanisms Are Central to Speciation

While geographical separation is the most important factor in getting speciation going, it cannot bring about speciation by itself. Following geographical separation, two populations of the same species must then undergo physical or behavioral changes that will *keep* them from interbreeding, should they ever be reunited. Allopatric speciation thus operates through a one-two process: first the geographic separation, then the development of differing characteristics in the two resulting populations. These are characteristics that will *isolate* them from each other in terms of reproduction.

Thus arises the concept of **reproductive isolating mechanisms**, which can be defined as any factor that, in nature, prevents interbreeding between individuals of the same species or of closely related species. Geographic separation is itself a reproductive isolating mechanism, because it is a factor that prevents interbreeding. But, because the mountains or rivers that are the actual barriers to interbreeding are outside of or *extrinsic to* the organisms in question, geographic separation is called an **extrinsic isolating mechanism**. For allopatric speciation to take place, what also must occur is the second in the one-two series of events: The evolution of *internal* characteristics that keep organisms from interbreeding. Such characteristics are referred to as **intrinsic isolating mechanisms**: evolved differences in anatomy, physiology, or behavior that prevent interbreeding between individuals of the same species or of closely related species.

Six Intrinsic Reproductive Isolating Mechanisms

What are these intrinsic mechanisms? A list of the most important ones is presented in **Table 18.1**. An easy way to remember them is to think about what sequence of events would be required for fertile offspring to be produced in any sexually reproducing organism. First, organisms that live in the same area must encounter one another; if they don't, the "ecological isolation" mechanism is in place. If they do encounter one another, they must then mate in the same time frame; if they don't, then temporal isolation is in place, and so on.

Ecological Isolation

Two closely related species of animals may overlap in their ranges and yet feed, mate, and grow in separate areas, which are called *habitats*. If they use different habitats, this means they may rarely meet up. If so, gene flow will be greatly restricted between them. Lions and tigers *can* interbreed, but they never have in nature, even when their ranges overlapped in the past. One reason for this is their largely separate habitats: Lions prefer the open grasslands, tigers the deep forests.

Temporal Isolation

Even if two populations share the same habitat, if they do not mate within the same time frame, gene flow will be limited between them. Two populations of the same species of flowering plant may begin releasing pollen at slightly different times of the year. Should their reproductive periods cease to overlap altogether, gene flow would be cut off between them.

Figure 18.3
Geographical Separation—Leading to Speciation?
These two varieties of squirrel had a common ancestor that at one time lived in a single range of territory. Then the Grand Canyon was carved out of land in Northern Arizona, leaving populations of this species separated from one another, about 10,000 years ago. In the area of the Grand Canyon, **(a)** the Abert squirrel lives only on the Canyon's south rim, while **(b)** the Kaibab squirrel lives on the north rim. It's unclear whether these two varieties of squirrel would be reproductively isolated if reunited today—that is, whether they are now separate species—but the geographical separation they have experienced is the first step on the way to allopatric speciation.

(a) Abert squirrel, south rim of Grand Canyon

(b) Kaibab squirrel, north rim of Grand Canyon

Behavioral Isolation

Even if populations are in contact and breed at the same time, they must choose to mate with one another for interbreeding to occur. Such choice is often based on specific courtship and mating displays, which can be thought of as passwords between members of the same species. Birds must hear the proper song, spiders must perform the proper dance, and fiddler crabs must wave their claws in the proper way for mating to occur.

Mechanical Isolation

Reproductive organs may come to differ in size or shape or some other feature, such that organisms of the same or closely related species can no longer mate. Different species of alpine butterfly look very similar, but their genital organs are different enough that one species cannot mate with another.

Gametic Isolation

Even if mating occurs, offspring may not result if there are incompatibilities between sperm and egg or between sperm and the female reproductive tract. In plants, the sperm borne by pollen may be unable to reach the egg lying within the plant's ovary. In animals, sperm may be killed by the chemical nature of a given reproductive tract, or may be unable to bind with receptors on the egg.

There is one form of gametic isolation, called *polyploidy*, that is very important in plants and that may have played a part in giving rise to the vertebrate evolutionary line that we humans are part of. You can read about it in "New Species through Genetic Accidents" on the next page.

Hybrid Inviability or Infertility

Even if offspring result, they may develop poorly—they may be stunted or malformed in some way—or they may be infertile, meaning unable to bear offspring of their own. A well-known example of such infertility is the mule, which is the infertile offspring of a female horse and a male donkey (**see Figure 18.4**).

Sympatric Speciation

Thus far, you have looked at the development of intrinsic reproductive isolating mechanisms strictly as a second step in speciation—one that follows a geographic

Figure 18.4
Mules Are Infertile Hybrids
Even though mules cannot themselves reproduce, humans frequently cross horses and donkeys to produce mules in order to take advantage of their exceptional strength and endurance. Chromosomal incompatibilities between horses and donkeys leave the mules sterile.

Table 18.1 Reproductive Isolating Mechanisms

Extrinsic isolating mechanism		**Geographic isolation** Individuals of two populations cannot interbreed if they live in different places (the first step in allopatric speciation).
Intrinsic isolating mechanisms		**Ecological isolation** Even if they live in the same place, they can't mate if they don't come in contact with one another.
		Temporal isolation Even if they come in contact, they can't mate if they breed at different times.
		Behavioral isolation Even if they breed at the same time, they will not mate if they are not attracted to one another.
		Mechanical isolation Even if they attract one another, they cannot mate if they are not physically compatible.
		Gametic isolation Even if they are physically compatible, an embryo will not form if the egg and sperm do not fuse properly.
		Hybrid inviability or infertility Even if fertilization occurs successfully, the offspring may not survive, or if it survives, may not reproduce (e.g., mule).

New Species through Genetic Accidents: Polyploidy

In the chapter so far, speciation has been portrayed as something that takes place over many generations. But plants have a means of speciating in a single generation. It is called *polyploidy*, and it is very important in the plant world; more than 100,000 species of flowering plants in existence today are thought to have been brought about through it. Beyond this, polyploidy may have had a great importance in the evolution of animals. Evolutionary biologists currently are debating whether a polyploidy "event" helped speed the rise of animals with backbones—the vertebrates, such as ourselves. As you'll see, whenever polyploidy comes about, it helps speed up evolution for reasons we'll get to.

How does polyploidy work? Here is one of the ways. From the genetics unit, you may recall that human beings have 23 pairs of chromosomes, *Drosophila* flies four pairs, and Mendel's peas seven. Whatever the *number* of chromosomes, the commonality among all these species is that their chromosomes come in *pairs*, which is the general rule for species that reproduce sexually.

As noted at the start of the chapter, plants are more adept than animals at producing "hybrids," with the gametes from two *separate* species coming together to create an offspring. Often these offspring are sterile, however, because when it comes time for them to produce *their* gametes (the counterparts to human sperm or eggs), the sets of chromosomes they inherited from their different parental species may not "pair up" correctly in meiosis, owing to differences in chromosome number or structure.

Now comes the accident. Suppose that, back when it was first created as a single-celled zygote, the hybrid offspring carried out the usual practice of doubling its chromosome number in preparing for its initial cell division. Now, however, the cell fails to actually divide; whereas it was supposed to put half its complement of chromosomes into one daughter cell and half into another, it doesn't do this. It doubles the chromosomes, but keeps them all in one cell. Equipped with *twice* the usual number of chromosomes, this single cell then proceeds to undergo regular cell division, meaning that every cell in the plant that follows will have this doubled number of chromosomes. Critically, this plant will have doubled both its *sets* of chromosomes—the set it got from parental species A and the set it got from species B. If you double a set, by definition every member of that set now has a partner to pair with in gamete formation. Thus the roadblock to a hybrid producing offspring has been removed; the chromosomes can all pair up.

This pairing up yields gametes (eggs and sperm), which can then come together and fuse thanks to another capability of plants. Recall that many plants can *self-fertilize*. A single plant contains male gametes that can fertilize that same plant's female gametes. Thus the plant with the doubled set of chromosomes can fertilize itself, theoretically beginning an unending line of fertile offspring. This line of organisms is "reproductively isolated" from either of its parental species because it has a different number of chromosomes than either parental species. With reproductive isolation, we have a *separate species*; and with self-fertilization, we have a species that can perpetuate itself.

Did polyploidy speed the evolution of our own vertebrate line?

A multiplication of the normal two sets of chromosomes to some other set number is known as **polyploidy**; here we have speciation by polyploidy, which is one type of sympatric speciation. The importance of this in the plant world is immense. Many of our most important food crops are polyploid, including oats, wheat, cotton, potatoes, and coffee. Indeed, the type of polyploid speciation we've looked at—which begins with a hybrid offspring—often produces bigger, healthier plants. As a result, breeders have developed ways to artificially induce polyploidy.

But what about animals? The self-fertilization that some plants are capable of does not exist among more complex animals. As a result, polyploidy that can be maintained over gener-

separation of populations. As it happens, however, intrinsic reproductive isolating mechanisms can develop between two populations in the *absence* of any geographic separation of them. If these isolating mechanisms reduce interbreeding between the populations sufficiently, speciation can take place. What occurs in this situation is not allopatric speciation, however; it is **sympatric speciation**, which can be defined as any speciation that does not involve geographic separation. (*Sympatric* literally means "of the same country.")

ations is extremely rare in the animal kingdom. But in the course of hundreds of millions of years of evolution, it clearly has been maintained several times. An extremely important polyploidy event may have occurred about 500 million years ago, just as vertebrate animals were evolving from invertebrates. A controversial hypothesis holds that this event doubled the chromosomal set of our invertebrate ancestors, thus greatly accelerating the pace of their evolution.

How can polyploidy speed up evolution? Think of it this way. If you added an extra room to a house, that would give you the flexibility to transform, say, an existing bedroom into a den. Now, if you add genes to a genome, as polyploidy does, that provides the *genome* with a kind of flexibility: the flexibility to have one kind of gene transformed into another. Where once an organism would have but two copies of a given gene—one allele from each parent—with polyploidy it has four. With this change, one pair of these alleles can mutate without causing harm to the organism because the *other* pair of alleles can carry out their function all by themselves. And in a few cases, alleles that mutate will produce new proteins that can help transform an organism—that can help it evolve down different lines. Polyploidy, then, provides for a tremendous increase in the genetic capacity to evolve. And it provides this capacity in a single generation.

Polyploidy in Wheat

Two different species of wheat exist in nature, with slightly different "genomes" or complements of DNA.

1. Gametes (eggs and sperm) are formed in the different species.

2. These gametes fuse, in fertilization, to form a zygote—a single cell that will develop into a new plant. Such a mixed-species or "hybrid" zygote generally develops into a sterile plant, because its chromosomes cannot pair up correctly when the plant produces its own gametes during meiosis.

3. In this case, however, the zygote doubles its chromosomes in preparation for cell division, but then fails to divide. With this doubling, each chromosome now has a compatible homologous chromosome to pair with during meiosis. This is the polyploidy event.

4. Gamete formation then takes place in the plant.

5. These gametes from the same plant then fuse, because this is a self-fertilizing plant. With this, a new generation of wheat plant has been produced—one that is a different species from either parent generation, because each of the parent species has two pairs of chromosomes, while this new hybrid species has four pairs.

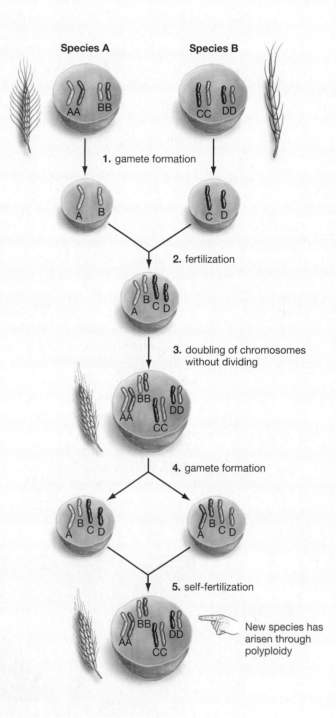

Species A   Species B

1. gamete formation

2. fertilization

3. doubling of chromosomes without dividing

4. gamete formation

5. self-fertilization

New species has arisen through polyploidy

Sympatric speciation has been a contentious subject in biology for years. One form of sympatric speciation is the polyploidy reviewed in "New Species through Genetic Accidents." But setting polyploidy aside, most biologists thought for years that if sympatric speciation took place at all, it was of trivial importance. By the mid 1990s, however, new evidence had convinced some of the most prominent researchers in evolutionary biology that sympatric speciation does have significance in the broader picture of speciation.

Sympatric Speciation in a Fruit Fly

To see how sympatric speciation works, consider one of its best-studied examples, a species of fruit fly named *Rhagoletis pomonella*. Prior to the European colonization of North America, *R. pomonella* existed solely on the small, red fruit of hawthorn trees (**see Figure 18.5**). The Europeans brought *apple* trees with them, however, and by 1862 some *R. pomonella* had moved over to them. It seemed to at least one mid-nineteenth-century observer, however, that with the introduction of apples there had arisen separate varieties of the flies—what are sometimes called apple flies and haw flies—with each variety courting, mating, and laying eggs almost exclusively on its own type of tree. A modern researcher named Guy Bush led the way, beginning in the early 1960s, in investigating whether this was the case, and the answer turned out to be yes. The two varieties of flies are separated from each other in all these ways. In one study undertaken on them, only 6 percent of the apple and haw flies interbred with one another. The apple and haw *R. pomonella* are not separate species yet, but they certainly give indication of being in transition to that status.

For our purposes, the first thing to note is that this separation has not come about because of *geographic* division. The apple and hawthorn trees that the two varieties of flies live on may scarcely be separated in space at all. Given this, how did this single species move toward becoming two separate species? Bush offers a likely scenario, based on a critical difference between the hosts the two species live on. Apples tend to ripen in August and September, while hawthorn fruit ripens in September and October. In the sum-

mer, all fruit flies emerge as adults after wintering underground as larvae, after which they fly to their host tree to mate and lay eggs in the fruit.

Bush believes that about 150 years ago (when hawthorns were hosts to all the flies), some individuals in a population of *R. pomonella* experienced either a mutation or perhaps a chance combination of rare existing alleles. In either event, this change did two things: It caused these flies to emerge slightly *earlier* from their underground state than did most flies; and it drew these flies to the smell of *apples* as well as hawthorns. Because apples mature slightly earlier than the hawthorn fruit, these flies had a suitable host waiting for them. More important, these flies would have bred *among themselves* to a high degree. Recall that the other flies were emerging later, and the adult fly only lives for about a month. Thus, the variant alleles were passed on to a selected population in the next generation. Today, the apple *R. pomonella* flies indisputably do emerge earlier than the hawthorn flies. This is one of the things that ensures reproductive isolation between the two groups; their periods of mating don't fully overlap.

Such a lack of overlapping mating periods may sound familiar, because it is one of the intrinsic reproductive isolating mechanisms you looked at earlier. (It is temporal isolation.) You also can see ecological isolation in operation with these flies. Though the two types of flies live in the same area, they meet up with relatively little frequency, because they have different habitats (hawthorn vs. apple trees). In short, these populations have developed intrinsic reproductive isolating mechanisms without ever having been separated from one another geographically. They are headed toward speciation, but in this case it is sympatric speciation.

Speciation through Hybridization

Sympatric speciation can also take place, at least in plants, through the hybridization mentioned earlier, wherein egg and sperm from different species come together to produce offspring. As you can imagine, if plant species A occupies one habitat, and a closely related plant species B occupies another nearby habitat, there will be a so-called "hybrid zone" running between these two habitats where both plants exist and fertilize each other by means of wind-borne pollen. In some instances, healthy hybrid offspring can be produced from this mixing of different species' gametes. However, most hybrid offspring are infertile; they grow, but like mules, they will be incapable of giving rise to a *succeeding* generation of plants.

In rare instances, however, hybrids will be able to have fertile offspring of their own, not by mating

Figure 18.5
A Species Undergoing Sympatric Speciation?
The fruit-fly species *Rhagoletis pomonella*, pictured here on the skin of a green apple, may be undergoing speciation.

(a) Hybrid species

(b) One parental species

(c) Another parental species

Figure 18.6
Evolving from Several Species
Louisiana's *Iris nelsonii* plant on the left is thought to have evolved through the hybridization of three other *Iris* species. One of these is *Iris fulva* in the middle and another is *Iris hexagona* on the right.

with hybrids of their type, but by mating with one or more of the *parental* species that gave rise to them. Should hybrids get established in this way, their genetic complement will change over time. As they exchange alleles with several parental species, they may develop a genome that isolates them from any one of these species, but that allows them to produce fertile offspring by mating with hybrids of their own kind. Once this happens, a new species has come into being. If you look at **Figure 18.6**, you can see *Iris nelsonii*, a hybrid species that grows in Louisiana, and two of the three parental species it came from.

To sum up, the most common form of speciation, allopatric speciation, takes place when a geographic separation is followed by the development of intrinsic reproductive isolating mechanisms. A less common form of speciation, sympatric speciation, occurs when these isolating mechanisms develop in the absence of geographic separation.

Two special means of sympatric speciation are speciation through polyploidy and speciation through hybridization.

18.3 When Is Speciation Likely to Occur?

The horseshoe crab (**see Figure 18.7**) is not really a crab at all, but instead is distantly related to land-dwelling arthropods such as spiders and scorpions. (It lacks the antennae that real crabs possess, but it *has* the pincers around the mouth that all spiders and scorpions do.) Horseshoe crabs have been around in something like their modern form for more than 300 million years. They exist today in a scant four or five species, each one of which is pretty much like the others. The crabs live in the shallow oceans off North America and Asia, pushing through sand or mud to feed on everything from algae to small-bodied invertebrate animals.

(a) Modern horseshoe crab

(b) Fossilized horseshoe crab

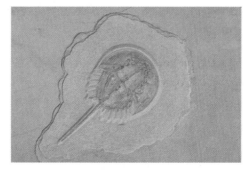

(c) Galapagos finch

Figure 18.7
Different Groups of Organisms Undergo Different Rates of Speciation

Because horseshoe crabs look essentially the same today as they did 300 million years ago, some scientists have called them "living fossils." Compare the modern organism, on the left, with the fossil at center, which is at least 145 million years old. By contrast, the Galapagos finches have diversified tremendously, forming 13 species in as little as 100,000 years.

Now consider the Galapagos (or Darwin's) finches you looked at in Chapter 17. There are *thirteen* of these species on the small Galapagos archipelago, all of them derived from a single species of South American finch that arrived on one of the islands perhaps 100,000 years ago.

Think of the difference between horseshoe crabs and Darwin's finches. The former have remained almost unchanged for more than 300 million years throughout the world, while the latter diverged into 13 different species in the last 100,000 years in a small cluster of islands off South America.

Specialists and Generalists

Why the great differences in rate of speciation here? Two general principles are at work. First, as Niles Eldredge has pointed out, horseshoe crabs are *generalists*: Their diet is extremely diverse; they will eat plants and small animals but will also scavenge for debris on the ocean bottom. By contrast, you saw in Chapter 17 how *specialized* some of the Galapagos finches are in their feeding behavior—particularly when food has become scarce because of drought conditions, as happens on the islands. In such times, species might exist on a single variety of plant seed. Species that are tied in this way to a particular food or environmental condition must adapt in connection with changes in them or face extinction. Think of how quickly the bills of the *Geospiza fortis* population on Daphne Major evolved toward greater depth when having a deeper bill meant the difference between life and death (see Chapter 17, page 336). By definition, this kind of adaptation means change—and change is what speciation is all about. By contrast, the horseshoe crab shifts from one food to another, depending on what is available, not adapting greatly in response to changes in any one food source.

New Environments: Adaptive Radiation

The second lesson offered by horseshoe crabs and finches concerns the kinds of environments that induce speciation. The Galapagos islands were formed by volcanic eruptions that brought the islands above the ocean's surface only about 5 million years ago. At that time, these volcanic outcroppings gave new meaning to the word *barren*. For a brief time, the islands were utterly sterile, with no life on them. Very quickly, however, life did come to the islands, with bacteria, fungi, plant seeds, and tiny animals landing on the islands, all being borne by air or ocean currents. The South American finches obviously did not arrive until much later; but by then, with lots of large plant species well established, what these birds encountered was an environment rich with possibility, for there were *no birds of their kind* on the islands. Imagine that you are a graphic artist, working in a big city with lots of other graphic artists, most of whom specialize in this or that (magazines, Internet websites, etc.). Now you and a few other graphic artists move to a new city in which there are few graphic artists, but a good number of *possibilities* for graphic arts work. You would thus be able to specialize fairly easily—filling a *niche* or working role in this new environment—because many of these niches would not yet have been taken. Just so did Darwin's finches rush in to fill previously unoccupied niches on the Galapagos Islands—this plant seed, this insect, this mating environment.

Such a situation is ripe with possibilities for change (meaning speciation) because, while niches are in flux, there is a good deal of shaping of species to environment. But more of this occurred on the Galapagos, because this niche-filling was taking place on 25 separate *islands*. The birds can fly from one island to another, but the water between the islands nevertheless represents a geographic barrier to bird interbreeding, and you know what follows from this: allopatric speciation.

The Galapagos finches exemplify something known as **adaptive radiation**: the rapid emergence of many species from a single species that has been introduced to a new environment. The finches radiated out to fill new niches on the islands, with populations adapting to the environments over time. In summary, two conditions that are conducive to speciation are specialization (of food source or environment) and migration to a new environment, particularly when there are no closely similar species in that environment.

Is Speciation Smooth or Jerky?

Charles Darwin knew that in looking at large-scale evolutionary changes, what is striking in the fossil record is the lack of "transitional forms," meaning fossils that demonstrate the incremental changes that took place as one species branched from another, in a line going from A to B to C to D. Darwin thought that this was simply a problem of not having collected enough fossils in the right places; that in time, the transitional forms would show themselves. To a certain extent, Darwin was right. In recent years, scientists have been unearthing fossils that show clear transitional forms—for example, in the evolution of whales from a medium-sized, four-legged land mammal.

These kinds of advances notwithstanding, when we look at the fossil record as a whole, there are still more gaps than Darwin would have been comfortable with. Indeed, the fossil record seems to speak against the mode of evolution as Darwin had envisioned it. He imagined evolution as a series of infinitesimally small changes accumulating in a slow, steady way in populations of a species until in time one species could be seen to have diverged into two.

What the fossil record exhibits again and again, however, is not evidence of slow, steady change in a species. Rather, there are indications of enormous lengths of time in which so-called *stasis* occurs—in which species stay exactly the same or undergo some minute modifications. After this long stasis, there appears to be an *abrupt* change to a new species.

In 1972, two young scientists proposed that it was time to take the fossil record at face value. Why isn't it plausible, they said, that species experience stasis for long periods of time, after which speciation then takes place in relatively brief bursts? This is the **theory of punctuated equilibria** proposed by Stephen Jay Gould and Niles Eldredge. (Periods of stasis are the equilibrium that species generally live in, with these then being "punctuated" by the speciation events.)

So, how rapid is speciation under this view? It can take place in thousands of years rather than the millions during which a species is in stasis. If that is the case, it would explain why there are so few transitional forms in the fossil record: They aren't around for long enough to leave many fossils (**see Figure 18.8**).

There has been great debate about the theory of punctuated equilibria ever since it was proposed. Some scientists have supported it, others have opposed it. Its detractors (sometimes called "gradualists") have noted that only the "hard parts" of species (bone and so forth) are preserved in the fossil record, while much evolutionary change must concern "soft parts." Other critics allege that the theory proposes nothing new, but merely puts terminology to ideas that already existed (in population genetics) about how rapidly species can evolve.

Some recent experimental evidence is consistent, however, with the idea of punctuated equilibria. Relying on the ability of bacteria to reproduce rapidly, scientists looked at 10,000 generations of them over a period of 4 years—freezing samples at various points in time. They found that these cells tended to stay the same size for hundreds of generations and then evolve to a larger average size very quickly. Other researchers have asked how much *genetic* change is required to bring about speciation and found that, in comparing two species of flowering plant at least, the answer is very little. They found that changes in as few as eight genetic locations may account for the differences seen between two species of monkey flower that differ in color and according-

ly attract different pollinators. If so few changes are required for speciation, then it could proceed more rapidly than if many changes were necessary.

One possibility is that speciation is both jerky (punctuated) and smooth (gradual), depending on the species and the characteristics being examined. Changes in "morphology" or physical structure are based on genetic changes, as you have seen. It could be that genetic changes are gradual, while the resulting morphological changes come about in spurts.

18.4 The Categorization of Earth's Living Things

This chapter began by noting the importance of the species concept—of being able to say that this organism is fundamentally separate from that. To have a species concept means, of course, that there has to be some means of *naming* separate species. You have probably noticed that most of the species considered in this chapter have been referred to by their scientific names; you didn't just look at a fruit fly, but rather *Rhagoletis pomonella*. To the average person, such names may be regarded as evidence that scientists are awfully exacting—or just plain fussy. Why the two names? Why the Latin? Can't we just say fruit fly?

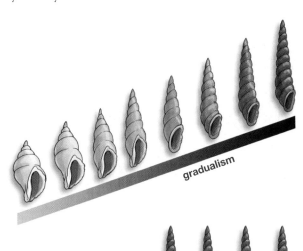

According to the gradualism model, speciation occurs through gradual change over long periods of time.

According to the punctuated equilibria model, long periods of stasis may periodically be "punctuated" by rapid bursts of speciation.

Figure 18.8
Is Speciation Smooth or Jerky?

To take these questions one by one, scientists can't just say "fruit fly" for the same reason a person can't say "the guy in the white shirt" while trying to identify a player at a tennis tournament. There are lots of guys in white shirts at tennis tournaments, and there are lots of different kinds of flies in the world. It is important to be able to say *which* fly we are talking about. The importance of this can be seen with *Rhagoletis pomonella* itself, which is a major pest in the apple industry. Tell some apple growers that you have spotted flies on their apples and you may get a shrug of the shoulders; tell them you have spotted *R. pomonella* and you may get a very different reaction.

The Latin that is used stems from the fact that this naming convention was standardized by the Swedish scientist Carl von Linné in the eighteenth century, at a time when Latin was still used in the Western world in scientific naming. (In fact, Linné is better known by the Latinized form of his name—Carolus Linnaeus.)

Linnaeus recognized the confusion that can result from having several common names for the same creature, and thus devoted himself to giving specific names to some 4,200 species of animals and 7,700 species of plants—all that were known to exist in his time. Many of the names he conferred are still in use today.

The fact that there are *two* parts to scientific names—a **binomial nomenclature**—points to the central question of groupings of organisms on Earth. Consider the domestic cat, which has the scientific name *Felis domestica*. The first part of its name is its genus, *Felis*, which designates a group of closely related but still separate species. It turns out that, worldwide, there are five other species of small cat within the *Felis* genus, such as the small cat *Felis nigripes* ("black-footed cat") found in Southern Africa.

Taxonomic Classification and the Degree of Relatedness

The practical importance of the genus classification is that, if we know that two organisms are part of the same genus, then to know something about one of them is to know a good deal about the other. But what is the basis for placing organisms in such a category? Modern science classifies organisms largely in accordance with how closely they are *related*. In this context, "related" has the same meaning as it does when the subject is extended families. If people have the same mother, they are more closely related than if they shared only a common *grandmother*, which in turn makes them more closely related than if they had only a common great-great grandmother, and so on.

In the same way, species can be thought of as being related. Domestic dogs (*Canis familiaris*) are very closely related to gray wolves (*Canis lupus*), with all dogs being descended from these wolves. Indeed, by some reckonings dogs and gray wolves are the same species (*Canis lupus*), because they will interbreed in nature. Such differences as exist between them are the product of a mere 15,000 years or so of the human domestication of dogs. It may be, then, that 15,000 years ago there was but *one* species (the gray wolf) that gave rise to both today's wolf and today's dog. A close relation indeed.

Domestic dogs are also related, however, to domestic *cats*; if we look far back enough in time, we can find a single group of animals that gave rise to both the dog and cat lines. This does not mean going back 10,000 years, however; it means going back perhaps 60 million years. So there is a big difference in how closely related dogs and wolves are, as opposed to dogs and cats. Establishing such *degrees* of related-

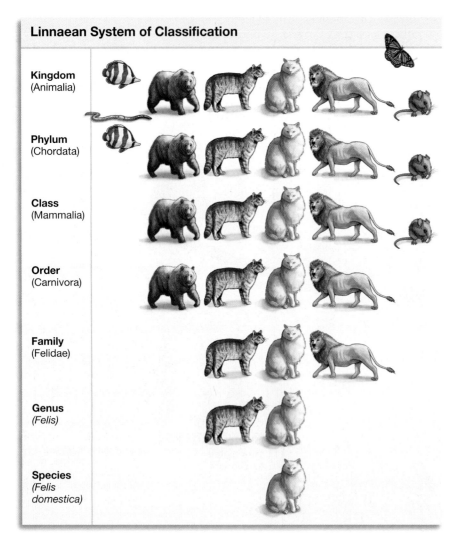

Linnaean System of Classification

Kingdom (Animalia)

Phylum (Chordata)

Class (Mammalia)

Order (Carnivora)

Family (Felidae)

Genus (Felis)

Species (Felis domestica)

Figure 18.9
Classifying Living Things
The classification of the house cat, *Felis domestica*, based on the Linnaean system.

ness is the most important task of the scientific classification system. There is a field of biology, called **systematics**, that is concerned with the diversity and relatedness of organisms; part of what systematists do is try to establish the truth about who is more closely related to whom. They study the evolutionary history of groups of organisms.

Setting aside for the moment the difficulty in *determining* what such evolutionary histories are, there obviously is a tremendous cataloguing job here, given the number of living and extinct species mentioned earlier. Given this diversity, a method of classifying organisms, called a *taxonomic system*, is employed in order to classify every species of living thing on Earth. There are eight basic categories in use in the modern taxonomic system: **species, genus, family, order, class, phylum, kingdom**, and **domain**. The organisms in any of these categories make up a grouping of living things, a **taxon**.

A Taxonomic Example: The Common House Cat

If you look at **Figure 18.9**, you can see how this taxonomic system works in connection with the domestic cat. As noted, this cat is only one species in a genus (*Felis*) that has five other living species in it. The genus then is a small part of a family (Felidae) that has 17 other genera in it (panthers, snow leopards, and others). The family is then part of an order (Carnivora) that includes not only big and small "cats," but other carnivores, such as bears and dogs. On up the taxa we go, with each taxon being more inclusive than the one beneath it until we get to the highest category in this figure, the kingdom Animalia, which includes all animals. (Later you'll look at the "supercategory" above kingdom, called domain; but for simplicity's sake, kingdom is the highest-level taxon considered here.)

Constructing Evolutionary Histories

So how do systematists go about putting organisms into these various groups? What evidence do they use in constructing their evolutionary histories? They rely on radiometric dating, the fossil record, DNA sequence comparisons—all the things reviewed in Chapter 16 that are used to chart the history of life on Earth.

If you look at **Figure 18.10**, you can see the outcome of some of this work, an evolutionary "tree" for one group of organisms, in this case one of the major groups of mammalian carnivores. You can see that the tree is "rooted," about 60 million years ago, with an ancestral carnivore whose lineage then split two ways: to dogs on the one hand and everything else on the other. One interesting facet of this evolutionary history is how closely related bears are to the aquatic carnivores. All such histories are hypotheses about evolutionary relationships, with each such hypothesis known as a **phylogeny**.

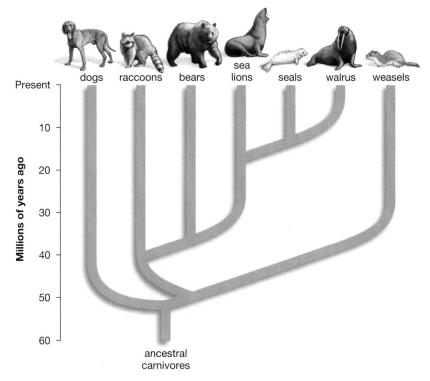

Figure 18.10
A Family Tree for a Selected Group of Mammalian Carnivores

Obscuring the Trail: Convergent Evolution

One of the things that makes the interpretation of evolutionary evidence difficult is that similar features may arise *independently* in several evolutionary lines. A bedrock of systematic classification is the existence of **homologies**, which can be defined as common structures in different organisms that result from a shared ancestry. In Chapter 16, a strong homology was noted in forelimb structure that exists in organisms as different as a gorilla, a bat, and a whale. If you look at **Figure 18.11**, you can see the gorilla and the bat again.

Now, however, consider an extinct group, the litopterns, that once lived in what is now Argentina. Like the modern-day horse, they roamed on open grasslands. Over the course of evolutionary time they developed legs that were extraordinarily similar to the horse's, right down to having an undivided hoof. In fact, litopterns are only very distantly related to the horse; but the leg similarities were enough to fool at least one nineteenth-century expert who thought he had found in litopterns the ancestor to the modern horse.

What is exhibited in the legs of the litoptern and horse is an **analogy**: a feature in different organisms that is the same in function and superficial appearance. When nature has shaped two separate evolutionary lines in analogous ways, what has occurred is **convergent evolution**. (Both the litoptern and horse lines converged in their development of similar legs.) But analogous features have nothing to do with common descent; they merely show that the same kinds of environmental pressures lead to the same kinds of designs. (There are only so many leg structures that work well for grass-eating mammals that roam over long distances, so it's not surprising that natural selection would have pushed two separate evolutionary lines toward a similar leg shape.) The problem for systematics

(a) Homology: Common structures in different organisms that result from common ancestry

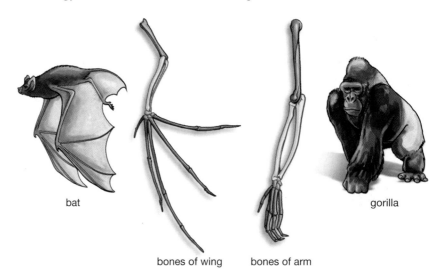

bat

gorilla

bones of wing   bones of arm

(b) Analogy: Characters of similar function and superficial structure that have *not* arisen from common ancestry

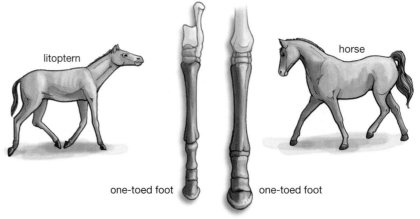

litoptern

horse

one-toed foot   one-toed foot

Figure 18.11
Related, as Opposed to Similar

(a) Bats and gorillas have the same bone composition in their forelimbs—one upper bone, joined to two intermediate bones, joined to five digits—because they share a common ancestor. Of course, these forelimbs are used today for very different functions.

(b) Horses and the extinct litopterns have long legs and one-toed feet that serve the same function, but these features were derived independently in the two creatures—in separate lines of descent.

is that analogy can be confused with homology; analogies can make us believe that organisms share common ancestry when in fact they do not.

18.5 Classical Taxonomy and Cladistics

Given difficulties such as convergent evolution and a less-than-perfect fossil record, systematists over the years have been driven to devise several methodologies for producing phylogenies. The essential question they have asked is: What means of *interpreting* the evidence will give us the most accurate picture of how Earth's living and extinct organisms are related?

One of the methodologies for interpreting evidence actually predates the field of systematics and most of the techniques used in it. Handed down from the time of Darwin, this methodology is sometimes called *classical taxonomy* (or descriptive taxonomy). In thinking about animals in Earth's family tree, classical taxonomists would look first and foremost at the physical form or "morphology" of the animals in question, as preserved in the fossil record, and compare it to the morphology of modern animals. Skull shape, "dentition" (teeth patterns), limb structure, and much more would be considered. They would also look at *where* the ancient forms existed, compared to modern forms. More recently, they have been using molecular techniques, such as comparing DNA or protein sequences in different living species to determine the relatedness among them. In essence, classical taxonomists would look at how many similarities one group has with another and, on that basis, try to judge relatedness among them. The word *judge* is used advisedly here, because at the end of the day in the classical system, subjective judgments usually need to be made about who is more closely related to whom. It is a matter of weighing one piece of evidence against another, but who's to say which piece of evidence is more important? In addition, as you'll see, classical taxonomy sometimes makes a distinction between the phylogenies it establishes and the taxonomic categories into which it puts different organisms.

Another System for Interpreting the Evidence: Cladistics

The subjectivity inherent in classical taxonomy helped bring about the formation of another system for establishing relatedness. Developed in the 1950s by the German biologist Willi Henning, it is called **cladistics** (from the Greek *klados*, meaning "branch"). In practice, cladistics has become the core of most of the phylogenetic work going on today.

In **Figure 18.12** you can see a **cladogram**, which is an evolutionary tree constructed within the cladistic system. Note that there is a very simple branched line and that no time scale is attached to it, as was the case with the evolutionary tree of carnivores. Cladistics concerns itself first with lines of descent—with the *order* of branching events. Once this order has been established, efforts can be made to fix events in time, but that is a secondary concern. First and foremost, this cladogram is a proposed answer to the question: Among the animals lizard, deer, lion, and seal, which two groups of animals have the most recent common ancestor? Then, which other two have the *next* most recent common ancestor, and so on. By extension, the question is: Who is more closely related to whom? Note that these questions are being asked about only four of the five animals listed. You'll get to the role of the hagfish in a moment.

Cladistics Employs Shared Ancestral and Derived Characters

The starting point for cladistic analysis is the difference between so-called ancestral characters and derived characters. If you look at any group of species (any taxon), its **ancestral characters** are those that existed in an ancestor common to them all. There are about 50,000 species of animals on Earth that possess a dorsal vertebral column (better known as a backbone). The ancestor to all fishes, reptiles, and mammals had such a feature, which makes the vertebral column an ancestral character for all these groups. On the other hand, only *some* vertebrates then went

Figure 18.12
A Simple Cladogram
Among the lizard, deer, lion and seal, which two are most closely related? The tentative answer displayed in this cladogram is the lion and the seal. The basis for this assertion is the number of unique or "derived" characters these two creatures share compared to any other two organisms under consideration. (They share three characters: tetrapod structure, mammary glands, and carnivorous feeding.) Such a system pays no attention to superficial similarities among organisms. Like fish, seals swim in the water; but they share far more derived characters with lions than they do with fish—enough to persuade scientists that seals have returned to the water in relatively recent times, following a period in which their ancestors lived on land.

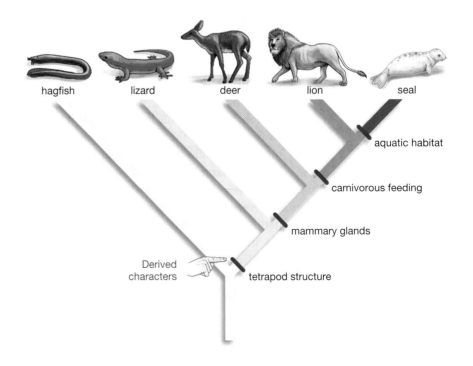

sympatric speciation always entails the development of intrinsic reproductive isolating mechanisms.

- There are two special forms of sympatric speciation. One is speciation through hybridization—a relatively rare phenomenon that is most important in plants. The other is polyploidy: a doubling of the number of chromosomes in a species that can bring about speciation in one generation. Polyploidy has been especially important in plant speciation, but it has played a part in animal evolution as well.
Web Tutorial 18.1: Speciation

18.3 When Is Speciation Likely to Occur?

- Speciation is more likely to occur when a species is highly specialized with respect to a food source or other environmental condition. Speciation is also more likely to occur when a species is introduced to an environment in which few other species of its kind exist. Such an environment has many available working roles or niches available to the new species—a situation that leads to rapid specialization, which in turn leads to speciation. This series of events is known as adaptive radiation: the rapid emergence of many species from a single species that has been introduced to a new environment.

- There is considerable debate among evolutionary biologists regarding the pace of evolution. Under one view, called the gradualist model, evolution proceeds at a slow, steady rate; under another view, called the punctuated equilibria model, organisms undergo long periods of "stasis" or little change, followed by relatively brief episodes of rapid speciation.

18.4 The Categorization of Earth's Living Things

- In science, a two-name or "binomial" nomenclature is used for each of Earth's species. The first name designates the genus, or group of closely related organisms, that the species is a member of; the second name is specific to the species. Genus and species fit into a larger framework of *taxonomy*, meaning the classification of species. Going from least to most inclusive, the eight commonly used groupings in this taxonomy are species, genus, family, order, class, phylum, kingdom, and domain.

- Species are put into these categories largely on the basis of relatedness—species that are closely related are in the same genus; species that are distantly related may only be in the same phylum or kingdom. The biological discipline of systematics is concerned with establishing degrees of relatedness among both living and extinct species.

- Systematists establish evolutionary histories or "phylogenies" by reviewing various kinds of evidence, including radiometric dating, the fossil record, and DNA sequence comparisons. Based on this evidence, they determine the evolutionary history of a given species.

- In establishing phylogenies, one of the things systematists look for is homologous structures: common structures in different organisms that result from a shared ancestry. One problem with their use is that they can be confused with analogous structures: similar features that developed independently in separate lines of organisms (as with the legs of modern horses and extinct litopterns).

18.5 Classical Taxonomy and Cladistics

- One method of determining phylogeny, classical taxonomy, establishes evolutionary relationships among living and extinct organisms in accordance with such factors as physical form, distribution, and molecular similarities; but classical taxonomy employs subjective judgments in deciding on what weight to give one piece of evidence as opposed to another. The phylogenetic system known as cladistics does away with this element of subjectivity by following a firm rule for inferring relatedness. It counts the number of shared derived characters two organisms have, meaning the features they uniquely share that are derived from another group of organisms.

- Cladistics is centrally concerned with establishing lines of descent, though species can be put into taxonomic categories that are derived from cladistic analysis. Classical taxonomy holds that, in putting organisms into various categories, factors other than phylogeny ought to be taken into account. Cladistics holds that the only criterion for taxonomic placement is phylogeny—a group must contain all descendants stemming from a given branch point.

Key Terms

adaptive radiation   358
allopatric speciation   351
analogy   362
ancestral character   363
binomial nomenclature   360
biological species concept   350
cladistics   363
cladogram   363
class   361
convergent evolution   362
derived character   364
domain   361
extrinsic isolating mechanism   352
family   361
genus   361
homology   362

intrinsic isolating mechanism   352
kingdom   361
order   361
phylogeny   361
phylum   361
polyploidy   354
reproductive isolating mechanisms   352
speciation   351
species   361
sympatric speciation   354
systematics   361
taxon   361
theory of punctuated equilibria   359

Understanding the Basics

Multiple-Choice Questions (answers in the back of the book)

1. In the biological species concept, the factor that defines a species is
   a. geographical separation
   b. reproductive isolation
   c. hybridization
   d. cladogenesis
   e. its behavior

2. In allopatric speciation (select all that apply):
   a. Allele frequency changes are shared between populations.
   b. Populations are geographically isolated.

    c. Gene flow between populations is greatly reduced by physical barriers.

    d. Geographically isolated populations accumulate genetic changes over time that eventually block successful interbreeding.

    e. Populations are directly adjacent.

3. Factors that can be problems in determining the true relationships among organisms may include (select all that apply):
    a. homologous traits
    b. analogous traits
    c. convergent evolution
    d. poor fossil record
    e. shared derived traits

4. Possible intrinsic reproductive isolating mechanisms include (select all that apply):
    a. mating at different times of year or times of day
    b. geographical isolation
    c. incompatibility between eggs and sperm that prevent fertilization
    d. singing a different mating song than another species
    e. producing fertile hybrids

5. Match the reproductive isolating mechanism with the means by which it works:
    a. temporal isolation — populations live in different environments
    b. ecological isolation — populations are separated by physical barriers
    c. geographical isolation — mating occurs at different times of day or year
    d. gametic isolation — egg and sperm do not fuse
    e. hybrid sterility — progeny of a cross are unable to reproduce

6. In ocean-dwelling creatures such as sea urchins or corals, sperm and eggs are released into the water by males and females, and fertilization occurs externally. If several closely related species of coral live in the same location, what reproductive isolating mechanisms are likely to be effective at preventing interbreeding? (Select all that apply.)
    a. behavioral isolation
    b. gametic isolation
    c. temporal isolation
    d. ecological isolation
    e. hybrid infertility

7. Order the following categories into the appropriate biological hierarchy, with the least inclusive category first.
    a. family
    b. species
    c. phylum
    d. genus
    e. class

8. Adaptive radiations (select all that apply):
    a. are most likely to occur on mainlands
    b. are examples of rapid speciation that occurs when a number of ecological niches have not been filled
    c. concern the evolution of heat loss in vertebrates experiencing stressful climates
    d. are exemplified by the Galapagos finches
    e. often occur on islands

9. Which pair of animals is probably most closely related evolutionarily?
    a. animals in the same genus
    b. animals in the same order
    c. animals in the same family
    d. animals in the same phylum
    e. animals in the same class

Brief Review

1. What is convergent evolution, and why is it a problem for phylogenetic analyses?

2. Contrast the gradual and punctuated equilibria models of long-term evolution. What type of evidence has been used to support the punctuated equilibria theory?

3. Describe one major difference between allopatric and sympatric speciation. What is one thing that they have in common?

4. State the biological species concept.

5. Why is the evolution of intrinsic reproductive isolation mechanisms required for two groups to be called separate species? Why isn't simple geographical isolation sufficient?

Applying Your Knowledge

1. Living things are classified into a set of hierarchical categories. Why is this more useful to biologists than simply giving everything a one-word name (or a number) and eliminating some of the categories?

2. Populations of some kinds of organisms may be isolated by a barrier as small as a roadway, while other organisms require a much larger physical barrier to block gene flow. What features of organisms might determine how readily they become geographically isolated? What impact might these features have on how often or how rapidly speciation is likely to occur in these groups?

3. In an adaptive radiation, one species may colonize a previously empty environment, such as an island, and diversify evolutionarily into a set of closely related species that occupy a very wide range of habitats. For example, in the Darwin's finches on the Galapagos Islands, one species fills the role of woodpecker by using a stick to tap on trees and dig out insects. Do you think that such a species would have been likely to evolve in an area that already had woodpeckers? Why or why not?

19 *The History of Life on Earth*

The history of life written in layers of rock.
(Section 19.1, page 371)

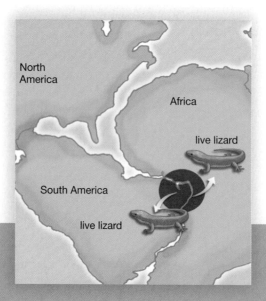

Parting company
(Essay, page 373)

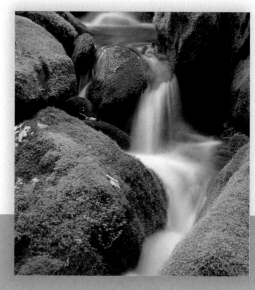

One of the earliest-evolving groups of plants.
(Section 19.6, page 381)

Within 5 minutes of its birth, a newborn wildebeest on the plains of central Africa can be off and running with its herd (**see Figure 19.1**). By contrast, human infants probably will not be able to take a step by themselves until they are about a year old. Female lions reach sexual maturity at about 3 years, and killer whales at about 7, but in human hunter-gatherer societies, reproduction generally begins at age 18 or 19. Even in comparison to our fellow mammals, humans mature at a notably slow rate.

At first glance, it might seem puzzling that our evolution took this turn. From what we have learned of natural selection, wouldn't it make sense that survival would *decrease* for offspring that developed more slowly? What is the fitness payoff in having young that are essentially helpless for several years and then immature for many more? Here is one possible answer. As a species, we have survived not so much because of our physical capabilities, but because of our wits. Our success is the product of our remarkable capacity to *learn*. But learning takes time, and it is arguably best undertaken by minds that remain for a long time in what might be called a state of flexible immaturity. (Think of how easily young children acquire a second language in comparison with adults.) Among our human ancestors, it may be that those who survived tended to be those who could learn the most. And who could learn the most? Those who took the longest to mature.

Under this view, we carry our evolutionary heritage with us in the form of delayed maturity. We did not leave this heritage behind in the African savanna, where human beings first evolved; we can see it every time we look at a 2-year-old. In the same way, we can see evolutionary vestiges in us from a much earlier time as well. Note the way we walk, our *left arm* swinging forward as our *right leg* does the

Figure 19.1
No Time to Lose
A newborn wildebeest struggles to its feet immediately after its birth on the plains of Africa. Wildebeest predators, such as hyenas, often concentrate their efforts on the newborn.

On the way to flowering plants.
(Section 19.6, page 383)

What Is a Primate?
(Section 19.7, page 386)

The artistic flowering of early humans.
(Section 19.8, page 394)

Figure 19.2
Our Evolutionary Heritage
The gait exhibited by almost all four-limbed animals, including human beings, stems from the reversing S-pattern that our evolutionary ancestors used in getting around when they were making the transition from aquatic to terrestrial life. Here a salamander employs this same motion in getting from one place to another.

same. Almost all four-limbed animals walk this way. One view is that we inherited this from the elongated, four-finned fish we evolved from. They came onto the muddy land by slithering in an S-pattern, the left-front fin going back as the left-rear fin went forward, then the reverse (**see Figure 19.2**).

Note the sweep of evolution in all this. Maturation is with us from one period, walking style from another. If we look at the genetic code—*this* DNA sequence specifying *that* amino acid—we have within

Figure 19.3
A Timescale for Earth and Its Living Things

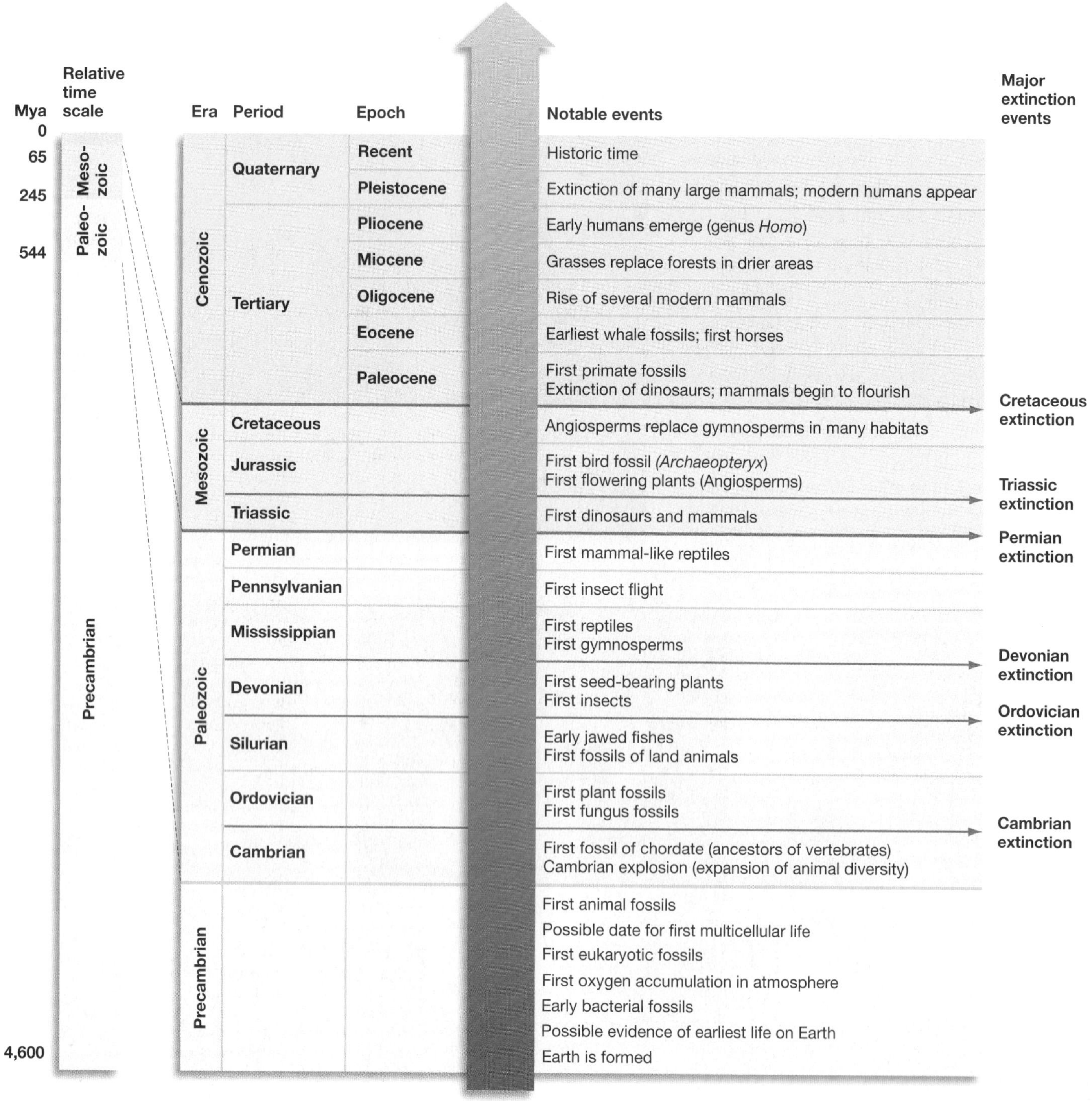

Mya	Relative time scale	Era	Period	Epoch	Notable events	Major extinction events
0				Recent	Historic time	
65	Meso-zoic	Cenozoic	Quaternary	Pleistocene	Extinction of many large mammals; modern humans appear	
245				Pliocene	Early humans emerge (genus *Homo*)	
544	Paleo-zoic		Tertiary	Miocene	Grasses replace forests in drier areas	
				Oligocene	Rise of several modern mammals	
				Eocene	Earliest whale fossils; first horses	
				Paleocene	First primate fossils Extinction of dinosaurs; mammals begin to flourish	
		Mesozoic	Cretaceous		Angiosperms replace gymnosperms in many habitats	Cretaceous extinction
			Jurassic		First bird fossil (*Archaeopteryx*) First flowering plants (Angiosperms)	Triassic extinction
			Triassic		First dinosaurs and mammals	
			Permian		First mammal-like reptiles	Permian extinction
			Pennsylvanian		First insect flight	
		Paleozoic	Mississippian		First reptiles First gymnosperms	
	Precambrian		Devonian		First seed-bearing plants First insects	Devonian extinction
			Silurian		Early jawed fishes First fossils of land animals	Ordovician extinction
			Ordovician		First plant fossils First fungus fossils	
			Cambrian		First fossil of chordate (ancestors of vertebrates) Cambrian explosion (expansion of animal diversity)	Cambrian extinction
			Precambrian		First animal fossils	
					Possible date for first multicellular life	
					First eukaryotic fossils	
					First oxygen accumulation in atmosphere	
					Early bacterial fossils	
					Possible evidence of earliest life on Earth	
4,600					Earth is formed	

us a heritage that stretches back to the beginning of life on Earth, because this code is shared by nearly every living thing. Under this view, evolution is not some abstract process that lies separate from us; on the contrary, it is with us in every step we take. But how has evolution proceeded in its long sweep down the ages? Starting with the spark of life itself, how did it produce the range of creatures that inhabit the Earth? In this chapter, you'll begin to find out.

19.1 The Geological Timescale: Life Marks Earth's Ages

We can start this inquiry by getting a sense of life's timeline. The Earth, which is home to all the life we know of, came into being about 4.6 billion years ago. Given the human life span of 80-plus years, 4.6 billion years is an unimaginably long time. Indeed, it's a long period of time relative to any time frame we know of. The universe as a whole is thought to be 14 billion years old at most, so the Earth's history stretches back a third of the way to the beginning of time.

In measuring something as long as 4.6 billion years, it obviously won't do to think in terms of individual years, so scientists use something called the geological timescale, which you can see in **Figure 19.3**. It divides earthly history into broad *eras*, shorter *periods*, and shorter-still *epochs*. The scale begins with the formation of the Earth (at the bottom of the figure), and runs to "historic time," meaning time in the last 10,000 years.

But what do the demarcations in this timescale indicate? What, for example, is the distinction between the Permian period you can see at the middle left of Figure 19.3, and the Triassic period that followed it? Remember that back in Charles Darwin's time, there was no radiometric dating of materials—no using the decay of uranium, for example, to provide absolute dates for any fossils that might be found. Hence, no one could say whether a fossil was 10 million or 100 million years old. Geologists of the time did know, however, that sedimentary rocks had been put down in layers (or strata), with the oldest strata on the bottom and the youngest strata on top (**see Figure 19.4**). And as it turned out, each stratum tended to have within it a group (or "assemblage") of fossils that was unique to it. The result was that each stratum and its fossils could be assigned a *relative* date—each was older than this stratum, but younger than that one. Since fossils helped define the geological strata, it's not surprising that scientists marked off the geological timescale in accordance with the *transitions* they saw in fossil forms. The Permian period is considered different from the Triassic period because, with move-ment from one to the other, a different set of life-forms appears, as recorded in the fossil record.

Features of the Timescale

Now a couple of notes on the timescale. First, it may surprise you to learn what the basis is for many of the transitions in it: death on a grand scale, in the form of "major extinction events." Six of these events are noted in Figure 19.3 (over on the right), with all of them defining the end of an era or period. The most famous extinction event, the **Cretaceous Extinction**, occurred at the boundary between the Cretaceous and Tertiary periods. This was the extinction, aided by the impact of one or more giant asteroids, that brought about the end of the dinosaurs, along with many other life-forms. The greatest extinction event of all, however, occurred earlier, in the boundary between the Permian and Triassic periods. This was the **Permian Extinction**, in which as many as 96 percent of all species on Earth were wiped out (see "Physical Forces That Have Shaped Evolution" on the next page).

This is not to say, however, that all Earth's eras or periods are marked off by extinctions. The transition between the Precambrian and Paleozoic eras marks not a major extinction, but a seeming explosion in the diversity of animal forms. Nevertheless, this so-called Cambrian Explosion, which you'll read about later, is just a different kind of transition in living things.

Web Tutorial 19.1
Evolutionary Timescales

Figure 19.4
Revealing Layers
The history of life can be traced through fossilized life-forms found in layers of sediment, seen here in America's Grand Canyon.

Now, when did some of these transitions come about? You might think that a transition such as a sudden increase in animal forms would have occurred very early in life's history, but the Cambrian Explosion took place 544 million years ago. This is a long time ago, to be sure, but life is thought to have begun 3.8 *billion* years ago. (Another way of writing this is 3,800 million years ago or 3,800 Mya). There may not have been an animal on the planet for 3.2 billion years after life got going—nor any plants either. Life was single-celled to begin with and, 2 billion years later, it still consisted of nothing but microbes. In the Precambrian era, life evolved at such a slow pace that we scarcely have words for it. ("Glacial" greatly overstates things.) Conversely, if the fossil record is correct, all the birds, reptiles, fish, plants, and mammals that exist today came about in the last 16 percent of evolutionary time—the last 600 million years of life's 3.8-billion-year span. As you'll see, our own species, *Homo sapiens*, is an extreme latecomer in this series of events. (For a look at why various "arrival dates" are uncertain, see "Fossils and Molecular Clocks: Dating Life's Passages" on page 376.)

What Is Notable in Evolution Hinges on Values

One final word about the timescale and the path you will follow in this chapter. Picking out the "notable events" in evolution is inevitably an exercise in making value judgments. The notable events covered in this chapter will lead you along a line that begins with microscopic sea creatures and ends with human beings. The value judgment guiding this path is that students have a great interest in knowing about their own evolution, and that the path to humans arguably presents the broadest sweep of evolutionary development. Such a course of study may leave the impression, however, that all of evolution amounts to a march toward the development of human beings, who then get to occupy the highest branch of an evolutionary tree.

In reality, we occupy one ordinary branch among a multitude of branches. Under notable events in the Cenozoic era, the table could just as easily have listed milestones in the evolution of fish or birds, or the emergence of social insects such as bees. These creatures have continued evolving in the modern era, along with our own species, but in lines separate from us. Focusing on a line that leads to humans is like focusing on a railroad route that leads from, say, Baltimore to Denver, while ignoring the multitude of lines that intersect with it or that are separate from it. We regard the Baltimore–Denver route as special because we are *interested* in it, but that does not mean

that it is fundamentally *different* from any other line. Some special qualities have evolved in human beings, but the same could be said for almost any species. In phylogenetic terms, we simply lie at the tip of one evolutionary branch, while bees lie at another, and cactus plants lie at still another.

The Kingdoms of the Living World Fit into Three Domains

In tracing the history of life, you'll be looking to some extent at all the major life-forms that have evolved on Earth. Thus it will be helpful to have at the outset a sense of what these major life-forms are. This book employs a system of three *domains* of life, which then have within them "kingdoms," as noted in Chapter 18. The three domains are Archaea, Bacteria, and Eukarya.

Archaea is a domain of life populated entirely by microscopic, single-celled organisms, many of which live in "extreme" environments, such as boiling-hot vents on the ocean floor. **Bacteria** is likewise a domain of single-celled microscopic creatures, but these are the bacteria that are more familiar to us. The third domain, **Eukarya**, then includes all the organisms that are most familiar to us—plants, animals, and fungi—along with another group that is less familiar, the protists. (The protist kingdom has within it lots of single-celled water dwellers, but also includes some large organisms, such as giant sea kelp.)

19.2 How Did Life Begin?

Taking a historical approach to tracing life on Earth, the first question that arises is one of the toughest: How did life begin? Darwin himself thought about this and imagined that life began in what he called a "warm little pond" in the early Earth. Well, maybe; but the *very* early Earth had no warm little ponds. What it had was an environment more akin to hell on Earth.

You have seen that Earth was formed about 4.6 billion years ago. This took place in a process of "accretion," in which ever-larger particles clumped together: cosmic dust to gravel, gravel to larger balls, larger balls to objects the size of tiny planets. One consequence of such development was that the accretion now called Earth was periodically being slammed into by *other* large accretions—meteorites and comets and proto-planets. One of these objects was as large as Mars, and the result of its impact on the Earth, 100 million years after Earth formed, may have been the creation of our moon. There were no oceans at this time, and Earth was so hot that it was covered with a layer of molten rock. Here are biologists Christopher Wills and Jeffrey Bada writing in *The Spark of Life* about the young planet:

> Earth at this stage must have been a spectacular planet, its thick atmosphere swirling with color as Jupiter's atmosphere does today. The day side would have been dazzling, and the darkness of the night side would have been dispelled periodically by the flashes of huge lightning storms and the fitful, red glow of mighty volcanic eruptions. Occasionally, bright explosions would blaze up out of the haze surrounding the planet, as [objects] the size of Mount Everest plunged down to strike the surface.

The bombardment of Earth seems to have slowed by about 3.8 billion years ago. With this change, we are left with a large question about what the planet was like. Scientists believe that the sun shone some 20–30 percent less brightly then than it does now. So one of two things is possible. If Earth's atmosphere contained a sufficient concentration of "greenhouse" gases, such as methane and carbon dioxide, then it would have remained a warm place—perhaps even a broilingly hot place. Conversely, if the concentration of these gases was low, then the Earth was frozen over in a global glaciation. Even the seas may have been frozen to a depth of a few hundred meters.

These two views lead to varying ideas about where life could have arisen. It may have originated in a *cold* little pond—one that was under a glacier—but then again, it may have gotten started in just the opposite environment: a hot-water environment, as we'll see. A third possibility is that something like Darwin's warm little pond did exist, in the form of tide pools at the edge of an ancient ocean. As different as all of these environments are, note that all of them are aqueous environments. Life got going in water.

A requirement of any of these environments is that they had to have in them the chemical raw materials that could be used in forming life's critical molecules. You saw in Chapter 3 that life's informational molecules (DNA and RNA) are composed of certain kinds of building blocks (nucleotides), while its proteins are composed of other kinds of building blocks (amino acids). The raw materials for such building blocks could have been the gases methane and ammonia. These substances certainly would have been spewed out (or "outgassed") by Earth's early volcanoes. The question is whether they would have been quickly broken down in the ultraviolet light, coming from the sun, that bathed the early Earth's atmosphere.

Figure 19.5
Space Travelers
Fragments of the Murchison meteorite, which landed on the township of Murchison in Australia in 1969. The meteorite contained water and organic molecules such as amino acids. Some scientists hypothesize that organic materials like these could have seeded life on Earth.

Some scientists believe that building blocks such as amino acids could have arrived ready-made, delivered by the meteorites and comets that smashed into the young Earth. It's clear, from meteorites that have landed on Earth in recent decades, that such objects *can* carry organic molecules (**see Figure 19.5**). It appears unlikely, however, that such accidental "seeding" from outer space could have brought a sufficient *quantity* of organic materials to get life going.

Life May Have Begun in Very Hot Water

A third possibility for the supply of organic materials is that they came from the methane and hydrogen sulfide that gush out even today from deep-sea vents on the floors of the world's oceans. Interestingly enough, there are creatures living near these vents today. Some of these are microscopic archaea and bacteria that thrive at temperatures of 105° Celsius (247° Fahrenheit), which is above the temperature at which water boils (**see Figure 19.6**). To judge by some of the molecular sequencing that has produced life's family tree, the very oldest organisms on Earth may be "hyperthermophiles" such as these. Various lines of evidence, including the fact that these organisms live around the

Figure 19.6
Hot Habitat
Material pours forth from a hot-water vent on the floor of the Atlantic Ocean. The fluid being emitted is mineral-rich enough to support bacteria and archaea that in turn form the basis for a local food web.

Fossils and Molecular Clocks: Dating Life's Passages

The geological timescale in Figure 19.3 has a lot of "firsts" in it—first reptiles, first dinosaurs, and so forth. But how can scientists assign even a rough date to the first appearance of a life-form? Through radiometric dating, they can assign dates to *fossils* of life-forms (see Chapter 16, page 318). But that's not the same thing as assigning a date for the *appearance* of a life-form, meaning the date at which it first branched off from an existing group, in the way that reptiles did from amphibians. Imagine that an ancient fossil can be dated radiometrically. What are the chances that this creature was one of the *earliest* organisms of its kind? The answer is, slim to none. Because if an organism was one of the first of its kind to exist, by definition it was one of only a few that existed. And a small "starting set" of this sort is unlikely to leave any fossils. (Think of postage stamps. If only 10 copies of a stamp were originally made a century ago, the chances of one of them existing today are very small. If 10,000 were made, conversely, the chances of one existing go up considerably.) What fossils generally indicate is an *abundance* of a given species—an abundance that can only come about long after the species first appeared.

Beyond this, fossils themselves are a limiting factor in determining when different life-forms first appeared. Recall that one of the ways that relatedness among species is determined is by looking for similarities in their physical characteristics, as detailed in their fossils. But when it comes to Earth's early one-celled life-forms (such as bacteria), fossils can be few to begin with, and those that exist may not differ much in appearance. So, how can scientists begin to construct family trees for Earth's most ancient living things?

One answer, which scientists hit upon in the 1960s, is to use a different kind of dating method than the molecular clocks reviewed in Chapter 16 (see page 319). By comparing molecular sequence differences between living creature A and living creature B, scientists could begin to calculate how long it had been since they shared a common ancestor—how long it was since they branched off into separate species, in other words. How is this possible? Imagine knowing that two streams you are near result from the branching of one stream at some point off in the distance. Further, imagine that you know the *rate* at which these streams are diverging from each other—one stream going in this direction, the other going in that.

You may not be able to travel the whole distance to see the branching point itself, but because you know the rate at which the streams are parting company, you can predict the point at which they diverge. Molecular clocks give scientists a rate of divergence among groups of species, which can be used to infer their branching points.

Molecular dating work was aided tremendously beginning in the 1960s by a scientist named Carl Woese of the University of Illinois who realized that one particular molecule, ribosomal RNA (rRNA), exists in all living things in a standard form. This meant that rRNA sequences could be used as a common molecular clock to measure very ancient lineages. Today, it is not just rRNA, but DNA and proteins as well that are used as molecular clocks to judge when various species arose. The result has been nothing less than a revolution in constructing life's family trees. One of the most robust scientific enterprises of today is the use of molecular clocks to try to "infer phylogenies"—to develop family trees, complete with the dates of the branching points on them.

The problem with this is that the "molecular dates" that are derived often differ drastically from the "fossil dates." The oldest primate fossils we have are about 55 million years old. Meanwhile, molecular dating tells us the earliest primates arose about 90 Mya. You would *expect* the molecular date to be older than the fossil date, because the oldest primate fossil is not likely to be the oldest primate. But should the dates differ by 35 million years? Is it reasonable to expect that primates left no fossils at all for the first 35 million years of their existence? Closer to home, the earliest undisputed dog fossils we have date from 14,000 years ago, but molecular sequencing says dogs arose from wolves 135,000 years ago.

In these cases, as in many others, fossil and molecular scientists disagree about where the truth lies. The molecular scientists think the fossil scientists need to look for earlier fossils; meanwhile, the fossil scientists think the molecular scientists need to recalibrate their molecular clocks. Of course, no matter what method is used, we could never hope to develop precise dates for events that occurred so far in the past; estimates and uncertainty simply come with the territory in developing the history of life. Hopefully in the future, however, science will undergo its own convergence, giving us a clearer picture of how old the branches are on life's family tree.

right sorts of organic materials, leads to an idea that has a good deal of support among origin-of-life researchers. It is that life began in the "prebiotic soup" of hot-water systems—in the deep-sea vents or the kind of hot-spring pools that exist in Yellowstone Park (**see Figure 19.7**).

On the other hand, there are respected researchers who feel that, whatever the deep-sea vents may have had in the way of life-inducing qualities, their scalding temperatures would have obliterated any early self-replicating molecules, which were bound to be fragile. A much more likely first home for life, these researchers feel, are sheltered stretches of ancient ocean beaches, where the combination of tides and the composition of shoreline rocks would have "sorted" organic materials in accordance with their adhesiveness and other qualities. In this environment, the right initial combination of materials could have come together and then been joined by fresh materials, delivered continuously by the ocean waves.

Wherever life began, it cannot have been created in a single step from basic compounds such as methane and hydrogen sulfide. There is a huge gap between these materials and organisms as complex as the archaea. If organic molecules did not come from outer space, early Earth had to be a kind of natural chemical lab in which ever more complex organic molecules were produced from simpler ones. A major portion of origin-of-life research involves creating, in the laboratory, simulations of early Earth environments and then watching to see if life's building blocks will come together in them. Scientists now have scenarios that can account for the creation of most, but not all, of these compounds.

The RNA World

Any living thing must have at least one critical feature: the ability to reproduce itself. Today, a key player in reproduction is DNA. In Chapter 14, you saw that our "genome," or inventory of DNA, can be thought of as a kind of cookbook containing recipes (DNA sequences) that lead to products (proteins) that undertake a wide variety of tasks. Yet, as noted there, though DNA is the cookbook, it is not the cook. When it comes time for DNA to be copied, the double helix cannot unwind itself nor pair its bases up nor do anything else. Instead, all these tasks are carried out by the *proteins* known as enzymes. Hence there is a chicken-and-egg dilemma: Enzymes can't be produced without the information contained in DNA, but DNA can't copy itself without the activity of enzymes. So, how did life get going if neither thing can function without the other?

In the late 1960s, researchers hit on a solution that has come to have a good deal of support. In early life-forms, a *single* molecule performed both the DNA and enzyme roles. This molecule was RNA (ribonucleic acid), which was portrayed back in Chapter 14 as a kind of genetic middleman, ferrying DNA's information to the sites where proteins are put together. A critical piece of evidence about RNA's role in early life came in 1983, when researchers discovered the existence of enzymes that are composed of RNA instead of protein: **ribozymes**. These molecules can encode in-

formation *and* act as enzymes (by, for example, facilitating bonds that bind RNA units together).

This evidence leads researchers to the presumed existence of the "RNA world"—an early living world that consisted solely of self-replicating RNA molecules. But there are questions about this RNA world. One of them is how it came into being in the first place. RNA is so complex, and so fragile, that many researchers believe it had to have taken over the role of a simpler, sturdier precursor. There are several candidates for this role, though none has proven itself yet.

The Step at Which Life Begins

The critical step in life's development comes in imagining RNA or its precursor beginning to carry out this replication. *This* is the step at which life can be said to have begun. Given a molecule that can make copies of itself, a line of such molecules can come into existence. Given the *mistakes* that are bound to take place in such copying, molecules can be produced that differ slightly from one another in being, say, more resistant to the elements. Molecules that have such an advantage will leave more daughter molecules, which might just as well be called offspring. What happens, in short, is natural selection among self-replicating molecules. This selection, of course, leads to evolution among these molecules, and with evolution life begins to diversify into the innumerable forms we see today.

Web Tutorial 19.2
RNA-like Self Replication

Figure 19.7
Only Seemingly Inhospitable
The hot-water Grand Prismatic Pool in Yellowstone National Park, shown here in an aerial view, gets its blue color from several species of heat-tolerant cyanobacteria. The colors at the edge of the pool come from mineral deposits. The "road" at the top of the pool is a walkway for visitors.

Once replication got going, we were still a long way from life as we know it today. Even the most primitive contemporary living things are encased in the protective linings; and they can get rid of wastes, react to their environments, and so forth. Whatever the sequence of events, once life's elaborations are joined to replication, we move from simple molecules to the cellular ancestors of today's organisms.

19.3 The Tree of Life

Once life was established, how did it evolve? If you look at **Figure 19.8**, you can see the biggest picture of all, which scientists produced by doing the kind of ribosomal RNA sequences note in Dating Life's passages on page 376. At the bottom you see a "universal ancestor," which is the organism—or perhaps a co-operative group of organisms—that gave rise to all current life. The evolutionary line that leads from the universal ancestor then branches out to yield, on the one hand, the domain Bacteria and, on the other, a line that leads to both the domain Archaea and the domain Eukarya. Archaea were long thought of as simply a different kind of bacteria, but recent research has shown these organisms to be distinct from bacteria in fundamental ways. Nevertheless, there are great similarities. Bacteria and archaea are

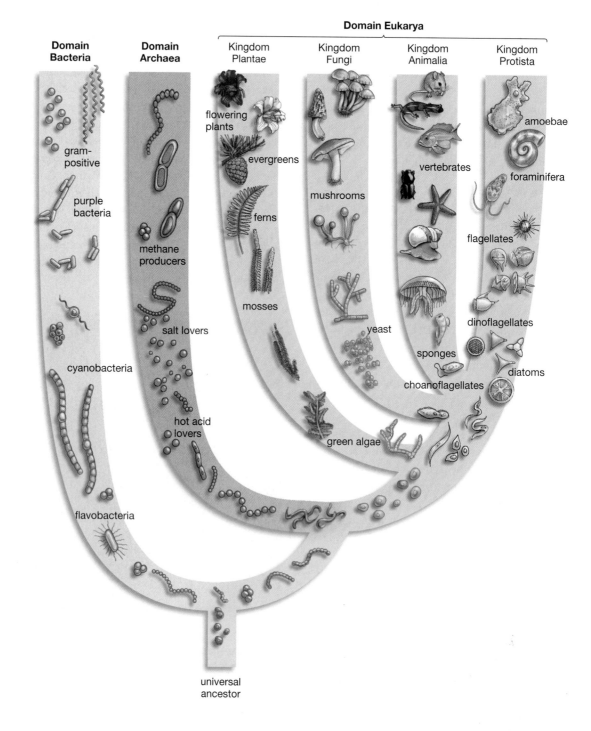

Figure 19.8
The Universal Tree of Life
The universal ancestor at the base of the tree gave rise to two domains of organisms whose cells lack nuclei—Bacteria and Archaea. The third domain of life is the Eukarya, whose earliest kingdom was the protists. From these protists all plants, animals, and fungi evolved.

essentially single-celled, and no archaeic or bacterial cell has a nucleus. Conversely, *all* the organisms in the domain Eukarya have nucleated cells, and are thus "eukaryotes."

As noted earlier, domain Eukarya has four "kingdoms" within it that you'll be learning about. Looking at the tree, you can see that the earliest kingdom in Eukarya is Protista—mostly one-celled, water-dwelling creatures—and it is from Protista that the other three eukaryotic kingdoms evolve. An unknown species within Protista gave rise to all the fungi that exist on Earth today; another grouping of protists called the choanoflagellates probably gave rise to all of the animal kingdom; and the protists we call green algae gave rise to all of today's plants. Note that, though it's common to think of plants and fungi as being alike, fungi and animals actually are more closely related than are fungi and plants.

19.4 A Long First Period: The Precambrian

With this picture in mind, let's start tracing life from its cellular beginnings. As noted, living things started small and stayed that way for a long, long time. Recall that the Earth was formed about 4,600 Mya. The earliest evidence we have of life is not a fossil but instead a kind of chemical signature of life left in the barren, frozen Isua rocks of Greenland (**see Figure 19.9**). Carbon trapped in these 3,700 to 3,800-million-year-old rocks is of a type produced only by living things. This evidence is controversial, but carbon signatures indicating photosynthesis are abundant beginning at 3,500 Mya, and microfossils found in Western Australia indicate that by 3,200 Mya, filament-type bacteria existed there. Most authorities believe that life existed by 3,800 Mya, or at the latest by 3,500 Mya. All of this life was of two types, bacteria or archaea, and all of it seems to have existed in the ancient oceans.

Once the bacteria and archaea got going, the long, slow Precambrian era had began. It was probably almost 2 billion years before any other life-form came into existence. When the first *multi*celled organisms appeared isn't clear. Certainly they exist by the time we have the first fossil evidence for animals, which is 600 Mya. But molecular clock studies put their evolution much further back, perhaps about 1,500 Mya. Even supposing this latter figure is right, note the big picture here. It took about 900 million years for life to appear after the Earth was formed; but then it took another 2.3 billion years to get from these single-celled life-forms to multicelled life-forms. Some authorities

Figure 19.9
Looking at the Oldest Rocks
The rocks at Isua, Greenland are believed to be the oldest in the world. In examining them, scientists have found possible chemical signatures of life that have been dated to nearly 3.8 billion years ago.

say this is evidence that the real hurdle in evolution was not the initiation of life, but rather the initiation of complex, multicellular life. What is certain is that evolution took its time in the Precambrian. The years rolled by in the millions, the land was barren, and the oceans were populated mostly by creatures too small to see with the naked eye.

Notable Precambrian Events

None of this is to say, however, that "nothing happened" in the Precambrian era. Those of you who have read Chapter 8 know how critical photosynthesis is for life on Earth. This capability first came about in the Precambrian era—and fairly early in the era at that.

Photosynthesis began in bacteria at least by the time the first bacterial fossils left their imprint, about 3,200 Mya. This event was a turning point. Had this capacity not been developed, evolution would have been severely limited, for the simple reason that there would have been so little to eat on Earth. The earliest organisms subsisted mostly on organic material in their surroundings, but the supply of this material was limited. In photosynthesis, the sun's energy is used to *produce* organic material—in our own era, the leaves and grasses and grains on which all animal life depends. Large organisms require more energy and, through photosynthesis, a massive quantity of energy-rich food was made available.

One particular kind of photosynthesis, again beginning in the Precambrian, had a second dramatic impact on evolution. One very early type of bacteria, the *cyanobacteria*, were the first organisms to produce *oxygen* as a by-product of photosynthesis. For us the word oxygen seems nearly synonymous with

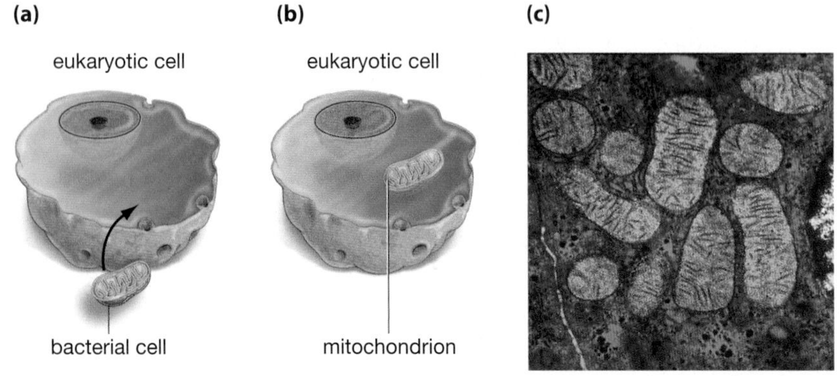

Figure 19.10
Adapting to Oxygen
The "powerhouses" of our cells—the organelles called mitochondria—have several characteristics of free-standing bacteria.

(a) There is general agreement among scientists that our mitochondria originally were bacteria that invaded eukaryotic cells many millions of years ago. These bacteria benefited from the eukaryote's cellular machinery, and the cell benefited from the bacteria's ability to metabolize oxygen.

(b) Over time, the bacteria came to be integrated into the host cells, replicating along with them. Two symbiotic organisms had now become a single organism.

(c) Today, the mitochondria in your cells turn the energy from food into a form that allows you to run, think, and turn the pages of this book. These artificially colored mitochondria functioned in a liver cell of a mouse.

the word life, but until about 2,300 Mya there was almost no oxygen in the atmosphere. When it finally did arrive, through the work of the cyanobacteria, its effect was anything but life-giving. For most organisms it was a deadly gas, producing what has been termed an "oxygen holocaust" and establishing a firm rule for life on Earth: Adapt to oxygen, stay away from it, or die.

In the creatures that adapted to oxygen, we find one of the most interesting stories in evolution. Those of you who went through Chapter 7, on cellular energy harvesting, will recall that most of the energy that eukaryotes get from food is extracted in special structures within their cells, called *mitochondria*. Though mitochondria are tiny organelles within eukaryotic cells, they have characteristics of free-living *cells*, specifically free-living bacterial cells. As it turns out, that is probably what they once were. Ancient bacteria that could metabolize oxygen took up residence in early eukaryotic cells and eventually struck up a mutually beneficial relationship with them (**see Figure 19.10**). The bacteria benefited from the eukaryotic cellular machinery, and the eukaryotes got to survive in an oxygen-rich world. In a second bacterial invasion, algae and their plant descendants got not only mitochondria, but the chloroplasts in which photosynthesis is carried out (see "Endosymbiosis" in Chapter 4, page 84).

The oxygen revolution then had one more momentous consequence. Molecules of oxygen came together to form the gas called *ozone*, which rose through the atmosphere to form the ozone layer, giving Earth, for the first time ever, protection against the ultraviolet radiation that comes from the sun. Marine creatures had been shielded from this radiation by the ocean itself, but prior to the formation of the ozone layer, there was no chance for life to develop on *land*.

19.5 The Cambrian Explosion

As the Precambrian was coming to a close, life consisted of bacteria, archaea, and many kinds of protists, most of which lived in the ocean. By about 600 Mya, however, we get the first fossil evidence of *animals* in the seas. The term "animal" here fits our common sense definition: They all are multicelled organisms, and they get their nutrition from other organisms or organic material.

There currently is a debate about whether the earliest animal fossils represent a line of creatures that died out, or are instead related to today's animals. Whatever the case, these early animals can be seen as a kind of early stirring; a first blip on the screen bringing news of something big to come. To judge by the fossil record, about 60 million years after the emergence of these animals, we get it. A tidal wave of new animal forms, the like of which has never been seen before or since in evolution. This is the **Cambrian Explosion**, which began about 544 Mya.

Recalling the taxonomic system you went over in Chapter 18, you may remember that just below the "domain" and "kingdom" categories there is the **phylum**, which can be defined informally as a group of organisms that share the same body plan. There are at least 36 phyla in the animal kingdom today, some of these being Porifera (sponges), Arthropoda (insects, crabs), and Chordata (ourselves, salamanders). With one exception, the fossil record indicates, every single one of these came into being in the Cambrian Explosion. This seeming riot of sea-floor evolution was actually more extensive than even this implies, in that a good number of phyla that *don't* exist today—phyla that have become extinct—also seem to have first appeared in the Cambrian. Some of these creatures are so bizarre, it's surprising that Hollywood hasn't mined the Cambrian archives for new ideas on monsters (**see Figure 19.11**). As if all this weren't enough, the Cambrian Explosion appears to have taken place in a very short period of time relative to evolution's normal pace—perhaps as little as 5 million years.

But did it? The Cambrian Explosion is so extreme that one of the primary challenges to evolutionary biology has been to explain why such a thing came about—or why it came about only once. Note the series of events: only a few animal forms before the Cambrian, then the explosion, then almost no new animal forms since. In the 1990s, several lines of evidence suggested something else: that animal forms actually began to diverge well back in the *Precambrian*. Under this view, all the Cambrian Explosion amounted to was an explosion of forms big and hard enough to leave *fossils*. The forms themselves appeared earlier and more gradually, this idea goes, but were too small and fragile to leave imprints behind. And indeed, molecular clock studies suggest that animals appeared as far back as 1.5 billion years ago.

(a)

eukaryotic cell

bacterial cell

(b)

eukaryotic cell

mitochondrion

(c)

If the Cambrian Explosion did in fact occur, what could have sparked it? Lots of ideas have been proposed; one of the best-received is that the explosion was triggered by the rise in atmospheric oxygen. To get bigger creatures, you need more oxygen, and levels of atmospheric oxygen may have reached a critical threshold about this time. Once larger forms *could* appear, adaptive radiation took place on a grand scale. So many niches were available that there was an explosion of forms. Once all the basic niches were taken, however, this frenzy of new forms not only came to a stop, it was pruned back (in the sense that some Cambrian phyla became extinct).

19.6 The Movement onto the Land: Plants First

The teeming seas of the early Cambrian period stand in sharp contrast to what existed on land at the time, which was no life at all except for some hardy bacteria. Earth was simply barren—no greenery, no birds, no insects. When multicelled life did come to the land, the first intrepid travelers were plant-fungi combinations. (Even today, most plants have a symbiotic relationship with fungi, the plants providing fungi with food through photosynthesis, the fungi providing plants with water and mineral absorption through their filament extensions.) Exactly when this transition onto land took place is a matter of debate. The earliest undisputed fossil evidence we have of plants and fungi on land dates from 460 Mya, meaning these organisms had become abundant by that time. But when did they first appear? New molecular work estimates that land (or "terrestrial") fungi have been around for 1.3 billion years. These fungi couldn't have existed without plants—since they need them to live—which would mean that plants were on land at this time too. The general transition here began with marine algae in the ocean, then continued with freshwater algae, then freshwater algae that came to exist in *shallow* water, living partially above the waterline. When the transition to full-time living on land came, it was to damp environments.

Adaptations of Plants to the Land

Such a change required a lot of adaptation. Aquatic algae did not have to deal with water loss or the crushing effects of gravity, but land plants did. One of the plants' responses to the water problem was to evolve a waxy outer covering, called a *cuticle*, that could retain moisture. Meanwhile, an initial response to gravity was to stay low. Some of the most primitive land plants are mosses that often hug the ground like so much green carpet (**see Figure 19.12**).

Figure 19.11
Life on the Ancient Sea Floor
The Cambrian Explosion resulted in a multitude of new animal forms. In this artist's interpretation of the fossil evidence, a number of now-extinct Cambrian animals are shown in their ancient sea-floor habitat. Two of these are Anomalocaris, raising up with the hooked claws, and Hallucigenia, on the sea floor with the spikes extending from it.

Then there was the problem of reproduction. When green algae reproduce sexually, their gametes (eggs and sperm) float off from them as individual cells, after which they are brought together by ocean currents or the cells' own movement. Once this happens, the resulting embryos develop completely on their own. An embryo left to mature by itself on land, however, would dry out and perish. Plants adapted to the land by developing a protective housing for embryos; in a plant, embryos mature *within* a parent.

Figure 19.12
Lying Low
Moss covering rocks along a stream in Tennessee's Great Smoky Mountains National Park. Mosses are members of the earliest-evolving group of plants, the bryophytes.

Another Plant Innovation: A Vascular System

If you look at **Figure 19.13**, you can see how the major divisions of plants evolved. The most primitive land plants, the **bryophytes** (represented by today's mosses), had no vascular structure—meaning a system of tubes that transports water and nutrients. Plants without such a system have very limited structural possibilities. They can grow out, but they lack ability to grow *up* very far, against gravity. By contrast, the ancestors of today's ferns developed a vascular system; with this, over the ensuing 100 million years, a variety of **seedless vascular plants** evolved, including huge seedless trees such as club mosses, some of them 40 meters tall, which is about 130 feet.

Plants with Seeds: The Gymnosperms and Angiosperms

Even as the seedless vascular plants were reaching their apex, a revolution in plants was well under way as some of them developed *seeds*. With the first seed plants, offspring no longer had to develop within a delicate plant on the forest floor. Instead they developed within a seed, which can be thought of as a reproductive package that includes not only an embryo (brought about when sperm fertilized egg), but food for this embryo, and a tough outer coat.

The living descendants of the first seed plants are **gymnosperms**—today's pine and fir trees, for example—whose seeds are visible in the well-known pinecone. Gymnosperms also represent the final liberation of plants from their ocean past. Mosses and ferns had sperm that could make the journey to eggs only through water. Not much water was needed; a thin layer would do, as with the last of a morning's dew on a fern leaf. With seed plants, however, the water requirement is gone entirely. Sperm are encased inside pollen grains, which can be carried by the *wind*. Think how far a windblown pollen grain could disperse compared to a sperm moving across a watery leaf. Several types of gymnosperms developed beginning about 350 Mya. Today, gymnosperm trees cover huge stretches of the Northern Hemisphere.

The Last Plant Revolution So Far: The Angiosperms

At the time the dinosaurs reigned supreme among land animals, the first flowering on Earth occurs, with the development of flowering plants, also known as **angiosperms**. Evolving about 165 Mya, the angiosperms eventually succeed the gymnosperms as the most dominant plants on Earth. Today there are about 700 gymnosperm species, but some 260,000 angiosperm species, with more being identified all the time. Angiosperms are not just more numerous than gymnosperms, they are vastly more diverse as well. They include not only magnolias and roses, but oak trees and cactus, wheat and rice, lima beans and sunflowers. If you look at **Figure 19.14**, you can see the fossilized impression of what may be an early flowering plant, dating from 125 Mya, along with an artist's conception of what it looked like.

19.7 Animals Follow Plants onto the Land

The movement of plants onto the land made it possible for animals to follow. With plants came food and shelter from the sun's rays. Recalling the division of the animal kingdom into various phyla, it was the phylum of arthropods that first moved to

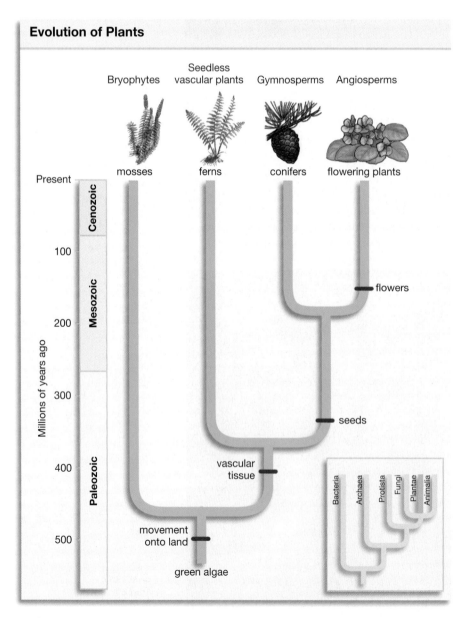

Figure 19.13
How Plants Evolved
The tree of life for bryophytes, seedless vascular plants, gymnosperms, and angiosperms.

land. About 440 Mya—20 million years after the first plant fossils were laid down—a creature similar to a modern centipede laid down the oldest set of terrestrial animal tracks we know of. It makes sense that arthropods were the first animals to come onto land, because the hallmark of all of them is a tough external skeleton (an "exoskeleton"), which can prevent water loss and guard against the sun's rays.

Other kinds of arthropods, the insects, came on land some 50 million years after the first arthropod immigrants. The insect pioneers were wingless, perhaps being something like modern silverfish. Insect flight would not develop for tens of millions of years more, but when it did insects had the skies to themselves for 100 million years. The more primitive flying insects are represented today by dragonflies, whose wings cannot be folded back on their body (**see Figure 19.15**). The compactness that came with foldable wings gave later insects the advantage of being able to get into small spaces. (Think of a common housefly.) In both winged and wingless forms, the insects took to the land with a vengeance. About 1.4 million species of living things have been identified on Earth to date, of which *half* are insects. For millions of years, the insects and their fellow arthropods were the only animals on land—but this would change.

Vertebrates Move onto Land

You'll recall that back at the time of the Cambrian Explosion, almost all animals existed on the ancient ocean floor. Within just a few million years, however, animals with backbones, known as vertebrates, were moving throughout the ocean; several of them were primitive, now extinct, orders of fishes. About 450 Mya, we get the rise of the family of fishes, the gnathostomes, that are the ancestors of nearly all the fish species still alive today, from sharks to gars to goldfish. Critically, gnathostomes are the first creatures to have jaws. With the development of jaws, these vertebrates and their descendants (one of whom is us) could securely grasp all kinds of things; big chunks of food, to be sure, but also offspring and inanimate objects. The new foods that became available to those with jaws allowed the gnathostomes to outgrow their jawless vertebrate relatives.

Primitive Fish First

It was a particular, later-arriving type of gnathostome, the "lobe-finned" fishes, that represent the transitional organisms between sea life and land life. The lobed *fins* these creatures possessed are the precursors to the

Figure 19.14
A Possible Early Flowering Plant
In 2002, a team of Chinese and American scientists announced the discovery of *Archaefructus sinensis*, whose fossilized, 125-million-year-old impression can be seen at left. The plant clearly had angiosperm-like traits—for example, the male reproductive structures known as anthers. Yet it had no petals. One hypothesis is that it represents a transitional form in the evolution of angiosperms from lake-dwelling herbaceous plants. If so, it stands to overturn the long-held hypothesis that angiosperms evolved from woody gymnosperms. On the right is an artist's conception of what *A. sinensis* looked like.

Figure 19.15
The First to Take Flight
Dragonflies, such as this dew-covered individual, are descendants of the insects that first developed the ability to fly. Flying insects that evolved later could fold their wings back, as with today's housefly.

Figure 19.19
Primate Characteristics
These modern-day lemurs in Madagascar exhibit several characteristics common to most primates: Opposable digits that enable grasping, front-facing eyes that allow binocular vision, and a tree-dwelling existence.

brought the mammals out of hiding. They radiated into niches far and wide, becoming the dominant form of large land animal, a status they retain to this day. Some of their members even returned to the sea, as with seals and whales. (Who is the whale's closest living relative? It turns out to be the hippopotamus.) An obvious question here is: Why did the mammals survive the Cretaceous Extinction while the dinosaurs died out? Alas, we have no clear answers. Pure size may have had something to do with it; small creatures seem to survive extinction events better than large ones.

The Primate Mammals

A species of placental mammals gave rise to the order of mammals called *primates*, though exactly when this happened is not clear. The oldest primate fossils we have date from 55 Mya, but it will not surprise you to learn that molecular work places their origins much further back—at about 90 Mya. If you look at **Figure 19.19**, you can see, in the lemur of Madagascar, a modern-day descendant of these earliest primates. Three things characteristic of most primates are apparent in this picture: large, front-facing eyes that allow for binocular vision (which enhances depth perception); limbs that have an opposable first digit, like our thumb (which makes grasping possible); and a tree-dwelling existence. There are about 230 species of primates living today (**see Figure 19.20**), which is a small portion of the 4,600 species of mammals. It's a tiny number of species indeed compared to, say, the 60,000 species of molluscs, or the 750,000 species of insects. **Figure 19.21** sets forth a possible primate family tree.

19.8 The Evolution of Human Beings

If you look to the upper right in Figure 19.21, you can see a fork that leads to chimpanzees on the one hand and human beings on the other. Down at the branching point of this fork is a presumed "com-

(a) A prosimian

(b) A New World monkey

(c) A great ape

Figure 19.20
Several Types of Primates

(a) A loris from India (*Loris tardigradus*) is representative of an early primate lineage, the prosimians, that includes llemurs.

(b) A woolly spider monkey (*Brachyteles arachnoides*) from the Atlantic rain forest in Brazil. Note the tail, which is capable of grasping objects. Only New World monkeys have such tails.

(c) A lowland silverback gorilla (*Gorilla gorilla*), native to Africa.

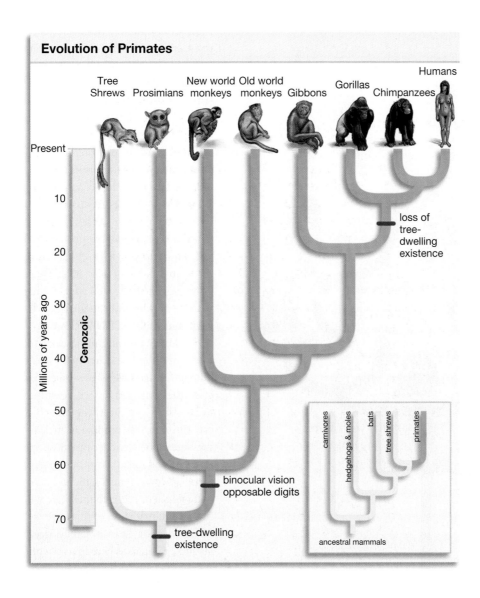

Evolution of Primates

Tree Shrews — Prosimians — New world monkeys — Old world monkeys — Gibbons — Gorillas — Chimpanzees — Humans

Present

Millions of years ago

Cenozoic

10
20
30
40
50
60
70

loss of tree-dwelling existence

binocular vision opposable digits

tree-dwelling existence

carnivores — hedgehogs & moles — bats — tree shrews — primates

ancestral mammals

**Figure 19.21
How Primates May Have
Evolved**

mon primate ancestor"—a single species that gave rise to both the chimpanzee family tree and the human family tree. By working backward through a set of primate molecular clocks, geneticists have determined that this branching probably occurred between 6 and 7 million years ago.

This split, of course, raises the question: Who was the common primate ancestor? We know that it was not the equivalent of a modern-day chimpanzee; today's chimpanzees did not arise until about the same time modern humans did. All we can say is that it was an ape-like creature whose closest living relatives include not only ourselves and chimpanzees, but gorillas and the chimp-like primates known as bonobos.

Figure 19.21 might leave the impression that it only took a couple of speciation events to go straight from the primate ancestor to modern human beings and chimpanzees. But the line extending up to hu-

mans is something like a satellite image of a river: it shows only the river's general outline, instead of its turns and branches. For a kind of close-up of the line leading to human beings, look at **Figure 19.22** on the next page. In it, you can see one possible way in which the human lineage might have evolved, from the common primate ancestor to modern *Homo sapiens*.

It's important to note at the outset that the human family tree in Figure 19.22 represents just one hypothesis, among many that have been proposed, for how human beings evolved. The problem in constructing such a tree is not with the dates you see that are lined up with the species; most of them are solid. And the problem is not with the individual species; we have fossils of all of them. The problem comes with *connecting* the species: who arose from whom? There is scarcely a branching point or line of connection in Figure 19.22 that all authorities agree to, and even many of the species and genus names in it are

Evolution of Human Beings

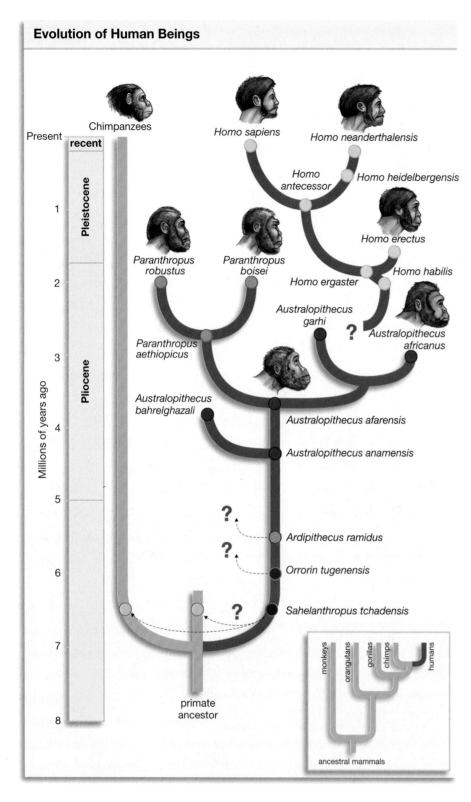

Figure 19.22
A Possible Family Tree for Human Beings

eage. Without family trees, human evolution can become an incomprehensible set of Latin names.

Debates about where an animal species belongs in a family tree are normally carried out in scientific papers and conferences, away from the public spotlight. When it comes to the human lineage, however, the public *cares* about who goes where, for the same reason that any of us would care about whether our family ancestors came from Ireland or Peru. Hence, major fossil finds are the subjects of major newspaper stories. And the scientists who make these finds are pressed for details about the meaning of their discoveries, as if they were private detectives who had just shown up with a load of family secrets.

Who are these researchers? There are archaeologists and geneticists who study human origins, along with biological anthropologists, who sequence ancient and modern DNA. However, the central evidence in human-origins research has always been fossils, and the researchers who seek out and analyze human-lineage fossils are called paleoanthropologists. These scientists go wherever the fossils are likely to be, which can mean the French countryside, but can also mean the middle of Africa's Sahel desert or the bottom of a sweltering pit in Spain. There, they can be found turning over handfuls of earth in an effort to uncover bone fragments and rock impressions that will tell us more about where we came from.

The ultimate goal of this work is the construction of an accurate human family tree, with the word "family" being used here in its scientific sense. All the species in Figure 19.22 are members of the taxonomic family *Hominidae*, which is to say the family of man-like primates. Every member of the *Hominidae* family is referred to as a **hominid**, including human beings.

What are some of the notable features of this tree as it's been constructed so far? Well, if you look at the root of Figure 19.22, you'll notice it that starts with one species, the primate ancestor. Up at the top is our species, *Homo sapiens*. In between these two points, there is a kind of bush-like branching of the various genera and species. Now here's the interesting thing: Every species in this tree is extinct except for our own. Numerous species arose in the human family tree over the past 6 or 7 million years, but all of them are gone except us. If you could take a time machine back 50,000 years, you would find, in Africa and elsewhere, fully modern human beings, which is to say hominids who physically are indistinguishable from us. But in Europe you would also find stocky Neanderthals (species *Homo neanderthalensis*), and over in the Far East you would find the dwindling remnants of a once far-flung species, *Homo erectus*. Jump forward another 10,000 years and all the *H. erectus* are

matters of dispute. (Is the *Paranthropus* genus at center left really its own genus, or should the species in it be included in the nearby *Australopithecus* genus?) Given these uncertainties, you might ask, why produce such trees at all? Scientists have use for them because they provide a starting point for analyzing the human lineage. We have use for them because they provide a starting point for *understanding* this lin-

gone; jump forward another 13,000 and all the Neanderthals are gone. Today, only we *Homo sapiens* remain among the hominids.

Another way to think of this tree is in terms of relative time frames. An "ancient" civilization, such as that of Egypt, stretches back perhaps 5,000 years. Yet the Neanderthals disappeared from the face of the Earth 22,000 years before Egyptian civilization began, and they had been in existence for 200,000 years. Even so, the Neanderthals were late-comers among the hominids. Recall that the *Hominidae* line stretches back as far as 7 million years. Thus, human evolution took a great deal of time, even when dated just from the last primate ancestor. On the other hand, when we consider the entire sweep of evolution, all hominids are late-comers. If we allow that life on Earth began 3.8 billion years ago, and think of all of evolution as taking place in one 24-hour day, then the most ancient of the hominids evolved in the last $2\frac{1}{2}$ minutes of that day. And modern human beings? We arose at most 200,000 years ago, which means we have only been around for the last 5 seconds of Earth's evolutionary day.

Yet another way to think of human evolution is in terms of *where* it took place. The message here is quite straightforward: For perhaps 70 percent of the time we hominids have been around, we lived and died solely in Africa. This is clear because the earliest hominid fossils found *outside* of Africa date to only 1.75 Mya. Meanwhile, all the early and mid-period hominid fossils have been unearthed on the African continent, as you can see in **Figure 19.23**.

As little as 10 years, ago, a finer point could have been put on this: It was not just in Africa, but in *East* Africa that most hominid evolution took place. Until the mid-1990s, all early- and middle-period hominid fossils were found in a swath of Africa that runs from the eastern portion of South Africa up through the Great Rift Valley that traverses modern Kenya, Tanzania, and Ethiopia. Then, in 1995, a French researcher named Michel Brunet announced that he had turned up hominid fossils in Chad, some 2,500 kilometers west of the Great Rift Valley. In 2002, Brunet and his colleagues stunned the research world when they announced that they had found a very primitive hominid in Chad, a creature they dubbed *Sahelanthropus tchadensis* but who they informally refer to as Toumaï. If Toumaï's status as a hominid is confirmed, it pushes the geographical range of early human ancestors across much of the African continent.

The Toumaï fossils actually were remarkable in another way. They capped a period of two years in which the root of the hominid family tree was pushed back by more than 2 million years. Prior to

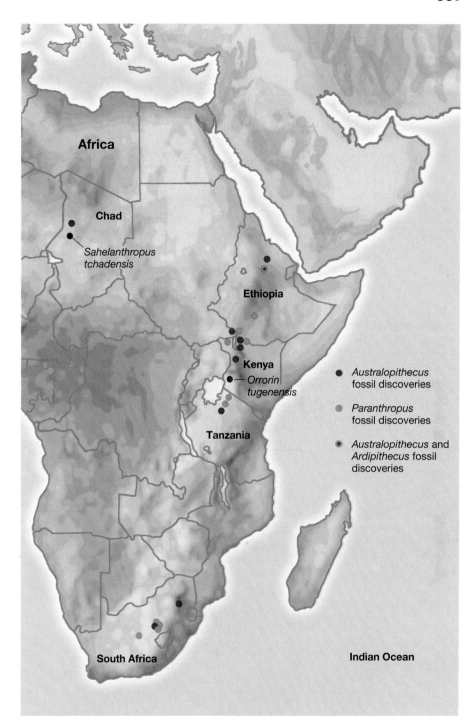

Figure 19.23
Hominid Fossil Sites in Africa
All of the oldest hominid fossils have been found on the African continent.

2000, the oldest putative hominid fossil was dated at 4.4 million years old. Then, in rapid succession, all three of the species you can see at the base of the tree in Figure 19.22 were announced by different teams of researchers. With the Toumaï find, hominid fossils were pushed back to between 6 and 7 Mya. As you may recall, molecular anthropologists have calculated that the human and chimpanzee lines only split at about 7 Mya at most. One possibility, then, is that Brunet's

Toumaï is a creature that lies right at the base of the hominid family tree. Another possibility, however, is that Toumaï is not a hominid, but belongs in the line of primate ancestors (which is why it's marked with dotted lines and a question mark in Figure 19.22). How do scientists determine placements like these? Let's see.

Interpreting the Evidence

If you look at **Figure 19.24**, you can see part of the *S. tchadensis*/Toumaï fossils that Brunet and his team unearthed in the Chadian desert. This remarkably complete cranium of Toumaï was complemented by additional artifacts the researchers found: two lower jaw segments and some individual teeth. Fragmentary as this evidence may seem, it actually is more extensive than some fossil finds. The claim for the primitive *Orrorin tugenensis* species seen in Figure 19.22 originally was based mostly on three thigh bones and several teeth.

So, what are paleoanthropologists looking for in such fossils? What features help them decide whether a given species should be placed in the hominid family tree or be put over with the apes? To Brunet, three features stood as evidence that Toumaï was a hominid. One was its canine teeth; they are smaller and less sharp than those of the apes. Another was the enamel on its teeth—it is thicker than the enamel apes have. A third was what might be called the slope of the face; the lower face doesn't jut out beyond the

eyes in the way ape faces do. Against this, Toumaï clearly has some ape-like features. Note its widely spaced eyes and massive brow, for example.

Brunet's views, however, are *interpretations* of evidence, and interpretations can vary. He may see, in Toumaï's teeth, the canines of a male hominid, but a French colleague of his, Brigitte Senut, sees in these teeth the canines of a female *ape*. It may not surprise you to learn that Senut has her own "earliest hominid" candidate—the *Orrorin tugenensis* specimen, which she and her colleagues unearthed in Kenya.

Tracing Human Evolution over Time

Bearing such considerations in mind, we can begin to trace the hominid lineage that leads to our own species. When we ask whether the ancient *S. tchadensis* belongs at the base of this lineage, there is a critical piece of evidence that is missing. Brunet needs to find some pelvic or leg fossils. Along with canine teeth structure, the most important criterion for placing a creature in *Hominidae* is evidence of *bipedalism*, meaning the tendency to walk on two limbs rather than four. But the evidence for *S. tchandensis* thus far only includes fossils from the head, so who knows how it walked?

Bipedalism is critical to hominid evidence because this trait eventually became characteristic of all hominids. By contrast, our closest non-hominid relatives—the chimpanzees and gorillas—are "knuckle walkers" that can rise up on two legs, but normally get around on four. We can trace our split with the apes by tracing the split between four-legged and two-legged walking. Senut and her colleagues claim to have good evidence that the six-million-year-old *O. tugenensis* walked upright, and we are fairly certain that the later *Australopithecus anamensis*—near the middle of Figure 19.22—was bipedal. By the time we get to the *Australopithecus afarensis* species, there is no doubt that the human line was bipedal.

The name of this last species, *Australopithecus afarensis*, is not likely to ring a bell but one particular *A. afarensis* is perhaps the most famous hominid ever found. She is "Lucy," who lived 3.18 million years ago in what is now Ethiopia. In her, we actually can see a transition to bipedalism. Lucy had the pelvis of a biped, but nevertheless possessed long arms, short legs, and feet that were built for grasping—features consistent with a species that came down from life in the trees, but that probably still spent considerable time in them. To get some idea of Lucy's stature compared with ours, see **Figure 19.25**.

The Australopithecines are generally agreed to be ancestral to our own genus, *Homo*; but which Australopithecines stand in this line of descent? If you

Figure 19.24
An Ancient Human Ancestor?
This cranium of *Sahelanthropus tchadensis*, or Toumaï, was discovered in the deserts of Chad, and has been determined to be 6 or 7 million years old. Toumaï's discoverers believe it is a very ancient hominid, but other researchers believe it should be grouped with apes.

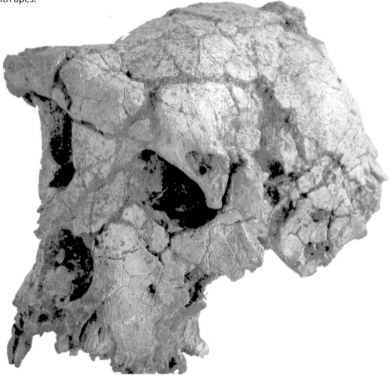

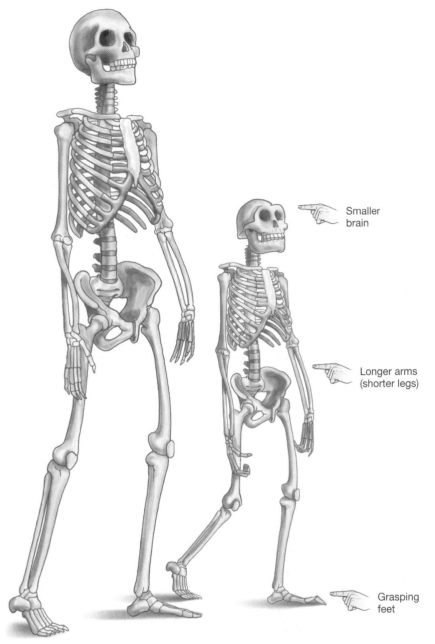

Smaller
brain

Longer arms
(shorter legs)

Grasping
feet

Figure 19.25
"Lucy" and Modern Humans
Compared
What was the hominid Lucy like compared to a modern human female? The figure gives an idea. Lucy stood about three-foot-seven and had a much smaller brain than modern humans, even allowing for her smaller stature. Her brain size or "cranial volume" was about that of a modern chimpanzee—450 cubic centimeters, compared to 1,400 cubic centimeters for a modern *H. sapiens*. Lucy's hip and pelvic bones make clear that she was bipedal, but note the longer arms and grasping feet common to tree-dwelling primates. Lucy's combination of features have prompted some researchers to label her a "bipedal chimpanzee." Her species, *Australopithecus afarensis*, is important because our own species may be descended from it. (Adapted from Boyd, R. and Silk, J.B., *How Humans Evolved*. New York: W.W. Norton & Company, 2000, p. 329.)

look at Figure 19.22, you can see two candidate species, *A. africanus* and the more recently discovered 2.5-million-year-old *A. garhi* from Ethiopia. *A. garhi* has the right mix of physical characteristics and may also have been the first hominid to use tools to butcher animals. Stone tools—actually little more than rocks with sharpened edges—were found with the *A. garhi* fossils, along with fossil remains of animals that had cut marks consistent with butchering and extraction of bone marrow. Before this discovery, it was thought that the first toolmaker was *Homo habilis*.

Into the Genus *Homo: Habilis* and *Ergaster*

Homo habilis—"handy man" in honor of its tool-making abilities—gets us to our own genus, *Homo*. But there is significant disagreement about whether *H. habilis* was *enough* like us that it deserves to be placed in our genus. The main reason that *H. habilis* gets the *Homo* designation is its use of tools; in physical form it is arguably an Australopithecine.

Changes to physical forms that are more like ours come at several steps in the hominid line, but the most dramatic change by far comes with the rise of *Homo ergaster*, whose best-preserved remains come from a boy who died 1.6 million years ago on the shores of Lake Turkana in Kenya. Experts estimate that, had this 9-year-old "Turkana Boy" grown to maturity, he would have reached a height of 2 meters (6 feet). His brain was 30 percent larger than that of *Homo habilis* and more than half the size of the average modern *Homo sapiens* brain. He had a much more modern face, long limbs typical of those humans who dwell today in arid climates in Africa, and advanced tool technology. If you look at **Figure 19.26**, you can see Turkana Boy's skeleton, which was found in amazingly complete form. Indeed, the only features missing from it were hands and feet. But what did Turkana Boy's species actually look like? **Figure 19.27** provides an artist's interpretation of it and several others.

Migration from Africa

All the hominid species we've reviewed so far lived solely in Africa. Then, about 2 Mya, the initial human migration out of Africa took place. Who were the first intercontinental travelers? As usual, we can pinpoint some individuals; what is more difficult is connecting the dots. There is no doubt that fossils from the same tall, modern-looking, brainy species, *Homo erectus*, have been found very far afield of Africa, on the Pacific island of Java and the Chinese mainland. There's also no doubt that these specimens are extremely old; some of the Javanese fossils seem to date from about 1.6 Mya. On the other hand, the *very* oldest non-African fossils we have are from a hominid found near the Black Sea who dates back to 1.75 Mya. We might expect this hominid to look like the *H. erectus* found in Java, but, alas for clarity's sake, its features sit somewhere between those of *H. erectus* and the much more primitive *Homo ha-*

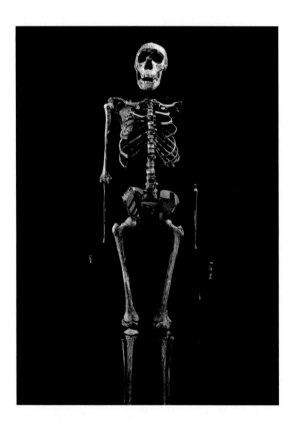

Figure 19.26
Much More Like Us
In the remarkably complete skeleton of "Turkana Boy," we can see the evolution of hominids to a form more like our own. This member of the species *Homo ergaster* was tall and had a much larger brain capacity than the earlier hominid "Lucy." His remains, dating from 1.6 million years ago, were found near Kenya's Lake Turkana.

bilis. To add to this confusion, a skull dating from 1 Mya was recently found in Ethiopia that has characteristics of both the Far East *H. erectus* and the earlier, purely African *H. ergaster*.

If you look at Figure 19.22 again, you can see what seems to be (just barely) the conventional wisdom about all this: That *Homo ergaster* gave rise not only to *Homo erectus*, but to our own line of descent. *Homo erectus* then went on to see the world, as it were, but was an evolutionary dead end; the last of its line died out about 40,000 years ago in Java. Note, however, that this species had a long run—almost 2 million years, which is a period of existence at least 10 times as long as the one our species has had so far. Also, keep your eye on the *H. erectus* extinction date of 40,000 years ago; it's interestingly close to another extinction date we'll be looking at later.

Into Europe

H. erectus was in the Far East more than 1.6 million years ago, but we have to come all the way up to 800,000 years ago to find evidence of hominids in Europe. The first European hominid we know of was a member of the species called *Homo antecessor*, whose remains were found at the Gran Dolina excavation site in northern Spain in 1994. Coming forward another 100,000 years, we find a second European hominid, *H. heidelbergensis*, first found at a town near Heidelberg, Germany. Then there is the best known of all the ancient hominids, the Neanderthals (*H. neanderthalensis*), whose earliest remains date from 200,000 years ago, and whose range was extremely broad—not only most of Europe, but parts of the Middle East and Asia as well. Coming very late to Europe was our own species, *Homo sapiens*, which arrived about 40,000 years ago.

Connecting the dots here is difficult. If you look at Figure 19.22, you can see the story as the discoverers of the Spanish *H. antecessor* fossil believe it unfolded. While still in Africa, *H. antecessor* gave rise to both *H. heidelbergensis* and to *Homo sapiens*. Then *H. heidelbergensis* went on to give rise to the Neanderthals. One piece of evidence that makes plausible the idea that *H. antecessor* gave rise to modern humans is that *H. antecessor* was in some ways strikingly like modern humans, particularly for a creature that was so ancient.

The Neanderthals

Of all the lost members of the hominid family, it is the Neanderthals that have most strongly captured the modern human imagination. This is so partly because the Neanderthals were the first extinct hominids to be discovered. When Neanderthal remains were uncovered in 1856, at a limestone quarry in Germany's Neander Valley, humanity got a wake-up call about human evolution. Though the notion was resisted at first, the Neanderthal bones made clear that there once had been a species on Earth that was very much like humans, but that was not *exactly* human. In the years that followed, the Neanderthals became the best known of the extinct hominids, simply because they left so many traces of themselves behind. In Europe alone, there are perhaps 20 major Neanderthal excavation sites, and others have been found as far away as Uzbekistan.

So, what were these near humans like? **Figure 19.27** shows you an artist's interpretation of their physical stature: short, stocky bodies—the average man was about five-foot-six—powerfully built, with a heavy "double arched" brow and a receding chin; big-boned, with large joints to match. It's easy to see how such features were turned into a *caricature* that has been with us since the nineteenth century: that of the caveman. It is the Neanderthals that provided us with the image of dumb, prehuman brutes, grunting their way through the Stone Age. But does this image fit with the reality? Well, consider that the Neanderthals had a cranial capacity as large as ours; that some put up shelters in their campsites; and that one reason we have so many artifacts from them is that they took the trouble to bury their dead.

This is not to say, however, that they were "just like us." Whenever direct comparisons are made, the Neanderthals do indeed look primitive in comparison with ancient *Homo sapiens*. As paleoanthropologist Ian Tattersall has pointed out, Neanderthals thrust spears at prey, but humans developed the valu-

able technique of *throwing* spears from a distance. The Neanderthals used nothing but stone for their relatively primitive tools, while *Homo sapiens* used bone and antler as well as stone in constructing their much finer implements. And the Neanderthals were "foragers" while ancient humans were "collectors." What's the difference? Humans monitored their environments and used "forward planning" by, for example, placing their campsites near animal migration paths. By contrast, Neanderthals do not appear to have timed their migrations in this way.

Given all this, it's not difficult to see why the Neanderthals died out: They encountered a species they could not compete with. They had been the sole hominids in Europe for nearly 200,000 years, but this isolation ended when modern humans arrived there about 40,000 years ago. Within another 13,000 years, all the Neanderthals were gone. There is a temptation, of course, to think that they died at the hands of the newly arrived *Homo sapiens*, but we have no evidence of violent conflict between the two species. What does seem clear is that modern humans could out-compete the Neanderthals. Humans knew where the antelope were going to be; they used better tools; they organized their campsites more efficiently. As a result, it's reasonable to assume that they got the lion's share of resources.

One fascinating idea stemming from this scenario is that it may have occurred elsewhere. Do you remember when it was that *Homo erectus* died out in the Far East? About 40,000 years ago. Perhaps, like the Neanderthals, they died because they could not compete with a newer, cleverer species that had moved into their territory.

Figure 19.27
Members of the Hominid Family
What did now-extinct members of the hominid family look like? Working from fossil and other evidence, artist Jay Matternes has produced drawings that provide some idea of their appearance.

(a) *Paranthropus boisei* lived about 2 million years ago. Though it was not a member of our genus (*Homo*), and not an ancestor of modern humans, it was a hominid.

(b) *Homo ergaster* is the species represented by Turkana boy. Its fossils date from a time only slightly later than *P. boisei*, but its physical form clearly is more human-like—not surprising since it probably was a human ancestor.

(c) *Homo neanderthalensis* lived in Europe and elsewhere in a period running from about 200,000 years ago to 27,000 years ago. These were the Neanderthals, a species that lived in proximity to modern humans for thousands of years before becoming extinct.

(a) *Paranthropus boisei*   **(b)** *Homo ergaster*   **(c)** *Homo neanderthalensis*

Modern *Homo Sapiens*

The newcomers to Europe and the rest of the world were us—modern human beings. Where had we come from? The short answer is Africa, though there is some question about the routes we took to get to Europe. Critically, we fully evolved *in Africa* before migrating to Europe and the rest of the world. This has been a contentious issue for years in anthropology, but it now seems to have been settled, mostly by the testing of ancient and modern DNA. This same kind of molecular testing has yielded approximate dates for the first appearance of modern human beings—between 100,000 and 200,000 years ago, with 130,000 years ago being a best estimate. If the Gran Dolina site researchers are correct, modern humans evolved in Africa during this time period as direct descendents of the *Homo antecessor* species. (The *H. antecessors* that came to Spain 800,000 years ago could be thought of as a hominid migratory expedition that didn't pan out.)

By the time modern humans arrived in Europe, they already possessed many of the capacities that set them apart from the Neanderthals. But there was more to come. The new Europeans de-veloped lamps to light caves; they made clay kilns; they dug pits in the permafrost that served as meat freezers. Most impressive of all, perhaps, they were artists. If you look at **Figure 19.28**, you can see a flowering of their creative minds: a portion of the 17,000-year-old cave paintings at Lascaux France—the product of a team of artists working by artificial light in the deep recesses of the cave.

These advances seem to have come about very rapidly. Almost none of them are seen before the arrival of *Homo sapiens* in Europe; but all of them are seen very soon after it. For years, researchers marveled that there was so little in the African record to indicate that African *H. sapiens* possessed the kind of mental capacities that European *H. sapiens* did. This lead to the idea that human beings may have undergone a genetic change that resulted in what has been termed the "creative explosion"— the flowering of human culture in Europe. Under this view, human mental capacity did not evolve in a slow, steady way. Instead, while still in Africa, our species went through a period of rapid evolution, perhaps 50,000 years ago, that made the European advances possible.

Figure 19.28
Flowering of Human Culture
Paintings done by a team of artists some 17,000 years ago on cave walls near Lascaux, France. Other cave paintings in France date from about 30,000 years ago.

Figure 19.29
The Earliest Art?
This piece of ocher, found in South Africa, has been dated to 77,000 years ago. The researchers who discovered it believe it represents the world's oldest known work of art. The date of the piece is important, as it predates the supposed creative explosion that began about 40,000 years ago, when modern humans migrated to Europe.

Against this, a more recent school of thought holds that there was no creative explosion. Instead, there was only the appearance of one, caused by the fact that modern human beings left so many well-preserved sites in Europe. When the full record from Africa and elsewhere is uncovered, these researchers believe, it will tell the story of a human mental evolution that was slow and steady.

Most scientists believe that the essence of what sets modern humans apart from other species is symbolic thinking. Human beings can design tools before they make them; they can plan for the future; they can represent three-dimensional animals on a two-dimensional cave surface. Until recently, however, almost all the earliest manifestations of symbolic thinking were found in Europe. But this may have begun to change. If you look at **Figure 19.29**, you can see a piece of red ocher stone found recently in South Africa and dated to 77,000 years ago. The researcher who uncovered this stone believes it represents the earliest-known example of human art. Even if it is not art, other researchers say, the geometric pattern on it provides clear evidence of symbolic thinking.

So, did we modern human beings become *fully* modern only in the last 50,000 years? Or did the last major event in human evolution occur 130,000 years ago, with our emergence as a species? Fortunately, our long evolution has given us the capacity not only to ask such questions, but to answer them.

On to the Diversity of Life

In reviewing the history of life in this chapter, you necessarily looked, in some small way, at almost all of Earth's major life-forms—bacteria, animals, plants, and so forth. In the three chapters coming up, you'll take a more detailed look at each of these life-forms. This is a tour that will make plain the astounding diversity that exists in the living world.

Chapter Review

Summary

19.1 The Geological Timescale: Life Marks Earth's Ages

- Earth's 4.6 billion years of history are measured in the geological timescale, which is divided into broad eras, shorter periods, and shorter-yet epochs. These time frames were defined in the nineteenth century, based in part on the different life-forms that existed within each of them, as reflected in the fossil record. Many of the major transitions in the fossil record have come about because of major extinction events.

- Life is thought to have begun on Earth about 3.8 billion years ago; from that time until perhaps 2 billion years ago, it was strictly microscopic and single-celled. According to the fossil record, all the animals and plants that exist today came about in the last 16 percent of this 3.8-billion-year period.

- Evolution can be conceptualized as a branching tree, with humans sitting on one ordinary branch. It is not shaped as a pyramid with humans at the apex. The story of evolution recounted in the chapter leads to human beings, but it could just as easily have led to any other modern organism.

- Every living thing on Earth belongs to one of three domains of life: Archaea, Bacteria, or Eukarya. Both bacteria and the archaea are single-celled, microscopic life-forms; the domain Eukarya is composed of four kingdoms: plants, animals, fungi, and protists.
Web Tutorial 19.1: Evolutionary Timescales

19.2 How Did Life Begin?

- Life began with the origin of self-replicating molecules. There currently is a debate regarding not only the kind of environment in which life is likely to have arisen but also how this environment could have accumulated the necessary chemical raw materials to allow the construction of life's complex organic molecules. One hypothesis is that life began in a "prebiotic soup" of hot-water systems—in the deep-sea vents or the kind of hot-spring pools that exist in Yellowstone Park today. Other researchers feel, however, that early replicating molecules could not have withstood these temperatures, and that a temperate location near a shoreline is a more likely environment for life to have gotten started.

- A critical question for origin-of-life researchers had been to account for the existence in living things of both the worker molecules called enzymes and the information-bearing molecules DNA and RNA, since neither kind of molecule can be synthesized without the other. The discovery of RNA molecules, called ribozymes, that have enzymatic capabilities helped solve this dilemma. Ribozymes provided evidence for an ancient "RNA world" in which the only living things were simple RNA molecules that could bring about their own replication.

- Replication is the step at which life could be said to begin. This is because, through mistakes, replicating molecules will begin to differ, meaning that some will leave more offspring than others, which is the beginning of Darwinian evolution and the diversification of living things.
Web Tutorial 19.2: RNA-like Self Replication

19.3 The Tree of Life

- Life today can be traced to a universal ancestor that gave rise to the domains Archaea, Bacteria, and Eukarya. The earliest kingdom in Eukarya, the protist kingdom, went on to give rise to the other kingdoms in Eukarya: animals, plants, and fungi.

19.4 A Long First Period: The Precambrian

- The earliest evidence we have for life is a chemical signature of it, carbon utilization, found in rocks in Greenland dated to about 3.8 billion years ago. Carbon signatures indicating photosynthesis (and hence life) are abundant beginning 3.5 billion years ago. All of this life was either bacterial or archaeal, and all of it seems to have existed in the oceans.

- Oxygen came to exist in the Earth's atmosphere through the activity of photosynthesizing organisms, originally photosynthesizing bacteria. Photosynthesis meant an abundance of food on the Earth. However, atmospheric oxygen was toxic to many existing species, and its production set a strict condition for all life: Adapt to oxygen, stay separate from it, or perish. Some early eukaryotes adapted to oxygen by ingesting, and then continuing to live in symbiosis with, bacteria that were able to metabolize oxygen. Algal cells struck up a similar relationship with other bacterial cells that could perform photosynthesis.

19.5 The Cambrian Explosion

- The fossil record indicates a tremendous, rapid expansion in the number of animal forms in a "Cambrian Explosion" that began about 544 Mya, but this evidence has been challenged. It may be that the Cambrian Explosion was merely an explosion in the number of animal forms big and hard enough to leave fossils, and that the surge in animal diversity took place over a relatively long period of time beginning well back in the Precambrian era. Molecular evidence indicates that animals arose much earlier than the Cambrian Explosion, perhaps 1.5 billion years ago.

19.6 The Movement onto the Land: Plants First

- Plants, which evolved from green algae, made a gradual transition to land in tandem with their symbiotic partners, the fungi. Plant-fungi fossils exist from about 460 Mya, but molecular studies indicate they were on land 1.3 billion years ago. Major transitions in plant life came with the development of a water-retaining covering (the cuticle), embryos that matured internally, and a "vascular" or fluid-transport system.

- Later plants went on to develop seeds, which can be thought of as packages containing an embryo and food for it, encased in a tough outer covering. The living ancestors of the first seed plants are the gymnosperms, represented by today's conifers. The flowering plants, called angiosperms, developed about 165 Mya and eventually succeeded the gymnosperms as the most dominant plants on Earth. Today, angiosperms include many food crops, cactus, and tree varieties.

19.7 Animals Follow Plants onto the Land

- The first land animals were arthropods; a centipede-like creature laid down the oldest terrestrial animal tracks we know of. Insects soon followed in great abundance. The arthropods were the only land animals for millions of years thereafter.

- Four-limbed vertebrates (tetrapods) moved onto land in the form of lobe-finned fishes, which gave rise to amphibians (represented by today's frogs and salamanders). Reptiles later branched off from amphibians, and mammals branched off from reptiles. Birds are the living descendents of the reptile branch that included the dinosaurs.

- Amphibians inhabit two worlds, water and land, in keeping with their evolutionary transition from one to the other. Reptiles evolved a protective amniotic egg that allowed their offspring to develop away from water, freeing reptiles to migrate inland.

- Mammals appear as a group of small, insect-eating animals about 210 million years ago. They began to evolve into the varied forms seen today shortly before the reign of the dinosaurs was ended with an asteroid impact 65 million years ago; after this they radiated into many niches, becoming the dominant land animals.

- The order of mammals called primates, whose fossils date from 55 Mya, is characterized by large, front-facing eyes, limbs with an opposable first digit (thumbs in today's human beings), and a tree-dwelling existence (which some later-arriving primates lost).

19.8 The Evolution of Human Beings

- A common primate ancestor is believed to have given rise to both the chimpanzee and the human family trees between 6 and 7 million years ago. The structure of the human family tree is a matter of considerable debate among researchers. Human evolution is studied by several kinds of scientists, but the core evidence in the field is provided by fossils, which are found and interpreted by scientists called paleoanthropologists.

- Human evolution is the study of the family *Hominidae*—the family of man-like primates. Every member of this family is referred to as a hominid, including human beings. All the members of the hominid family tree are extinct except for *Homo sapiens*, the human species. All early and mid-period hominid evolution took place in Africa. Until recently, it was believed that all such hominid evolution took place in East Africa, but several fossil discoveries in Chad may have extended the range of early human evolution across much of the African continent. One of these fossils, dubbed *Sahelanthropus tchandensis*, may be as much as 7 million years old, meaning it may lie at the base of the hominid family tree. Conversely, it may belong to the primate line.

- Paleoanthropologists interpret features of fossils—such as characteristics of teeth and skulls—in order to make judgments about where a given fossil-form lies in the hominid family tree. Interpretations can differ, however. Two features critical to inclusion in the hominid tree are tooth structure and evidence of bipedalism or upright walking.

- Bipedalism is clearly seen in the hominid genus *Australopithecus*, which is thought to be ancestral to our own genus, *Homo*. The most famous Australopithicene, "Lucy," lived 3.18 million years ago. A change to a physical form and mental capacity much closer to ours comes with the evolution of *Homo ergaster*, exemplified by "Turkana Boy," who lived 1.6 Mya.

- Hominids are thought to have first migrated from Africa about 2 Mya. A species called *Homo erectus* has been found at sites dating from 1.6 Mya in Java and China. Hominid fossils in Europe date from 800,000 years ago in the form of *Homo antecessor*, believed by some researchers to be the direct ancestor to modern human beings.

- The best known of the extinct hominids are the Neanderthals (*Homo neanderthalensis*), who populated Europe as well as parts of Asia and North Africa for nearly 200,000 years. The last of them disappeared 27,000 years ago—13,000 years after *Homo sapiens* first migrated to Europe. It seems likely that the Neanderthals died out because they could not compete with the more advanced humans.

- Modern human beings fully evolved in Africa between 200,000 and 100,000 years ago, with a best-estimate date being 130,000 years ago. Their arrival in Europe 40,000 years ago coincides with a remarkable cultural flowering that occurred among them. There is a debate as to whether humans may have undergone a genetic change shortly before leaving Africa that made this "creative explosion" possible. Some researchers feel there was only the appearance of such an explosion, and that the African human record eventually will reveal a picture of slow, steady human evolution.

Key Terms

amniotic egg   384
angiosperm   382
Archaea   374
Bacteria   374
bryophytes   382
Cambrian Explosion   380
continental drift   373
Cretaceous Extinction   371

Eukarya   374
gymnosperm   382
hominid   388
Permian Extinction   371
phylum   380
ribozyme   377
seedless vascular plant   382

Understanding the Basics

Multiple-Choice Questions (answers in the back of the book)

1. Put the following in the proper temporal sequence, earliest first.
   a. first dinosaurs
   b. first bacteria
   c. first amphibians
   d. first primates
   e. free oxygen accumulation in the atmosphere
   f. first land plants
   g. first birds

2. The time frames in the geological timescale generally denote _____.
   a. 10,000-year increments
   b. 10-million-year increments
   c. the reappearance of certain life-forms
   d. transitions in life-forms
   e. the appearance of larger life-forms

3. The first life on Earth may have been (select all that apply)

   a. eukaryotic cells
   b. ribozyme-like molecules
   c. bacteria
   d. formed on land
   e. formed at extreme temperatures

4. Which of the following characteristics of animals and/or plants are adaptations to life on land? (Select all that apply.)
   a. multicellularity
   b. pollen and seeds
   c. amniotic egg
   d. muscles
   e. internal fertilization

5. What is the order of evolution among the following organisms, earliest first?
   a. amphibians
   b. lobe-finned fish
   c. mammals
   d. reptiles
   e. primates

6. True or False: Dinosaurs were often hunted by humans.

7. Photosynthesis was a huge innovation in the history of life because (select all that apply)
   a. It allowed RNA to replicate itself.
   b. It put huge quantities of oxygen into the atmosphere.
   c. It resulted in the production of more food than had been previously available.
   d. It resulted in blockage of UV light, increasing habitability of the land for terrestrial organisms.
   e. It helped the Earth to thaw out.

8. Mitochondria and chloroplasts (select all that apply)
   a. are thought to have arisen as symbiotes
   b. are very similar to bacteria
   c. are very similar to algae
   d. are organelles within cells
   e. were early protists

9. Which of the following organisms have an amniotic egg? (Select all that apply.)
   a. grasshopper
   b. lizard
   c. salamander
   d. fish
   e. bird
   f. turtle

10. The family tree of the hominids is rooted in _____ and is thought to extend back _____ years.
    a. Asia; 3 to 4 million
    b. Europe; 3 to 4 million
    c. Asia; 6 to 7 million
    d. Africa; 3 to 4 million
    e. Africa; 6 to 7 million

Brief Review

1. What are two differences between domain Bacteria and domain Eukarya?

2. What is the amniotic egg, and why was it a breakthrough for terrestrial animals? In which vertebrate class did the amniotic egg evolve?

3. What are two characteristics of primates that are not found in other mammals? Why are they important for the evolution of hominids?

4. Which two groups are more closely related, humans and chimps or humans and gibbons?

5. Were Neanderthals the direct ancestors of humans?

6. When did life begin, and how much time passed between the origin of the first bacterial cells and the colonization of land? What was the land like during that time?

7. Describe three characteristics that evolved in mammals that had not been present in earlier organisms.

Applying Your Knowledge

1. How would Earth be different if photosynthesis had never developed in any organism?

2. Why does it make sense that life started so small? Why does it make sense that it began in water?

3. Given what the text says about the characteristics of Neanderthals in comparison with human beings, what would the world be like if Neanderthals had survived into modern times?

20 *The Diversity of Life 1*

Creatures large and small.
(page 402)

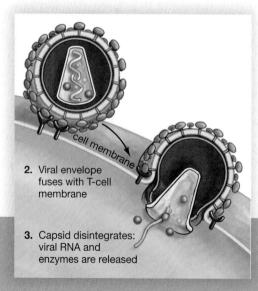

2. Viral envelope fuses with T-cell membrane

cell membrane

3. Capsid disintegrates: viral RNA and enzymes are released

The cause of AIDS.
(Section 20.2, page 404)

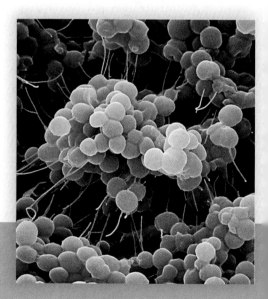

One kind of bacteria.
(Section 20.3, page 409)

The smallest living thing discovered to date is a type of bacteria that measures about two-tenths of a micrometer in diameter—that is, two-tenths of one-millionth of a meter in diameter. When we turn to the *biggest* living thing, elephants or whales might come to mind, but it turns out that animals aren't even in the running for this title. At a maximum length of 27 meters (or about 89 feet), the blue whale is the largest animal that has ever existed, but it's small compared to California's coastal redwood trees, which might reach a height of 100 meters or about 330 feet. This is large indeed, but consider that a giant sea kelp once was found that was 274 meters in length, meaning it was about 900 feet long. It may be, though, that in terms of size, both kelp and redwoods have to take a back seat to a life-form normally thought of as quite small. In 1992, researchers reported finding a fungus growing in Washington State that runs underground through an area of 1,500 *acres* and that arguably is a single organism.

Now, how about the *deepest*-living things? The current record-holders are bacteria that were found in the early 1990s—in a combination of oil drilling and scientific exploration—living almost 3 kilometers beneath the Earth's surface, which is about 1.9 miles down. Food and water are so scarce at such depths that these bacterial cells may live in a kind of suspended animation, dividing perhaps once a year, or even once a *century*. Turning to the highest living things, 12 species of bacteria have been found to be living in the Himalayas at 8,300 meters or 27,000 feet above sea level, which is just a couple thousand feet lower than the peak of Mount Everest. A species of chickweed survives in the Himalayas at over 20,000 feet.

All kinds of life-forms—clams, tube worms, crabs—live kilometers beneath the surface of the oceans, near the "hot-water vents" that spew out water and minerals from the Earth's interior onto the ocean floor. Microbes called archaea, living at these deep-sea vents, set the pace for life in a hot environment. An average, midday temperature in the Sahara Desert would be so *cold* it would end the reproduction of the archaean *Pyrolobus fumarii*, which lives within hot-water vents at temperatures of up to 113° Celsius, which is hotter than the temperature at which water boils. At the other extreme, the well-named bacterium *Phormidium frigidum* carries out photosynthesis in Antarctic lakes whose temperatures sit right at the freezing point of water.

Finally, what about the oldest living things? Humans have been known to live to 120, but this is an eyeblink compared to the lives of some plants. In California there is a Sequoia tree that was growing 3,500 years ago, when the pharaohs ruled Egypt, and

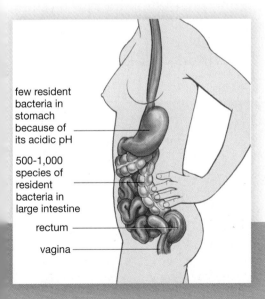

At home within us.
(Section 20.4, page 411)

few resident bacteria in stomach because of its acidic pH

500-1,000 species of resident bacteria in large intestine

rectum

vagina

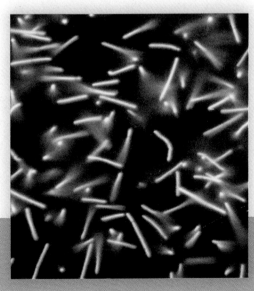

Hot-water lover.
(Section 20.6, page 413)

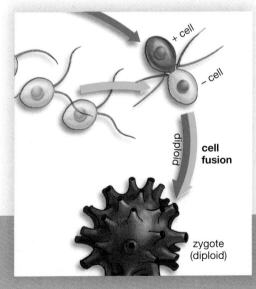

+ cell

– cell

diploid

cell fusion

zygote (diploid)

An early version of male and female.
(Section 20.8, page 415)

there is a bristlecone pine tree (*Pinus longaeva*) that has been living in the dry slopes of eastern California for 4,900 years. These trees are mere youngsters, however, compared to some microbes, which can survive in a form called spores for immensely long periods of time. In the 1990s, bacterial spores were found living within a bee that had been encased in amber for 25 to 40 million years. Even this eye-popping number did not prepare scientists for what came next, however. In 2000, researchers announced that they had found, encased within salt crystals at the bottom of a New Mexico air shaft, archaean spores that date from 250 million years ago. These spores were revived and billions of their offspring now exist. This claim is controversial—one possibility is that the material encasing the spores wasn't really 250 million years old—but if borne out, it raises the question of whether some microbes are "effectively immortal," as one researcher has put it.

Looking at life like this—at its biggest or deepest or oldest extremes—is one way to get at its incredible diversity. But it does not begin to do justice to life's variety, because life is diverse in so many ways. Honeybees do a "dance" to let their hive-mates know the location of a food supply. Corn plants, when attacked by army worms, can call in an air force: the plants release an air-borne substance that attracts parasitic wasps, which then prey on the worms. Pacific salmon live years in the ocean only to make one arduous, upstream journey to spawn in the waters where they were born, after which they die. Look closely at almost any creature, and you are likely to find something only slightly less dramatic (**see Figure 20.1**).

20.1 Life's Categories and the Importance of Microbes

The purpose of this chapter and the two that follow is to introduce you to life as it exists across all its large-scale categories. Those readers who went through Chapter 19 know that all living things can be placed into one of three "domains." Two of these are Archaea and Bacteria, whose members are all single-celled and microscopic. The third domain, Eukarya, encompass-

(a) Mouse beneath the foot of an elephant

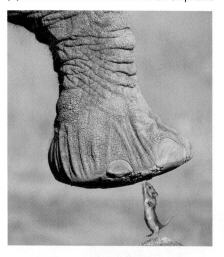

(b) Aquatic snails moving across kelp

(c) Fungus growing on a fallen log

(d) Water flea swimming near green algae

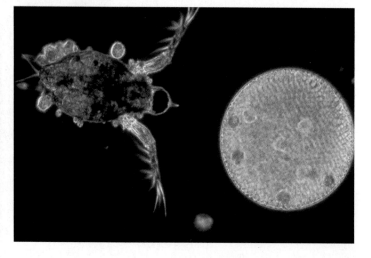

Figure 20.1
An Amazing Diversity in the Living World

es such incredible diversity that it is further divided into four separate "kingdoms." These are plants, animals, fungi, and a group of mostly microscopic organisms called protists. We will begin our tour of the living world by starting small, looking in this chapter at Domains Bacteria and Archaea and at the protists within Domain Eukarya. Then in Chapter 21, fungi and plants will be covered; in Chapter 22 animals will be reviewed. **Figure 20.2** shows you the evolutionary relationships between all these categories of life.

Because this chapter is devoted to mostly unfamiliar *microbes*—meaning living things so small they can't be seen with the naked eye—it might be helpful to say a word at the start about their significance. Microbes are something like the foundation of a house: seldom thought of but critically important. To the extent that they cross our minds at all, we are aware of them mostly for the diseases they cause. This is an important topic—one we'll go into extensively—but thinking of microbes strictly as disease-causing organisms is like thinking of cars strictly as wreck-causing machines. Where does the oxygen we breathe come from? If your answer is "plants," that's partly correct. But the photosynthesis carried out by plants supplies less than half our oxygen; the balance comes from microscopic algae and bacteria—mostly drifters on the ocean's surface. How about the nitrogen that is indispensable to all plants and animals? Among living things, only bacteria are capable of plucking nitrogen out of the air and turning it into a form that plants can use (see page 713). If there were no bacteria, there would be no plants—and consequently no animals. And how is it that dead tree branches or orange peels or the bones of animals are broken down and recycled into the Earth? Almost entirely through the work of bacteria and fungi. Without these organisms, Earth would long ago have become a garbage heap of dead, organic matter. Animals cannot decompose a dead tree branch, but bacteria and fungi can.

The upshot of all this is that life on Earth would grind to a halt without microbes; they are an essential underpinning to all forms of life. They come to this status in part because of the environments they live in, which is to say all environments on Earth that support life of any kind. It probably didn't escape your attention that the coldest- and hottest- and highest- and deepest-living things were all bacteria or archaea. Wherever a plant or animal can live, microbes can be found too. Just as impressive is the sheer number and tonnage of these tiny organisms. A common denominator for measuring the amount of living material is something called *biomass*, defined as the total dry weight of material produced by a given set of organisms. Each bacterium or microscopic alga has an infin-

itesimally small weight, of course—but then again, there are so many of them! It has been estimated that there are more bacteria living in your mouth now than the number of people who have ever lived, and there seem to be more bacterial cells in the human body than there are human cells (a topic we'll return to). A typical gram of fertile soil—an amount about the size of a sugar cube—has about 100 million bacteria living in it, and a mere drop of water near the ocean's surface contains thousands of the tiny algae called phytoplankton. All this adds up. When the calculations are done, we find that microbes probably make up more than half of Earth's biomass. Put another way, the microbes of the world outweigh all the plants and animals and visible fungi of the world combined.

In sum, microbes matter, and not just in ways that are harmful. We'll now start to look at the portion of the living world that is made up of them. Our first stop on this tour, however, will be a category of tiny replicating entities that most biologists would say lie just outside of life, though they have a great effect on living things.

20.2 Viruses: Making a Living by Hijacking Cells

If a living thing is too small to be seen and can cause disease, most people would put it in the vague category known as "germs." But there is a fundamental distinction to be made between *two* kinds of infectious organisms. On the one hand there are bacteria, which may be very small but nevertheless are *cells*, complete with a protein-producing apparatus, a mechanism for extracting energy from the environment, a means of getting rid of waste—in short, complete with everything it takes to be a self-contained living thing. On the other hand there are viruses, which by themselves possess

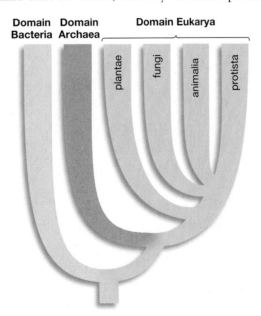

**Figure 20.2
The Tree of Life**
Life is presumed to have developed from a single common ancestor (the base of the tree), which gave rise to the three domains of the living world: Bacteria, Archaea, and Eukarya. Domain Eukarya is then further divided into four "kingdoms," plants, fungi, animals, and protists.

none of these features. Indeed, viruses can be likened to a thief who arrives at a factory he intends to rob possessing only two things: the tools to get inside and some software that will make the factory turn out items he can use. The factory being broken into is a living cell, and the software that the virus brings is its DNA or RNA, which it puts inside this "host" cell. Once this is accomplished, viruses employ different tactics, as you'll see, but in every case the result is that viruses make more copies of themselves. Viruses are an integral part of the living world, and not all of them cause harm. But

unlike the case with bacteria, it's right to think of them first and foremost in terms of the diseases they cause, not only in humans, but in all kinds of living things. **Viruses** are non-cellular replicating entities that must invade living cells to carry out their replication. To put this in the scientific terminology, they are obligate intracellular parasites.

HIV: The AIDS Virus

If you look at **Figure 20.3a**, you can see a simplified rendering of a particular virus, the human immuno-

Figure 20.3
The Virus that Causes AIDS
The anatomy and life-cycle of HIV, the human immunodeficiency virus.

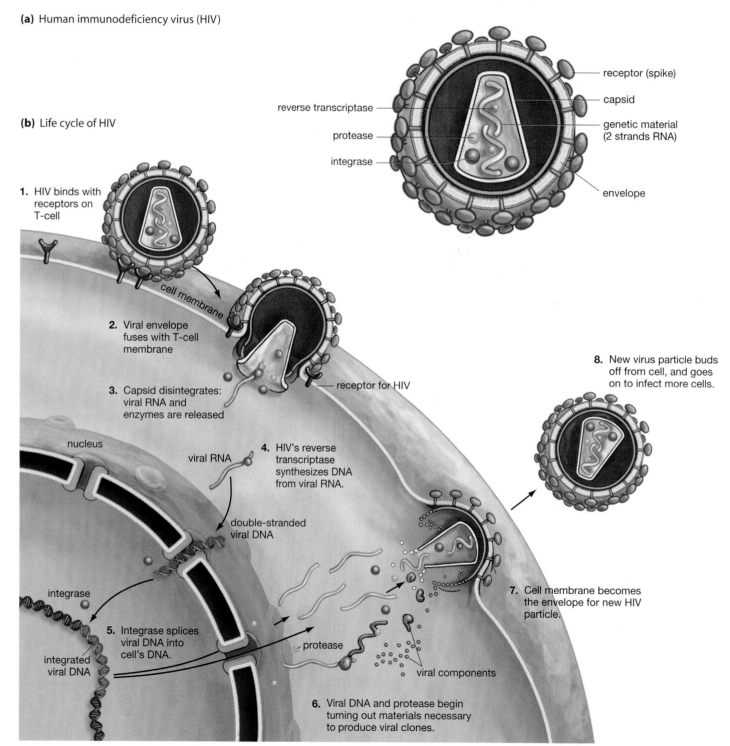

(a) Human immunodeficiency virus (HIV)

receptor (spike)
capsid
reverse transcriptase
genetic material (2 strands RNA)
protease
integrase
envelope

(b) Life cycle of HIV

1. HIV binds with receptors on T-cell

cell membrane

2. Viral envelope fuses with T-cell membrane

receptor for HIV

3. Capsid disintegrates: viral RNA and enzymes are released

nucleus

viral RNA

4. HIV's reverse transcriptase synthesizes DNA from viral RNA.

double-stranded viral DNA

integrase

5. Integrase splices viral DNA into cell's DNA.

integrated viral DNA

protease

viral components

6. Viral DNA and protease begin turning out materials necessary to produce viral clones.

7. Cell membrane becomes the envelope for new HIV particle.

8. New virus particle buds off from cell, and goes on to infect more cells.

deficiency virus (HIV), which is the cause of AIDS. Almost all viruses have two of the three large-scale structures you can see in HIV: the genetic material at the core (in HIV's case, two strands of RNA), and a protein coat, called a **capsid**, that surrounds the viral genetic material. Many viruses then go on to have a third major element—a fatty membrane, called an *envelope*, which you can see surrounding the HIV capsid. Protruding from the HIV envelope are a series of receptors, often referred to as spikes, which are proteins capped with carbohydrate chains. These serve the critical role of binding with receptors that protrude from the target cell, thus giving the virus a way to get in.

If you look at **Figure 20.3b**, you can see how the life cycle of HIV proceeds from this initial binding. The target cell in this case is the one most commonly invaded by HIV, an immune-system cell called a helper T-cell. As you can see, once HIV binds with two receptors on the T-cell's surface, the viral envelope fuses with the T-cell membrane. With this, HIV's capsid, with the genetic material enclosed, is inside the cell. Once there, the capsid disintegrates. Now two HIV enzymes that had been enclosed in the capsid get to work. If you look at Figure 20.3a, you can see these enzymes, integrase and reverse transcriptase. The reverse transcriptase gets busy first, turning HIV's RNA strands into double-stranded DNA. Then integrase does just what its name implies: It integrates the viral DNA into the *cell's* DNA through a cut-and-paste operation.

At this point, the viral DNA might simply stay integrated in the T-cell's DNA, causing no great harm but getting copied, along with the cell's DNA, each time the cell divides. The effect of this, however, is that every "daughter" cell of this infected cell will be infected too; each new cell will have viral DNA spliced into its own DNA. Each of these cells thus has a time bomb ticking within it, because at some point the HIV DNA will change course. It will no longer simply go on replicating within the cell's DNA. Instead, it will start turning out the materials necessary to make whole new copies of the virus. As you can see in the figure, virus construction takes place just inside the cell's outer membrane (helped along by a third enzyme that HIV brought with it, protease). This location is important, because the cell's membrane ends up serving as the final structural part of each new HIV "particle," as the viral clones are called. When the capsid is completed, it buds off from the cell, taking with it part of the cell's membrane, which now becomes the capsid's envelope. When enough viral particles have budded off from the cell in this way, it dies. The viral particles that emerge from the cell then go on to infect other cells.

Viral Diversity

In the HIV life cycle, you can see four steps that are common to almost all viruses: get genetic material inside the host cell, turn out viral component parts, construct new particles from these parts, and move these particles out of the cell. HIV uses RNA as its genetic material, but many viruses use DNA. Viruses that lack an envelope don't get their genetic material inside the cell by fusing with the cell's membrane in the way HIV does. If you look at **Figure 20.4**, you can see a spacecraft-shaped virus called T-4 that infects bacteria through an alternate means. T-4 keeps its capsid outside the target cell, but gets its DNA inside by injecting it through the bacterium's cell wall and membrane.

(a)

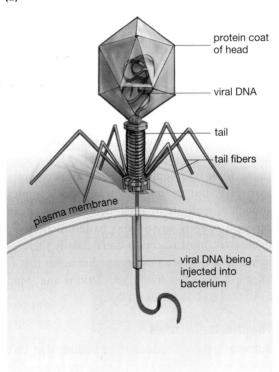

protein coat
of head

viral DNA

tail

tail fibers

plasma membrane

viral DNA being
injected into
bacterium

(b)

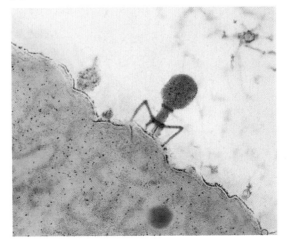

Figure 20.4
Another Variety of Virus

(a) The T4 virus looks like a spacecraft that has landed on the surface of a bacterium.

(b) A viral invader lands. An artificially colored T4 virus injects its DNA into an *E. coli* bacterium (in blue).

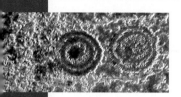

Unwanted Guest: The Persistence of Herpes

The first outbreak usually is the worst, but the real problem is that the first outbreak usually isn't the last. Herpes comes back, as herpes sufferers well know. A few weeks or months after the initial outbreak, everything seems fine, but then there may be a tingling—nothing that can be seen yet, but instead just a feeling. Not much later, however, it's back: the same unsightly "cold sore" on the mouth or lesions on the genitals, in the same place that the first outbreak occurred.

This series of events is familiar to lots of people. An estimated one in five Americans over the age of 12 is infected with the herpes virus, though this statistic overstates the trouble caused by it. The vast majority of herpes carriers do not even realize they harbor the disease, because such outbreaks as they have are so mild as to go unnoticed. This very thing facilitates the spread of herpes, however. People who have no symptoms can nevertheless be "silent shedders": they can pass herpes along to others, who may get the full-blown form of the affliction.

But why does herpes come back in the way it does? Where does it go between outbreaks and why can't we just get rid of it permanently? The term "herpes" actually is short-hand for *herpes simplex viruses*, a pair of viruses that belong to a group of 100 known viruses in the herpes family. (Others in the family cause chicken pox, shingles, and the Epstein-Barr disease.) Herpes simplex viruses come in "type 1" and "type 2" varieties. The type 1 variety supposedly is "oral," while the type 2 is "genital." But either variety can end up in either location, because of oral sex.

The herpes sores that can be so distressing are the result of newly formed virus particles breaking out of skin cells, thereby destroying them. Fever, swollen glands, and other symptoms often accompany these lesions, particularly in an initial infection. Within a couple of weeks or so, however, the immune system has killed all the virus particles in and around the skin cells and the symptoms are gone. The problem is that, by this time, the virus has also infected *nerve* cells in the area. Such nerve cells (or neurons) have long extensions of themselves called axons. With the initial infection, herpes virus particles start making a slow migration up these axons to the *body* of the nerve cells, specifically to the nucleus inside the cell body. In the case of genital herpes, this means a journey from the vagina or penis to cell bodies that sit right at the base of the spinal cord. And these cell bodies are where the virus stays—forever. Once there, it does next to nothing; it's not causing any nerve damage or replicating. The nerve cells it is in never divide, so it isn't passed on to a larger group of cells. But neither is it eliminated. It just resides in the nerve cell nuclei until one day—often a stressful day—it gets activated by an unknown mechanism and begins a journey back up the very axons it came in on. Once it reaches the original nerve-cell endings it spreads to skin cells, and the result is new lesions in the same old place.

As noted, some people with herpes can shed virus particles from time to time even when they have no lesions, meaning they are infectious during these periods. This, of course, puts these carriers at terrible risk of passing the virus on to their uninfected partners. The good news about herpes is that there is a drug, called Valacyclovir, that has shown significant ability to suppress the virus at all times. In a large clinical trial, herpes carriers who took one Valacyclovir tablet a day transmitted the virus to their partners at less than half the rate of carriers who took only a placebo.

Figure 20.6
Small Is a Relative Thing
The bacteria, protist, and viral clones shown in the figure are all microscopic. Nevertheless, they differ greatly among themselves in terms of size. A typical human cell might be about a third the size of the *Paramecium*.

- *No membrane-bound organelles.* A nucleus is an example of an *organelle*: a highly organized structure within a cell that carries out specific cellular functions. You may remember from Chapter 4 that, in addition to a nucleus, eukaryotic cells have such organelles as mitochondria and lysosomes, which are surrounded by a membrane. Bacteria have only a single kind of organelle, the ribosome, which differs from other organelles in that it is not surrounded by a membrane.

- *Single-celled organisms with small cell size.* Bacteria often come together in groupings that have an organization to them, but each bacterial cell is a self-contained living thing—each one can live and reproduce without the aid of other cells. Bacterial cells are also very small, relative to eukaryotic cells; thousands of them could typically fit into a single eukaryotic cell. (If you look at **Figure 20.6**, you can see a size comparison of bacteria, a T4 virus, and a protist you'll be looking at called a paramecium.)

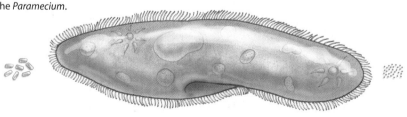

bacteria (*E. coli*)
2 µm long

protist (*Paramecium*)
75 µm long

viral clones (T4)
0.2 µm long

- *Asexual reproduction.* Bacteria reproduce by a simple cell-splitting or **binary fission**: one cell splits into two, with both "daughter" cells being exact replicas of the parental cell. In contrast, most eukaryotes are capable of sexual reproduction, in which the genetic material of two separate organisms produces offspring, usually through the fusion of egg and sperm.

If you look at **Figure 20.7**, you can see micrographs of the three forms bacterial cells most often take: round, rod-shaped, and spiral-shaped. A round bacterium is known as a *coccus*; a rod-shaped bacterium is a *bacillus*; and a spiral-shaped bacterium goes under a variety of names, the best known of which is a *spirochete*. Different species of bacteria fit into all three groups.

Almost all bacteria have a structure called a cell wall at their periphery that is largely composed of a material called peptidoglycan. This may seem like a technical detail, but the fact that bacteria have cell walls is of great importance to humans because it represents a *difference* between bacterial cells and human cells. As you'll see, we humans have exploited this difference in our development of the medical drugs known as antibiotics.

Bacterial Diversity

Scientists have taken small soil samples, looked for DNA sequence differences among the bacteria in these samples, and concluded that about 10,000 different species of bacteria exist in such handfuls of earth. Extrapolating from these samples, it's safe to say that millions of species of bacteria exist. Surprisingly, perhaps, only 4,500 species of bacteria and archaea have been identified to date.

An incredible diversity exists among the various kinds of bacteria. It may seem strange to think of these tiny creatures as "diverse," given that they are all microscopic and single-celled. But physical form is only one kind of diversity. Imagine that only some human beings got their nutrition in the conventional way—by ingesting food—while others could *make* their own food by carrying out photosynthesis. Bacteria do things both ways. Imagine that only some human beings needed oxygen to live, while others could take it or leave it, and still others were poisoned by it. Bacteria fit into all three categories. Animals and plants have diverse forms, but bacteria have diverse metabolisms. If you look at "Modes of Nutrition" on the next page, you can see the various ways that life's creatures get the nutrition they need. Note that bacteria utilize every one of these modes, while humans utilize only one.

(a) Round bacteria (cocci)

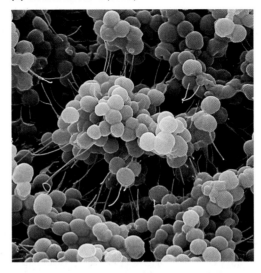

(b) Rod-shaped bacteria (bacilli)

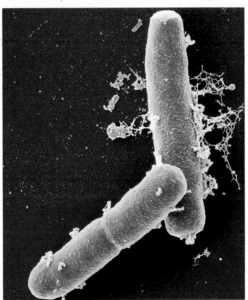

(c) Spiral-shaped bacterium (spirochete)

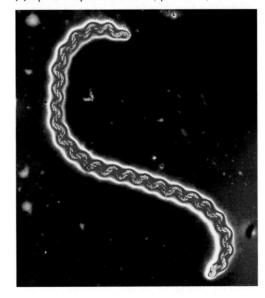

Figure 20.7
Different Shapes of Bacteria

(a) The round bacterium *Staphylococcus epidermis* is a normal part of the bacteria that cover our skin.

(b) The rod-shaped bacterium *Bacillus anthracis* is the cause of the deadly disease anthrax.

(c) The spiral-shaped bacterium *Leptospira interrogans* often infects rodents, sometimes infects dogs, and occasionally infects human beings.

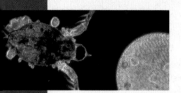

Modes of Nutrition: How Organisms Get What They Need to Survive

All living things need energy and nutrients. The question is, how do they get them? What is their "nutritional mode," as biologists would put it?

The most fundamental distinction in nutritional mode separates groups known as autotrophs from those called heterotrophs. "Autotroph" means "self-feeding." **Autotrophs** are organisms that can manufacture their own food, defined as some form of organic (carbon-containing) molecule that can be broken down to yield energy. All organisms that are not autotrophs are **heterotrophs**, meaning "other feeders;" they cannot manufacture their own food, but must get it from elsewhere, as animals do.

The idea of heterotrophs is not strange to us—it's what we are, after all—but the notion of autotrophs may be a little more exotic. Those who have gone through Chapter 8 on photosynthesis will recall that almost all plants are autotrophs in that they manufacture their own food, initially a sugar. To do this, they need a carbon source, because carbon is the "backbone" of the organic molecules we know as food, and they need an energy source that can drive the complex process of photosynthesis. The carbon source for plants turns out to be carbon dioxide (CO_2), obtained from the atmosphere, while the energy source is sunlight. Here we see the *two-part* requirement for nutrition that holds for all organisms: a source of carbon on the one hand, and an energy source on the other.

Within this framework there are four nutritional modes, which you can review in **Table 1**. The two most important of these modes are *photoautotrophy*, which is the nutritional mode of plants, and *chemoheterotrophy*, in which organic materials (better known as food) act as both carbon supplier *and* energy source. This is another way of saying that the cereal you ate this morning supplied you, as a chemoheterotroph, with both carbon and high-energy electrons.

There is also *chemoautotrophy*, in which organisms—some bacteria and archaea—get their carbon from carbon dioxide, like plants do, but power the production of food from this carbon by oxidizing (pulling electrons from) such inorganic materials as hydrogen sulfide and ammonia. A small number of bacteria and archaea practice *photoheterotrophy*, in which the sun supplies the energy but the carbon comes from surrounding organic material.

Table 1

Nutritional mode	Carbon source	Energy source	Practiced by
Autotrophy			
Photoautotrophy	Carbon dioxide (CO_2)	The sun's rays	Almost all plants, some bacteria, and many protists
Chemoautotrophy	Carbon dioxide	Inorganic compounds such as hydrogen sulfide and ammonia	Some bacteria and archaea
Heterotrophy			
Photoheterotrophy	Organic material	The sun's rays	A few bacteria and archaea
Chemoheterotrophy	Organic material	Organic material	Almost all animals, all fungi, most bacteria, many protists, and a few plants

20.4 Intimate Strangers: Humans and Bacteria

In the chapter introduction, you saw how bacteria are indispensable to life in general because they can capture atmospheric nitrogen, decompose dead organic material, and so forth. As it turns out, however, bacteria may be indispensable to *human* life in an even more direct way. From the time we travel down our mother's birth canal, we enter into a lifelong interdependence with these microorganisms. If you look at **Figure 20.8**, you can see the places bacteria are found in quantity outside and inside the human body, along with a micrograph showing what one class of these bacteria looks like. Outside,

(a)

scalp

nasal
passages

about 200
species of
resident
bacteria in
mouth

skin

digestive
tract

few resident
bacteria in
stomach
because of
its acidic pH

armpit

500–1,000
species of
resident
bacteria in
large intestine

rectum

vagina

(b)

Figure 20.8
At Home in Us

(a) We human beings live with trillions of "resident" bacteria, meaning bacteria that are always present on us, or within us. (Bacteria on the scalp or skin might temporarily be washed away, but they quickly return.) The greatest concentration of these bacteria is found in the digestive tract, particularly the large intestine.

(b) An example of resident bacteria. Pictured is a kind of bacterium commonly found in the human nose. The little yellow balls are *Staphylococcus aureus*, seen here among the hair-like cilia that extend from the cells lining the nasal passages. Many of these bacteria can be seen adhering to mucus (in blue).

we are covered in bacteria from head to toe, with heavy concentrations in the armpits and scalp. Inside, they exist in nasal passages, in the vagina in females, and, most especially, in the digestive tract. Perhaps 100 trillion bacterial cells live within the length of the tract, from mouth to rectum. The number of *human* cells in a human body is thought to be about 100 trillion as well, so when we count bacterial cells not just in the digestive tract but elsewhere, we find that bacterial cells outnumber our own. (The reason we are not bloated with them is that they are such small cells.) This is not to say that bacteria are everywhere within us; most of our tissues are kept bacteria-free, except for occasional invaders. But about half the contents of our colon are bacteria, as are perhaps a quarter of our feces by weight. These are not transient bacteria who come for brief periods and then are gone. They are so-called "resident" bacteria whose permanent home is the human body.

This account may make you feel like you've been colonized, but the relationship between us and many of the bacteria we harbor actually is one of **mutualism**, which is to say a relationship between two organisms that benefits both of them. If we could shrink ourselves down to the size of a bacterium and enter, say, the human small intestine, we would see a community of

competing but *interdependent* organisms, most of which are bacteria, but one of which is us. What bacteria get from participating in this community is food and habitat. What humans get is a fully functional digestive system. How do bacteria help bring this about? Consider that rats whose bacteria have been wiped out through laboratory techniques require 30 percent more calories to maintain their body weight than do normal rats. Without bacteria, the structure of these rats' intestines is altered in a way that's unhealthy: they have low numbers of the cells that line the digestive tract and move nutrients from the tract to the bloodstream. Researchers have made mice "germ-free," then reintroduced bacteria into them, and then watched as mouse genes were turned on that help metabolize sugars and fats. Bacteria even metabolize some sugars that we cannot digest, and they produce some vitamins as well. The essential message here is that, within the digestive tract, mammals and bacteria have had such a long working relationship that each has become dependent on the other. We humans don't just tolerate the bacteria in our gut; we need them.

20.5 Bacteria and Human Disease

It goes without saying, of course, that not all bacteria are beneficial to human beings. Even within our digestive tract, there are bacteria that are "commensal" with us—they benefit from the relationship, while we are unaffected by it. And, of course, there are bacteria that are **pathogenic**, which is to say disease-causing. These are not bacteria that routinely inhabit any part of our body in large numbers. Instead, they are opportunists that invade specific tissues. Among the diseases they cause are tuberculosis, syphilis, gonorrhea, cholera, tetanus, botulism, leprosy, typhoid fever, and diphtheria, not to mention all manner of food poisonings and blood-borne infections. In the fourteenth century, the bacterially caused bubonic plague wiped out a *third* of Europe's population in just four years. Closer to home, you may recall a substance referred to as "anthrax" that was maliciously mailed to various offices after the September 11 attacks. Anthrax actually is a bacterium (*Bacillus anthracis*) that, when not in the hands of terrorists, occasionally infects farm animals.

How do pathogenic bacteria cause illness? A few invade human cells and reproduce there. More often, however, bacterial damage comes from substances bacteria either secrete or leave behind when they die. These compounds are *toxins*, which is to say substances that harmfully alter living tissue or interfere with biological processes. In the case of respiratory anthrax, *Bacillus anthracis* secretes a trio of toxins that cause blood vessels in the lungs and brain to "hemorrhage" or leak. The botulism bacterium, *Clostridium botulinum*, secretes a toxin that blocks the signals that go from nerves to muscles. The end result can be a paralysis of the muscles that control the diaphragm, in which case victims lose the ability to breathe. Of course, all pathogenic bacteria must have a way to get into the body in the first place. The bacterium that causes bubonic plague, *Yersinia pestis*, enters on the bite of a flea, while the tetanus bacterium, *Clostridium tetani*, may hitch a ride in on a rusty nail. Meanwhile, the tuberculosis bacterium, *Mycobacterium tuberculosis*, can simply be inhaled.

Killing Pathogenic Bacteria: Antibiotics

Prior to the 1930s, doctors were nearly powerless to stop any of these bacterial invaders. Some progress was made in the late 1930s with the development of the synthetic chemicals called sulfa drugs. But modern medicine was born in the 1940s with the development of **antibiotics**, meaning chemical compounds produced by one microorganism that are toxic to another microorganism. The first antibiotic, penicillin, was a substance produced by a fungus. Like all antibiotics, it had two critical qualities: it could kill microorganisms within human beings while leaving the humans themselves relatively unharmed. (For an account of penicillin's development, see "The Discovery of Penicillin" on page 419.)

How can antibiotics be toxic to bacteria but not to people? They exploit the differences between bacterial cells and human cells. Remember earlier when we noted that almost all bacteria have cell walls, while human cells don't? Penicillin and several other antibiotics work by blocking cell-wall construction, which bacteria have to undertake when they divide. Penicillin confiscates a construction tool; it binds with a bacterial enzyme that helps stitch the cell wall together. A bacterium without a cell wall is doomed: it ruptures, floods, and dies. But the *human* cells around it—lacking cell walls—are left unharmed.

It is hard to overstate how much of human health hinges today on antibiotic processes such as these. If you have grandparents who were alive in the 1930s, they may well have had friends or relatives who died of tuberculosis or blood poisoning or tetanus. But do you know of anyone who, during your lifetime, has died of such an illness? Probably not. Antibiotics have made all the difference.

The Threat of Antibiotic Resistance

This power of antibiotics ought to spark alarm in us about something that is happening today, which is that antibiotics are steadily *losing* their ability to kill bacteria. How could this happen? An estimated 30 percent of antibiotic prescriptions written today are thought to be needless, in the sense that the person taking the antibiotic doesn't even have a bacterial infection, but rather is suffering from a viral or other illness. (So why do people get such prescriptions? They ask their doctors for them and the doctors go along.) Even when people do have bacterial infections, they often take their antibiotics only until they start feeling better, rather than following through with the full course of recommended treatment, which can last weeks or even months.

Those of you who have read through the chapters on evolution may be able to see what this leads to. Normal bacteria may be wiped out by a needlessly short course of antibiotics. But a few bacteria will survive these inadequate treatments because they have sets of genes that allow them to do so. In the *absence* of antibiotics, these survivors might not have had anything special going for them, relative to others in their population. (They aren't *generally* superior bacteria; they just resist antibiotics better.) But with antibiotics these survivors have been left without bacterial competitors for food and habitat. As a result, they flourish, giving rise to antibiotic-resistant *lines* of bacteria. Moreover, they will do this rapidly, since one generation of bacteria can give rise to another in as little as 20 minutes.

Other factors also enter into this picture. It may surprise you to learn that nearly half the antibiotics used in the United States don't go into people at all; they are put into animal feeds because they serve as growth stimulants. In a similar vein, farmers spray huge tracts of fruit trees with antibiotics to ward off bacterial infections. Hand soaps now contain antimicrobial agents whose residues stay on bathroom and kitchen surfaces, wiping out some bacteria while leaving the resistant variety to multiply in the absence of competitors. (Such hand soaps are unnecessary in the first place because plain soap and water will work just as well to wash the germs down the drain instead of killing them.)

The upshot of this massive over-use of antibiotics is that bacterial infectious diseases are making a comeback. The problem is most pressing in hospitals, where physicians are seeing one formerly useful drug after another fail to cure infections. If we cannot stay ahead of bacterial evolution by developing new antibiotics—while cutting back on the needless use of existing antibiotics—we may find ourselves returning to the days of our grandparents, when a simple cut could be life-threatening.

20.6 Archaea: From Marginal Player to Center Stage

As little as 10 years ago, any biology textbook you picked up would likely have referred to the archaea as archae*bacteria*, and most would have mentioned the archaea as a kind of marginal life-form, existing in only a few extreme habitats, such as hot springs or salty lakes. Today, the archaea are recognized as constituting their own domain in life—standing alongside Domain Bacteria and Domain Eukarya—

and they have belatedly been recognized as existing in abundance in lots of environments. The importance of the archaea stretches beyond this, however, in that, of all the organisms living today, archaea may have been the first to exist. Whether this means that archaea are ancestral to all other existing life-forms remains to be seen—many researchers have their doubts—but it is certain that archaea lie near the trunk of life's family tree. In particular, the *thermophiles*, or heat-loving archaea, appear to be a very old branch on life's family tree, which is one piece of evidence leading many scientists to suspect that life began in or near hot-water vents of the type seen on the ocean floor today.

It took a long time to recognize the unique qualities of archaea because, superficially, they are very similar to bacteria. They are microscopic, single-celled, and their cells lack a nucleus (making them prokaryotes, along with the bacteria). In addition, it is hard to learn much about archaea because it is so difficult to "culture" them, meaning to grow populations of them in a laboratory. The lab environment is so different from the extreme environments inhabited by many archaea that cultures of them are quickly taken over by common bacteria.

Beyond these things, however, there was a long-term resistance among scientists to seeing archaea as anything but an odd type of bacteria. Indeed, a single individual, Carl Woese of the University of Illinois, had to wage a lonely battle lasting years before his claims about the uniqueness of the archaea were generally accepted. In 1996 there came a triumphant confirmation of his ideas about the archaea with the sequencing of the entire genome of an archaean species, *Methanococcus jannaschii*. This work revealed that an amazing 56 percent of *M. jannaschii*'s genes were completely unknown to science—they were unlike anything seen in either bacteria or in eukaryotes. As for the remaining 44 percent of the genome, some *M. jannaschii* genes worked like those

(a)

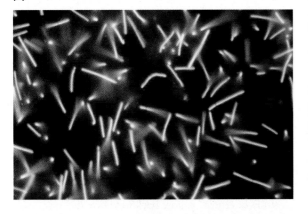

(b)

Figure 20.9
Life in Extreme Environments

(a) Deep-sea archaeans *Methanopyrus kandleri* is an archaean that lives in the ocean at temperatures near those of boiling water. It has been found in geothermal vents on the sea-floor at depths of 2,000 meters, or about 1.2 miles down. One of many archaeans that produce the gas methane as part of their metabolism, it can only live in environments in which there is no oxygen.

(b) Prospecting for extremeophiles Researchers looking for archaean extremeophiles in the Obsidian Pool in Yellowstone National Park.

of eukaryotes, while others worked like those of bacteria. It makes sense, then, that archaea lie between bacteria and eukaryotes on life's family tree, as you can see in Figure 20.2 (page 403).

All this speaks to the uniqueness and evolutionary place of the archaea, but what about the widespread distribution of them? In the 1980s, scientists developed a technique that allowed them to do a kind of census taking of archaea out in nature. This method revealed that archaeans live in extreme environments, such as the hot springs in Yellowstone National Park, in more numbers than had been thought. But the real surprise came when more expansive environments were tested; archaea turn out to account for almost a third of the microscopic organisms living in the surface waters off Antarctica, for example, and huge populations of them inhabit deeper-ocean waters.

Prospecting for "Extremophiles"

The species of archaea that inhabit extreme environments have in recent years become the target of a kind of new-age prospecting, with "miners" being scientists from chemical and biotechnology firms. The archaea they are looking for are **extremophiles**—archaea or bacteria that flourish in conditions that would kill most organisms, such as high heat, high pressure, high salt, or extreme pH (**see Figure 20.9**). One of the most extreme of the extremophiles was reported on in 2000, when researchers revealed the existence of an archaean they found thriving in the acidic wastes of an abandoned copper mine. Dubbed *Ferroplasma acidarmanus*, this hardy microbe can live in liquid that has a pH of zero—a habitat more acidic than battery acid. There are also archaea that not only live in, but seem to require, the crushing water pressure that comes with life at the bottom of the ocean; and there are archaeans that thrive in habitats so salty that no plant or animal could survive in them.

To live in such environments, extremophiles ("extreme-lovers") must produce *enzymes* that function in them. And, through modern biotechnology processes, enzymes can be isolated from an organism and then turned out in quantity in factories. These enzymes can then be put to use in *human-made* extreme environments, such as the inside of a washing machine or the confines of a laboratory container. One U.S. company has already introduced a clothing detergent containing a cleaning additive that is simply an enzyme from a heat-loving extremophile.

20.7 Protists: Pioneers in Diversifying Life

We've thus far looked at two of life's domains, Bacteria and Archaea. As noted, the third domain, Eukarya, is so diverse we will look at only one of its four kingdoms in this chapter, the mostly microscopic organisms known as protists.

What's a protist? No one can provide a satisfactory definition. They are something like the objects in a drawer marked "miscellaneous": the primary thing they have in common is that they don't belong in any other drawer. Domain Eukarya has within it three well-defined kingdoms—plants, animals, and fungi—but protists are defined in terms of what they are *not* relative to these life-forms: Protists are eukaryotic organisms that do not have all the defining characteristics of either a plant or an animal or a fungus.

You may recall from Chapter 19 that animals, plants, and fungi *evolved* from protists—animals from protists called choanoflagellates, plants from green-algae protists, and fungi from an unknown variety of protist. Not surprisingly, then, there are protists that are *like* plants, animals, and fungi. Some protists ingest food, as animals do; others make their own food, as plants do; and still others send out slender extensions of themselves and digest their food externally, as fungi do. These similarities will provide us with a way to conceptualize the protist kingdom, but the protists themselves don't actually stay confined within these categories. To give one well-known example of a category-breaker, there is a microscopic protist called *Euglena gracilis* that is bright green and, when at rest, takes on a cigar shape (see **Figure 20.10**). The green color of *Euglena* comes from the chloroplasts within it, which you might remember are the organelles that perform photosynthesis. So, when swimming around its home, which tends to be a freshwater pond, *Euglena* is making its own food through photosynthesis, which would presumably place it with the plant-like protists. Yet when sunlight becomes scarce, *Euglena* simply switches over to an animal-like nutritional mode—it ingests organic matter. Then, when sunlight returns, it goes back to photosynthesis.

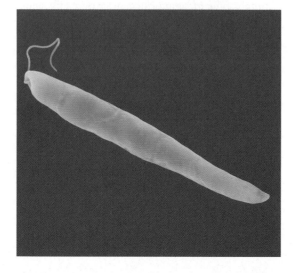

Figure 20.10
More Like a Plant or an Animal?
The protist *Euglena gracilis* switches back and forth between the plant-like behavior of making its own food and the animal-like behavior of capturing food from its external environment.

Protists, then, are difficult to categorize. The only things we can say about all of them is that they are eukaryotes and that they live in environments that are at least moist, if not fully aquatic—damp forest floors as well as oceans and lakes. Almost all are microscopic, though, as you'll see, some are very large. About 100,000 species are known to exist. Though only a small proportion are parasites that affect humans, the rogue's gallery of protists is a formidable one. A microscopic protist, *Plasmodium falciparum*, causes one of the world's most widespread diseases, malaria, by passing between human and mosquito hosts. Campers in the United States are aware of the intestinal parasite *Giardia*, which contaminates water. Other protists cause sleeping sickness and amoebic dysentery. Pathogenic protists also affect humans in less direct ways. The cause of the great Irish Potato Famine of the 1840s was a fungus-like protist, *Phytophthora infestans*, that devastated the Irish potato fields. A close relative of this pest, *Phytophthora ramorum*, is now wiping out stands of California oak trees and bay laurels.

20.8 Protists and Sexual Reproduction

When we look at the living world as a whole, the protists in it are fascinating because they are the organisms that made the break with prokaryotic life—with life as nothing but single-celled bacteria and archaea. After life first appeared on Earth, it consisted solely of bacteria and archaea for more than 2 billion years. The first life-forms to appear other than bacteria and archaea were the protists. And in these organisms, we can see how the living world branched out; we can see living things making transitions to life as we know it today.

One such transition has to do with reproduction. You may recall that bacteria reproduce through a simple splitting of one cell into two. As it turns out, most protists reproduce this way as well. But many protists can reproduce by mixing genetic material from *two* individuals; they give rise to offspring through *sexual* reproduction, in other words. This change was terrifically important to evolution because it meant that offspring would be different from their parents. Bacterial daughter cells are clones of their parent cell—they are genetically identical to it. But the offspring of sexually reproducing organisms have a mixture of genes from both parents, meaning these offspring will not be identical to *either* parent. Once variation started appearing simply as a consequence of reproduction, evolution was off to the races in terms of creating new life-forms, because so much variation was being produced.

This does not mean, however, that life went in one step from reproduction by simple cell splitting to reproduction as we now know it today in, say, human beings. Remember, the protists show us transitions. Let's look at what sexual reproduction is like in an ancient, early-evolving protist.

The protist in question is a type of single-celled algae called *Chlamydomonas*. Most of the time, *Chlamydomonas* reproduces through cell splitting. In times of little nutrition, however, *Chlamydomonas* switches over to sexual reproduction. This begins with two normal *Chlamydomonas* cells, which means two cells that are haploid—that contain only a single set of chromosomes. You may recall that bacteria and archaea are always haploid. It makes sense, then, that the usual state for *Chlamydomonas* is haploid, since, on life's family tree, it lies so close to the bacteria and archaea. Yet Chlamydomonas has made a switch, as you can tell by looking at **Figure 20.11**. Note that at one point, two haploid *Chlamydomonas* cells fuse to create a zygote—the same thing that takes place with human sperm and egg cells. The result is a cell with *two* sets of chromosomes, one set coming from each parent. This cell then undertakes the meiosis reviewed in Chapter 10. You may recall that meiosis always entails a swapping of pieces between the chromosomes that came from parent A and those that came from parent B. Further, these chromosomes then assort randomly, such that any "daughter" cell will get a different mix of chromosomes from each parent. In *Chlamydomonas*, four new offspring cells are produced from the fusion of the original two

Figure 20.11
Sexual Reproduction in a Protist
The simple alga *Chlamydomonas* normally is a haploid organism, meaning one that possesses a single set of chromosomes. In reproduction, however, *Chlamydomonas* cells of opposite mating types (+ and −) can fuse to create a zygote or fertilized egg that is diploid—that has a paired set of chromosomes. Meiosis then ensues in the zygote and the result is four haploid *Chlamydomonas* cells, each of which differs genetically from the others, thanks to the mixing of chromosomes that took place in meiosis.

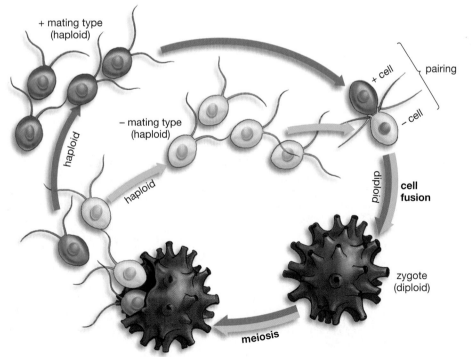

cells. The critical thing is that, because of the shuffling done in meiosis, each of these cells has a unique mixture of chromosomes from each parent. The difference between bacteria and *Chlamydomonas* is the difference between clones and diverse offspring.

One more detail about this process is important. You may wonder about the little plus (+) and minus (−) signs next to the *Chlamydomonas* cells in Figure 20.11. This is an early protist version of what it means to be male and female. There are so-called + and − "mating types" of *Chlamydomonas*, and only these opposites can come together in sexual reproduction. (A + cell cannot fuse with another + cell.) The whip-like flagella that extend from + and − mating types differ in their micro-structure. As a result, only opposite types of flagella can adhere to one another—something that is necessary to get the fusion process going. What we think of as male-female differences are later-arriving variations on this theme.

With this as background, let's now look at a few representatives of the protists, within the categories of plant-like, animal-like, and fungus-like.

20.9 Plant-Like Protists

Plant-like protists are perhaps the only large group of protists known to the average person. The catchall name for them is **algae**, defined as protists that perform photosynthesis. If you look at Figure 20.13, you can look at three varieties we'll be reviewing.

Figure 20.12a shows a microscopic "golden" alga called *Synura scenedesmus* that can be found swimming in freshwater ponds in the United States. In this species, we can see a transition between the single-celled life carried out by bacteria and the multicelled life carried out by more complex organisms. All varieties of golden algae come in species that exhibit **colonial multicellularity**: a form of life in which individual cells form stable associations with one another but do not take on specialized roles. Each of the three

objects you see in the picture of *Synura* is a cluster of *Synura* cells—a colony of them. Each member of the colony keeps its two whip-like flagella pointed to the outside of the group, the flagella beat back and forth, and the colony literally rolls through the water. None of these cells, however, performs a function different from any other cell, which is why this *Synura* is referred to as colonial and not truly multicellular.

If you look at **Figure 20.12b**, you can see a form of *green* algae, called *Volvox*, that represents an evolutionary step beyond *Synura*. The beautiful, translucent circles that outline *Volvox* are a gelatinous material that initially encloses a single layer of anywhere from 500 to 60,000 *Volvox* cells, all of them spaced around the periphery of the sphere. Almost all of these cells have flagella that they move back and forth in unison, so that *Volvox* moves slowly through the water in a given direction, spinning along an axis as it goes (something like a torpedo). For our purposes, the key thing about *Volvox* is that not all of the cells in this colony will reproduce. Only a select group of cells without flagella, generally at the rear pole of the axis, will divide to produce new colonies. These colonies in turn give rise to new spheres that exist inside the original parent sphere. Eventually the gelatinous material around the parent sphere ruptures and the daughter spheres emerge, each of them then continuing the process. The important point is that *Volvox* has cells that specialize. As such, *Volvox* has arguably achieved **true multicellularity**: a form of life in which individual cells exist in stable groups, with different cells in a group specializing in different functions. In human beings, of course, we can see a great specialization in, for example, nerve and muscle cells. *Volvox* shows us specialization in a rudimentary form.

Finally, if you look at **Figure 20.12c**, you can see some representative *brown* algae—the giants of the protist world. One variety of brown algae, the sea kelp, has individuals that may be 100 meters long,

Figure 20.12
Transitions Visible in Algae

(a) Three colonies of the golden algae *Synura*, each of which is composed of a cluster of individual cells. *Synura* is a colonial protist, as none of the cells in a colony takes on any specialized function.

(b) Colonies of the green algae *Volvox* with daughter colonies growing inside. *Volvox* colonies can be regarded as truly multicellular organisms, in that they have specialized reproductive cells. The largest *Volvox* colonies can be seen with the naked eye.

(c) A frond of the brown algae known as sea kelp. These unusually large protists have assemblages of cells that conduct food throughout the organism.

(a) Golden algae

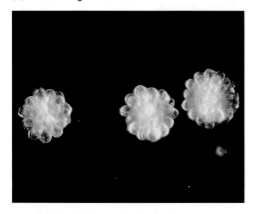

(b) Green algae

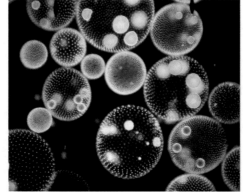

(c) Brown algae

and that often grow in enormous "kelp forests." For our purposes, brown algae are notable because their leaf-like "blades" and stem-like "stalks" represent another leap up in complexity: they contain organized assemblages of cells that conduct food throughout the organism.

One other thing is important about the algae. Note that the golden algae and *Volvox* share three characteristics: they are microscopic, aquatic, and carry out photosynthesis. If we were to go through all the algae, we would find these same qualities in countless species. Creatures in this category, including some bacteria, are known as **phytoplankton**: small photosynthesizing organisms that float near the surface of water. Such organisms are extremely important to life on Earth for two reasons. First, they produce most of the Earth's oxygen. Second, they sit right at the base of most aquatic food chains. All large life-forms on Earth ultimately depend on photosynthesizers for their food. On land, lions may eat zebras, but zebras eat grass. On the ocean, whales may eat the shrimp-like creatures called krill, but krill eat phytoplankton.

20.10 Animal-Like Protists

Once referred to as protozoa, the animal-like protists are alike in that they do not get their nutrients from photosynthesis, as algae do, but instead get them from consuming other organisms or bits of organic matter that they find. Furthermore, they digest this material internally, and they have evolved various means of moving toward their prey. If you look at Figure 20.13, you can see three varieties of animal-like protists.

Figure 20.13a shows a common animal-like protist, *Paramecium*, which is known as a ciliate because of all the hair-like cilia that cover its exterior. These cilia beat in unison, allowing paramecia to move forward or back in an exacting fashion—toward or away from objects in its environment. *Paramecium* may be

single-celled, but it is complex. It takes food in through a mouth-like passageway called a gullet, digests it in a special organelle called a food vacuole, and then empties the resulting waste into the outside world through a special pore on its exterior. Some species even have tiny dart-like structures (called trichocysts) that they can shoot out when threatened.

Figure 20.13b shows a so-called ameboid protist, meaning a protist that gets around by means of a pseudopod or "false foot." The amoeba sends out a slender extension of itself (the foot) and then the rest of its body flows into this extension. It takes in bits of food by encircling them with extensions and then fusing these extensions together. What once had been outside the amoeba is now inside, being digested. The ameboid protists are where we get the idea of a life-form as a "blob," since many of them have no fixed shape, but instead are constantly changing form. Many ameboid protists are parasites, and one of them, *Entamoeba histolytica*, is a serious human parasite, causing an estimated 100,000 deaths per year, mostly in the tropics, by entering the human digestive system through contaminated water. When a parasite invades the human intestinal wall, the resulting condition is known as dysentery. When that parasite is an amoeba, the result is amoebic dysentery.

Figure 20.13c shows another human parasite that enters human beings through contaminated water. So widespread is this "camper's parasite," *Giardia lamblia*, that health authorities have advised Americans never to drink stream or lake water without both filtering and boiling it, or filtering and treating it (with iodine or chlorine). *Giardia* lives in the small intestines of humans and other animals, where it causes symptoms that include nausea, diarrhea, and vomiting. It gets around not by the false foot of the amoeba, nor by the cilia of *Paramecium*, but by pairs of whip-like flagella, as we saw in the algae protists.

**Figure 20.13
Animal-like Protists**

(a) A green *Paramecium*, whose many cilia both help it move and bring food to its gullet, visible in the center.

(b) An amoeba bends itself around food, which it will soon ingest.

(c) The parasite *Giardia lamblia* within the human small intestine. Seen here as the pear-shaped green and white objects, *Giardia* use sucking discs to attach themselves to finger-like extensions of the intestine's inner lining.

(a) A *Paramecium*

(b) An amoeba

(c) The *Giardia* parasite

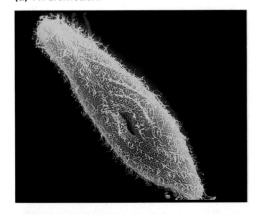

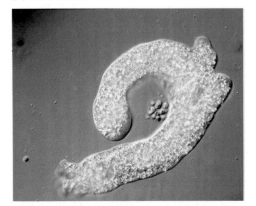

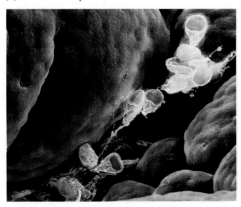

20.11 Fungus-Like Protists

As you'll see next chapter, all fungi get nutrients through a two-step process: They send out slender filaments to a food source, and then release enzymes that digest the food externally. Once this is done, the food is taken back up into the fungus. One class of fungus-like protists, called the oomycetes, operates this same way—so much so that many of these protists can be mistaken for fungi, even by experts. The protist noted earlier that caused the Irish Potato Famine, *Phytophthora infestans*, is in this category. If you look at **Figure 20.14a**, you can see filaments of an oomycete called a water mold extending from a goldfish. In general, the water molds and their terrestrial relatives are **saprobes**, meaning organisms that obtain their nutrition from *dead* organic matter. Clearly, however, they attack living organisms as well.

Figure 20.14b shows another kind of fungus-like protist, this one known as a plasmodial slime mold. This organism assumes several forms during its lifecycle. In the form you see in the picture, its name is appropriate, because it is essentially moving slime. If you looked at it under a microscope, you would see lots of cell nuclei, but almost none of the *boundaries*—the plasma membranes—that normally separate one cell from another. The result is a freeflowing, but slowly moving mass of cytoplasm; the interior of a single cell spread out over a large area, with only one plasma membrane at its periphery. This material moves by the same "cytoplasmic streaming" we saw in the ameboids. As it rolls over rocks or forest floor, it consumes underlying bacteria, fungi, and bits of organic material. Should an object stand in its path, it simply flows around it, or even through it, if the object is porous enough. In short, this is life as a blob again, only this is a large blob. The plasmodial slime mold's similarity with fungi is that, when nutrition is scarce, it changes shape, forming a structure that releases fungus-like spores.

By now, you may be getting the impression that protists could be star contestants in a show called "Can You Top This for Strangeness?" If you look at **Figure 20.14c**, you can see another contestant. Pictured is a much-studied organism, *Dictyostelium discoideum*, a so-called cellular slime mold that has a multiple personality. At one point in its life cycle, it exists in the form of individual microscopic cells that crawl along the forest floor feeding on bacteria. Each "Dicty" is on its own at this stage, but if the bacterial food runs low, these individual cells begin "aggregating" or coming together in an organized group. Eventually, they develop into a migrating "slug" of up to 100,000 cells that has front and back ends and that migrates to a more fertile area of the forest. Then at a certain point, they change form again. They arrange themselves into a stalk-like reproductive structure that has, at its top, spores that will be dispersed and develop into new individual cells that once again crawl along the forest floor. The whole aggregation process will not take place unless starvation is imminent, however; without hunger as a motivator, the individual cells stay separate.

Now, once you get past the shock of realizing that such a thing could happen—that individual cells could just collect themselves and create a larger, moving organism—you can begin to see what interesting questions it raises. What's the signaling mechanism that brings 100,000 separate cells together? How do they begin to cooperate to form these various structures? (To put this another way, how does multicellularity begin to work?) And, perhaps most intriguingly, why should some cells sacrifice themselves by serving on the *stalk* of the reproductive structure—from which location they will not reproduce themselves—while others get to be at the top and disperse to a new life in better hunting grounds? If evolution is all about passing on genes, why should some cells give up the chance to pass on theirs? Scientists study Dicty in hopes of learning more about all these questions.

Figure 20.14
The Fungi-like Protists

(a) The wispy white outgrowths stemming from the goldfish are the filaments of a parasitic protist known as a water mold.

(b) This plasmodial slime mold, *Physarium polycephalum*, is growing on and around a tree trunk in Pennsylvania.

(c) A *Dictyostelium* in its slug stage, composed of tens of thousands of individual cells that have come together to produce an organism capable of moving across the forest floor.

(a) Water mold growing on fish

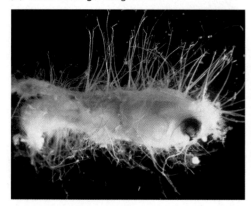

(b) Plasmodial slime mold

(c) *Dictyostelium* slug

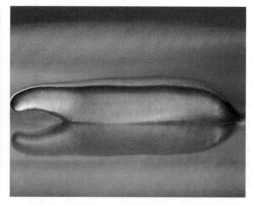

How Did We Learn?
The Discovery of Penicillin

In wartime England, late in 1940, as the German bombings raged, a 43-year-old policeman lay dying in the Radcliffe Infirmary in Oxford. Albert Alexander was, however, not a victim of the German air raids. His enemy was smaller and more subtle. A few months earlier, he had scratched his cheek on a rosebush thorn. This minor wound had then become infected, and from that point on he steadily went downhill as bacteria worked away inside him. By December his face was a mess of festering wounds, one of which cost him an eye.

Incomprehensible as it may seem, in 1940 a scratch from a rosebush thorn could do this to a person. The notable thing about Albert Alexander was not that he became deathly ill from such a trivial injury, but that he came close to beating his tiny enemies. He happened to live near an Oxford University laboratory where a handful of researchers were hurriedly trying to extract, from a common fungus, a product that the world eventually came to know as penicillin. When Albert Alexander became the first human being to receive a therapeutic dose of penicillin in February 1941, the results were startling. His fever dropped, his infection began to clear; even his appetite returned. But there was to be no happy ending for him, for the researchers had only a five-day supply of the drug. Trying everything they could think of, they even extracted precious micrograms of penicillin from Alexander's own urine, to put back into him. But it was not enough; he died days later.

Nevertheless, Alexander's case had shown that penicillin could work in human beings, and clear-cut successes with the drug were not long in coming. A 15-year-old boy with an infected hip underwent a remarkable turn-around when injected with it; children had their eyesight saved when blindness would have been the expected outcome. These early trials made clear that, if ever there was a "miracle drug," penicillin was it. Prior to its advent, doctors could only watch and hope for the best as their patients were wracked by infections that ranged from pneumonia to meningitis to diphtheria to blood poisoning. With the advent of penicillin, at last doctors could do something.

How did it happen that a small band of researchers in wartime England ended up realizing the wondrous potential of this substance? The story begins not in the 1940s, but in 1928, when another scientist in England, Alexander Fleming, discovered the existence of penicillin through a combination of perceptiveness and luck. Returning to his hospital laboratory from a summer vacation, he noticed spot of mold (a type of fungus) in one of the petri dishes in which he was growing or "culturing" bacteria. Such bacterial cultures can easily become contaminated; this dish probably became tainted by means of fungal spores wafting in from another lab in the building. Something about this particular contamination caught Fleming's eye, however: the bacteria in the vicinity of the mold had clearly burst—they were dead. Realizing that chance had brought to him a substance that had the power to kill bacteria, Fleming grew more of the fungal mold, prepared a broth from it, tested its power against several kinds of deadly bacteria, and even went so far as to inject it into healthy mice to see if it had adverse effects (which it didn't). This latter finding was crucial. There are lots of substances that can kill bacteria; the trick is to find one that can kill bacteria while leaving human beings unharmed.

The team had to brew up liters of fungal fluid to get mere millionths of a gram of active penicillin

Having advanced his work with penicillin to this point, however, Fleming then abandoned it. He had clearly recognized its therapeutic potential right from the start, but his subsequent research seemed to convince him that penicillin was just one bacteria killer among many that had no future in medicine. He found that it was unstable; that it appeared very difficult to purify from the stew of fungus chemicals it existed in; and that it quickly lost its power to kill bacteria, once they had been exposed to it. After 1931, Fleming didn't publish anything on penicillin nor did he encourage his students to work with it. He seemed to regard it as a substance that was useful mostly for killing off unwanted bacteria in petri dishes. A great deal of credit goes to him for being perceptive enough to recognize what had been laid in front of him. But had research on penicillin ended with him, there never would have been a drug called penicillin. By 1935, as Trevor Williams has written, "there was not one person in the world who believed in penicillin as a practical aid to medicine."

But this was soon to change, thanks to the work of two immigrants to Britain. One was Howard Florey, an Australian trained as a medical doctor who was then head of a school of pathology at Oxford University. The other was Ernst Chain, a refugee from Hitler's Germany who had been hired by Florey to develop a biochemistry unit in the school. The critical decision the two made in 1938 was to investigate substances in the same class as penicillin—antibiotics, which is to say substances made by one microorganism that are toxic to another. Critically, one of the compounds Florey and Chain decided to investigate was penicillin, whose existence they knew of from reading the earlier papers of Fleming and others.

continued on next page

From the beginning, the team's research on penicillin proceeded on two tracks. One was simply seeing if the substance worked. The other was finding out what the substance *was* and trying to get enough of it to work with. Remember that penicillin is the product of a living organism—originally the fungus *Penicillium notatum*—and as such it existed within a mixture of thousands of other compounds the fungus produces. The trick was to isolate and characterize the germ-killing substance within the fungal soup. This biochemical work was the job of Chain. While this was going on, the team had to brew up liters of fungal fluid to get mere millionths of a gram of active penicillin.

Finally armed with enough of the drug to put it to the test, the group carried out the critical experiment. On the morning of May 25, 1940 they injected eight mice with deadly doses of streptococcal bacteria, and an hour later they injected four of the mice with penicillin. Then a round-the-clock watch began. Normally, the dose given would have killed all the mice within 24 hours, and indeed by 3:30 a.m. the next day the four unprotected mice were dead. But the four mice given the penicillin were all fine! A long road lay ahead with human subjects such as Albert Alexander, but penicillin had done in mice what had so long been sought: It killed bacterial invaders while leaving their "host" unharmed.

This effort was being carried out against the backdrop of the outbreak of World War II. By the time the mouse experiments were started, much of Europe had fallen to Hitler and it was feared that England might be next. (The wartime importance of a drug such as penicillin was clear to the Oxford team, which had contingency plans to destroy all its lab equipment and notes in the case of a successful German invasion. Only the precious *P. notatum* spores would be retained, by means of the researchers rubbing them into their clothing.) In this chaotic environment, the scientists were forced to go outside England to get the help they needed to ramp up the production of penicillin to an industrial scale.

Howard Florey thus went to America to enlist aid there. Oxford researchers eventually found themselves working with chemical engineers from the U.S. Department of Agriculture in the unlikely location of Peoria, Illinois. Where once the English scientists had fermented penicillin in milk bottles, eventually they and their American colleagues were brewing it up in 25,000-gallon tanks.

Getting to this point meant coming up not only with better fermenting methods, but with better starting ingredients. To this end, military pilots collected soil samples from around the world in an effort to find a *Penicillium* species that would yield more of the precious drug. As luck would have it, a species that did the trick was found on the project's doorstep. In 1943, lab worker Mary Hunt bought a moldy cantaloupe at a Peoria market. The melon contained a fungus, *Penicillium chrysogenum*, that doubled the amount of penicillin produced; its daughter spores are the ones used to this day to produce penicillin.

Despite some detours, this crash program was a smashing success. The drug that once was restricted to mice, then to a few human subjects, was put into commercial production early in 1944. By the time of the allied invasion of Europe, in June 1944, enough was being produced that every soldier who needed it could have it. Not much later, all restrictions on it were lifted; penicillin was available by prescription from the corner drugstore.

In 1945, Alexander Fleming, Howard Florey, and Ernst Chain were awarded the Nobel Prize in Physiology or Medicine for their work in discovering this wonder drug. Bestowing the prize just after the war ended, the Nobel committee noted:

In a time when annihilation and destruction through the inventions of man have been greater than ever before in history, the introduction of penicillin is a brilliant demonstration that human genius is just as well able to save life and combat disease.

Penicillin's Discoverers
The three scientists who were awarded the Nobel Prize for their work on the development of penicillin. From left to right, Alexander Fleming, Howard Florey, and Ernst Chain.

How Did We Learn? Penicillin

Until the 1940s, human beings had few defenses against infections by the pathogenic bacteria we looked at earlier. What happened in the 1940s to change this? A group of researchers in England discovered the first antibiotic, penicillin. To learn more about how they did it, see "The Discovery of Penicillin" on page 419.

On to Fungi and Plants

The bacteria, archaea, and protists reviewed in this chapter represent only a portion of the living world. Coming up next are two life-forms that most people think of as related and similar, though in fact they are only distantly related and are quite dissimilar. These are fungi and plants.

Chapter Review

Summary

20.1 Life's Categories and the Importance of Microbes

- All living things on Earth can be classified as falling into one of three domains of life: Bacteria or Archaea, whose members are all single-celled and microscopic; or Eukarya, whose diversity is so broad it is further divided into four kingdoms: plants, animals, fungi, and protists.

- Microbes—living things so small they cannot be seen with the naked eye—are indispensable to all life on Earth. They produce more than half of Earth's atmospheric oxygen; the bacteria among them are responsible for putting nitrogen into a form plants can use; and bacteria and fungi are the most important "decomposers" of the natural world—they break down dead organic matter, such as tree branches, and recycle the resulting elements back into the Earth.

- Microbes live in all environments in which larger life-forms exist. They are present in numbers so immense that the weight or "biomass" of all microbes on Earth exceeds the biomass of all larger life-forms.

20.2 Viruses: Making a Living by Hijacking Cells

- Viruses are non-cellular replicating entities that invade living cells to carry out their replication. Because viruses cannot carry out replication by themselves, most scientists do not classify them as living things.

- The Human immunodeficiency virus (HIV), which causes AIDS, has two structures common to almost all viruses: genetic material and a protein coat, called a capsid, surrounding this material. HIV also has one other structural element that many viruses possess: a fatty membrane, called an envelope, which surrounds the capsid.

- Almost all viruses carry out four steps in their life cycle. They get their genetic material inside a "host" cell; turn out viral component parts, construct new virus clones or "particles" from these parts; and move the new particles out of the cell, at which point the particles go on to infect more cells.

- Viruses cause a host of human illness. New viruses harmful to human beings periodically come along by means of viruses that infect other animals mutating and then "jumping" to human hosts. All forms of life are vulnerable to viral infections.
Web Tutorial 20.1: Life Cycles of Viruses

20.3 Bacteria: Masters of Every Environment

- Bacteria are microscopic, single-celled organisms that are prokaryotes: organisms whose genetic material is not contained within a nucleus. Other defining features of bacteria are that they have only a single organelle (the ribosome) and reproduce asexually through a simple cell splitting called binary fission. Millions of species of bacteria exist. Bacteria are metabolically far more diverse than plants or animals.

20.4 Intimate Strangers: Humans and Bacteria

- Bacteria live on and in human beings in great numbers. In the digestive tract, the relationship between humans and many bacteria is one of mutualism: a relationship between two organisms that benefits both of them. Bacteria get food and habitat from this relationship; human beings get an efficiently functioning digestive system.

20.5 Bacteria and Human Disease

- Only a small number of bacteria are pathogenic or disease causing, but these bacteria are responsible for some of humanity's worst diseases. A few pathogenic bacteria cause harm by invading human cells, but bacteria generally do their damage by releasing or leaving behind harmful substances called toxins.

- The primary human defense against pathogenic bacteria is the class of drugs known as antibiotics, defined as substances produced by one microorganism that are toxic to another. The first antibiotic, penicillin, was developed in the 1940s. Antibiotics work by exploiting the differences between bacterial and human cells, such that they kill bacteria while leaving human cells relatively unharmed.

- The power of antibiotics is being threatened by the emergence of antibiotic-resistant strains of bacteria. These bacteria are evolving in greater numbers because of an overuse of antibiotics in medicine and agriculture.

20.6 Archaea: From Marginal Player to Center Stage

- Archaea were once thought to be a form of bacteria, but have been shown to differ greatly from bacteria in their genetic makeup and metabolic functioning. Like bacteria, they are single-celled, microscopic, and are prokaryotes (cells without nuclei). They appear to be among the earliest evolving organisms on Earth and exist in large numbers both in extreme environments, such as hot springs, and in the world's oceans.

- The extreme environments that so many archaea live in have prompted biotechnology firms to "prospect" for novel archaeans in their home environments, with the intent of developing commercial products from the enzymes these "extremophiles" produce.

20.7 Protists: Pioneers in Diversifying Life

- A protist is a eukaryotic organism that does not have all the defining features of a plant, an animal, or a fungus. Protists are mostly single-celled, and all of them live in environments that are at least moist, if not aquatic. Plants, animals, and fungi evolved from them. Thus there are protists that are plant-like, animal-like, and fungi-like, but many protists do not fit neatly into any of these categories.

20.8 Protists and Sexual Reproduction

- For the first two billion years after life appeared, it consisted solely of bacteria and archaea. Protists were the first life-form to evolve other than bacteria or archaea; they were the organisms that made transitions to many of the capabilities and forms seen in larger organisms today. Among these transitions was the change to sexual reproduction, which protists were the first to practice.

20.9 Plant-Like Protists

- Protists that get their nutrition by performing photosynthesis are known as algae. Some algal species provide examples of colonial multicellularity, defined as a form of life in which individual cells form stable associations with one another but do not take on specialized roles. Other algal protists provide examples of true multicellularity: a form of life in which individual cells exist in stable groups, with different cells specializing in different functions.

- Most algae are microscopic but some varieties, particularly the brown algae, grow to very large sizes. Microscopic algae are important members of the group of organisms known as phytoplankton: small photosynthesizing organisms that float near the surface of water. Phytoplankton are very important to life in general, because they produce most of the Earth's oxygen and form the base of so many aquatic food chains. All phytoplankton are either algae or bacteria.

20.10 Animal-Like Protists

- Once referred to as protozoa, the animal-like protists are alike in that they do not get their nutrients by performing photosynthesis, but instead get them from consuming either other organisms or bits of organic matter. They digest this material internally and have evolved various means of moving toward their prey. The so-called ciliate protists move through use of hair-like cilia surrounding their bodies; amoeboid protists move by means of extending pseudopodia or "false feet." Many amoeboid protists are parasites, and some of them are human parasites that live in the digestive system.

20.11 Fungus-Like Protists

- Some protists get their nutrients by extending slender filaments to a food source, after which they digest that food source externally, in the way fungi do. Protists in this category include the oomycetes, one group of which is known as the water molds. Many oomycetes are saprobes, meaning organisms that obtain their nutrition from dead organic matter.

- Fungus-like protists exist in the strange forms known as plasmodial slime molds and cellular slime molds. The latter category includes the organism *Dictyostelium discoideum*, which exists for part of its life cycle as a collection of separate cells that crawl over the forest floor feeding on bacteria. In times of starvation, however, these cells collect into a single, mobile organism composed of up to 100,000 cells that moves to a more favorable location and distributes individual spores that begin the life cycle again.

Key Terms

algae 416	**heterotroph** 410
antibiotics 412	**mutualism** 411
autotroph 410	**pathogenic** 411
binary fission 409	**phytoplankton** 417
capsid 405	**prokaryote** 406
colonial multicellularity 416	**saprobe** 418
eukaryote 406	**true multicellularity** 416
extremophile 413	**virus** 404

Understanding the Basics

Multiple-Choice Questions (answers in the back of the book)

1. Microbes
   a. produce most of the world's oxygen
   b. make nitrogen available to plants
   c. recycle dead organic matter
   d. account for most of the world's biomass
   e. all of the above

2. Viruses
   a. are the smallest living things
   b. have only one organelle
   c. are beneficial to most organisms
   d. replicate only by invading living cells
   e. have circular chromosomes

3. All viruses possess _____ which they _____.
   a. toxins; kill cells with
   b. cell walls; protect themselves with
   c. DNA or RNA; copy inside themselves
   d. many cells; insert inside a host cell
   e. DNA or RNA; transfer into a host cell

4. Bacteria
   a. are single-celled eukaryotes
   b. live outside cells
   c. reproduce sexually or asexually

d. are multicelled prokaryotes

e. are single-celled prokaryotes

5. Within the human body, bacteria exist in greatest numbers in the _____, with many of these bacteria having a _____ relationship with their human host.
   a. lungs; commensal
   b. feet; antagonistic
   c. digestive tract; mutualistic
   d. digestive tract; antagonistic
   e. kidneys; mutualistic

6. Damage done by pathogenic bacteria most often comes from
   a. bacterial use of bodily resources
   b. bacterial invasion of human cells
   c. the access bacteria give to other microbes
   d. bacterial blockage of blood vessels
   e. bacterial release of toxins

7. Archaea are
   a. single-celled and often found in extreme environments
   b. multi-celled but restricted to marine environments
   c. multi-celled and found only at higher elevations
   d. superficially similar to fungi
   e. a later-evolving, single-celled life-form

8. Protists were the first life-forms to
   a. reproduce sexually
   b. exhibit true multicellularity
   c. exhibit colonial multicellularity
   d. evolve following the bacteria and archaea
   e. all of the above

9. Phytoplankton are important because they
   a. produce antibiotics
   b. purify water systems
   c. sit at the base of terrestrial food chains
   d. sit at the base of aquatic food chains
   e. fight viruses

10. The hallmark of true multicellularity is
    a. the number of cells involved
    b. the stability of the association between cells

c. specialization among cells in the association

d. the existence of nervous tissue

e. sexual reproduction

Brief Review

1. What is an extremophile?

2. Why do most biologists classify viruses as nonliving?

3. List two reasons why phytoplankton are so important to life on Earth.

4. Antibiotics used in humans must have at least two qualities. What are they?

5. What is an amoeba? What class of diseases is associated with them?

6. What is the probable cause of Mad Cow disease? Why is it such an unusual infectious agent?

7. Which of the following classes of living things evolved from protists? Animals, fungi, plants.

8. How do plants get the nitrogen they need?

Applying Your Knowledge

1. What does it mean to be a "successful" organism? Bacteria are certainly the most numerous organisms on Earth, they live in nearly all environments, and they are the least susceptible to being eliminated through environmental catastrophe. Are they Earth's most successful organisms?

2. What does it mean to be an "organism"? The cellular slime mold *Dictyostelium discoideum* exists initially as a collection of separate cells moving over the forest floor. In times of starvation, these cells come together to form a single "slug" that moves to more fertile territory. Eventually it will change into a new structure that will throw off separate cells again. Is "Dicty" an organism or a group of temporarily cooperating cells?

3. What would life on Earth be like if there were no decomposers?

21

The Diversity of Life 2

Breaking down what remains.
(Section 21.1, page 426)

dikaryotic

Life cycle of a fungus.
(Section 21.3, page 429)

Reproducing a club fungus.
(Section 21.4, page 431)

Common houseflies can suffer a hostile takeover by a fungus called *Entomophthora muscae*. It's not unusual for one organism to live off another, but in the case of *E. muscae*, the story has a bizarre twist. This fungus gets inside flies either by drilling through their tough outer skeleton or by squeezing between some cracks in it. The strange thing is that, upon entry, the first thing *E. muscae* does is send out extensions of itself to the fly's brain—specifically to parts of the brain that control crawling behavior. Eventually, the fly begins to crawl *upward*. This usually means upward on a plant stem or leaf, but it may mean upward on the wall of a house, toward the ceiling. Like any fungus, *E. muscae* lives by digesting organic material, and in this case the organic material is the fly's body. By the time the fly begins making its upward trek, it is nearly dead from the internal damage *E. muscae* has done.

If you look at **Figure 21.1**, you can see the end result of all this: a fly that has crawled up on a conifer needle, and, in one of its last acts, has put its mouthparts on the needle, to which they are now glued by the fungus. The white you see around the fly's abdomen are fungal spores. Soon they will be showered over everything *beneath* the fly—including, perhaps, some more flies, which then can be parasitized as well. Fungi have various ways of getting their spores up off the ground for dispersal. The structure we call a mushroom is essentially a spore-dispersal tower. But *Entomophthora muscae* has taken another tack. It gets its *prey* to lift it to higher ground; then it sticks its prey in place; then it uses the prey as a launching pad for a shower of fungal spores. *E. muscae* shows us once again that, while nature is certainly diverse, it is not always pretty.

Figure 21.1
Doing a Parasite's Bidding
Pictured is a root maggot fly (genus *Delia*) that has been killed by an *Entomophthora* fungus, whose white spores can be seen bursting out from the fly's abdomen. Just before death, the fly made its way up to the conifer needles pictured and stuck its feeding organ to them. From the height the fly reached, the *Entomophthora* spores shower down on any additional flies that may lie below. *Entomophthora* brings about a behavioral change in its prey that causes them to climb or fly to heights from which its spores can be dispersed.

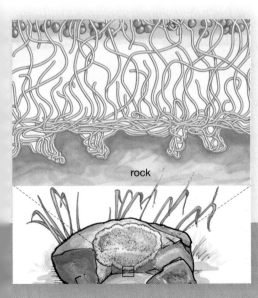

A mutually beneficial arrangement.
(Section 21.5, page 432)

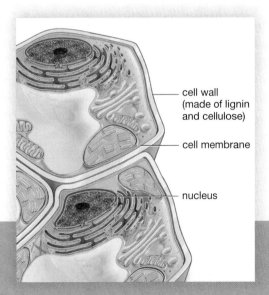

cell wall (made of lignin and cellulose)

cell membrane

nucleus

An important type of cell.
(Section 21.6, page 433)

A bull's eye for bees.
(Section 21.8, page 439)

21.1 The Fungi: Life as a Web of Slender Threads

Entomophthora muscae is a member of one of the "kingdoms" of the living world, the fungi. There are lots of fungi that make their living as parasites, as *E. muscae* does, but then again there are fungi that live in a beneficial association with plants and fungi that do us the favor of decomposing fallen trees (**see Figure 21.2**). Fungi are a large and diverse kingdom, and hence the subject of the first half of this chapter. Once we've reviewed them, we'll look at a better-known kingdom, the plants.

At first glance, plants and fungi may seem to be alike. After all, they're fixed in one spot and tend to grow in the ground. But as you'll see, the main connection between plants and fungi is that many of them have struck up underground partnerships. When it comes to evolution, the surprising thing is that fungi and *animals* are more closely related than fungi and plants. There was a single evolutionary line of protists that gave rise to fungi, plants, and animals, but plants branched off from this line first.

So if a fungus is not a plant, what is it? As the fly example shows, fungi do not make their own food as plants do. Instead, fungi are *heterotrophs*—they consume existing organic material in order to live. They make a living by sending out webs of slender, tube-like threads to a food source. This food source could be a fly, as in *E. muscae*'s case, but it could just as easily be bits of organic matter below ground or a fallen log in a forest. Whatever the case, fungi cannot immediately take in or "ingest" their food in the way that animals do. Fungal cells have cell walls, which means that only relatively small molecules can pass into the fungal filaments. The fungus' solution to this problem is to digest its food *externally*, through digestive enzymes it releases. When the food has been broken down sufficiently by these enzymes, the resulting molecules are taken up into the filaments. Almost all fungi obtain nutrition in this manner. Whether we are talk-

ing about a microscopic bread mold or extensive webs of fungi that spread underneath the forest floor, this is a defining characteristic of fungi. This characteristic and others can be stated in a more formal way.

- Fungi largely consist of slender, tube-like filaments called **hyphae**.

- Fungi get their nutrition by dissolving their food externally and then absorbing it into their hyphae.

- Collectively, the hyphae make up a branching web, called a **mycelium**. If you look at **Figure 21.3**, you can see the nature of this structure. You may have thought of a mushroom as the main part of a fungus, but in reality it is merely a reproductive structure, called a *fruiting body*, that sprouts up from the larger mycelium below. A fruiting body amounts to a large, folded up collection of hyphae.

- Fungi are **sessile** or fixed in one spot, but their hyphae grow. The fungal mycelium does not move, but its individual hyphae grow toward a food source. This growth comes, however, only at the tips of the hyphae, not throughout their length.

- Fungi are almost always multicellular. The exception to this is the yeasts, which are unicellular and thus do not form mycelia.

This last point leads to a more general concept about the living world. You may recall from last chapter that all bacteria and archaea are single-celled, and that protists are usually single-celled. But protists are the limit of single-celled life. With fungi, plants, and animals, life becomes almost completely multicellular. There is no such thing as a single-celled plant or animal, and the one group of single-celled fungi, the yeasts, seem to have evolved into this state from an earlier multi-celled form. This is not to say that all life becomes *big* once we get past the protists. As you'll see, there are plenty of microscopic fungi (and even some microscopic animals). But from here on out, all the life-forms we'll be looking at are made up of more than one cell.

**Figure 21.2
Fungal Diversity**

(a) Fungi serve with bacteria as the major "decomposers" in nature. Here a honey fungus, *Armillaria mellea*, is breaking down a dead tree in a forest in Poland.

(b) Fungi are also human pests. Pictured is a human nail with fungal spores on it that are the cause of athlete's foot.

(c) Fungi distribute their spores in various ways. Pictured are spores being expelled from some puffballs in Costa Rica. What brings about the expulsion? Drops of rain falling on the puffballs.

(a)

(b)

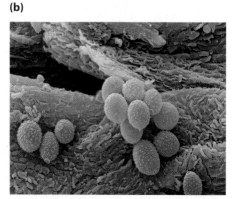

(c)

21.2 Roles of Fungi in Society and Nature

Like bacteria, fungi are mostly hidden from us—and mostly known to us by the trouble they cause. This trouble is considerable, as fungi are responsible for mildew, dry rot, vaginal yeast infections, bread molds, general food spoilage, toenail and fingernail infections, and a variety of related skin afflictions, among them athlete's foot, "jock itch," and ringworm. The agricultural blights of corn smut and wheat rust are fungal infections. Indeed, fungi probably are the single worst destroyers of crop plants. One survey in Ohio indicated that only 50 plant diseases there were caused by bacteria and 100 by viruses, but 1,000 by fungi. If you look at **Figure 21.4**, you can see an example of what happened to American elm trees in cities across the United States beginning in the 1930s because of a fungus called *Ophiostoma ulmi*—the cause of Dutch elm disease. In recent years, fungi called molds have caused great damage to housing in such states as Texas and California. In Texas alone, insurance companies paid more than $1 billion in mold insurance claims in 2001–2002. (To learn more about the strange effects of a fungus that infects rye, see "A Psychedelic Drug from an Ancient Source" on page 430.)

Even given all this fungal pestilence, we could say of the fungi what we said of bacteria: While we could do without some of them, we could not live without others. Mold in houses may be a huge problem, but there is another mold that is the source of penicillin. If you know someone who has high cholesterol, chances are he or she controls it with one of the so-called "statin" drugs (Pravastatin, Lovastatin, or Simvastatin). All these drugs contain a compound produced by the fungus *Aspergillus terreus*. In the realm of food production, brewer's yeast makes bread rise and beer ferment, and blue cheese is blue because of the veins of mold within it. Most soft drinks contain citric acid that is produced from a fungus, and we simply eat fungi outright in the form of mushrooms. Out in nature, fungi join bacteria as the major "decomposers" of the living world, breaking down organic material such as garbage and fallen logs and turning it into inorganic compounds that are recycled into the soil. In fact, the final breakdown of woody material is almost entirely the work of fungi. And, as noted, fungi are involved in a critical association with plants, about which you'll learn more later.

Some 77,000 species of fungi have been identified so far, but this is just a fraction of the number that actually exist. Because fungi tend to grow in inaccessible places, most of them are unknown to us. By one estimate, the total number of fungal species may be about 1.6 million.

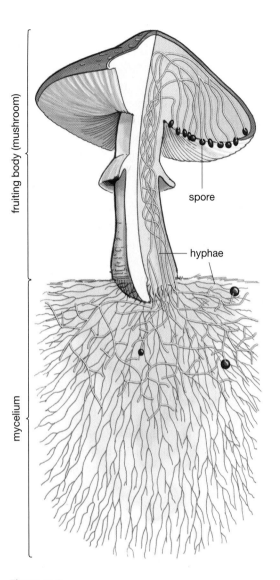

fruiting body (mushroom)

mycelium

spore

hyphae

Figure 21.3
Structure of a Fungus
Fungi are composed of tiny slender tubes called *hyphae*. The hyphae form an elaborate network called a *mycelium*. The same hyphae also form a reproductive structure called a fruiting body, which in this case is the familiar mushroom.

(a) Gillet Avenue, Waukegan Illinois, 1962   **(b)** Gillet Avenue, Waukegan Illinois, 1972

Figure 21.4
The Effects of Dutch Elm Disease

21.3 Structure and Reproduction in Fungi

Let's now look at just one type of fungus as a means of learning something about fungi in general. In **Figure 21.5**, you can see a so-called club fungus that we know as the common mushroom. If you look down at the figure's lower-right, you can see that the mushroom's hyphae amount to thin lines of cells—something that is true of all hyphae. In many cases, individual cells in the hyphae are separated from one another by dividers called septa. But as it happens, septa are somewhat porous dividers; they have tiny openings that allow for a fairly free flow of cellular material between one cell and the next. This allows for rapid movement of cellular resources right to the tip of the hyphae—the site of hyphal growth. The result is that hyphae can grow very quickly, whether they are growing toward a food source or organizing themselves into a mushroom cap. Did it ever seem to you that a mushroom sprouted in your yard overnight? It probably did.

If you look at the upper left of the figure, you can see how the mushroom cap is structured. The underside of each cap is made up of a profusion of accordion-like folds called gills. Each gill is simply a collection of hyphae. And right at the tip of some of these hyphae is a reproductive structure called a basidium. Reproductive cells called spores are produced within the basidium and then ejected from it. Caught by the wind, these spores are then carried away to new ground on which they can "germinate" or sprout new hyphae. Fungal spores are, however, something like lottery tickets: a multitude are made, but only a few will be winners. Such a tiny fraction of spores end up germinating that huge numbers must be released to ensure that all nearby environments receive some. Tens of millions will be ejected each hour from the underside of a fairly small mushroom cap. We noted earlier that the cap's *height* allows spores to be caught by the wind, rather than just falling to the ground below. It turns out that the cap's *shape* allows them to be shielded from the rain. Water plays a part in the ejection of spores from the gills, but it wouldn't do for the spores simply to be rained on.

The mushroom cap is one example of a large fungal reproductive structure, but most fungi don't have this kind of size. Think of a flat circle of blue-gray mold on some spoiled bread. Reproductive structures are present, topped by spores, but magnification is necessary to make out any of this detail, as you can see in **Figure 21.6** on page 430. To a great extent, fungi can be regarded as microbes.

The Life Cycle of a Fungus

We've noted the role that fruiting bodies play in releasing mushroom spores, but how do these spores get produced in the first place? For that matter, what's a spore? These questions can best be understood within the context of the life cycle of a fungus, which is set forth in Figure 21.5. To understand it, we need to go over three tricky concepts related to fungal reproduction. One of these can be called the dikaryotic phase of life; the others are the nature of sex in fungi and the nature of spores.

The Dikaryotic Phase of Life

Taking the life cycle in Figure 21.5 from the beginning, over at step 1, you can see that it starts with spores that have been released from fruiting bodies. These spores germinate, generate hyphae and, in step 2, two of these hyphae fuse—two hyphal cells come together and become one cell, from which a single hypha grows. Note, however, that the cells in this new hypha have *two* nuclei each (symbolized by the little black and white dots within them). This is an unusual condition for a cell; in the normal case, one cell has one nucleus. This condition has come about because, while two cells fused in step 2, their nuclei remained separate. Thus begins a phase of life that is unique to fungi: the **dikaryotic phase**, in which cells in a fungal mycelium have two nuclei. And what is the nature of the genetic material in these nuclei? Each nucleus exists in what is known as a haploid state—there is but a *single* set of chromosomes in each one. This contrasts with the diploid state in almost all human cells. Our nuclei have *paired* sets of chromosomes within them. The haploid state of these nuclei may seem like a technical detail, but it's important for what's coming up. Once a mushroom mycelium enters dikaryotic phase it can stay there for a long time. The phase can go on for years, in fact, and ends only when the most visible manifestation of the mycelium, the mushroom cap, sprouts above ground as seen in step 3.

Sex in Fungi

The dikaryotic state initially exists in all the cells in the mushroom—even the cells that make up the spore-releasing basidia that line the gills. But in some of these cells, right at the tips of the basidia, this will change. The two nuclei in these cells will fuse into one nucleus as seen in step 4. This change means these cells have become diploid; they now have a single nucleus that contains *two* sets of chromosomes. Now think about where each of these sets of chromosomes came from. One of them came originally from

Web Tutorial 21.1
Life Cycles of Fungi

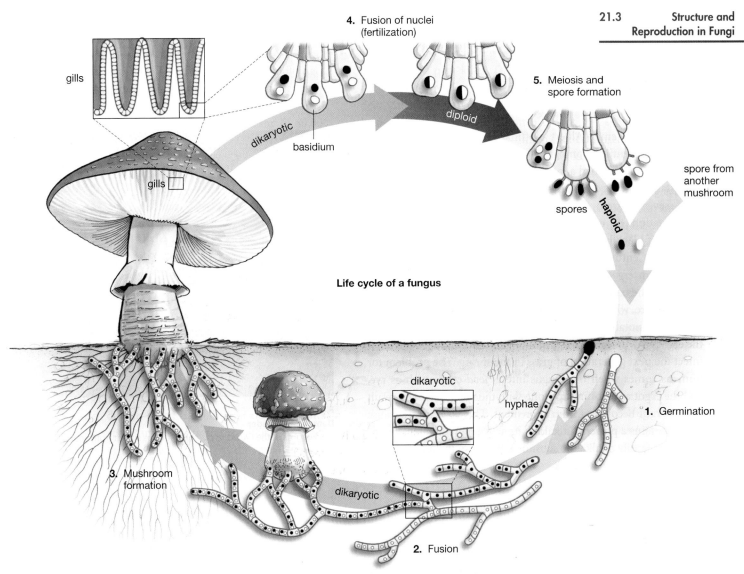

**Figure 21.5
The Life Cycle of a Fungus**

the black spore over at step 1, while the other came from the white spore. They existed separately in a single cell for a while, but now they have fused. With this, we have had a fusion of genetic material from two separate organisms—sex has taken place. It's not the merger of egg and sperm we're used to in human reproduction, but it's sex nevertheless.

Next, as you can see in step 5, each of the cells in which this fusion has taken place undergoes the meiosis that was reviewed in Chapter 10. This means a mixing of chromosomes from the two parents and two rounds of cell division. The result is four haploid spores attached like little bags to the tip of a basidium; these will be among the millions of spores released from the mushroom cap. Each of these spores will be of one "mating type" or another—symbolized by the black and white colors in the figure. Just as only opposite *sexes* can come to-

gether to create offspring in humans, so only opposite *mating types* can come together (in hyphae fusion) to create offspring in fungi when they reproduce sexually.

The Nature of Spores

This gets us to the nature of a **spore**, which can be defined as a reproductive cell that can develop into a new organism without fusing with another reproductive cell. Note the difference between the spores of the mushroom and the *gametes* of a human being—eggs or sperm. A human being cannot develop from an egg or a sperm; a fusion of both cells is required. Meanwhile, the spores that mushrooms develop from do not undergo fusion. A fungal spore is produced in meiosis and, once it comes into being, it is dispersed and can develop into a whole new mycelium by itself.

21.5 Fungal Associations: Lichens and Mycorrhizae

Lichens

Everyone has seen the thin, sometimes colorful coverings called *lichens* that seemingly can grow almost anywhere—on rocks as well as on trees; in Antarctica as well as in a lush forest (**see Figure 21.8**). As it turns out, lichens are not a single organism. A **lichen** is a composite organism composed of a fungus and either algae or photosynthesizing bacteria. If you look at **Figure 21.9**, you can see that this association is structured as a kind of sandwich: an upper layer of densely packed fungal hyphae on top; then a zone of less dense hyphae that includes a layer of algal or bacterial cells; then another layer of densely packed

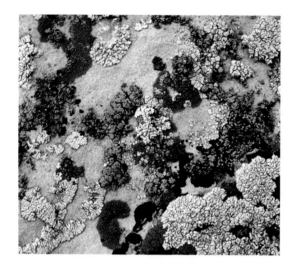

Figure 21.8
Life on the Rocks
Lichens growing on sandstone rocks, near the Montana-Wyoming border.

Figure 21.9
The Structure of a Lichen
A lichen is not a single organism. It is composed of a fungus and an alga (or sometimes a photosynthesizing bacterium) living in a mutually beneficial arrangement. The fungal hyphae form a dense layer on the top and the bottom and a loose layer in the middle, within which the alga is nestled. The fungus may provide a moist, protective environment for the alga, and the alga, which performs photosynthesis, provides food for the fungus.

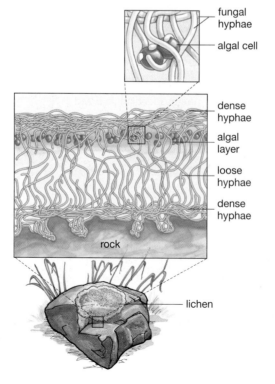

fungal hyphae
algal cell
dense hyphae
algal layer
loose hyphae
dense hyphae
rock
lichen

hyphae on the bottom that sprout extensions down into the material the lichen is growing on (such as a rock). The example here is an association between a fungus and an alga, which is the case in about 90 percent of lichens. In such a relationship, the fungal hyphae either wrap tightly around the algal cells or actually extend into them.

A lichen is mostly a fungus, then, but it is a fungus that could not grow without its algal partner. Why? Because the alga makes its own food through photosynthesis and then proceeds to supply the fungus with most of the nutrients it needs. For years, it was assumed that what fungi do for algae is keep them from drying out. Increasingly, however, scientists have come to believe that most fungi are doing nothing at all for the algae, in which case the fungi should be considered an algal parasite, rather than a partner.

In nature, the general significance of lichens comes from their role as pioneers at sites where life is just getting established, such as barren rocks. Because the algae in lichens make their own food, and because the fungi in them can dig into rock itself, lichens can make a living in such habitats, whereas most other organisms cannot. The lives of lichens, however, bear some comparison with French movies: they tend to be long and slow. Some lichens have been alive for thousands of years, but at most they spread out at a rate of about 3 centimeters a year, which is a little over an inch. Indeed, they have been observed to grow as little as 1 millimeter per year, at which rate they would grow 4 inches in a century.

Mycorrhizae

By some estimates, 90 percent of seed plants live in a cooperative relationship with fungi. The linkage here is one of plant roots and fungal hyphae, with both the plants and the fungi benefiting. What the fungus gets from the association is food, in the form of carbohydrates that come from the photosynthesizing plant; what the plant gets is minerals and water, absorbed by the fungal hyphae. So important is this relationship to plants that some species of trees, such as pine and oak, cannot grow without fungal partners. Other plants will have stunted growth without a fungal partner.

Fungal hyphae generally grow into the plant roots, but in some instances they wrap around the root without penetrating it. In either case, the root-hyphae associations of plants and fungi are known as **mycorrhizae** (**see Figure 21.10**). The mycorrhizal relationship is one of the oldest and most important in nature. It appears that at least 460 million years ago, when the very first plants made the transition to life on land, they did so with fungal partners.

21.6 Plants: The Foundation for Much of Life

In your review of the living world so far, you've seen that human beings have reason to feel ambivalent about bacteria, protists, and fungi. While there is much to appreciate in these organisms, there is a good deal to fear as well. When it comes to plants, however, it's hard to find much to fear—but easy to find much to appreciate. For starters, we humans are utterly dependent upon plants for food. Try naming something you eat that isn't a plant or that didn't itself eat plants, and you will end up with a very short list. Then there is the atmospheric oxygen that plants produce as a by-product of photosynthesis. Beyond this, plants stabilize soil, provide habitat for animals, and lock away the "greenhouse" gas carbon dioxide. They yield lumber, medicines, and certain varieties of them are so beautiful that human beings spend hours tending them. Against all this, what's the worst thing we could say of plants? That a few varieties grow a little too much for our satisfaction? What other form of life helps us so much and harms us so little?

The Characteristics of Plants

We all have an intuitive sense of what plants are, though not all plants match our intuitions. A few are parasites (on other plants) and some, like the Venus flytrap, consume animals. Allowing for such exceptions, plants share the following characteristics.

- *Plants are photosynthesizing organisms that are fixed in place, multi-celled, and mostly land-dwelling.* We

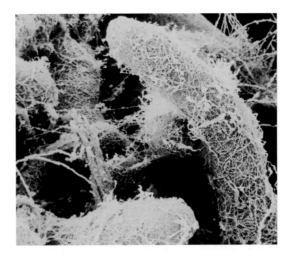

know that plants are fixed in one spot (sessile), and that they make their own food through photosynthesis. That they are multi-celled follows from the fact that they develop from embryos, which are multi-celled by definition. During the course of their evolution from green algae, plants made a transition from water to land, and land is where most of them are found today—though some have made the transition *back* to water, as with water lilies.

- *Plant cells have cell walls and organelles specialized for photosynthesis.* Both plant and animal cells are enclosed by a plasma membrane, but plant cells have something just outside the membrane: a **cell wall**, defined as a relatively thick layer of material that forms the periphery of plant, bacterial, and fungal cells (**see Figure 21.11**). The

Figure 21.10
Underground Partners
Mycorrhizae are associations between plant roots and fungal hyphae that benefit both the plant and the fungus. Shown is an association between Aspen tree roots and thread-like mushroom hyphae that are wrapped around them. The hyphae help bring water and minerals to the trees, while the tree provides the fungus with food.

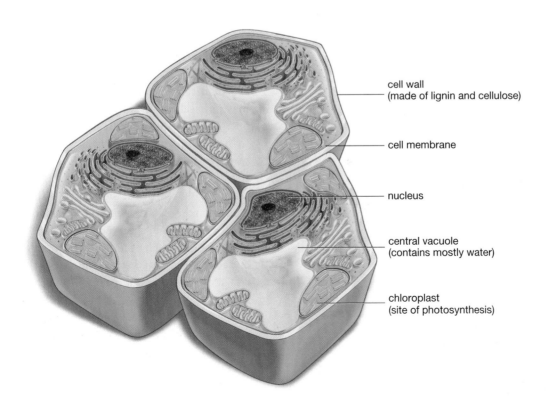

cell wall
(made of lignin and cellulose)

cell membrane

nucleus

central vacuole
(contains mostly water)

chloroplast
(site of photosynthesis)

Figure 21.11
Characteristics of Plants
All cells have an outer membrane, but plant cells have a wall external to this membrane. The compounds cellulose and lignin, which help make up the cell wall, impart strength to it. Plant cells have a higher proportion of water in them than do animal cells, with much of this water located in an organelle called a central vacuole. The sites of photosynthesis in plants are the organelles called chloroplasts.

plant cell wall is composed in large part of a tough, complex compound called cellulose. In woody plants, cellulose is joined by a strengthening compound called lignin. Together, cellulose and lignin allow plant cells to support the massive weight of trees. Plant (and algae) cells also have organelles called **chloroplasts** that are the sites of photosynthesis. If there is one feature that has fundamentally shaped plant evolution, it is the chloroplast. Why are plants *able* to be fixed in one spot? Because they make their own food in chloroplasts and thus don't need to move toward food, as animals do. Given this, it makes sense that plants would evolve into relatively inflexible organisms with protective exteriors—the very thing provided by cell walls.

- *Successive generations of plants go through what is known as an alternation of generations.*

One way to conceptualize this characteristic is to compare plant reproduction with human reproduction. The eggs and sperm that human beings produce are so-called gametes that are *haploid* cells, meaning they have but a single set of chromosomes in them. When these haploid cells fuse in the moment of conception, what's produced is a *diploid* fertilized egg, meaning an egg with two sets of chromosomes—one set from the father and one from the mother. This egg gives rise to more diploid cells through cell division, and the result is a whole new human being (**see Figure 21.12a**).

A typical *plant* in its diploid phase will, like human beings, produce a specialized set of haploid reproductive cells. Instead of these cells being egg or sperm, however, what's produced is a *spore*: a reproductive cell that can develop into a new organism without fusing with another reproductive cell. Like the spores we saw in fungi, plant spores have the ability to grow into a new generation of plant all on their own. In some plants, this spore lands on the ground, starts dividing, and after a time a mature plant exists. When *this* plant reproduces, however, it does not do so through more spores. It produces gametes—eggs and sperm. These go on to fuse, and the result is a diploid plant just like the one we started with.

The upshot of all this, as you can see in **Figure 21.12b**, is that plants move back and forth between two different kinds of generations, one of them spore-producing, the other gamete-producing. In this cycle, the generation of plant that produces spores is known as the **sporophyte generation**, while the generation of plant that produces the gametes is the **gametophyte generation**. The fern shown in Figure 21.12b gives you some idea of how physically different these generations can be, but the disparities actually can be much greater than this. In a massive organism such as a redwood tree, the tree that is familiar to us is the sporophyte generation. But what do its spores develop into? On the male side, barely visible pollen grains. On the female side, small collections of cells, hidden deep within a redwood "pinecone." The fact that these gametophyte individuals are tiny, however, doesn't mean that they are unnecessary. The eggs and sperm they produce are required to bring about a whole new redwood tree. In sum, all plants go through an **alternation of generations**: a life cycle practiced by plants, in which successive plant generations alternate between the diploid sporophyte condition and the haploid gametophyte condition.

Figure 21.12
The Alternation of Generations in Plants, Compared to the Human Life Cycle
(a) Humans Almost all cells in human beings are diploid or 2*n*, meaning they have paired sets of chromosomes in them. The exception to this is human gametes (eggs and sperm), produced through meiosis, which are haploid or 1*n*, meaning they have a single set of chromosomes. In the moment of conception, a haploid sperm fuses with a haploid egg to produce a diploid zygote that grows into a complete human being through mitosis.
(b) Plants In plants, conversely, diploid (2*n*) plants—the multicellular sporophyte fern in the figure—go through meiosis and produce individual haploid (1*n*) spores that, without fusing with any other cells, develop into a separate generation of the plant. This is the multicellular gametophyte shown. This gametophyte-generation plant then produces its own gametes, which are eggs and sperm. Sperm from one plant fertilizes an egg from another, and the result is a diploid zygote that develops into the mature sporophyte generation. The alternation between the sporophyte and gametophyte forms is called the alternation of generations.

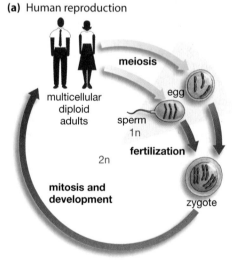

(a) Human reproduction

(b) Plant alternation of generations

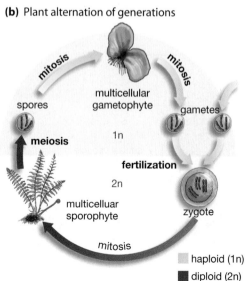

haploid (1*n*)
diploid (2*n*)

21.7 Types of Plants

For our purposes, it is convenient to separate the members of the plant kingdom into a mere four types (**see Figure 21.13**). These are the *bryophytes*, which include mosses; the *seedless vascular plants*, which include ferns; the *gymnosperms*, which include coniferous ("cone bearing") trees; and the *angiosperms*, a vast division of flowering plants—by far the most dominant on Earth today—that includes not only flowers such as orchids, but also oak trees, rice, and cactus. We'll look at all four types briefly

here. Chapters 23 and 24 provide detailed descriptions of the workings of flowering plants.

Bryophytes: Amphibians of the Plant World

If you look in low-lying, wet terrain, you are likely to see a carpet-like covering of moss. In countries such as Ireland, whole fields of peat moss grow and are harvested to be used as fuel for heating and cooking. Mosses are the most familiar example of a primitive type of plant that falls under the classification of **bryophyte**, which can be defined as a type of plant lacking a true vascular system (**see Figure 21.14**).

Figure 21.13
Four Main Varieties of Plants
The most primitive type of plant **(a)** the bryophytes (moss in the figure), lack a fluid-transporting vascular system. **(b)** Ferns, representing the seedless vascular plants, have a vascular system but do not have seeds. **(c)** Gymnosperms (conifers in the picture) do utilize seeds. **(d)** Flowering plants (blossoming pear trees in the picture) produce seeds and are responsible for several other plant innovations, among them fruit.

(a) Moss **(b)** Ferns **(c)** Conifers **(d)** Flowering plants

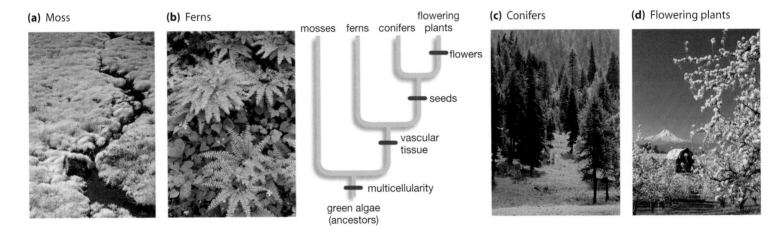

(a) Mosses

(b) Liverworts

(c) Hornworts

Figure 21.14
Three Kinds of Bryophytes
(a) Mosses, **(b)** liverworts, and **(c)** hornworts. These plants have no vascular tissue and thus tend to be small. The sperm of bryophytes must travel through water to get to eggs. For this reason bryophytes are found most often in moist environments.

What's a vascular system? A network of tubes within an organism that serves to transport fluids. We can get a good idea of what the bryophytes are like by looking at the mosses.

Mosses are representatives of some of the earliest plants that made the transition from water to land. As such, they can be conceptualized as plants that made only a partial break with the aquatic living of their evolutionary ancestors, the algae. In making this transition, they had to deal with something the algae didn't, which is the effect of gravity. Lacking a vascular system, mosses cannot transport water and other substances very *far* against the force of gravity; instead they must lie low, hugging the surface to which they are attached while spreading out horizontally to maximize their exposure to sunshine. Nevertheless, the mosses are hearty competitors in some tough environments, such as the arctic tundra and the cracks in sidewalks. In keeping with their aquatic origins, they have sperm that can get to eggs only by swimming through water. Not much water is necessary; a thin film left over from the morning dew will suffice. But some water must be present.

Not surprisingly, mosses and other bryophytes usually are found in moist environments. Bryophytes have no roots at all, but instead use single-celled extensions called *rhizoids* to anchor themselves to their underlying material, which may be soil or rock or wood. You might expect that rhizoids would serve to absorb water, as roots do, but this is true only to a very limited extent. Bryophytes take in water almost entirely through their above-ground exterior surface.

Seedless Vascular Plants: Ferns and Their Relatives

As noted, all plants except the bryophytes have a *vascular system*, or network of fluid-conducting tubes, which transport both food and water. When we begin to look at vascular plants, we find that the most primitive variety of them are the so-called **seedless vascular plants**: plants that have a vascular system, but that do not produce seeds as part of reproduction.

Easily the most familiar representatives of the seedless vascular plants are ferns, with their often beautifully shaped leaves, called fronds (**see Figure 21.15**).

(a) Ferns

(b) Horsetails

(c) Club mosses

Figure 21.15
Three Kinds of Seedless Vascular Plants

(a) Fall-colored ferns in New Hampshire, **(b)** horsetails, and **(c)** club mosses. Because these plants have vascular tissue, they are able to grow taller than most bryophytes. Like bryophytes, however, they do not produce seeds and are tied to moist environments.

These plants have moved a step further in the direction of separation from an aquatic environment. Their vascular system allows them to grow *up* as well as *out*, and it allows for roots that extend into the ground, where they serve their absorptive function. Despite this evolutionary innovation, the sperm of the seedless vascular plants is like that of the bryophytes: it needs to move through water to fertilize eggs.

(a) Spruce tree

(b) *Ginkgo biloba*

Figure 21.16
Gymnosperms
(a) This spruce tree is a member of the grouping of gymnosperms known as conifers, which account for about three-quarters of all gymnosperm species. The conifers also include redwood, pine, juniper, and cypress trees.
(b) There are other types of gymnosperms as well. Pictured are the leaves and seeds of the maidenhair tree, *Ginkgo biloba*, which are used today in herbal preparations.

The First Seed Plants: The Gymnosperms

There are two kinds of seed plants, the gymnosperms and the angiosperms, but putting things this way makes it sound as though these plants are on a kind of equal footing with the bryophytes and seedless vascular plants. In reality, the gymnosperms replaced seedless vascular plants as the most dominant plants on Earth, though this event took place a long time ago—about the time the dinosaurs came to dominance. Then the gymnosperms lost this status to the angiosperms, which began flourishing about 80 million years ago. Today there are only about 700 gymnosperm species compared to 260,000 angiosperm species. Nevertheless, the gymnosperm presence is considerable, especially in the northern latitudes, where they exist as vast bands of coniferous trees, including pine, fir, and spruce. Conifers such as these provide most of the world's lumber (**see Figure 21.16**).

Reproduction through Pollen and Seeds

Why were the gymnosperms able to outcompete so many of the seedless vascular plants back in the dinosaur days? Part of the answer lies in what might be called sperm transport. In gymnosperms, sperm are contained in tiny spheres called pollen grains. Critically, these grains are transported to the female eggs through the *air*, rather than being limited to *swimming* to them, as is the case with the seedless plants. Given this innovation, the gymnosperms could propagate over great distances, which gave them a competitive advantage over the seedless plants.

Another advantage the gymnosperms had were their seeds, which can be thought of as tiny packages of food and protection. If you look at **Figure 21.17**, you can see one example of a gymnosperm seed, this

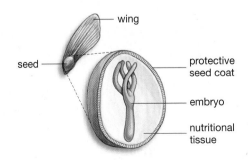

Figure 21.17
A Gymnosperm Seed
Seeds are tiny packages of food and protection. They come in many shapes and sizes, but they all contain an embryo, some food, and a protective seed coat.

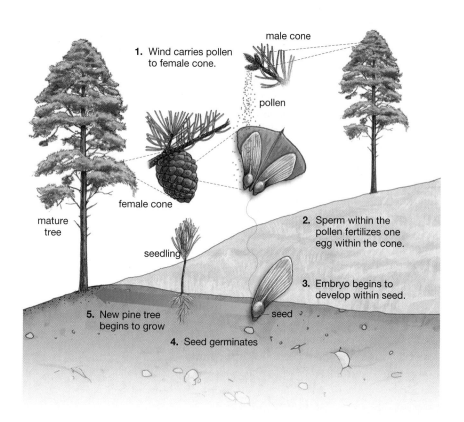

1. Wind carries pollen to female cone.

male cone

pollen

female cone

mature tree

seedling

2. Sperm within the pollen fertilizes one egg within the cone.

3. Embryo begins to develop within seed.

5. New pine tree begins to grow

seed

4. Seed germinates

Figure 21.18
The Life Cycle of a Gymnosperm
Pine trees have two kinds of cones. Pollen is produced within the smaller male cones, while eggs are produced within the larger female cones. When the wind carries pollen onto the female cone, the sperm within the pollen fertilizes one of the eggs within the cone. An embryo then begins to develop inside a seed, which falls to the ground. Once conditions are suitable, the seed germinates and a whole new pine tree begins to grow.

Figure 21.19
Angiosperm Variety

ture, and the result is a fertilized egg that begins to develop as an embryo. This is no different in principle than human egg and sperm coming together to create a human embryo. In the pine tree, however, the embryo is developing inside a seed, a structure that is unique to plants. Seeds have a tough coat, inside of which is not only the embryo but a food supply for it—a kind of sack lunch of stored carbohydrates, proteins, and fats. In sum, a **seed** is a plant structure that includes a plant embryo, its food supply, and a tough, protective casing. When seeds reach the ground, they contain all the starting materials necessary to bring about germination of a new generation of plant.

Gymnosperm seeds turn out to differ from the seeds produced by angiosperms. Angiosperm seeds come wrapped in a layer of tissue—called fruit—that gymnosperm seeds do not have. The details of this anatomical feature are presented in Chapter 24. For now, just be mindful that a **gymnosperm** can be defined as a seed plant whose seeds are not surrounded by fruit. The very name *gymnosperm* comes from the Greek words *gymnos*, meaning "naked," and *sperma*, meaning "seed."

one from a pine tree. **Figure 21.18** then shows you how this seed fits into the life cycle of the pine. As you can see, pollen grains are carried on the wind to the female reproductive structure, which resides in the familiar pinecone. Then the sperm in the pollen grain comes together with an egg in the reproductive struc-

Angiosperms: Nature's Grand Win-Win Invention

As noted, there are about 260,000 known species of the flowering plants known as angiosperms. The term *flowering plants* may bring to mind roses or tulips and, indeed, these flowers are angiosperms. But the angiosperm grouping includes all manner of other plants as well—almost all trees except for the conifers, all our important food crops, cactus, shrubs, common grass: the list is very long (**see Figure 21.19**). As noted, **angiosperms** are defined by an aspect of their anatomy; their seeds are surrounded by the tissue called fruit. The details of angiosperm anatomy and physiology

(a) Calla lilies on the California coast

(b) Cholla cactus in Arizona

(c) Corn in a field

21.8 Angiosperm-Animal Interactions

For an angiosperm such as a new honeysuckle plant to come into existence, a two-part process must take place. First, sperm—developing once again inside a pollen grain—must get to an egg and fertilize it, which produces the embryo encased in a seed coat. Next, this seed must land on a patch of soil somewhere and begin to germinate or sprout.

Now, in the angiosperms, how is it that sperm gets to egg in the first place? Remember that with gymnosperms, pollen grains were carried by the *wind*. By contrast, the angiosperms have induced *animals* to carry pollen from one plant to another—to pollinate them. Not all angiosperms are pollinated in this way; some are pollinated by wind, and a few aquatic species rely on water currents. But most are pollinated with the help of animals, be they insects, birds, mammals, or even snails (**see Figure 21.20**).

Angiosperms use various attractants to encourage animal pollination. The most important of these is nectar, which is essentially sugar-water, but angiosperms don't stop there. So important is pollination to them that they have developed a host of pollination marketing strategies. How about a *fragrance* that, say, bees are sensitive to? How about *colors* that they're attracted to? Insect-pollinated flowers generally have both sweet fragrances and coloration patterns—nature's equivalent of homing signals and landing lights. They serve to get an animal to the flower in the first place and then into the right *position* for feeding and pollination (**see Figure 21.21**).

Why do plants spend so much energy attracting animal pollinators while gymnosperms let the wind do the work? Windborne pollination in gymnosperms can be thought of as a kind of scatter-shot approach: If the wind *happens* to blow a pollen grain onto the fe-

(a) Pollen-covered honeybee on a dandelion

(b) Carib martinique pollinating a flower

(c) Lesser long-nosed bat pollinating a cactus

Figure 21.20
Animals Help Pollen Get from Here to There
Flowering plants were the first to take advantage of the mobility of animals to transport pollen from one plant to another. Relationships between flowers and pollinators often are species-specific, which helps to ensure that the pollen will be delivered to the right address.

male reproductive structure, then pollination occurs. Think how much more directed things are with animal pollination. It is essentially door-to-door service. Animal pollination is one of the reasons that angiosperms are by far the most dominant plants on Earth.

(a) This is what we see (normal sunlight).

(b) This is what the bee sees (ultraviolet light).

(c) This is what the bee "thinks."

Figure 21.21
Food Lies This Way
Animals may be attracted to a particular flower by its fragrance, color, or pattern of visual elements. Insects are guided into the nutrients in the center of the flower by color patterns called nectar guides, visible to an insect—because it perceives ultraviolet light—but not to us.

Seed Endosperm: More Animal Food from Angiosperms

Once pollination has occurred, the resulting seed, with the embryo inside, has to reach the ground and then germinate in it. All seeds contain food reserves for the growing embryo, but angiosperm seeds develop a special kind of nutritive tissue, called **endosperm**, that often surrounds the embryo (**see Figure 21.22**). This endosperm supplies much of the food for human beings throughout the world. Rice and wheat grains are seeds, consisting in large part of the endosperm meant to sustain the plant embryo. It's worth noting that *whole-grain* wheat is wheat that still has its embryo (usually called "wheat germ") and its seed coat (usually called "bran"). By contrast, white bread, regular pasta, and so forth are made from wheat whose wheat germ and bran have been removed through processing.

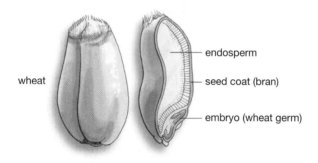

wheat — endosperm — seed coat (bran) — embryo (wheat germ)

Figure 21.22
Food for the Seedling, Food for Humans
Endosperm is a food reserve in seeds. It feeds the embryo before it can make its own food. It is this same endosperm that nourishes humans throughout the world in the form of wheat, corn, rice, and even coconut.

Fruit: An Inducement for Seed Dispersal

Some flowering plants also add an inducement for the *dispersal* of their seeds: the tissue called fruit. This is not fruit in the technical sense, but fruit as that term is commonly understood—the flesh of an apricot or cherry, for example. As such, it represents one more piece of bounty that plants provide to animals. To understand how fruit benefits plants as well, imagine a bear who consumes a wild berry, which consists not only of the fruit flesh, but of the seeds inside this flesh. The fruit is digested by the bear as food, but the *seeds* are tough; they are passed through its system intact to be deposited, with bear feces as fertilizer, at a location that may be very remote from the place where the bear *ate* the berry. The result is seed dispersal at what may be a promising new location for a berry plant (**see Figure 21.23**).

Given pollination and the varieties of food that angiosperms produce, it's easy to see that, with the rise of the flowering plants, animals and plants entered into a much more interdependent relationship than had existed before, one that is still evolving today.

On to a Look at Animals

In this chapter and the one that preceded it, you've looked at bacteria and archaea, and at three of the kingdoms within the eukaryotic domain—protists, fungi, and plants. This leaves one more eukaryotic kingdom to go. It is a kingdom that is arguably the most familiar of all, because we human beings are part of it. Yet it is a kingdom so diverse that it holds a good many surprises as well. It is the kingdom of animals.

Figure 21.23
Seed Carriers
Angiosperms have taken advantage of the mobility of animals not only to transfer pollen, but also to disperse seeds.
(a) Some seeds are wrapped in tasty fruit that is consumed by animals, such as bears.
(b) Other seeds come wrapped in burrs and spines that stick to the fur of animals and are carried away.

(a) Wild berries induce seed dispersal

(b) Burrs help seeds travel

Chapter Review

Summary

21.1 The Fungi: Life as a Web of Slender Threads

- Fungi do not make their own food, as plants do, but instead are heterotrophs—they consume existing organic material in order to live. Fungi consist largely of webs of slender tubes, called hyphae, that grow toward food sources. Fungi cannot immediately ingest the food the hyphae reach, since only small molecules can pass through the cell walls of fungal cells. Fungi thus digest their food externally, through release of digestive enzymes, and then bring the resulting small molecules into the hyphae.

- Collectively, hyphae make up a branching web, called a mycelium, that forms the bulk of most fungi. Many fungi have a reproductive structure called a fruiting body that produces and releases reproductive cells called spores. All fruiting bodies are organized collections of hyphae.

- Fungi are sessile or fixed in one spot, though their hyphae grow toward food sources. Such growth comes only at the tips of the hyphae, not throughout their length.

- Almost all fungi are multicellular. The sole exception to this, the single-celled yeasts, are thought to have evolved to the single-celled state from an earlier multi-celled condition.

21.2 Roles of Fungi in Society and Nature

- Fungi are the cause of many crop diseases, human infections, and forms of damage to buildings and homes. They are also sources of medicines, and are used extensively in food processing. In nature, fungi join bacteria as major "decomposers" of the living world, breaking down organic material such as fallen trees and turning it into inorganic compounds that are recycled into the soil.

21.3 Structure and Reproduction in Fungi

- Individual cells that form hyphae often are separated from one another by dividers called septa. These septa are porous enough, however, that they allow for a fairly free flow of cellular material between one hyphal cell and the next—so much so that hyphae can bring resources very quickly to the point of growth at the hyphal tips. As a result, fungi can grow very quickly.

- The fruiting body of the mushroom fungi, the mushroom cap, is made up on its underside of accordion-like gills that both produce spores and release them into the wind, which carries them to new locations. Only a tiny fraction of fungal spores will successfully germinate, or sprout new mycelia. Because of this, fruiting bodies release huge quantities of spores as a means of ensuring new fungal growth.

- Within their life cycle, many fungi have a so-called dikaryotic phase of life, in which the fusion of cells from different hyphae produces a single hypha whose cells each have two nuclei. Each

of these nuclei is haploid, meaning each contains only a single set of chromosomes.

- When the below-ground mycelium in a mushroom fungus sprouts into the above-ground mushroom cap, all the cells in the cap initially are dikaryotic, including those in the reproductive structure called the basidium. The nuclei in certain cells of the basidia then undergo fusion, which is the first step in carrying out sexual reproduction. These cells are now diploid—they have a single nucleus that contains a paired set of chromosomes. These same cells then undergo meiosis, and the result is four haploid spores attached to tip of the basidium. These spores are then released and blown by the wind to new locations, where they may germinate and result in new mycelia.

- A spore is a reproductive cell that can develop into a new organism without fusing with another reproductive cell. Unlike human eggs and sperm, which fuse to make a fertilized egg, fungal spores do not need to fuse with any other cell to give rise to a new generation of fungi.

  Web Tutorial 21.1: Life Cycles of Fungi

21.4 Categories of Fungi

- There are three major categories or "phyla" of fungi: basidiomycetes (also known as club fungi), ascomycetes (also known as cup fungi), and zygomycetes (also known as bread molds). Fungi are placed in these categories in accordance with the structures they use in sexual reproduction.

- Fungi that have not yet been observed to reproduce sexually are placed in a holding category, called the mitosporic (or imperfect) fungi, until such time as they are seen reproducing sexually; at that point they are placed in the appropriate phylum. The single-celled yeasts are ascomycetes, but often are studied as a unified group.

21.5 Fungal Associations: Lichens and Mycorrhizae

- Lichens are composite organisms, made up of both fungi and algae (or fungi and photosynthesizing bacteria). Within a lichen, the relationship between fungi and algae may be mutually beneficial to both kinds of organisms: Fungi derive food from the photosynthesizing algae, and the algae are shielded by the fungi from forces that would dry them out. Many scientists have concluded, however, that algae derive nothing from the relationship, in which case the fungi actually are parasites.

- Up to 90 percent of seed plants live in a cooperative association with fungi that links plant roots with fungal hyphae. In this relationship, plants supply fungi with food produced in photosynthesis, while the fungi supply plants with minerals and water, gathered by the web of fungal hyphae. Associations of plant roots and fungal hyphae are called mycorrhizae.

21.6 Plants: The Foundation for Much of Life

- Plants are the foundation for much of life on Earth, because they are responsible for much of the living world's production of food and oxygen. In addition, they stabilize soil, provide habitat for animals, and lock up carbon dioxide. All plants are multi-celled, and almost all are fixed in one spot and carry out photosynthesis. All plant cells have a cell wall and contain organelles called chloroplasts, which are the sites of photosynthesis. Plants reproduce through an alternation of generations: a life cycle in which successive plant generations produce either spores (the sporophyte generation) or gametes (the gametophyte generation). Within a given species, these two generations can differ greatly in size and structure.

21.7 Types of Plants

- The four principal categories of plants are bryophytes, seedless vascular plants, gymnosperms, and angiosperms. Bryophytes include mosses; seedless vascular plants include ferns; gymnosperms include coniferous trees; and angiosperms include a wide array of plants, such as orchids, oak trees, rice, and cactus.

- Bryophytes are representative of the earliest plants that made the transition from water to land. They lack a fluid-transport or vascular system and thus tend to be low-lying. Bryophyte sperm can get to eggs only by swimming through water. Thus, bryophytes are most commonly found in damp environments.

- Seedless vascular plants have a vascular system but do not produce seeds in reproduction. Their sperm must move through water to fertilize eggs.

- Gymnosperms are seed-bearing plants whose seeds are not encased in tissue called fruit. There are only about 700 gymnosperm species, but their presence is considerable, particularly in northern latitudes, where gymnosperm trees, such as pine and spruce, often dominate landscapes. The sperm of gymnosperms is encased in pollen grains that are carried to female reproductive structures by the wind. Gymnosperms produce seeds in carrying out reproduction. Seeds are structures that include a plant embryo, its food supply, and a tough, protective casing.

- Angiosperms, or flowering plants, are seed plants whose seeds are encased in tissue called fruit. Angiosperms are easily the most dominant group of plants on Earth, with some 260,000 species having been identified to date.

21.8 Angiosperm-Animal Interactions

- Angiosperm pollen grains generally are transferred from one plant to another by animals, such as insects and birds. Such animal-assisted pollination is unique to angiosperms. To induce animals to carry out this pollination, angiosperms produce nectar and have developed attractive colorations and fragrances.

- Angiosperm seeds contain tissue called endosperm, which functions as food for the growing embryo. Endosperm supplies much of the food that human beings eat. Rice and wheat grains consist largely of endosperm.

- Angiosperm seeds are unique in the plant world in being wrapped in a layer of tissue called fruit. Fruit that is edible functions in angiosperm seed dispersal, because animals will eat and digest the fruit but then excrete the tough seeds inside, often in a different location.

Key Terms

alternation of generations	434	**hyphae**	426
angiosperm	438	**lichen**	432
bryophyte	435	**mycelium**	426
cell wall	433	**mycorrhizae**	432
chloroplast	434	**seed**	438
dikaryotic	428	**seedless vascular plant**	436
endosperm	440	**sessile**	426
gametophyte generation	434	**spore**	429
gymnosperm	438	**sporophyte generation**	434

Understanding the Basics

Multiple-Choice Questions (answers in the back of the book)

1. Most fungi are composed almost entirely of
   a. ball-shaped objects called hypha
   b. leaves and stems
   c. reproductive cells called spores
   d. slender filaments called hyphae
   e. slender filaments called septa

2. With respect to food, animals _____ while fungi _____
   a. digest, then ingest ... ingest, then digest
   b. consume existing organic material ... make their own
   c. ingest only ... digest only
   d. digest externally; digest internally
   e. ingest, then digest ... digest, then ingest

3. Fungi reproduce through cells called spores, which
   a. have a single set of chromosomes
   b. can germinate without fusing with another cell
   c. often are released from structures called fruiting bodies
   d. usually are released in huge numbers
   e. all of the above

4. Only one of these organisms grows toward new sources of food. Which one is it?
   a. flowering plant
   b. bacterium
   c. fungus
   d. moss
   e. algae

5. Fungi cause (select all that apply)
   a. tetanus
   b. athlete's foot
   c. common colds
   d. dry rot
   e. wheat rust

6. Defining characteristics of plants include (select all that apply)
   a. alternation of generations
   b. hyphae
   c. cell walls
   d. reproduction through fruiting bodies
   e. nutrition from photosynthesis

7. You are walking through a botanical garden and you see a 6-meter-tall plant that shows no evidence of flowering, but seeds are present. To which group of plants does it belong?
   a. gymnosperms
   b. angiosperms
   c. mosses
   d. ferns
   e. cacti

8. Only angiosperms (select all that apply)
   a. are pollinated by animals
   b. produce seeds
   c. encase seeds in fruit
   d. have seeds containing endosperm
   e. have sperm encased in pollen grains

9. In the alternation of generations that plants employ (select all that apply)
   a. all generations produce eggs and sperm
   b. one generation produces spores, the next produces eggs and sperm
   c. succeeding generations tend to be about the same size
   d. succeeding generations tend to differ in size
   e. all generations produce spores

10. Plants (select all that apply)
   a. produce oxygen
   b. decompose organic matter
   c. destabilize soil
   d. lock up carbon dioxide
   e. usually form associations with underground fungi

Brief Review

1. Fungi could be said to take a "lottery" approach to reproduction. In what respect?

2. How does a spore differ from a gamete?

3. Animals have evolved various means of moving toward the food they eat. What is the approach that fungi take?

4. Describe two means by which angiosperm reproduction is aided by animals.

5. Describe two common associations that fungi make with other organisms. What does each party in each of these associations do for the other, and what do they gain themselves?

Applying Your Knowledge

1. Animals move because they need to—they have to get to prey and they must evade predators. These dual needs led to the evolution of the animal nervous system, which in turn led to the development of varying levels of intelligence in animals. Meanwhile plants never had a need to move, because they made their own food through photosynthesis. Do the separate evolutionary paths of animals and plants mean that there is an inherent value to struggle in the living world—in the animals' case, the struggle to obtain food?

2. Like all living things, fungi require water to begin growing. The recent scourge of mold fungi in American homes has primarily affected new homes. Why should they be affected more than old homes?

3. The *Entomophthora muscae* fungal parasite that invades flies has no consciousness, yet it manages to undertake actions that affect the behavior of the flies it attacks. (It sends hyphae into the flies' brains and, as a result, the flies crawl upward.) Almost all parasitism is disturbing to human sensibilities, but this brand is particularly unsettling. Why should this be so?

22 *The Diversity of Life 3*

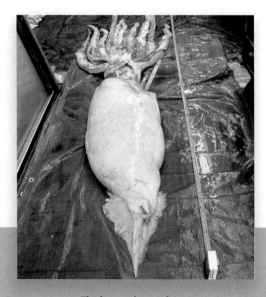

The largest invertebrate.
(Section 22.7, page 458)

Shedding an old skeleton.
(Section 22.9, page 461)

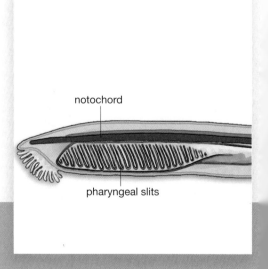

notochord

pharyngeal slits

One of our chordate relatives.
(Section 22.11, page 467)

The diversity of animals is one of the wonders of the natural world.

We human beings are, in a sense, intimate strangers with the bacteria and fungi that were reviewed in the last two chapters. We are intimate with these organisms in that we carry out our lives right along beside them: we walk through clouds of them, wash off layers of them, and have trillions of them living inside us. That said, except for the occasional outbreak of athlete's foot or stomach flu, they never cross our minds. So small are bacteria and most fungi that, while they may be everywhere, for us they are effectively nowhere. It's a different story with the animals that are the subject of this chapter. Our cats and dogs and neighborhood birds are part of our conscious lives. We pay attention to a new spider web on our porch or to the lizard on our hiking path. We wonder about the circling hawk overhead.

Animals come from the long evolutionary line of *other* feeders: the heterotrophs, who cannot manufacture their own food but must get their nutrition from outside themselves. Many bacteria fit into this category, as do lots of protists and all of the fungi. But animals, in their quest for food, added something that is unique to them: A nervous system; a system for transmitting complex messages rapidly over long pathways in the body. Once this happened, the living world was off to the races in terms of adding novel abilities. Think of sight, hearing, smell, flight, walking, singing, and reading. Then there are such things as the echo-location of bats and the waggle-dance of honey bees. For the average person, to be alive means to sense, to investigate, to respond, to move. In all these areas, animals reign supreme in the living world.

This chapter is about the broad diversity that exists in animals (**see Figure 22.1** on the next page). Given that animals are part of our daily lives, some of what follows will be familiar. But there are likely to be many surprises as well.

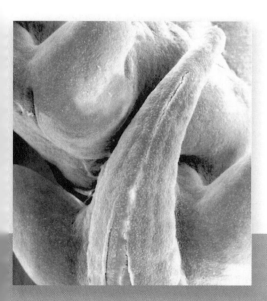

A human tail.
(Section 22.11, page 467)

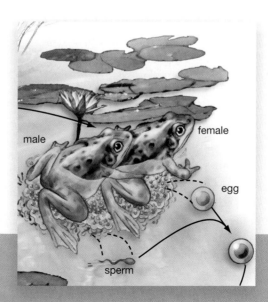

male   female

egg

sperm

A double life.
(Section 22.11, page 470)

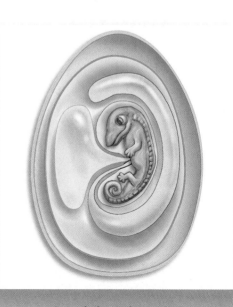

An evolutionary innovation.
(Section 22.11, page 470)

445

Figure 22.1
Animal Diversity
An octopus, a feather star, and a chimpanzee are very different creatures, yet all are members of the animal kingdom.

(b) Feather star

(a) Octopus

(c) Chimpanzee

(a) Octopuses often are found crawling over ocean rocks, but they swim as well, by expelling a jet of water from a hose-like siphon, often propelling themselves "backwards" as with this one.

(b) Though they look like plants, ocean-dwelling feather stars such as this one are animals that are part of the same phylum as sea stars. They extend their arms to catch bits of food drifting in the ocean currents.

(c) Chimpanzees are the closest living relatives of human beings.

22.1 What Is an Animal?

In defining animals, the question that needs to be answered is: What characteristics do *all* animals have that other organisms *don't* have? It turns out there is a single, rather technical feature that is sufficient to set animals apart from all other living things.

- Animals pass through something called a blastula stage in their embryonic development. A *blastula* is a hollow, fluid-filled ball of cells that forms soon after an egg is fertilized by sperm (see Chapter 29 for details). All animals go through a blastula stage, but no other living things do.

Three other characteristics are found in all animals, but they're found in other kinds of organisms as well. All animals:

- Are multicelled; there are no single-celled animals.
- Are heterotrophs; they must get their nutrition from outside themselves.
- Are composed of cells that do not have cell walls. (The outer lining of animal cells is the plasma membrane. By contrast, all plants and most other creatures have a relatively thick additional lining—the cell wall—outside their plasma membrane.)

Apart from these universals, animals *usually* go on to share other characteristics. Animals tend to move, though there are animals that, for at least part of their lives, are as *sessile*, or fixed in one spot, as any plant. Except for the animal group that includes sponges, animal bodies are organized into **tissues**, each tissue being a collection of similar cells that serves a common function.

Then there are characteristics of animals that don't fit our preconceived notions. While it's true that animals generally are large relative to bacteria or protists, some animals are microscopic. We often think of animals as creatures with vertebral columns or "backbones," but there are far more **invertebrates**, or animals without vertebral columns, than animals with them. Indeed, 99 percent of all animal species are invertebrates. We generally think of animals as land creatures and, with a huge assist from insects, there certainly are a lot of land animals. However, in terms of animal *diversity*, as measured by basic body plans, there are more kinds of ocean-going animals than land animals, as you'll see. This makes sense because animals existed in the sea for at least 100 million years before any of them came onto land.

22.2 Animal Types: The Family Tree

Basic body plans turn out to provide a sensible way of dividing up the animal kingdom. Depending on who is counting, there are between 36 and 41 animal phyla, with each phylum being informally defined as a group of organisms that share a basic body plan. The classification scheme for the living world introduced in Chapter 18—that of kingdom, phylum, class, order, family, genus, and species—provides the basis for a more formal definition of phylum. A **phylum** is a category of living things, directly subordinate to the category of kingdom, whose members share traits as a result of shared ancestry. (For more on classification, see Chapter 18, page 359.)

So, in what ways are the various animal phyla related to one another? To put this another way, what does the animal family tree look like? At the moment, this question is a matter of great debate among **zoologists**, meaning biologists who study animals. By measuring similarities in DNA and RNA sequences among living animals, these scientists have come up with entirely new conceptions of who is more closely related to whom. A consensus seems to have emerged about this, however, and you can see it laid out for you in **Figure 22.2**. Not all the 36–41 animal phyla are shown; some of the phyla at the top represent groups of phyla. For more on how DNA and RNA evidence has changed our view of animal evolution, see "Redrawing the Animal Family Tree" on page 449.

One way to conceptualize this tree is that, over time, animals became more complex through a series of *additions* to the characteristics found in more primitive animals. The twist is that only some varieties of animals evolved to get these additions, while others retained the primitive, "ancestral" condition. Let's take a tour of the tree now, focusing on features that were added to the animal kingdom through evolution.

Web Tutorial 22.1
The Animal Family Tree

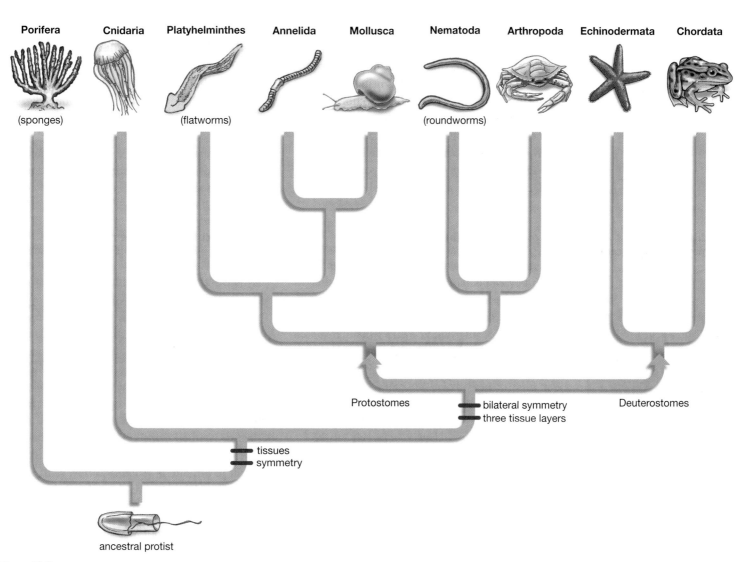

Figure 22.2
A Possible Family Tree for Animals
The animal kingdom is divided into groups called phyla. The members of each phylum have in common physical features that are evidence of shared ancestry. Nine of the estimated 36–41 animal phyla are shown here. All these phyla are regarded as having evolved from an ancestral species of protist, shown at lower left in the family tree. The red horizontal lines on the tree mark the points in animal evolution at which certain structural innovations appeared, such as tissues and bilateral symmetry. Human beings are members of phylum Chordata, at upper right.

coelom as one tube that encircles another; the coelom is generally tube-shaped, and it surrounds the tube that is the digestive tract. (There are, however, lots of variations on this general principle.) **Figure 22.4** displays what it means to have no coelom, as in flatworms; a *pseudocoel*, as in roundworms; and a *true coelom*, as in earthworms.

The concept of a coelom is intimately linked to that of tissue layers. Almost all animal embryos have what are known as germ layers, defined as layers of cells that *become* various types of tissue in adult ani-

mals. Most animals have three types of germ layers: endoderm, mesoderm, and ectoderm. Endoderm is initially an inner layer of cells in the embryo, mesoderm a middle layer, and ectoderm an outer layer. The general importance of these tissue layers will be reviewed later. For now, just note the linkage, displayed in Figure 22.4, between mesoderm and a coelom. This linkage provides a way of formally defining a **coelom**. It is an enclosed body cavity in an animal, that is lined with cells of mesodermal origin. You may wonder why there are no red lines on Figure 22.2 showing where the coelom was added in animal evolution. So valuable is this internal space that it seems to have evolved independently several times among animals—at least once each among the two large groupings you can see toward the base of the tree, the protostomes and the deuterostomes. But what are these two large groupings? Let's see.

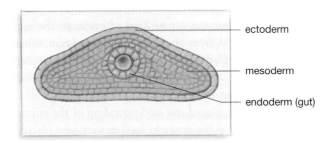

— ectoderm
— mesoderm
— endoderm (gut)

flatworm

No coelom (acoelomate). Phylum Platyhelminthes, composed of flatworms, is one of the phyla that has no coelom. Note that from its gut to its exterior, the flatworm is composed of uninterrupted tissue–endoderm (inner tissue), mesoderm (middle tissue), and ectoderm (outer tissue).

A Split in the Animal Kingdom: Protostomes and Deuterostomes

The protostome/deuterostome categories you can see toward the bottom of the animal family tree simply represent an early split among animals along two evolutionary paths. The difference between the two types of animals is grounded in how each type develops in its early embryonic stages. Remember how you saw earlier that the chief defining characteristic of animals is that they all go through something called the blastula stage in development? Well, in **Figure 22.5a**, you can see what happens to the hollow ball of cells called a blastula: It develops an invagination whose opening is called a blastopore. In protostomes, this blastopore becomes the mouth, but in deuterostomes it becomes the anus. (*Protostome* means "mouth first," while *deuterostome* means "mouth second.")

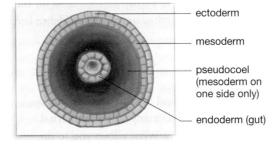

— ectoderm
— mesoderm
— pseudocoel (mesoderm on one side only)
— endoderm (gut)

roundworm

Pseudocoel. Roundworms, by contrast, have an enclosed cavity, known as a "pseudocoel," that lies between their gut and mesodermal tissue layers. The pseudocoel develops as a continuation of the space that exists when an embryonic animal is in the blastula or hollow ball-of-cells stage.

If you look at Figure 22.5b, you can see the effect that this developmental difference and others have on adult protostomes and deuterostomes. Note that the protostome has a dorsal heart and a ventral nerve cord, while in the deuterostome this arrangement is reversed.

Having reviewed the broad-scale features of the animal kingdom, we will now take a brief tour of it by walking through each of the nine phyla shown in Figure 22.2.

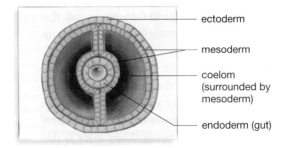

— ectoderm
— mesoderm
— coelom (surrounded by mesoderm)
— endoderm (gut)

earthworm

Coelom. The earthworm is one of the many animals that has a true coelom (or just-plain coelom). Note that both the coelom and the gut of the earthworm are lined with mesodermal tissue layers (whereas in the roundworm, only the pseudocoel is lined by mesoderm). The true coelom develops from an animal's mesodermal tissue.

Figure 22.4
An Important Space
Most animal bodies have an enclosed cavity or coelom—an internal space that surrounds their digestive tract or other internal structures. A coelom gives an animal flexibility, protects its organs from external blows, and provides space for the expansion of such organs as the stomach. Only three of the phyla covered in the chapter lack a coelom.

22.3 Phylum Porifera: The Sponges

Back in the 1780s, the United States was just that: a group of states that had just united into a single entity. The sponges that make up the phylum Porifera bear some comparison to this. Each sponge is a single, unified entity, but not by much. Sponges have no

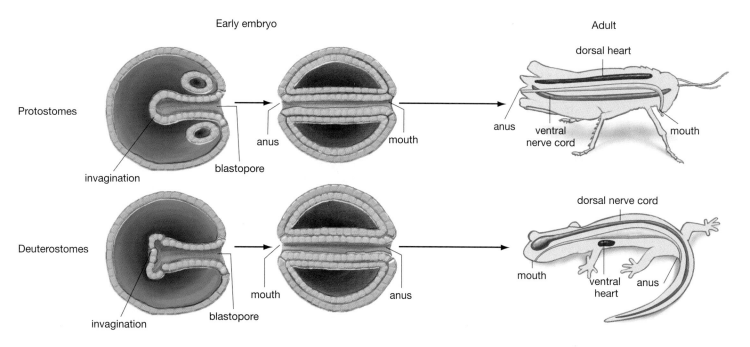

Early embryo

Adult

Protostomes

invagination

blastopore

anus

mouth

dorsal heart

anus

ventral nerve cord

mouth

Deuterostomes

invagination

blastopore

mouth

anus

dorsal nerve cord

mouth

ventral heart

anus

At one point in their embryonic lives, all animals pass through a so-called blastula stage of development, the blastula being a hollow ball of cells. In most animals, this blastula then invaginates to form a structure that develops into the animal's gut. The opening to this invagination is called a blastopore. In protostomes, the blastopore becomes the mouth. In deuterostomes, by contrast, the blastopore becomes the anus.

Developmental differences such as these have consequences in the adult. Protostomes, such as the insect shown, have their heart on their top or dorsal side and nerve cord on their bottom or ventral side. In deuterostomes, such as the reptile shown, this placement is reversed. The portions of the embryos colored yellow, red, and blue signify the endodermal, mesodermal, and ectodermal tissues noted in the discussion of the animal coelom.

Figure 22.5
The Protostome-Deuterostome Divide

organs; no stomach, no heart. They don't really even have tissues. Each of their cells acquires its own oxygen and eliminates its own wastes. In experiments, scientists have strained some sponge species through a filter, "disaggregating" them into the individual cells they are made of. The scientists then watched as these cells came back together to make up a single sponge once again. This is possible only in an organism in which each cell functions with a great deal of independence. If you were disaggregated into your individual cells, there is no chance they would come back together to re-form you. On the other hand, the cells in sponges clearly work together in a coordinated way. Sponges may not show much organization relative to fish, but they show a great deal of organization relative to, say, bacteria.

So, what manner of organism is a sponge? The simplest varieties have a layer of outer cells that is pockmarked with thousands of microscopic pores. Water flows in through these pores in the outer-layer cells and then is expelled out through a single large opening at the top of the sponge (**see Figure 22.6**). Because sponges are fixed in one spot (generally on the sea floor), everything they need in the way of food and oxygen must come *to* them, while everything they need to get rid of must be washed away *from* them. The water that flows through them takes care of both needs. Food and oxygen wash in through the micro-

scopic pores, and wastes wash out through the single large opening (called an *osculum*). Most sponges have elaborations on this theme, but the essential concept is the same: Move the water in, capture the food that comes with it, and move the water out, expelling wastes in the process. Millions of cells, called collar cells, wave tiny, whip-like flagella that keep the water flowing through.

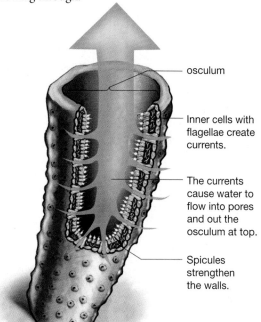

osculum

Inner cells with flagellae create currents.

The currents cause water to flow into pores and out the osculum at top.

Spicules strengthen the walls.

Figure 22.6
Sponges: A Body Plan for a Simple Lifestyle
Sponges filter water through themselves in order to live. Cells on their interior surface have hair-like extensions, called flagella, whose rapid back-and-forth movement draws water in through pores in the sponge's side walls. After filtering this water for nutrients, the sponge then expels it through a large opening at the top of its body, called an osculum. Sponges lack true tissue layers, but possess many specialized cell types. The cells are held together in a stable arrangement by tiny, hard structures called spicules and by the flexible protein called collagen.

Most sponges have skeletons composed of small, barbed structures called *spicules* that keep the sponge pores open and generally give the sponge structure. Some sponges have no spicules, but instead have skeletons made of a special variety of the protein collagen, which can be a soft, pliable material. (Think of the collagen *injections* that people sometimes get to improve their looks.) Sponges with collagen skeletons are the sponges from which we get our everyday notion of a sponge, meaning a soft, absorbent object used in cleaning. Put another way, the sponges that people once used were the skeletal remains of ocean-dwelling animals (which were harvested by divers). Today our sponges generally are made of synthetic materials.

Depending on whose count you accept, between 5,000 and 10,000 species of sponge have been recognized so far, with probably lots more remaining to be identified. Almost all these species dwell in ocean water, though there are a few freshwater varieties (**see Figure 22.7**).

Sponges can reproduce merely by budding. A group of cells breaks off from an existing sponge and then develops into a new sponge. Sexual reproduction exists too, however, and here again the sponge puts water currents to use. Sperm are released from one sponge and make their way via water currents to another sponge, where they are taken up and transported to egg cells. An offspring that comes from sex-

ual reproduction has a moving or "motile" stage. It swims for a time as a larva before landing on a solid surface—perhaps a rock or the body of a living animal—whereupon its outer layer of cells secretes a substance that will allow it to attach to the material beneath it. Then it begins its life of filtering water.

22.4 Phylum Cnidaria: Jellyfish and Others

If the signature activity of sponges is filtering water to get food, the signature activity of cnidarians is stinging prey to get it. In the main, cnidarians (knee-DAR-ee-uns) harpoon their prey with extensions that are not only barbed, but that may release poisons. These extensions are *tiny*, since each one springs from inside a single cell, but they are numerous enough that a single jellyfish may be able to immobilize and eat animals the size of small fish. Only a few species have a sting potent enough to cause real harm to human beings, but it makes sense to give all jellyfish a wide berth.

Jellyfish are undoubtedly the best-known cnidarians, but the phylum also includes corals, whose calcified remains make up coral reefs; the delicate sea anemones, whose petal-like tentacles close up when touched; and hydrozoans, which can look more like sea plants than animals (**see Figure 22.8**). One type of hydrozoan is the hydra that you may see in your biolo-

Figure 22.7
Sponge Diversity
Sponges differ greatly in size and shape, as you can see from these examples.

(a) A leaf sponge

(b) A yellow, warm-water tube sponge

(a) Jellyfish

(b) Sea anemone

(c) Coral polyps

Figure 22.8
Cnidarian Diversity
One characteristic that unifies cnidarians as diverse as jellyfish, sea anemones, and corals is their ability to immobilize prey with harpoon-like stinging cells.
(a) Pictured are some Pacific Ocean jellyfish, *Chrysaora fuscescens*, whose sting, while painful, generally is not harmful to human beings. One of its relatives, *Chrysaora quinquecirrha*, is a major irritant during the summer in the Chesapeake Bay.

(b) The large sea anemone *Urticina piscivora* is found off the shores of British Columbia in Canada. At about 20 centimeters or 8 inches across, it is so big it can eat small fish.

(c) What builds a coral reef? Coral polyps, such as these seen in a coral reef in Western Australia. Note the mouths and extended tentacles of the polyps, which help form the living outer layer of coral reefs.

gy lab. There are about 9,000 identified species of cnidarians, nearly all of them ocean dwellers, with a few living in freshwater.

The basic body plan of a cnidarian is that of a sack, though it often is a sack that is turned upside down. If you look at **Figure 22.9**, you can see how this plays out in connection with a hypothetical jellyfish. In the familiar adult or "medusa" stage of a jellyfish, the sack is upside down; note that the mouth is on the *underside* of the body in this stage. Conversely, look at this jellyfish's immature "polyp" stage, when it is fixed to the ocean floor; now the mouth is on top. Medusa and polyp stages are present in many cnidarians, though corals, sea anemones, and hydra have

only the polyp stage. The medusa stage predominates in jellyfish—misnamed, because they are not fish—while the polyp stage predominates in most hydrozoans. In all medusa and in many polyps, however, the basic body plan is the same: a single opening to the outside, serving as both mouth and anus, is surrounded by tentacles that both sting prey and bring the prey to the mouth. Note also the material that lies between the jellyfish's gastrovascular cavity and its exterior. This is the *mesoglea*, a secreted, gelatinous material that makes up most of the medusa-stage jellyfish. Its consistency accounts for the name jellyfish. The gastrovascular cavity functions in both digestion (*gastro–*) and in circulation (*vascular*).

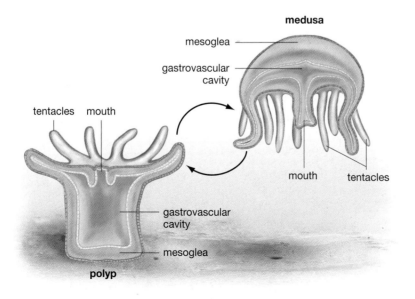

medusa

mesoglea

gastrovascular cavity

tentacles   mouth

mouth   tentacles

gastrovascular cavity

mesoglea

polyp

Figure 22.9
Two Stages of Life for Many Cnidarians
Many cnidarians go through both a polyp stage, in which they are attached to a solid surface beneath them, and a medusa stage, in which they swim freely through the water. Note that in the polyp stage, the cnidarian's mouth and tentacles face upward, while in the medusa stage, they face downward. The gastrovascular cavity of the cnidarian functions in both digestion and circulation. The mesoglea noted in the figure is a secreted, gelatinous material that makes up the bulk of the medusa stage in jellyfish cnidarians. All cnidarians use stinging tentacles to capture prey.

Figure 22.10 shows you the life cycle of a cnidarian, this one a hydrozoan known as *Obelia*. Medusa-stage male *Obelia* release sperm into the water while females release eggs; these unite, eventually producing a larval stage offspring that settles to the ocean-floor bottom, now becoming a polyp. This starting polyp can give rise to an entire colony of *Obelia* polyps through the process of *budding*: Groups of cells break off from one polyp and form another polyp. Note that in this polyp stage, some of the branch-like outgrowths are "feeding polyps" that have the familiar sack-like appearance of cnidaria (tentacles, mouth), while other outgrowths are entirely given over to reproduction, containing "medusa buds." These buds are immature medusa-stage *Obelia*, which eventually will be released to become free-swimming individual medusae. The *Obelia* life cycle is one among many that exist in cnidarians.

Cnidarians don't have any true organs, but they are a long way from sponges in that they have a nervous system, and cells that function like muscle cells. This gives medusa-stage cnidarians the ability to swim. (Polyp-stage cnidarians sometimes can slide slowly along the ocean floor.) The typical jellyfish medusa contracts cells that ring its mesoglea, thus expelling a jet of water from its underside. This action both moves the animal in the opposite direction from the jet and pushes in the mesoglea. Like a compressed cylinder of foam rubber, the mesoglea then snaps back into shape, ready for the next contraction.

Special mention should be made of the reef-building corals, which have no medusa stage but live almost entirely as polyp-stage cnidarians. The fantastic ocean cities known as coral reefs take their names from the coral animals, which actually are close relatives of sea anemones. Coral reefs have a rock-like appearance and, indeed, they are composed mostly of calcium carbonate, better known as limestone. But this limestone comes from living organisms. The corals secrete it, a practice they share with a type of red algae with whom they share their habitat. The limestone forms an external skeleton for the corals, and when each coral polyp dies, its limestone skeleton remains. Thus coral reefs are composed largely of the stacked-up remains of countless generations of coral polyps, along with the remains of their neighboring red algae. But the thin outer veneer of each reef is composed of *living* algae and the latest generation of pinhead-sized polyps. If you looked up close at any one of these coral animals, you would see the basic features of a cnidarian polyp: stinging tentacles pointed upward, surrounding a mouth.

The Protostomes

22.5 Phylum Platyhelminthes: Flatworms

With the flatworms, you've arrived at animals that fit our everyday notions of what animals look like, in that they have something like a head and have the same

Figure 22.10
Cnidarian Reproduction

This hydrozoan cnidarian, known as *Obelia*, reproduces through both sexual and asexual means. Sexual reproduction can be observed at the right of the figure, where male and female medusae release sperm and eggs, respectively, that fuse to form the new generation of *Obelia*. The polyp that results from this fusion, however, can reproduce by asexual means. Cells bud off from the first polyp (center) and grow into another polyp, shown at left, which may be the first of many that will develop this way. Polyps within these larger structures specialize. Some are involved in feeding, while others function in reproduction. The reproductive polyps have medusa developing inside them (the medusa buds), which are released to the surrounding water, starting the life cycle over again. The different stages of the life cycle are drawn at different scales.

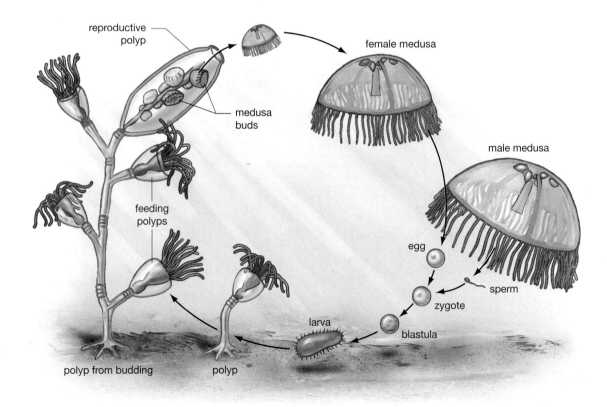

side-to-side (bilateral) symmetry as, say, a mouse. Further, flatworms go beyond the tissues of the cnidarians and have **organs**, meaning highly organized structures formed of several *kinds* of tissues. (Your stomach is an organ that includes nerve and muscle tissues.)

Despite this complexity, flatworms lack a good many features that exist in almost all the phyla further up the complexity ladder. They have no coelom, or enclosed internal cavity. Thus, except for the tube of their digestive tract, they have very little internal space in them. This may sound like a good thing, but it actually limits their flexibility. Further, they have no system of blood circulation. Therefore, every cell in a flatworm must get its oxygen directly from the environment around it, through diffusion from water or air outside the body. This is why flatworms are flat—they have to maximize the number of cells that are near an exterior surface, as opposed to being buried deep inside (**see Figure 22.11**). Those flatworms that are on land tend to be found under stones or rotting wood; many live at the bottom of the ocean, and a few live in freshwater. Some 25,000 species of flatworm have been identified.

Humans are unlikely to notice flatworms since, of the minority that live around us, most are very small. When they do come into our consciousness, it is often in the unpleasant role of parasites. Two of the three traditional classes of flatworms live strictly as **parasites**, defined as organisms that feed off their prey but do not kill them, at least immediately. Some of these worms can infect human beings. There are, for example, the tapeworms that can enter the human body in undercooked meat or fish. Tapeworms can be small, like most flatworms, but some reach up to 20 meters in length, which is about 66 feet. Then there are several species of another variety of flatworm, the flukes, that cause the disease schistosomiasis, which affects more than 200 million people in tropical areas around the world, often causing serious damage to the human bladder, liver, or spleen. With flukes, the means of entry is not food but unprotected human skin, as with people who are standing in fishing areas or flooded rice paddies.

The flukes are in one class of flatworms (Trematoda), while the tapeworms are in another (Cestoidea). There is then a third class of flatworms, the Turbellaria, whose members generally are not parasitic, but free-living. If you look at **Figure 22.12**, you can see something of the anatomy of one variety of turbellarian, the flatworm *Dugesia*, which is likely to

Figure 22.12
The Flatworm *Dugesia*

(a) Flatworm anatomy

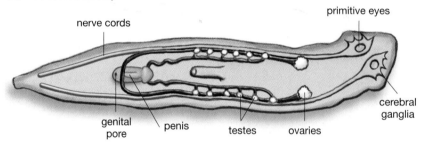

Dugesia's nervous system includes primitive eyes and two collections of nerve cells, the cerebral ganglia, that connect to nerve cords that run the length of the animal. In reproduction, *Dugesia* is hermaphroditic, meaning it possesses male sex organs (testes and penis) as well as female sex organs (ovaries and other structures). When two *Dugesia* copulate, each projects its penis and inserts it in the genital pore of the other.

(b) Feeding and digestion

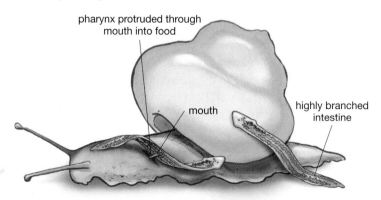

Dugesia feeds by turning its muscular pharynx inside out and projecting it into the food source. Food taken up through the pharynx is distributed through the body by *Dugesia's* highly branched intestine. Flatworms lack a true circulatory system and thus depend on their digestive system for the distribution of nutrients. Their digestive system lacks an anus, meaning *Dugesia* has only one digestive-tract opening, its mouth.

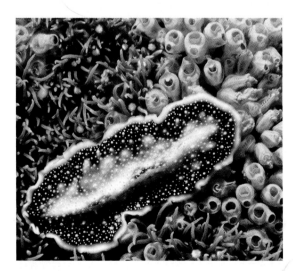

Figure 22.11
Flatworms
The flatworm in this picture, *Pseudoceros ferrugineus*, was photographed in the oceans off the coast of the Philippines. Most flatworms are free-living like this one, but there are also many parasitic flatworms that live for at least part of their lives inside a host.

be found crawling under rocks at the bottom of freshwater streams. About 1 centimeter or less than half an inch long, *Dugesia* has a "head," in that one end of it has not only a pair of primitive eyes (called ocelli) but also a concentration of nerve cells (the cerebral ganglia) that are connected to nerve trunks that run the length of the body.

Given this concentration of features in *Dugesia*'s head, the location of its mouth may come as a surprise. It's located midway down its body, on its ventral (bottom) side. To feed, it shoots a muscular structure called a pharynx outside its mouth, coats the prey with enzymes to break down its tissues, penetrates the prey with the pharynx, and then sucks the resulting semi-liquid material back in. (Your pharynx is the area at the back of your throat where your windpipe and digestive tract meet. Imagine being able to shoot it outside your mouth.) Food goes from the pharynx into an intestine, where it is digested and then simply diffuses into nearby tissues. Remember that flatworms have no blood circulation system. Thus, *Dugesia*'s intestine has to be highly branched because it is serving as both a food digestion and a food delivery system. Flatworms have no anus; their mouth is the single opening to their digestive tract, and waste material is expelled out of it.

Like most flatworms, each *Dugesia* has both ovaries and testes (the organs noted earlier). It is thus **hermaphroditic,** meaning a state in which one animal possesses both male and female sex organs. When two *Dugesia* copulate, each projects its penis and inserts it in the genital pore of the other. Most turbellarians can also reproduce asexually by a straightforward means: They break themselves in half. The posterior ("tail") part of the worm grasps the material beneath it while the anterior portion moves forward. The separated halves then grow whatever tissues they need to make themselves complete worms.

Members of phylum Platyhelminthes, such as *Dugesia*, have organs thanks to a critical embryonic addition found in flatworms, and in all animals more complex than flatworms. Recall from the earlier discussion of the body cavity (the coelom) that all animal embryos have what are known as germ layers, meaning layers of cells that *become* various types of tissue in adult animals. These layers are the endoderm, mesoderm, and ectoderm. If you looked at the embryos of jellyfish just past the blastula stage, you would see that they have only the endoderm and ectoderm. But Platyhelminthes and all the other more complex phyla also have a third layer, the mesoderm, literally in the middle of the other two. (Endoderm is initially an inner layer of cells and ectoderm an outer layer; mesoderm comes in the middle.) Animals with

the three germ layers are said to be **triploblastic**, while the cnidarians, with two layers, are **diploblastic**. The addition of mesoderm was an important event in animal evolution because the interactions between germ layers in embryos allow for the development of more complex structures. (For details on germ layers, see Chapter 29 on animal development.)

22.6 Phylum Annelida: Segmented Worms

More worms? Actually, three of the nine phyla covered in this chapter are worms. (Still to come are the roundworms in Nematoda.) Evolution resulted in a lot of worm or worm-like phyla—perhaps a dozen altogether—but the reason so many are being covered here is that they occupy interesting places in the spectrum of animal diversity.

Annelid worms occupy such a position, in that they provide the clearest example of a feature that exists in a lot of animals, including ourselves: **body segmentation**, which can be simply defined as a repetition of body parts in an animal. Just as a set of identical Lego blocks can be snapped together to make a larger structure, so can a repeated set of segments be connected to make a larger body. Think of the vertebrae that make up your backbone; 24 of them run in a line from your neck to the base of your back, each of them having a similar basic structure and a similar function (support and protection).

Segmentation is so useful in animals that it evolved independently in them several times. It happened at least once in the protostome lineage that you've just been going over and then once again in the deuterostome lineage—more specifically in our own phylum, Chordata.

Though segmentation is widespread in the animal kingdom, it is most visible in the annelid worms. Indeed, the name Annelida comes from the Latin word *anellus*, or "little ring." That pretty much describes what we see in the best-known annelid, the common earthworm, whose body seems to be composed of a series of little rings that have been joined together (**see Figure 22.13**).

Segmentation is valuable in earthworms for the same reason that individual rooms are valuable in a house: The activities being carried out in one room can *differ* from those being carried out in another. Each of the earthworm's 100–150 segments is separated from adjoining segments by a partition, and each segment can function independently of the others to some degree. The muscles in individual segments (or groups of segments) can contract independently. This, in turn, makes for independent control of needle-like bristles that the worm can extend from each segment.

These features account for the mesmerizing way in which earthworms move. One set of muscles toward the rear of the earthworm contracts, making the worm thick and compact in the area, even as the bristles in these segments are being pushed into the ground like so many stakes. Now, with its hindquarters fixed in place, the worm uses a different set of muscles to make the mid-portion segments *lengthen*, thinning out as they do, meaning this part of the worm has moved forward. Then the worm plants bristles in the front, anchors itself, and literally brings up its rear. Note that this movement is possible only because of segmentation. If the *whole worm* lengthened, without part of it being compacted and fixed in one spot, it would get nowhere.

Segmentation is important in annelids, then, and is thought to have evolved in them precisely because of its power to help them move. (Earthworms use this power extensively, because they burrow in the dirt for hours at a time, swallowing soil for the nutrients it contains and eating leaves or other decaying vegetation.) In more general terms, segmentation is valuable to living things because it can confer flexibility while maintaining strength. We can see this in our own "backbones." Imagine how inflexible you would be if, instead of 24 moveable vertebrae, you had a single bone running from your skull to your posterior.

Earthworms generally are helpful to human beings—and not just as fish bait. They churn up soil, aerating it and depositing their cylindrical "worm casting" waste, which enriches the soil. Earthworms have no lungs; they take in oxygen right through their skin. Their skin needs to be moist for this to happen,

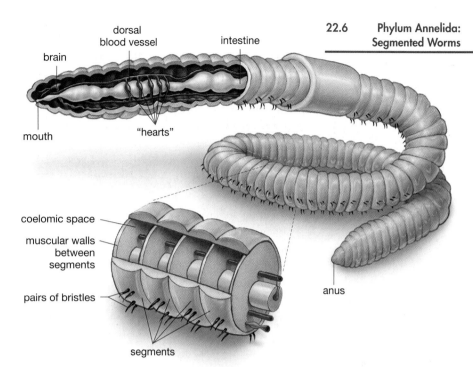

however. The great threat to an earthworm is drying out, which is why these "nightcrawlers" generally stay underground during the daytime. Like the flatworms, earthworms are hermaphrodites. Each member of a copulating pair inseminates the other.

Though we've focused on earthworms, they actually represent only one small part of phylum Annelida, which contains about 17,000 named species. Most annelids actually are ocean-dwelling, and some of these colorful creatures do not fit in with our idea of drab worms (**see Figure 22.14**). There also are freshwater

Figure 22.13
The Body Plan of an Earthworm
The segmentation of body parts—a widespread feature in the animal kingdom—is clearly seen in earthworms, whose segments are repeated hundreds of times. In general, such segmentation provides flexibility in combination with strength. In the earthworm, it also allows the actions being carried out in one segment to differ from those being carried out in another. Note, in the blowup of the earthworm's body, the muscular walls between segments; these give the earthworm independent control of the segments and the bristles that can be extended from them. The earthworm also expands our notion of what a "heart" amounts to. It has a dorsal blood vessel that contracts, as do our human hearts, giving the main impetus to the circulation of red blood that moves through the worm's system. But it also has five pulsating vessels on each side of its intestine that serve as accessory hearts.

(a) Hawaiian Christmas tree worm

(b) Medicinal leech

Figure 22.14
Annelid Diversity
Not all annelid worms are drab, and not all live underground.
(a) Polychaetes are ocean-dwelling worms. Pictured is a Hawaiian Christmas tree worm in the genus *Spirobranchus*. These polychaetes stay in one spot, burrowed into tubes they themselves have constructed, generally within a coral. The bushy "Christmas trees" that extend from the worm's head are structures called radioles that it uses to filter-feed from the water.
(b) Leeches are a smaller class of mostly parasitic annelid worms. Pictured is a leech that is feeding from a human arm. This particular leech, *Hirudo medicinalis*, serves a therapeutic function. These leeches are used today in connection with human reattachment and plastic surgery. Enzymes they secrete serve as anticoagulants—they keep blood from clotting—and help reduce pain at the site of surgery. A leech might be attached at the site of surgery for half an hour or so and then be removed, after which its enzymes will continue working for hours longer.

annelids, who are part of the same class as the earthworms but who tend to be much smaller. Then there is the smallest class of the annelids, which is composed entirely of leeches. Most leeches live up to their reputation of being blood-sucking parasites. Though most live in freshwater, some live in the ocean while others dwell in damp-land habitats.

22.7 Phylum Mollusca: Snails, Oysters, Squid, and More

With the molluscs, we reach a phylum that has lots of members who are familiar to us. Everyone has seen a snail, many of us have eaten clams or mussels, and squid and octopus are probably known to us from television or books. But molluscs are even more varied than these examples would suggest. Some molluscs are slugs and others worm-like; some have highly developed brains and others no brains at all; some are filter feeders and others fierce ocean predators. Some are very small, but one is a behemoth that is one of the enduring mysteries of the animal world. The giant squid, *Architeuthis dux*, reaches a length of about 20 meters or 60 feet and weighs up to a ton (**see Figure 22.15b**).

Depending on the expert consulted, there are between 50,000 and 100,000 described species of molluscs. They live in freshwater, saltwater, and on land. There actually are eight classes of molluscs, but here we will look only at the three most familiar classes, the **gastropods** (snails, slugs), the **bivalves** (oysters, clams, mussels), and the **cephalopods** (octopus, squid, nautilus).

What features unite this huge, diverse group as a single phylum? Outside the realm of technical anatomical details, there really is only one feature common to all molluscs. If you look at **Figure 22.16a**, you can see an aquatic snail that displays this feature along with others. Note a layer near the dorsal (top) surface of the snail called a *mantle*. This is a fold of skin-like tissue that surrounds the upper body of the mollusc and that usually secretes material that forms a shell (though there are lots of molluscs without shells).

Past this universal feature, the snail displays a number of structures that are common to *most* molluscs. The mantle generally drapes over part of the mollusc body like a tablecloth, meaning there is a mantle cavity—a space underneath the mantle. This cavity generally is the site of gills, by which all aquatic molluscs obtain oxygen, and by which the filter-feeding bivalves (clams, mussels, oysters) obtain not only oxygen but food. In land-dwelling molluscs, such as garden snails, much of the cavity is a primitive lung that evolved from the gill.

Then there is the single or "unitary" foot as shown on the snail's ventral (bottom) side. Through wave-like muscular contractions, this foot can propel the mollusc along the surface underneath it, be it a leaf or the ocean bottom. Some sessile bivalves, such as oysters, have lost the foot altogether, however; and as you'll see, in squid and octopus it has evolved into something else entirely. Note, in the blowup of the snail's mouth, there is a tooth-covered membrane called the *radula*. Present in most molluscs except the bivalves, this organ can be extended outside the mouth, where it serves as a kind of rasping file. Mol-

Figure 22.15
Mollusc Diversity
Pictured are species in three of the best-known classes of molluscs:

(a) Snails are gastropods. Pictured is a colorful terrestrial snail in the *Liguus* genus. These snails live in trees in southern Florida and through much of the Caribbean, though their numbers are dwindling in Florida.

(b) The largest cephalopod of all is the giant squid *Architeuthis dux*, shown here, which can reach a length of about 20 meters or 60 feet. This squid, prepared for an exhibit at New York's American Museum of Natural History, measured a mere 25 feet in length and weighed 250 pounds. It was netted off the coast of New Zealand in 1997.

(c) The bivalve *Argopecten irradians* is better known as the bay scallop—a part of many seafood dishes. Note the small blue dots around the edge of the scallop. These are a few of the animal's many primitive eyes.

(b) Cephalopod
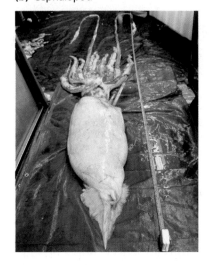

(a) Gastropod

(c) Bivalve

luscs use it for scraping a surface or for cutting small prey; then they retract it back into the mouth, thus moving food into their digestive tracts.

The snail has some other features worth pointing to now (though we could have looked at variations on them in the annelids as well). These are a heart and a stomach. These organs obviously aren't unique to snails (or even molluscs), and that's just the point. Along with gills and lungs, these organs are signs of the molluscs' advanced position in terms of structural specialization.

Why does something like a heart exist? Because creatures as large as molluscs need a *delivery system* for food and oxygen. Simple diffusion of food from a gut won't work anymore, because some tissues are buried so deep they would never be served. The delivery system's vehicle is blood, but that blood must be propelled by some force so that it can make a set of rounds through the animal, picking up oxygen from the gills and food from the digestive tract. The propelling force for this work is the beating of the heart.

The circulation of the cephalopods (squid, octopus) is like ours in that it is a **closed circulation system**—a circulation system in which the blood *stays within* the blood vessels, while the oxygen and nutrients diffuse out of the vessels to the surrounding tissues. Most molluscs, however, have an **open circulation system**: a circulation system in which blood flows out of blood vessels altogether, into spaces or "sinuses" where it *bathes* the surrounding tissues. Veins collect the blood in these sinuses for a return trip to the gills and the heart. Gastropods such as snails can be either single-sex or hermaphroditic, and this is the case for the bivalves as well. All cephalopods, however, are single-sex.

Figure 22.16b gives you some idea of one set of variations evolution has wrought with the basic body plan of the mollusc. Remember how the unitary foot is a usual mollusc feature? Look at what has become of it in the squid; it has been divided into eight arms (with suckers) and two longer tentacles. The radula is still there, but now it lies next to a set of jaws that operate something like a parrot's beak. Meanwhile, the mantle cavity now has another function; the siphon you can see at one end of it acts like a jetting water hose that not only helps propel squid movement—they expel water out of it—but also directs this movement by being aimed first one direction and then another.

Cephalopods deserve mention for another reason, which is that they are not only the biggest invertebrates in the world, they are the smartest. One researcher who worked for decades with octopus said that they had the intelligence of dogs, which is very smart indeed in the animal world. Many researchers

(a) Aquatic snail

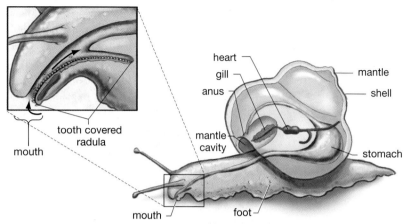

Aquatic snails such as this one have features that are common to many molluscs. All molluscs possess a skin-like tissue called a mantle, which secretes material that can form a shell, whether it is an external shell (as in the snail) or an internal shell (as in a squid). Snails also possess a tooth-covered rasping structure for feeding, called a radula, and they are complex enough to have evolved some specialized organs that are familiar to us—the heart and the stomach. The unitary "foot" of the snail, which works through wave-like contractions, is another common mollusc feature.

(b) Squid

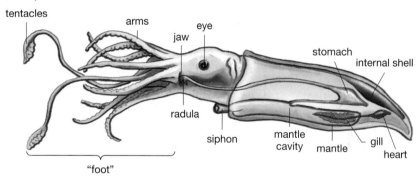

In the squid, the mollusc's unitary foot has evolved into a multisegmented appendage of eight arms and two tentacles. Note how other mollusc features have been modified in the squid. The mantle cavity houses not only gills, but a siphon the squid jets water out of, thus propelling itself. The radula now exists near jaws that function like a parrot's beak.

Figure 22.16
The Body Plans of Molluscs

would go further, however, saying that octopus and squid seem to plan and calculate in very sophisticated ways. These animals also are fast swimmers, have keen eyesight, can change color in an instant (to hide from predators), and as a last resort in defending themselves, have the ability to shoot out black "ink" from a special sac. Given their remarkable capabilities, the curious thing about these molluscs is that they live only a year or two; most animals this sophisticated live much longer lives.

22.8 Phylum Nematoda: Roundworms

Nematodes, commonly referred to as roundworms, exist in such enormous numbers that it is impossible to imagine life without them. They fit into numerous ecological roles, such as that of "detritivore," meaning an organism that feeds on dead or cast-off organic material. Nevertheless, many farmers in the United States would *like* to imagine life without them, because several species of these animals are a major cause of damage to such crops as soybeans and corn. Farmers and research scientists have been united for decades in an entrenched battle against them. In addition, some are human parasites. The disease trichinosis, usually brought on by eating undercooked pork, is caused by a roundworm; likewise there is hookworm, which affects hundreds of millions of people in warmer climates throughout the world (**see Figure 22.17**).

An acre of prime farmland might contain several hundred billion roundworms, but you don't have to go to farmland to find these animals. They exist in desert sands, polar ice, ocean bottoms, lakes—name the place where life is going on, and the nematode worms are likely to be among the living. Indeed, scientists have only the vaguest idea of how many *species* there are; about 15,000 have been catalogued, but 100,000 seems to be the minimum number that actually exist.

Most roundworm species are very small—the crop pests, which feed on roots, generally are microscopic—but some parasitic roundworms are several meters long. The roundworms that have been studied most intensely are those that cause disease and crop damage, and these have a characteristic appearance: transparent, smooth, C-shaped, and cylindrical with tapered ends (hence the term *roundworm*). Across the breadth of roundworm species, however, there is a great diversity in form.

Roundworms generally are of one sex or the other, meaning they are not hermaphroditic, as the flatworms are. Reproduction is always sexual; unlike flatworms, roundworms can't reproduce by pulling themselves apart. Roundworms are the first of two "molting" animal phyla we will look at; they have an external skeleton, called a cuticle, that they shed or "molt" four separate times during their lifetime. We'll take a more detailed look at molting, however, in the animals coming right up, the arthropods.

22.9 Phylum Arthropoda: Insects, Lobsters, Spiders, and More

Phylum Arthropoda is large and varied indeed. It is so large that there are more described species in one arthropod group, the insects, than there are in all the other animal phyla combined. It is so varied that its members range from legless barnacles to hundred-legged millipedes (**see Figure 22.18**). There are lots of ways to group the animals in this phylum, but for our purposes, it will be convenient to think about them batched into three "subphyla." These are:

- Subphylum Uniramia, which includes insects, millipedes, and centipedes, among other organisms
- Subphylum Crustacea, which includes shrimp, lobsters, crabs, and barnacles, among other organisms
- Subphylum Chelicerata (chel-is-er-AH-ta), which includes spiders, ticks, mites, and horseshoe crabs, among other organisms

Before getting to these subphyla, it will be useful to review the characteristics that all arthropods share. First, all arthropod animals have an **exoskeleton**, defined as an external material covering the body, providing support and protection. You can see an example of an exoskeleton in the grasshopper pictured in **Figure 22.19**. Second, all arthropods have what are known as **paired, jointed appendages**, which are just what they sound like: appendages, such as legs, that come in pairs and that have joints. (The word *arthropod* actually means "jointed leg.")

Taking the exoskeleton first, it is made of a tough carbohydrate called chitin that comes embedded in protein. The crustacean arthropods (crabs, lobsters, and so forth) have, in addition, a calcified component to their skeleton that makes it very hard, as anyone who has ever "cracked" a lobster knows. Such a skeleton is not only protective but also serves to anchor arthropod muscles very securely. Your own muscles work by pulling against your *endo*skeleton—your internal skeleton. Arthropod muscles pull against a skeleton too, but it's a skeleton that surrounds their bodies.

Strong though it may be, the exoskeleton only works because it is also flexible. Like a knight's armor, the exoskeleton has plates that overlap and thin, flexible sections between these plates that can bend as necessary. Even with this flexibility, however, the exoskeleton presents a problem: The arthropod body can only grow so much before it expands right into it.

Figure 22.17
Roundworms
The tiny worms called nematodes are a normal part of many ecosystems, including soil ecosystems. However, a small number of roundworm species are harmful to humans, either directly as parasites, or indirectly as crop pests. Pictured is an artificially colored micrograph of the nematode *Toxocara canis*, a parasite that infects dogs. Humans can become infected with *T. canis* eggs if they have contact with an infected dog or contaminated soil. The worm is magnified × 450.

(a) Subphylum Chelicerata

(b) Subphylum Crustacea

(c) Subphylum Uniramia

Figure 22.18
Big Numbers, Much Diversity
The arthropods represent the largest animal phylum, in sheer numbers as well as in diversity. Shown are representatives of the three arthropod subphyla:
(a) A web-building *Argiope* spider, this one found on the Snake River at the Idaho-Oregon border. The Chelicerata grouping to which it belongs is the only arthropod subphylum whose members do not have antennae.
(b) A flame lobster from Hawaii. Note the well-defined arthropod characteristics in this crustacean: an exoskeleton and the legs that exist as paired, jointed appendages.
(c) The arthropod subphylum Uniramia includes insects. Pictured is a short-horned grasshopper native to western North America.

So how does the arthropod solve this problem? Its solution is **molting**: a periodic shedding of an old skeleton followed by growth of a new one. A crustacean such as a crab will retreat to a relatively safe place and begin to slip out of its old exoskeleton, as if it were Houdini trying to get out of a straightjacket (**see Figure 22.20**). Through chemical processes, the old skeleton has begun to split by this time. Since the new, developing exoskeleton lies just underneath the old one, you might think this process would leave the crab just as hemmed in as before. But at this stage the new skeleton can be *stretched*, and the crab proceeds to do just that by ingesting a large amount of water, thus inflating both its body and the skeleton to a new size. The new skeleton then hardens in its inflated size, the crab loses the extra water, and shrinks to its previous size. The result? It now has room to grow. Insects pull off this same feat, but stretch themselves with air instead of water.

The jointed appendages that arthropods possess come in many forms. There are legs for walking, claws for predation and defense, and wings for flying. These appendages are flexible and often come with sensory attachments, such as sensory "hairs," that keep the arthropod in touch with its environment.

Figure 22.20
Old and New
Arthropods have an outer or exoskeleton of a fixed size. They are able to grow by periodically shedding, or molting, this stiff skeleton. The crab in the photograph has just finished molting; its old exoskeleton lies nearby.

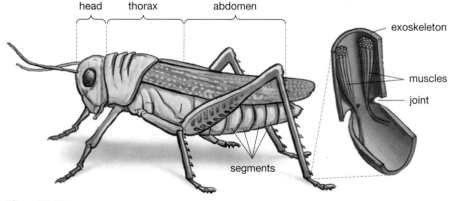

Figure 22.19
Arthropod Features
Arthropods have paired, jointed appendages, as with the legs of this grasshopper. The insects among the arthropods have bodies made up of three large-scale sections: head, thorax, and abdomen. Like the annelid worms, arthropods are segmented. This segmentation is clearly visible in the abdomen region of the grasshopper, but is less pronounced in the head and thorax because there the segments have fused. The entire arthropod body is covered in a rigid exoskeleton that is jointed to allow movement.

Together, jointed appendages and an exoskeleton make for an animal that is often nimble and well defended—a combination that is one of the reasons arthropods are so numerous. In addition, arthropods often have keen senses, such as sight and chemical perception (as with ants following a "pheromone" or chemical scent trail).

Subphylum Uniramia: Insects First

Recall that the arthropod phylum is divided into three subphyla. The first of these subphyla, Uniramia, has within it a group of animals we all have some familiarity with, the insects.

Who can comprehend how many insects there are in the world? There are more species of them— about a million have been counted so far—than there are species in all the other groupings of animals combined. If we could get an accurate count of their numbers, we would probably find that there are *tens* of millions of insect species. To look at their numbers another way, for every person on Earth, there may be 200 million insects. The vertebrates in the animal world have been the biggest creatures on land for the last 350 million years. But size counts for only so much in the face of the numbers the insects can put up. Here is the lead sentence from a September 2000 article in the Australian *Financial Review*:

> Extensive spraying could start as early as next week to try to control an impending locust plague which threatens to cause up to $500 million damage to Australia's forecast big grain harvest.

Had the *Financial Review* existed thousands of years ago, it could have made this same kind of report. Consider this passage on locusts from the Bible's Book of Joel, which dates from about 2,500 years ago:

> For a nation has invaded my land, powerful and innumerable, ... It has laid waste my vines, and splintered my fig trees ...

The battles that insects and human beings have been engaged in for so long have involved not only plagues of locusts, but the bubonic plague (carried by fleas), malaria (transmitted by mosquitoes), and the everyday predations of lice, termites, and ants. On the other hand, we have insects to *thank* for most of the plant growth we see in the world, in that insects pollinate flowering plants. In addition, insects provide us with honey, they help break down organic matter, and many of them are beneficial to humans in that they feed on the very insect pests that are so troublesome to humans.

If you look again at Figure 22.19, you can see the body parts of a typical insect. The three larger body regions featured in this grasshopper—head, thorax, and abdomen—are the rule in insects. In the abdomen, you can see a regular segmentation reminiscent of that you saw with the annelid worms.

The main cavities in the insect are the sinuses, or spaces, that its blood flows into. This blood bathes tissues in an "open" circulation system that almost all arthropods have. Sexes are separate in the insects, and reproduction is generally sexual.

Other Uniramians: Millipedes and Centipedes

In addition to insects, the uniramian subphylum also includes millipedes and centipedes (**see Figure 22.21**). Millipedes don't really have a thousand legs, despite their name. But hundreds of legs are possible, because these creatures have up to 100 body segments, each of

(a) Millipede

(b) Centipede

**Figure 22.21
Millipedes and Centipedes**
Insects constitute only one of the three classes of arthropods in the arthropod subphylum Uniramia. Millipedes and centipedes make up the other two classes.

(a) Millipedes have two pairs of legs per segment and live mostly on decaying plant matter.

(b) Centipedes have only one pair of legs per segment and are carnivores. Shown is the Texas giant centipede, *Scolopendra heros*.

which has two pairs of legs. This structure makes for movement that is slow but relatively powerful as millipedes make their way across the damp, dark environments they favor, which often means the forest floor.

Centipedes look much like millipedes, but are a different class of arthropod. They have only one pair of legs per segment and a pair of more prominent antennae extending from their head. Moreover, centipedes are carnivores, whereas millipedes feed largely on decaying plants. Centipedes come equipped for their predatory work, as their front two legs have become modified into venom-injecting fangs. This may sound ominous, but only a few tropical species pose a threat to human beings.

Subphylum Crustacea: Shrimp, Lobsters, Crabs, Barnacles and More

The second arthropod subphylum, Crustacea, includes several kinds of animals that don't seem to have much in common at first glance. How are a crab and a barnacle alike, for example? But remember that in putting living things together in a category, biologists are first and foremost concerned with shared *ancestry*, not with whether one creature has the same general appearance as another. At a crime scene, convincing evidence is likely to be found in small details, such as hair samples and fingerprints. The same thing holds true for evidence about the relatedness of groups of animals: the telling details are likely to involve small features.

This turns out to be the case with crustaceans. What all of them have in common is five pairs of appendages extending from their heads—two pairs of antennae and three pairs of feeding appendages. If you look at **Figure 22.22**, however, you can see the fantastic diversity that evolution wrought in crustaceans.

One of the crustaceans pictured in Figure 22.22, the water flea in the genus *Daphnia*, certainly is not familiar in the way that crabs or barnacles are. But it is representative of a variety of small, aquatic crustaceans that exist in huge numbers, mostly in the oceans, but also in freshwater bodies. As a whole, the crustaceans are mostly ocean-dwelling, but many live in freshwater, while a few live in damp-land environments. Crustaceans are known to humans mostly as a food source, such as shrimp and lobsters, or as a marine pest, such as the barnacles that attach themselves to ships. It is not just humans who use crustaceans as a food source, however. Closely related to shrimp are the crustaceans known as krill. Generally no more than an inch long, these animals feed on the plant life that floats in the icy ocean waters off Antarctica. Traveling in huge, dense swarms, they then become an important part of the diet of whales, seals, and penguins.

(a) Barnacles   **(b)** Water flea   **(c)** Shrimp

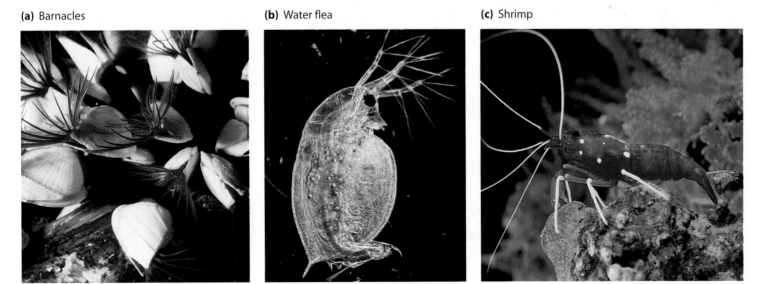

Figure 22.22
Crustacean Diversity
Diversity among the arthropod subphylum Crustacea is stunning.

(a) One class of crustaceans includes barnacles. When seen on a pier or ship, barnacles may seem lifeless to us, but notice the feathery extensions coming from these barnacles in the south Pacific. These are cirri, which barnacles extend from their shell and then use to catch drifting plankton.

(b) Water fleas in the genus *Daphnia*, such as the one shown here, live in huge numbers in ponds and lakes. Though barely visible with the naked eye, they are complex animals. Under microscopes, their tiny hearts can be seen beating a furious rate. As they are not insects, they are not really fleas. They get their name from the jumping, jerky motions they make in the water. Note the clearly visible digestive tract in this water flea.

(c) The crustacean class that includes shrimp also includes lobsters, crabs, and the pill bugs so often seen when we turn over rocks. Shown is the colorful fire shrimp, *Lysmata debelius*.

Subphylum Chelicerata: Spiders, Ticks, Mites, Horseshoe Crabs, and More

The third arthropod subphylum, Chelicerata, provides another example of small physical features yielding critical clues about relatedness. You might think that horseshoe crabs should be categorized as crustaceans, along with the other crabs, but it turns out that horseshoe crabs aren't really crabs at all. If you looked at the underside of a horseshoe crab, near its mouth, you would see two small appendages called *chelicerae*, which have pincers and which the crab uses to bring food to its mouth. If you looked near the mouth of a common spider, you would again see two chelicerae, only now each chelicera has become modified to become a venom-injecting *puncturing* appendage, rather than a grasping appendage. Ticks, mites, and all the other chelicerates likewise have paired chelicerae, from which this subphylum gets its name (**see Figure 22.23**).

Horseshoe crabs may be familiar to anyone who has spent time on the beaches of the Atlantic seaboard in May or June. Fearsome though they may look, they are harmless creatures that feed on worms, clams, or whatever else they can scavenge from the ocean bottom.

It is another grouping of chelicerates, however, that really inspires human fear: the arachnids, a class that includes mites, ticks, and scorpions, along with the more familiar spiders. Why "arachnophobia" should be so widespread in human beings is something of a mystery since, of the 34,000 identified species of spiders, only a handful cause any real harm to people. Part of the problem may be that spiders are blamed for wounds actually inflicted by other creatures. In one study of 600 suspected "spider bite" occurrences in southern California, 80 percent turned out to be the work of other arthropods—mostly ticks and some of the true bugs of the insect world (meaning bedbugs and other members of their order). Spiders evolved one of the world's great prey-snaring abilities, which is the spinning of their remarkable silk webs.

If we want to look for true causes of human misery, a better place to start would be with another arachnid, the microscopic dust mite, which is now known to be a major cause of asthma and other human allergic reactions. In the last decade, the mite's fellow arachnid, the *Ixodes* tick, has become feared because of its role as a carrier of the bacterium that spreads Lyme disease (named after the town of Lyme, Connecticut, where the disease was first identified).

(a) Horseshoe crabs

(b) Tarantula spider

(c) Dust mite

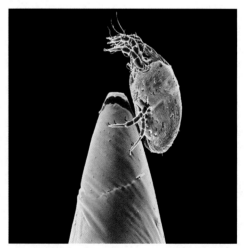

Figure 22.23
Chelicerate Diversity
All members of the arthropod subphylum Chelicerata possess paired chelicerae, which are pincer-like appendages used for feeding. These appendages stand in contrast to the mandibles used by the other two arthropod subphyla.

(a) In the springtime, East Coast beaches often have lots of horseshoe crabs such as these. In this picture, a group of males is competing for access to a female.

(b) The much-feared but generally harmless tarantula, this one a dry-weather Chilean rose tarantula, *Grammostola spatulata*, that is often kept as a pet. Like all spiders, the tarantula is a member of the Chelicerata class Arachnida, from which we get the term *arachnophobia*.

(c) Dust mites actually are eight-legged arachnids as well. The dust mite shown is standing on the head of a sewing needle. Such mites exist in almost all homes in the millions. Body fragments and excrement from them actually make up a large portion of the "dust" that is visible when sunlight shines into a room. This material sets off allergic reactions in many people.

(a) Sea star

(b) Sea cucumber

(c) Sea urchins

Figure 22.24
Echinoderm Diversity
There are five classes in the echinoderm phylum, three of which are represented here.

(a) Sea stars such as this one are fierce predators. Here a sea star is managing to open the shell of a bivalve known as a rock cockle.

(b) True to its name, this sea cucumber does look something like the vegetable, but sea cucumbers come in many shapes and colors. Like most echinoderms, sea cucumbers move slowly along the ocean floor.

(c) Kelp is a favorite food of many sea urchins. Here a group of sea urchins is shown foraging on some kelp.

The Deuterostomes

22.10 Phylum Echinodermata: Sea Stars, Sea Urchins, and More

With the echinoderms, we cross into the animal grouping noted earlier, the deuterostomes, that also includes our own phylum, Chordata. You might expect that, as our relatives, the echinoderms would have the kind of sophisticated features we've recently been seeing—brains, well-developed sense organs, strictly sexual reproduction—but none of this is true in the best-known echinoderm, the sea star (**see Figure 22.24a**). It has no brain, its only sensory or-gans are the primitive eyespots found on the tips of its arms, and a whole new sea star can be regenerated from the arm and part of a central disk of an existing sea star. Even the bilateral symmetry that you've seen in every animal since the jellyfish is absent in the adult sea star: It has *radial* symmetry, like the jellyfish.

This is not to say, however, that the sea star and its echinoderm relatives are completely prim-itive. Most have a remarkable system for moving that involves forcing water into a series of suction-tipped tube feet that then extend, like so many in-flated water balloons, from the underside of the sea star's arms (**see Figure 22.25**). Moreover, the

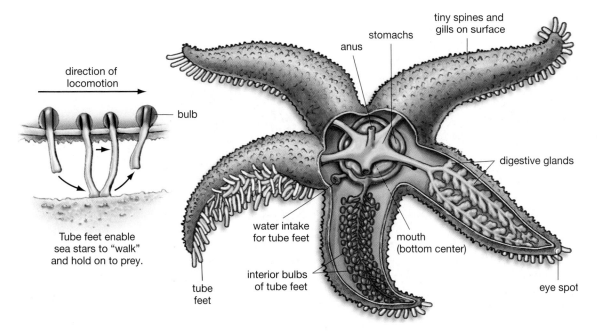

direction of
locomotion →

bulb

Tube feet enable
sea stars to "walk"
and hold on to prey.

tube
feet

interior bulbs
of tube feet

water intake
for tube feet

anus

stomachs

tiny spines and
gills on surface

digestive glands

mouth
(bottom center)

eye spot

Figure 22.25
The Body Plan of a Sea Star
The sea star is radially symmetrical in its adult form, and it has no brain. Its only sensory organs are the primitive eyespots found on the tips of its arms. It has, however, evolved a remarkable system for moving along the sea floor and grasping prey. It first takes in water from the surrounding sea and then channels this water into a series of water bulbs. When in-flated with the water, these bulbs lengthen, functioning as suction-tipped "tube feet."

sea star evolved into radial symmetry *from* bilateral symmetry. How do we know? Sea star larvae have bilateral symmetry, a quality that is lost when the larvae develop into adults. Looked at together, these echinoderm features add up to a phylum that is something of a mystery to zoologists—a mixture of characteristics from all over the animal map.

Sea stars aren't the only familiar echinoderms. Sea urchins, with their spine-covered spherical bodies, are echinoderms as well, as are sand dollars. Then there are the less familiar but well-named sea cucumbers (**see Figure 22.24b**). There are about 7,000 identified species of echinoderms, all of them ocean-dwelling and nearly all of them inhabiting the ocean floor, though "floor" in this case may just mean the bottom of a tide pool. Some are sessile, but most move across the ocean floor at a slow pace. Their shared habitat and measured pace make sense when we realize that the entire phylum evolved from a group of sessile filter feeders.

Despite their slow movement, echinoderms can be formidable predators. Sea stars often feed on molluscs, such as oysters and mussels. In some cases they use the suction tubes on their feet to pry open a mollusc's shell just a little. Then they "evert" their stomach—turn it inside out—and slide it in through the narrow gap they have created between shell halves. Juices secreted by their stomach then start to digest the prey's tissues. Sea urchins, meanwhile, are some of the animal kingdom's most voracious algae eaters.

22.11 Phylum Chordata: Mostly Animals with Backbones

Say the word "animal," and most people think of a *vertebrate* animal—be it tiger, elephant, deer, lizard, or some other large and probably toothy creature. One thing that ought to be clear by now, however, is what a small portion of the animal world vertebrates make up. Nevertheless, there are a lot of vertebrate species. The 50,000 or so that have been identified are nothing compared to the 1 million species of insects, but there are more species of vertebrates than there are of sponges, or cnidarians (jellyfish, etc.), or flatworms, and the vertebrate numbers may match those of molluscs.

In other ways, too, the vertebrates have been a big success. All the largest and swiftest animals are vertebrates, whether in water or in air or on land. And the sheer variety of sensory capabilities that vertebrates have—from echo-location to reading—is unrivaled anywhere in the living world.

Large and important as they are, the vertebrates don't even constitute a phylum; instead they make up only a subphylum, Vertebrata. The larger phylum into which vertebrates fit, called Chordata, actually has two additional subphyla in it. One of these (Cephalochordata) is made up of only a single kind of animal, a small, eel-like creature called a lancelet, represented by only 25 or so species (**see Figure 22.26a**). The other subphylum (Urochordata) has about 3,000 ocean-going species in it, but is represented by only three classes of animals. The largest of these classes is made up of the tunicates, whose practice of expelling water earns them their alternative name of sea squirt.

Figure 22.26
Chordate Diversity

(a) A single type of animal, the lancelet, is the sole representative of one of the three subphyla of chordates, Cephalochordata. Lancelets spend much of their time burrowed into the sand or mud of the shallow ocean floor, filter-feeding on small organisms or passing food particles. Only a few centimeters long, the diminutive lancelets are chordates, but not vertebrates.

(b) The African cheetah shown here is a vertebrate, and thus a representative of another of the chordate subphyla, Vertebrata.

(c) Urochordates, representing the third chordate subphylum Urochordata, come in about 3,000 species, most of them similar to the vase-like tunicates shown here. These animals are filter feeders who spend their adult lives attached to rocks. Their common name, sea squirts, comes from their practice of suddenly ejecting a spout of water from an opening near the top of their bodies. This group was photographed in the ocean off the coast of Indonesia.

(b) Subphylum Vertebrata

(a) Subphylum Cephalochordata

(c) Subphylum Urochordata

What Is a Vertebrate?

What *distinguishes* vertebrates from these other chordates? Only the vertebrates have just what you'd expect: a **vertebral column**. Better known as a backbone, this flexible column of bones extends from the anterior to posterior end of an animal. Conversely, what *unites* all the chordates are four features, which you can see pictured in a lancelet in **Figure 22.27**. At some point in their lives, all chordates have

- A **notochord**: a stiff, rod-shaped support structure, composed of cells and fluid and surrounded by a lining of fibrous tissue, running from the chordate's head to its tail
- A **dorsal nerve cord**: a rod-shaped structure consisting of nerve cells, running from the chordate's head to its tail
- A series of **pharyngeal slits**: openings to the pharyngeal cavity
- A **post-anal tail**: a tail located posterior to the anus

At this point, as a chordate, you may be wondering where your *own* post-anal tail or pharyngeal slits went, so a little explanation of these features is in order. Although the word "notochord" sounds like "nerve cord," the notochord is a flexible but tough *support* structure, not a nervous-system feature. A lancelet retains its notochord throughout life, but most vertebrates lose theirs in embryonic life. (Its outer portion develops into the backbone.)

The dorsal nerve cord may not seem like a feature unique to chordates—you've seen nerve cords in lots of other phyla—but it is the qualifier *dorsal* that makes the difference. The nerve cords you saw in the annelids and arthropods were ventral, meaning they were on the underside of these animals. The verte-

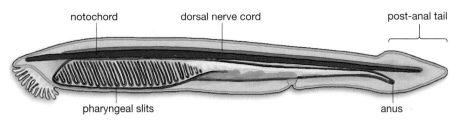

Figure 22.27
Four Universal Chordate Features
All chordates possess four structures at some point in their lives: pharyngeal slits, a support structure called a notochord, a dorsal nerve cord, and a post-anal tail. In the lancelet, all these features are present in the adult. In humans, only the dorsal nerve cord is clearly evident in adults.

brates among the chordates get not only the dorsal nerve cord, but a protective structure that develops around it, the vertebral column. The anterior (top) part of the vertebrate nerve cord undergoes tremendous development, becoming the brain.

The pharyngeal slits you see in the lancelet in Figure 22.27 take on more meaning when you remember that, in both humans and lancelets, the pharynx is a passageway just behind the mouth that can be thought of as the first chamber of the digestive tract. Pharyngeal *slits*, then, are perforations of the pharynx, and these serve different functions in different chordates. In the filter-feeding lancelets, they serve to trap food. Meanwhile, in embryonic fish, the slits *develop into* gills, which fish use to obtain oxygen. And in people? If you look at a month-old human embryo, you can see a series of pharyngeal clefts that for a time have openings in them. These pharyngeal slits, however, develop not into gills but into various other structures, including the space inside the middle ear.

Finally, what about that tail? If you look at **Figure 22.28**, you can see an 8-week-old human embryo with a very clearly developed tail. By the

(a) Human embryo tail at eight weeks

tail

(b) Human embryo tail at ten weeks

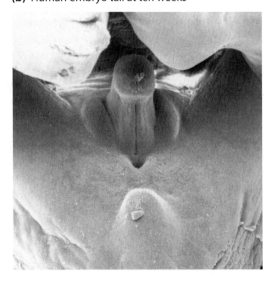

Figure 22.28
The Human Tail
(a) Eight weeks after conception, the human embryo has a very clearly developed tail.
(b) Two weeks later, the tail has almost disappeared. In adult humans the only vestige of our embryonic tail is a small bone at the base of the vertebral column, the coccyx.

time the embryo is 10 weeks old, however, the tail has almost disappeared. All chordates have a tail at some point. In our oceangoing ancestors, it was used to help move them through the water, as is the case with modern fish. In human adults, though, all that remains of our embryonic tails is a small bone, called the coccyx, located at the base of our vertebral column.

Diversity among the Vertebrates

For the remainder of the chapter, we'll concentrate on the vertebrates among the chordates. The evolutionary relationships among these vertebrates were covered in Chapter 19. For a summary of which animals gave rise to which others, see Figure 19.17 on page 385.

A signal shift in vertebrate evolution came with the development of a very familiar feature: jaws. There is general agreement that ancient, jawless vertebrates gave rise to the jawed variety. The development of jaws is considered perhaps the single most important event in vertebrate history, for it gave the vertebrates a great power to capture prey, carry offspring, and generally interact with the world in more effective ways. So useful was this feature that today all true vertebrates have jaws—with one exception.

A Jawless Vertebrate

The lamprey is a long, thin animal that looks something like an eel, but it should not be confused with one. Eels are fish, complete with jaws; lampreys, by contrast, have a sucking disk at their anterior end with which they attach themselves to their prey, which are usually fish (**see Figure 22.29**). The lamprey's oral disk has numerous teeth around its edge. It combines these teeth with a rasping tongue to pierce the flesh of its prey. It then hangs on to the victim through suction, continuing to rasp away, ingesting the fluids and flesh that it obtains in this way.

Lampreys are best known in the United States as a destructive force in the Great Lakes. Once confined to the eastern Great Lakes, parasitic lampreys were able to move west by the mid-twentieth century thanks to human alteration of a canal around Niagara Falls. By the 1950s, they had decimated a once-thriving trout fishing industry in the western Great Lakes. Human efforts to kill off lampreys while reintroducing trout have had only limited success. The main culprit seems to be human pollution, which is keeping stocked trout from reproducing.

Fish: Cartilaginous and Bony

All the early jawed vertebrates were fish, which today are a big success story in the vertebrate world. Of the 50,000 vertebrate species, more than half are fish. And of the 25,000 or so fish species, more than 24,000 are so-called bony fishes, whose name tells their story: with a couple of exceptions, their skeletons are made of bone. Another 750 fish species are cartilaginous fishes, meaning their skeletons are made of the more pliable connective tissue, called cartilage, that also gives shape to human ears.

By far the best-known cartilaginous fish are sharks, but not far behind are the graceful rays that seem to fly through the ocean (**see Figure 22.30**). The sharks' reputation for being an ancient life-form is well deserved. Their lineage dates back more than 350 million years; hence, they had been around for 130 million years by the time the first dinosaurs ap-

Figure 22.30
Cartilaginous Fish
Sharks and rays, such as the giant manta ray shown here, have cartilaginous, rather than bony skeletons. The ancient cartilaginous fishes represent only about 3 percent of all fish species. This ray and its accompanying diver were photographed off the coast of Mexico.

Figure 22.29
No Jaws, Powerful Predator
Lampreys, such as the one shown here, are jawless fish. Note the lamprey's sucking oral disk, which it uses to latch onto and then parasitize prey. This sea lamprey, *Petromyzon marinus*, is the species that, thanks to human intervention, was able to invade the western Great Lakes in the mid-twentieth century. It remains there today, although in smaller numbers than before.

peared. Unlike most of their bony fish relatives, sharks have no **swim bladder**, which is an inflatable organ in a fish that the fish can fill with gas for optimum buoyancy. Fish that do have swim bladders can inflate them as needed to maintain a neutral buoyancy—that is, to float in the water at a given depth without expending any energy. By contrast, a shark must keep swimming or it will sink.

Sharks, rays, and their close kin aside, all the rest of the fish world is made up of bony fish, from goldfish to tuna to herring (**see Figure 22.31**). What accounts for their great numbers today? Their basic body plan seems to have allowed them to adapt to every kind of underwater environment. Some have evolved a slender shape (eels), others a fantastic camouflage (anglerfish), and still others a biological "antifreeze" for frigid waters (the Antarctic icefish).

Almost all bony fish are so-called "ray-finned" fishes, so named because their dorsal fins are supported by straight-line structures that look like rays (see Figure 22.31). Four surviving species of bony fishes belong to another category, however, the lobed-finned fishes. These fish are important because it is their ancestors who brought vertebrates onto the land by struggling out of the swamps some 375 million years ago, using their muscular, lobed fins to propel themselves. Their lobed *fins* evolved into the four *limbs* of the **tetrapods**, meaning the four-limbed vertebrates (see Figure 19.16 on page 384 for this transition). All land vertebrates are tetrapods: amphibians, reptiles, birds, and mammals. Even snakes and whales are tetrapods; these animals simply lost the limbs their ancestors had. If you look at **Figure 22.32**, you can see a picture of one of the rare, living lobe-finned fish, the coelacanth (SEE-la-kanth).

Figure 22.32
Lobe-finned Fish
One of the few living lobe-finned (rather than ray-finned) fish is the coelacanth, which was first discovered in modern times in the Indian Ocean near Madagascar. A fish similar to this one is thought to be the ancestor of all land vertebrates. Note its four fleshy, lobed fins. Fins such as these evolved, in the lobe-finned ancestors, into the four limbs of land vertebrates.

Amphibians: At Home in Two Worlds

Amphibians were the first truly terrestrial vertebrates, having evolved from lobe-finned fishes, but this statement needs qualification. Are amphibians really land animals? Most of them actually live in two worlds—land and water. (Their very name means "double life.") If you look at **Figure 22.33** on the next page, you can see the life cycle of the best-known amphibian, the frog.

Even adult frogs must live in environments that are at least moist, because water can evaporate right through their skin. But frog offspring are fully aquatic for a time. They hatch from eggs that must be laid in water lest they dry out, and most young frogs spend the first part of their lives as swimming tadpoles. There are exceptions to the rule that amphibians must remain near standing water. Some frogs have become fully terrestrial, but even they must inhabit moist environments.

Apart from frogs, there are two other varieties of amphibians: salamanders and newts, collectively known as the tailed amphibians; and some wormlike, tropical amphibians known as caecilians. Salamanders may look like lizards, but their watery amphibian heritage betrays them. With some exceptions, they must live where it is moist, a fact that often keeps them hidden from human view.

Amphibians once were the dominant land animals; indeed, they were the only land vertebrates for some 75 million years. But about 250 million years ago they ceded their dominant status to the very group that evolved from them, the reptiles.

Reptiles and Birds

It may seem strange to lump reptiles and birds together, but remember that scientific classification

Figure 22.31
Bony, Ray-finned Fish
About 97 percent of fish species have bony (rather than cartilaginous) skeletons, and of the bony fish, almost all are "ray-finned," meaning that their dorsal fins are supported by stiff rays such as the ones visible on this yellowtail parrotfish, *Sparisoma rubripinne*.

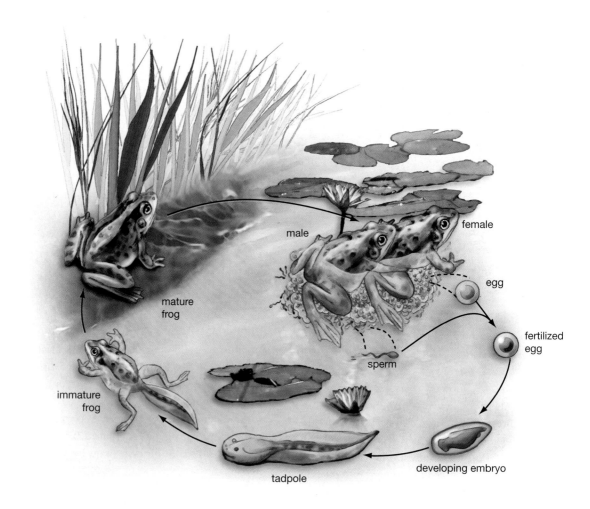

Figure 22.33
Amphibian Life Cycle
Amphibians (such as frogs, salamanders, and newts) require an aquatic, or at least moist, environment for at least part of their life cycle. Note the female frog in the figure is laying her eggs into the water. The male then releases sperm that falls on the eggs, fertilizing them. Most frogs dwell solely in the water through much of their development, first as embryos, then as tadpoles.

Figure 22.34
An Egg for Many Environments
The amniotic egg was an evolutionary development that allowed vertebrates to move inland from the water. The egg's features meant that, in dry environments, the embryo would not "desiccate" or perish through a loss of fluids. The egg's tough shell is the first line of defense; it limits evaporation of fluids. The amnion is one of several membranes that provide some cushioning protection for the embryo. The allantois serves as a repository site for the embryo's waste products. The chorion provides cushioning protection and works with the allantois in gas exchange, meaning the movement of oxygen to the embryo and the movement of carbon dioxide away from it. The yolk sac provides the nutrients the embryo will need as it develops.

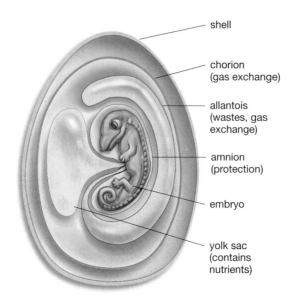

has mostly to do with shared ancestry. There is general agreement that birds are the direct descendants of the reptiles we know as dinosaurs. If so, then the separate categories of "bird" and "reptile" make sense mostly in terms of describing *features* of the two kinds of animals.

One seemingly innocuous feature unites all reptiles and birds, but it is a feature that had far-reaching consequences. If you look at **Figure 22.34**, you can see an illustration of something known as the **amniotic egg**, named after a membrane (called an amnion) that aids the embryo inside of the egg. Recall that most amphibians must lay their eggs in water, lest the eggs dry out and the embryos inside them perish. The amniotic egg of reptiles and birds is different; it can be laid in environments ranging from moist forest floor to desert. Its tough shell keeps the egg from drying out, and the various membranes within it protect the embryo. The amniotic egg therefore gave reptiles a tremendous advantage over the amphibians in the kinds of locations the reptiles could call home; unlike the amphibians, reptiles could settle away from the water.

Characteristics of Reptiles

Several other things distinguish reptiles from amphibians. Looking back at the frog life cycle in Figure 22.33, you can see that the female frog lays her eggs in the water, after which the male spreads his sperm on top of them. This is external fertilization. By contrast, all reptiles employ internal fertil-

(a) Green turtle

(b) Tree python

(c) American crocodile

Figure 22.35
Reptile Diversity
The three main groups of modern-day reptiles are the turtles, the lizards and snakes, and the crocodiles and alligators. Although some of these animals live in water, their scaly skin and amniotic eggs free them from dependency on it.

(a) This green turtle (*Chelonia mydas*) is swimming in the Red Sea.

(b) A newborn green tree python (*Chondropython viridis*) in New Guinea. It will change into its adult color in 6 to 8 months.

(c) American crocodiles such as this one (*Crocodylus acutus*) are found in far-southern Florida, through the Caribbean, Central America, and northern South America.

ization—eggs are fertilized inside the female's body. A requirement for this, of course, is that males possess a copulatory organ that allows sperm to be deposited inside the female. Another difference between amphibians and reptiles is that reptiles have a tough, scaly skin that conserves water, as opposed to the thin amphibian skin that allows water to escape. Reptiles also have a stronger skeleton than amphibians, more efficient lungs, and a better-developed nervous system.

There are three main varieties of modern-day reptiles: lizards and snakes; crocodiles and alligators; and turtles (**see Figure 22.35**). Then there is an ancient, extinct variety of reptiles that is as well known as the living examples. The dinosaurs first appeared about 220 million years ago and were certainly the most fearsome creatures ever to have walked on land. The last of them died out 65 million years ago.

Characteristics of Birds

Surprising as it may seem, one view is that "dinosaurs," properly defined, never did die out, but live on in their descendants, the birds. Indeed, many evolutionary specialists routinely think of birds as "avian dinosaurs." Two lines of evidence have convinced most scientists that birds descended from a line of bipedal (two-footed), meat-eating dinosaurs. One line concerns bone similarity. Long, S-shaped necks, types of hip bones, relative leg sizes. Only dinosaurs and birds share this list of characteristics. Second, there is the famous crow-sized fossil *Archaeopteryx*—a clear "transitional form" between dinosaurs and birds that had the dinosaur's teeth and claws, but the bird's feathers (**see Figure 22.36**). In the 1990s, scientists began filling in the transitional gaps on the other side, unearthing fossils of dinosaurs that were flightless but nevertheless had feathers.

Figure 22.36
Dinosaur and Bird
Scientists believe that this fossil, called *Archaeopteryx*, represents a transitional form between dinosaurs and birds. It has the teeth and claws of a dinosaur, but the feathers of a bird. The drawing at right is an artist's interpretation of the fossil at left.

Birds are a vertebrate success. They exist all over the globe in great numbers, and their 9,100 recognized species exceeds the number of reptile species (7,000) or amphibian species (4,300). Though there are obvious differences between an eagle and a hummingbird, all flying birds have a very similar appearance. By contrast, think how different snakes are from crocodiles among the reptiles, or frogs are from salamanders among the amphibians. The similarity among birds is a consequence of the strict requirements of vertebrate flight: light, hollow bones, wings, and powerful flight muscles that attach to a breastbone (**see Figure 22.37**). All birds use their own metabolism to maintain a relatively stable internal temperature—a quality called *endothermy* that we'll review shortly in connection with our last major grouping of animals.

Mammals: Small Numbers, Big Impact

The story of mammals could be titled "Big Effects from a Small Group." There are only about 4,400 species of mammals—not even half the number of bird species, and scarcely more than the number of amphibian species. The total number of *individual* mammals is infinitesimal compared to the number of insects or roundworms. Yet it's the mammals that are the biggest creatures not only on land (the elephants) but in the sea (the whales). It's the mammals that are the fiercest creatures from the savanna (lions) to the arctic ice (polar bears). And, of course, it's the mammals, in the form of humans, who control the fate of so much of the rest of the living world.

But what is a mammal? Here are two mammal universals and two near-universals. All mammals have mammary glands and maintain a near-constant internal temperature. Most mammals have hair and have eggs that develop inside their mother's body. Here's some detail on each of these characteristics.

Mammals are named for the **mammary glands** that all of them possess: a set of glands that, in females, provide milk for the young. Next, almost all mammals have hair, while no other animals have it. This feature is related to another universal mammal characteristic, which is that mammals are **endothermic**: their internal body temperature is relatively stable and their body heat is generated internally—by their own metabolism. Conversely, amphibians and reptiles are **ectothermic**, meaning their internal temperature is controlled largely by the temperature of their environment.

Endothermy, sometimes misleadingly called warm-bloodedness, is rare in the animal world. In fact, other than mammals, the only creatures to possess it are birds. Endothermy and ectothermy are very important in determining which creature will live where in the animal world. After a cold night in the desert, the muscles of a lizard cannot function at full capacity until that lizard has warmed itself in the sun. By contrast, a desert rat is as functional at sunrise as it is at sunset. But there is a price to be paid for its readiness, and that price is energy expenditure. It takes a great deal of energy to maintain the constant, relatively high body temperatures of mammals and birds. Pound per pound, the "basal" or resting metabolic rate of a rat may be four times that of a lizard. And it takes food to fuel this energy expenditure. The upshot is that, pound per pound, mammals and birds simply have to eat more than reptiles or amphibians.

The very quality of being able to maintain a stable body temperature has, however, allowed mammals to live where amphibians and reptiles

(a) Heron

(b) Bald eagle

Figure 22.37
Bird Similarity
Despite the obvious differences between the heron on the left and the bald eagle on the right, there are great similarities in the birds as well. Indeed, all flying birds display a similarity in body form. This is so because animals as large as birds must meet a strict set of requirements in order to fly. All birds have thin, hollow bones, tails to stabilize their flight, and powerful flight muscles that originate on their breastbone or sternum.

cannot—in cold climates. Consider the fact that there are arctic hares, arctic foxes, and arctic birds, but there are no arctic lizards or frogs. Past a certain point, north or south on the globe, there is no place a lizard can go to warm its body enough to live. Mammals have their limits as well, but the range of climates they can adapt to is greater. Their hair is a great aid in maintaining temperature, as is the thick layer of fat that lies just beneath the skin of such creatures as whales and seals.

With two important exceptions, mammals are **viviparous**: a condition in which fertilized eggs develop inside a mother's body. Contrast this with the **oviparous** condition, seen in all birds and many reptiles, in which fertilized eggs are laid outside the mother's body and then develop there. All embryos need nutrients and protection. In egg-laying (oviparous) animals, both things are provided by the egg itself (though parents also may provide protection). Conversely, a human egg (and then embryo) is protected by the mother's body and draws its nutrients directly from the mother in ways we'll look at shortly.

Reproduction turns out to be the defining feature of the three evolutionary lines of mammals: the monotremes, the marsupials, and the placental mammals (**see Figure 22.38**). **Monotremes** are egg-laying mammals, represented by the duck-billed platypus and spiny anteaters found in Australia. The monotremes do nourish their young on milk from mammary glands, but they have no nipples; thus the young must lap up the milk that diffuses onto the hair of the mother.

Marsupials are mammals in which the young develop within the mother to a limited extent, inside an egg that has a membranous shell. Marsupials are represented by several Australian animals, including the kangaroo, and by several animals in the Western Hemisphere, including the North American opossum. Early in development, the egg's membrane disappears, and in a few days time, a developmentally immature but active marsupial is delivered from the mother. In the case of a kangaroo, the tiny youngster has just enough capacity to climb up its mother's body, into her "pouch," and begin suckling from a nipple there. We can think of the kangaroo as developing partly inside its mother's reproductive tract, and partly inside her pouch.

All other mammals are known as **placental mammals**. They are mammals nurtured before birth by the **placenta**; a network of maternal and embryonic blood vessels and membranes that allows nutrients and oxygen to diffuse *to* the embryo from the mother while allowing embryonic wastes to diffuse *from* the embryo to the mother. (For more on the placenta, see Chapter 30, page 665.) Thus, in placental mammals, the embryonic young derive their nutrition not from food stored in an egg, but directly from the circulation of the mother. Further, the embryos of placental mammals can develop for a very long time inside the mother. The human gestation period—the time from conception to birth—is about nine months, and in elephants it is almost two years. By contrast, the gestation time for a kangaroo is about a month, after which its "pouch life" might last six months.

The diversity of the placental mammals is impressive. There are flying mammals (bats), swimming ocean behemoths (whales, dolphins, and so forth), all manner of large herbivores (horses, deer,

Figure 22.38
Mammal Diversity
All mammals have mammary glands that provide milk for their young. However, mammals differ in their reproductive strategies.

(a) This duck-billed platypus is a monotreme—a mammal that lays eggs.

(b) Kangaroos are marsupials—they give birth to physically immature young, which then develop further within the mother's pouch. Pictured are a mother and her fairly mature "joey" in eastern Australia.

(c) The young of placental mammals, such as this grizzly bear (*Ursus arctos*), develop to a relatively advanced state inside the mother, with nutrients supplied not by an egg, but by a network of blood vessels and membranes, the placenta.

(b) Marsupial

(a) Monotreme

(c) Placental mammal

Table 22.1 Animal Phyla

Phylum	Members include	Live in	Characteristics
Porifera	Sponges	Ocean water (marine), with a few living in fresh water	Sessile (fixed in one spot) as adults, no symmetry. System of pores through which water flows serves to capture nutrients. Bodies do not have organs or tissues. Sexual and asexual reproduction.
Cnidaria	Jellyfish, hydrozoans, sea anemones, coral animals	Almost all marine; a few in fresh water	Radial symmetry, medusa (swimming) and polyp (largely sessile) stages of life. Use stinging tentacles to capture prey. Have tissues. Can reproduce sexually or by budding.
Platyhelminthes	Flatworms (flukes, tapeworms, turbellarians)	Marine, fresh water, or moist terrestrial environments	Bilateral symmetry, hermaphroditic, possess organs, no central cavity, most primitive animals with triploblastic tissue structure. Many are parasites. Can reproduce by dividing themselves.
Annelida	Segmented worms	Most are marine, some fresh water and moist terrestrial	Distinct body segmentation in varieties such as earthworms. Leeches are annelids.
Mollusca	Gastropods (snails, slugs), bivalves (oysters, clams, mussels), cephalopods (octopus, squid, nautilus)	Marine, fresh water, moist terrestrial	Mantle that can secrete material that becomes shell; mantle cavity housing gills or lungs; unitary foot (lost in some sessile species, evolved into arms and tentacles in cephalopods).
Nematoda	Roundworms	Almost all environments	Small size in most, but some are several meters long. Many are crop pests, others animal parasites. Possess pseudocoelom. Reproduction always sexual.
Arthropoda	Vast group that includes insects, spiders, mites, crabs, shrimp, and centipedes	Many aquatic and terrestrial environments	External skeleton made of chitin; jointed, bilateral appendages; molting in many species; segmented bodies; generally an open circulation system.
Echinodermata	Sea stars, sea urchins, sea cucumbers	Always marine, most on ocean floor	Generally have radial symmetry as adults; tube feet in some species, slow ocean floor movement in most.
Chordata	Vertebrates, lancelets, sea squirts	Great variety of aquatic and terrestrial environments	At some point, all possess notochord, dorsal nerve cord, pharyngeal slits, post-anal tail. Vertebrates have vertebral column.

cows) and smaller herbivores (rodents, which make up more than half of all mammal species). Then, of course, there are the primates such as ourselves, defined by our forward-facing eyes, grasping hands, and arboreal or tree-dwelling habitat. Several species of primates, including our own, have come down from the trees to dwell on the ground (see Chapter 19, page 386). **Table 22.1** summarizes all of the animal phyla reviewed in this chapter.

On to Plants

In life's family tree, the animals you've just looked at are far removed from plants. Yet animals and plants are alike in one sense, in that they sit atop two great evolutionary lines. Plants occupy the higher branch-

es of the line of "self-feeders": the *autotrophs* who make their own food by harnessing the power of the sun. The first creatures who performed this feat were bacteria; then came the algae (which are protists). And then, evolving from the algae, came the plants, moving onto land and constructing their green edifice wherever they took root, thereby providing food and shade for all animals that came after them. In a sense, this sequence of events describes the relationship that animals and plants have: Plants first, then animals. In a world without plants, there would be no land animals, but in a world without land animals, there would still be plants. The two chapters coming up describe the nature of the indispensable, green members of the living world, the plants.

Chapter Review

Summary

22.1 What Is an Animal?

- A single physical feature is sufficient to define animals. All animals pass through a blastula stage in embryonic development, but no other living things do. A blastula is a hollow, fluid-filled ball of cells that forms once an egg is fertilized by sperm.

- Three other features are characteristic of all animals, but also are shared, to varying degrees, by members of other kingdoms in the living world. All animals are multicelled, are heterotrophs (they cannot make their own food), and are composed of cells that do not have cell walls.

22.2 Animal Types: The Family Tree

- The animal kingdom is divided into large-scale categories called phyla. There are between 36 and 41 animal phyla, each phylum being a group of organisms that share a set of physical characteristics that result from shared ancestry.

- Animals can be thought of as having evolved from simpler to more complex forms through a series of additions to the characteristics found in more primitive animals. These additions are tissues, radial symmetry, bilateral symmetry, and an enclosed body cavity or coelom.

- A fundamental split in animal evolution came with the divergence of the protostome and deuterostome animal lines. Protostome animals include flatworms, roundworms, molluscs, annelid worms, and arthropods; deuterostome animals include echinoderms and chordates.
Web Tutorial 22.1: The Animal Family Tree

22.3 Phylum Porifera: The Sponges

- Sponges are simple animals, lacking organs, tissues, or symmetry. Nearly all sponges live in marine environments, though there are a few fresh water varieties.

- Sponges live by drawing water into themselves through a series of tiny pores on their exterior, and then filtering this water for food.

- Sponges can reproduce sexually, with eggs and sperm carried from one sponge to another by water currents; or they can produce asexually, through budding.

22.4 Phylum Cnidaria: Jellyfish and Others

- The defining characteristic of members of phylum Cnidaria is their use of stinging tentacles to capture prey.

- Cnidarians include jellyfish, sea anemones, hydras, and the reef-building coral animals whose skeletons make up the bulk of coral reefs.

- Many cnidarians have both an adult, medusa stage of life, which swims in the water; and an immature, polyp stage of life, which generally remains fixed to rocks, animals, or other solid surfaces. Some cnidarians have only the polyp stage. Cnidarians have no organs, but they do have tissues.

- Cnidarians have radial symmetry, and reproduction can be sexual or asexual.

22.5 Phylum Platyhelminthes: Flatworms

- The flatworms of phylum Platyhelminthes are mostly small creatures, dwelling either in aquatic or moist terrestrial environments. Some parasitic flatworm species can, however, reach enormous lengths.

- Flatworms have bilateral symmetry, but have no coelom or system of blood circulation. They have no anus, meaning the flatworm's mouth is the single opening to its digestive system.

- Flatworms are the least complex animals to be triploblastic. They have three embryonic germ layers—endoderm, mesoderm, and ectoderm—whereas cnidarians lack a mesoderm and are therefore diploblastic.

- Most flatworms are hermaphroditic—a single flatworm is likely to possess both male and female sex organs. Many flatworms can reproduce asexually by breaking themselves in half.

- Parasitic tapeworms, which can infect human beings, are flatworms, as are the flukes that cause the disease schistosomiasis.

22.6 Phylum Annelida: Segmented Worms

- The worms of phylum Annelida provide a clear example of a physical feature that is widespread in the animal kingdom, body segmentation, meaning a repetition of body parts in an animal. Segmentation is generally valuable because it provides both flexibility and strength.

- Earthworms are representative of the annelid or segmented worms; earthworm bodies display a clear segmentation, allowing the worm to exercise a degree of independent control over each segment or group of segments.

- Most annelids are marine, and some dwell in freshwater. All annelids exist in environments that are at least moist. Leeches are parasitic annelid worms.

22.7 Phylum Mollusca: Snails, Oysters, Squid, and More

- Phylum Mollusca is an extremely varied group of animals whose members range from sessile, brainless mussels to agile, intelligent squid. Mollusc habitats range from marine and freshwater to moist terrestrial.

- Three important classes of molluscs are the gastropods, which include snails and slugs; the bivalves, which include oysters, clams, and mussels; and the cephalopods, which include octopus, squid, and nautilus.

- All molluscs possess a fold of skin-like tissue, called a mantle, that usually secretes material that forms a shell, though many molluscs do not have shells. Molluscs tend to have a mantle cavity that houses either gills (if the molluscs are aquatic) or lungs (if they are terrestrial); many have a unitary foot, whose wave-like contractions allow movement. Many molluscs have a tooth-lined membrane called a radula that can be extended outside the body for scraping and retrieving food.

- Cephalopods have a closed circulation system, in which blood stays within blood vessels, while oxygen and nutrients diffuse out of the vessels to the surrounding tissues. Most molluscs have an open circulation system, in which blood flows out of blood vessels into openings called sinuses, where the blood bathes tissues.

22.8 Phylum Nematoda: Roundworms

- The roundworms of phylum Nematoda exist in enormous numbers in all kinds of habitats on Earth. Most are microscopic, though some are large.

- A number of roundworms are agricultural pests, and some are human parasites. The disease trichinosis is caused by roundworms, and the parasites known as hookworms are roundworms.

- Roundworms are the least complex animals to possess an enclosed cavity—technically a pseudocoel instead of a true coelom. Their reproduction is always sexual, and they are not hermaphroditic.

22.9 Phylum Arthropoda: Insects, Lobsters, Spiders, and More

- Phylum Arthropoda is enormous and extremely varied. There are more species in one class of arthropods, the insects, than there are in all the other groupings of animals combined.

- Arthropods can be divided into three subphyla: Uniramia, which includes millipedes and centipedes along with insects; Chelicerata, which includes spiders, ticks, mites, and horseshoe crabs; and Crustacea, which includes shrimp, lobsters, crabs, and barnacles. All arthropods have an external skeleton or exoskeleton, and paired, jointed appendages. Arthropods go through molting, meaning the periodic shedding of an old exoskeleton.

- Members of the subphylum Uniramia include the insects, which have bodies divided into three sections: head, thorax, and abdomen. They have an open circulation system, sexes are separate, and reproduction is generally sexual. Centipedes and millipedes make up the two other classes of Uniramians.

- All members of subphylum Crustacea have five pairs of appendages extending from their heads—two pairs of antennae and three pairs of feeding appendages. Crustaceans range from large lobsters to microscopic water fleas.

- All members of subphylum Chelicerata have in common a pair of appendages, called chelicerae, that serve various feeding functions—grasping in horseshoe crabs, for example, and puncturing in spiders.

22.10 Phylum Echinodermata: Sea Stars, Sea Urchins, and More

- Echinoderms include sea stars, sea urchins, sand dollars, and sea cucumbers, among others. All members of phylum Echinodermata are marine, and most inhabit the ocean floor.

- Echinoderms have a radial symmetry as adults. The sea stars among them evolved to this condition from bilateral symmetry; sea star larvae have a bilateral symmetry that is lost as they develop into adults. Sea stars have no brain and only one sensory organ, the primitive eyespots on the tips of their arms. A new sea star can be generated asexually from the arm and part of the central disk of an existing sea star.

- Echinoderms tend to move slowly across the sea floor; but many are formidable predators, often feeding on such molluscs as oysters or mussels.

22.11 Phylum Chordata: Mostly Animals with Backbones

- Phylum Chordata is made up of three subphyla: Vertebrata, which includes all the vertebrates, including human beings; Cephalochordata, made up of creatures called lancelets; and Urochordata, made up of three classes of animals, including the tunicates or sea squirts. Only the vertebrates have a vertebral column, meaning a flexible column of bones running from the anterior to posterior ends of an animal.

- All chordates possess, at some point in their lives, a rod-shaped support structure called a notochord; a nerve cord on their dorsal side; a post-anal tail; and a series of pharyngeal slits that develop into various structures, depending on the type of chordate.

- With the exception of lampreys, all true vertebrates have jaws. Today's jawed vertebrates developed from ancient jawless vertebrates.

- Fish account for more than half the 50,000 species of vertebrates, with the vast majority of fish being bony fish, as opposed to the more ancient line of cartilaginous fishes, such as sharks, whose skeletons are made of the connective tissue cartilage.

- Most bony fish are ray-finned fishes, named after the ray-like structures in their dorsal fins. Four species of fish belong to another category of fish, the lobe-finned fishes. These fish are important because an ancient variety of them is thought to have given rise to all tetrapods, meaning four-limbed vertebrates. All land vertebrates are tetrapods: amphibians, reptiles, birds, and mammals.

- Amphibians were the first terrestrial vertebrates. Today amphibians include frogs; the tailed amphibians, salamanders and newts; and some wormlike amphibians known as caecilians. All amphibians must live in environments that are at least moist, most employ external fertilization of eggs, and most live in aquatic environments for part of their lives.

- Reptiles evolved from amphibians. An important feature in their evolution was the development of the amniotic egg, which can keep embryos moist even in dry environments. The amniotic egg allowed the reptiles to move inland from water sources. Reptiles employ internal fertilization and have a tough, scaly skin that conserves water. There are three main varieties of modern-day reptiles: turtles, lizards and snakes, and crocodiles and alligators. Dinosaurs were reptiles.

- Most scientists believe that birds are the direct descendants of dinosaurs. Birds exist all over the globe in great numbers. The similar appearance of all flying birds stems from the strict requirements of flight for creatures as large as vertebrates: light bones, wings, and powerful flight muscles that attach to a breastbone. All birds are endothermic: their internal body temperature is relatively stable and body heat is generated internally, by their own metabolism. Conversely, amphibians and reptiles are ectothermic: their internal temperature is controlled largely by the temperature of their environment. Birds and mammals are the only true endothermic animals.

- Though there are relatively few mammal species, mammals have had a great impact on the natural world. They account for the largest creatures on land and sea, and one species of mammal, human beings, controls the fate of much of the living world.

- All mammals have mammary glands, meaning glands that deliver milk to the young. Most mammals have hair, and their fertilized eggs develop inside the mother's body (making them viviparous). There are three evolutionary lines of mammals: the monotremes, which are egg-laying (or oviparous) mammals, represented by the platypus; marsupials, in which the young develop within the mother only to a limited extent, represented by kangaroos; and placentals, in which young are nourished within the mother through a placenta, represented by most of world's mammals.

Key Terms

amniotic egg 470
bilateral symmetry 448
bivalves 458
body segmentation 456
cephalopods 458
closed circulation system 459
coelom 450
diploblastic 456
dorsal nerve cord 467
ectothermic 472
endothermic 472
exoskeleton 460
gastropods 458
hermaphroditic 456
invertebrate 446
mammary glands 472
marsupial 473
molting 461
monotreme 473
notochord 467

organ 455
open circulation 459
oviparous 473
paired, jointed appendages 460
parasite 455
pharyngeal slits 467
phylum 447
placenta 473
placental mammal 473
post-anal tail 467
radial symmetry 448
swim bladder 469
symmetry 448
tetrapods 469
tissue 446
triploblastic 456
vertebral column 467
viviparous 473
zoologist 447

Understanding the Basics

Multiple-Choice Questions (answers in the back of the book)

1. All animals (select all that apply)
   a. go through a blastula stage in development
   b. have cells that lack cell walls
   c. are multicelled
   d. are heterotrophs
   e. all of the above

2. Bilateral symmetry is a condition in which
   a. no one animal species can overpower another
   b. the top and bottom halves of an animal are symmetrical
   c. two animal species evolve along parallel lines
   d. one side of an animal is symmetrical to the other
   e. one side of an animal is as strong as the other

3. A coelom is
   a. an enclosed body cavity, not found in sponges, cnidarians, or flatworms
   b. an enclosed body cavity, found only in deuterostomes
   c. a digestive organ, found in all animals except sponges
   d. an evolutionary offshoot of a species
   e. an offspring that arises from asexual reproduction

4. An animal has bilateral symmetry, but no coelom. What phylum is it in?
   a. Mollusca
   b. Cnidaria
   c. Chordata
   d. Porifera
   e. Platyhelminthes

5. Only _____ are endothermic, a physiological condition whose chief direct cost is _____.
   a. cephalopods and mammals; a short life-span
   b. cephalopods and echinoderms; smaller size
   c. mammals and birds; increased energy expenditure
   d. reptiles and amphibians; cold-induced lethargy
   e. mammals and birds; fewer offspring

6. A marine animal with radial symmetry goes through both medusa and polyp stages in its life cycle. Which of the following features is it certain not to have? (Select all that apply.)
   a. a brain
   b. a heart
   c. a coelom
   d. legs
   e. a mouth

7. The world's largest invertebrates, the _____, always _____.
   a. crustaceans; have exoskeletons
   b. echinoderms; have radial symmetry
   c. cephalopods; are single-sex
   d. chordates; have a post-anal tail
   e. cephalopods; are filter feeders

8. In earthworms, we can see a clear example of the widespread animal feature known as
   a. filter feeding
   b. body segmentation
   c. viviparous reproduction
   d. a post-anal tail
   e. a radula

9. Which of the following is not a physical feature of all vertebrates at some point in their lives? (Select all that apply.)
   a. pharyngeal slits
   b. a vertebral column
   c. a ventral nerve cord
   d. a post-anal tail
   e. a mantle cavity

10. If an animal has an exoskeleton, paired jointed appendages, and goes through the process of molting, it is:
    a. a bivalve
    b. a gastropod
    c. a tetrapod
    d. an arthropod
    e. a chordate

Brief Review

1. A coelom amounts to internal space in an animal. Why is such a feature valuable?

2. Do all animals have a single ancestor, or do the different phyla of animals have different ancestors? What evidence supports your answer?

3. What are the benefits and the costs of endothermy? What animals are endothermic?

4. Why is it fair to characterize sponges as "primitive" animals?

5. What is a coral reef chiefly composed of?

6. What feature common to molluscs was modified through evolution to become the arms and tentacles of the squid?

7. Which animal is a spider more closely related to, a dust mite or a grasshopper?

8. A mother provides milk for her offspring, and yet those offspring are hatched from eggs. What kind of animal is the mother?

Applying Your Knowledge

1. The text notes the tremendous number of insects in the world—both the number of insect species and the number of individual insects. One of the most spectacular examples of insect numbers can be seen in the so-called periodical cicadas, which emerge in the Eastern and Central United States in groups called "broods" after completing an underground larval stage that lasts either 13 or 17 years. One 17-year brood that emerged in 1956 near Chicago was found in densities of 1,500,000 per acre in lowland forests—about 533 tons of cicadas per square mile over their entire range. Biologists often talk of the reproductive "strategies" of various species. These are the traits of a species that help ensure its continued reproduction, generation after generation. What is the reproductive strategy of the periodical cicadas? Does their long underground immature stage aid in this strategy?

2. One of the distinguishing features of animals is that their cells do not have the thick outer lining known as a cell wall. Plant cells have a cell wall, as do fungi cells. Why would such a feature be advantageous for plants, but disadvantageous for animals?

3. Plants can have elaborate defense systems, and they can respond in sophisticated ways to their environment—for example, in preserving resources, they can go into a dormant state in winter. Given this, can plants have intelligence? Or is it only animals that have this trait, to one degree or another?

23 *An Introduction to Flowering Plants*

Heart medicine.
(Section 23.1, page 482)

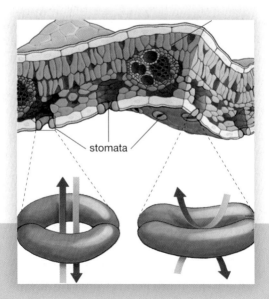

stomata

Guard duty.
(Section 23.2, page 485)

A stem for support, storage, and transport.
(Section 23.2, page 487)

The optometrist has almost completed the examination but would like a better look at the interior of the patient's eyes. Would she agree to have her pupils dilated with the application of a couple of drops to each eye? The patient tilts her head back, and in her pose, it's possible to imagine her as a lady of the Italian Renaissance, lifting her eyes upward and applying to them a few drops of a substance she believes will make her a *bella donna*—a beautiful lady. And what does this substance do? It dilates her eyes just slightly, making her pupils large and entrancing. Spin back further in time to the eleventh century in Scotland where, as legend has it, the Scottish leader Earl Macbeth poisons an invading army of Danish soldiers by lacing their drinks with juice from a plant known as sleepy nightshade. Only later will the English-speaking world call this plant by the name used today: *Atropa belladonna* (**Figure 23.1**). Skip forward to 1831, when scientists isolate from the belladonna plant one of its active ingredients, a compound called atropine. This is the substance that actually dilated the eyes of the Italian ladies and that, at one time, dilated eyes in optometry clinics. Come forward then to the fall of 2002. As the United States is getting ready for war with Iraq, it re-

ceives word that Iraq has ordered a million doses of atropine, along with "auto-injectors" that can be used to administer the drug into the legs. The U.S. is deeply suspicious of this move. Atropine works as an antidote to poison gases; why would Iraq be ordering an antidote to these gases unless it intended to use them as a weapon in the coming war? Doctors note that atropine is routinely employed in medicine, but not in connection with poison gas. In ambulances and emergency rooms, it is used to revive heart-attack victims. An injection of atropine can make an ailing heart beat faster.

Figure 23.1
A Plant that has Served Many Purposes
The flowering plant *Atropa belladonna*

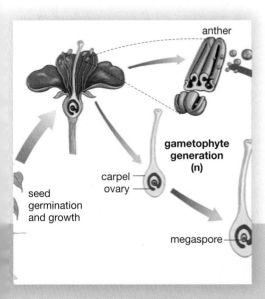

Part of a life cycle.
(Section 23.3, page 489)

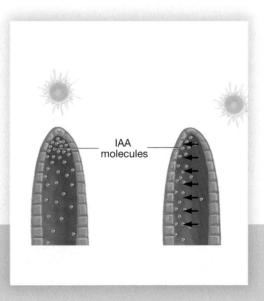

The power of light.
(Section 23.4, page 496)

Climbing up by circling around.
(Section 23.4, page 497)

23.1 The Importance of Plants

Like a talented employee, belladonna keeps getting called back to work by human beings. If you were keeping track, we humans have derived, from just this one plant, a poison, an antidote to poison, a cosmetic, an aid to visual diagnosis, and a heart medicine. But that's the way with plants. The number of things they do for human beings is enormous. The best-known thing they do, of course, is provide us with food. But even in this role, their significance is not fully appreciated. In the world today, up to 90 percent of the calories that human beings consume come directly from plants. Indeed, as some botanists have noted, there are about a dozen plants that stand between humanity and starvation. (The most important of these plants is rice, with wheat coming in second.) Plants are able to serve this function because they make their own food through photosynthesis. The bounty they produce through this process has made Earth a planet of *surplus*—a planet of grains and leaves and roots that are available for the taking. Along these same lines, plants also produce much of the oxygen that most living things require. Then there are the products that human beings have learned to derive from plants, among them lumber, medicines, fabrics, fragrances, and dyes (**see Figure 23.2**).

Apart from these uses, plants are a kind of anchoring environmental force. Their roots prevent soil erosion, while their leaves absorb such pollutants as sulfur dioxide and ozone. One of the main greenhouse gases thought to be warming Earth is carbon dioxide, and plants absorb carbon dioxide when they perform photosynthesis. Thus, there is not only a *production* side to atmospheric carbon dioxide—in significant part the pollution human beings produce—there is also a *consumption* side to it, in the amount plants absorb. One proposed way to fight greenhouse warming, therefore, is to plant trees. A single mature maple tree will absorb about 450 kilograms (about 1,000 pounds) of carbon dioxide from the atmosphere over the course of a single summer.

A Focus on Flowering Plants

Though the plant kingdom is vast and varied, it turns out that there are just four principal types of plants in it. These are the bryophytes, represented by mosses; the seedless vascular plants, represented by ferns; the gymnosperms, represented by coniferous (evergreen) trees; and the flowering plants, also known as angiosperms. Chapter 21, which introduced plants, contained information on the bryophytes, the seedless vascular plants, and the gymnosperms. In this chapter, we will focus strictly on flowering plants. Why pay so much attention to just one type of plant? Because of the overwhelming dominance of flowering plants within the plant kingdom. There are about

Figure 23.2
Uses Galore
Plants are used by human beings in innumerable ways.

(a) Food, here in the form of Concord grapes.

(b) Wood, being harvested from a forest in British Columbia.

(c) A foxglove plant, which yields the heart medicine digitalis.

(a)

(b)

(c)

(a)

(b)

(c)

Figure 23.3
Angiosperm Variety
(a) A trumpet vine (*Clytostoma callistegiodes*) spreading on a bush.
(b) A Senita cactus, on the left, stands next to a Saguaro cactus in the Arizona desert.
(c) The cereal grain rye (*Secale*) growing in a field.

16,000 species of bryophytes, 13,000 species of seedless vascular plants, and 700 species of gymnosperms, but there are an estimated 260,000 species of flowering plant. The term "flowering plant" may bring to mind roses or orchids, and these plants are indeed angiosperms. But food crops such as rice and wheat are flowering plants, as are all cacti, almost all the leafy trees, innumerable bushes, pineapple plants, cotton plants and ice plant—the list is very long (**see Figure 23.3**). In this chapter, you'll get an overview of how flowering plants are structured, of how they function internally, and of how they respond to signals from the outside world. Chapter 24 will go into greater detail on angiosperm structure and function. A formal definition of angiosperms will be provided once you've learned a little about them.

23.2 The Structure of Flowering Plants

Let's begin our tour of the angiosperms by looking at their component parts.

The Basic Division: Roots and Shoots

We'll first look at the larger-scale structures of the angiosperms, starting with a simple, two-part division that rhymes: roots and shoots. Plants live in

two worlds, air and soil, with their root system below ground and their shoot system above (**see Figure 23.4**).

The function of the root system is straightforward: Grow to water and minerals and absorb them from the soil and then begin transporting them up

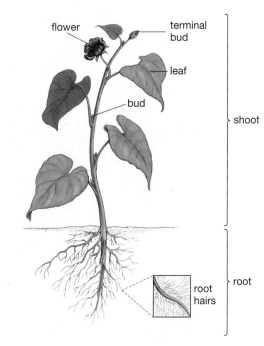

Figure 23.4
A Life-Form in Two Parts
Flowering plants live in two worlds, with their shoots in the air and their roots in the soil. Although flowering plants differ enormously in size and shape, they generally possess the external features shown.

through the rest of the plant. (For more on the minerals plants absorb, see "What Is Plant Food?" on page 486.) Most roots also serve as anchoring devices for plants, and some act as storage sites for food reserves.

The function of the shoot system is more complex. Photosynthesis takes place in this system, primarily in leaves, which means leaves must be positioned to absorb sunlight. Plants are in *competition* with one another for sunlight, which is a primary reason so many plants are tall. The food that is produced in photosynthesis must be distributed throughout the plant. It is not just food production that is centered in the shoot system, however; it is plant reproduction as well. Within flowers and their derivatives, we find all the components of reproduction—seeds, pollen, and so forth. Now let's look in more detail at the components of the root and shoot systems.

Roots: Absorbing the Vital Water

If you look at **Figure 23.5**, you can see pictures of the two basic types of plant root systems, one a **taproot system** consisting of a large central root and a number of smaller lateral roots, and the second a **fibrous root system** consisting of many roots that are all about the same size. **Figure 23.6** then shows you not only the taproot of a young sweetcorn plant but also another feature common to root systems. These are **root hairs**: threadlike extensions of roots that greatly increase their absorptive surface area. Each root hair actually is an elongation of a single outer cell of the root. Be-

tween roots and root hairs, the root structure of a given plant can be extensive. One famous analysis, conducted in the 1930s on a rye plant, concluded that its taproot and lateral roots alone, if laid end to end, would have totaled some 622 kilometers, which is about 386 miles. Meanwhile, the root hairs collectively were 10,620 kilometers long, meaning the plant had almost 6,600 miles of root hairs! This from a plant whose *shoot* stood 8 centimeters (about 3 inches) off the ground.

Why do plants lavish such resources on their roots? Essentially because plants have such a great need to take up water, and roots are the structures that allow them to do this. To perform photosynthesis, plants must have a constant supply of carbon dioxide, which enters the plants through microscopic pores, called *stomata*, that exist mostly on the underside of plant leaves. Open stomata, however, don't just let carbon dioxide in; they let the plant's water *out*, as water vapor. To accommodate this loss while still keeping vital tissues moist, plants continually pass water through themselves, from roots, up through the stem, and into the leaves (at which point it exits as water vapor). The evaporation of water from a plant's shoot is known as **transpiration**; through it, more than 90 percent of the water that enters a plant evaporates into the atmosphere. Because the roots of a single tall maple tree can absorb about 220 liters (or nearly 60 gallons) of water per *hour* on a hot summer day, the scale of transpiration is immense. Thus do we see the importance of root development.

Web Tutorial 23.1
Flowering Plants

Figure 23.5
Two Root Strategies

(a) Dandelions such as this one employ a taproot system—a large central root and a number of smaller lateral roots.

(b) The French marigold employs a fibrous root system—a collection of roots that are all about the same size.

(a) Taproot system

(b) Fibrous root system

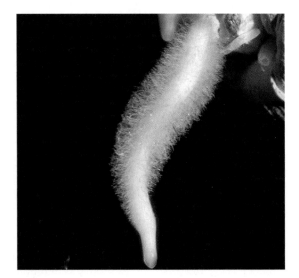

Figure 23.6
Root Hairs
Root hairs enormously increase the surface area of a root. The greater the surface area, the more fluid absorption can occur. Shown here are root hairs on the taproot of a sweetcorn plant.

The food storage aspect of roots has a meaning for human beings. Sweet potatoes and carrots, which are roots, can represent stored food for people as well as for plants.

Shoots: Leaves, Stems, and Flowers

Now let's turn to the shoot system, looking first at the leaves within it.

Leaves: Sites of Food Production

The primary business of leaves is to absorb the sunlight that drives photosynthesis, which is why most leaves are thin and flat. This leaf shape maximizes the surface area that can be devoted to absorbing sunlight while minimizing the number of cells in leaves that are irrelevant to photosynthesis. Beyond this, through their tiny stomata, leaves serve as the plant's primary entry and exit points for gases. As noted, the most important gas that's entering is carbon dioxide, which is one of the starting ingredients for photosynthesis. What's exiting is the by-product of photosynthesis, oxygen, and the water vapor.

The broad, flat leaves that are so common in nature have in essence a two-part structure—a **blade** (which we usually think of as the leaf itself) and a **petiole**, more commonly referred to as the leaf stem. If you look at the idealized cross section of this leaf in **Figure 23.7b**, you can see that the blade can be likened to a kind of cellular sandwich, with layers of cuticle, or waxy outer covering, on the outside, and a layer of epidermal cells just inside them. In the leaf's interior there are *vascular bundles*, which bear some relation to animal veins in that they are part of a transport system. Then there are several layers of mesophyll cells. It is these cells that are the sites of most photosynthesis.

Now note, on the underside of the blade, the openings called **stomata** (singular, *stoma*). These are the pores, noted earlier, that let water vapor out and carbon dioxide in—but for most plants only during the day. When the sun goes down photosynthesis can no longer be performed, and it is not cost-effective for a plant to lose water without gaining carbon dioxide. As such, the stomata close up until photosynthesis begins again the following day. If you look at **Figure 23.7c**, you can see how this opening and closing of stomata is achieved: Two "guard cells," juxtaposed like the sides of a coin purse, are arranged around the stomata. When sunlight strikes the leaf, these cells engorge with water, which makes them

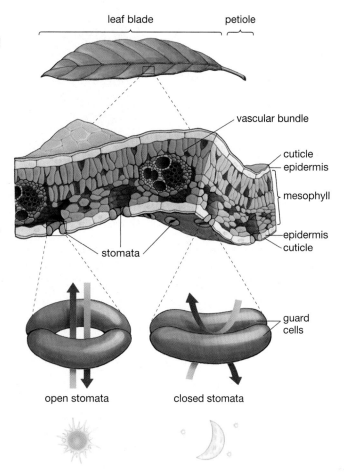

(a) Leaves tend to be broad and flat to increase the surface area exposed to sunlight.

(b) Photosynthesis occurs primarily within the mesophyll cells in the interior of the leaf. The vascular bundles carry water to the leaves and carry the product of photosynthesis, sugar, to other parts of the plant. Pores called stomata, mostly on the underside of leaves, allow for the passage of gases in and out of the leaf.

(c) Guard cells of the stomata control the opening and closing of the stomata. When sunlight shines on the leaf during the day, the guard cells engorge with water that makes them bow apart, thus opening the stomata. When sunlight is reduced, water flows out of the guard cells, causing the stomata to close.

leaf blade   petiole

vascular bundle

cuticle
epidermis

mesophyll

epidermis
cuticle

stomata

guard cells

open stomata   closed stomata

Figure 23.7
Site of Photosynthesis

What Is Plant Food?

Plants are able to make their own food through photosynthesis if they have just four things: water, sunlight, carbon dioxide, and a few nutrients. Water, sunlight, and probably even carbon dioxide are no mystery to you by now, but what exactly are these nutrients you've been reading about?

A **nutrient** is simply a chemical element that is used by living things to sustain life. Recall from Chapter 2 that "elements" are the most basic building blocks of the chemical world. Silver is an element, as is uranium, but these are not nutrients because living things do not need them to live. In the broadest sense, carbon, oxygen, and hydrogen are nutrients. And it turns out that almost 96 percent of the weight of the average plant is accounted for by these three elements, which come to plants primarily from water and air.

Plants need at least another 13 elements to live, however, and not all of these are supplied by water and air. When we think of common "plant food," we are generally referring to those nutrients that plants can *use up* in their soil and that thus must be replenished to assure continued growth and reproduction. Of these, the most important three are nitrogen, phosphorus, and potassium, whose chemical symbols are N, P, and K. When you look at a package of an average plant fertilizer, you will see three numbers in sequence on it (for example, 10-20-10). What these refer to are the percentage of the fertilizer's weight accounted for by these nutrients (see **Figure 1**). The growth of plants usually is enhanced by a fertilizer that has equal ratios of the three elements, but if increased *blooming* is your goal, an increased phosphorus ratio generally is recommended, as with the 10-20-10 example.

It is not just houseplants or lawns that benefit from fertilizer, but farm crops as well. Indeed, the planting of a crop on a parcel of land generally requires a large investment in fertilizer because a crop such as wheat removes a great deal of N, P, and K from the soil in a single season. Historically, fertilizers used on farms were *organic*, meaning they came from decayed living things, such as the fish that Native Americans taught the Pilgrims to use when planting corn. Nowadays, however, commercial fertilizers generally are *inorganic*, meaning they are mixtures of pure elements within binding materials, produced by chemical processes.

Nature is, of course, perfectly capable of producing a green bounty in such places as forests and marshes without the aid of any human-made fertilizer. Decaying plant and animal matter put nutrients back into the soil, and bacteria are able to fix nitrogen, meaning to absorb it from the air and transform it into a form that plants can take up. Looked at one way, however, houseplants and farm crops are *isolated* plants whose participation in this web of life is very limited, and thus they require a helping hand from humans in order to remain robust.

Figure 1
Needed Nutrients
Plants make their own food from sunlight, water, carbon dioxide, and a few nutrients, some of which come from the soil. The most important of these nutrients are nitrogen (N), phosphorus (P), and potassium (K). Because garden plants and houseplants are isolated from natural ecosystems, they must often get these nutrients in the form of fertilizer.

bow apart. Then, in the absence of light, water flows out and the door closes again. A given square centimeter of a plant leaf may contain from 1,000 to 100,000 stomata.

Figure 23.8 gives you some idea of the varied forms of leaves. In addition to the so-called simple leaf we've been looking at, there are compound leaves in which individual blades are divided into a series of "leaflets." Pine needles and cactus spines are likewise actually leaves. The cactus, in particular, has reduced the amount of leaf surface it has as a means of conserving water. So how does it perform enough photosynthesis to get along? It carries out most of its photosynthesis in its stem.

Stems: Structure and Storage

We all generally understand what the stem of a plant is, though it is less generally appreciated that the trunk of a tree is simply one kind of plant stem. The main functions of stems are to give structure to the plant as a whole and to act as storage sites for food reserves. In addition, water, minerals, hormones, and food are constantly shuttling through (and to) the stem, with water on its way up from the roots and food on its way down from the leaves to the rest of the plant.

If you look at **Figure 23.9**, you can see a cross section of one type of plant stem, showing the vascular bundles or veins you first saw in the leaves and the outer, or *epidermal, tissue* just as the leaves had. You can also see so-called *ground tissue* of two types: an outer cortex and an inner pith, both of which can play a part in food storage and wound repair and provide structural strength to the plant.

Flowers: Many Parts in Service of Reproduction

Flowers are the reproductive structures of plants. A single flower generally has both male and female reproductive structures on it, which might make you think that a given plant would fertilize *itself*. This is indeed the case with some plants, such as Gregor Mendel's pea plants, reviewed in Chapter 11. However, because the evolutionary benefit of sexual reproduction is to get the genetic diversity that comes with *mixing* genetic material from different individual organisms, natural selection has worked against self-fertilization by endowing many flowering plants

with ways to reduce the incidence of it. For example, the male pollen of a given plant might be genetically incompatible with that plant's female reproductive structures. Some of the illustrations you'll be seeing show a plant fertilizing itself, but this is done only for visual simplicity.

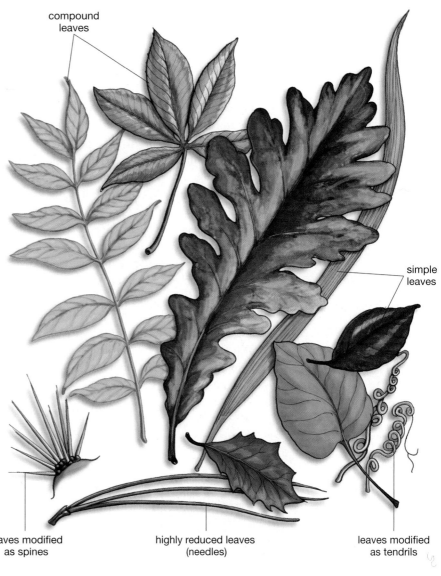

Figure 23.8
Leaves Come in Many Sizes, Shapes, and Colors
Simple leaves have just one blade. In compound leaves, the blade is divided into little leaflets. Some leaves are so modified that the average person can hardly recognize them as leaves—for example, the spines of a cactus.

compound leaves

simple leaves

leaves modified as spines

highly reduced leaves (needles)

leaves modified as tendrils

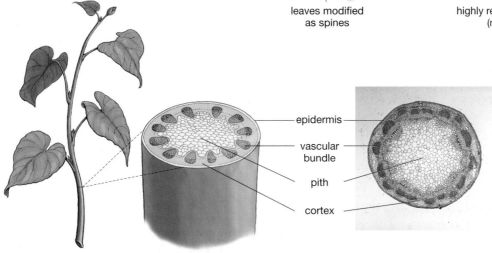

epidermis

vascular bundle

pith

cortex

Figure 23.9
The Stem and Its Parts
Stems provide support to the rest of the plant, act as storage sites, and conduct fluids. Vascular plants have a "plumbing" system that transports food, water, minerals, and hormones. The vascular bundles in the figure are groups of tubes, running in parallel, that serve this transport function.

If you look at **Figure 23.10**, you can see the components of a typical flower. Taking things from the bottom, there is a modified stem, called a pedicel, which widens into a base called a receptacle, from which the flowers emerge. Flowers themselves can be thought of as consisting of four parts: sepals, petals, stamens, and a carpel. The **sepals** are the leaf-like structures that protect the flower before it opens. (Drying out is a problem, as are hungry animals.) The function of the colorful **petals** is to announce "food here" to pollinating animals.

The heart of the flower's reproductive structures consists of the stamens and the carpel. If you look at Figure 23.10, you can see that the **stamens** consist of a long, slender **filament** topped by an **anther**. These anthers contain cells that ultimately will yield sperm-bearing pollen grains. At maturity, these pollen grains will consist of three cells surrounded by an outer coat.

It is these pollen grains that will be released and then carried—perhaps by a pollinating bee or bird—to the carpel of another plant (or perhaps of the same plant). As Figure 23.10 shows, a **carpel** is a composite structure, composed of three main parts: the **stigma**, which is the tip end of the carpel, on which pollen grains are deposited; the **style**, a slender tube that raises the stigma to such a prominent height that it can easily catch the pollen; and the **ovary**, the area in which fertilization of the female egg and then early development of the plant embryo take place.

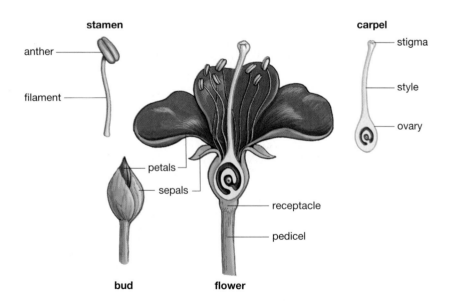

Figure 23.10
Parts of the Flower
Flowers are composed of four main parts: the sepals, petals, stamens, and a carpel. The sepals protect the young bud until it is ready to bloom. The petals attract pollinators. The stamens are the reproductive structures that produce pollen grains (which contain sperm cells). The carpel is the female reproductive structure; it includes an ovary that contains one or more eggs.

23.3 Basic Functions in Flowering Plants

Having looked at the structure, or anatomy, of the angiosperms, let's now go over the activities of some of these components. We'll think in terms of systems here: the reproductive system, the transport system, the hormonal system, the communication system, and the defense system. In addition, you'll look briefly at the nature of plant growth. Because you've just reviewed the parts of a flower, let's start with reproduction.

Reproduction in Angiosperms

The function of a pollen grain is the fertilization of an egg. There are cells in the anthers of the original flower that undergo the type of cell division reviewed in Chapter 10, meiosis, thereby producing **microspores** that are haploid—that contain only a single set of chromosomes. As you can see in **Figure 23.11**, each pollen grain develops from one of the microspores inside the anther. In the type of plant pictured in the figure, the grain will consist, by the time it leaves the anther, of a tough outer coat and three cells: one **tube cell** and two **sperm cells**. Though the pollen grain never gets much more complex than this, it is not just a component of the original plant; it is its alternate generation on the male side. (For more on the alternation of generations in plants, see page 434 in Chapter 21.)

You can see in Figure 23.11 what the activities of the pollen grain's cells are in the context of the angiosperm life cycle. When the pollen grain leaves the anther, it is bound for the stigma of another plant (or its own plant). But for a grain to merely *land* on a stigma doesn't mean that anything has been fertilized. The tube cell in the grain must then begin to *germinate* on the stigma, forming a **pollen tube** that grows down through the style. Once this has taken place, one of the sperm cells in the grain travels through the tube, gets to the female egg, and only *then* is there fertilization. (The other sperm moves down along with the first, but then spurs the growth of food reserves for the growing embryo, about which you'll learn more in Chapter 24.)

And how does the egg arise that is fertilized by the sperm? Together with its supporting cells, the egg is the plant's alternate generation on the female side. Inside the plant's ovary, there is a cell that also goes through meiosis, thus producing a haploid **megaspore** (*mega*, because it's bigger than the male microspore). It in turn gives rise to several kinds of cells, one of which is the egg that the sperm from the pollen grain will fertilize.

Watch closely now, because once this happens— once sperm (from pollen grain) has fertilized egg (from megaspore)—it is the beginning of the kind of flowering plant we started with. We have alternated back to this generation of plant. Many a step remains before arriving at something that *looks* like the original plant; the fertilized egg (now called a zygote) must first have a tough covering develop around it. The combination of embryo, its food supply, and the covering is called a **seed**. This seed must be released and then land on a suitable patch of earth, there to germinate and grow to a full flowering plant. But the step at which this new generation appears is the point at which sperm fertilizes egg.

To put some terms to these generations, note that both the megaspore and pollen grain are haploid spores. Because *spores* are what the original, flowering plant produced, this flowering plant is the **sporophyte generation**. Meanwhile, the pollen and megaspore generation produced *gametes*—sperm or egg—and are thus the **gametophyte generation**. When the sperm and egg came together, they fused their haploid sets of chromosomes and thus produced a sporophyte embryo with a doubled, or diploid, set of chromosomes. A detailed account of angiosperm reproduction can be found in Chapter 24, beginning on page 520.

Recall that the angiosperm egg is fertilized and develops inside the structure called the ovary (see Figure 23.10). Also recall that as the fertilized egg develops into an embryo, it becomes enclosed within a seed. As this is taking place, the ovary that surrounds the seed is also developing into something: a tissue called fruit. We are all familiar with fruit, of course, and sometimes the tissue surrounding the seed is fruit as we commonly understand that term—the flesh of an apricot, for example. But ovaries can mature into fruit in other forms. The pod that surrounds peas is fruit in this scientific sense, as is the outer covering of a kernel of corn. A **fruit** is simply the mature ovary of a flowering plant. All angiosperms have fruit in this sense, and it provides us with our definition of the flowering plants. An **angiosperm** is a flowering seed plant whose seeds are enclosed within the tissue called fruit.

Asexual Reproduction in Angiosperms: Vegetative Reproduction

We are used to the human mode of reproduction, which—pending the arrival of cloned human beings—is always *sexual* reproduction: At some point, sperm must fertilize egg to produce a new embryo. But plants, including angiosperms, have another option—**vegetative reproduction**, which is *asexual* reproduction. This is familiar to us in the form of "cuttings" that can be taken from one houseplant to start another. A growing cutting represents a new plant that is an

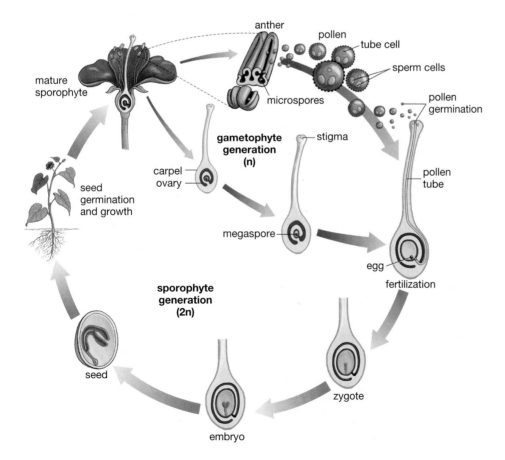

Figure 23.11
The Angiosperm Life Cycle
The mature sporophyte flower produces many male microspores within the anther, and a female megaspore within the ovary. The male microspores then develop into pollen grains that contain the male gamete, sperm. Meanwhile the female megaspore produces the female gamete, the egg. Pollen grains, with their sperm inside, move to the stigma of a plant. There the tube cell of the pollen grain sprouts a pollen tube that grows down toward the egg. The sperm cells then move down through the pollen tube, and one of them fertilizes the egg. This results in a zygote that develops into an embryo protected inside a seed. This seed leaves the parent plant and sprouts or "germinates" in the earth, growing eventually into the type of sporophyte plant that began the cycle.

exact genetic replica of the original plant; no alternate generation or fertilization is required. One sporophyte plant has been grown from another.

Human beings are motivated to put such vegetative reproduction to use when nature produces a particularly *valuable* plant, generally through mutation. Navel oranges, for example, seem to have arisen through mutation on a conventional orange tree that was grown in Brazil in the nineteenth century. If pollen from that tree had helped fertilize an egg from another tree, we may well have *lost* the advantage this unique tree provided—fat, seedless navel oranges—through the shuffling of genetic material that comes with sexual reproduction. The trick was to use cuttings (or "grafts") from the original tree to produce one *identical* generation of orange tree after another. Vegetative reproduction is also common in nature. The roots of aspen trees, for example, produce a form of shoot known as a sucker which, if physically separated from the parent plant, will grow into a new aspen tree. Oftentimes, then, a stand of aspen trees amounts to a group of clones of one another (**Figure 23.12**).

Plant Plumbing: The Transport System

In looking at leaves and stems earlier, you saw that they have something called *vascular bundles* running through them. This term refers to collections of tubes through which fluid materials move from one part of the plant to another. This transport system bears obvious similarities with the circulation systems of animals, but there are differences as well. Many animals employ blood as a transport medium, and animals such as ourselves have a transport *pumping* device, called a heart. Plants do not have blood, and they have no pumping system. (In fact, plants have almost no "moving parts" at all.) Yet think of the transport job plants have to do. Water that is transpired from the top of a redwood tree may have made a journey of more than 100 meters straight up—more length than a football field! What kind of power is at work here?

There are two essential components to the plant transport system. First there is **xylem**, which can be defined as the tissue through which water and dissolved minerals flow in vascular plants. (**Tissue** means a group of cells that perform a common function.) Second there is **phloem**, the tissue through which the *food* produced in photosynthesis—mostly sucrose—is conducted, along with some hormones and other compounds. As noted before, water is making a directional journey from root through leaf, but the plant must be able to transport food and hormones everywhere within itself.

If you look at **Figure 23.13**, you can see an idealized view of a plant's transport system as it would appear within a stem. You'll also see that the vascular bundles noted earlier are composed of bundles of linked xylem and phloem tubes running in parallel.

Xylem is composed of two types of fluid-conducting cells, *vessel elements* and *tracheids*, which have different shapes as you can see. Upon reaching maturity, these cells do the rest of the plant the favor of dying. The cells' content is cleared out, leaving strands of empty cells stacked one on top of another—leaving tubes, in other words. The walls of these cells remain, however, and indeed are reinforced with the materials cellulose and lignin, which provide the *load-bearing* capacity that allows something as massive as a redwood tree to be so tall. (For more on the role of xylem in everyday houseplants, see "Keeping Cut Flowers Fresh," on page 492.)

Phloem is also composed of two types of cells, but the arrangement here is very different. One type of phloem cell—a *sieve element*, which does the actual nutrient conducting—doesn't undergo cell death upon maturity, but it does lose its cell nucleus, meaning nearly all its DNA. Those of you who went through the genetics unit may wonder how any cell could function without its DNA information center, but there is an answer. Each sieve element cell has associated with it one or more *companion cells* that retain their DNA and that seem to take care of all the housekeeping needs of their sister sieve elements. So, why do sieve elements lose their nuclei? Apparently to make room for the rapid flow of food through them. In Chapter 24 you'll look at the means by which phloem and xylem conduct their respective materials.

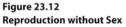

Figure 23.12
Reproduction without Sex
Individual aspen trees often are clones of one another, reproducing without the fusion of eggs and sperm from different individuals. This stand of aspen trees is in Colorado.

Communication: Hormones Affect Many Aspects of Plant Functioning

We're so used to associating hormones with *animal* functioning that it may come as a surprise to learn that plants also have hormones. Taken as a whole, plant hormones do many of the same things that animal hormones do. Their most important roles are to regulate growth and development and to integrate the functioning of the various plant parts. In less abstract terms, hormones help buds grow and leaves fall and fruit ripen. This goes hand in hand with helping plants respond to their environments—to heat and cold, munching goats, and voracious insects.

Though most people have an intuitive sense of what hormones are, a definition might be helpful. **Hormones** are chemical messengers; they are substances that, when released in one part of an organism, go on to prompt physiological activity in another part of that organism. In animals, hormones are generally synthesized in well-defined organs, called *glands*, whose main function is to produce hormones (for example, the thyroid or adrenal glands). In plants, however, hormone production is a more diffuse process, taking place not in glands, but in collections of cells that carry out a range of functions.

If you look at **Table 23.1** you can see a list of five of the most important hormones that function in plants. Three of these actually are *classes* of hormones, and two are individual hormones. You can read about the hormone ethylene in "Ripening Fruit Is a Gas." Let's go over just one other plant hormone, a member of the family of hormones known as *auxins*, to give you some idea of what plant hormones do.

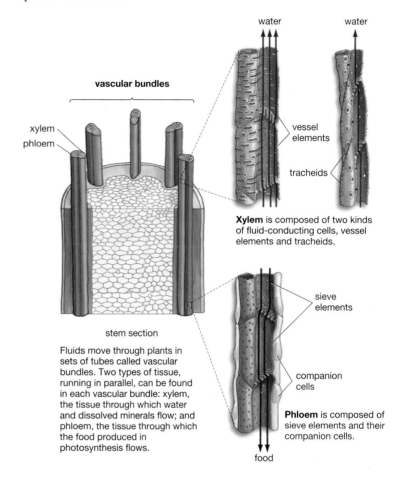

Xylem is composed of two kinds of fluid-conducting cells, vessel elements and tracheids.

Phloem is composed of sieve elements and their companion cells.

Fluids move through plants in sets of tubes called vascular bundles. Two types of tissue, running in parallel, can be found in each vascular bundle: xylem, the tissue through which water and dissolved minerals flow; and phloem, the tissue through which the food produced in photosynthesis flows.

Figure 23.13
Fluid-Transport Structure of Plants

Table 23.1 Plant Hormones		
Hormone	**Major functions**	**Where found or produced in plant**
Auxins	Suppression of lateral buds; elongation of stems; growth and abscission (falling off) of leaves; differentiation of xylem and phloem tissue	Root and shoot tips; young leaves
Cytokinins	Stimulate cell division; active in the development of plant tissues from undifferentiated cells	Roots
Gibberellins	Stem elongation, growth of fruit, promotion of seed germination	Seeds, apical meristem tissue, young leaves
Ethylene	Ripening of fruit, retardation of lateral bud growth, promotion of leaf abscission	Nearly all plant tissue
Abscisic acid	Induces closing of leaf pores (stomata) in drought; promotes dormancy in seeds; counteracts growth hormones	Young fruit; leaves, roots

- In temperate climates, deciduous trees exhibit a coordinated, seasonal loss of leaves and enter into a state of dormancy, existing on stored nutrient reserves in colder months.

- Plants can sense the passage of seasons and time their reproductive activities accordingly. One mechanism that assists in this process is photoperiodism, which is the ability of a plant to respond to changes it is experiencing in the daily duration of darkness, as opposed to light. Some plants that exhibit photoperiodism are long-night plants, meaning those whose flowering comes only with an increased amount of darkness—in late summer or early fall. Others are short-night plants, meaning those that flower only with a decreased amount of darkness—in early to midsummer.

Web Tutorial 23.2: Response to External Signals

Key Terms

angiosperm   489
anther   488
apical dominance   492
blade   485
carpel   488
deciduous   497
dormancy   497
fibrous root system   484
filament   488
fruit   489
gametophyte generation   489
gravitropism   496
hormone   491
megaspore   488
microspore   488
mycorrhizae   495
nutrient   486
ovary   488
petal   488
petiole   485

phloem   490
photoperiodism   498
phototropism   496
pollen tube   488
root hair   484
seed   489
sepal   488
sperm cell   488
sporophyte generation   489
stamen   488
stigma   488
stomata   485
style   488
taproot system   484
thigmotropism   497
tissue   490
transpiration   484
tube cell   488
vegetative reproduction   489
xylem   490

Understanding the Basics

Multiple-Choice Questions (answers in the back of the book)

1. Which of the following is not a benefit that plants provide to other living things?
   a. act to decompose dead organic material
   b. build up food supply
   c. lock up carbon dioxide from the atmosphere
   d. produce oxygen
   e. prevent soil erosion

2. Plants take in _____ and _____ through their roots, but take in _____ primarily through tiny pores on their leaves.
   a. minerals; carbon dioxide; water
   b. oxygen; carbon dioxide; water
   c. water; carbon dioxide; minerals
   d. oxygen; minerals; carbon dioxide
   e. minerals; water; carbon dioxide

3. Which of the following are not functions of roots? (Select all that apply.)
   a. nutrient storage
   b. loss of water through transpiration
   c. loss of oxygen in photosynthesis
   d. grow to water supplies
   e. absorption of water

4. You decide that you want your spruce tree in the front yard to be rounded instead of conical. What do you need to do to ensure that the plant no longer grows taller?
   a. Treat the top of the tree with growth hormones.
   b. Cut off the top of the tree.
   c. Trim the sides of the tree, but leave the top alone.
   d. Cover the top of the tree, so light cannot reach it.
   e. none of the above

5. Pollen grains are
   a. hormonal messengers that travel from one plant to another
   b. tiny food packets that move from one plant to another
   c. sperm-bearing sporophyte plants
   d. toxin-bearing packets that stunt the growth of other trees
   e. sperm-bearing gametophyte plants

6. Plant growth (select all that apply)
   a. occurs only in the midsection of any leaf, stem, or root
   b. serves almost solely to lengthen a plant, rather than widen it
   c. occurs, in vertical growth, only at the tips of roots or shoots
   d. must come to an end after a fixed number of months or years
   e. can go on as long as the plant is alive

7. The sucrose produced by photosynthesis is transported from the leaf to other parts of the plant through
   a. sieve elements
   b. vessel elements
   c. tracheids
   d. rhizoids
   e. guard cells

8. Fertilization in angiosperms occurs when
   a. a collection of seeds from a plant reaches the ground
   b. the megaspore in an ovary goes through meiosis
   c. pollen from one plant reaches the stigma of another
   d. a sperm cell from a pollen grain fuses with an egg cell in an ovary
   e. a seed begins to germinate in the ground

9. Phototropism and gravitropism lead to the bending of shoots or roots toward or away from a stimulus (light or gravity, respectively). Such bending is brought about by
   a. increased growth over the general surface of the root or shoot
   b. differential growth on opposite sides of the root or shoot
   c. decreased growth at the root or shoot apex, and increased growth at the base
   d. muscular contractions, leading to a curvature in the root or shoot
   e. all of the above

10. Photoperiodism allows some plants to respond to _____ by timing such things as _____ in accordance with the amount of _____ they are exposed to.
    a. predators; dormancy; moisture
    b. competition; growth; moisture
    c. the seasons; growth; toxins
    d. other plants; flowering; ethylene
    e. the seasons; flowering; darkness

Brief Review

1. The stomata that exist mostly on leaves allow important gas exchanges to take place in plants, but these exchanges also present plants with a problem they must solve. What gases are exchanged through the stomata, and in which direction? What problem must plants solve because of these exchanges, and how do they solve it?

2. What functions do roots perform? What are the main types of roots produced by flowering plants?

3. List the main parts of a flower, and describe the functions performed by each part.

4. Contrast the way people grow, and explain how plants grow and develop over their lifetimes.

5. Give examples of mutually beneficial relationships between plants and bacteria, and plants and fungi.

6. Distinguish between the gametophyte and sporophyte plant generations.

Applying Your Knowledge

1. Gardeners often "pinch off" the terminal shoot apex to stimulate bushiness in young plants. Explain the physiological basis for this common horticultural practice.

2. The text makes clear that plants respond to their environments in sophisticated ways, even to the point of signaling one another during attacks by predators. Given this, can plants be said to be conscious beings, or are animals the only conscious beings?

3. The "race for sunlight" in plants led to the evolution of the tallest living things in existence, trees. But sunlight is not the only thing that stands to make a tree successful or not in reproducing. Think about how plants function and then answer this question: What are the *costs* of being taller, as opposed to shorter?

24 *Form and Function in Flowering Plants*

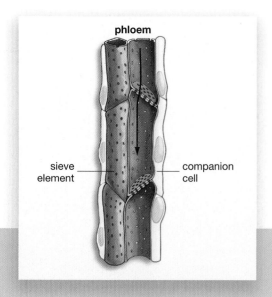

Plant tubes.
(Section 24.3, page 509)

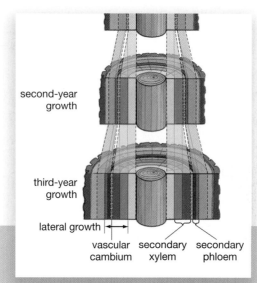

How woody plants grow wider.
(Section 24.5, page 513)

Pass the flapjacks.
(Essay, page 518)

Insects that eat plants are not much of a surprise, but plants that eat insects? About 500 species of flowering plants are carnivorous, with at least one variety, the tropical pitcher plant *Nepenthes*, able to capture animals up to the size of a small bird. But the most famous plant predator of them all undoubtedly is the Venus flytrap (*Dionaea muscipula*), which is native to the wetlands of North and South Carolina. So how does this plant, which has no muscles and is fixed in one spot, end up consuming insects?

Each trap is a single leaf folded in two, with the interlocking "fingers" of the two halves able to snap together in less than half a second, imprisoning any small wanderer (**see Figure 24.1**). But what keeps the leaf halves from wasting their energy by snapping together in pursuit of wind-borne blades of grass or other objects? The chemical reaction that puts the trap in motion is set off by a series of trigger hairs on the inside of each of the leaf's two sections. The trick is that *two* trigger hairs must be tripped in order for the halves of the leaf to snap shut. Insects wandering around inside the leaf are likely to trip two hairs, but blades of grass are not. And how does the plant slam its "jaws" shut without muscles? When the trigger hairs are tripped, they set off a chemical reaction that engorges cells at the base of the leaf with water. This rapid movement causes the leaf halves to come together. Enzymes are then released that begin to digest the struggling insect.

Figure 24.1
Fatal Entry
A fly enters the leaf of a Venus flytrap. Once the fly pulls on two trigger hairs within the leaf, its opposing halves will snap shut, trapping the fly. Then the plant will begin to digest its prey.

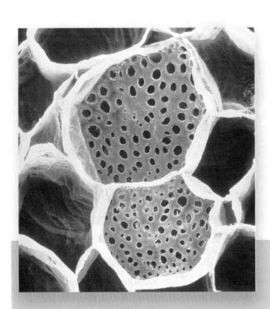

Food carriers.
(Section 24.6, page 519)

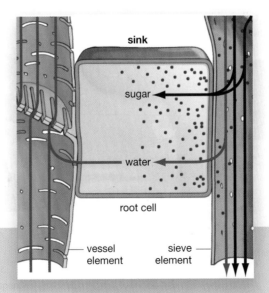

sink

sugar

water

root cell

vessel element

sieve element

Food and water take a trip.
(Section 24.6, page 520)

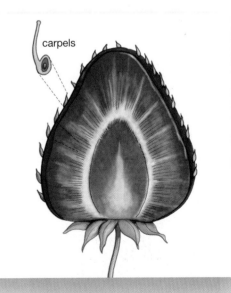

carpels

Where are the strawberry seeds?
(Section 24.8, page 526)

two small collections of darker-colored cells on each side of the apical meristem dome. These are collections of meristem tissue that have recently been left behind; they will become lateral buds.)

Figure 24.15
Tissue Development from Apical Meristems
Apical meristems give rise to the primary tissues. In development, cells first are mostly engaged in division, just above the apical meristem, then elongation, and finally differentiation. The result is three fully formed kinds of tissue.

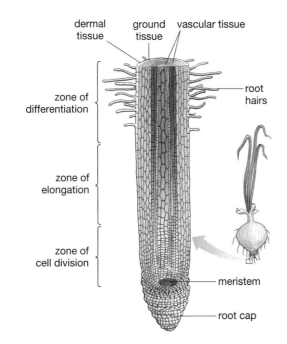

Figure 24.16
Growth by Cell Elongation
Much of the elongation of the shoot and root in a young plant results from cell elongation rather than from cell division. Thus it is cells just above the root tip, in the zone of elongation, that provide most of the root's growth, as shown here. The tip of the root (where cell division occurs) has elongated very little, and the cells in the upper root have completed most of their elongation. This same pattern holds true for the shoot.

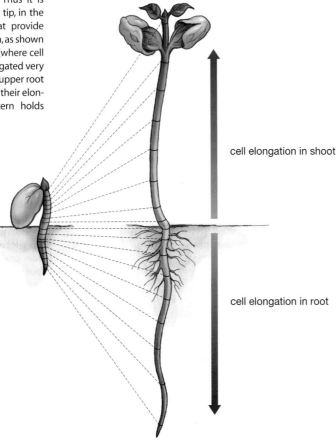

Root Meristems

Because roots are pushing their way into the ground, root apical meristems are located not at the very tip of the root, but just up from the tip. The root apical meristems are adjacent to a collection of cells called a **root cap** that both protects the meristematic cells and secretes a lubricant that helps ease the way for the root in its progress. Toward the middle of the apical meristem, there is a collection of cells called the quiescent center. These slowly dividing cells amount to another kind of insurance policy for the plant. Should the meristematic cells become damaged, quiescent center cells will start dividing more rapidly, producing new meristematic cells. Cells in the quiescent center are, in short, a reserve that goes into action in times of trouble. They are well suited for this because, in their dormancy, they are more resistant than apical meristem cells are to injury from such influences as toxic chemicals.

How the Primary Tissue Types Develop

You've looked so far at ground, dermal, and vascular tissues. And you know that the fourth type of primary tissue, meristematic tissue, gives rise to all these. So how do plants get from meristem to the others? Not in a single step, it turns out—not directly from apical meristem cells to, say, fully functioning vascular tissue. Instead, there is a transition involving three main changes: *More* cells are produced through cell division; cells *elongate* through cell growth; and cells *differentiate* into the three kinds of tissue. These three changes are overlapping—cells are elongating even as they are differentiating—but events do take place in roughly this order. You can see this by looking at **Figure 24.15**. In a gradual transition with no sharp boundaries, the apical meristem gives way to a **zone of cell division**, then there is a **zone of elongation**, and finally a **zone of differentiation**, which is where the ground, dermal, and vascular tissues fully take shape. Note that root hairs don't appear until the zone of differentiation. Though the three kinds of tissues *finish* taking shape in the zone of differentiation, they actually begin this process just above the apical meristem (in the zone of cell division). It's accurate to think of the tissues in the zone of cell division as being precursors to the three tissue types.

If you look at **Figure 24.16**, you can see the practical effect of a zone of elongation: It is where most of the primary plant growth occurs. There is some growth with the addition of more cells down near the apical meristem, but most of the vertical expansion of the plant takes place because of cell elongation. Such elongation has the effect of extending the roots into the soil and the stem into the air.

Insects that eat plants are not much of a surprise, but plants that eat insects? About 500 species of flowering plants are carnivorous, with at least one variety, the tropical pitcher plant *Nepenthes*, able to capture animals up to the size of a small bird. But the most famous plant predator of them all undoubtedly is the Venus flytrap (*Dionaea muscipula*), which is native to the wetlands of North and South Carolina. So how does this plant, which has no muscles and is fixed in one spot, end up consuming insects?

Each trap is a single leaf folded in two, with the interlocking "fingers" of the two halves able to snap together in less than half a second, imprisoning any small wanderer (**see Figure 24.1**). But what keeps the leaf halves from wasting their energy by snapping together in pursuit of wind-borne blades of grass or other objects? The chemical reaction that puts the trap in motion is set off by a series of trigger hairs on the inside of each of the leaf's two sections. The trick is that *two* trigger hairs must be tripped in order for the halves of the leaf to snap shut. Insects wandering around inside the leaf are likely to trip two hairs, but blades of grass are not. And how does

the plant slam its "jaws" shut without muscles? When the trigger hairs are tripped, they set off a chemical reaction that engorges cells at the base of the leaf with water. This rapid movement causes the leaf halves to come together. Enzymes are then released that begin to digest the struggling insect.

Figure 24.1
Fatal Entry
A fly enters the leaf of a Venus flytrap. Once the fly pulls on two trigger hairs within the leaf, its opposing halves will snap shut, trapping the fly. Then the plant will begin to digest its prey.

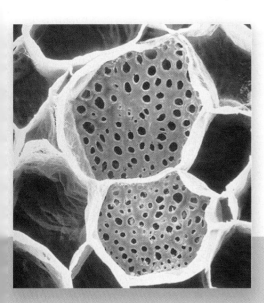

Food carriers.
(Section 24.6, page 519)

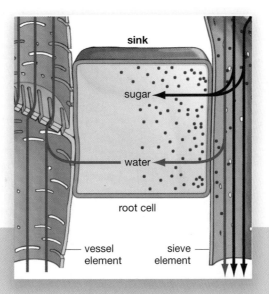

sink

sugar

water

root cell

vessel
element

sieve
element

Food and water take a trip.
(Section 24.6, page 520)

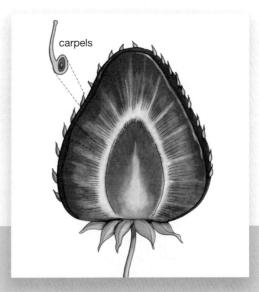

carpels

Where are the strawberry seeds?
(Section 24.8, page 526)

It may surprise you to learn that Venus flytraps make their own food through photosynthesis, just like any other plant. So why do they need these side orders of insects? For nutritional supplements. Like many other carnivorous plants, Venus flytraps grow naturally in mineral-poor soil, and the insects they catch are a good source of nitrogen and phosphate.

The world of plants is filled with species as unique in their own way as the Venus flytrap is. Queen Victoria water lilies (*Victoria regia*) have circular leaves that can be nearly 2 meters (about 6 feet) in diameter and that can serve as a floating platform for a person if pressed into service (**see Figure 24.2**). The seeds of orchids are so tiny that they resemble dust more than seeds, while the seeds of the double coconut come wrapped in a fruit that might be 0.6 meter (about 2 feet) across.

Yet there is a unity to plant life, particularly in the case of the flowering plants, or angiosperms, that you'll be looking at in this chapter. All angiosperms have a common set of cell types, for example, and all of them reproduce in a similar way. The goal in this chapter is to cover common angiosperm features in four areas: cell and tissue types, growth, fluid transport, and reproduction. Before we begin, it might be helpful to revisit the representation of the whole plant that you first saw in Chapter 23, so that you can "find your place" when reading about a plant's component parts. If you look at **Figure 24.3**, you can see a diagram of a typical plant, with its two-system division—roots and shoots—and the various parts of the plant that lie within these two systems. The function of each of these parts will become clear to you as you go through the details on them.

Figure 24.2
Floating Platforms
A woman plays a violin while standing atop a Victoria water lily at the Missouri Botanical Garden in St. Louis at the turn of the 20th century. The lilies are native to South America. Some South American Indians call the lilies *Yrupe,* which can be translated as "big water tray."

24.1 Two Ways of Categorizing Flowering Plants

The constituent parts of plants are often organized one way in a given kind of plant, but another way in a different type. So what are these types? Here are a couple of ways of *categorizing* plants, the significance of which will become apparent as you go along.

The Life Spans of Angiosperms: Annuals, Biennials, and Perennials

One important question that might be asked about any plant is: How long does it take for it to go through its entire life cycle—from being a seed germinating in the ground, through growth, flowering, seed dispersal, and then death? Plants that go through this cycle in one year or less are known as **annuals**. Some of these are food crops, such as tomatoes and the commercial grains. Plants that go through the cycle in about two years are **biennials**, which include carrots and cabbage. (Have you ever seen the flowers of a carrot plant? Probably not; we *pick* the taproots called carrots after one year, so their shoots don't get a chance to flower.) Plants that live for many years are known as **perennials**, a category that includes trees, woody shrubs such as roses, and many grasses. **Figure 24.4** shows some representative annuals, biennials, and perennials.

A Basic Difference among Flowering Plants: Monocotyledons and Dicotyledons

A second distinction among the flowering plants has to do with their anatomy, meaning the arrangement

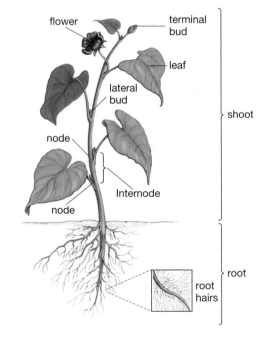

Figure 24.3
Anatomy of a Flowering Plant (An Angiosperm)

of the structures that make them up. With respect to anatomy, there are two broad classes of flowering plants: monocotyledons and dicotyledons, which are almost always referred to as monocots and dicots. A cotyledon is an embryonic leaf, present in the seed, as you can see in **Figure 24.5**. **Monocotyledons** are plants that have one embryonic leaf; **dicotyledons** are plants that have two.

This distinction in the embryos is only the start of how monocots and dicots differ. Their roots are different, their leaves are different, their transport tubes, or vascular bundles, are arranged differently (Figure 24.5). More than 75 percent of all flowering plants are the broad-leafed dicots. But given the

(a) One-year lifespan

(b) Two-year lifespan

(c) Variable lifespan

Figure 24.4
Categorizing Plants by Life Span

(a) Annuals such as tomato plants live for only one year.

(b) Biennials such as carrot plants have a two-year life span.

(c) Perennials such as this white oak tree live for many years, some for many hundreds of years. Plants with long life spans tend to be woody and larger than plants with short life spans.

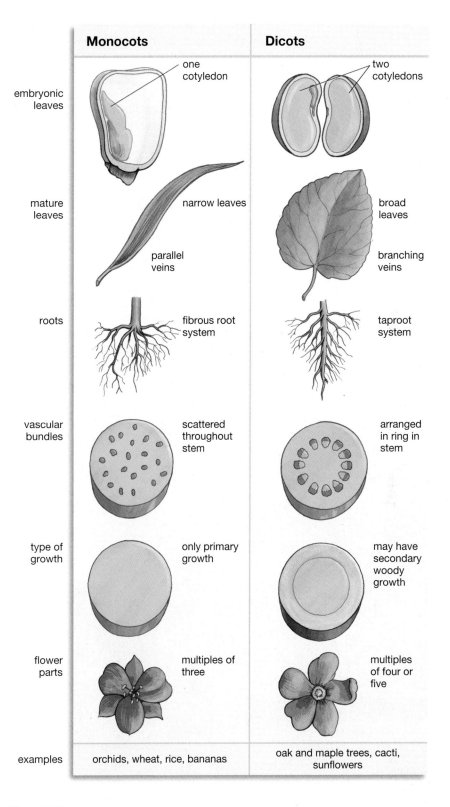

	Monocots	Dicots
embryonic leaves	one cotyledon	two cotyledons
mature leaves	narrow leaves / parallel veins	broad leaves / branching veins
roots	fibrous root system	taproot system
vascular bundles	scattered throughout stem	arranged in ring in stem
type of growth	only primary growth	may have secondary woody growth
flower parts	multiples of three	multiples of four or five
examples	orchids, wheat, rice, bananas	oak and maple trees, cacti, sunflowers

Figure 24.5
A Cotyledon Is an Embryonic Leaf
Plants with one embryonic leaf (the monocots) are structured differently than plants with two embryonic leaves (the dicots), as shown in the figure.

260,000 known species of flowering plants, this still leaves more than 50,000 species of narrow-leafed monocots, which include most of the important food crops (corn, wheat, rice). **Figure 24.6** shows you an example of both a monocot and a dicot.

With these distinctions in mind, you're now ready to start looking at the basic elements that go into making up flowering plants.

24.2 There Are Three Fundamental Types of Plant Cells

You saw in Chapter 4 that all living things are made up of cells, and plants are no exception to this rule. As it happens, there are three fundamental types of plant cells that, alone or in combination with each other, go on to make up most of the plant's *tissues*, which are *groups* of cells that carry out a common function.

Parenchyma Cells

First, there are **parenchyma cells**, which have thin cell walls and are easily the most abundant type of cell in the plant, being found almost everywhere in it. Accordingly, these cells have a lot of different functions. They form the flesh of many fruits, the outer surface of most young plants, parts of leaves that are active in photosynthesis, and much of the "ground" tissue you'll be hearing about shortly. Further, parenchyma are alive at maturity, a quality that might hardly seem worth mentioning except that some plant cells are dead in their mature, functioning state.

Sclerenchyma Cells

This dead-at-maturity condition characterizes a second type of plant cell, **sclerenchyma cells**, which have thick cell walls that help the plant return to its original shape when it has been deformed by some force, such as the push of wind or animals. Sclerenchyma cells are found in parts of the plant that are mature—that will not be growing further, and that need the support that sclerenchyma can provide. Like all plant cells, sclerenchyma have, running through their cell walls, a compound called **cellulose** that can be likened to the steel bars in reinforced concrete. Most sclerenchyma cell walls also are infused with a tough compound called *lignin*. The combination of lignin and cellulose results in cells that can bear the crushing weight of massive trees (**see Figure 24.7**).

Collenchyma Cells

The third type of plant cell, **collenchyma cells**, can be thought of as support cells, like sclerenchyma cells, but with the function of stretching or elongating in parts of the plant that are growing, such as young leaves and stems. Collenchyma cells combine the properties of parenchyma and sclerenchyma cells, performing some of the functions of each of these cell types. They provide mechanical strength in areas that are actively growing, and can also participate to some extent in photosynthesis and wound repair.

Parenchyma as Starting-State Cells

Both collenchyma and sclerenchyma are derived from parenchyma cells; that is, some parenchyma cells differentiate *into* collenchyma or sclerenchyma cells. Parenchyma are thus a kind of starting-state cell, but their capabilities are greater than this. They can be transformed "backward," in a sense, into the type of embryonic cells that give rise to the whole plant. You may have wondered in Chapter 23 how we can get a new plant by taking a cutting from an existing plant and placing it in the ground. The answer is that when a plant is cut, parenchyma cells lying close to the cut are transformed into *growth* cells of a type you'll look at shortly, and these can sprout a new root. This obviously gives plants a flexibility not possessed by animals such as ourselves. If one of our fingers is cut off, we do not grow a new finger, to say nothing of a whole new person.

(a) Monocot plant   **(b)** Dicot plant

Figure 24.6
Categorizing Plants by Physical Features

(a) Corn is one of several monocot plants that are food crops.

(b) Geraniums are an example of a dicot. See Figure 24.5 for a list of the anatomical differences between monocots and dicots.

24.3 The Plant Body and Its Tissue Types

You saw earlier that tissues are groups of cells that carry out a common function. So what kinds of tissues are there in angiosperms? The short answer is dermal, ground, vascular, and meristematic, with each of these four types of tissue generally being composed of one or more of the cell types you just looked at. Dermal tissue can be thought of as the plant's outer covering, vascular tissue as its transport or "plumbing" tissue, meristematic tissue as its growth tissue, and ground tissue as almost everything else in the plant.

First: A Distinction Between Primary and Secondary Growth Tissue

Before proceeding further with this topic, we need to take note of a basic distinction in plant tissue that has to do with the way plants grow. Some plants are capable only of what is called **primary growth**, meaning growth at the tips of their roots and shoots that principally increases their *length*. In contrast, other plants exhibit not only primary growth but also something known as **secondary growth**, which can be thought of as the *lateral* growth, or thickening, that occurs in **woody plants**. Secondary growth occurs in trees, for example, but not in orchids or strawberry plants. Non-woody plants such as an orchid are known as **herbaceous plants**, meaning those that never develop wood (or bark) and that thus contain only primary tissue. It is this *primary* tissue that you will look at first, with a review of secondary tissue to come shortly.

If you look at **Figure 24.8**, you can see the location of all four kinds of primary tissue in a dicot. Looking at the stem in cross section, you can see that ground tissue is well named, because visually it forms a kind of background against which you can see the tubes of the vascular tissue. Dermal tissue then is at the periphery of the plant, forming a "skin" layer around the ground tissue. The placement of the meristematic tissue is discussed shortly. First, let's look in a little more detail at dermal, ground, and vascular tissue.

Dermal Tissue Is the Plant's Interface with the Outside World

Plants can't move, they can't bite, and, as you saw in Chapter 23, they must carefully control their water supply. All this makes their **dermal tissue**, or *epidermis*, very important to them. This outer coat is generally only one layer of cells thick, but it serves to protect the plant and control its interaction with the outside world

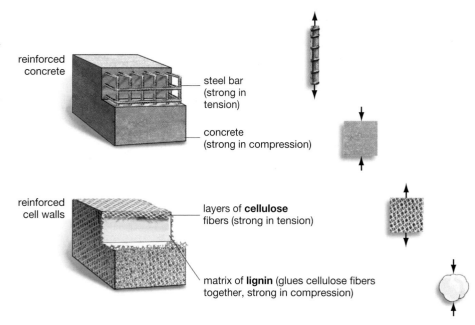

Figure 24.7
Plant-Cell Walls Are Composed of Composite Materials, as Is Reinforced Concrete
The cellulose fibers of plant-cell walls are very strong in tension, as are the steel bars in reinforced concrete. Lignin is similar to the matrix of concrete in that it binds the fibers together and resists compression. Both plant-cell walls and reinforced concrete are important in providing support for structures that are tall and heavy.

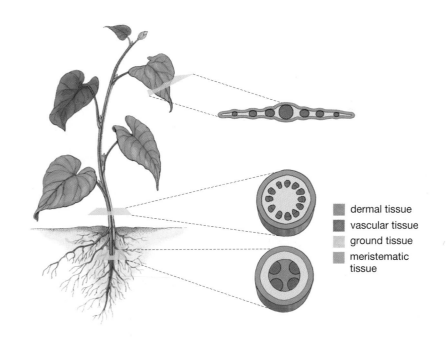

Figure 24.8
Four Types of Tissue in Primary Plant Growth
Dermal tissue (dark green) is found on the outside of the plant and is the interface between the plant and its environment. Ground tissue (pale green) is found throughout the plant and gives the plant its shape, stores food, and is active in photosynthesis. Vascular tissue (purple) transports water, nutrients, and sugar throughout the plant. Meristematic tissue (blue) occurs at the tips of shoots and roots and at leaf bases, and gives rise to the entire plant, through the production of cells that develop into the other three tissue types.

(**see Figure 24.9**). A plant's first line of defense against outside predators is a waxy coating called the *cuticle* that covers the epidermis of the plant's above-ground parts (its shoot). The cuticle is a kind of waterproofing, serving to keep water in, and an invader-proofing, helping to keep infecting bacteria and fungi out.

Plants need to exchange gases with their environments, however. They take in carbon dioxide and emit oxygen and water vapor through the microscopic pores called stomata, found mostly on the underside of their leaves. (You went over the functioning of stomata in Chapter 23, starting on page 485.) Stomata also represent a kind of weak point for plants, however, because they provide a passageway for microbial invaders and a conduit for excessive loss of moisture.

Dermal tissue also forms several kinds of extensions, called trichomes, that protrude from the plant's surface. You learned about one variety of them in Chapter 23: root hairs, meaning the threadlike outgrowths of individual epidermal root cells. As we noted there (page 484), these wispy outgrowths greatly increase the plant's ability to absorb water and minerals. Other types of trichomes serve not to absorb substances, but to secrete them. In some plants, trichomes secrete toxic chemicals that ward off plant-eating animals.

Ground Tissue Forms the Bulk of the Primary Plant

Most of the primary plant is ground tissue, which can play a role in photosynthesis, storage, and structure (**see Figure 24.10**). Large parts of ground tissue may be simple parenchyma, but other parts are combinations of cell types.

Vascular Tissue Forms the Plant's Transport System

As you saw Chapter 23, the plant's vascular system has two main functions. First, it must transport water and minerals; second, it must transport the food made in photosynthesis. In line with this, there are two main components to the plant vascular system. First there is **xylem**, the tissue through which water and dissolved minerals flow. Second there is **phloem**, the tissue that conducts the food produced in photosynthesis, along with some hormones and other compounds (**see Figure 24.11**). The movement of water through xylem is directional—from root up through stem and then out through leaves as water vapor. Meanwhile, the food made in photosynthesis must travel through phloem to every part of the plant, though, as a

Figure 24.9
Where Plant Meets Environment
Dermal tissue serves as the interface between a plant and the environment around it. The waxy cuticle and the single layer of cells in the epidermis work together to protect the plant and to control interactions with the outside world.
(a) Specialized epidermal cells called guard cells regulate the opening of the plant pores called stomata, thus controlling gas exchange between the plant and its environment. Here, guard cells have opened two stomata. In most plants, the majority of stomata are found on the leaves.
(b) Some epidermal cells have hair-like projections, called trichomes, that serve various functions. Two kinds of trichomes are visible in this rose plant. The larger of the two varieties, with the bulbous tips, help secrete chemicals that guard against plant-eating predators.

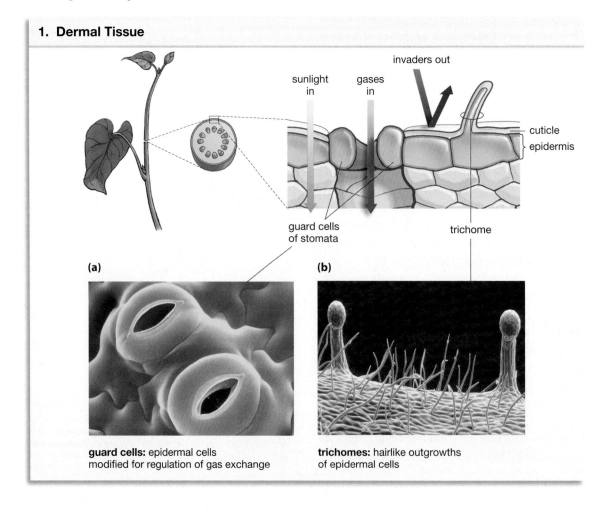

1. Dermal Tissue

invaders out

sunlight in   gases in

cuticle
epidermis

guard cells of stomata

trichome

(a)

(b)

guard cells: epidermal cells modified for regulation of gas exchange

trichomes: hairlike outgrowths of epidermal cells

2. Ground Tissue

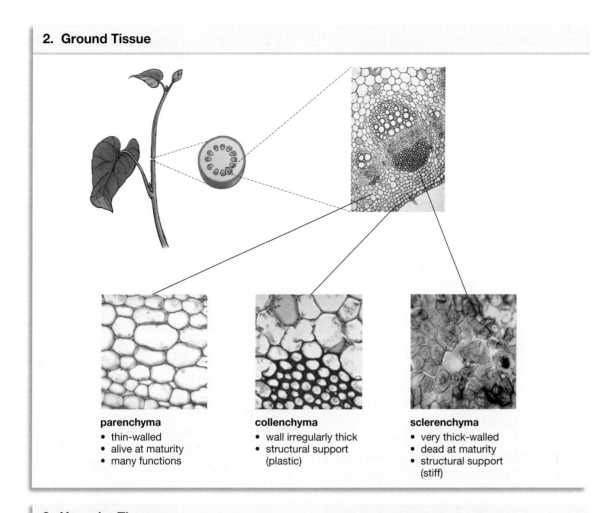

parenchyma
- thin-walled
- alive at maturity
- many functions

collenchyma
- wall irregularly thick
- structural support (plastic)

sclerenchyma
- very thick-walled
- dead at maturity
- structural support (stiff)

Figure 24.10
Ground Tissue
Most of a plant is made of ground tissue, which is composed of parenchyma, collenchyma, and sclerenchyma cells. Examples of these cell types shown here come from several different plants. In a cross section of a stem, thin-walled parenchyma can be found in most of the interior. Collenchyma—at the bottom of the collenchyma picture, surrounded by parenchyma—are support cells that can elongate in growing parts of the plant. Very thick-walled sclerenchyma, such as these from a pear, can provide strength to tissues.

3. Vascular Tissue

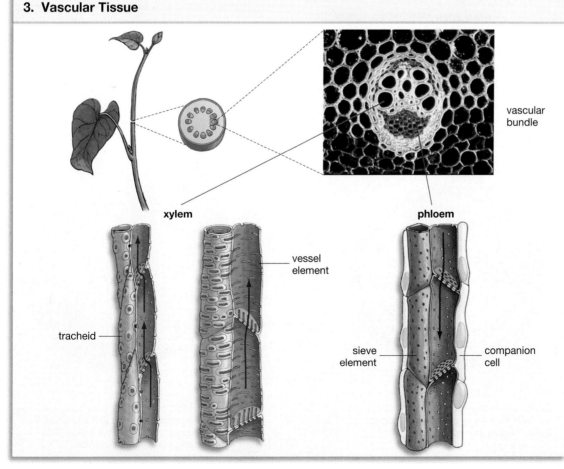

vascular bundle

xylem

vessel element

tracheid

phloem

sieve element

companion cell

Figure 24.11
Materials Movers
Vascular tissue transports fluids throughout the plant. Xylem is composed of cells called tracheids and vessel elements that transport water and dissolved minerals. Phloem is composed of sieve elements (and their companion cells), which transport the food produced during photosynthesis. These cell types are stacked upon one another to form long tubes.

practical matter, in temperate climates the net flow is downward in summer (from food-producing leaves) but upward in early spring (from root and stem storage sites).

Xylem and phloem are arranged in *vascular bundles*, which is to say collections of xylem and phloem tubes that run together in parallel in the stem—xylem tubes toward the inside of the stem, phloem tubes toward the outside. The bundles are arranged in different ways within a stem, depending on whether the plant is a monocot or a dicot; the dicot bundles are configured in a circle, the monocot in an irregular pattern. This is why dicot ground tissue often is conceptualized as existing in two regions, which you can see in **Figure 24.12**: a *cortex* that is outside a ring of the vascular bundles, and a *pith* that is inside it, with the pith cells being specialized for storage.

Meristematic Tissue and Primary Plant Growth

So you've seen three types of primary plant tissues—dermal, ground, and vascular. The question now is: How do these arise? How is it that a plant gets its vertical growth? The short answer is that the fourth type of plant tissue, meristematic tissue, gives rise to all the other tissue types. In considering the role of meristematic tissue, recall from Chapter 23 that, in their vertical growth, plants do not grow the way people do—throughout their entire length—but instead grow only at the tips of their roots and shoots. In addition, a plant's growth is *indeterminate*; in most plants, it can go on indefinitely at the tips of roots and shoots, as opposed to animal growth, which generally comes to an end when the animal matures.

Web Tutorial 24.1
Plant Tissue and Growth

24.4 How a Plant Grows: Apical Meristems Give Rise to the Entire Plant

Putting indeterminate growth together with the concept of growth at the tips, it follows that there must be plant cells at root and shoot tips that remain perpetually young—or more accurately, perpetually embryonic. That is, these cells are able to keep dividing, thereby giving rise to cells that then differentiate into all the primary cell and tissue types you've been reading about. Indeed, everything in plants (both their primary and secondary tissues) develops ultimately from these cells, which collectively form the **apical meristems** of plants.

If you look at **Figure 24.13**, you can see the apical meristem locations in a typical plant. Note that in this plant with a taproot, each lateral root tip has its own apical meristematic tissue that is capable of giving rise to yet more roots. In the shoot, plants confine their growth to the shoot apices that lie at the tip of each stem. Why? The better to compete for precious sunlight. It is the **shoot apical meristem** that gives rise to all the cells that allow this vertical growth. Note, however, that there is a second location for meristematic tissue in the shoot—the area nestled between leaf and stem. It's here that we find lateral buds (sometimes called *axillary buds*). Any **bud** is an undeveloped shoot, composed mostly of meristematic tissue. A **lateral bud** is meristematic tissue that may give rise to a branch or a flower (which obviously ends up growing laterally from the stem). Lateral buds also serve, however, as a kind of insurance policy for the plant in that they can switch roles. Should the plant's apical meristem be-

(a) Monocot stem

(b) Dicot stem

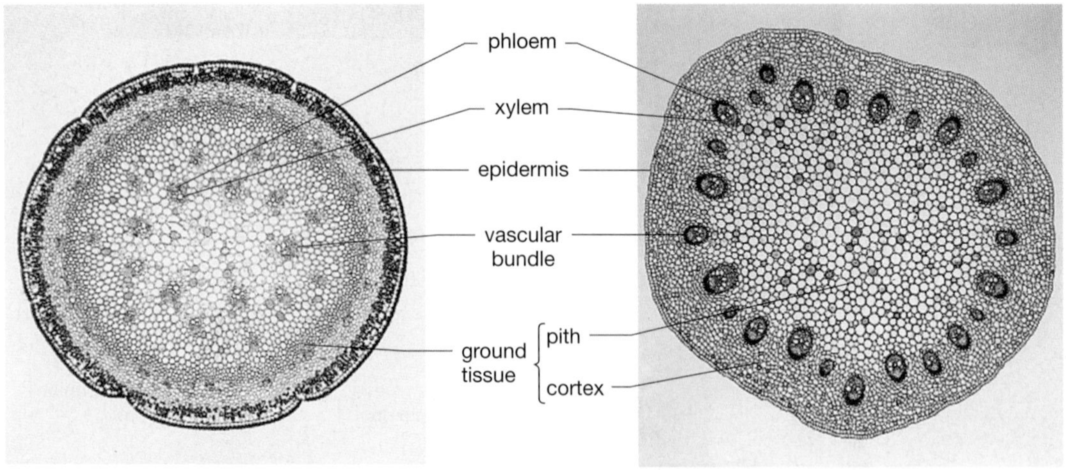

phloem

xylem

epidermis

vascular
bundle

ground
tissue { pith
cortex

Figure 24.12
Two Types of Flowering Plants
These cross sections show the organization of primary tissues in a monocot and a dicot.

Monocots (in this case an onion) have vascular bundles arranged in an irregular pattern within ground tissue.

Dicots (in this case a buttercup) have vascular bundles arranged in a ring that separates the ground tissue into pith on the inside of the bundles and cortex on the outside.

4. Meristematic Tissue

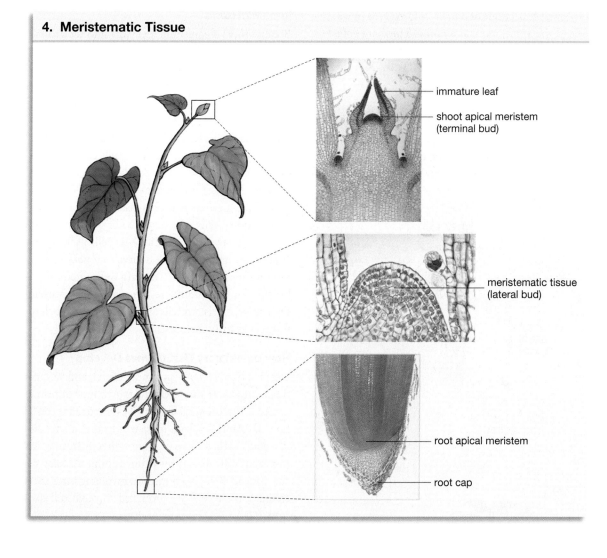

immature leaf

shoot apical meristem (terminal bud)

meristematic tissue (lateral bud)

root apical meristem

root cap

Figure 24.13
Meristematic Tissue
Growth originates from meristematic tissue. The shoot and root elongate at the apical meristems, while new branches originate at the lateral buds. Meristematic tissue gives rise to more dermal, ground, and vascular tissue, as well as more of itself.

come damaged, one of the lateral buds steps in to assume the role of the shoot apex, thus allowing the plant to maintain its vertical growth. As long as the original apical meristem is intact, however, it generally will produce hormones that *suppress* the growth of the lateral buds near it, leaving them dormant. The shoot apex is itself sometimes referred to as a **terminal bud**, particularly when *it* is dormant, as with trees in winter.

We can think of plant growth in terms of growth modules. Each module consists of an internode (the stem between leaves), a node (the area where a leaf attaches to stem), and one or more leaves (**see Figure 24.14**).

A Closer Look at Root and Shoot Apical Meristems

Now let's look a little closer at these important collections of cells from which all else comes, the shoot and root apical meristems. The right-hand pictures in Figure 24.13 show you what shoot apical meristem, lateral bud, and root apical meristem look like up close.

Shoot Meristems

As you can see in the topmost picture, the shoot apical meristem lies atop a dome-shaped collection of cells. Overlapping it are two budding leaves that serve a protective function. Next, in the middle picture, you can see a second shoot location for meristematic tissue—the lateral bud, nestled between leaf and stem. These buds are collections of meristematic tissue that are left behind, in a sense, by the apical meristem tissue as it continues its growth. (If you look again at the topmost picture, you can see

Figure 24.14
Vertical Growth in Plants Occurs in Modules
Each growth module (indicated here by alternating dark green and light green segments) consists of an internode, a node, and one or more leaves. This plant progresses from one module in week one to four modules in week four.

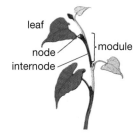

leaf

node

internode

module

week one

week two

week three

week four

two small collections of darker-colored cells on each side of the apical meristem dome. These are collections of meristem tissue that have recently been left behind; they will become lateral buds.)

Figure 24.15
Tissue Development from Apical Meristems
Apical meristems give rise to the primary tissues. In development, cells first are mostly engaged in division, just above the apical meristem, then elongation, and finally differentiation. The result is three fully formed kinds of tissue.

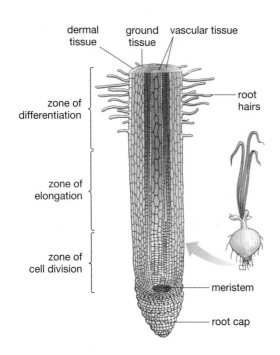

Figure 24.16
Growth by Cell Elongation
Much of the elongation of the shoot and root in a young plant results from cell elongation rather than from cell division. Thus it is cells just above the root tip, in the zone of elongation, that provide most of the root's growth, as shown here. The tip of the root (where cell division occurs) has elongated very little, and the cells in the upper root have completed most of their elongation. This same pattern holds true for the shoot.

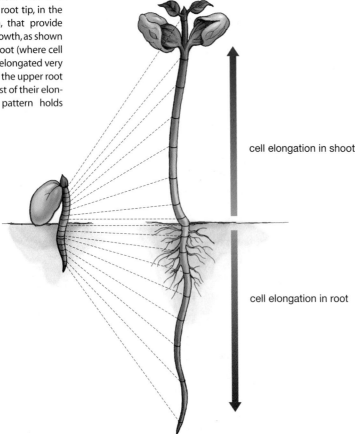

cell elongation in shoot

cell elongation in root

Root Meristems

Because roots are pushing their way into the ground, root apical meristems are located not at the very tip of the root, but just up from the tip. The root apical meristems are adjacent to a collection of cells called a **root cap** that both protects the meristematic cells and secretes a lubricant that helps ease the way for the root in its progress. Toward the middle of the apical meristem, there is a collection of cells called the quiescent center. These slowly dividing cells amount to another kind of insurance policy for the plant. Should the meristematic cells become damaged, quiescent center cells will start dividing more rapidly, producing new meristematic cells. Cells in the quiescent center are, in short, a reserve that goes into action in times of trouble. They are well suited for this because, in their dormancy, they are more resistant than apical meristem cells are to injury from such influences as toxic chemicals.

How the Primary Tissue Types Develop

You've looked so far at ground, dermal, and vascular tissues. And you know that the fourth type of primary tissue, meristematic tissue, gives rise to all these. So how do plants get from meristem to the others? Not in a single step, it turns out—not directly from apical meristem cells to, say, fully functioning vascular tissue. Instead, there is a transition involving three main changes: *More* cells are produced through cell division; cells *elongate* through cell growth; and cells *differentiate* into the three kinds of tissue. These three changes are overlapping—cells are elongating even as they are differentiating—but events do take place in roughly this order. You can see this by looking at **Figure 24.15**. In a gradual transition with no sharp boundaries, the apical meristem gives way to a **zone of cell division**, then there is a **zone of elongation**, and finally a **zone of differentiation**, which is where the ground, dermal, and vascular tissues fully take shape. Note that root hairs don't appear until the zone of differentiation. Though the three kinds of tissues *finish* taking shape in the zone of differentiation, they actually begin this process just above the apical meristem (in the zone of cell division). It's accurate to think of the tissues in the zone of cell division as being precursors to the three tissue types.

If you look at **Figure 24.16**, you can see the practical effect of a zone of elongation: It is where most of the primary plant growth occurs. There is some growth with the addition of more cells down near the apical meristem, but most of the vertical expansion of the plant takes place because of cell elongation. Such elongation has the effect of extending the roots into the soil and the stem into the air.

513

24.5   Secondary Growth
Comes from a Thickening
of Two Types of Tissues

Intercalary Meristems Keep Grasses Growing

There is one important variation on primary growth through apical meristems. What about plants such as grasses that are constantly losing their tops through such means as grazing or mowing? Grasses get their growth from tissue known as **intercalary meristems**, which are found at the base of *each node*. There thus exists a series of vertical growth tissues that are intercalated, or interspersed, between regions of nondividing cells as we go up the plant. One practical effect of this structure is that when we mow the top half of a blade of grass, it grows back. A second is that, because these plants are growing at *many* points along their length (rather than just their apex), they can manifest very rapid growth. We can easily see nodes in the giant grass known as bamboo, which exhibits intercalary growth (**see Figure 24.17**).

24.5 Secondary Growth Comes from a Thickening of Two Types of Tissues

So far, you've been looking at a plant's vertical or primary growth. In the herbaceous plants that's all the growth there is. But you've seen that some plants—the woody plants—grow out as well as up and down; they have primary *and* secondary growth, in other words (**see Figure 24.18**). So how does this secondary growth come about? Once again, through cells that are meristematic. In essence, secondary growth yields three new kinds of tissue: a different type of xylem and phloem—secondary xylem and secondary phloem—and several layers of outer tissue that collectively go under the familiar name of *bark*.

At the start, it's worth noting that secondary growth in plants is related to two of the plant characteristics that we reviewed at the start of the chapter. First,

secondary growth almost always takes place in perennials, rather than annuals (though plenty of perennials exhibit only primary growth, as with grasses). Second, with very few exceptions, the monocots described earlier have only primary growth. Meanwhile, dicots are more often woody, though many are herbaceous.

Secondary Growth through the Vascular Cambium: Secondary Xylem and Phloem

You saw earlier that the plant's vascular tissues are arranged into bundles of tube-like vessels, with the food-carrying phloem tubes toward the outside of the stem and the water-carrying xylem tubes toward the inside. In a cross section of a dicot plant stem, these tubes form concentric rings. If you look at the plant another way—through its entire *length*—you can see that these rings take the shape of cylinders that are nested, as if these layers of tissue were a series of open-ended cans, one inside the next (**see Figure 24.19**).

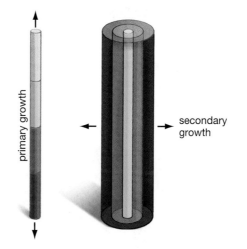

Figure 24.18
Growing Up versus Growing Out
Herbaceous plants grow at their tips, mostly vertically, through primary growth. Woody plants, however, grow not only vertically but also laterally, through what is called secondary growth. Dicots can be either woody or herbaceous, but almost all monocots are herbaceous.

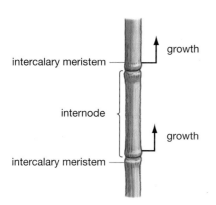

Figure 24.17
Intercalary Meristems Keep Grasses Growing
One reason bamboo grows so rapidly is that growth occurs at several points along its length at once. The intercalary meristems responsible for this growth are located at the base of each internode.

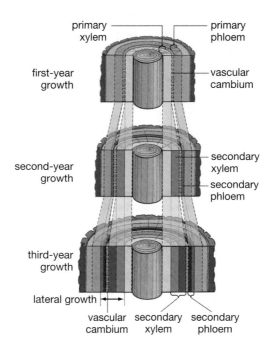

Figure 24.19
Vascular Cambium and Secondary Growth
Vascular cambium is meristematic tissue that produces secondary xylem tissue to its interior and secondary phloem tissue to its exterior. Over the years, the thickening of a tree comes about primarily because of the growth of secondary xylem.

encountering—one that gives rise to the plant's outer tissues (**see Figure 24.22**).

The cork cambium itself is a product of other tissue in the stem—in a mature tree, the secondary phloem tissue you just looked at. Cork cambium is once again secondary meristematic tissue, which is like a cell factory that is continually pushing out different kinds of cells to either side of itself. The main product of this activity goes to the *outside* of the cambium in the form of **cork** cells. As these cells mature, they go on to infuse their cell walls with a waxy substance that acts as nature's own waterproofing and invader-proofing. But cork cells are born to die, in the sense that their genetic blueprint brings about a so-called programmed cell death. It is in this dead, mature state that they serve their protective function. In a single growing season, the cork cambium can produce layer upon layer of cork cells, the result being a well-protected tree. To its interior, the cork cambium can produce, in a few species, a layer of parenchyma cells called phelloderm.

What Is Bark?

So, woody plants have cork cambium and its two products (cork and phelloderm). Then *inside* this three-part structure, there is the secondary phloem. Put together, all four parts constitute a region of a woody plant that goes by a familiar name **bark**. Put another way, bark is everything outside the vascular cambium. Figure 24.22 shows all the layers of tissues in secondary growth.

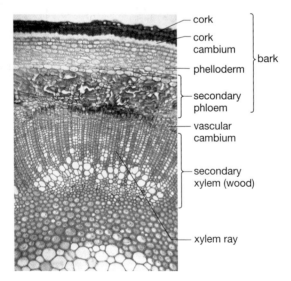

Figure 24.22
Secondary Growth Tissues
The secondary xylem, the wood with which we are all familiar, is responsible for water transport and the structural support of the plant. Xylem rays are lateral conduits for water. Bark is composed of four layers of tissue: secondary phloem, phelloderm, cork cambium, and cork.

cork
cork cambium
phelloderm ⎫ bark
secondary phloem ⎭
vascular cambium
secondary xylem (wood)
xylem ray

Figure 24.23
Death from Lack of Nutrients
This tree has been girdled by someone who intends to kill it. The bark was removed all the way down to the secondary phloem, the tissue layer that transports food from one part of the plant to another.

If a tree is "ringed" by having a strip removed around its entire circumference—as hungry animals might do in winter—it might live without its cork, and the cork cambium interior to it, and the phelloderm inside that. But if this cutting continues into the younger, food-conducting secondary phloem, that would be the end of the tree, because it could no longer move its energy stores to where they're needed. Indeed, this kind of cutting describes the "girdling" that is a common way to kill unwanted trees (**see Figure 24.23**).

To conclude this section on plant growth, **Figure 24.24** shows you how the entire plant grows, with the starting point for both secondary and primary growth being apical meristem tissue.

24.6 How the Plant's Vascular System Functions

You have had ample opportunity already to look at the plant's vascular or "plumbing" system as it relates to plant growth, but now it's time to take a closer look at how the vascular system itself works, starting with the water-conducting xylem and then continuing to the food-conducting phloem.

How the Xylem Conducts Water

Remember the statement at the top of the chapter that a good many plant cells are dead in their mature, functioning state? You saw this once already with cork cells; here is another variety of these cells, which form most of the xylem. Two types of cells make up the water-conducting portions of both the primary and secondary xylem. These are **tracheids** and **vessel elements**. Upon reaching maturity, these cells die, and their contents are then cleared out. What's left is a strand of empty, thick-walled cells stacked one on top of another—tubes, in other words (**Figure 24.25**). The walls of these xylem cells remain, however; indeed, *these* are some of the sclerenchyma cells noted earlier that have the strong lignin reinforcement in their cell walls. The upshot is not only a transport system but a *load-bearing* system, which is why something as massive as a tree can grow so tall. As noted earlier, this load-bearing function remains in a given group of xylem cells long after they have ceased to act as a conduit for the water and dissolved nutrients that are sometimes called **xylem sap**. (To get an idea of one of the uses human beings have for this sap, see "The Syrup for Your Pancakes Comes from Xylem" on the next page.)

A tube metaphor is more accurate for the vessel elements than for the slender, tapered tracheids, because the vessel elements not only lose their internal contents, but may also lose part or all of the cell-wall barriers between adjoining cells. The result is a great capacity to transport water relative to the tracheids, whose cell walls remain intact. Any two tracheid cells will have matched pairs of perforations that line up with one another, allowing water to move from one tracheid to another.

The Value of Having Tracheids and Vessel Elements

You may remember from Chapter 21 that, through evolutionary time, flowering plants took over as the dominant variety of plant from the gymnosperms (which include coniferous trees). One reason this happened was the development of vessel elements; almost all angiosperms have vessel elements *and* tracheids, while almost all woody gymnosperms possess only tracheids. As a result, angiosperms can simply transport more water than can gymnosperms. You may wonder, though: If vessel elements are so much more efficient than tracheids at transporting water, why has evolution preserved tracheids in angiosperms? As it turns out, compared with vessel elements, tracheids are less susceptible to trapping air bubbles that can arise in flowing water. Given this, tracheids may be another kind of insurance policy for plants as they face various environmental changes.

Suction Power Moves Water through Xylem

In some trees, water coming from the roots goes through a vertical journey of up to 100 meters, or about 300 feet, before it gets to the topmost leaves of the tree. What kind of force can move water such a long way? Plants spend no energy at all in this process. Instead, it is **transpiration** itself—the loss of water from a plant, mostly through the leaves—that moves water through xylem. As water vapor evaporates into the air, it creates a region of low pressure, and this pulls a continuous column of water upward

Web Tutorial 24.2
The Vascular System

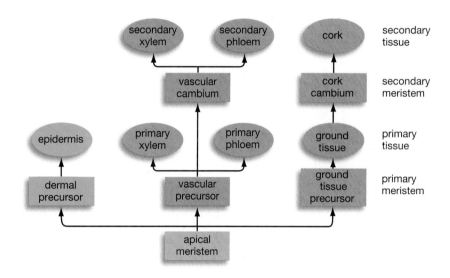

Figure 24.24
Summary of Primary and Secondary Growth in the Stem of a Vascular Plant
The "precursor" tissues that arise from the apical meristem are those that take shape in the zones of cell division, elongation, and differentiation noted earlier in the chapter.

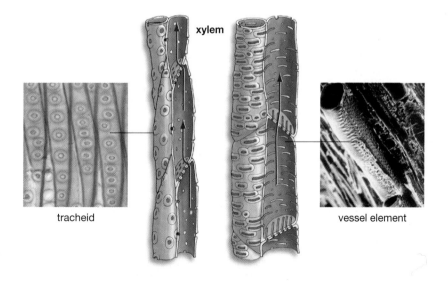

Figure 24.25
Water Carriers
The fluid-conducting cells of the xylem are tracheids and vessel elements. Both kinds of cells stack up to form tubes.

The Syrup for Your Pancakes Comes from Xylem

In early March, when winter is winding down in New England and Eastern Canada, farmers who own sugar maple trees are kept busy in a game of drilling and watching. What they are drilling is a series of two-to-three-inch holes, about three of them per maple tree. What they are watching is the weather, because they need a series of freezing nights followed by warmer days to get what they want. Such nights prompt the maple trees to convert nutrients they have stored into sugar; the warmer days then bring about a flow of this sugar, upward in sap through the tree's xylem. This material can then be "tapped" through spigots inserted into the drilled holes. The clear white sap that drips out—as much as 40 gallons from a single tree—is the starting material for pure maple syrup. Vermont is the chief maple-syrup-producing state in the United States, with an output of about 275,000 gallons in 2001. Canada, however, is the world's largest maple syrup producer.

The "maple syrup" that most Americans buy in stores actually has very little maple syrup in it (perhaps 2–3 percent, the bulk being made up of corn syrup and syrup from regular sugar). This is understandable, because pure maple syrup is expensive; the best varieties cost anywhere between $30 and $45 per gallon. One reason for this expense is that harvesting the syrup is a labor-intensive business; each tap must be put in by hand, just as it was hundreds of years ago (**see Figure 1**). Beyond this, it takes about 50 gallons of sap to make a single gallon of maple syrup. The sap must be boiled down, until its sugar concentration reaches 66.5 percent.

Too many taps can kill a tree, but experienced syrup harvesters know how to tap the same tree year after year—potentially for hundreds of years—without harming it. The 40 gallons of sap a tree can yield up sounds like a lot, but it may account for only 10 percent of a tree's total supply of sugar.

In the text, you've read much about the plant's food supply flowing through *phloem*, not xylem. But in sugar maples and some other deciduous trees, the food produced in summer's photosynthesis is stored as starch in xylem cells, to be broken down into sugar and released to the growing upper portions of the tree in early spring.

Figure 1
Source of Maple Syrup
Maple trees in Vermont being tapped for their xylem sap, which will become the key ingredient in maple syrup.

Water Transport by Transpiration

H₂O

1. Water evaporates from stomata on underside of leaves.

2. Water from stem is pulled up through xylem to replace water lost from leaves.

plant's energy *not* required

3. Water is pulled out of soil into roots to replace water lost from stem.

H₂O

(**see Figure 24.26**). Ultimately, the energy source for this movement is the sun, whose warm rays power the evaporation of water at the leaf surface.

Food the Plant Makes Is Conducted through Phloem

Now let's turn from water-and-nutrient transport to food transport. What is the nature of the food that flows through phloem? It's mostly sucrose—better known as table sugar—that is dissolved in water. Hormones, amino acids, and other compounds move through phloem as well, but sucrose and water are the main ingredients of this material, which is known as phloem sap.

Figure 24.26
Suction Moves Water through Xylem
When water evaporates from leaves, a low pressure is created that pulls more water up to fill the void. Water is so cohesive that it moves up in a continuous column. The sun's energy, rather than the plant's energy, fuels this process.

The phloem cells that this sap moves through are called **sieve elements**. Unlike xylem cells, sieve elements are alive at maturity, only in a curiously altered state. Each one has lost its nucleus, which you may remember is the DNA-containing information center of a cell. Sieve elements continue to function as living cells, however, because each one has associated with it one or more **companion cells** that retain their DNA and that seem to take care of all the housekeeping needs of their sister sieve elements (**see Figure 24.27**). The association between the two kinds of phloem cells is so intimate that when sieve elements die out, as they often do at the end of a growing season, their companion cells die as well. The reason sieve elements lose their nuclei seems to be to make room for the rapid flow of nutrients through them. The reason they are called *sieve* elements is that the ends of adjoining cells are studded with small holes, much like a sieve. Stacked on top of each other, these cells form the plant's food pipes, the **sieve tubes**.

Sugar Pumping and Water Pressure Move Nutrients through the Phloem

The motive power for the movement of food through phloem is different from that of water through xylem; with food, the plant does not get a free ride on transportation (via transpiration). Instead, to load the sucrose it makes into the phloem, the plant must expend its own energy (**see Figure 24.28**). After this, a principal player in moving fluid through the phloem is a phenomenon you've encountered before—osmosis.

Osmosis and the Movement of Phloem Sap

You can review the process of osmosis in Chapter 5 (page 103), but here's a brief summary. Any time there is a membrane, such as the lining of a cell, that is immersed in water and is semipermeable—so that some molecules can pass through it but others can't—this creates the conditions for osmosis. There is a net movement of water across the membrane to the *side* in which there is a greater concentration of solutes, or substances that can be dissolved in liquid (**see Figure 24.29**, on the next page). In plants, the solute is sucrose, which arrives at the cell walls of the sieve elements and is then *pumped* into them through special membrane channels. (The pumping is what requires energy from the plant.) The result is a higher concentration of sucrose inside the cells than before, resulting in a net movement of *water* into the cells

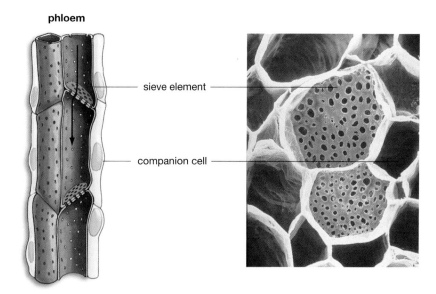

phloem

sieve element

companion cell

Figure 24.27
Food Carriers
The micrograph at right shows a cross section through some sieve elements (looking at them from above), surrounded by a number of companion cells. A plant's phloem sap moves from one sieve element to another through the perforations visible in the cells.

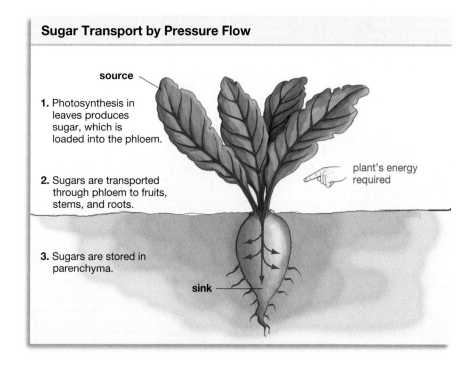

Sugar Transport by Pressure Flow

source

1. Photosynthesis in leaves produces sugar, which is loaded into the phloem.

plant's energy required

2. Sugars are transported through phloem to fruits, stems, and roots.

3. Sugars are stored in parenchyma.

sink

Figure 24.28
How Food Moves through the Plant
Sucrose is produced in the leaves (source) by photosynthesis, after which the plant expends its own energy to load the sucrose into the phloem. Osmosis is then a key factor in transporting the sucrose through the plant for use and storage (sink).

because of osmosis. The effect is the same as water coming into a garden hose from a spigot: The water moves through its path of least resistance—through the hose and out the nozzle. Just so is this solution of water and sucrose moved through the sieve tubes under the power of this pressure flow

Figure 24.29
Osmosis in Action

(a) An aqueous solution divided by a semipermeable membrane has a solute—in this case salt—poured into its right chamber.

(b) As a result, though water continues to flow in both directions through the membrane, there is a net movement of water toward the side with the greater concentration of solutes in it.

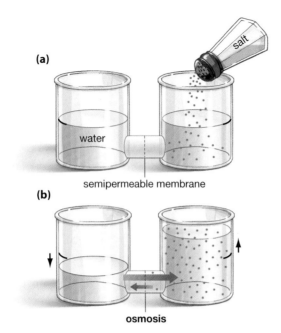

(a)

water

semipermeable membrane

(b)

osmosis

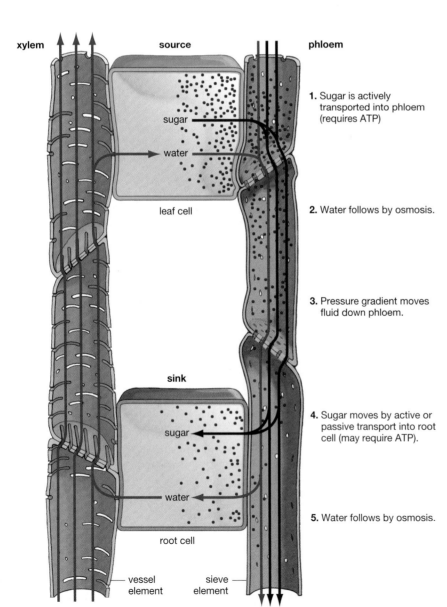

xylem   source   phloem

1. Sugar is actively transported into phloem (requires ATP)

sugar

water

leaf cell

2. Water follows by osmosis.

3. Pressure gradient moves fluid down phloem.

sink

sugar

4. Sugar moves by active or passive transport into root cell (may require ATP).

water

5. Water follows by osmosis.

root cell

vessel element   sieve element

(see **Figure 24.30**). The name given to this hypothesis about phloem sap movement is the **pressure-flow model**, first developed in the 1920s by German plant researcher Ernst Münch.

The loading site of this system is called a "source" and the unloading site is called a "sink," but these sites could also be called the "producer" and the "consumer," or perhaps "producer" and "bank." Sources tend to be sugar-producing leaves, while sink sites tend to be storage tissue such as roots, along with growing fruits and stems. At the sink, sucrose either moves passively or is actively pumped from sieve element cells to sink cells (the roots, for example). This movement triggers osmosis once again; only this time the net water flow is *out* of the sieve tube, because there has been an increase in the solute concentration outside the sieve tube. This flow creates a pull that, like the push at the source, helps move the phloem sap along.

Water Flows from Xylem to Phloem and Back Again

You saw earlier that phloem and xylem tubes are stacked together in bundles. These transport systems are not sealed off from each other, however; instead, they are closely interconnected. Remember the phloem sources, where water flowed into the sieve tube? Well, where did that water come from? From the adjacent xylem. And in the phloem sinks, when water flows out of the sieve tubes, most of it flows into adjacent xylem, where it joins water coming from the roots and moves on up with it (see Figure 24.30). With this, let's take our leave of transport in plants in order to cover another critical topic, reproduction and development.

24.7 Sexual Reproduction in Flowering Plants

To begin, it may be helpful to look again briefly at the reproductive parts of plants, their flowers, which you can see in **Figure 24.31**. You may recall from Chapter 23 that the heart of the flowers' re-

Figure 24.30
Phloem Sap Transport and Phloem-Xylem Linkage
In food transport, on the right, (1) sugar produced within leaves is actively pumped by the plant into the sieve elements of the phloem. (2) Water then "follows" the sugar in, under the power of osmosis. This creates (3) a pressure gradient that moves the sugar-water compound through the phloem. At the sink, (4) sugar moves either passively or through pumping, this time out of phloem cells, with (5) water following, again through osmosis. The phloem and xylem systems are linked. Water returns to the xylem from the phloem (bottom of the figure). Meanwhile, water is also transported to the *phloem* from the *xylem* (top left of the figure).

productive structures are the stamens and the carpel. A *stamen* consists of a long, slender filament that is topped by an *anther*, whose chambers contain cells that give rise to the male pollen. The carpel, meanwhile, is a composite structure, composed of three parts: the *stigma*, which is the tip end of the carpel, on which pollen grains are deposited; the *style*, a slender tube that raises the stigma to such a prominent height that it can easily catch the pollen; and the *ovary*, the area in which fertilization of the female egg and then development of the next generation of plant will take place.

Flowering Plants Reproduce through an Alternation of Generations

If flowering plants reproduced the way people do, then maple trees, as an example, would produce gametes—eggs and sperm—that would come together and result in a new generation of maple trees. But flowering plants don't reproduce this way. What a generation of maple trees produces is not gametes but spores. A **spore** is a reproductive cell that can develop into a new organism without fusing with another reproductive cell. Maple tree spores do exactly this: they develop into an alternate generation of the maple tree plant. To say the least, however, this alternate generation does not look like a maple tree. On the male side, the alternate generation amounts to pollen grains, each one of which is microscopic and consists of three cells and an outer coat, at most. On the female side, the alternate generation is a small collection of cells, called an embryo sac, that is hidden away deep inside the reproductive parts of the maple tree. Though these entities are tiny, they have a critical role to play: *They* will produce the eggs and sperm that, once fused, will result in more maple trees.

As may be apparent from this account, plants exhibit a back-and-forth between generations. In their reproduction, they go through an *alternation of generations* (see page 434). Because the generation that is familiar to us—the maple tree—produces spores, it is known as the sporophyte generation. Because the generation that is less familiar to us produces gametes, it is called the gametophyte generation. Given how small the gametophyte generation is in angiosperms, most students have trouble seeing it *as* a generation. From a human perspective, pollen grains and embryo sacs look like small component parts of the sporophyte generation. Nevertheless, gametophytes are a separate generation of plant, with their own critical role to play in reproduction. For

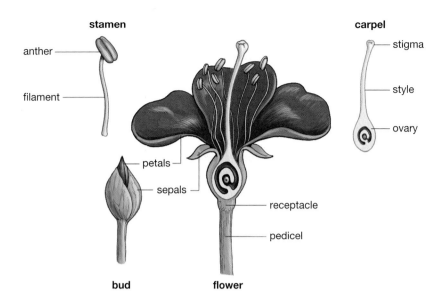

Figure 24.31
Anatomy of a Flower: The Reproductive Structure of an Angiosperm

a view of how gametophyte and sporophyte generations differ not just in angiosperms but in all types of plants, see **Figure 24.32**. In the more primitive plants, such as mosses, it is the gametophyte generation that is larger and more dominant, but the reverse is true with the later-evolving angiosperms.

Web Tutorial 24.3
Plant Reproduction

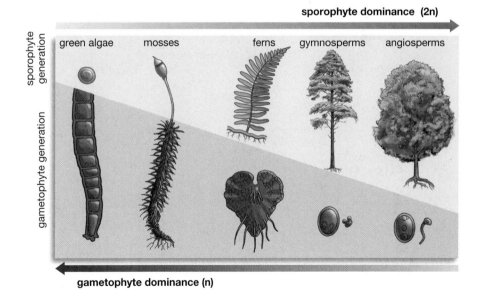

Figure 24.32
Generational Dominance Varies with the Type of Plant
In more primitive plants, the gametophyte generation is larger and more dominant. In mosses, for example, the gametophyte generation houses the sporophyte generation and supplies it with nutrition. By contrast, in the most recently evolved plants, the angiosperms, the gametophyte generation has been reduced to a microscopic entity that is almost wholly dependent on the sporophyte generation (in this example a tree). Though not plants, green algae are included for comparative purposes.

Figure 24.33
Overview of Angiosperm Reproduction
1. The mature sporophyte flower produces many male microspores within the anther and female megaspore within the ovary. The male microspores then develop into pollen grains that contain the male gamete, sperm. **2.** Meanwhile, the female megaspore produces the female gamete, the egg. **3.** Pollen grains, with their sperm inside, move to the stigma of a plant. Moving through a pollen tube, the sperm reaches the egg and fertilizes it. **4.** This results in a zygote that develops into an embryo protected inside a seed. **5.** This seed leaves the parent plant and sprouts or "germinates" in the earth, growing eventually into the type of sporophyte plant that began the cycle. In this diagram, the plant is self-fertilizing: a sperm from the plant is fertilizing an egg in this same plant. This does occur in nature, but it is the exception rather than the rule. More often, pollen from one plant will fertilize one or more eggs in a different plant.

Look now at **Figure 24.33** to get an overview of angiosperm reproduction. One of the things you'll see in the figure is how the gametophyte generation starts out within the sporophyte generation. As you proceed through the story of angiosperm reproduction, you'll be looking in greater detail at each of the stages shown in this figure.

The Sporophyte and Gametophyte Generations Are Genetically Different

Apart from being different in size and appearance, the sporophyte and gametophyte generations are genetically very different. Let's look at this genetic difference first on the male side, starting with the sporophyte of a typical flowering plant (see the left side of **Figure 24.34**). Students who have been through the genetics section will recall that the normal state for the cells of most complex organisms is diploid, meaning that cells have chromosomes that come in matched *pairs*—a state sometimes referred to as $2n$. All the cells in the sporophyte generation are diploid, or $2n$. But special cells called **microspore mother cells**, which lie within the anthers of the sporophyte plant, will undergo the special kind of cell division called meiosis (reviewed in Chapter 10). Meiosis in the microspore mother cells will result in *microspores* that have only a single set of chromosomes each

(making them haploid or n). It is these microspores that will develop into pollen grains. Keeping an eye on the forest as well as the trees, the line between one generation and the next is crossed after the microspore mother cell has undergone meiosis. Once the microspores have been produced, the result is the gametophyte generation.

On the female side, both meiosis and development of the female gametophyte take place entirely inside the ovary lying at the base of the plant's carpel (see the right side of **Figure 24.34**). In this case, a single diploid cell—called a **megaspore mother cell**—undergoes meiosis, yielding female *megaspores* that are haploid. (Why is the female spore "mega" compared to the male's "micro"? Because the female spore is bigger.) Once the megaspores have been produced, the female gametophyte generation has come into existence.

Both the male and female gametophyte generations are multicelled structures (though just barely), with these cells arising by mitosis, or common cell division, from the original single haploid spores. Because mitosis makes exact copies of cells, this means that *every cell* in the gametophyte generation is haploid. In sum, the gametophyte generation is haploid (n), while the sporophyte generation is diploid ($2n$).

Development of the Male and Female Gametophyte Generation

The male and female gametophytes we've been talking about are going to produce the sperm and eggs that will come together and result in a new generation of the flowering plant we started with. But how do male and female gametophytes develop, such that they can do this? Let's see.

The Male Gametophyte

On the male side, things are pretty simple. As you can see on the left side of Figure 24.34, meiosis in each microspore mother cell yields four haploid microspores. Each spore develops into a pollen grain that initially consists of three things: an outer coat, a tube cell, and a generative cell. This pollen grain can be thought of as an *immature* male gametophyte plant. At some point, the generative cell divides once, resulting in two *sperm* cells. (In the type of plant you're looking at, this division comes before the pollen grain is released from the sporophyte plant, but in other species this division comes later.) Now having three cells, the pollen grain is released from the sporophyte plant and makes its way, by wind or bird or insect, to the stigma of another plant. This is what pollination is. More formally, **pollination** is the transfer of pollen from the anther of a flowering plant to the stigma of a flowering plant. You'll get to its story in a second.

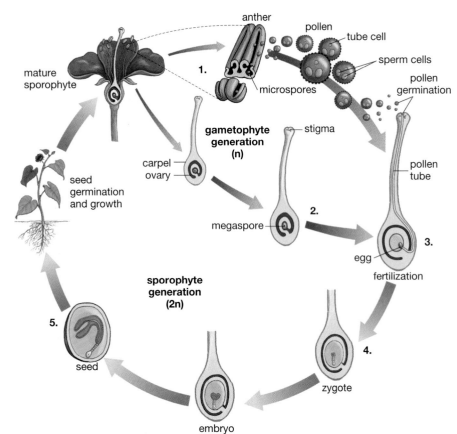

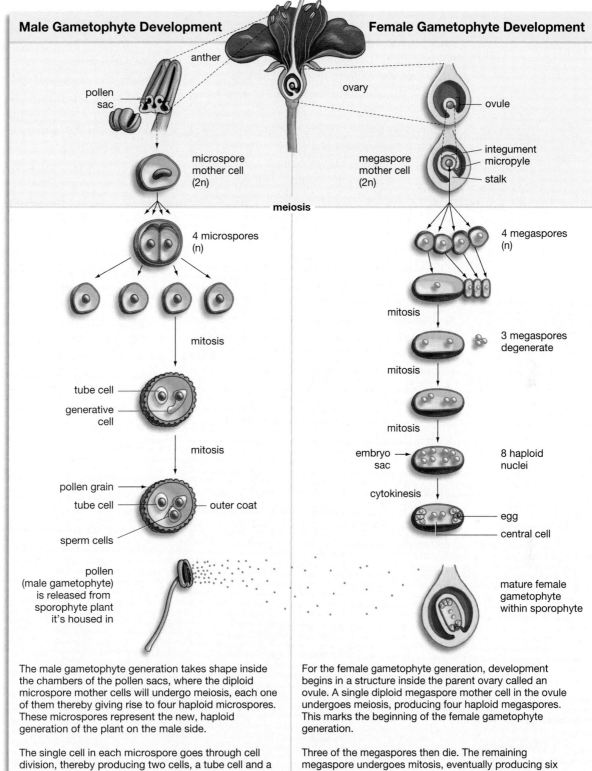

Male Gametophyte Development

Female Gametophyte Development

anther

pollen sac

ovary

ovule

microspore mother cell (2n)

megaspore mother cell (2n)

integument micropyle

stalk

meiosis

4 microspores (n)

4 megaspores (n)

mitosis

3 megaspores degenerate

mitosis

mitosis

tube cell

generative cell

mitosis

mitosis

embryo sac

8 haploid nuclei

cytokinesis

pollen grain

tube cell

outer coat

egg

central cell

sperm cells

pollen (male gametophyte) is released from sporophyte plant it's housed in

mature female gametophyte within sporophyte

The male gametophyte generation takes shape inside the chambers of the pollen sacs, where the diploid microspore mother cells will undergo meiosis, each one of them thereby giving rise to four haploid microspores. These microspores represent the new, haploid generation of the plant on the male side.

The single cell in each microspore goes through cell division, thereby producing two cells, a tube cell and a generative cell. Before or during this time, a protective coat develops around the microspore. The combination of the cells and protective coat is the pollen grain. At some point, the generative cell in the grain divides into two sperm cells. With this cell division, a mature male gametophyte has developed. The pollen grain is then released from the anthers to make its way to a stigma.

For the female gametophyte generation, development begins in a structure inside the parent ovary called an ovule. A single diploid megaspore mother cell in the ovule undergoes meiosis, producing four haploid megaspores. This marks the beginning of the female gametophyte generation.

Three of the megaspores then die. The remaining megaspore undergoes mitosis, eventually producing six cells with a single nucleus each and one central cell with two nuclei. These seven cells form the embryo sac, which is the mature female gametophyte. One of the seven cells in the embryo sac is the egg that will undergo fertilization by one of the sperm in the pollen grain.

Figure 24.34
Development of the Male and Female Gametophyte Generation

The Female Gametophyte

On the female side, note in Figure 24.34 that there develops from the ovary wall a structure called an *ovule*—a part of the sporophyte plant—that will enclose the female gametophyte generation. The ovule consists of the protruding stalk, a central core of cells, and a couple layers of tissue called *integuments*. The integuments wrap around the central cells, almost coming together except for a tiny opening they leave (called a micropyle). From the ovule's central mass of diploid cells, a single cell—the megaspore mother cell—undergoes meiosis, thereby producing four megaspores. Then three of the megaspores will die, leaving the one that remains as an immature female gameto-phyte plant.

The surviving megaspore now undergoes three rounds of mitosis, but of a curious sort: There are several rounds of doubling of the cell's DNA (in the nucleus), but no subsequent cell division, as is usu-ally the case. The result is a single cell that has eight haploid nuclei. This cell eventually does undergo cell division, but once again there is a twist. Six of the cells that result will each have a single nucleus, but one will have two nuclei. You can see, in Figure 24.34, the most common arrangement of these seven cells. Collectively, they form the **embryo sac**, which can be thought of as the mature female ga-metophyte plant. At one end is the cell of greatest importance, the egg, flanked by a couple of cells on either side. On the opposite end there are three other cells; and in the middle, a very large central cell, which is the cell that has two nuclei.

Fertilization of Two Sorts: A New Zygote and Food for It

The stage is now set for male to come into contact with female. The pollen grain, with its two sperm cells and one tube cell, now lands on the stigma of a plant with an egg and embryo sac matured down in the ovule (**see Figure 24.35**). The tube cell then sprouts something: a pollen tube, which tunnels its way down through the style, eventually gaining access to the female reproductive cells within the ovule. The two sperm cells travel down this tube to achieve fertilization. (Why *two* sperm cells? Stay tuned.) In sum, in its most mature state, the male gametophyte generation consists of three cells, one of which grows a tube and two of which are sperm cells—quite a different thing from the sporophyte generation with its billions of cells.

The pollen tube grows through the ovule's mi-cropyle and ejects its two sperm cells into the em-bryo sac. One of these sperm fuses with the egg, yielding the zygote that is the beginning cell in the new generation of plant—another sporophyte generation like the flowering plant we started with. This is no different in principle than what takes place with *human* life, where egg and sperm fuse, producing a single-celled zygote that devel-ops into a whole human being. And, as with human egg and sperm, the plant egg and sperm were both haploid, containing a single set of chro-mosomes each. So their fusion produces a diploid zygote—one with a *paired* set of chromosomes—that will divide over and over, resulting in a diploid sporophyte plant.

**Figure 24.35
Double Fertilization in
Angiosperms**
When the male pollen (shown greatly enlarged) lands on the stigma, the tube cell within it sprouts a pollen tube, through which the two sperm pass in mov-ing to the ovule. One sperm cell fertilizes the egg, forming a zy-gote—the new sporophyte plant. The other sperm cell fuses with the two nuclei of the central cell, forming the endosperm that will later nourish the embryonic plant.

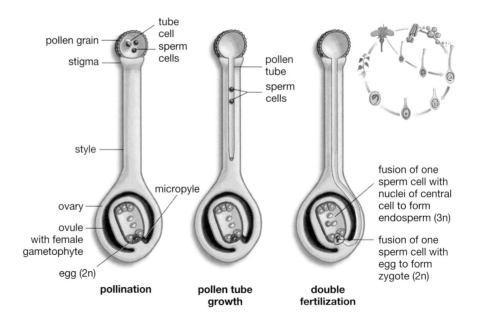

And what of the other sperm? Here is one of the beauties of flowering plants. This sperm enters the central cell with its two nuclei, producing a cell with *three* nuclei (making it triploid or *3n*.) This too is a fertilization, only what it sets in motion is the development of *food* for the embryo—endosperm, which is tissue rich in nutrients that can be used by the embryo that will now develop from the zygote. Thus two fertilizations have taken place. This is known as **double fertilization** in angiosperms—a fusion of gametes and of nutritive cells at the same time. Double fertilization is an innovation that belongs almost solely to the flowering plants. As it happens, it has been a boon to all kinds of *animals*, including people. It is endosperm that supplies much of the food we human beings consume. The rice and wheat grains we're familiar with consist in large part of endosperm meant to sustain the plant embryo (**see Figure 24.36**).

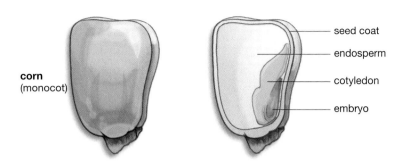

Figure 24.36
Endosperm Is Food for the Embryo
In monocots such as corn, endosperm persists in the mature seed, providing nutrition for the embryo inside. The cotyledon in the figure is the corn plant's embryonic leaf.

24.8 Embryo, Seed, and Fruit: The Developing Plant

The Development of a Seed

Egg fertilization produces a single-celled zygote. The question is, how does the plant develop from this zygote into an embryo-containing *seed* that falls to the ground, eventually giving rise to a growing sporophyte plant? Many steps are involved here; what follows are only the highlights. First, recall that the female gametophyte, the embryo sac, was nearly surrounded by ovule tissue called *integuments*. These will now develop into the **seed coat** that will protect the embryonic plant until it germinates in the ground. What once was an ovule with a gametophyte inside is now a seed: a reproductive structure of a plant that includes an outer seed coat, an embryo, and nutritional tissue for the embryo (the endosperm). This seed is a reproductive structure that bears similarities to one employed by some animals, the egg (**see Figure 24.37**).

The Development of Fruit

Something more will now develop *outside* the seed coat. Recall that the ovule was surrounded by the ovary of the flower. The ovary wall now develops into fruit. Strictly speaking, **fruit** is the mature ovary of any flowering plant. Given this, *all* flowering plants develop fruit, but it's clear that most of the 260,000 species of flowering plants are not fruits as we commonly understand that term. It follows that

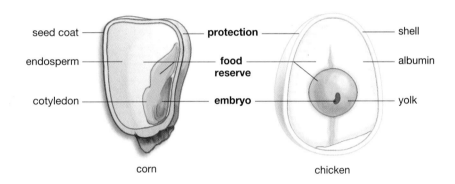

Figure 24.37
A Kernel of Corn and a Chicken Egg Are Similar in Some Ways
Both are home to an embryo, both contain plenty of food for development, and both offer protection for the embryo in dry, land environments.

the botanical definition of fruit differs from the common definition; under the botanical definition, fruits can take on lots of forms. The pod of a pea plant is a fruit, for example, as is the outer covering of a kernel of corn (though here fruit has fused with the corn seed coat, leaving scarcely anything separate that can be recognized as fruit). On the other hand, fruits can take exactly the form we're used to, as with the flesh of an apricot or cherry. To add another twist, where is the fruit in a strawberry? The answer, shown in **Figure 24.38**, may surprise you. Fruit develops in all angiosperms, and it develops *only* in angiosperms. Thus, angiosperms can be defined by this anatomical feature. An **angiosperm** is a plant whose seeds develop inside the tissue called fruit.

Figure 24.38
Fruits Come in Many Forms
A fruit, the mature ovary of a flowering plant, surrounds a seed or seeds. Thus **(a)** the flesh of an apricot fits this definition of fruit, but so does **(b)** the pod of a pea, which has several seeds inside. The structure we commonly think of as **(c)** the strawberry fruit actually is the strawberry plant receptacle, with each receptacle having many fruits on its surface. What are commonly thought of as strawberry "seeds" are tiny strawberry plant carpels, each complete with its own fruit and seed. Fruits are important not only because they protect the seeds but also because they attract animals that disperse the seeds.

(a) Apricot

one carpel, one seed

(b) Pea

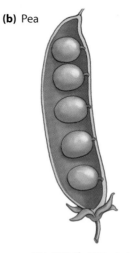

one carpel, many seeds

(c) Strawberry

carpels

many carpels, many seeds, one receptacle

receptacle

Figure 24.39
Development of a Plant Embryo, Germination of a Plant Seed
Egg and sperm have come together in this dicot to produce a single-celled zygote, which divides repeatedly in developing as the plant embryo. The zygote is surrounded by the plant ovule, which develops into a seed.

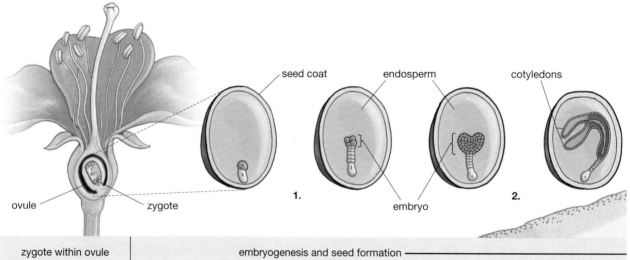

seed coat   endosperm   cotyledons

embryo

ovule   zygote

1.   2.

| zygote within ovule | embryogenesis and seed formation |

1. The zygote divides into two cells, one of which is the embryo (green in the figure). The other cell (yellow in the figure) develops into a structure that will push the embryo into the endosperm.

2. The embryo starts to develop its cotyledons or embryonic leaves. Most of the endosperm nutrients eventually are taken up by these cotyledons. (Because this plant is a dicot, it has two cotyledons.)

Fruits Serve in Protection and in Seed Dispersal

What is common to many fruits is that they serve a protective function for the underlying seed, and many act as *attractants* for seed dispersal. As noted in Chapter 21 (page 439), the wild berry that a bear eats consists not only of the fruit flesh, but of the seeds it surrounds. The fruit is digested by the bear as food, but the seeds pass through the bear's digestive tract. Thus has this plant engineered seed dispersal at what may be a promising new location for a berry plant. Plants with such attractants *time* their fruit ripening, so that the fruit is not attractive before the seed is ready for dispersal.

Development of the Embryo and Germination of the Seed

While fruit and seed coat are maturing, the embryo inside them is also developing. **Figure 24.39** shows you the outline of the development of a plant embryo, this one a dicot plant. Recall that the process starts with a single cell, the zygote, encased within the structure that will become the seed, the ovule. The zygote divides, yielding two cells, one of which becomes the embryo. The other cell gives rise to a paddle-shaped structure that pushes the growing embryo into the endosperm.

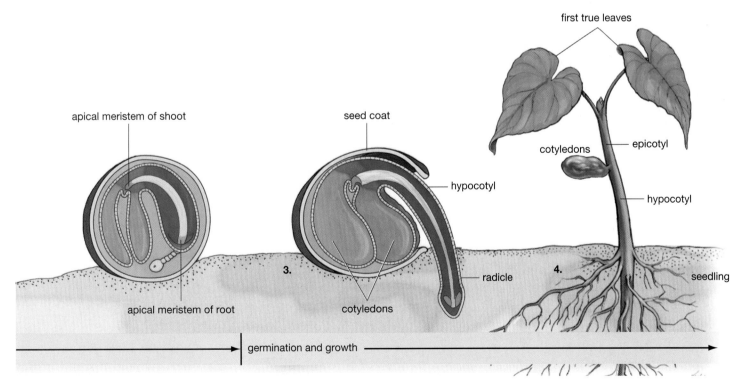

germination and growth

3. Having made its way to the ground, the seed is ready to "germinate" or start sprouting. The seed coat splits, allowing the emergence of the hypocotyl (tissue below the cotyledons), which includes the plant's radicle, or early root structure.

4. Root and shoot have both sprouted. The plant's cotyledons, having completed their function, will wither away. The plant sprouts its first true leaves and begins the process of photosynthesis. The reproductive cycle is ready to begin again.

Eventually, the embryonic tissue develops two **cotyledons**, or embryonic leaves. When the embryo is still tightly encased in the seed coat, the cotyledons can take on an important role you've seen before, which is to provide nutrients to the growing embryo. You may say: But isn't this the function of the endosperm? In most monocots (for example, the grains) the endosperm retains this function, but in most dicots the nutrients stored in endosperm tissue are *taken up* by the cotyledons as they develop. The foods we know as beans are comprised of embryos with large cotyledons (**see Figure 24.40**).

The seed that encases the embryo eventually separates from the sporophyte parent plant and makes its way into a suitable patch of earth, there to germinate, or to sprout both roots and shoots. First to emerge from the seed is the root structure or radicle, which is attached to **hypocotyl**, which can be thought of as all the plant tissue below the cotyledons. Then there emerges the **epicotyl**, meaning the tissue above the cotyledons. (In Figure 24.40, you can see the location of hypocotyl and epicotyl at an earlier stage within the bean seed.) The epicotyl gives rise to the first true leaves of the plant. The seed has sprouted into a new sporophyte plant.

Figure 24.40
Cotyledons Serve a Nutritive Function

In a dicot such as this bean plant, the endosperm food reserves are taken up by the embryonic leaves or cotyledons of the embryo, which then serve as the mature seed's food reserves. The hypocotyl in the figure is tissue that eventually will lie below the cotyledons in the sprouting plant. Epicotyl is tissue that will lie above the cotyledons.

bean
(dicot)

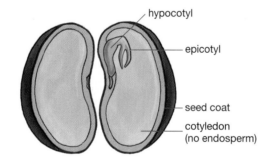

hypocotyl

epicotyl

seed coat

cotyledon
(no endosperm)

Seed Dormancy Can Be Used to a Plant's Advantage

One of the survival strategies of plants is the use of dormancy, or a prolonged low metabolic level in seeds. Seeds can *postpone* germinating until they have favorable conditions around them, such as the proper temperature or amount of light. What generally sets seed germination in motion is an uptake of water from the surrounding environment. It is, however, the triggering mechanisms for this water uptake that tell the tale in dormancy. Some seeds require a scraping or abrasion before they will take up water; others simply need to dry out, once they've shed the fruit from around them; still others require the action of enzymes from an animal's gut, or even charring by fire.

Once the plant has germinated and matured a little, you could take a cross section of its stem and find concentric circles of the tissue types mentioned near the start of this chapter: dermal, vascular, and ground. Meanwhile, meristematic tissue could be found at the tips of the root and shoot, allowing the plant to grow further. With this, we have come full cycle: Another generation of plants is maturing.

On to Animals

In Chapters 23 and 24, you have looked at a group of living things, the plants, whose members are silent, fixed in one spot, and have almost no moving parts. Now it's time to shift gears. There are animals that are silent and animals that are fixed in place, but you'll also find with animals nature's only sound-makers and its supreme travelers. The next five chapters are concerned with the functioning of animals in general, but they focus on one animal that in some ways is very familiar to us—the human animal.

Chapter Review

Summary

24.1 Two Ways of Categorizing Flowering Plants

- Plants can be categorized by how long it takes them to go through a cycle from germination to death. Those that go through this cycle in a year or less are annuals; those that go through in about two years are biennials; those that live for many years are perennials.

- A cotyledon is an embryonic leaf, present in the seed. Angiosperms are classified according to how many cotyledons they have—one in the case of monocotyledons, two in the case of dicotyledons. Monocot and dicot plants are structured in different ways.

24.2 There are Three Fundamental Types of Plant Cells

- There are three fundamental types of cells in plants that, alone or in combination, make up most of the plant's tissues, meaning groups of cells that carry out a common function. These three cell types are parenchyma, sclerenchyma, and collenchyma.

24.3 The Plant Body and Its Tissue Types

- Some plants are capable only of primary growth, meaning growth at the tips of their roots and shoots that primarily increases their length. Plants that exhibit only primary growth are herbaceous plants, composed solely of primary tissue. Other plants exhibit both vertical growth and lateral, or secondary growth. These are the woody plants, composed of primary and secondary tissue.

- There are four tissue types in the primary plant body: dermal, vascular, meristematic, and ground. Dermal tissue can be thought of as the plant's outer covering, vascular tissue as its "plumbing," meristematic tissue as its growth tissue, and ground tissue as almost everything else in the plant.
  Web Tutorial 24.1: Plant Tissue and Growth

24.4 How a Plant Grows: Apical Meristems Give Rise to the Entire Plant

- The entire plant develops from meristematic cells in regions called apical meristems. Meristematic cells remain perpetually embryonic, able to give rise to cells that differentiate into all the plant's tissue types.

- Shoot apical meristems give rise to the entire shoot of the plant. In addition to providing for vertical growth, shoot apical meristems produce meristematic tissue called lateral buds at the base of leaves that can give rise to a branch or flower. Root apical meristems are located just behind a collection of cells at the very tip of the root, called the root cap.

- The plant's tissue types develop in stages from meristematic cells. This development takes place in a series of regions adjacent to the apical meristem. In a gradual transition, the apical meristem gives way to a zone of cell division, followed by a zone of elongation (in which developing cells lengthen) followed by a zone of differentiation (in which cells fully differentiate into different tissue types).

24.5 Secondary Growth Comes from a Thickening of Two Types of Tissues

- Secondary growth in plants takes place through the division of cells in two varieties of meristematic tissue: the vascular cambium, which gives rise to secondary phloem and xylem; and cork cambium, which gives rise to the outer tissues of woody plants. Secondary xylem, also known as wood, is responsible for most of a tree's widening.

- Looking at a tree from the secondary phloem outward to the tree's periphery, four tissues constitute the tree's bark: secondary phloem, phelloderm, cork cambium, and cork. The cork cells are dead in their mature state and provide layers of protection for the tree.

24.6 How the Plant's Vascular System Functions

- Two types of cells make up the water-conducting portions of xylem tissue. These are tracheids and vessel elements, both of which are dead in their mature, working state. Vessel elements, which exist almost solely in angiosperms, conduct more water than tracheids. Their existence in angiosperms is one of the reasons for the angiosperms' dominance in the plant world.

- Water movement through xylem is driven by transpiration, meaning the loss of water from a plant, mostly through the leaves. As water evaporates into the air, it pulls a continuous column of water upward through the plant.

- Sucrose is the main product that flows through phloem. The fluid-conducting cells in phloem are called sieve elements, which lack cell nuclei in maturity. Each sieve element has associated with it one or more companion cells, which retain their nuclei and seem to take care of the housekeeping needs of their related sieve elements.

- Plants must expend their own energy to load the sucrose they produce into the phloem. Dissolved sucrose then moves through the plant through the power of osmosis and the fluid pressure that results from it.
  Web Tutorial 24.2: The Vascular System

24.7 Sexual Reproduction in Flowering Plants

- All plants, including angiosperms, reproduce through an alternation of generations. A sporophyte generation (the familiar tree or flower) produces haploid spores that develop into their own generation of plant, the gametophyte generation. The male gametophyte is the pollen grain, consisting in maturity of an outer coat, two sperm cells, and one tube cell. The female gametophyte consists at maturity of an embryo sac composed of seven cells, one of which is the egg. The female gametophyte is housed inside a structure of the parent sporophyte plant called an ovule.

- Fertilization of the egg by sperm requires that a pollen grain land on the stigma of a plant. The tube cell of the pollen grain then germinates, sprouting a pollen tube that grows down through the sporophyte plant's stigma and style, eventually reaching the female reproductive cells inside the ovule. The sperm cells inside the pollen grain travel down through the pollen tube, and one of the sperm cells fertilizes the egg in the ovule, producing a zygote. With this, the new sporophyte generation of plant has come into being.

- The second sperm cell in the pollen grain enters the central cell in the embryo sac, setting in motion the development of food for the embryo, endosperm tissue. This second fertilization completes the process of double fertilization—a fusion of gametes and of cells producing nutritive tissue.
Web Tutorial 24.3: Plant Reproduction

24.8 Embryo, Seed, and Fruit: The Developing Plant

- With fertilization, the ovule integuments that surrounded the embryo sac begin to develop into the seed coat that will surround the growing sporophyte embryo. The ovary that surrounded the ovule then starts to develop into a layer of tissue that will surround the seed—fruit, which is defined as the mature ovary of a flowering plant. Under this definition the pod of a pea plant is fruit, as is the flesh of an apricot.

- The seed with its fruit covering eventually separates from the sporophyte parent plant and then germinates in the earth, sprouting both roots and shoots.

Key Terms

angiosperm 525	epicotyl 528
annual 504	fruit 525
apical meristem 510	herbaceous plant 507
bark 516	hypocotyl 528
biennial 504	intercalary meristem 513
bud 510	lateral bud 510
cellulose 506	megaspore mother cell 522
collenchyma cell 506	microspore mother cell 522
companion cell 519	monocotyledon (monocot) 505
cork 516	parenchyma cell 506
cork cambium 515	perennial 504
cotyledon 528	phloem 508
dermal tissue 507	pollination 522
dicotyledon (dicot) 505	pressure-flow model 519
double fertilization 525	primary growth 507
embryo sac 524	root cap 512

sclerenchyma cell 506	tracheid 517
secondary growth 507	transpiration 517
secondary phloem 514	vascular cambium 514
secondary xylem 514	vessel element 517
seed coat 525	wood 514
shoot apical meristem 510	woody plant 507
sieve element 519	xylem 508
sieve tube 519	xylem sap 517
spore 521	zone of cell division 512
terminal bud 511	zone of differentiation 512
tissue 506	zone of elongation 512

Understanding the Basics

Multiple-Choice Questions (answers in the back of the book)

1. Dermal tissue _____, while vascular tissue _____.
   a. protects the plant; conducts fluids
   b. provides strength; facilitates growth
   c. facilitates reproduction; protects the plant
   d. facilitates growth; provides strength
   e. protects the plant; facilitates reproduction

2. Biennials, like cauliflower, radishes, and beets,
   a. go through their entire life cycle in one season
   b. go through their entire life cycle in about two years
   c. grow and bloom each season, for many years
   d. have the shortest life span, among annuals, biennials, and perennials
   e. go through their entire life cycle twice each year

3. Sclerenchyma cells
   a. lack cellulose
   b. function in metabolic processes such as photosynthesis
   c. are generally dead at maturity
   d. are commonly found in rapidly growing parts of the plant
   e. are found exclusively in roots

4. Secondary growth
   a. leads to increased height of the plant
   b. is more widespread in herbaceous plants than in woody shrubs and trees
   c. leads to an increase in the girth of the plant
   d. is brought about through cell divisions in the cuticle
   e. is growth that occurs only in a plant's second year

5. Which of the following statements about the ascent of water in tall trees is true?
   a. Plants must expend metabolic energy to power the ascent of water into treetops.
   b. The evaporation of water from the leaf surface is required to lift water from the roots to the canopy of a tall tree.
   c. The ascent of water in the xylem is fastest when the stomata are closed.

d. The formation of embolisms, or air bubbles, in xylem elements increases the speed of ascent of water.

e. All of the above are true.

6. Phloem sieve tubes develop a positive pressure because
   a. Osmotic uptake of water occurs when sugars are loaded into the sieve elements.
   b. Sieve elements lack a nucleus.
   c. Sieve elements are dead at maturity.
   d. Sucrose is a bulky molecule that exerts a pressure against the sieve tube walls.
   e. All of the above are true.

7. Endosperm is formed when
   a. A sperm nucleus from the pollen fuses with one haploid nucleus from the embryo sac.
   b. Two nuclei from the embryo sac fuse with each other.
   c. A sperm nucleus from the pollen fuses with the egg nucleus in the embryo sac.
   d. An egg nucleus from the pollen fuses with a diploid nucleus in the embryo sac.
   e. A sperm nucleus from the pollen fuses with two central cell nuclei in the embryo sac.

8. Which of the following statements about spores is true?
   a. Spores are diploid structures that give rise to gametes.
   b. Spores give rise to the gametophyte generation of plants.
   c. Spores are part of the sporophyte generation.
   d. Spores give rise to multicellular haploid gametes.
   e. Spores give rise to single-celled diploid gametes.

9. Which of these is a gametophyte plant? (Select all that apply.)
   a. petal
   b. stigma
   c. embryo sac
   d. pollen grain
   e. sepal

10. The difference between a seed and a fruit is that
    a. Fruit is composed of triploid tissue; seeds are diploid.
    b. Seeds are always hard; fruits are always soft.
    c. Seeds develop from ovules while fruit tissue is derived from ovary walls.
    d. Annuals produce seeds while perennials produce fruits.
    e. Fruits are annuals while seeds are perennial.

Brief Review

1. Compare monocots and dicots with respect to their anatomy, morphology, and seed structure.

2. What are the three main cell types in a plant, and what distinct functions are performed by each?

3. What is meristematic tissue? Describe the organization of the meristem.

4. Compare the structure and function of the water-conducting (xylem) and food-conducting (phloem) cells in a plant.

5. What does a growth ring on a tree trunk represent? Why can you tell the age of a tree by counting its growth rings?

6. Describe the process of double fertilization in plants.

Applying Your Knowledge

1. If you were to regularly shear a lilac bush down to a stub, you would probably kill it. Yet, mowing after mowing, the grass in your lawn continues to grow. Is there a fundamental difference in the location of growing points in the lilac and the grass? Explain.

2. Beavers often kill young trees by girdling them—that is, removing a strip of bark from the circumference of a tree trunk. Why does this bark removal kill a tree?

3. Gardeners often speak of outdoor plants having become "established," meaning the plants have reached a point where they need less human care. What threshold does a plant have to cross in order to become established?

MediaLab

Why Do We Need Plants Anyway? The Importance of Plant Diversity

It is speculated that a meteorite struck the Earth near the Gulf of Mexico about 65 million years ago. After that event, there is no fossil record of any animal species over 65 pounds surviving. Most animals were probably not killed by rocks or tidal waves, but died when Earth's plant life could no longer support their needs. A significant meteor impact would send millions of tons of dust and debris into the atmosphere, blocking the sun's rays from reaching Earth's surface, significantly reducing photosynthesis, and ultimately leaving little food to support animal life. This event illustrates the essential role of plants in creating an environment that supports animal life. We simply couldn't survive without them. To learn more, the *Web Tutorial* will introduce you to the basics of plant structure and growth. In the *Web Investigation*, you will see how plants have adapted these basic structures to specific habitats and conditions. Finally, in the *Communicate Your Results* section, you will synthesize and apply your knowledge to the plant life around you.

This *MediaLab* can be found in Chapter 24 on your Companion Website (www.prenhall.com/krogh3).

WEB TUTORIAL

Plants, because of their ability to trap solar energy in a chemical form, are a food source for all other living organisms. They are pivotal members of every ecosystem and have adapted their basic body parts to suit these various environmental conditions. In this *Web Tutorial*, you will complete the following activity.

Activity

1. First, review the four primary plant tissues—dermal, vascular, meristematic, and ground.

2. Then, cover the process of by which plants transport water and nutrients.

3. Finally, explore plant reproduction through the life cycle of a typical sporophyte plant.

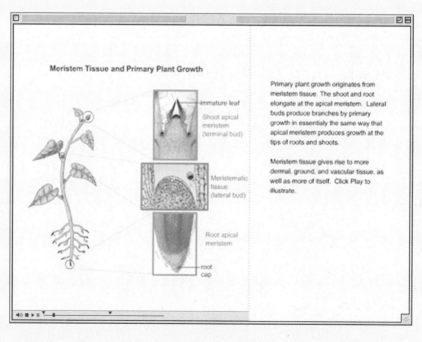

Investigation 1
Estimated time for completion = 15 minutes

There are two major groups of flowering plants, the monocots and the dicots, whose lineages diverged early in their evolutionary history. To contrast how these two groups of plants are constructed, select the keywords **DICOT** and **MONOCOT** on your Website for this *MediaLab*. Then compare the architecture and anatomy of the tomato plant, a dicot, and the rice plant, a monocot. How many differences can you observe in these two plants? Examine the roots, stems, leaves, and flowers.

Investigation 2
Estimated time for completion = 5 minutes

Flowering plants appeared on Earth about 150 million years ago. Since then, they have moved into an extraordinary number of habitats. Over genera-tions, plant species have altered their form and function to become better adapted to living in varying habitats. Select the keyword **SILVERSWORD** on your Website for this *MediaLab* to view a striking example of adaptation by a plant species found in Hawaii. How do the different forms of these plants suit their environment?

Investigation 3
Estimated time for completion = 5 minutes

The structures of flowering plants are the culmination of millions of years of plant evolution. Features such as vascular tissue and enclosed seeds have appeared in various plant groups at various times in the evolutionary history of plants. Select the keyword **FOSSIL PLANTS** on your Website for this *MediaLab* to determine when these plant features first appeared in the fossil record.

Now that you've gained some understanding of plant structures and functions by doing the preceding activities, consider these questions.

Exercise 1
Estimated time for completion = 30 minutes

Use your observations from *Web Investigation 1* to construct a chart illustrating the differences between the architecture and anatomy of dicot and monocot plants. You could start with something like **Table 22.1**, but you should be able to greatly expand on the differences. To do this exercise most efficiently, you may want to systematically work your way from flowers to roots, or follow any other order you would prefer.

Exercise 2
Estimated time for completion = 30 minutes

Using the adaptations you observed of the silversword species to environments in the Hawaiian Islands, answer this question: How would you construct a plant to be well adapted to the following environment?

- *Poor-quality soil, frequently disturbed by rock slides*
- *Growing season about 8 months long*
- *Annual rainfall of about 18 inches, most of it during the first third of the growing season*
- *About 8 hours of sunlight at the start of the growing season; about 16 hours by the last half*

After deciding how to construct this plant so that it will be adapted to its environment, draw your hypothetical plant. Then, list how its adaptations help the plant cope with its environmental hardships.

Exercise 3
Estimated time for completion = 30 minutes

Using the information you obtained from *Web Investigation 3*, construct a graph illustrating when the various plant features first appeared in the fossil record. Your chart should extend from the time when the first land plants appeared to the present day.

25 The Integumentary, Skeletal, and Muscular Systems

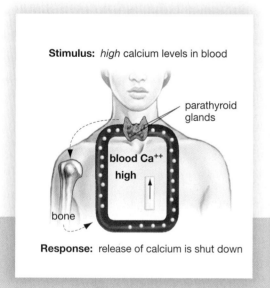

Negative feedback is key.
(Section 25.2, page 536)

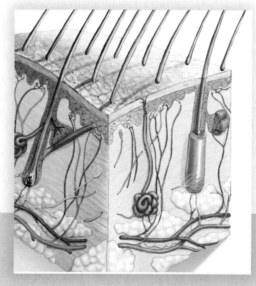

Our outer wrap.
(Section 25.7, page 543)

Some of our 206 bones.
(Section 25.8, page 549)

> Just as several individuals may make up an office, and several offices a department, so our bodies are organized into a number of small working units that go on to make up a functioning human being. Three of the larger units involve skin, bone, and muscle.

There is an old country saying that goes, "You don't miss your water until your well runs dry." Doesn't this seem applicable to the way we think of our bodies? We go along taking for granted our ability to eat or talk or run—until the day comes when we can't do one of these things. If you've ever, say, broken a leg, and ended up with a cast and crutches, you know that just walking around with ease can look like the most wonderful thing in the world. Broken bones heal, of course, and that reassures us—until we run up against a condition that won't heal. While still in his twenties, the actor Michael J. Fox woke up one morning and noticed, as he later remembered, that "the pinky in my left hand was twitching. It was a curiosity as much as anything because I couldn't stop it, no matter how I tried to." About a year later, after many doctor's visits, Fox got definitive word that he had Parkinson's disease, which is incurable. By then, it was not just his little finger that was twitching; it was both hands. Imagine if being able to hold your hands steady seemed as beautiful as the moon and just as far away.

We take our bodies for granted partly because they work so well most of the time, but partly because what goes on inside them is hidden from us and so complicated that our physical functioning just seems mysterious. The next six chapters of this book are aimed at clearing up some of this mystery. At their conclusion, you will have a basic understanding of how your heart beats, for example, and of how your immune system fights off invaders, and of how your eyes and brain provide you with a picture of the world.

If you think about just the first of these things—the workings of the heart—you can get a sense of what's in store. The human heart beats about 100,000 times *each day*, pumping about 8,800 quarts of blood through the body in the process. Moreover, it keeps this rhythmic contraction up without a second's rest from the third week of our embryonic life to the day we die. As if this weren't enough, it is carrying out this feat mostly with its original equipment. Very few of the muscle cells in the heart ever divide; as a result, the cells that are working for us at age 8 are pretty much the cells that are working for us at 80.

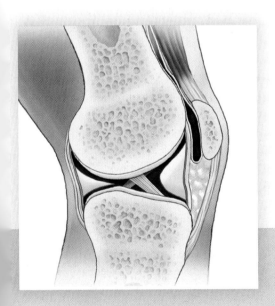

Comes with shock absorbers.
(Section 25.8, page 550)

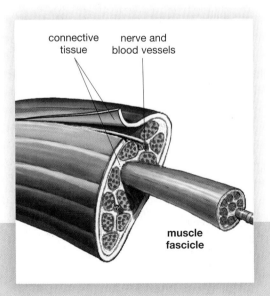

connective tissue

nerve and blood vessels

muscle fascicle

Muscles help us move.
(Section 25.9, page 550)

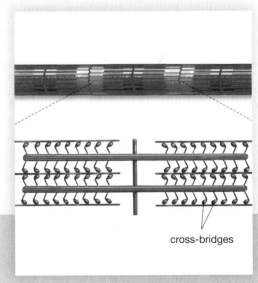

cross-bridges

How does a muscle contract?
(Section 25.9, page 551)

An engineer who could invent such a machine would be considered a genius. But the truly remarkable thing is that there are lots of organs and systems in the body that are just as impressive as the heart. Once you've learned how these things work, it's hard to take your body for granted in the same way, because you've come to understand what an incredible machine the human body is.

25.1 The Disciplines of Anatomy and Physiology

When we study the makeup or functioning of any complex organism, we have entered the realms of two overlapping disciplines in biology: anatomy and physiology. All organisms have an anatomy and a physiology. But in these chapters, our primary concern is with human beings. Within this framework, human anatomy is the study of the body's structure and the relationships between the body's constituent parts. Human physiology is the study of how these parts work. Thus anatomy tells us where the heart is, for example, how it is structured, and how it is situated in relation to other organs. Physiology describes to us *how* the heart works.

25.2 How Does the Body Regulate Itself?

Before we get to specifics about the human body, it will be helpful to go over a couple of general concepts about its structure and functioning. One of these concepts concerns the question: Who's in charge here? We are in charge of our conscious actions, of course; human beings get to decide what they are going to do from one moment to the next. But what about the huge number of unconscious actions that the body undertakes each second—getting oxygen to cells, getting rid of waste, being prompted to breathe while asleep? You might think the brain would be in charge of all this, and indeed

messages from the brain do prompt a lot of physical activity. But there are plenty of physical processes that don't involve brain signals, and the brain *gets* lots of messages that say, in effect, "get this process going now." Human society is hierarchical to a great degree; we're used to bosses and employees who work in a straight line of authority. But the body doesn't work this way. Instead, it functions through a process of *circular* causation in which A prompts B and B prompts C, but in which C may then prompt A. It is a system of self-regulation in which all parts work together to influence each other.

In understanding self-regulation, it's useful to think about what kind of internal environment the body needs in order to function properly. The answer is a *stable* environment. The human body must guard against being too hot or too cold; too dehydrated or too hydrated; too stimulated or too relaxed. If it is to exist at all, it must avoid extremes. To put this another way, it must seek **homeostasis**, meaning the maintenance of a relatively stable internal environment. And it turns out that a single process is almost solely responsible for bringing about homeostasis. Let's look at this process by way of an analogy.

Most people are aware of how a home heating system works. Falling temperature causes a thermostat to turn on a furnace. To look at this another way, there is a stimulus (cold air) that brings about a response (furnace operation). The *product* of this response is hot air. When enough hot air circulates to raise the temperature, the thermostat senses this and shuts the furnace down.

Now let's think of your body which has, as an example, a certain amount of calcium circulating through it in the bloodstream. When levels of calcium in the blood fall too low (stimulus), organs called the parathyroid glands sense this and secrete a hormone that causes your bones to release calcium (response). Thus, the product of *this* stimulus-response chain is released calcium; when enough of it is circulating, the parathyroid glands sense this and stop releasing their calcium-liberating hormone (**see Figure 25.1**).

With both the thermostat and the parathyroid glands, we can see that their responses to a stimulus bring about a decrease in their own activity. In other words, their responses feed back on their activity in a negative way—they reduce it. This is **negative feedback**, which can be defined as a process in which the elements that bring about a response have their activity reduced by that response. The calcium example is

Figure 25.1
Calcium Levels: An Example of Negative Feedback
Calcium levels in the bloodstream are regulated in part by the parathyroid glands. These glands can sense when calcium levels are too low (the stimulus in the drawing at left). In response, the parathyroids secrete a hormone that causes the bones to release calcium. This brings about high calcium levels, which the glands also can sense (the stimulus in the drawing at right). High calcium levels cause the parathyroids to halt their production of the calcium-liberating hormone.

Stimulus: *low* calcium levels in blood

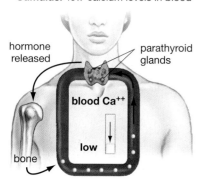

Response: release of calcium into blood

Stimulus: *high* calcium levels in blood

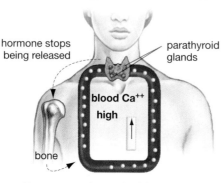

Response: release of calcium is shut down

only one of the countless negative feedback loops that exist in the body. In it, we can see not only how the body maintains homeostasis, but how the body is self-regulating. There was no central boss giving commands in this process. Instead two elements (calcium levels and the parathyroid glands) influenced each other to bring about stability. This is the way the body generally works.

Large-Scale Features of the Body

Many of the large-scale features of a human being—four limbs, one trunk, and so forth—are obvious, but a couple features are hiding in plain sight, we might say. The first of these is an internal body cavity. All animals more complex than a flatworm have such a cavity. Humans and other mammals actually have two primary body cavities. The first of these is a dorsal (back) cavity that is further divided into cranial and spinal cavities, into which our brain and spinal cord fit (**see Figure 25.2**). The second is a ventral (front) cavity that includes a thoracic or "chest" portion containing the lungs and heart as well as a lower abdominopelvic cavity containing such organs as the stomach, liver, and intestines. A muscular sheet called the diaphragm separates the thoracic and abdominopelvic cavities. Cavities provide flexibility and protection. With the space provided by a cavity, an organ like the stomach can expand as we eat and is protected from external blows to some extent.

Humans also have an internal skeleton that supports the body. We take for granted a skeleton that exists inside our skin, but it stands in contrast to what most animals have, which is an *external* skeleton (as with crabs and grasshoppers) or no skeleton at all (as with jellyfish).

25.3 Levels of Physical Organization

Features of the human body, such as a skeleton, fit into a larger organizational framework that is common to all living things. In our everyday world, several individuals may make up an office, several offices a department, and several departments a working company. In a similar fashion, each living thing is built up from a series of individual units that go on to make up a working organism. The organization of these units runs as follows. Recall from Chapter 4 that the basic unit of all life is the cell. Cells and cellular products that work together to perform a common function are known as **tissues**. There is, for example, muscle tissue—a group of cells that perform the common

function of contracting. Tissues in turn are arranged into various types of **organs**, meaning complexes of several kinds of tissues that perform a special bodily function. The stomach is an organ that has both muscle and nerve tissue within it. Organs then go on to form **organ systems**, which are groups of interrelated organs and tissues that serve a particular function. Contractions of the heart push blood into a network of blood vessels. The heart, blood, and blood vessels form the cardiovascular system, which is one of 11 systems in the human body.

Tissues, organs, and organ systems are three levels of organization within animals, but you may remember from Chapter 1 that the levels of organization in living things do not *start* with tissues (see page 12). Tissues are made of cells, cells have within them tiny working structures called organelles, and organelles are in turn made of the building blocks called molecules. At the bottom of this chain is the fundamental unit of matter, the atom. In this chapter, however, the focus is on the

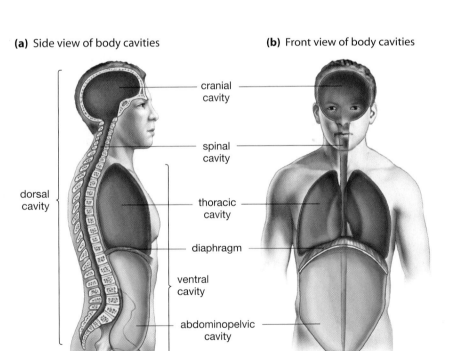

(a) Side view of body cavities   **(b)** Front view of body cavities

- cranial cavity
- spinal cavity
- dorsal cavity
- thoracic cavity
- diaphragm
- ventral cavity
- abdominopelvic cavity

Figure 25.2
Body Cavities

(a) A side or "lateral" view of the two main cavities in the human body—the dorsal (back) and ventral (front) cavities—and the smaller cavities within them. The dorsal body cavity is bounded by the bones of the skull and vertebral column and includes both the cranial and spinal cavities. The ventral cavity includes a thoracic (chest) cavity and an abdominopelvic cavity, which is separated from the thoracic cavity by a muscular diaphragm.

(b) A front or "ventral" view of the body's cavities.

levels of organization that run from tissues through organ systems. **Figure 25.3** shows the various levels of structural organization within the human body, using the cardiovascular system as an example.

25.4 The Human Body Has Four Basic Tissue Types

Looking at the smallest level of organization we're interested in, the tissue, we can then ask: what kinds of tissues are there in humans and other animals? It turns out there are only four fundamental tissue types: epithelial, connective, muscle, and nervous. All other tissue varieties are subsets of these fundamental types. Let's look briefly at the characteristics of these four varieties.

Epithelial Tissue

The key word in epithelial tissue is *surfaces*, because **epithelial tissue** is tissue that *covers* surfaces exposed to an external environment. The outside of our body is exposed to the external environment, and as such, the outer layer of our skin is composed of epithelial tissue. But there are other surfaces as well. Think of a pipe; it has not only an external surface but also an internal surface that comes into contact with something that is external to it—the fluid that runs through it. Similarly, your body has arteries and veins that have the fluid called blood running through them and the tissues that line the internal blood-vessel surfaces are epithelial tissues. Likewise, the lining of your stomach and the lining of your lungs are surfaces that come into contact with something from outside you; once again, this contact is made by epithelial tissues. Given their location, epithelial tissues often serve as a barrier, as in the case of our skin; but many epithelial tissues are "transport" tissues that materials are moved across. (The food that enters your bloodstream does so by being transported across the epithelial tissue that lines your small intestines.) Epithelial tissues can be several layers of cells thick, or as thin as a single layer, depending on their function (**see Figure 25.4**).

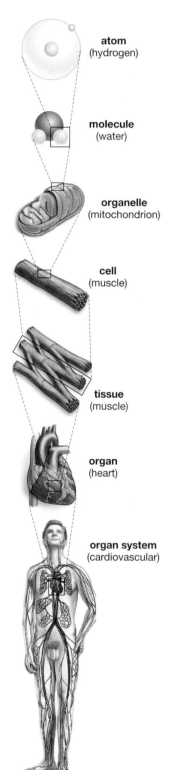

atom
(hydrogen)

molecule
(water)

organelle
(mitochondrion)

cell
(muscle)

tissue
(muscle)

organ
(heart)

organ system
(cardiovascular)

Figure 25.3
Levels of Organization in the Human Body
Atoms form molecules, which in turn combine to form organelles, which combine to form cells, such as heart muscle cells. A group of cells that performs a common function is a tissue; here cells have formed heart muscle tissue (whose common function is heart muscle contraction). Two or more tissues combine to form an organ such as the heart. The heart is one component of the cardiovascular system; other parts of the system are blood and blood vessels. All the organ systems combine to create an organism, in this case a human being.

Epithelial tissues often produce substances that aid in various bodily processes. The skin has epithelial tissue that produces the water-resistant protein keratin, for example, while the stomach has epithelial tissue that produces digestive juices. This substance-producing function can be carried out by isolated epithelial cells, but it is sometimes undertaken by concentrations of cells. Such concentrations are known as **glands**: organs or groups of cells specialized to secrete one or more substances. Later we'll look at two basic kinds of glands.

Connective Tissue

Connective tissue, defined as tissue that stabilizes and supports other tissue, is very different from epithelial tissue. Whereas epithelial tissue is always in contact with an external environment, connective tissue never is. Whereas epithelial tissue is almost completely composed of cells, connective tissue usually is composed of cells that are separated from each other by an extracellular material—a material that is secreted by the connective tissue cells themselves. Indeed, the prime function of many connective tissues is to produce this kind of material.

Bone and cartilage are clear examples of connective tissue. They are connecting something, after all (bodily structures), and their support and stabilization role is clear. They likewise display another hallmark of connective tissue, which is that their cells lie within a so-called *ground substance*. The ground substance of bone, which the bone cells themselves secrete, happens to be a mix of closely packed protein fibers and calcified material. These are the substances that largely make up the hard bones we're familiar with.

But ground substances need not be hard. Adipose or fat tissue is connective tissue, and its ground substance is soft. Indeed, a ground substance can even be fluid, as is the case with blood and a bodily fluid called lymph. How can a ground tissue be a fluid? Well, to take blood as an example, the ground substance that surrounds blood cells is mostly water, with electrolytes, proteins, and other materials in the mix.

Muscle Tissue

Muscle tissue is tissue that is specialized in its ability to contract, or shorten. There are three kinds of muscle tissue—skeletal, smooth, and cardiac—which differ in their structure and in the way they are prompted to contract. **Skeletal muscle** is the ordinary muscle that is attached to bone and is contained in, for example, our biceps. It is under our conscious control and has a striped or "striated" appearance when looked at under a microscope, owing to the parallel orientation of the

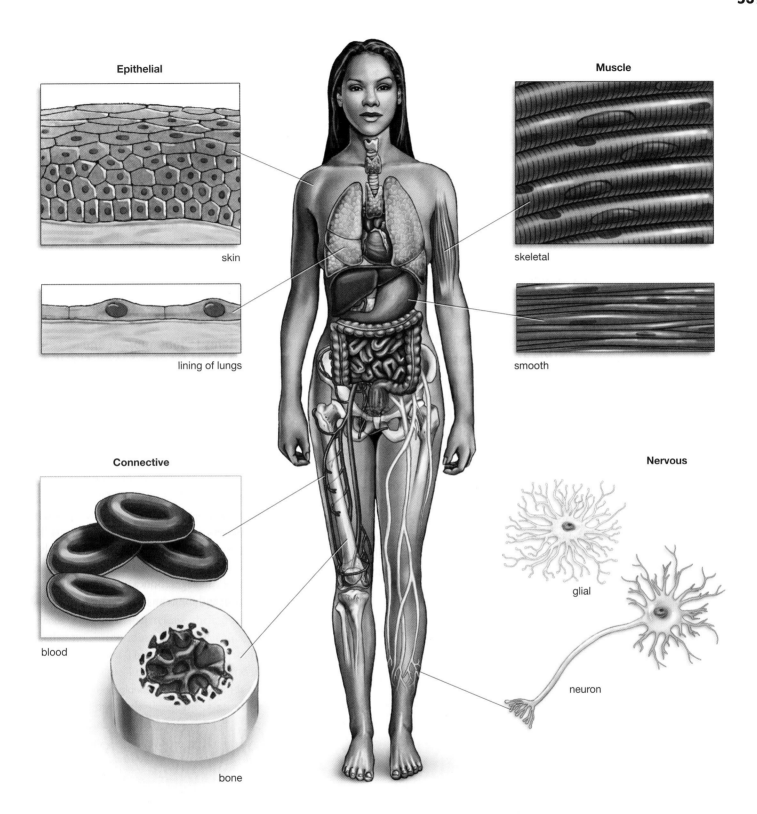

Epithelial

skin

lining of lungs

Muscle

skeletal

smooth

Connective

blood

bone

Nervous

glial

neuron

Figure 25.4
Four Types of Tissue
There are four basic tissue types in the human body: epithelial, connective, muscle, and nervous. Pictured are examples of each of the four types in some representative locations. Epithelial tissue, which always covers a surface exposed to an external environment, can take the form of many layers of cells (as with the skin) or a single layer of cells (as with the lining of lungs). Connective tissue, which supports and stabilizes other tissue, can take such diverse forms as bone and blood. Muscle tissue, which is specialized for contraction, comes in a striped or "striated" form (in both skeletal and cardiac muscle) and in a smooth form that lines the digestive tract, blood vessels, and other structures. Nervous tissue is specialized for electrical signal transmission. The two types of nerve cells are neurons, which conduct nervous system signals, and glial cells, which support the neurons.

(g) The cardiovascular system

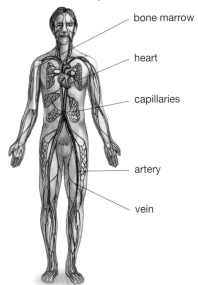

- bone marrow
- heart
- capillaries
- artery
- vein

(h) The respiratory system

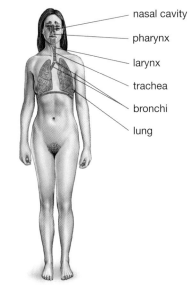

- nasal cavity
- pharynx
- larynx
- trachea
- bronchi
- lung

(i) The digestive system

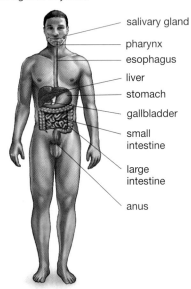

- salivary gland
- pharynx
- esophagus
- liver
- stomach
- gallbladder
- small intestine
- large intestine
- anus

(j) The urinary system

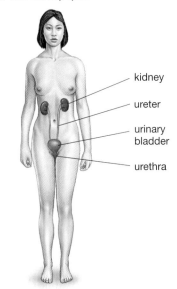

- kidney
- ureter
- urinary bladder
- urethra

(k) The male reproductive system

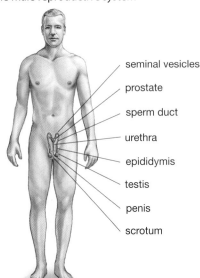

- seminal vesicles
- prostate
- sperm duct
- urethra
- epididymis
- testis
- penis
- scrotum

(l) The female reproductive system

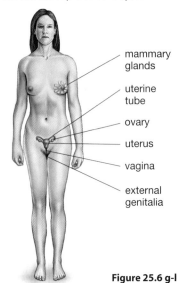

- mammary glands
- uterine tube
- ovary
- uterus
- vagina
- external genitalia

Figure 25.6 g-l

as it passes through the system in an effort to identify invading microorganisms. Lymphoid organs include the well-known lymph nodes located under the arms and elsewhere, along with the thymus gland, tonsils, and spleen.

Organ Systems 3: Transport and Exchange with the Environment—The Cardiovascular, Respiratory, Digestive, and Urinary Systems

- The **cardiovascular system** (**Figure 25.6g**) consists of the heart, blood, and blood vessels and an inner, "marrow" portion of bones where red blood cells are formed. The cardiovascular system is the body's mass transit system. It carries nutrients, dissolved gases, and hormones *to* tissues throughout the body, and it carries waste products *from* tissues to the sites where they are removed: the kidneys and lungs.

- The **respiratory system** (**Figure 25.6h**) includes the lungs and the passageways that carry air to and from the lungs. Through this system, oxygen comes into the body and carbon dioxide is expelled from it.

- The central feature of the **digestive system** (**Figure 25.6i**) is the digestive tract, a long tube that begins at the mouth and ends at the anus. Along its course, there are the digestive organs of the system—the stomach, small intestines, and large intestines—and accessory glands, such as the pancreas and liver.

- The **urinary system** (**Figure 25.6j**) has two major functions, one of them well known, the other not so much appreciated. The well-known function is the elimination of waste products from the blood through the production of urine. The underappreciated function is conservation. The system is very good at retaining what is useful to the body—water, proteins, and so forth—even as it eliminates what is useless. The primary organs that carry this selection process out are the kidneys, but the system also includes various other parts, such as the bladder.

- Males and females have different **reproductive systems** (**Figures 25.6k and 25.6l**). The two are linked, of course, in that together they produce offspring. Both systems are discussed in detail in Chapter 30.

With this brief review of the organ systems completed, you're ready to look in more detail at the first three organ systems noted earlier—the integumentary, skeletal, and muscular systems. They are batched together in this chapter because each of them has to do with body support and movement. We'll start with the integumentary system.

25.7 The Integumentary System: Skin and Its Accessories

The primary component of the integumentary system is the organ known as skin. It may seem strange to think of skin as an organ, since it is not a compact, clearly defined structure like the heart or the liver. But remember the definition of an organ: a complex of several kinds of tissues that performs a special bodily function. Clearly, that's what skin is. Joining skin to make up the integumentary system are the related structures of hair, nails, and a variety of exocrine glands.

The primary function of skin is easy to see: It covers the body and seals it off from the outside world. Skin also protects underlying tissues and organs, however, and helps regulate our temperature. In addition, it stores fat and makes vitamin D. Meanwhile, exocrine glands that are accessories to the skin excrete materials such as sweat, water, oils, and milk; and specialized nerve endings in the skin detect such sensations as pressure, pain, and temperature.

The Structure of Skin

Skin is an organ that is organized in two parts: a thin outer covering, the epidermis, and a thicker underlying layer, the dermis, that is composed mostly of connective tissue. Beneath the dermis—and not part of the skin proper—is another layer, called the hypodermis, made up of connective tissue that attaches the skin to deeper structures such as muscles or bones. You can see a cross section of human skin in **Figure 25.7**.

The Outermost Layer of Skin, the Epidermis

The outer layer of our skin, the epithelial tissue called the **epidermis**, would seem to be facing an impossible task. On the one hand, it has to serve as a permanent, protective barrier, keeping out everything from water to invading microorganisms; on the other it has to continually renew itself, given that it is cut, scraped, and simply worn away. How does it manage to do both things? The answer lies in the many *layers* of cells the epidermis is organized into. If we could take a micro-elevator down to the innermost layer of the epidermis, we would see a group of rounded cells rapidly dividing and, in the process, pushing the cells on top of them *up*, toward the surface of the skin. As we move toward the surface, we notice that the epidermal cells are flattening out. Furthermore, they're getting less active; at a certain point they cease dividing altogether. What is happening is that they are now so far from the underlying blood vessels (down in the dermis) that they are losing their blood supply. They are still carrying out a task, however, which is the production of **keratin**—a flexible, water-resistant protein, abundant in the outer layers of skin, that also makes up hair and fingernails. By the time epidermal cells

Figure 25.7
Our Outer Wrap: What Is Skin Made Of?

(a) A micrograph of human skin, showing the epidermal and dermal layers. Note how the cells of the epidermis flatten out about half-way toward the top.

(b) A cross-section of the skin's epidermis and dermis layers, and the hypodermis layer beneath them (which is not part of the skin).

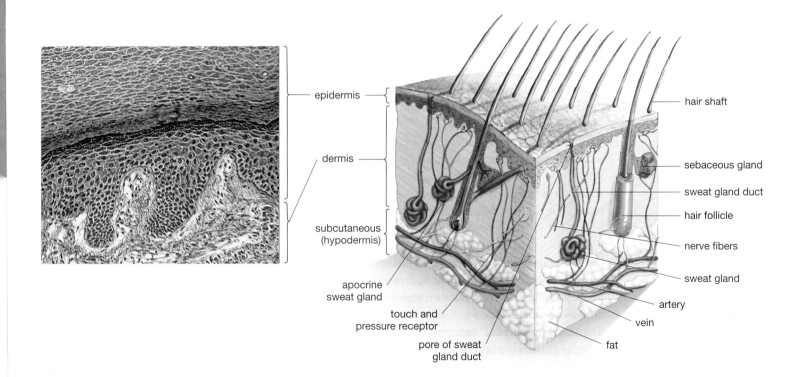

epidermis

dermis

subcutaneous (hypodermis)

apocrine sweat gland

touch and pressure receptor

pore of sweat gland duct

hair shaft

sebaceous gland

sweat gland duct

hair follicle

nerve fibers

sweat gland

artery

vein

fat

which produce an oily substance called sebum that is secreted into hair follicles and that then moves through them onto the skin (see Figure 25.7). Once on the skin, sebum lubricates the hair and inhibits bacterial growth in the surrounding area.

The problem with this process is that, under hormonal influence, sebaceous glands can be *too* active—they can produce too much sebum, particularly in teenagers. When this excess sebum mixes with skin cells that are being shed from the sebaceous follicle, the result can be a blocked follicle that swells with sebum—a whitehead. Once this blockage occurs, bacteria normally found on the skin begin to multiply in the follicle, producing irritating substances that can cause general inflammation and pimples. Taken together, these various skin afflictions are known as acne. No matter what you may have heard, acne is not caused by eating fatty foods, nor is it caused by stress, nor by a lack of face scrubbing. As with baldness, the hormones called androgens are the culprits. They prompt the increased sebaceous activity; since males at puberty have more androgens in them than females, it is males who suffer more from acne.

The second type of exocrine gland in the skin is the **sweat gland**, simply defined as a gland that produces the fluid sweat. This can mean a couple of different things, however, because sweat glands come in two varieties. One type is a *merocrine* (or *eccrine*) gland that produces the clear, saltwater-like fluid we normally think of as sweat. We have, at a minimum, 2 million merocrine sweat glands on our bodies. They exist on almost all skin surfaces, but we have an especially dense concentration of them on our feet, forehead, and palms (hence "sweaty palms"). The primary function of the merocrine sweat glands is to bring sweat to the surface of the skin, so that it can

cool us by evaporating. These glands have their own ducts and follicles, which begin deep in the dermis.

The other sweat glands in the body, called *apocrine* glands, are very different from the merocrine glands. First, they only exist in a few places in the body—notably the groin, the anal area, and the armpits. As such, they have little effect on body cooling. Second, they secrete their products into hair follicles, not up through their own ducts. Most important, apocrine gland secretions contain not only regular sweat, but a fatty, milky fluid whose function in humans is unknown. The *effect* of this fluid is very well known, however. Bacteria feed on it, and the resulting bacterial byproducts are the cause of body odor. Apocrine glands only begin functioning at puberty, which is why children seldom have body odor.

Nails

Our final accessory structure of the integumentary system is the nails found on our fingers and toes. Like hair, nails are made of a variety of the protein keratin. They are clearly useful for protection and for prying, scratching, and picking up objects. Nails allow us to look, in a sense, at the blood vessels underlying the dermis, because it's the reddish color of these vessels that gives nails their pink hue.

25.8 The Skeletal System

We now shift gears, going from a consideration of our outer covering to a consideration of the framework that underlies it—our skeletal system. This system has only three structural elements to it: bone, ligaments, and cartilage. Most of us are quite familiar with **bone**, which can be defined as a connective tissue that provides support and storage capacity, and that is the site of blood cell production. Ligaments are fairly inaccessible to us; but if you look at **Figure 25.8**, you can see some typical ligaments carrying out a typical task—linking bones together. You can also see the third element of the skeletal system, cartilage. Note that cartilage is located *in between* bones in the foot, and this turns out to be one of its common functions. **Cartilage** is a connective tissue that serves as padding in most joints. It is more familiar to us, however, in other parts of the body. It forms our larynx (voice box), the C-shaped "rings" on our trachea (windpipe), our nose-tip and outer ears, and it links each of our ribs to our breastbone.

It's a good idea to look at Figure 25.8 to get a sense not only of the difference between bone, ligaments, and cartilage but also of the difference between all of these skeletal elements and another type

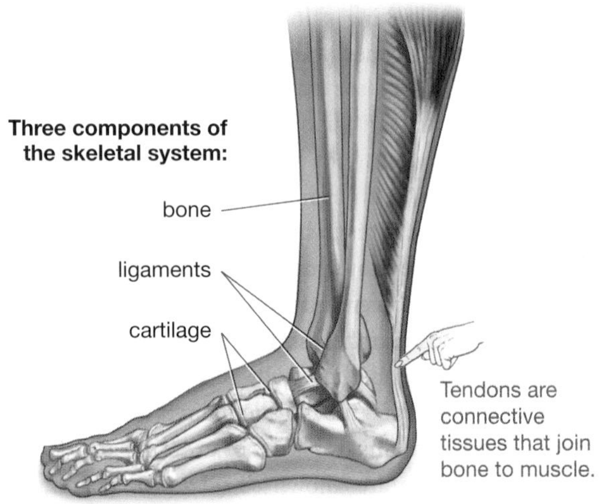

Three components of the skeletal system:

bone

ligaments

cartilage

Tendons are connective tissues that join bone to muscle.

**Figure 25.8
Elements of the Skeletal System**
Bones are the basic element of the skeletal system. Ligaments link one bone to another, and cartilage often serves as padding between bones. Tendons, while not part of the skeletal system, always link bone to muscle.

25.7 The Integumentary System: Skin and Its Accessories

The primary component of the integumentary system is the organ known as skin. It may seem strange to think of skin as an organ, since it is not a compact, clearly defined structure like the heart or the liver. But remember the definition of an organ: a complex of several kinds of tissues that performs a special bodily function. Clearly, that's what skin is. Joining skin to make up the integumentary system are the related structures of hair, nails, and a variety of exocrine glands.

The primary function of skin is easy to see: It covers the body and seals it off from the outside world. Skin also protects underlying tissues and organs, however, and helps regulate our temperature. In addition, it stores fat and makes vitamin D. Meanwhile, exocrine glands that are accessories to the skin excrete materials such as sweat, water, oils, and milk; and specialized nerve endings in the skin detect such sensations as pressure, pain, and temperature.

The Structure of Skin

Skin is an organ that is organized in two parts: a thin outer covering, the epidermis, and a thicker underlying layer, the dermis, that is composed mostly of connective tissue. Beneath the dermis—and not part of the skin proper—is another layer, called the hypodermis, made up of connective tissue that attaches the skin to deeper structures such as muscles or bones. You can see a cross section of human skin in **Figure 25.7**.

The Outermost Layer of Skin, the Epidermis

The outer layer of our skin, the epithelial tissue called the **epidermis**, would seem to be facing an impossible task. On the one hand, it has to serve as a permanent, protective barrier, keeping out everything from water to invading microorganisms; on the other it has to continually renew itself, given that it is cut, scraped, and simply worn away. How does it manage to do both things? The answer lies in the many *layers* of cells the epidermis is organized into. If we could take a micro-elevator down to the innermost layer of the epidermis, we would see a group of rounded cells rapidly dividing and, in the process, pushing the cells on top of them *up*, toward the surface of the skin. As we move toward the surface, we notice that the epidermal cells are flattening out. Furthermore, they're getting less active; at a certain point they cease dividing altogether. What is happening is that they are now so far from the underlying blood vessels (down in the dermis) that they are losing their blood supply. They are still carrying out a task, however, which is the production of **keratin**—a flexible, water-resistant protein, abundant in the outer layers of skin, that also makes up hair and fingernails. By the time epidermal cells

Figure 25.7
Our Outer Wrap: What Is Skin Made Of?

(a) A micrograph of human skin, showing the epidermal and dermal layers. Note how the cells of the epidermis flatten out about half-way toward the top.

(b) A cross-section of the skin's epidermis and dermis layers, and the hypodermis layer beneath them (which is not part of the skin).

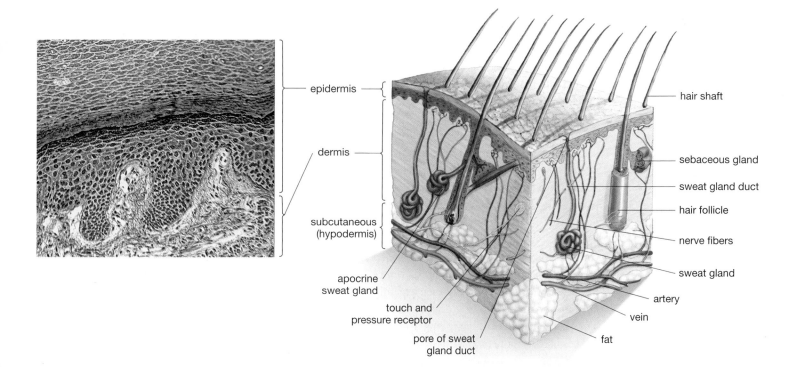

epidermis

dermis

subcutaneous (hypodermis)

apocrine sweat gland

touch and pressure receptor

pore of sweat gland duct

hair shaft

sebaceous gland

sweat gland duct

hair follicle

nerve fibers

sweat gland

artery

vein

fat

reach the surface, they are scarcely cells at all anymore. They are dead and don't even have the remnants of internal organelles within them. Instead, they have become a series of tightly interlocked keratin sacs. In another two weeks they will be scraped or washed away, replaced by the next cells that have pushed up from the inner epidermis.

Beneath the Epidermis: The Dermis and Hypodermis

Whereas the epidermis is epithelial tissue, the inner layer of skin, the **dermis,** is mostly connective tissue. But if you look again at Figure 25.7, you can see that nerves and muscles exist in the dermis as well, along with sweat glands, hair follicles, and more. This is just another example of how all the various kinds of tissues tend to exist side by side in a typical organ.

Many of the structures in the dermis are serving the epidermis. For example, the nervous system touch and pressure receptors shown in Figure 25.7 allow us to know when something makes contact with our skin, and the blood vessels pictured provide a blood supply to the dividing lower cells of the epidermis. But the dermal blood supply takes on another role as well. It plays a big part in controlling our temperature. If you get too cold, the dermal blood vessels will constrict, thus moving blood away from your surface and retaining the heat that's in the blood. If you get too hot, the dermal blood vessels expand, thereby moving blood toward your surface so that the heat that's in blood can radiate to the outside world (making your skin feel "flushed"). One more thing to note about the dermis is that its cells are relatively fixed, unlike those of the outer epidermis. Ever wonder why a tattoo must be made with a needle sunk into the skin? The tattoo's pigment must be deposited into the fixed cells in the dermis, rather than the ever-changing cells of the epidermis.

As noted earlier, the "hypodermis," the layer beneath the dermis, is not actually part of the skin. But it deserves special mention here because of one of the aspects of it you can see in Figure 25.7—its abundant fat cells. The body has fat cells in other places, but when we talk about how much "fat" we have, what we basically mean is how much fat we have in the hypodermis. This is a layer of fat whose composition changes throughout the life span. It is relatively thick when we are infants, then gets thinner when we are children, and then expands steadily around our mid-sections and elsewhere as we age. The term "hypodermis" may ring a bell because it sounds like the more familiar term hypoder*mic,* as in needles. The fatty tissue of the hypodermis doesn't have much of a blood supply and thus is a good place to inject medicines that need a relatively slow, steady entry into circulation.

Here's one final thing about the skin. If you want to make it age fast, one sure way to do it is to spend a lot of time in the sun. You can read more about this in "There Is No Such Thing as a Fabulous Tan" on the next page.

Accessory Structures of the Integumentary System

As noted, skin has a number of accessory structures associated with it, including glands, nails, and hair.

Hair

Five million or so hairs exist on the human body, though only about 100,000 of them are on the human head. Hair clearly serves a major cosmetic role in modern society, but why does it exist on human beings in the first place? You might think that heat retention for our ancestors would have been its main function, but maybe not. Rub your hand with just the lightest stroke over the hair on your head and you can get a sense of how exquisitely sensitive hair is to our sense of *touch.* Like a cat's whiskers, hair may have served primarily to fine-tune our sense of contact with the world around us.

How can hair have this quality since it is not alive, but merely a shaft of the protein keratin, which is a product of underlying cells? If you look at Figure 25.7, you can see that each hair grows out of a tube-like structure called a hair follicle that begins fairly deep in the dermis. Each follicle is a tissue complex that has not only its own blood vessels and muscles, but its own nerve endings as well. Thus, when you touch a hair, you stimulate a hair follicle's nerve endings.

True to what we noted at the start of the chapter, the body's hair is perhaps more missed when it is gone than appreciated when it is here. Many men start losing the hair on their heads in their early twenties, a trend that accelerates as they get older. Women can lose hair too, though their loss tends to come later in life and usually involves a general thinning of the hair, rather than receding hairlines or bald spots. What causes hair loss?

Follicles go through growth and resting phases, with a given hair on our heads growing for anywhere from two to six years until a short resting phase is reached. At that point the hair drops out, and its follicle's growth phase starts again. Hair loss generally is a condition in which an ever-growing proportion of follicles start to shrink, resulting in a

There Is No Such Thing as a Fabulous Tan

The French clothing designer Gabriel Chanel once remarked that "fashion changes, but style remains." Decades ago, however, Chanel herself helped launch a fashion that seems stubbornly resistant to change. The fashion is that of getting a suntan, but this practice might as well be called the fashion of damaging your skin so that in a couple of decades it will be saggy, wrinkled, and spotted.

Throughout the early part of the twentieth century, it was chic to have pale skin, rather than a tan, as a tan was a sign of someone who *had* to be outdoors—working a job. According to one popular account, Gabriel Chanel (of Chanel No. 5 fame) got some sun on an ocean cruise in the 1920s, stepped off the boat tanned, and conveyed the message that it was chic to look that way. People in industrial countries were ready to hear this message because it came at a moment when their leisure time was increasing, meaning they had more opportunities to be out in the sun playing, rather than working.

These days, the constant message from health professionals is to *avoid* the sun, but this is a message society has not really taken to heart. Consider that in a British survey done in 2001, more than 70 percent of the respondents said that people who get tanned are taking a health risk. So far, so good; but this same percentage of respondents believed that "people with sun-tans look healthy." There are an estimated 15,000 businesses in the United States devoted solely to *giving* people tans through the same exposure to ultraviolet light that the sun delivers.

But in what ways does sun exposure harm the skin? The most serious thing it does is help cause skin cancer—three varieties of it, actually—but let's set this aside and focus here on what it does to the appearance of the skin.

The clearest way to get a sense of the cosmetic effects of the sun is to put your forearm up against that of a fair-skinned 50-year-old who has spent a lot of time outdoors. By some estimates, 90 percent of the difference you will see is attributable to sun damage. See the brown marks commonly known as "liver spots?" They don't have anything at all to do with the liver; they're purely caused by sun exposure. How about the white spots that seem to be patches of skin where the pigment has just given up, leaving small splotches of pure white behind? Sun exposure again. Now what about the sagging that can be seen on the arm and the wrinkling on the hand? It turns out that just below the surface of the skin, in the layer called the dermis, there is a protein called elastin that does pretty much what it sounds like: It springs back to its original shape after having been bent or twisted in some way. Unfortunately, ultraviolet light breaks down elastin, so that after enough sun exposure, the skin loses that full, robust look—it sags and wrinkles. These aging effects are so well understood among dermatologists that they simply refer to sun-caused skin changes as "photo-aging."

But what about sunblocks? They are indeed recommended by dermatologists, with the understanding that they need to be SPF 15 or higher, applied in nice thick layers, and reapplied after activities such as swimming. And there is general agreement that they can block most of the harmful effects of the sun. There is a debate going on within the research community, however, about whether sun blocks are lulling people into a false sense of security about how much sun damage they are getting. The concern is that people will stay out in the sun until they get the general "flushed" feeling telling them they have had enough. It takes longer to get this feeling when sunscreen has been applied, but the critical thing is that sun damage has occurred *whenever* this point has been reached. What difference does it make if a person blocks the sun's rays for three hours before allowing them in? The general advice of dermatologists is simply to avoid lengthy exposure to direct sunlight, particularly between the hours of 10 a.m. and 2 p.m. Sunblocks and covering up help, but they have their limitations. If you avoid the sun, you may not end up with the most fashionable-looking skin over the next two weeks but over the next two decades, you will be the clear winner.

briefer growing phase and hair that comes out either short and fine (like a baby's) or doesn't come out at all. Genes clearly play a part in this process although, contrary to popular belief, it is not just genes from the mother's side of the family that are involved. Hormones called androgens clearly are involved as well; men have higher levels of androgens circulating in their bodies than women do, and thus experience more hair loss.

Glands

You may remember that when we went over the endocrine system, you saw that glands come in two varieties: those that secrete their substances directly into the bloodstream or surrounding tissue (endocrine glands) and those that secrete their substances through the tubes called ducts (exocrine glands). Skin contains two types of exocrine glands. The first of these are known as **sebaceous glands,** which produce an oily substance called sebum that is secreted into hair follicles and that then moves through them onto the skin (see Figure 25.7). Once on the skin, sebum lubricates the hair and inhibits bacterial growth in the surrounding area.

The problem with this process is that, under hormonal influence, sebaceous glands can be *too* active—they can produce too much sebum, particularly in teenagers. When this excess sebum mixes with skin cells that are being shed from the sebaceous follicle, the result can be a blocked follicle that swells with sebum—a whitehead. Once this blockage occurs, bacteria normally found on the skin begin to multiply in the follicle, producing irritating substances that can cause general inflammation and pimples. Taken together, these various skin afflictions are known as acne. No matter what you may have heard, acne is not caused by eating fatty foods, nor is it caused by stress, nor by a lack of face scrubbing. As with baldness, the hormones called androgens are the culprits. They prompt the increased sebaceous activity; since males at puberty have more androgens in them than females, it is males who suffer more from acne.

The second type of exocrine gland in the skin is the **sweat gland**, simply defined as a gland that produces the fluid sweat. This can mean a couple of different things, however, because sweat glands come in two varieties. One type is a *merocrine* (or eccrine) gland that produces the clear, saltwater-like fluid we normally think of as sweat. We have, at a minimum, 2 million merocrine sweat glands on our bodies. They exist on almost all skin surfaces, but we have an especially dense concentration of them on our feet, forehead, and palms (hence "sweaty palms"). The primary function of the merocrine sweat glands is to bring sweat to the surface of the skin, so that it can

cool us by evaporating. These glands have their own ducts and follicles, which begin deep in the dermis.

The other sweat glands in the body, called *apocrine* glands, are very different from the merocrine glands. First, they only exist in a few places in the body—notably the groin, the anal area, and the armpits. As such, they have little effect on body cooling. Second, they secrete their products into hair follicles, not up through their own ducts. Most important, apocrine gland secretions contain not only regular sweat, but a fatty, milky fluid whose function in humans is unknown. The *effect* of this fluid is very well known, however. Bacteria feed on it, and the resulting bacterial byproducts are the cause of body odor. Apocrine glands only begin functioning at puberty, which is why children seldom have body odor.

Nails

Our final accessory structure of the integumentary system is the nails found on our fingers and toes. Like hair, nails are made of a variety of the protein keratin. They are clearly useful for protection and for prying, scratching, and picking up objects. Nails allow us to look, in a sense, at the blood vessels underlying the dermis, because it's the reddish color of these vessels that gives nails their pink hue.

25.8 The Skeletal System

We now shift gears, going from a consideration of our outer covering to a consideration of the framework that underlies it—our skeletal system. This system has only three structural elements to it: bone, ligaments, and cartilage. Most of us are quite familiar with **bone**, which can be defined as a connective tissue that provides support and storage capacity, and that is the site of blood cell production. Ligaments are fairly inaccessible to us; but if you look at **Figure 25.8**, you can see some typical ligaments carrying out a typical task—linking bones together. You can also see the third element of the skeletal system, cartilage. Note that cartilage is located *in between* bones in the foot, and this turns out to be one of its common functions. **Cartilage** is a connective tissue that serves as padding in most joints. It is more familiar to us, however, in other parts of the body. It forms our larynx (voice box), the C-shaped "rings" on our trachea (windpipe), our nose-tip and outer ears, and it links each of our ribs to our breastbone.

It's a good idea to look at Figure 25.8 to get a sense not only of the difference between bone, ligaments, and cartilage but also of the difference between all of these skeletal elements and another type

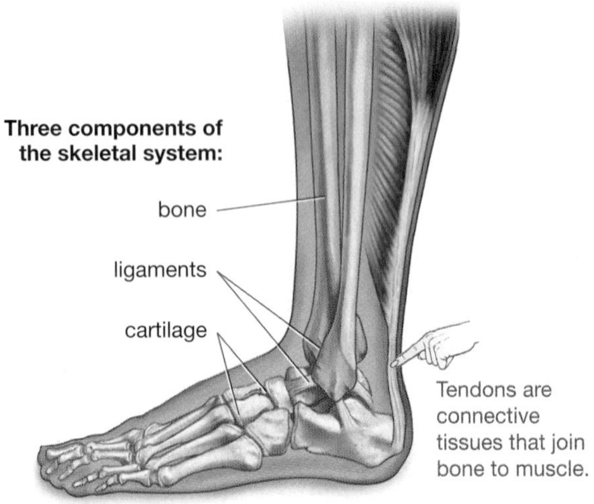

Three components of the skeletal system:

bone

ligaments

cartilage

Tendons are connective tissues that join bone to muscle.

Figure 25.8
Elements of the Skeletal System
Bones are the basic element of the skeletal system. Ligaments link one bone to another, and cartilage often serves as padding between bones. Tendons, while not part of the skeletal system, always link bone to muscle.

of tissue that is pictured, tendons. Though not part of the skeletal system, tendons are always associated with the bones of the system. Looking at all four elements in the figure, here's one way to think of them. Bones are the basic framework; cartilage often serves as padding between bones; **ligaments** are tissues that join bone to bone; and **tendons** are tissues that join bone to muscle.

Function and Structure of Bones

It's tempting to think of bones as being nothing more than support beams, but this conception is wide of the mark in two ways. First, support beams are not alive, but bone is very much a living, dynamic tissue, as you'll see. Second, bones do more than just support us. Once we are past infancy, all of our blood cells (both red and white) are created in the interior of our bones. Some bones store fat, in the form of yellow marrow, as a kind of energy reserve, and bones serve as the storage sites for important minerals such as calcium and phosphate. Beyond this, the only reason we can move at all is that our muscles are attached to bones. When we contract our bicep, for example, our forearm is raised by means of the bicep pulling on the bones of the forearm.

The Structure of Bone

Each bone in our body actually is considered a separate organ, because each bone is composed of several kinds of tissue. There is the calcified or "osseous" tissue we think of as bone, to be sure, but each bone also has its own blood vessels and nerves. (Cartilage, meanwhile, is composed of a single kind of protein fiber—a collagen that can be up to 80 percent water—and it has no nerves and no blood vessels.)

Bones are a great example of connective tissue, in that the *cells* in bones may account for as little as 2 percent of bone mass. What is the other 98 percent of bone made of? It is overwhelmingly a ground substance, secreted by bone cells, that is composed of hard, calcium-containing crystals and tough, yet flexible collagen fibers. This combination means that bones are hard without being brittle—they can support our weight and yet can bend or twist a little without breaking.

Large-Scale Features of Bone

The typical features of a long bone such as the humerus (in the arm) are shown in **Figure 25.9**. A long bone has a central shaft, or diaphysis, and expanded ends, or epiphyses (singular, epiphysis). There are two types of bone. **Compact bone**, which

Figure 25.9
Large- and Small-Scale Features of a Typical Bone
Note that in (c), the building units of compact bone, the osteons, run parallel to the long axis of the bone. Spongy bone does not contain osteons, but instead is composed of an open network of calcified rods or plates.

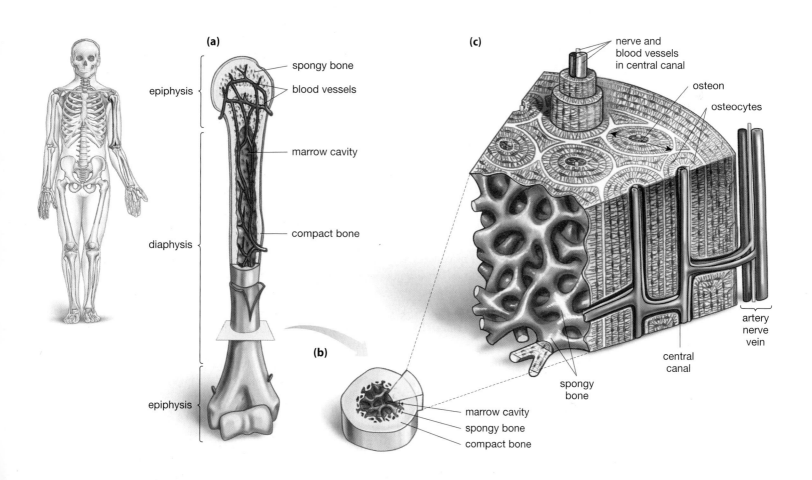

is relatively solid, forms the outer portion of the bone (all the way around it, as you can see in Figure 25.9b). **Spongy bone** then fills the epiphyses, and is well named. It is porous enough that it contains another type of tissue, called marrow, that comes in two forms: **red marrow**, which gives rise to blood cells; and **yellow marrow**, made up of energy-storing fat cells. Yellow marrow also exists in the so-called *marrow cavity* of a long bone, which you can see in Figure 25.9.

Small-Scale Features of Bone

Dropping down to the microscopic level and looking at bone *cells*, we find that three different types are involved in the growth and maintenance of bone. **Osteoblasts** are immature bone cells that are responsible for the production of new bone. They secrete the material that becomes the bone ground substance. Once osteoblasts are surrounded by this material, they reduce their production of it, and this marks their transition to osteocytes, the second type of bone cell. **Osteocytes** are mature bone cells; they *maintain* the structure and density of normal bone by continually recycling the calcium compounds around themselves. The third type of bone cell, the osteoclasts, could be thought of as the demolition team of bone tissue. **Osteoclasts** are cells that move along the outside of bones, releasing enzymes that eat away at bone tissue, thus liberating minerals stored in the bone. This may sound harmful, but it actually is essential for bone growth and for the regulation of blood levels of calcium. (This is the calcium-releasing function we reviewed at the start of the chapter.) The upshot of all this is that, at any given moment, osteoblasts are adding to bone, osteocytes are maintaining it, and osteoclasts are removing it.

Within compact bone, the basic functional unit is the osteon seen in Figure 25.9c (which is also known as the Haversian system). The essence of bone growth is that, within each osteon, osteoblasts produce layer after layer of concentric cylinders of bone—much like layers of insulation wrapped around a pipe. These layers surround what could be thought of as the hole in the pipe, which is a central canal (also known as a Haversian canal). These canals parallel the long axis of the bone and have blood vessels and nerves running through them—a kind of support system for the bone.

Unlike compact bone, spongy bone has no osteons. It consists of an open network of interconnecting calcified rods or plates. The red and yellow marrow fills the spaces within this network.

Practical Consequences of Bone Dynamics

Now, let's think about the practical consequences of this structure. Remember the earlier observation that bone is a living tissue? With its three kinds of cells constantly adding, maintaining, and breaking down the tissue, you can see the truth of this. But bone is more dynamic than even this would indicate, in that bones are responsive to our activities. As physiologist Frederic Martini has noted, when you undertake an exercise like jogging, which stresses your bones, the calcium-containing crystals of the affected bones create small electrical fields that apparently attract osteoblasts. Upon arrival, these osteoblasts begin producing bone. The result? If you take up jogging, you get denser leg bones. But there's a flip side to this coin: People who are using crutches to take the weight off a broken leg bone may temporarily lose up to a third of the mass in this bone because it is now scarcely being stressed at all.

Interestingly, though we develop almost all our height by late adolescence, we are still gaining in bone density up until about the age of 30. Then, beginning in middle age, the body may start removing more bone than it adds. This phenomenon particularly afflicts women, in the form of a condition known as *osteoporosis*, meaning a thinning of bone tissue that can result in bone breakage and deformation. One way we can guard against osteoporosis is to reach the age of 30 with as much bone density as we can muster—the equivalent of putting money in the bank as a hedge against a later withdrawal. And there are a couple of proven ways to get more bone density before 30: Take up moderate programs of "weight-bearing" exercise such as jogging or walking, and modestly increase calcium intake.

The Human Skeleton

Taking a step back from the microscopic world, you can see in **Figure 25.10** that there are a lot of bones in the skeletal system—206 of them in all. (Only the major bones are visible in the figure.) There are two main divisions to the skeletal system: the **axial skeleton**, whose 80 bones include the skull, the vertebral column, and the rib cage that attaches to it; and the **appendicular skeleton**, whose 126 bones include those of our paired appendages—the arms and the legs, along with the pelvic and pectoral "girdles" to which they are attached. Also note the blue-colored tissue joining ribs to our breastbone and lying in between each of the vertebrae. These are but a few of the cartilages in the body.

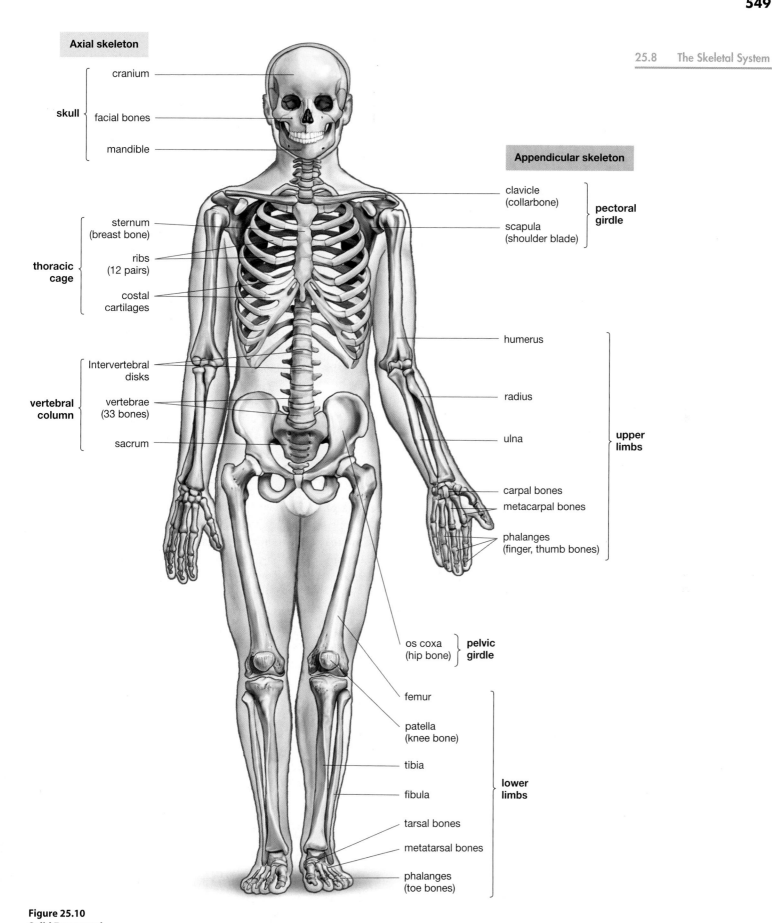

Figure 25.10
Solid Framework
A view of the human skeleton, which has two divisions, axial and appendicular. The major bones of the axial division are listed on the left, and the major bones of the appendicular division on the right.

Joints

Joints, or articulations, exist wherever two bones meet. What qualities are required of a joint? Well, some only need strength. We think of our skull as being a single bone; but in fact, it's composed of several bones that interlock with each other in joints that are immovable. For other joints, flexibility is the key requirement. The ball-and-socket joint at our shoulder permits a range of motion so extensive that movement of our upper arms is limited more by our shoulder muscles than by our bones.

Highly movable joints are typically found at the ends of long bones, such as those in the legs as well as the shoulder. You can see a view of such a joint—the human knee joint—in **Figure 25.11**. A joint such as this must serve two competing functions. On the one hand, it must keep the massive femur (thigh bone) and large tibia (shin bone) in a stable arrangement. On the other, it serves as a hinge that allows the tibia to swing as free as a screen door off the femur. How does it do both things? Well, the two bones are partly held in place by four main ligaments, some of which are visible in the figure. Recall that ligaments link bone to bone, but there are also tendons, which attach bone to muscle; the knee joint has these in place as well.

With respect to the motion that such a joint allows, a critical element is that one bone does not actually touch another. Instead, an extensive network of padding and lubricating tissues in between the bones performs the same function as the "gel" found in running shoes. Note that the top of the tibia and bottom of the femur are surrounded by cartilage, and that more cartilage—in the form of the meniscus tissue—lies on each side of the joint. The yellow tissue shown is fat padding that fills in spaces that are created when the bones move, and the space in between the bones is filled with a thick, viscous fluid that not only lubricates joints, but aids in their metabolism.

Joints can get injured, of course, and as we age they can wear down altogether. If you follow sports, the name of one of the ligaments shown in Figure 25.11 may ring a bell. The anterior cruciate ligament (ACL), which runs from the back of the femur to the front of the tibia, is a frequent site of injury for athletes who participate in sports that involve rapid "cutting," such as volleyball, soccer, and basketball. An athlete will come to a quick stop, hear a pop, and be down with a torn or ruptured ACL that usually will have to be replaced by grafting tissue onto it from another part of the body. For older people, the usual problem is not a sudden injury, but rather a slow, steady degeneration of hip, knuckle, or knee joints in the disease known as osteoarthritis. Such "wear and tear" arthritis results in a breakdown of the cartilage in between bones, such that joints become swollen and misaligned, with bone sometimes grinding against bone.

25.9 The Muscular System

Bones may provide a scaffolding for the body, but how is it that this scaffolding is capable of movement? The answer is muscles, specifically the skeletal muscles that were introduced earlier, which are numerous and large enough that they account for about 40 percent of our body weight. (When we talk about the "meat" in, say, cattle, we are mostly talking about skeletal muscle.) Like bones, skeletal muscles are organs in that they contain a variety of tissues. At each end of a muscle, fibers of the outer muscle layer come together to form the tendons that attach muscle to bone.

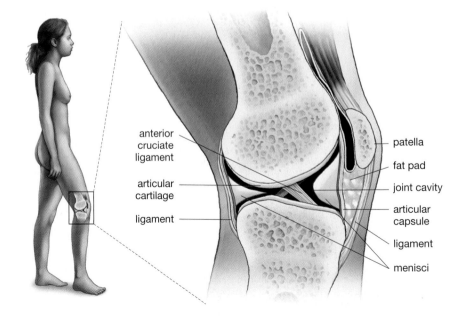

Figure 25.11
A Highly Moveable Joint
A simplified view of the human knee joint. Note that the two bones never touch each other because they are covered with cartilage. The joint has shock-absorbing cartilage in the form of the menisci and a fat pad that protects the cartilage.

anterior cruciate ligament

articular cartilage

ligament

patella

fat pad

joint cavity

articular capsule

ligament

menisci

The Makeup of Muscle

As you can see in **Figure 25.12**, each muscle has within it a number of oval-shaped bundles called fascicles. Inside each fascicle there is a collection of muscle cells, but the term cell here may be misleading, because any one of these cells can be as long as the muscle itself—a gigantic length relative to the strictly microscopic dimensions of most cells. Because skeletal muscle cells are elongated, they are referred to as **muscle fibers**. Along their length, muscle fibers are divided into a set of about 10,000 repeating units, called **sarcomeres**, that are the fundamental units of muscle contraction. But what is it that's contracting in a muscle? A look inside a single muscle fiber reveals that it is composed of more strands—perhaps a thousand long, thin structures called myofibrils that run the length of the cell. Each myofibril, in turn, has a large collection of two kinds of strands inside it that alternate with one another; these are thin filaments made of the protein *actin* and thick filaments made of the protein *myosin*. It is these filaments that contract, as you'll now see.

How Muscles Work

To understand contraction, note first that the thick myosin filaments lie in the *center* of a given sarcomere. Meanwhile, the thin actin filaments are attached to either *end* of the sarcomere and extend toward the center, where they overlap with the thick filaments (**see Figure 25.13** on the next page). In contraction, the thin filaments slide toward the center, causing the unit to shorten, which is to say, causing the muscle to contract. To understand this, place just the tips of the fingers of your right hand in between the tips the fingers of your left hand. Now slide your right hand as far as it will go to the left. Note that neither set of *fingers* shortened in length, but that the total distance made up by both sets of fingers did shorten—exactly what happens within a sarcomere.

But what enables the thin filaments to slide? The thick myosin filaments have numerous club-shaped "heads" extending from them. These myosin heads are capable of alternately binding with or detaching from the adjacent thin filaments, creating so-called *cross-bridges*. The myosin heads pivot at their base, as if they were on a hinge. Here's the order of their activity: Attach to the actin filament, pull it toward the center of the sarcomere, detach from the actin, reattach to it, and pull again. In actual muscle contraction, this process happens many times in the blink of an eye; at any one point in a contraction, some myosin heads will be pulling an actin filament while others are detaching from it and still others binding to it.

This whole process gets going only when a signal arrives from a nerve telling a muscle to contract—often when a person has decided to move. This is why we speak of skeletal muscles as being under voluntary control, although it's easy to find exceptions to this rule. (The diaphragm that allows you to breathe is composed of skeletal muscles, but normally there is little that's voluntary in using them.)

When a nerve signal arrives, the actual message doesn't go directly from nerve to muscle. Instead, a kind of ferry-boat operation takes place in which the nerve end secretes a chemical, called a

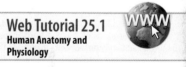

Web Tutorial 25.1
Human Anatomy and Physiology

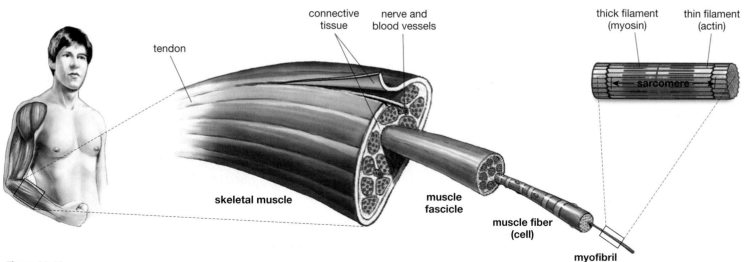

Figure 25.12
Structure of Skeletal Muscle
Skeletal muscles have oval-shaped bundles within them, called fascicles, that are composed of a number of muscle cells, each one of which is so elongated it is referred to as a "fiber." Each fiber is composed of many thin myofibrils, and each myofibril has within it a collection of alternating thick and thin protein strands called filaments. The thin filaments are made of the protein actin, the thick filaments of the protein myosin. Each myofibril is divided up lengthwise into a series of repeating functional units called sarcomeres.

neurotransmitter, that travels across a tiny gap that lies between nerve and muscle. Once the neurotransmitter arrives at the muscle, it binds with it, setting in motion a chemical reaction that results in calcium ions (Ca^{2+}) being released inside a sarcomere. These calcium ions have the effect of allowing the myosin heads to bind to the actin filaments, which of course has to happen before the myosin heads can *pull* the actin filaments.

Now, what about the practical consequences of this physiology? The cycle that runs from myosin binding, through pulling, and then detaching is known as a *twitch*. The muscle fibers that go through this cycle come in two varieties—fast-twitch and slow-twitch—whose names describe how long it takes them to complete a cycle. Fast-twitch fibers are not just faster than slow-twitch fibers, however; they also can generate more power. But there is a catch to this, in that fast-twitch fibers can generate power only for brief periods. In contrast, slow-twitch fibers generate less power but can do it over a sustained period of time. Thus, fast- and

slow-twitch fibers are something like the hare and tortoise of muscle tissue.

Because slow-twitch fibers are infused with an oxygen-binding molecule called myoglobin, they have a red coloration that true fast-twitch fibers lack. The "dark meat" of a chicken leg is dark because it is primarily made of myoglobin-infused slow-twitch fibers. The chicken constantly uses its legs, but doesn't generate much power in doing so—the very thing that slow-twitch fibers are good at. The chicken will occasionally engage its breast muscles in a flurry of contraction so that it can flap its wings. Not surprisingly, chicken breasts are white-meat, which is to say muscle that is primarily composed of fast-twitch fibers. Most muscles contain a mixture of fast- and slow-twitch fibers. We simply utilize different proportions of each kind of fiber, depending on what we are doing. As you might guess, a long-distance runner primarily utilizes slow-twitch fibers while a sprinter primarily uses fast-twitch.

Finally, recall that any muscle contraction begins with a nerve ending sending a chemical packet called a

Figure 25.13
How a Skeletal Muscle Contracts

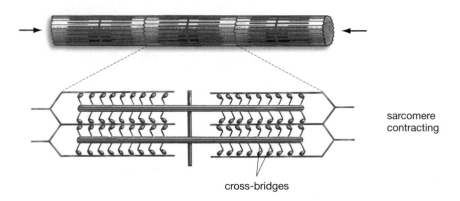

(a) Muscle at rest

The myofibril contains several sarcomeres, one of which is shown at rest. The thin filaments within the sarcomere, made of actin, overlap with adjacent thick filaments, made of myosin. The thick filaments have club-shaped outgrowths, called heads. The myofibril is also shown in cross section, at a location where thick and thin filaments overlap.

myofibril in cross section

thick filaments

thin filaments

myofibril

sarcomere at rest

thin filaments (actin)

thick filaments (myosin)

myosin "heads"

(b) Muscle contraction

Muscle contraction is a matter of the thin actin filaments sliding together within a sarcomere. This comes about when the myosin heads attach to the actin filaments and pull them toward the center of the sarcomere. The myosin heads pivot at their base and are capable of alternately binding with, or detaching from, the thin filaments.

sarcomere contracting

cross-bridges

neurotransmitter over to a muscle. It turns out that the only neurotransmitter that operates in muscle contraction is one called acetylcholine. Now imagine what would happen if something *interfered* with acetylcholine's release: Muscles could no longer get a message telling them to contract. This would be bad enough in the case of an arm muscle, but disastrous in the case of the muscles that allow us to breathe. This kind of muscle paralysis does take place, however, in connection with the puncture-related infection tetanus, and with a kind of food poisoning that's not so well known, botulism. In both cases, bacteria that enter the body release toxins that impede acetylcholine release, and in both cases the result can be death from suffocation. Thanks to vaccines and good hygiene, both afflictions are very rare in the United States—50 cases a year of tetanus, 110 of botulism—but they do point up that our survival depends on all these microscopic processes remaining in working order.

Given all this, it may seem strange that one of the best-selling drugs in the United States in recent years has been a form of … botulism toxin! In 2002, the U.S. Food and Drug Administration (FDA) approved the cosmetic use of Botulinum Toxin Type A—better known as Botox—which is intended to temporarily get rid of forehead creases and "frown lines." Botox is sterile, purified, and stays localized in the area where a doctor injects it, which is the forehead. Nevertheless, like germ-borne botulism toxin, its effect is to paralyze muscles.

On to the Nervous and Endocrine Systems

Having looked at the basic characteristics of the body and at three of its organ systems, we're now ready to move on to two other systems, both of which function in communication. These are the nervous and endocrine systems.

Chapter Review

Summary

25.1 The Disciplines of Anatomy and Physiology

- Anatomy is the study of the body's internal and external structure and the physical relationships between the body's constituent parts. Physiology is the study of how these parts work.

25.2 How Does the Body Regulate Itself?

- The body regulates itself not through a hierarchical command structure, but rather through a system of circular control, in which A prompts B and B prompts C, but in which C may then prompt A. By self-regulating in this way, the body seeks homeostasis or the maintenance of a relatively stable internal environment. A single mechanism is almost singly responsible for bringing about homeostasis. This mechanism is negative feedback: a process in which the elements that bring about a response have their activity reduced by that response.

- Humans have two primary body cavities: a dorsal cavity that is further divided into cranial and spinal cavities; and a ventral cavity that is further divided into a thoracic cavity that contains the lungs and heart, and an abdominopelvic cavity that contains such organs as the stomach and liver.

25.3 Levels of Physical Organization

- Animal bodies function through a series of smaller-to-larger working units whose order runs as follows: cells, tissues, organs, and organ systems. Tissues are groups of cells that have a common function. Organs are complexes of several kinds of tissues that perform a special bodily function. Organ systems are groups of organs and tissues that perform a given function.
  Web Tutorial 25.1: Human Anatomy and Physiology

25.4 The Human Body Has Four Basic Tissue Types

- There are four fundamental tissue types in the human body: epithelial, connective, muscle, and nervous. Epithelial tissue covers all body surfaces exposed to an external environment and forms glands, which are organs that secrete one or more substances.

- Connective tissue supports and protects other tissues. Connective-tissue cells are embedded in a "ground" material, which they themselves generally secrete. A prime function of most connective-tissue cells is to produce this extracellular material, which varies in consistency from a fluid to a solid. For example, bone cells secrete a connective ground material that largely forms the solid bones we are familiar with.

- Muscle tissue is specialized in its ability to contract. There are three primary types of muscle tissue: skeletal, cardiac, and smooth. Skeletal muscle is always attached to bone and is under voluntary control. Cardiac muscle exists only in the heart and beats under the control of its own pacemaker cells. Smooth muscle is "smooth" because it lacks the striations that skeletal muscle has; it is not under voluntary control and is responsible for contractions in blood vessels, air passages, and hollow organs such as the uterus.

- Nervous tissue is specialized for the rapid conduction of electrical impulses. It is made up of two basic types of cells: neurons, which actually transmit the impulses; and glial cells, which are support cells for neurons.

25.5 Organs Are Made of Several Kinds of Tissues

- The stomach provides a case in point for how various tissue types come together to form a working organ. A typical cross section of the stomach is made up of muscle, nervous, connective, and epithelial tissue.

25.6 Organs and Tissues Make Up Organ Systems

- There are 11 organ systems in the body. The first three of these—the integumentary, skeletal, and muscular systems—function in body support and movement. The integumentary system is made up of skin and such associated structures as glands, hair, and nails. The skeletal system is made up of bones, ligaments, and cartilages. The muscular system includes all the skeletal muscles of the body, but not smooth or cardiac muscles.

- The nervous, endocrine, and lymphatic systems function in communication, regulation, and defense of the body. The nervous system is a rapid communication system that includes the brain, spinal cord, all the nerves, and sense organs such as the eye and ear. The endocrine system is a communication system that works more slowly, through the substances called hormones, many of which are produced in specialized glands. (Glands that release their substances directly into the bloodstream or extracellular fluid are endocrine glands. Glands that release their substances through the tubes known as ducts are exocrine glands.) The lymphatic system overlaps with the immune functions of the body. It consists of a network of lymphatic vessels that collect extracellular fluid and deliver it to the blood vessels as a fluid called lymph. Lymphatic system organs include the lymph nodes, thymus gland, tonsils, and spleen. Cells within lymph glands inspect lymph as it passes through the system to identify invading microorganisms.

- The cardiovascular, respiratory, digestive, and urinary organ systems function in transport and in exchange with the environment. The cardiovascular system consists of the heart, blood, blood vessels, and the inner "marrow" portion of the bones where red blood cells are formed. The respiratory system includes the lungs and the passageways that carry air to them. The central feature of the digestive system is the digestive tract, a tube that begins at the mouth and ends at the anus. Along its course, there are digestive organs of the system: the stomach, small intestines, and large intestines and such accessory glands as the liver and pancreas. The urinary system functions to eliminate waste products from the blood through the formation of urine and to retain substances that are useful to the body. It is made up of the kidneys and various other structures, such as the bladder.

- Humans have separate reproductive systems for females and males.

25.7 The Integumentary System: Skin and Its Accessories

- The primary component of the integumentary system is the organ called skin. Integumentary structures associated with skin are hair, nails, and a variety of exocrine glands.

- In addition to covering the body, skin protects underlying tissues and organs, controls the evaporation of body fluids, regulates heat loss, stores fat, and makes vitamin D. Exocrine glands associated with skin excrete such materials as water, oils, and milk. Specialized nerve endings in the skin detect touch, pressure, pain, and temperature.

- Skin has a thin outer epithelial covering, the epidermis; and a thicker underlying dermis, composed mostly of connective tissue. Beneath the dermis—and not part of the skin—is a third layer of tissues, the hypodermis. Epidermal skin cells are constantly worn away and replaced by new epidermal cells being pushed up from the inner epidermis. As they move toward the surface, these cells are transformed into a series of dead, tightly linked sacs made of the protein keratin. The dermis is filled with accessory structures, such as hair follicles and sweat glands, and contains blood vessels and nerves that support the surface of the skin. The blood supply of the dermis is important in controlling body temperature.

- The hypodermis, beneath the dermis, contains the layer of fat cells that largely determine how much "fat" we have on our bodies.

- Hairs originate in tiny organs called hair follicles and may have been important for our evolutionary ancestors in both heat retention and touch sensitivity.

- The skin contains two types of exocrine glands: sebaceous glands, that produce sebum, which lubricates hair and limits bacterial growth on the skin; and sweat glands, which produce sweat that cools the body by evaporating. Overactive sebaceous glands are the cause of acne. Sweat glands come in two varieties: the more-numerous merocrine glands, which produce the perspiration; and the less-numerous apocrine glands, which produce sweat mixed with a thick, cloudy secretion that moves through hair follicles onto the skin. Bacteria feed on this secretion, and the resulting bacterial by-products are the cause of body odor.

- Nails are made of the protein keratin and form over the tips of the fingers and toes, where they are useful in protection and in prying, scratching, and picking up objects.

25.8 The Skeletal System

- The human skeletal system is composed of bone, ligaments, and cartilage. Bone is a connective tissue that provides support and storage capacity, and that is the site of blood-cell production. Cartilage serves as padding in most joints, forms our larynx (voice box) and trachea (windpipe), and links each rib to the breastbone. Ligaments are tissues that join bone to bone. Tendons link bone to muscle but are not part of the skeletal system.

- Each bone of the skeleton is an organ that contains not only connective tissue—mostly the supporting connective tissue called osseous tissue—but blood vessels and nervous tissue as well.

- The typical bone features a long central shaft, called a diaphysis, and expanded ends, called epiphyses. There are two types of bone: compact bone, which is relatively solid and forms the outer portion of the bone; and spongy bone, which is less dense and fills the epiphyses. Spongy bone is filled with marrow, which comes in two forms: red marrow, which gives rise to blood cells; and yellow marrow, which is composed of energy-storing fat cells. Long bones have a central marrow cavity that is filled with yellow marrow.

- Three different types of cells are involved in bone growth and maintenance. Osteoblasts are responsible for the production of new bone; osteocytes are mature bone cells that maintain the structure and density of normal bone; and osteoclasts release enzymes that breaks down bone, thus releasing stored substances, such as calcium.

- In compact bone, the basic functional unit is the osteon. Osteoblasts produce layers of concentric cylinders of bone around a central canal of the osteon. The central canal, which parallels the long axis of the bone, has blood vessels and nerves running through it.

- Bone is responsive to our activities. The density of a bone will decrease to the extent that the bone is not used and will increase to the extent that the bone is stressed, as in some forms of exercise.

- The human skeleton has 206 bones in all, in two main divisions: an axial skeleton, whose 80 bones include the skull, the vertebral column, and the rib cage; and the appendicular skeleton, whose 126 bones include those of the arms and legs and the pelvic and pectoral girdles to which they are attached.

- Joints or articulations exist wherever two bones meet. Some joints provide a strong linkage between bones, while others provide great flexibility. Ligaments (joining bone to bone) and tendons (joining bone to muscle) help provide the stability that large-bone joints need. Normally, the bony surfaces of highly movable joints do not meet directly, but are instead covered with special cartilages and pads that provide shock absorption and lubrication.

25.9 The Muscular System

- A given skeletal muscle cell can be as long as the muscle itself; because of their elongation, these cells are called muscle fibers. Along their length, muscle fibers are divided into a set of repeating units called sarcomeres, which are the basic units of muscle contraction. Each muscle fiber is composed of thin structures, called myofibrils, that are in turn composed of assemblies of two kinds of protein strands that alternate with one another: thin filaments made of the protein actin, and thick filaments made of the protein myosin.

- Actin filaments attached to the end of each sarcomere overlap with the myosin filaments that lie in the center of each sarcomere. The myosin filaments bring about contraction by attaching to the actin filaments, pulling them toward the center of the sarcomeres, detaching, and then pulling again. As the actin filaments slide toward the center of the sarcomere, the sarcomere shortens.

- Skeletal muscles contract only when prompted to by the arrival of a signal from a nerve. Nerve signals can be blocked by, for example, bacterial toxins, in which case skeletal muscles can become paralyzed.

- Muscle fibers differ in how long it takes them to go through the cycle of myosin binding, pulling, and detaching—a cycle referred to as a twitch. Muscle fibers can be categorized as slow-twitch or fast-twitch. Slow-twitch fibers contract with less force, but can do so in a sustained manner; fast-twitch fibers contract with relatively more force, but can do so only for brief periods of time.

Web Tutorial 25.1: Human Anatomy and Physiology

Key Terms

appendicular skeleton	548	**muscular system**	541
axial skeleton	548	**negative feedback**	536
bone	546	**nervous system**	541
cardiac muscle	540	**nervous tissue**	540
cardiovascular system	542	**organ**	537
cartilage	546	**organ system**	537
compact bone	547	**osteoblast**	548
connective tissue	538	**osteoclast**	548
dermis	544	**osteocyte**	548
digestive system	542	**red marrow**	548
endocrine gland	541	**reproductive system**	542
endocrine system	541	**respiratory system**	542
epidermis	543	**sarcomere**	551
epithelial tissue	538	**sebaceous glands**	545
exocrine gland	541	**skeletal muscle**	538
gland	538	**skeletal system**	540
homeostasis	536	**skin**	543
hormone	541	**smooth muscle**	540
integumentary system	540	**spongy bone**	548
keratin	543	**sweat gland**	546
ligaments	547	**tendons**	547
lymphatic system	541	**tissue**	537
muscle fiber	551	**urinary system**	542
muscle tissue	538	**yellow marrow**	548

Understanding the Basics

Multiple-Choice Questions (answers in the back of the book)

1. To survive, the human body needs to maintain _____ and does so primarily through the means of self-regulation known as _____.
   a. high-energy output, positive feedback
   b. adequate fat layers, exocrine secretion
   c. bone density, muscle contraction
   d. homeostasis, negative feedback
   e. nerve signal transmission, hormonal secretion

2. The four basic types of tissue in the human body are
   a. epithelial, connective, muscle, nervous
   b. epithelial, blood, muscle, nervous
   c. skin, bone, muscle, nervous
   d. skin, connective, heart, nervous
   e. epithelial, bone, connective, muscle

3. Endocrine glands differ from exocrine glands in that endocrine glands
   a. have ducts
   b. are ductless
   c. produce only perspiration
   d. consist of epithelia and muscle tissue
   e. are found only in the liver

4. Muscle fibers that are striated and voluntary are found in
   a. cardiac muscle
   b. skeletal muscle
   c. smooth muscle
   d. rough muscle
   e. all four types of muscle

5. Which of the following drugs would seem most likely to cause muscle spasms (uncontrolled contractions)?
   a. one that prevents attachment of myosin heads to actin
   b. one that prevents the use of ATP
   c. one that blocks the detachment of myosin heads from actin
   d. one that elongates the sarcomere

6. Surface epidermal cells of the skin
   a. rapidly divide and are filled mostly with keratin
   b. never divide and remain in place through adult life
   c. rapidly divide and are in place only a short time
   d. are dead and filled with keratin
   e. rapidly divide and are filled with fat

7. Body odor comes primarily from
   a. the tightly linked nature of epidermal cells
   b. the odor of sweat
   c. bacteria that exist in the blood servicing the dermis
   d. byproducts of the bacteria that feed on apocrine sweat
   e. the layer of fat in the hypodermis

8. Cells called _____ build bone tissue _____ maintain it, and _____ break it down.
   a. osteoblasts, osteoclasts, osteocytes
   b. osteoclasts, osteons, osteoblasts
   c. osteoblasts, osteocytes, osteoclasts
   d. osteons, osteocytes, osteoclasts
   e. osteocytes, osteoblasts, octeoclasts

9. The essence of muscle contraction is that thick and thin muscle proteins
   a. slide past each other
   b. link irreversibly with each other
   c. are short enough to respond quickly
   d. stay completely separate
   e. pull in opposite directions

10. No skeletal muscle can contract without first
    a. receiving a hormone signal
    b. checking on the status of nearby skeletal muscles
    c. lining up its actin and myosin filaments
    d. receiving a nervous system signal
    e. linking to a bone

11. Which type of bone cell would you expect would be most active in a female with osteoporosis:
    a. osteocytes
    b. osteoclasts
    c. osteoblasts
    d. cartilage

Brief Review

1. Beginning with the cell, list in correct sequence the four levels of organization of the human body, from the simplest to the most complex.

2. What role does the lymphatic system play in immune function?

3. Why is our skin water-resistant?

4. List three functions of the skeletal system.

5. What is the general mechanism at work in hair loss?

6. A sample of bone shows an interconnecting network of calcified rods and plates filled in with both red and yellow marrow. Is this sample from the shaft (diaphysis) or the end (epiphysis) of a long bone?

7. People who have torn their anterior cruciate ligament report that their knee feels "wobbly" until they get it repaired. Why would this be?

8. Why does skeletal muscle appear striated when viewed under a microscope?

Applying Your Knowledge

1. Cells called melanocytes that lie deep in the epidermis produce a pigment, called melanin, that gives skin its color. Exposure to ultraviolet light—from the sun or a tanning lamp—causes melanocytes to produce more melanin, which they pass along to the skin cells above them, yielding a suntan. But why would sun exposure prompt this increased melanin production? What is the function of a tan, in other words?

2. The text notes that bone density increases in response to the stresses on bone that are provided by exercises such as running. Imagine two fairly inactive people who decide to start exercising. One becomes a runner, the other becomes a swimmer. Who experiences the greater increase in bone density: the swimmer (in the arm bones) or the runner (in the leg bones)?

3. If Botox works by temporarily paralyzing muscles, what does this say about the cause of age lines on the forehead?

26 *The Nervous and Endocrine Systems*

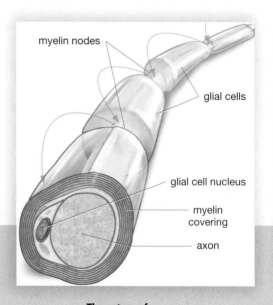

The nature of a nerve.
(Section 26.2, page 563)

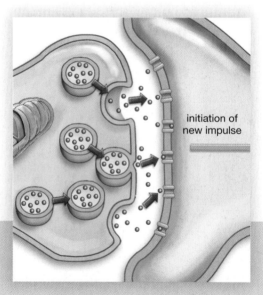

A nerve impulse is transmitted.
(Section 26.3, page 566)

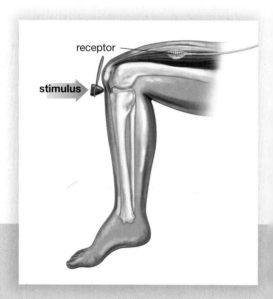

A knee-jerk reaction.
(Section 26.4, page 568)

To get through a day, our bodies must undertake a huge number of tasks each second, many of them coordinated with each other. How do we control such an enormous set of activities? Through our nervous and endocrine systems.

With its depiction of a man who can neither remember his present nor forget his past, the film *Memento* was a big hit in 2001. In it, we watch as the fictional Leonard Shelby relentlessly tracks down the murderer of his wife. He does so, however, while taking Polaroid pictures and scribbling notes to himself anywhere he can—some on his own body as tattoos. He does this because he has a big handicap to deal with. Having sustained a head injury in the same break-in in which his wife was killed, he can remember everything in his life up through the moment of his injury, but has lost the ability to create *new* long-term memories. Now, anything that happens more than 15 minutes in the past is lost to him. Hence the constant notes and Polaroids: Where am I staying? Who did I talk to this morning? What did he look like?

Though Leonard Shelby is a movie character, the condition he has actually exists in some people. A version of it was brilliantly described by the physician-writer Oliver Sacks in his book *The Man Who Mistook His Wife for a Hat*. Sacks introduces us to "Jimmy G.," who was admitted to Sacks's care in the mid-1970s. Though Jimmy G. is a gray-haired, 49-year-old ex-sailor, he perpetually believes himself to be the 19-year-old sailor he was back in 1945. About 1970, he apparently drank so heavily that he damaged his brain, losing not only his memory back to 1945, but his ability to make new memories as well. Hence, in his mind he is always a young man and it is always the end of World War II. This can lead to brief, horrible moments of disorientation, as Sacks describes.

The Importance of Memory
In the film *Memento*, the actor Guy Pearce played Leonard Shelby, a man who has lost his ability to create new long-term memories.

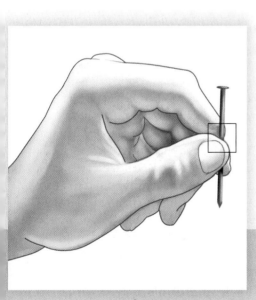

Stretch receptors and touch.
(Section 26.8, page 572)

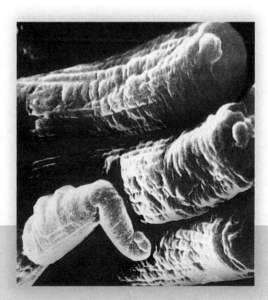

Seeing in color; seeing in black and white.
(Section 26.12, page 579)

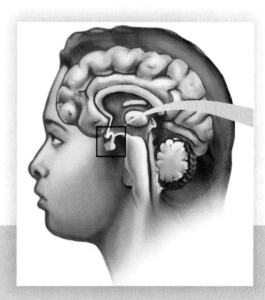

Controlling hormones.
(Section 26.15, page 583)

"Here," I said, and thrust a mirror toward him. "Look in the mirror and tell me what you see. Is that a nineteen-year-old looking out from the mirror?" He suddenly turned ashen and gripped the sides of the chair. "Jesus Christ," he whispered. "Christ, what's going on? What's happened to me?"

Sacks would not have done this but for his certainty about what would happen next. Two minutes later, Jimmy G. has no memory of the mirror or of Sacks. When Sacks briefly leaves the room and then reenters it, Jimmy G. greets him as a total stranger.

As Jimmy G. shows, the "self" that we take for granted largely amounts to a set of memories, which exist in the small physical space of the brain. We can think of memory as an agent of communication and control—one that allows us to act in the present based on our experience of the past. The human nervous system, which includes the brain, can be thought of as a wider-yet communication and control operation—one that constantly monitors our present, along with our past, and issues commands that allow us to function. Only some of these commands are conscious orders, however; most are actions that the brain takes without our knowledge.

Beyond the nervous system, the body has an additional communication and control operation in its hormonal or "endocrine" system. This system works ceaselessly, mostly outside our conscious awareness, to handle such things as our growth when we are children and our bodies' water-balance when we are adults.

Our goal in this chapter is to review both these communication and control operations: the nervous and endocrine systems. We'll start with the nervous system, which is such a wonder when healthy and, as Jimmy G. shows, such a heart-breaker when not.

26.1 Structure of the Nervous System

Taking the broadest view, the nervous system includes all the cells in the body that can be defined as nervous tissue, plus the sense organs we have, such as the eye and ear. Recall from Chapter 25 that two types of cells make up nervous tissue. These are neurons, which actually transmit nervous system messages; and glial cells, which support the neurons.

One helpful way to think about the nervous system is to consider three essential tasks it has to perform. It first has to *receive* information, both from outside and inside our bodies. (For example, what taste sensations are coming in?) It next has to *process* this information. (Do I like this?) Then it has to *send*

information out that allows our body to deal with this input. ("Lift your hand for another spoonful.")

The nervous system in which these activities take place can be divided into two fundamental parts. The first of these is the **central nervous system**: that portion of the nervous system consisting of the brain and the spinal cord. The second is the **peripheral nervous system**: that portion of the nervous system outside the brain and spinal cord, plus the sensory organs (**see Figure 26.1a**).

The central nervous system's brain and spinal cord are not physically separate entities; the spinal cord simply expands greatly at its top, and we call this flowering the brain. Because the brain and spinal cord are so different in terms of structure and function, however, they are thought of as individual units.

Meanwhile, the peripheral nervous system, or PNS, can be pictured as a group of nerves and related nerve cells that fan out from either the brain or spinal cord. One way to think about these peripheral nerves and related cells is to consider the direction of the messages they carry. Any nerves that help carry messages *to* the brain or spinal cord are said to be part of the **afferent division** of the peripheral nervous system. Any nerves that help carry messages *from* the brain or spinal cord are said to be part of the **efferent division** of the PNS. An easy way to remember the difference between these two terms is to remember that "efferent" sounds like "effect." And the efferent division *effects change* in various organs in the body.

What kinds of change? Many of the activities the nervous system controls are voluntary activities, which often means movement. You decide to lift your finger; you decide to stand up, and so forth. You may remember from Chapter 25 that all movement is handled by skeletal muscles—muscles, like the biceps, that are attached to bones and that are under voluntary control. So, one part of the PNS's efferent division is the **somatic nervous system**: that portion of the peripheral nervous system's efferent division that provides voluntary control over skeletal muscle.

But, there are lots of processes in the body that are not under voluntary control. When you walk from bright sunshine into a darkened movie theater, your pupils dilate, to let in more light, but you have no control over this dilation. Certain muscles allow our pupils to open up, but these are not skeletal muscles; they are the involuntary "smooth muscles." Likewise, the cardiac muscle of your heart beats in a rhythm that is largely out of your conscious control. And your glands release hormones in a way you have little control over. It is a *second* part of the PNS's efferent division that controls these operations, the

(a) The nervous system has two components

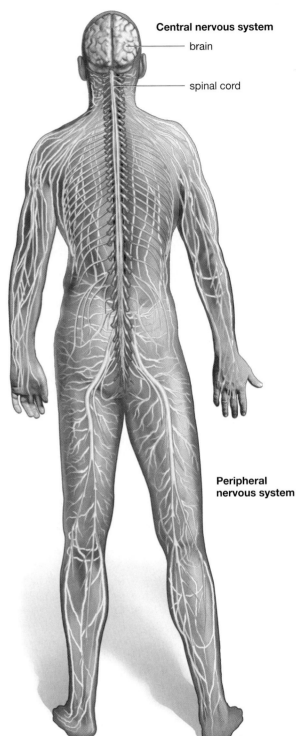

Central nervous system
- brain
- spinal cord

Peripheral nervous system

(b) How these two components interact

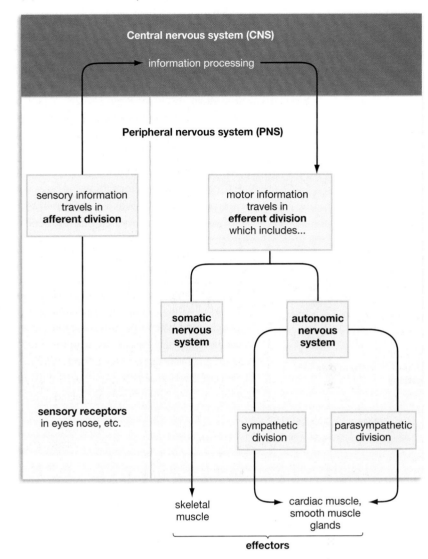

Figure 26.1
Divisions of the Nervous System

(a) The two central branches of the nervous system are the central nervous system (CNS), which consists of the brain and the spinal cord; and the peripheral nervous system (PNS), which consists of all the nervous tissue outside the brain and spinal cord, plus the sensory organs.

(b) Information about the body and its environment comes to the nervous system from its sensory receptors—for example, cells in the eyes—which are part of the peripheral nervous system. This information then goes through the afferent division of the peripheral nervous system to the brain and spinal cord. After processing this information, the brain and spinal cord issue motor commands through the peripheral nervous system's efferent division. These commands go to "effectors" such as skeletal muscles and glands. The peripheral nervous system's efferent division has two systems within it: The somatic nervous system, which provides voluntary control over skeletal muscles; and the autonomic system, which provides involuntary regulation of smooth muscle, cardiac muscle, and glands. The autonomic system is further divided into sympathetic ("fight-or-flight") and parasympathetic ("rest-and-digest") divisions.

autonomic nervous system: that part of the peripheral nervous system's efferent division that provides involuntary regulation of smooth muscle, cardiac muscle, and glands (**see Figure 26.1b**).

Now let's go down just one more level, this time strictly within the autonomic system. It turns out that the autonomic system is divided into two divisions, the *sympathetic* and the *parasympathetic*. These terms will be formally defined later; for now, just be aware of two characteristics they have. First, these divisions differ in that the nerves that make them up stem from different locations in the brain and spinal cord. Second, the sympathetic division generally has stimulatory effects on us—we get adrenaline going through this division, for example—while the parasympathetic division generally has relaxing effects.

26.2 Cells of the Nervous System

So, what about the cells that make up these systems? Remember first that the nervous-system cells that transmit signals are called neurons. These cells come in three varieties that neatly parallel the idea of a nervous system that receives, processes, and sends information (**see Figure 26.2a**). The first type of neuron is the **sensory neuron**, which does just what its name implies: It senses conditions both inside and outside the body and brings this received information to the central nervous system. (Given the direction of this information, sensory neurons are afferent neurons.) When someone brushes the top of your hand, sensory neurons just beneath the skin sense lots of things about this touch—what direction it came from, what shape the touching object was—and then convey a message containing this information that goes from your hand, into your spinal cord, then up into your brain.

The second type of neuron is the interneuron (or association neuron). Located solely in the central nervous system, **interneurons** interconnect other neurons. These can be very simple connections, but they can be complex as well. The memory we talked about at the start of the chapter amounts to a massive mobilization of interneurons. How do we remember? We process information in complex webs of interneurons.

The third type of neuron is the **motor neuron**: a peripheral-system neuron that sends instructions from the central nervous system (CNS) to such structures as muscles or glands. (Given the direction of this transmission, these are efferent neurons.) The key to understanding motor neurons is to think of them as neurons that transmit messages to organs or tissues *outside* the nervous system, thus prompting some kind of action. If you look at the right side of **Figure 26.2a**, you can see how this works. A message comes from the cell body of a motor neuron, travels down an extension of it, and then gets transferred to a muscle—causing the muscle to contract. Any organ or group of cells that responds to this kind of nervous-system signal is called an *effector*. Thus, a gland prompted to release hormones through a motor neuron signal is also an effector.

To sharpen the point about the difference between sensory and motor neurons, consider the newly popular drug Botox. Physicians inject this drug into the foreheads of older people who want to get rid of the "age lines" there. Botox works by

Figure 26.2
Types of Neurons and Neuron Anatomy

(a) Three types of neurons

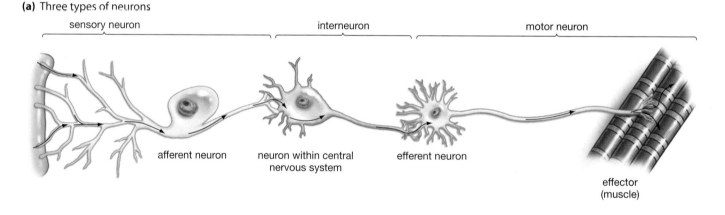

sensory neuron   interneuron   motor neuron

afferent neuron   neuron within central nervous system   efferent neuron

effector (muscle)

(b) Anatomy of a neuron

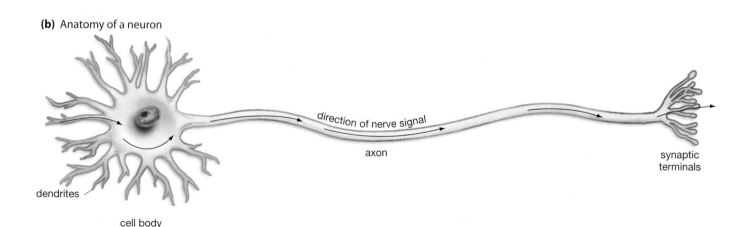

direction of nerve signal

axon

synaptic terminals

dendrites

cell body

blocking the signal between motor neurons and the muscles they stimulate. The linkage works as follows: Botox blocks the nerve signal going to forehead muscles; this means there will be no forehead muscle contraction; and this means there will be no age lines. Now, one of the frequently asked questions about Botox is: Will it make my forehead numb? The answer is no, because Botox is blocking *motor* neuron transmission, not sensory neuron transmission. The motor signals are going out, the sensory signals are coming in, and only the motor neurons are responsive to Botox.

Anatomy of a Neuron

Now let's look at how a neuron is structured. Like any other cell, a neuron has a nucleus, it makes proteins, it gets rid of waste, and so forth. If you look at Figure **26.2b**, you can see, however, that the neuron has some extensions sprouting from it that clearly separate it from other cells. Projecting from the cell body are a variable number of **dendrites**: extensions of neurons that carry signals *toward* the neuronal cell body. There is also a single, large **axon**: an extension of the neuron that carries signals *away* from the neuronal cell body. Though you can't see it, the other thing that separates the neuron from other cells is its outer or "plasma" membrane, which has an amazing ability to respond to various kinds of stimulation, as you'll see.

The Nature of Glial Cells

Now, what about the second kind of nervous-system cell, the glial cells (also known as glia)? Found in both the CNS and PNS, glia have no signal-processing ability of their own, but they provide all kinds of support to neurons. For example, some glia wrap their cell membranes around the axons of neurons in the CNS and PNS. The membranous covering that glia provide to neurons is called **myelin**, and an axon wrapped in this way is said to be myelinated. The importance of this is that axons that are myelinated carry nerve impulses faster than those that are not. (The impulse skips from one gap or "node" in the myelin covering to the next.) If this sounds like some technical detail, consider that the disease multiple sclerosis results from a dismantling of the myelin covering in the brain and spinal cord by the body's own immune system. Because myelin is fat-rich, areas of the brain and spinal cord containing myelinated axons are glossy white. Thus do we get the term *white matter* of the CNS, which contains mostly axons. By contrast, areas containing mostly neuron cell bodies are *gray matter*. If you look at **Figure 26.3a**, you can see what a myelinated axon looks like.

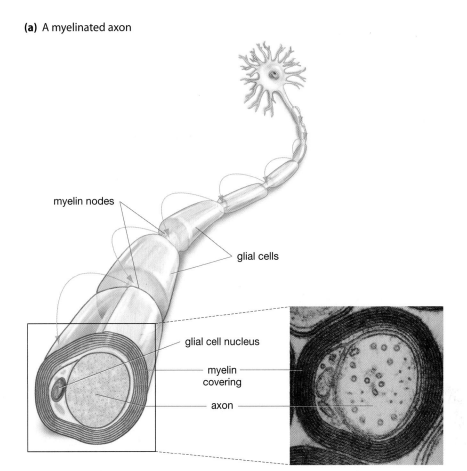

(a) A myelinated axon

myelin nodes

glial cells

glial cell nucleus

myelin covering

axon

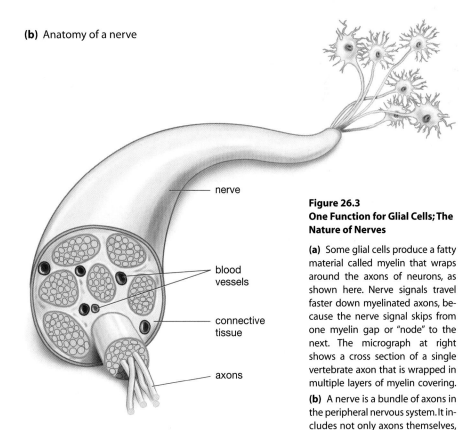

(b) Anatomy of a nerve

nerve

blood vessels

connective tissue

axons

Figure 26.3
One Function for Glial Cells; The Nature of Nerves

(a) Some glial cells produce a fatty material called myelin that wraps around the axons of neurons, as shown here. Nerve signals travel faster down myelinated axons, because the nerve signal skips from one myelin gap or "node" to the next. The micrograph at right shows a cross section of a single vertebrate axon that is wrapped in multiple layers of myelin covering.

(b) A nerve is a bundle of axons in the peripheral nervous system. It includes not only axons themselves, but supporting blood vessels and connective tissue.

Nerves

All this information about neurons and systems provides a way to understand a term that's been used extensively thus far, but that has not been defined. What is a **nerve**? It is a bundle of axons in the PNS that transmits information to or from the CNS. If you look at **Figure 26.3b**, you can see a representation of a nerve. Note that nerves have support tissue in the form of blood vessels and connective tissue.

26.3 How Nervous-System Communication Works

With this anatomy under your belt, you're ready to see how nervous-system signaling works. We'll look at a two-step process, which we can think of as (1) signal movement down a cell's axon and (2) signal movement from this axon over to a second cell, across what is known as a synapse.

Communication within an Axon

All the action in this first step is going to take place on either side of the thin plasma membrane that constitutes the outer border of any animal cell. Inside the plasma membrane is the cell and all its contents; outside is extracellular fluid (**see Figure 26.4**). Those of you who read Chapter 5 will recall that the plasma membrane regulates the passage of substances in and out of the cell. You may remember that some of these substances are the electrically charged particles called ions, meaning atoms that have gained or lost one or more electrons. Because ions are charged, they get the little + or – signs after their chemical names, as with Na^+ for the positively charged ion sodium. Ions cannot simply pass through the plasma membrane; rather, they need to move through special passageways, called protein channels. Some of these are so-called "leak" channels that are always open, but others are "gated" channels that can open or close.

It turns out that a "resting" neural cell—one not transmitting a signal—has a greater positive charge outside itself than inside. This is so mostly because there are more proteins inside the cell than outside, and proteins carry a *negative* charge. The charge difference also exists, however, because there are more of the positively charged Na^+ ions outside the membrane than there are positively charged potassium ions (K^+) inside, as you can see in Figure 26.4. As you know, in electricity, opposites attract, meaning the positively charged Na^+ ions outside the plasma membrane are attracted to the negatively charged interior of the cell. This amounts to a form of potential energy, in the same way that a rock perched at the top of a hill does. The rock would roll down the hill (releasing energy) if given a push, and the Na^+ ions would rush into the cell (releasing energy) if the proper channels were opened to them. The charge difference that exists from one side of the neuronal plasma membrane to the other is known as the **membrane potential**.

This potential exists in part because of what you've already seen: the greater abundance of Na^+ ions outside the cell. It's important to note that this abundance doesn't exist by accident. Every cell keeps this imbalance in place by constantly pumping three Na^+ ions out for every two (K^+) ions it pumps in. It takes a lot of energy to maintain this "sodium-potassium pump," but the cell does it for the same reason that you charge your cell-phone battery: in order to have ready access to a capability. In your case, the capability is talking on the phone; in the cell's case, the capability is that of having Na^+ ions available to rush into it. These ions can get the electrical impulse called a nerve signal going, as you'll now see.

When a neuron is stimulated by getting a chemical signal from another neuron, the effect of this stimulation is that, up near where the neuron's axon meets its cell body, Na^+ ion gates open up. Now the Na^+ ions can act on their attraction, and they do so, rushing into the cell. After a very brief time, this influx is great enough that the inside of the cell is more positive than the outside. This very state causes the *potassium* gates to open up more fully, however. Recall that potassium (K^+) ions exist in more abundance inside the cell than outside it. With the opening of their gates, they rush out of the cell, and the positive charge that has briefly existed on the inside of the cell begins to decline. Eventually, enough K^+ ions have moved out that the cell returns to its resting state, with a negative charge inside the membrane. This entire process takes place in a few milliseconds, or thousandths of a second.

Movement Down the Axon

All this alters the membrane potential at *one spot* along the neuron's axon. But how does a nerve signal get transmitted down the entire length of the axon? The key to this transmission is that an influx of Na^+ ions at an initial location on the membrane triggers reactions that cause the *neighboring* portion of the membrane to begin an Na^+ influx (see Figure 26.4). What occurs, in other words, is a chain reaction that moves down the entire length of the axon membrane. This is

Web Tutorial 26.1
The Nervous System

Figure 26.4
Nerve Signal Transmission within a Neuron

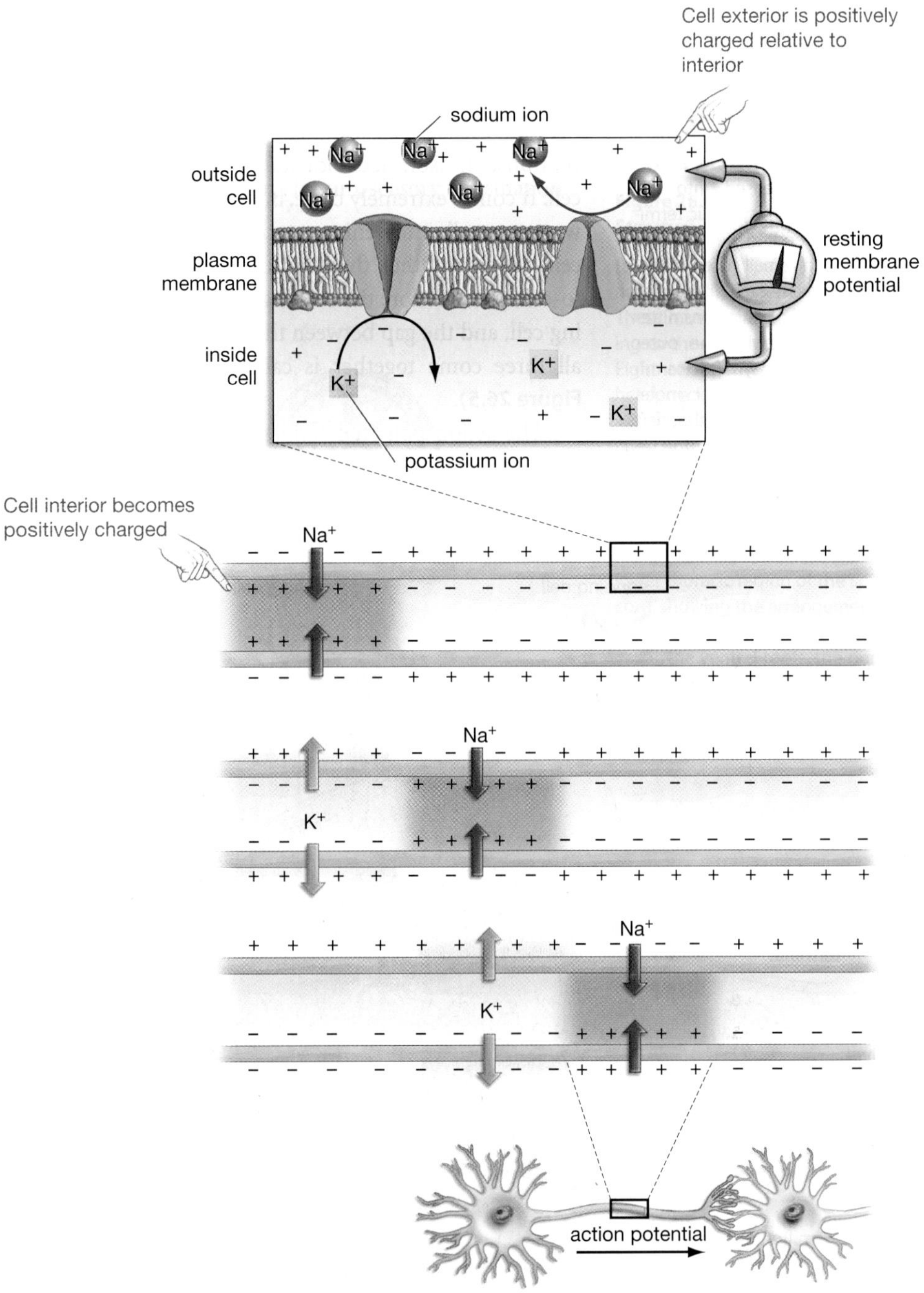

(a) Resting potential

Electrical energy is stored across the plasma membrane of a resting neuron. There are more negatively charged compounds just inside the membrane than outside of it. As a result, the inside of the cell is negatively charged relative to the outside. The charge difference creates a form of stored energy called a membrane potential. Protein channels (shown in green) that can allow the movement of electrically charged ions across the membrane remain closed in a resting cell, thus maintaining the membrane potential.

(b) Action potential

1. Nerve signal transmission begins when, upon stimulation, some protein channels open up, allowing a movement of positively charged sodium ions (Na$^+$) into the cell. For a brief time, the inside of the cell becomes positively charged.

2. The Na$^+$ gates close and the gates for positively charged potassium ions (K$^+$) open up, allowing a movement of K$^+$ out of the cell. With this, there is once again a net positive charge outside the membrane. The influx of Na$^+$ at one point in the cell membrane then triggers the same sequence of events in an adjacent portion of the membrane—note the influx of Na$^+$ next to the outflow of K$^+$.

3. The nerve signal continues to be propagated one way along the axon by means of this action potential.

Cell exterior is positively charged relative to interior

sodium ion

outside cell

plasma membrane

inside cell

resting membrane potential

potassium ion

Cell interior becomes positively charged

action potential

an **action potential**: a temporary reversal of cell-membrane potential that results in a conducted nerve impulse down an axon. (Temporary reversal? Remember the inside of the cell goes from negative to briefly positive, then back.) Action potentials have been compared to what happens to a lighted fuse: The heat of the spark causes the neighboring section of the fuse to catch fire, thus moving the spark along.

All action potentials are of the same strength. Once the original signal on the cell-body/axon border reaches sufficient strength to allow Na$^+$ ions to rush in, the action potential will get going. Thanks to its fuse-like quality, this potential will be the same strength all the way down the axon. How fast can this signal go? At best, about 120 meters per second—very fast indeed, but much slower than the electricity that comes from a wall socket.

576

(a) Anatomy of the ear

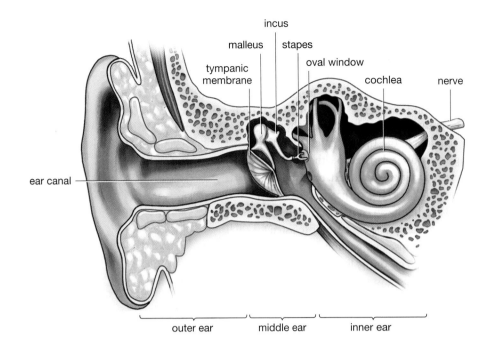

incus

malleus stapes

tympanic membrane

oval window

cochlea nerve

ear canal

outer ear | middle ear | inner ear

(b) From air vibration to nerve signal

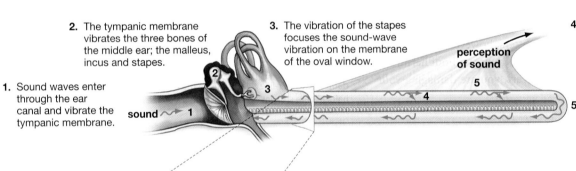

2. The tympanic membrane vibrates the three bones of the middle ear; the malleus, incus and stapes.

3. The vibration of the stapes focuses the sound-wave vibration on the membrane of the oval window.

4. The oval window's vibrations cause fluid vibrations within the coiled, tubular cochlea (shown elongated here for illustrative purposes).

perception of sound

1. Sound waves enter through the ear canal and vibrate the tympanic membrane.

sound

5. These fluid vibrations cause cells within the cochlea to release a neurotransmitter which triggers a nerve signal to the brain.

(c) How fluid triggers nerve signal

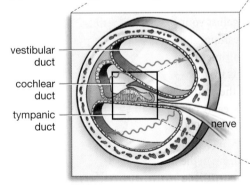
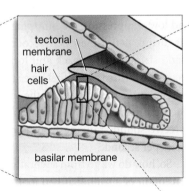
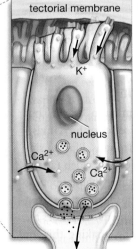

vestibular duct

cochlear duct

tympanic duct

nerve

tectorial membrane

hair cells

basilar membrane

tectorial membrane

K^+

nucleus

Ca^{2+} Ca^{2+}

3. As the hair cells contact the tectorial membrane, cilia on them bend. This change in position causes "trap door" channels in the hair cells to open, which allows potassium ions (K^+) to flow into them.

4. This influx triggers an influx of calcium ions (Ca^{2+}) at the base of the hair cells, which in turn causes the cells to release a neurotransmitter.

5. The neurotransmitter is received by adjacent dendrites and a nerve signal is sent to the brain.

1. Seen in cross-section, the cochlea has vestibular and tympanic ducts, in which fluid is vibrating.

2. This vibration shakes the basilar membrane, pushing hair cells on it up against the overlying tectorial membrane.

Figure 26.14
Our Sense of Hearing

three smallest bones in the body, the malleus, incus, and stapes (or "hammer, anvil, and stirrup"), which lie in the middle ear. In the same way that a lens serves to take light coming from a large area and focus it on a small one, these three bones take the vibrations coming from the relatively large tympanic membrane and focus them on the smaller oval window. This is important, because it *amplifies* the vibration signal for the next step. The oval window's vibrations now shake fluid that lies within the pea-sized structure called the **cochlea**: the coiled, membranous portion of the inner ear in which vibrations are transformed into the nervous system signals perceived as sound.

In the cross-section of the cochlea pictured in **Figure 26.14c**, you can see how vibrations of fluid in the cochlea are transformed into the sensation of sound. The cochlea can be seen to have a couple of ducts (the vestibular and tympanic) which are filled with fluid. In the middle, there is the cochlear duct, likewise filled with fluid, which has at its bottom a "basilar" membrane on which sit the elements that will produce sound. Note that, supported by the basilar membrane, there are a group of so-called "hair cells," so-named because each of them sprouts a group of hair-like cilia that come close to touching the tectorial membrane that folds over them in the cochlea.

Now, how does all of this yield sound? Think of it this way: If you rested your left hand on a water balloon and then pushed rhythmically on this same balloon with your right hand, your left hand would move up and down slightly. If there were something directly above your left hand, then your hand might make contact with it. In the same way, the vibrations of the fluid in the cochlea move the basilar membrane up and down, and this movement can push the hair cells up against the tectorial membrane that lies above them. If you look at the blow-up of Figure 26.14c, you can see what happens next. The cilia on the hair cells have microscopic "trap door" channels on them, and when these cilia bend from touching the tectorial membrane, the trap doors open. With this, potassium (K^+) ions flow in, leading to an influx of calcium (Ca^{2+}) ions at the base of the hair cell; and this leads the hair cells to release a neurotransmitter. Waiting to receive the neurotransmitter are adjacent dendrites, and, with this, vibration has been turned into a nerve signal.

The details of the hearing process actually are much more complicated than this account would indicate. While we need not go over these details, it's worthwhile to note just one of the ways this system is able to discriminate one sound from another. To lo-

cate where a sound is *coming from*, the brain in effect calculates the difference between how intense a sound is in one ear as opposed to the other, and what the time difference is between when sound arrives at one ear and then the other.

Finally, it is easy for hearing to be damaged, and this damage often is to the hair cells we've been looking at, as you can see in "Too Loud" on page 575.

26.12 Our Sense of Vision

Our visual system has to accomplish three central tasks. First, it has to capture light from the outside world and focus it at a very precise location within our eyes. Second, it has to take this focused light and convert it into a nervous-system signal. Third, it has to make sense of the visual information it receives. If you look at **Figure 26.15**, you can see some of the structures of the eye that function in the first two steps. Light comes into our eyes through the transparent cornea and passes through the opening known as the pupil. Surrounding the pupil is the iris, which is composed partly of smooth muscle and thus is capable of contracting or dilating, meaning it can let in less light in a bright environment, or more light in a darkened one. The incoming light is bound for the layer of cells at the back of the eye called the **retina**: an inner layer of tissue in the eye containing cells that transform light into nervous-system signals. But as the light rays come in, they are bent or "refracted"—first by the cornea

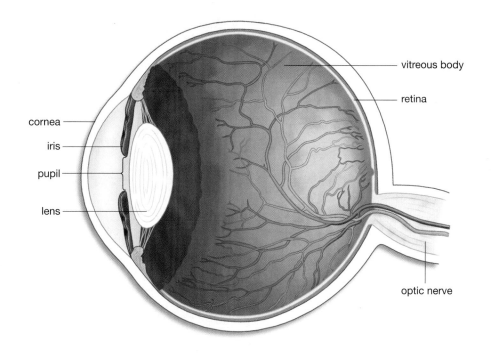

Figure 26.15
Anatomy of the Human Eye

Figure 26.16
Focused and Unfocused
Sharp vision requires that incoming light rays converge just as they reach the retina.

(a) In normal vision, the eye's cornea and lens bring about this convergence; they refract the incoming light at an angle that matches the length of the eye.

(b) People who are farsighted have a mismatch between the length of their eye and the amount of refraction provided by their cornea and lens: Their eye is too short for the amount of refraction provided. Thus, light rays converge behind their retina. As a result, these people cannot see nearby objects clearly.

(c) People who are nearsighted have the opposite problem: Their eye is too long for the amount of refraction provided, with the result that light rays converge in front of their retina. These people cannot see distant objects clearly. Both nearsighted and farsighted vision can be corrected.

and then by the lens that lies behind it—such that they converge on a small area of the retina, as you can see in **Figure 26.16**. This is no different in principle than what happens with a camera. We take a picture of a large object (such as a statue) that ends up as a small image on a frame of film. In a similar way, thanks to refraction, large objects we view with our eyes end up as *very* small images on our retina. How small? At fractions of a millimeter, some of them are microscopic. The eyes do not always focus light rays in such a way that they converge properly on the retina, however. You can see what it means to be nearsighted or farsighted in Figure 26.16.

Now, how do retinal cells convert rays of light into electrical signals? If you look at **Figure 26.17**, you can see that there are three layers of cells in the retina: ganglion cells, connecting cells, and photoreceptors. The **photoreceptors** are the sensory receptors for vision, and they come in two varieties: rods and cones. **Rods** are photoreceptors that function in low-light situations, but that provide only black-and-white vision. **Cones** are photoreceptors that respond best to bright light, and that provide color vision.

The blow-up of these cells shows that their posterior ends—the ends toward the back of the head—are filled with pigments, which lie embedded in a series of flattened, membranous discs. A pigment is simply a chemical compound that absorbs light. So, our light comes in and moves to, say, a rod in the retina. There it is absorbed by a pigment, and this absorption causes a normally bent part of the pigment molecule to … straighten out. That's it. Our entire sense of vision hinges on this tiny initial effect. Yet, as a shouted "hello" might cause an avalanche, so this small step puts big things in motion. In his book *What Makes You Tick?* vision researcher Thomas Czerner puts it this way:

> The infinitesimal energy of a photon [of light] changes the shape of a *single* molecule of pigment, which causes the release of *dozens* of molecules of an enzyme, which cause the rapid breakdown of *hundreds* of molecules of a chemical messenger, which cause the gates of *thousands* of ion channels in the cell membrane to slam shut, which blocks *millions* of sodium ions from entering the cell, which produces the grand finale, a local change in the resting membrane potential.

You might expect that this change in photoreceptor membrane potential would, like the others you've seen, result in the release of a neurotransmitter to an adjacent neuron. The twist in vision, however, is that in their *resting* state rods and cones are releasing neurotransmitter molecules in abundance to their connecting cells. When light strikes the rods and cones it has the effect of reducing this neurotransmitter release. Thus, the lack of neurotransmitter release from a rod or cone transmits the message "light stimulation here."

The signal that begins with a rod or cone goes first to connecting neurons, and then to the ganglion cells you can see in Figure 26.17. These cells have axons that are like tributaries coming together to form a river; only this river is the optic nerve, composed of about a million axons (see Figure 26.15). Because we have two eyes, we thus have a pair of million-fiber optic nerves carrying visual information from the eyes to the brain. The optic nerve takes the visual information to the thalamus, and from there it is transmitted, on other nerve fibers, to an area at the very back surface of the brain, the *primary visual cortex*, where most vision processing is done.

The details of this processing are much too complicated to go into here. But a central lesson of this complexity is that the human visual system does not work like a camera; it does not passively record bits of light and register a collection of these

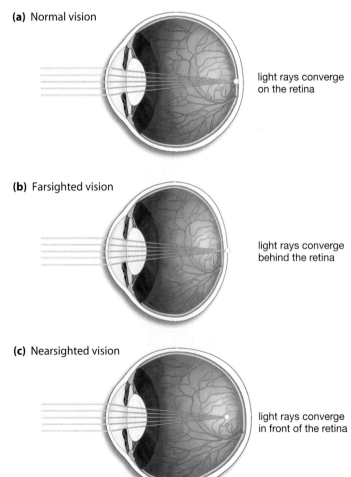

(a) Normal vision

light rays converge
on the retina

(b) Farsighted vision

light rays converge
behind the retina

(c) Nearsighted vision

light rays converge
in front of the retina

(a)

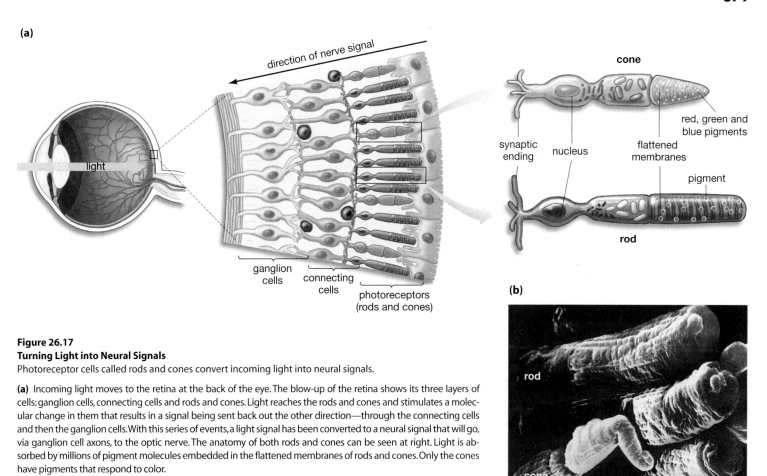

direction of nerve signal

light

cone

red, green and blue pigments

synaptic ending

nucleus

flattened membranes

pigment

rod

ganglion cells

connecting cells

photoreceptors (rods and cones)

(b)

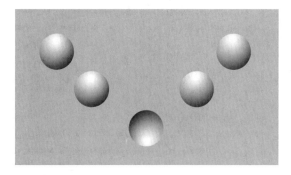

rod

cone

Figure 26.17
Turning Light into Neural Signals
Photoreceptor cells called rods and cones convert incoming light into neural signals.

(a) Incoming light moves to the retina at the back of the eye. The blow-up of the retina shows its three layers of cells: ganglion cells, connecting cells and rods and cones. Light reaches the rods and cones and stimulates a molecular change in them that results in a signal being sent back out the other direction—through the connecting cells and then the ganglion cells. With this series of events, a light signal has been converted to a neural signal that will go, via ganglion cell axons, to the optic nerve. The anatomy of both rods and cones can be seen at right. Light is absorbed by millions of pigment molecules embedded in the flattened membranes of rods and cones. Only the cones have pigments that respond to color.

(b) A color-enhanced micrograph of rods and cones. The light blue structure is a cone.

bits as images. Rather, the brain *constructs* images as much as it records them. To get an idea of this, look at **Figure 26.18**. In it, you see a V, formed out of bumps and a dot-like indentation. Now turn the book upside down and look again. The bumps have become indentations, and vice versa. Why? Note that when the book is right side up, most dots in the V are shaded in such a way that light seems to be coming from above them—there are "shadows" on the lower part of the dots. Light generally comes from above us, whether from a lamp or from the sun. This is so common that evolution has produced a genetically based "rule" in our visual system that says: "When shadows are in the lower part of an object, that object is raised above a surface."

Why would evolution produce rules such as these? After all, this rule is yielding an *inaccurate* interpretation of the dots on the page—they aren't really raised above the surface at all. Our visual system has such rules because what mattered in the evolution of our ancestors was not whether they perceived objects "correctly." What mattered was whether they perceived objects in ways that helped them survive and reproduce. Over the ages, individuals who could perceive, say, a

snake more quickly than other individuals survived longer and thus left more offspring. The many visual rules we have amount to rules for survival. They provide a series of "best bets" for quickly telling us what we are looking at. (If an object's lower half is shaded,

Web Tutorial 26.2
The Vertebrate Eye

Figure 26.18
Seeing Is Believing
Look at the figure and then turn the book upside down and look at it again. With the shift, dots that once appeared to be raised above the surface of the drawing now appear to be indented, and vice versa. Our visual system has a rule that causes us to perceive objects that are shaded at the bottom as protruding from a surface, while objects that are shaded at the top appear to be indented. This is but one example of how our visual system creates visual images, rather than simply recording them.

the best bet is that it's raised above a surface.) The idea that our visual system is often creating perceptions, rather than simply recording them, may seem strange, but that's the reality of visual perception.

26.13 The Endocrine System

We now leave the nervous system to focus on a different system in the body, the endocrine system. Like the nervous system, the endocrine system is in the communication and control business. And, like the nervous system with its neurotransmitters, the endocrine system works through a group of chemical messengers. The endocrine messengers are not neurotransmitters, however, but instead are substances called hormones. In this text, we will define **hormones** as substances secreted by one set of cells that travel through the bloodstream and affect the activities of other cells. Such a definition sets hormones apart from other kinds of signaling molecules in the body. There are signaling molecules, called *paracrines*, that do not travel through the bloodstream, but instead diffuse from one or more cells to a nearby group of cells, causing a metabolic change in them. Likewise, a cell can be affected by its own secreted chemical messenger, called an *autocrine*. Note that both paracrines and autocrines carry out strictly *local* communication between cells. The endocrine hormones we'll be looking at can have their effects over short distances, but they also have the

ability to carry chemical signals throughout the body, given their transportation through the bloodstream.

This very means of distribution provides a key for getting to the heart of what separates the endocrine and nervous systems. In the nervous system, signals go from neurons A to B to C, in well-defined lines of transmission—rather like a telephone call going through relay stations. By contrast, a typical hormone is "broadcast," in a sense, as it moves through the bloodstream. Like a television signal, it can be "picked up" by any cell that has the proper "receiver." What are these receivers? They are receptors that are shaped in such a way that they can latch onto the hormone. With this, we get to the concept of **target cells**: those cell types that can be affected by a given hormone (see **Figure 26.19**). The target cells for a hormone called anti-diuretic hormone are located primarily in the kidneys, while the target cells for the hormone insulin are located throughout the body. Thus, hormones differ greatly with respect to the number of target cells they affect and the location of these cells.

Hormonal production takes place to a significant extent within specialized organs called endocrine glands, and this represents another contrast with the nervous system. You may remember that a gland is any localized group of cells that work together to secrete a substance. **Endocrine glands** are glands that release their materials directly into the bloodstream or into surrounding tissues, without using ducts. Thus, the endocrine system has a set of large organs whose primary function is to secrete chemical messengers. In contrast, the nervous system has no such organs—only its individual neurons, secreting neurotransmitters. The major endocrine glands are shown in **Figure 26.20**. While most hormones are secreted by these glands, it's important to note that not *all* hormones are. The heart, kidneys, stomach, liver, small intestine, placenta, and fatty tissue secrete hormones related to their functioning, yet they are not glands.

In a final point of comparison, the endocrine system tends to work more slowly than the nervous system, but its effects tend to be more long-lasting. The fastest-acting hormones take several seconds to work, while the slowest-acting ones may take several hours. Contrast this with the almost instantaneous effects of nervous-system messages. The opposite side of this coin is that the longest-lasting hormones can keep exerting their effects for hours after they have been released. In contrast, nervous-system signals disappear as fast as they arise. Given the time scales in which hormones act, it's not surprising that they tend to regulate processes that unfold over minutes,

Web Tutorial 26.3
The Endocrine System

Figure 26.19
Hormones and Their Target Cells
For a hormone to affect a target cell, that cell must have receptors that can bind the hormone. The binding of hormone to receptor then initiates a change in the target cell's activity. The figure shows a peptide hormone that affects skeletal muscle tissue. The hormone does not affect the nerve cell, also shown in the figure, because this cell does not have the appropriate receptors to bind with the hormone.

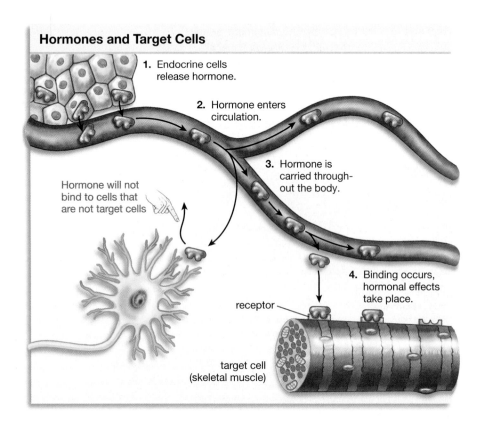

Hormones and Target Cells

1. Endocrine cells release hormone.

2. Hormone enters circulation.

Hormone will not bind to cells that are not target cells

3. Hormone is carried throughout the body.

4. Binding occurs, hormonal effects take place.

receptor

target cell (skeletal muscle)